Digital Design and Verilog HDL Fundamentals

Digital
Design
and
Verilog
HDL
Fundamentals

Joseph Cavanagh

Santa Clara University
California, USA

CRC Press
Taylor & Francis Group
Boca Raton   London   New York

CRC Press is an imprint of the
Taylor & Francis Group, an **Informa** business

CRC Press
Taylor & Francis Group
6000 Broken Sound Parkway NW, Suite 300
Boca Raton, FL 33487-2742

© 2008 by Taylor and Francis Group, LLC
CRC Press is an imprint of Taylor & Francis Group, an Informa business

No claim to original U.S. Government works

International Standard Book Number: 978-1-4200-7415-4 (Hardback)

Library of Congress Cataloging-in-Publication Data

Cavanagh, Joseph J. F.
    Digital design and Verilog HDL fundamentals / Joseph Cavanagh.
        p. cm.
    Includes bibliographical references and index.
    ISBN 978-1-4200-7415-4 (hardback : alk. paper) 1. Logic circuits--Computer-aided design. 2. Verilog (Computer hardware description language) 3. Digital electronics. I. Title.

TK7868.D5C3945 2008
621.39'5--dc22   2008012851

Visit the Taylor & Francis Web site at
http://www.taylorandfrancis.com

and the CRC Press Web site at
http://www.crcpress.com

By the same author:

DIGITAL COMPUTER ARITHMETIC: Design and Implementation

SEQUENTIAL LOGIC: Analysis and Synthesis

VERILOG HDL: Digital Design and Modeling

———————————————

The Computer Conspiracy — A novel

To my son, Brad, for his continued help and support.

PREFACE

The field of digital logic consists primarily of the analysis and synthesis of combinational logic circuits and sequential logic circuits, also referred to as finite-state machines. The principal characteristic of combinational logic is that the outputs are a function of the present inputs only, whereas the outputs of sequential logic are a function of the input sequence; that is, the input history. Sequential logic, therefore, requires storage elements which indicate the present state of the machine relative to a unique sequence of inputs.

Sequential logic is partitioned into synchronous and asynchronous sequential machines. Synchronous sequential machines are controlled by a system clock which provides the triggering mechanism to cause state changes. Asynchronous sequential machines have no clocking mechanism — the machines change state upon the application of input signals. The input signals provide the means to enable the sequential machines to proceed through a prescribed sequence of states.

The purpose of this book is to provide a thorough exposition of the analysis and synthesis of combinational and sequential logic circuits, where sequential logic consists of synchronous and asynchronous sequential machines. Emphasis is placed on structured and rigorous design principles that can be applied to practical applications. Each step of the analysis and synthesis procedures is clearly delineated. Each method that is presented is expounded in sufficient detail with several accompanying examples.

The Verilog hardware description language (HDL) is used extensively throughout the book for both combinational and sequential logic design. Verilog HDL is an Institute of Electrical and Electronics Engineers (IEEE) standard: 1364-1995. The book concentrates on combinational and sequential logic design with emphasis on the detailed design of various Verilog HDL projects. The examples are designed first using traditional design techniques, then implemented using Verilog HDL. This allows the reader to correlate and compare the two design methodologies.

The book is intended to be tutorial, and as such, is comprehensive and self contained. All designs are carried through to completion — nothing is left unfinished or partially designed. Each chapter includes numerous problems of varying complexity to be designed by the reader, including both traditional logic design techniques and Verilog HDL design techniques in appropriate chapters. The Verilog HDL designs include the design module, the test bench module which tests the design for correct functionality, the outputs obtained from the test bench, and the waveforms obtained from the test bench.

Chapter 1 covers the number systems of different radices such as binary, octal, decimal, and hexadecimal, including conversion between radices. The chapter also presents the number representations of sign magnitude, diminished-radix complement, and radix complement. Binary weighted and nonweighted codes are covered, including conversion to and from binary-coded decimal (BCD), plus the Gray code. Chapter 1 also introduces error detection and correction codes that are presented in more detail in a later chapter.

Chapter 2 presents Boolean algebra and illustrates methods to minimize switching functions. These methods include algebraic minimization, Karnaugh maps, Karnaugh maps using map-entered variables, the Quine-McCluskey algorithm, and the Petrick algorithm.

Chapter 3 presents the analysis and synthesis (design) of combinational logic circuits. Examples include sum-of-products and product-of-sums notation, disjunctive normal forms, and conjunctive normal forms. Logic macro functions are also covered, including multiplexers, decoders, encoders, and comparators.

Chapter 4 introduces Verilog HDL, which will be used in this chapter to design combinational logic. Verilog HDL is the state-of-the-art method for designing digital and computer systems and is ideally suited to describe both combinational and sequential logic. Verilog provides a clear relationship between the language syntax and the physical hardware. The Verilog simulator used in this book is easy to learn and use, yet powerful enough for any application. It is a logic simulator — called SILOS — developed by Silvaco International for use in the design and verification of digital systems. The SILOS simulation environment is a method to quickly prototype and debug any logic function. It is an intuitive environment that displays every variable and port from a module to a logic gate. SILOS allows single stepping through the Verilog source code, as well as drag-and-drop ability from the source code to a data analyzer for waveform generation and analysis. This chapter introduces the reader to the different modeling techniques, including built-in primitives for logic primitive gates and user-defined primitives for larger logic functions. The three main modeling methods of dataflow modeling, behavioral modeling, and structural modeling are introduced.

Chapter 5 presents a detailed exposition on the design of computer arithmetic circuits and includes topics in the following categories: fixed-point addition including ripple-carry and carry lookahead; fixed-point subtraction; fixed-point multiplication, including the sequential add-shift technique, the Booth algorithm, bit-pair recoding, and a high-speed array multiplier; decimal addition and subtraction; decimal multiplication and division; floating-point addition, subtraction, multiplication, and division.

Chapter 6 covers the Verilog HDL design of a variety of computer arithmetic circuits for fixed-point addition, subtraction, and multiplication and for decimal addition and subtraction. Fixed-point addition includes implementations of a high-speed full adder, a 4-bit ripple adder, and a carry lookahead adder. Fixed-point subtraction includes a unit that combines addition and subtraction. Fixed-point multiplication includes the implementation of a Booth algorithm circuit for signed operands in 2s complement representation and a high-speed array multiplier. Decimal addition and subtraction circuits are also designed, including a 9s complementer for subtraction.

Chapter 7 presents methods of analysis and synthesis for synchronous and asynchronous sequential machines. These techniques form the basic mechanisms for effective analysis and synthesis. Each method of analysis is accompanied by appropriate examples. The synthesis of synchronous sequential machines includes methods to design registers, counters, Moore machines, and Mealy machines. The synthesis procedure is outlined, then methods are described to determine state equivalence. If equivalent states can be identified, then redundant states can be eliminated, resulting in a machine with a minimal number of logic gates. The primary focus of this chapter is on the synthesis of deterministic synchronous sequential machines in which the next state is uniquely determined by the present state and the present inputs. This chapter also covers the analysis and synthesis of asynchronous sequential machines where state changes occur on the application of the input signals only — there is no machine clock. A final topic is the analysis and synthesis of pulse-mode asynchronous sequential machines in which each input variable is active in the form of a pulse. There is also no clock input in pulse-mode asynchronous sequential machines.

Chapter 8 applies the concepts given in Chapter 7 to design methodologies using Verilog HDL. Synchronous Moore and Mealy machines are designed using traditional methods and then implemented in Verilog HDL using various modeling constructs. A synchronous counter is designed that counts in a nonsequential pattern. Various asynchronous sequential machines are designed using dataflow modeling, behavioral modeling, structural modeling, and mixed design modeling, which incorporates two of the previous modeling constructs. Moore and Mealy pulse-mode asynchronous sequential machines are designed using different Verilog HDL modeling constructs.

Chapter 9 presents topics in programmable logic and discusses their use in both combinational and sequential logic circuits. The programmable devices include a programmable read-only memory (PROM), a programmable array logic (PAL), a programmable logic array (PLA), and a field-programmable gate array (FPGA).

Chapter 10 covers topics in digital-to-analog (D/A) conversion and analog-to-digital (A/D) conversion. Operational amplifiers are introduced, which are integral devices used in converting from digital to analog and from analog to digital. The digital-to-analog methods include a binary-weighted resistor network D/A converter and an $R - 2R$ resistor network DA converter. A special type of operational amplifier called a comparator is introduced, which is used in analog-to-digital conversion. The analog-to-digital methods include a counter A/D converter, a successive approximation A/D converter, and a high-speed simultaneous (of flash) A/D converter.

Chapter 11 presents magnetic recording fundamentals, which covers different techniques to encode digital data on a magnetic recording surface. The encoding concepts are applicable to disk drives, tape drives, and other magnetic systems. The following encoding methods are introduced: return to zero, nonreturn to zero, nonreturn to zero inverted, frequency modulation, phase encoding, modified frequency modulation, run-length limited, and group-coded recording. Peak shift and write precompensation are also covered plus a section on vertical recording.

Chapter 12 presents additional topics in digital design. Functional decomposition is a process of decomposing a function into smaller functions for the purpose of minimization; that is, to hierarchically decompose a system into its functional components. Functional decomposition examples are designed using traditional methods and then implemented using Verilog HDL. Iterative networks are one-dimensional or multi-dimensional arrays of identical cells in which the output of a cell depends on the input from previous cells. Typical applications are sequence detectors, shift registers, and array multipliers. Examples of iterative networks are designed using Verilog HDL. The section on Hamming code error detection and correction expands the concepts presented in Chapter 1 by providing the theory, logic design, and Verilog design of a Hamming code circuit. An overview of the cyclic redundancy check (CRC) code is presented. Residue checking and parity prediction techniques are discussed relative to detecting errors in arithmetic operations. An arithmetic and logic unit (ALU) is designed using Verilog HDL and a section on memories is discussed and a typical memory is designed using Verilog HDL.

Appendix A presents a brief discussion on event handling using the event queue. Operations that occur in a Verilog module are typically handled by an event queue. Appendix B presents a procedure to implement a Verilog project. Appendix C contains the solutions to select problems in each chapter.

The material presented in this book represents more than two decades of computer equipment design by the author. The book is intended as a text for a two-course sequence on combinational and sequential logic design. Chapter 1 through Chapter 6 can be used for combinational logic; Chapter 7 through Chapter 12 can be used for sequential logic. The book presents Verilog HDL with numerous design examples to help the reader thoroughly understand this popular hardware description language.

This book is designed for undergraduate students in electrical engineering, computer engineering, and computer science, for graduate students who require a noncredit course in logic design, and for practicing electrical engineers, computer engineers, and computer scientists.

A special thanks to Dr. Ivan Pesic, CEO of Silvaco International, for allowing use of the SILOS Simulation Environment software for the examples in this book. SILOS is an intuitive, easy-to-use, yet powerful Verilog HDL simulator for logic verification.

I would like to express my appreciation and thanks to the following people who gave generously of their time and expertise to review the manuscript and submit comments: Professor Daniel W. Lewis, Chair, Department of Computer Engineering, Santa Clara University, who continues to support me in all my endeavors; Dr. Geri Lamble; Steve Midford for his helpful suggestions and comments; and Ron Lewerenz. Thanks also to Nora Konopka and the staff at Taylor & Francis for their support.

Joseph Cavanagh

CONTENTS

Chapter 7   Sequential Logic 565

Chapter 8   Sequential Logic Design Using Verilog HDL 739

1

Number Systems, Number Representations, and Codes

Digital systems contain information that is represented as binary digits called *bits*. The alphabet of these bits is the set {0, 1}, which represents the logical value of the bits. The physical value is determined by the logic family being used. The transistor-transistor logic (TTL) family represents a logic 0 typically as +0.2 volts and a logic 1 typically as +3.4 volts using a +5 volt power supply; the emitter-coupled logic (ECL) 100K family represents a logic 0 typically as −1.7 volts and a logic 1 typically as −0.95 volts using a −4.5 volt power supply.

Thus, a signal can be asserted either positive (plus) or negative (minus), depending upon the active condition of the signal at that point. The word *positive*, as used here, does not necessarily mean a positive voltage level, but merely the more positive of two voltage levels, as is the case for ECL.

1.1 Number Systems

Numerical data are expressed in various positional number systems for each *radix* or *base*. A *positional number system* encodes a vector of n bits in which each bit is weighted according to its position in the vector. The encoded vector is also associated with a radix r, which is an integer greater than or equal to 2. A number system has exactly r digits in which each bit in the radix has a value in the range of 0 to $r − 1$, thus the highest digit value is one less than the radix. For example, the binary radix has two

1

digits which range from 0 to 1; the octal radix has eight digits which range from 0 to 7. An n-bit integer A is represented in a positional number system as follows:

$$A = (a_{n-1}a_{n-2}a_{n-3} \ldots a_1 a_0) \tag{1.1}$$

where $0 \le a_i \le r-1$. The high-order and low-order digits are a_{n-1} and a_0, respectively. The number in Equation 1.1 (also referred to as a vector or operand) can represent positive integer values in the range 0 to $r^n - 1$. Thus, a positive integer A is written as

$$A = a_{n-1}r^{n-1} + a_{n-2}r^{n-2} + a_{n-3}r^{n-3} + \ldots + a_1 r^1 + a_0 r^0 \tag{1.2}$$

The value for A can be represented more compactly as

$$A = \sum_{i=0}^{n-1} a_i r^i \tag{1.3}$$

The expression of Equation 1.2 can be extended to include fractions. For example,

$$A = a_{n-1}r^{n-1} + \ldots + a_1 r^1 + a_0 r^0 + a_{-1}r^{-1} + a_{-2}r^{-2} + \ldots + a_{-m}r^{-m} \tag{1.4}$$

Equation 1.4 can be represented as

$$A = \sum_{i=-m}^{n-1} a_i r^i \tag{1.5}$$

Adding 1 to the highest digit in a radix r number system produces a sum of 0 and a carry of 1 to the next higher-order column. Thus, counting in radix r produces the following sequence of numbers:

$$0, 1, 2, \ldots , (r-1), 10, 11, 12, \ldots, 1(r-1), \ldots.$$

Table 1.1 shows the counting sequence for different radices. The low-order digit will always be 0 in the set of r digits for the given radix. The set of r digits for various radices is given in Table 1.2. In order to maintain one character per digit, the numbers 10, 11, 12, 13, 14, and 15 are represented by the letters A, B, C, D, E, and F, respectively.

Table 1.1 Counting Sequence for Different Radices

Decimal	$r = 2$	$r = 4$	$r = 8$
0	0	0	0
1	1	1	1
2	10	2	2
3	11	3	3
4	100	10	4
5	101	11	5
6	110	12	6
7	111	13	7
8	1000	20	10
9	1001	21	11
10	1010	22	12
11	1011	23	13
12	1100	30	14
13	1101	31	15
14	1110	32	16
15	1111	33	17
16	10000	100	20
17	10001	101	21

Table 1.2 Character Sets for Different Radices

Radix (base)	Character Sets for Different Radices
2	{0, 1}
3	{0, 1, 2}
4	{0, 1, 2, 3}
5	{0, 1, 2, 3, 4}
6	{0, 1, 2, 3, 4, 5}
7	{0, 1, 2, 3, 4, 5, 6}
8	{0, 1, 2, 3, 4, 5, 6, 7}
9	{0, 1, 2, 3, 4, 5, 6, 7, 8}
10	{0, 1, 2, 3, 4, 5, 6, 7, 8, 9}

(Continued on next page)

Table 1.2   Character Sets for Different Radices

Radix (base)	Character Sets for Different Radices
11	{0, 1, 2, 3, 4, 5, 6, 7, 8, 9, A}
12	{0, 1, 2, 3, 4, 5, 6, 7, 8, 9, A, B}
13	{0, 1, 2, 3, 4, 5, 6, 7, 8, 9, A, B, C}
14	{0, 1, 2, 3, 4, 5, 6, 7, 8, 9, A, B, C, D}
15	{0, 1, 2, 3, 4, 5, 6, 7, 8, 9, A, B, C, D, E}
16	{0, 1, 2, 3, 4, 5, 6, 7, 8, 9, A, B, C, D, E, F}

Example 1.1   Count from decimal 0 to 25 in radix 5. Table 1.2 indicates that radix 5 contains the following set of four digits: {0, 1, 2, 3, 4}. The counting sequence in radix 5 is:

$$000, 001, 002, 003, 004 = (0 \times 5^2) + (0 \times 5^1) + (4 \times 5^0) = 4_{10}$$
$$010, 011, 012, 013, 014 = (0 \times 5^2) + (1 \times 5^1) + (4 \times 5^0) = 9_{10}$$
$$020, 021, 022, 023, 024 = (0 \times 5^2) + (2 \times 5^1) + (4 \times 5^0) = 14_{10}$$
$$030, 031, 032, 033, 034 = (0 \times 5^2) + (3 \times 5^1) + (4 \times 5^0) = 19_{10}$$
$$040, 041, 042, 043, 044 = (0 \times 5^2) + (4 \times 5^1) + (4 \times 5^0) = 24_{10}$$
$$100 = (1 \times 5^2) + (0 \times 5^1) + (0 \times 5^0) = 25_{10}$$

Example 1.2   Count from decimal 0 to 25 in radix 12. Table 1.2 indicates that radix 12 contains the following set of twelve digits: {0, 1, 2, 3, 4, 5, 6, 7, 8, 9, A, B}. The counting sequence in radix 12 is:

$$00, 01, 02, 03, 04, 05, 06, 07, 08, 09, 0A, 0B = (0 \times 12^1) + (11 \times 12^0) = 11_{10}$$
$$10, 11, 12, 13, 14, 15, 16, 17, 18, 19, 1A, 1B = (1 \times 12^1) + (11 \times 12^0) = 23_{10}$$
$$20, 21 = (2 \times 12^1) + (1 \times 12^0) = 25_{10}$$

1.1.1 Binary Number System

The radix is 2 in the *binary number system*; therefore, only two digits are used: 0 and 1. The low-value digit is 0 and the high-value digit is $(r - 1) = 1$. The binary number system is the most conventional and easily implemented system for internal use in a digital computer; therefore, most digital computers use the binary number system. There is a disadvantage when converting to and from the externally used decimal system; however, this is compensated for by the ease of implementation and the speed of execution in binary of the four basic operations: addition, subtraction, multiplication, and division. The radix point is implied within the internal structure of the computer; that is, there is no specific storage element assigned to contain the radix point.

The weight assigned to each position of a binary number is as follows:

$$2^{n-1} 2^{n-2} \ \ldots \ 2^3 \ 2^2 \ 2^1 \ 2^0 . \ 2^{-1} 2^{-2} 2^{-3} \ \ldots \ 2^{-m}$$

where the integer and fraction are separated by the radix point (binary point). The decimal value of the binary number 1011.101_2 is obtained by using Equation 1.4, where $r = 2$ and $a_i \in \{0,1\}$ for $-m \leq i \leq n-1$. Therefore,

$$
\begin{array}{cccccccc}
2^3 & 2^2 & 2^1 & 2^0 & . & 2^{-1} & 2^{-2} & 2^{-3} \\
1 & 0 & 1 & 1 & . & 1 & 0 & 1_2
\end{array}
$$

$$
\begin{aligned}
&= (1 \times 2^3) + (0 \times 2^2) + (1 \times 2^1) + (1 \times 2^0) + \\
&\quad (1 \times 2^{-1}) + (0 \times 2^{-2}) + (1 \times 2^{-3}) \\
&= 11.625_{10}
\end{aligned}
$$

Digital systems are designed using bistable storage devices that are either reset (logic 0) or set (logic 1). Therefore, the binary number system is ideally suited to represent numbers or states in a digital system, since radix 2 consists of the alphabet 0 and 1. These bistable devices can be concatenated to any length n to store binary data. For example, to store 1 byte (8 bits) of data, eight bistable storage devices are required as shown in Figure 1.1 for the value 0110 1011 (107_{10}). Counting in binary is shown in Table 1.3, which shows the weight associated with each of the four binary positions. Notice the alternating groups of 1s in Table 1.3. A binary number is a group of n bits that can assume 2^n different combinations of the n bits. The range for n bits is 0 to $2^n - 1$.

| 0 | 1 | 1 | 0 | 1 | 0 | 1 | 1 |

Figure 1.1   Concatenated 8-bit storage elements.

Table 1.3 Counting in Binary

Decimal	Binary			
	8	4	2	1
	2^3	2^2	2^1	2^0
0	0	0	0	0
1	0	0	0	1
2	0	0	1	0
3	0	0	1	1
4	0	1	0	0
5	0	1	0	1
6	0	1	1	0
7	0	1	1	1
8	1	0	0	0
9	1	0	0	1
10	1	0	1	0

(Continued on next page)

Table 1.3 Counting in Binary

Decimal	Binary			
	8	4	2	1
	2^3	2^2	2^1	2^0
11	1	0	1	1
12	1	1	0	0
13	1	1	0	1
14	1	1	1	0
15	1	1	1	1

The binary weights for the bit positions of an 8-bit integer are shown in Table 1.4; the binary weights for an 8-bit fraction are shown in Table 1.5.

Table 1.4 Binary Weights for an 8-Bit Integer

2^7	2^6	2^5	2^4	2^3	2^2	2^1	2^0
128	64	32	16	8	4	2	1

Table 1.5 Binary Weights for an 8-Bit Fraction

2^{-1}	2^{-2}	2^{-3}	2^{-4}	2^{-5}	2^{-6}	2^{-7}	2^{-8}
1/2	1/4	1/8	1/16	1/32	1/64	1/128	1/256
0.5	0.25	0.125	0.0625	0.03125	0.015625	0.0078125	0.00390625

Each 4-bit binary segment has a weight associated with the segment and is assigned the value represented by the low-order bit of the corresponding segment, as shown in the first row of Table 1.6. The 4-bit binary number in each segment is then multiplied by the value of the segment. Thus, the binary number 0010 1010 0111 1100 0111 is equal to the decimal number $59,335_{10}$ as shown below.

$$(2 \times 8192) + (10 \times 4096) + (7 \times 256) + (12 \times 16) + (7 \times 1) = 59,335_{10}$$

Table 1.6 Weight Associated with 4-Bit Binary Segments

8192	4096	256	16	1
0001	0001	0001	0001	0001
0010	1010	0111	1100	0111

1.1.2 Octal Number System

The radix is 8 in the *octal number system*; therefore, eight digits are used, 0 through 7. The low-value digit is 0 and the high-value digit is $(r-1) = 7$. The weight assigned to each position of an octal number is as follows:

$$8^{n-1} 8^{n-2} \ldots 8^3\, 8^2\, 8^1\, 8^0 . 8^{-1} 8^{-2} 8^{-3} \ldots 8^{-m}$$

where the integer and fraction are separated by the radix point (octal point). The decimal value of the octal number 217.6_8 is obtained by using Equation 1.4, where $r = 8$ and $a_i \in \{0,1,2,3,4,5,6,7\}$ for $-m \le i \le n-1$. Therefore,

$$
\begin{array}{cccccl}
8^2 & 8^1 & 8^0 & . & 8^{-1} & \\
2 & 1 & 7 & . & 6_8 & = (2 \times 8^2) + (1 \times 8^1) + (7 \times 8^0) + (6 \times 8^{-1}) \\
& & & & & = 143.75_{10}
\end{array}
$$

When a count of 1 is added to 7_8, the sum is zero and a carry of 1 is added to the next higher-order column on the left. Counting in octal is shown in Table 1.7, which shows the weight associated with each of the three octal positions.

Table 1.7 Counting in Octal

Decimal	Octal		
	64	8	1
	8^2	8^1	8^0
0	0	0	0
1	0	0	1
2	0	0	2
3	0	0	3
4	0	0	4
5	0	0	5
6	0	0	6
7	0	0	7
8	0	1	0
9	0	1	1
...	...		
14	0	1	6
15	0	1	7
16	0	2	0
17	0	2	1
...	...		

Continued on next page

Table 1.7 Counting in Octal

Decimal	Octal		
	64	8	1
	8^2	8^1	8^0
22	0	2	6
23	0	2	7
24	0	3	0
25	0	3	1
...		...	
30	0	3	6
31	0	3	7
...		...	
84	1	2	4
...		...	
242	3	6	2
...		...	
377	5	7	1

Binary-coded octal   Each octal digit can be encoded into a corresponding binary number. The highest-valued octal digit is 7; therefore, three binary digits are required to represent each octal digit. This is shown in Table 1.8, which lists the eight decimal digits (0 through 7) and indicates the corresponding octal and binary-coded octal (BCO) digits. Table 1.8 also shows octal numbers of more than one digit.

Table 1.8 Binary-Coded Octal Numbers

Decimal	Octal	Binary-Coded Octal	
0	0		000
1	1		001
2	2		010
3	3		011
4	4		100
5	5		101
6	6		110
7	7		111
8	10	001	000
9	11	001	001
10	12	001	010
(Continued on next page)			

Table 1.8 Binary-Coded Octal Numbers

Decimal	Octal	Binary-Coded Octal		
11	13		001	011
...	...		...	
20	24		010	100
21	25		010	101
...	...		...	
100	144	001	100	100
101	145	001	100	101
...	...		...	
267	413	100	001	011
...	...		...	
385	601	110	000	001

1.1.3 Decimal Number System

The radix is 10 in the *decimal number system*; therefore, ten digits are used, 0 through 9. The low-value digit is 0 and the high-value digit is $(r-1) = 9$. The weight assigned to each position of a decimal number is as follows:

$$10^{n-1} 10^{n-2} \ldots 10^3 \ 10^2 \ 10^1 \ 10^0. \ 10^{-1} 10^{-2} 10^{-3} \ldots 10^{-m}$$

where the integer and fraction are separated by the radix point (decimal point). The value of 6357_{10} is immediately apparent; however, the value is also obtained by using Equation 1.4, where $r = 10$ and $a_i \in \{0,1,2,3,4,5,6,7,8,9\}$ for $-m \le i \le n-1$. That is,

$$\begin{array}{cccc} 10^3 & 10^2 & 10^1 & 10^0 \\ 6 & 3 & 5 & 7_{10} \end{array} = (6 \times 10^3) + (3 \times 10^2) + (5 \times 10^1) + (7 \times 10^0)$$

When a count of 1 is added to decimal 9, the sum is zero and a carry of 1 is added to the next higher-order column on the left. The following example contains both an integer and a fraction:

$$\begin{array}{ccccc} 10^3 & 10^2 & 10^1 & 10^0 & .\ 10^{-1} \\ 5 & 4 & 3 & 6 & .\ 5 \end{array} = (5 \times 10^3) + (4 \times 10^2) + (3 \times 10^1) + (6 \times 10^0) + (5 \times 10^{-1})$$

Binary-coded decimal   Each decimal digit can be encoded into a corresponding binary number; however, only ten decimal digits are valid. The highest-valued decimal digit is 9, which requires four bits in the binary representation. Therefore, four binary digits are required to represent each decimal digit. This is shown in Table 1.9, which lists the ten decimal digits (0 through 9) and indicates the corresponding binary-coded decimal (BCD) digits. Table 1.9 also shows BCD numbers of more than one decimal digit.

Table 1.9 Binary-Coded Decimal Numbers

Decimal	Binary-Coded Decimal
0	0000
1	0001
2	0010
3	0011
4	0100
5	0101
6	0110
7	0111
8	1000
9	1001
10	0001 0000
11	0001 0001
12	0001 0010
...	...
124	0001 0010 0100
...	...
365	0011 0110 0101

1.1.4 Hexadecimal Number System

The radix is 16 in the *hexadecimal number system*; therefore, 16 digits are used, 0 through 9 and A through F, where by convention A, B, C, D, E, and F correspond to decimal 10, 11, 12, 13, 14, and 15, respectively. The low-value digit is 0 and the high-value digit is $(r - 1) = 15$ (F). The weight assigned to each position of a hexadecimal number is as follows:

$$16^{n-1} 16^{n-2} \ \dots \ 16^3 \ 16^2 \ 16^1 \ 16^0 . \ 16^{-1} 16^{-2} 16^{-3} \ \dots \ 16^{-m}$$

where the integer and fraction are separated by the radix point (hexadecimal point). The decimal value of the hexadecimal number $6A8C.D416_{16}$ is obtained by using Equation 1.4, where $r = 16$ and $a_i \in \{0,1,2,3,4,5,6,7,8,9,A,B,C,D,E,F\}$ for $-m \le i \le n - 1$. Therefore,

$$16^3 \ 16^2 \ 16^1 \ 16^0 \ . \ 16^{-1}16^{-2}16^{-3}16^{-4}$$

$$
\begin{aligned}
6 \ \ A \ \ 8 \ \ C \ . \ D \ \ 4 \ \ 1 \ \ 6 \ &= \ (6 \times 16^3) + (10 \times 16^2) + (8 \times 16^1) \\
&\quad + (12 \times 16^0) + (13 \times 16^{-1}) + (4 \times 16^{-2}) \\
&\quad + (1 \times 16^{-3}) + (6 \times 16^{-4}) \\
&= \ 27{,}276.828846_{10}
\end{aligned}
$$

When a count of 1 is added to hexadecimal F, the sum is zero and a carry of 1 is added to the next higher-order column on the left.

Binary-coded hexadecimal   Each hexadecimal digit corresponds to a 4-bit binary number as shown in Table 1.10. All 2^4 values of the four binary bits are used to represent the 16 hexadecimal digits. Table 1.10 also indicates hexadecimal numbers of more than one digit. Counting in hexadecimal is shown in Table 1.11. Table 1.12 summarizes the characters used in the four number systems: binary, octal, decimal, and hexadecimal.

Table 1.10 Binary-Coded Hexadecimal Numbers

Decimal	Hexadecimal	Binary-Coded Hexadecimal
0	0	0000
1	1	0001
2	2	0010
3	3	0011
4	4	0100
5	5	0101
6	6	0110
7	7	0111
8	8	1000
9	9	1001
10	A	1010
11	B	1011
12	C	1100
13	D	1101
14	E	1110
15	F	1111
...	...	...
124	7C	0111 1100
...	...	...
365	16D	0001 0110 1101

Table 1.11 Counting in Hexadecimal

Decimal	Hexadecimal		
	256	16	1
	16^2	16^1	16^0
0	0	0	0
1	0	0	1
2	0	0	2
3	0	0	3
4	0	0	4
5	0	0	5
6	0	0	6
7	0	0	7
8	0	0	8
9	0	0	9
10	0	0	A
11	0	0	B
12	0	0	C
13	0	0	D
14	0	0	E
15	0	0	F
16	0	1	0
17	0	1	1
…		…	
26	0	1	A
27	0	1	B
…		…	
30	0	1	E
31	0	1	F
…		…	
256	1	0	0
…		…	
285	1	1	D
…		…	
1214	4	B	E

1.1.5 Arithmetic Operations

The arithmetic operations of addition, subtraction, multiplication, and division in any radix can be performed using identical procedures to those used for decimal arithmetic. The operands for the four operations are shown in Table 1.13.

Table 1.12 Digits Used for Binary, Octal, Decimal, and Hexadecimal Number Systems

0 1 2 3 4 5 6 7 8 9 A B C D E F

Binary

Octal

Decimal

Hexadecimal

Table 1.13 Operands Used for Arithmetic Operations

	Addition		Subtraction		Multiplication		Division
	Augend		Minuend		Multiplicand		Dividend
+)	Addend	−)	Subtrahend	×)	Multiplier	÷)	Divisor
	Sum		Difference		Product		Quotient, Remainder

Radix 2 addition   Figure 1.2 illustrates binary addition of unsigned operands. The sum of column 1 is 2_{10} (10_2); therefore, the sum is 0 with a carry of 1 to column 2. The sum of column 2 is 4_{10} (100_2); therefore, the sum is 0 with a carry of 0 to column 3 and a carry of 1 to column 4. The sum of column 3 is 3_{10} (11_2); therefore, the sum is 1 with a carry of 1 to column 4. The sum of column 4 is 4_{10} (100_2); therefore, the sum is 0 with a carry of 0 to column 5 and a carry of 1 to column 6.

The unsigned values of the binary operands are shown in the rightmost column together with the resulting sum.

Column	6	5	4	3	2	1	Radix 10 values
			1	1	1	0	14
			0	1	1	1	7
			1	0	1	0	10
+)			0_{11}	1_0	0_1	1	5
		0	0	1	0	0	36

Figure 1.2   Example of binary addition.

Radix 2 subtraction   The rules for subtraction in radix 2 are as follows:

$0 - 0 = 0$

$0 - 1 = 1$ with a borrow from the next higher-order minuend

$1 - 0 = 1$

$1 - 1 = 0$

Figure 1.3 provides an example of binary subtraction using the above rules for unsigned operands. An alternative method for subtraction — used in computers — will be given in Section 1.2 when number representations are presented. In Figure 1.3 column 3, the difference is 1 with a borrow from the minuend in column 4, which changes the minuend in column 4 to 0.

Column	4	3	2	1	Radix 10 values
	1^0	0	1	1	11
−)	0	1	0	1	5
		1	1	0	6

Figure 1.3   Example of binary subtraction.

Radix 2 multiplication   Multiplying in binary is similar to multiplying in decimal. Two n-bit operands produce a $2n$-bit product. Figure 1.4 shows an example of binary multiplication using unsigned operands, where the multiplicand is 7_{10} and the multiplier is 14_{10}. The multiplicand is multiplied by the low-order multiplier bit (0) producing a partial product of all zeroes. Then the multiplicand is multiplied by the next higher-order multiplier bit (1) producing a left-shifted partial product of 0000 111. The process repeats until all bits of the multiplier have been used.

								Radix 10
				0	1	1	1	7
			×)	1	1	1	0	14
0	0	0	0	0	0	0	0	
0	0	0	0	1	1	1		
0	0	0	1	1	1			
0	0_1	1_{11}	1_0	1_1				
1	1	0		0	0	1	0	98

Figure 1.4   Example of binary multiplication.

Radix 2 division   The division process is shown in Figure 1.5, where the divisor is n bits and the dividend is $2n$ bits. The division procedure uses a sequential shift-subtract-restore technique. Figure 1.5 shows a divisor of 5_{10} (0101_2) and a dividend of 13_{10} ($0000\ 1101_2$), resulting in a quotient of 2_{10} (0010_2) and a remainder of 3_2 (0011_2).

The divisor is subtracted from the high-order four bits of the dividend. The result is a partial remainder that is negative — the leftmost bit is 1 — indicating that the divisor is greater than the four high-order bits of the dividend. Therefore, a 0 is placed in the high-order bit position of the quotient. The dividend bits are then restored to their previous values with the next lower-order bit (1) of the dividend being appended to the right of the partial product. The divisor is shifted right one bit position and again subtracted from the dividend bits.

						P1	P2	P3	P4	P5	P6	P7	P8	
						0	0	0	1	0				Quotient
	0	1	0	1		0	0	0	0	1	1	0	1	
Subtract						0	1	0	1					
						1	0	1	1					
Restore						0	0	0	0	1				
Shift-subtract							0	1	0	1				
							1	1	0	0				
Restore						0	0	0	0	1	1			
Shift-subtract								0	1	0	1			
								1	1	1	0			
Restore						0	0	0	0	1	1	0		
Shift-subtract									0	1	0	1		
									0	0	0	1		
No restore						0	0	0	0	0	0	1	1	
Shift-subtract										0	1	0	1	
										1	1	1	0	
Restore						0	0	0	0	0	0	1	1	Remainder

Figure 1.5   Example of binary division.

This restore-shift-subtract cycle repeats for a total of three cycles until the partial remainder is positive — the leftmost bit is 0, indicating that the divisor is less than the corresponding dividend bits. This results in a no-restore cycle in which the previous partial remainder (0001) is not restored. A 1 bit is placed in the next lower-order quotient bit and the next lower-order dividend bit is appended to the right of the partial remainder. The divisor is again subtracted, resulting in a negative partial remainder, which is again restored by adding the divisor. The 4-bit quotient is 0010 and the 4-bit remainder is 0011.

The results can be verified by multiplying the quotient (0010) by the divisor (0101) and adding the remainder (0011) to obtain the dividend. Thus, $0010 \times 0101 = 1010 + 0011 = 1101$.

Radix 8 addition   Figure 1.6 illustrates octal addition. The result of adding column 1 is 17_8, which is a sum of 1 with a carry of 2. The result of adding column 2 is 11_8, which is a sum of 3 with a carry of 1. The remaining columns are added in a similar manner, yielding a result of 21631_8 or 9113_{10}.

Column	4	3	2	1	Radix 10 value
	7	6	5	4	4012
	6	5	4	7	3431
+)	3_1	2_1	0_2	6	1670
		1 6	3	1	9113

Figure 1.6   Example of octal addition.

Radix 8 subtraction   Octal subtraction is slightly more complex than octal addition. Figure 1.7 provides an example of octal subtraction. In column 2 (8^1), a 1 is subtracted from minuend 5_8 leaving a value of 4_8; the 1 is then added to the minuend in column 1 (2_8). This results in a difference of 6_8 in column 1, as shown below.

$$(1 \times 8^1) + (2 \times 8^0) = 10_{10}$$
Therefore, $10 - 4 = 6$

In a similar manner, in column 4 (8^3), a 1 is subtracted from minuend 6_8 leaving a value of 5_8; the 1 is then added to the minuend in column 3 (1_8), leaving a difference of $9 - 5 = 4$, as shown below.

$$(1 \times 8^3) + (1 \times 8^2) = 1100_8$$
Consider only the 11 of 1100_8, where $(1 \times 8^1) + (1 \times 8^0) = 9_{10}$
Therefore, $9 - 5 = 4$

	8^3	8^2	8^1	8^0
Column	4	3	2	1
	6	1	5	2
–)	5	5	3	4
		4	1	6

Figure 1.7   Example of octal subtraction.

Consider another example of octal subtraction shown in Figure 1.8, which shows a slightly different approach. A 1 is subtracted from the minuend in column 4 and added to the minuend in column 3. This results in a value of 13_8 in column 3 or 001 011 in binary-coded octal (also radix 2). Therefore, 001 011 − 101 = 000 110 = 6.

Column	8^3	8^2	8^1	8^0
	4	3	2	1
	6	3	7	2
−)	4	5	0	1
	6	7	1	

Figure 1.8   Example of octal subtraction.

Radix 8 multiplication   An example of octal multiplication is shown in Figure 1.9. The multiplicand is multiplied by each multiplier digit in turn to obtain a partial product. Except for the first partial product, each successive partial product is shifted left one digit. The subscripts in partial products 3 and 4 represent carries obtained from multiplying the multiplicand by the multiplier digits. When all of the partial products are obtained, the partial products are added following the rules for octal addition.

					7	4	6	3
				×)	5	2	1	0
Partial product 1	0	0	0	0	0	0	0	0
Partial product 2	0	0	0	7	4	6	3	
Partial product 3	0	0_1	6_1	0_1	4	6		
Partial product 4	0_4	3_2	4_3	6_1	7			
Carries from addition	1	2	2	2	1			
		0	0	1	0	4	3	0

Figure 1.9   Example of octal multiplication.

Radix 8 division   An example of octal division is shown in Figure 1.10. The first quotient digit is 3_8 which, when multiplied by the divisor 17_8, yields a result of 55_8. This can be verified as shown Figure 1.11, where $3_8 \times 7_8 = 25_8$, resulting in a product of 5_8 and a carry of 2_8. Another approach is as follows: $3_{10} \times 7_{10} = 21_{10}$, which is 5 away from 2×16; that is, a product of 5_8 and a carry of 2_8. Subtraction of the partial remainder and multiplication of the quotient digit times the divisor are accomplished using the rules stated above for octal arithmetic.

```
                        2    3
   1   7 | 6    1    4    5
            5    5
                 4    4
                 3    6
                      6    5
                      5    5
                           0
```

Figure 1.10   Example of octal division.

```
             1    7
  ×)              3
        3₂ = 5    5
```

Figure 1.11   Example of octal multiplication for the first partial remainder of Figure 1.10.

The results of Figure 1.10 can be verified as follows:

$$
\begin{aligned}
\text{Dividend} \quad &= (\text{quotient} \times \text{divisor}) + \text{remainder} \\
&= (323_8 \times 17_8) + 10_8 \\
&= 6145_8
\end{aligned}
$$

Radix 10 addition   Arithmetic operations in the decimal number system are widely used and need no introduction; however, they are included here to add completeness to the number system topic. An example of decimal addition is shown in Figure 1.12. The carries between columns are indicated by subscripted numbers.

```
             4    5    2
             7    6    5
             1    8    9
  +)        2₃   9₂    7
             7    0    3
```

Figure 1.12   Example of decimal addition.

Radix 10 subtraction   An example of decimal subtraction is shown in Figure 1.13. The superscripted numbers indicate the minuend result after a borrow is subtracted.

7^6	6^5	3	9	4
$-)$ 5	9	7	2	2
	6	6	7	2

Figure 1.13   Example of decimal subtraction.

Radix 10 multiplication   An example of decimal multiplication is shown in Figure 1.14. The subscripted numbers indicate the carries.

			2	9	6
		$\times)$	5	4	3
0	0	0	8	8	8
0	1	1	8	4	
1	4_1	8_1	0_1		
	6	0	7	2	8

Figure 1.14   Example of decimal multiplication.

Radix 10 division   An example of decimal division is shown in Figure 1.15.

					8
7	2	1	3	4	9
			7	2	
			6	2	9
			5	7	6
					3

Figure 1.15   Example of decimal division.

Radix 16 addition   An example of hexadecimal addition is shown in Figure 1.16. The subscripted numbers indicate carries from the previous column. The decimal value of the hexadecimal addition of each column is also shown. To obtain the hexadecimal value of the column, a multiple of 16_{10} is subtracted from decimal value and the difference is the hexadecimal value and the multiple of 16_{10} is the carry. For example, the decimal sum of column 1 is 28. Therefore, $28 - 16 = 12$ (C_{16}) with a carry of 1 to column 2. In a similar manner, the decimal sum of column 2 is $40 + 1$ (carry) $= 41$. Therefore, $41 - 32 = 9$ (9_{16}) with a carry of 2 to column 3.

Column	4	3	2	1
	A	B	C	D
	9	8	7	6
	E	F	9	4
+)	9_2	A_2	C_1	5
Radix 10 =	44	46	41	28
	C	E	9	C

Figure 1.16   Example of hexadecimal addition.

Radix 16 subtraction   Hexadecimal subtraction is similar to subtraction in any other radix. An example of hexadecimal subtraction is shown in Figure 1.17.

Column	4	3	2	1
	C^1	2^1	8	D
-)	8	F	E	9
		2	A	4

Figure 1.17   Example of hexadecimal subtraction.

The superscripted numbers indicate borrows from the minuends. For example, the minuend in column 2 borrows a 1 from the minuend in column 3; therefore, column 2 becomes $18_{16} - E_{16} = A_{16}$. This is more readily apparent if the hexadecimal numbers are represented as binary numbers, as shown below.

	1	8			0	0	0	1		1	0	0	0
-)		E	$\rightarrow$	-)	0	0	0	0		1	1	1	0
					0	0	0	0		1	0	1	0

In a similar manner, column 3 becomes $11_{16} - F_{16} = 2_{16}$ with a borrow from column 4. Column 4 becomes $B_{16} - 8_{16} = 3_{16}$.

Radix 16 multiplication   Figure 1.18 shows an example of hexadecimal multiplication. Multiplication in radix 16 is slightly more complex than multiplication in other radices. Each multiplicand is multiplied by multiplier digit in turn to form a partial product. Except for the first partial product, each partial product is shifted left one digit position. The subscripted digits in Figure 1.18 indicate the carries formed when multiplying the multiplicand by the multiplier digits.

Consider the first row of Figure 1.18 — the row above partial product 1.

$$10_{10} \times 4_{10} = 40_{10} = 8_{16} \text{ with a carry of } 2_{16}$$
$$10_{10} \times 13_{10} = 130_{10} = 2_{16} \text{ with a carry of } 8_{16}$$
$$10_{10} \times 9_{10} = 90_{10} = A_{16} \text{ with a carry of } 5_{16}$$
$$10_{10} \times 12_{10} = 120_{10} = 8_{16} \text{ with a carry of } 7_{16}$$

In a similar manner, the remaining partial products are obtained. Each column of partial products is then added to obtain the product.

					C	9	D	4
				×)	7	8	B	A
				7	8_{51}	A_8	2_2	8
Partial product 1	0	0	0	7	E	2	4	8
			8	4_6	3_8	F_2	C	
Partial product 2	0	0	8	A	C	1	C	
		0	6	4	8_6	8_2	0	
Partial product 3	0	6	4	E	A	0		
	5	4_3	F_5	B_1	C			
Partial product 4	5	8	4	C	C			
Carries from addition		1	2	3		1		
Product	5	F	2	E	0	4	0	8

Figure 1.18   Example of hexadecimal multiplication.

Radix 16 division   Figure 1.19 (a) and Figure 1.19 (b) show two examples of hexadecimal division. The results of Figure 1.19 can be verified as follows:

Dividend = (quotient × divisor) + remainder

For Figure 1.19 (a): Dividend = $(F0F_{16} \times 11_{16}) + 0 = FFFF_{16}$
For Figure 1.19 (b): Dividend = $(787_{16} \times 22_{16}) + 11_{16} = FFFF_{16}$

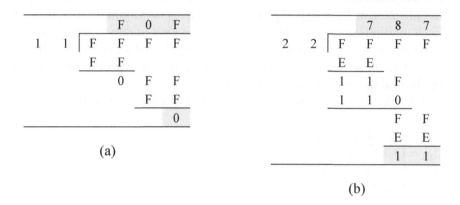

Figure 1.19   Examples of hexadecimal division.

1.1.6 Conversion Between Radices

Methods to convert a number in radix r_i to radix r_j will be presented in this section. The following conversion methods will be presented:

Binary	→	Decimal
Octal	→	Decimal
Hexadecimal	→	Decimal
Decimal	→	Binary
Decimal	→	Octal
Decimal	→	Hexadecimal
Binary	→	Octal
Binary	→	Hexadecimal
Octal	→	Binary
Octal	→	Hexadecimal
Hexadecimal	→	Binary
Hexadecimal	→	Octal
Octal	→	Binary-coded octal
Hexadecimal	→	Binary-coded hexadecimal
Decimal	→	Binary-coded decimal

Comparison between the following formats will also be examined:

octal → binary-coded octal and octal → binary
hexadecimal → binary-coded hexadecimal and hexadecimal → binary
decimal → binary-coded decimal and decimal → binary

There will also be an example to illustrate converting between two nonstandard radices and an example to determine the value of an unknown radix for a given radix 10 number.

Binary to decimal   Conversion from any radix r to radix 10 is easily accomplished by using Equation 1.2 or 1.3 for integers, or Equation 1.4 or 1.5 for numbers consisting of integers and fractions. The binary number 1111000.101_2 will be converted to an equivalent decimal number. The weight by position is as follows:

2^6	2^5	2^4	2^3	2^2	2^1	2^0		2^{-1}	2^{-2}	2^{-3}
1	1	1	1	0	0	0	.	1	0	1

Therefore,
$$
\begin{aligned}
1111000.101_2 &= (1 \times 2^6) + (1 \times 2^5) + (1 \times 2^4) + (1 \times 2^3) + \\
&\quad (0 \times 2^2) + (0 \times 2^1) + (0 \times 2^0) + \\
&\quad (1 \times 2^{-1}) + (0 \times 2^{-2}) + (1 \times 2^{-3}) \\
&= 64 + 32 + 16 + 8 + 0.5 + 0.125 \\
&= 120.625_{10}
\end{aligned}
$$

Octal to decimal   The octal number 217.65_8 will be converted to an equivalent decimal number. The weight by position is as follows:

8^2	8^1	8^0		8^{-1}	8^{-2}
2	1	7	.	6	5

Therefore,
$$
\begin{aligned}
217.65_8 &= (2 \times 8^2) + (1 \times 8^1) + (7 \times 8^0) + (6 \times 8^{-1})(5 \times 8^{-2}) \\
&= 128 + 8 + 7 + 0.75 + 0.078125 \\
&= 143.828125_{10}
\end{aligned}
$$

Hexadecimal to decimal   The hexadecimal number $5C2.4D_{16}$ will be converted to an equivalent decimal number. The weight by position is as follows:

16^2	16^1	16^0		16^{-1}	16^{-2}
5	C	2	.	4	D

Therefore,
$$
\begin{aligned}
5C2.4D_{16} &= (5 \times 16^2) + (12 \times 16^1) + (2 \times 16^0) + (4 \times 16^{-1}) + (13 \times 16^{-2}) \\
&= 1280 + 192 + 2 + 0.25 + 0.05078125 \\
&= 1474.300781_{10}
\end{aligned}
$$

Decimal to binary   To convert a number in radix 10 to any other radix r, repeatedly divide the integer by radix r, then repeatedly multiply the fraction by radix r. The first remainder obtained when dividing the integer is the low-order digit. The first integer obtained when multiplying the fraction is the high-order digit.

The decimal number 186.625_{10} will be converted to an equivalent binary number. The process is partitioned into two parts: divide the integer 186_{10} repeatedly by 2 until the quotient equals zero; multiply the fraction 0.625 repeatedly by 2 until a zero result is obtained or until a certain precision is reached.

$186 \div 2 =$	quotient = 93,	remainder = 0 (0 is the low-order digit)
$93 \div 2 =$	quotient = 46,	remainder = 1
$46 \div 2 =$	quotient = 23,	remainder = 0
$23 \div 2 =$	quotient = 11,	remainder = 1
$11 \div 2 =$	quotient = 5,	remainder = 1
$5 \div 2 =$	quotient = 2,	remainder = 1
$2 \div 2 =$	quotient = 1,	remainder = 0
$1 \div 2 =$	quotient = 0,	remainder = 1

$0.625 \times 2 =$	1.25	1	(1 is the high-order digit)
$0.25 \times 2 =$	0.5	0	
$0.5 \times 2 =$	1.0	1	

Therefore, $186.625_{10} = 10111010.101_2$.

The decimal number 267_{10} will be converted to binary-coded decimal (BCD) and binary. The binary bit configuration for BCD is

$$267_{10} = 0010 \ 0110 \ 0111_{BCD}$$

The bit configuration for binary is

$$267_{10} = 0001 \ 0000 \ 1011_2$$

The two results are not equal because the BCD number system does not use all sixteen combinations of four bits. BCD uses only ten combinations — 0 through 9.

Decimal to octal   The decimal number 219.62_{10} will be converted to an equivalent octal number. The integer 219_{10} is divided by 8 repeatedly and the fraction 0.62_{10} is multiplied by 8 repeatedly to a precision of three digits.

$219 \div 8 =$	quotient $= 27$,	remainder $= 3$	(3 is the low-order digit)
$27 \div 8 =$	quotient $= 3$,	remainder $= 3$	
$3 \div 8 =$	quotient $= 0$,	remainder $= 3$	

$0.62 \times 8 =$	4.96	4	(4 is the high-order digit)
$0.96 \times 8 =$	7.68	7	
$0.68 \times 8 =$	5.44	5	

Therefore, $219.62_{10} = 333.475_8$.

Decimal to hexadecimal   The decimal number 195.828125_{10} will be converted to an equivalent hexadecimal number. The integer is divided by 16 repeatedly and the fraction is multiplied by 16 repeatedly.

$195 \div 16 =$	quotient $= 12$,	remainder $= 3$	(3 is the low-order digit)
$12 \div 16 =$	quotient $= 0$,	remainder $= 12$ (C)	

$0.828125 \times 16 =$	13.250000	13 (D)	(D is the high-order digit)
$0.250000 \times 16 =$	4.000000	4	

Therefore, $195.828125_{10} = C3.D4_{16}$.

Binary to octal   When converting a binary number to octal, the binary number is partitioned into groups of three bits as the number is scanned right to left for integers and scanned left to right for fractions. If the leftmost group of the integer does not contain three bits, then leading zeroes are added to produce a 3-bit octal digit; if the rightmost group of the fraction does not contain three bits, then trailing zeroes are added to produce a 3-bit octal digit. The binary number 10110100011.11101_2 will be converted to an octal number as shown below.

0 1 0	1 1 0	1 0 0	0 1 1	.	1 1 1	0 1 0
2	6	4	3	.	7	2

Binary to hexadecimal   When converting a binary number to hexadecimal, the binary number is partitioned into groups of four bits as the number is scanned right to left for integers and scanned left to right for fractions. If the leftmost group of the integer does not contain four bits, then leading zeroes are added to produce a 4-bit hexadecimal digit; if the rightmost group of the fraction does not contain four bits, then trailing zeroes are added to produce a 4-bit hexadecimal digit. The binary number 11010101000.1111010111_2 will be converted to a hexadecimal number as shown below.

0 1 1 0	1 0 1 0	1 0 0 0	.	1 1 1 1	0 1 0 1	1 1 0 0
6	A	8	.	F	5	C

Octal to binary   When converting an octal number to binary, three binary digits are entered that correspond to each octal digit, as shown below.

2	7	5	4	.	3	6
0 1 0	1 1 1	1 0 1	1 0 0	.	0 1 1	1 1 0

When converting from octal to binary-coded octal (BCO) and from octal to binary, the binary bit configurations are identical. This is because the octal number system uses all eight combinations of three bits. An example is shown below in which the octal number 217_8 is converted to binary-coded octal and to binary.

The binary bit configuration for BCO is

$$217_8 = 010\ 001\ 111_{BCO}$$

The bit configuration for binary is

$$217_8 = (2 \times 8^2) + (1 \times 8^1) + (7 \times 8^0) = 143_{10}$$
$$143_{10} = 010\ 001\ 111_2$$

Octal to hexadecimal   To convert from octal to hexadecimal, the octal number is first converted to BCO then partitioned into 4-bit segments to form binary-coded hexadecimal (BCH). The BCH notation is then easily changed to hexadecimal, as shown below.

7	6	3	5	.	4	6	
1 1 1	1 1 0	0 1 1	1 0 1	.	1 0 0	1 1 0	0 0
F	9	D	.	9	8		

Hexadecimal to binary   To covert from hexadecimal to binary, substitute the four binary bits for the hexadecimal digits according to Table 1.10 as shown below.

F	A	9	7	.	B	6
1 1 1 1	1 0 1 0	1 0 0 1	0 1 1 1	.	1 0 1 1	0 1 1 0

When converting from hexadecimal to BCH and from hexadecimal to binary, the binary bit configurations are identical. This is because the hexadecimal number system uses all sixteen combinations of four bits. An example is shown below in which the hexadecimal number $1E2_{16}$ is converted to BCH and to binary.

The binary bit configuration for BCH is

$$1E2_8 = 0001 \ 1110 \ 0010_{BCH}$$

The bit configuration for binary is

$$1E2_{16} = (1 \times 16^2) + (14 \times 16^1) + (2 \times 16^0) = 482_{10}$$
$$482_{10} = 0001 \ 1110 \ 0010_2$$

Hexadecimal to octal   When converting from hexadecimal to octal, the hexadecimal digits are first converted to binary. Then the binary bits are partitioned into 3-bit segments to obtain the octal digits, as shown below.

B	8	E	.	4	D	
1 0 1 1	1 0 0 0	1 1 1 0	.	0 1 0 0	1 1 0 1	0
5	6	1	6	.	2	3	2

Conversion from a nonconventional radix to radix 10   Equation 1.4 will be used to convert the following radix 5 number to an equivalent radix 10 number: 2134.43_5.

$$
\begin{aligned}
2134.43_5 &= (2 \times 5^3) + (1 \times 5^2) + (3 \times 5^1) + (4 \times 5^0) . \ (4 \times 5^{-1}) + (3 \times 5^{-2}) \\
&= \ 250 + \quad\ 25 + \quad 15 + \quad 4 + \quad\ 0.8 + \quad\ 0.12 \\
&= \ 294.92_{10}
\end{aligned}
$$

Convert from radix *ri* to any other radix *rj*   To convert any nondecimal number A_{ri} in radix *ri* to another nondecimal number A_{rj} in radix *rj*, first convert the number A_{ri} to decimal using Equation 1.4, then convert the decimal number to radix *rj* by using repeated division and/or repeated multiplication. The radix 9 number 125_9 will be converted to an equivalent radix 7 number. First 125_9 is converted to radix 10.

$$125_9 = (1 \times 9^2) + (2 \times 9^1) + (5 \times 9^0)$$
$$= 104_{10}$$

Then, convert 104_{10} to radix 7.

$104 \div 7 =$	quotient = 14,	remainder = 6	(6 is the low-order digit)
$14 \div 7 =$	quotient = 2,	remainder = 0	
$2 \div 7 =$	quotient = 0,	remainder = 2	

Verify the answer.

$$125_9 = 206_7 = (2 \times 7^2) + (0 \times 7^1) + (6 \times 7^0)$$
$$= 104_{10}$$

Determine the value of an unknown radix   The equation shown below has an unknown radix a. This example will determine the value of radix a.

$$44_a{}^{0.5} = 6_{10}$$
$$44_a = 36_{10}$$
$$(4 \times a^1) + (4 \times a^0) = (3 \times 10^1) + (6 \times 10^0)$$
$$4a + 4 = 30 + 6$$
$$4a = 32$$
$$a = 8$$

Verify the answer.

$$44_8 = (4 \times 8^1) + (4 \times 8^0)$$
$$= 36_{10}$$

1.2 Number Representations

The material presented thus far covered only positive numbers. However, computers use both positive and negative numbers. Since a computer cannot recognize a plus (+) or a minus (−) sign, an encoding method must be established to represent the sign of a number in which both positive and negative numbers are distributed as evenly as possible.

There must also be a method to differentiate between positive and negative numbers; that is, there must be an easy way to test the sign of a number. Detection of a number with a zero value must be straightforward. The leftmost (high-order) digit is usually reserved for the sign of the number. Consider the following number A with radix r:

$$A = (a_{n-1}\, a_{n-2}\, a_{n-3}\, \cdots\, a_2\, a_1\, a_0)_r$$

where digit a_{n-1} has the following value:

$$A = \begin{cases} 0 & \text{if } A \geq 0 \\ r-1 & \text{if } A < 0 \end{cases} \tag{1.6}$$

The remaining digits of A indicate either the true magnitude or the magnitude in a complemented form. There are three conventional ways to represent positive and negative numbers in a positional number system: sign magnitude, diminished-radix complement, and radix complement. In all three number representations, the high-order digit is the sign of the number, according to Equation 1.6.

$$0 = \text{ positive}$$
$$r-1 = \text{ negative}$$

1.2.1 Sign Magnitude

In this representation, an integer has the following decimal range:

$$-(r^{n-1} - 1) \text{ to } + (r^{n-1} - 1) \tag{1.7}$$

where the number zero is considered to be positive. Thus, a positive number A is represented as

$$A = (0\, a_{n-2} a_{n-3}\, \cdots\, a_1 a_0)_r \tag{1.8}$$

and a negative number with the same absolute value as

$$A' = [(r-1)\, a_{n-2} a_{n-3}\ \cdots\ a_1 a_0]_r \tag{1.9}$$

In sign-magnitude notation, the positive version $+A$ differs from the negative version $-A$ only in the sign digit position. The magnitude portion $a_{n-2} a_{n-3}\ \cdots\ a_1 a_0$ is identical for both positive and negative numbers of the same absolute value.

There are two problems with sign-magnitude representation. First, there are two representations for the number 0; specifically $+0$ and -0; ideally there should be a unique representation for the number 0. Second, when adding two numbers of opposite signs, the magnitudes of the numbers must be compared to determine the sign of the result. This is not necessary in the other two methods that are presented in subsequent sections. Sign-magnitude notation is used primarily for representing fractions in floating-point notation.

Examples of sign-magnitude notation are shown below using 8-bit binary numbers and decimal numbers that represent both positive and negative values. Notice that the magnitude parts are identical for both positive and negative numbers for the same radix.

Radix 2

```
0   0 0 0   0 1 0 0   +4
1   0 0 0   0 1 0 0   -4
```

Magnitude

Sign

```
0   0 0 0   1 1 0 1 . 1 0 1   +13.625
1   0 0 0   1 1 0 1 . 1 0 1   -13.625
```

```
0   1 0 1   0 1 1 0 . 0 1 1   +86.375
1   1 0 1   0 1 1 0 . 0 1 1   -86.375
```

Radix 10

```
0   7 4 3   +743
9   7 4 3   -743
```

where 0 represents a positive number in radix 10, and 9 ($r-1$) represents a negative number in radix 10. Again, the magnitudes of both numbers are identical.

0 6 7 8 4   +6784
9 6 7 8 4   −6784

1.2.2 Diminished-Radix Complement

This is the *(r − 1) complement* in which the radix is diminished by 1 and an integer has the following decimal range:

$$-(r^{n-1} - 1) \text{ to } +(r^{n-1} - 1) \tag{1.10}$$

which is the same as the range for sign-magnitude integers, although the numbers are represented differently, and where the number zero is considered to be positive. Thus, a positive number A is represented as

$$A = (0\ a_{n-2}a_{n-3}\ \cdots\ a_1 a_0)_r \tag{1.11}$$

and a negative number as

$$A' = [(r-1)\ a_{n-2}'a_{n-3}'\ \cdots\ a_1'a_0']_r \tag{1.12}$$

where

$$a_i' = (r-1) - a_i \tag{1.13}$$

In binary notation ($r = 2$), the diminished-radix complement ($r - 1 = 2 - 1 = 1$) is the 1s complement. Positive and negative integers have the ranges shown below and are represented as shown in Equation 1.11 and Equation 1.12, respectively.

Positive integers: 0 to $2^{n-1} - 1$
Negative integers: 0 to $-(2^{n-1} - 1)$

To obtain the 1s complement of a binary number, simply complement (invert) all the bits. Thus, $0011\ 1100_2$ ($+60_{10}$) becomes $1100\ 0011_2$ (-60_{10}). To obtain the value of a positive binary number, the 1s are evaluated according to their weights in the positional number system, as shown below.

$$2^7 \ 2^6 \ 2^5 \ 2^4 \ 2^3 \ 2^2 \ 2^1 \ 2^0$$
$$0 \ \ 0 \ \ 1 \ \ 1 \ \ 1 \ \ 1 \ \ 0 \ \ 0 \qquad +60_{10}$$

To obtain the value of a negative binary number, the 0s are evaluated according to their weights in the positional number system, as shown below.

$$2^7 \ 2^6 \ 2^5 \ 2^4 \ 2^3 \ 2^2 \ 2^1 \ 2^0$$
$$1 \ \ 1 \ \ 0 \ \ 0 \ \ 0 \ \ 0 \ \ 1 \ \ 1 \qquad -60_{10}$$

When performing arithmetic operations on two operands, comparing the signs is straightforward, because the leftmost bit is a 0 for positive numbers and a 1 for negative numbers. There is, however, a problem when using the diminished-radix complement. There is a dual representation of the number zero, because a word of all 0s (+0) becomes a word of all 1s (−0) when complemented. This does not allow the requirement of having a unique representation for the number zero to be attained. The examples shown below represent the diminished-radix complement for different radices.

Example 1.3   The binary number 1101_2 will be 1s complemented. The number has a decimal value of -2. To obtain the 1s complement, subtract each digit in turn from 1 (the highest number in the radix), as shown below (Refer to Equation 1.12 and Equation 1.13). Or in the case of binary, simply invert each bit. Therefore, the 1s complement of 1101_2 is 0010_2, which has a decimal value of $+2$.

To verify the operation, add the negative and positive numbers to obtain 1111_2, which is zero in 1s complement notation.

$$
\begin{array}{r}
1 \ \ 1 \ \ 0 \ \ 1 \\
+) \ \underline{0 \ \ 0 \ \ 1 \ \ 0} \\
1 \ \ 1 \ \ 1 \ \ 1
\end{array}
$$

Example 1.4   Obtain the diminished-radix complement (9s complement) of 08752.43_{10}, where 0 is the sign digit indicating a positive number. The 9s complement is obtained by using Equation 1.12 and Equation 1.13. When a number is complemented in any form, the number is negated. Therefore, the sign of the complemented radix 10 number is $(r - 1) = 9$. The remaining digits of the number are obtained by using Equation 1.13, such that each digit in the complemented number is obtained by subtracting the given digit from 9. Therefore, the 9s complement of 08752.43_{10} is

$$\frac{9-0}{9} \quad \frac{9-8}{1} \quad \frac{9-7}{2} \quad \frac{9-5}{4} \quad \frac{9-2}{7} \cdot \frac{9-4}{5} \quad \frac{9-3}{6}$$

where the sign digit is $(r - 1) = 9$. If the above answer is negated, then the original number will be obtained. Thus, the 9s complement of $91247.56_{10} = 08752.43_{10}$; that is, the 9s complement of -1247.56_{10} is $+8752.43_{10}$, as written in conventional sign magnitude notation for radix 10.

Example 1.5 The diminished-radix complement of the positive decimal number 06784_{10} will be 9s complemented. To obtain the 9s complement, subtract each digit in turn from 9 (the highest number in the radix), as shown below to obtain the negative number with the same absolute value. The sign of the positive number is 0 and the sign of the negative number is 9 (refer to Equation 1.11 and Equation 1.12).

$$\frac{9-0}{9} \quad \frac{9-6}{3} \quad \frac{9-7}{2} \quad \frac{9-8}{1} \quad \frac{9-4}{5}$$

To verify the operation, add the negative and positive numbers to obtain 99999_{10}, which is zero in 9s complement notation.

$$
\begin{array}{r}
0\ 6\ 7\ 8\ 4 \\
+)\ 9\ 3\ 2\ 1\ 5 \\
\hline
9\ 9\ 9\ 9\ 9
\end{array}
$$

Example 1.6 The diminished-radix complement of the positive radix 8 number 05734_8 will be 7s complemented. To obtain the 7s complement, subtract each digit in turn from 7 (the highest number in the radix), as shown below to obtain the negative number with the same absolute value. The sign of the positive number is 0 and the sign of the negative number is 7 (refer to Equation 1.11 and Equation 1.12).

$$\frac{7-0}{7} \quad \frac{7-5}{2} \quad \frac{7-7}{0} \quad \frac{7-3}{4} \quad \frac{7-4}{3}$$

To verify the operation, add the negative and positive numbers to obtain 77777_8, which is zero in 7s complement notation.

$$
\begin{array}{r}
0\ 5\ 7\ 3\ 4 \\
+)\ 7\ 2\ 0\ 4\ 3 \\
\hline
7\ 7\ 7\ 7\ 7
\end{array}
$$

Example 1.7   The diminished-radix complement of the positive radix 16 number $0A7C4_{16}$ will be 15s complemented. To obtain the 15s complement, subtract each digit in turn from 15 (the highest number in the radix), as shown below to obtain the negative number with the same absolute value. The sign of the positive number is 0 and the sign of the negative number is F (refer to Equation 1.11 and Equation 1.12).

$$\frac{F-0}{F} \quad \frac{F-A}{5} \quad \frac{F-7}{8} \quad \frac{F-C}{3} \quad \frac{F-4}{B}$$

To verify the operation, add the negative and positive numbers to obtain $FFFFF_{16}$, which is zero in 15s complement notation.

$$
\begin{array}{r}
0 \ A \ 7 \ C \ 4 \\
+) \ F \ 5 \ 8 \ 3 \ B \\
\hline
F \ F \ F \ F \ F
\end{array}
$$

1.2.3 Radix Complement

This is the *r complement*, where an integer has the following decimal range:

$$-(r^{n-1}) \text{ to } +(r^{n-1}-1) \tag{1.14}$$

where the number zero is positive. A positive number A is represented as

$$A = (0 \ a_{n-2}a_{n-3} \ \cdots \ a_1a_0)_r \tag{1.15}$$

and a negative number as

$$(A')_{+1} = \{[(r-1) \ a_{n-2}'a_{n-3}' \ \cdots \ a_1'a_0'] + 1\}_r \tag{1.16}$$

where A' is the diminished-radix complement. Thus, the radix complement is obtained by adding 1 to the diminished-radix complement; that is, $(r-1) + 1 = r$. Note that all three number representations have the same format for positive numbers and differ only in the way that negative numbers are represented, as shown in Table 1.14.

Table 1.14 Number Representations for Positive and Negative Integers of the Same Absolute Value for Radix r

Number Representation	Positive Numbers	Negative Numbers
Sign magnitude	$0\ a_{n-2}a_{n-3}\ \cdots\ a_1 a_0$	$(r-1)\ a_{n-2}a_{n-3}\ \cdots\ a_1 a_0$
Diminished-radix complement	$0\ a_{n-2}a_{n-3}\ \cdots\ a_1 a_0$	$(r-1)\ a_{n-2}'a_{n-3}'\ \cdots\ a_1'a_0'$
Radix complement	$0\ a_{n-2}a_{n-3}\ \cdots\ a_1 a_0$	$(r-1)\ a_{n-2}'a_{n-3}'\ \cdots\ a_1'a_0' + 1$

Another way to define the radix complement of a number is shown in Equation 1.17, where n is the number of digits in A.

$$(A')_{+1} = r^n - A_r \tag{1.17}$$

For example, assume that $A = 0101\ 0100_2\ (+84_{10})$. Then, using Equation 1.17,

$$2^8 = 256_{10} = 10000\ 0000_2. \text{ Thus, } 256_{10} - 84_{10} = 172_{10}$$

$$(A')_{+1} = 2^8 - (0101\ 0100)$$

```
    1 0 0 0 0 0 0 0 0
-)    0 1 0 1 0 1 0 0
    ─────────────────
    1 0 1 0 1 1 0 0
```

As can be seen from the above example, to generate the radix complement for a radix 2 number, keep the low-order 0s and the first 1 unchanged and complement (invert) the remaining high-order bits. To obtain the value of a negative number in radix 2, the 0s are evaluated according to their weights in the positional number system, then add 1 to the value obtained.

$$
\begin{array}{cccc cccc}
2^7 & 2^6 & 2^5 & 2^4 & 2^3 & 2^2 & 2^1 & 2^0 \\
1 & 0 & 1 & 0 & 1 & 1 & 0 & 0 \\
\hline
\end{array}
$$

$$80 \qquad\qquad 3 + 1 = 4 \qquad -84_{10}$$

Table 1.15 and Table 1.16 show examples of the three number representations for positive and negative numbers in radix 2. Note that the positive numbers are identical for all three number representations; only the negative numbers change.

Table 1.15 Number Representations for Positive and Negative Integers in Radix 2

Number Representation	$+127_{10}$	-127_{10}
Sign magnitude	0 111 1111	1 111 1111
Diminished-radix complement (1s)	0 111 1111	1 000 0000
Radix complement (2s)	0 111 1111	1 000 0001

Table 1.16 Number Representations for Positive and Negative Integers in Radix 2

Number Representation	$+54_{10}$	-54_{10}
Sign magnitude	0 011 0110	1 011 0110
Diminished-radix complement (1s)	0 011 0110	1 100 1001
Radix complement (2s)	0 011 0110	1 100 1010

There is a unique zero for binary numbers in radix complement, as shown below. When the number zero is 2s complemented, the bit configuration does not change. The 2s complement is formed by adding 1 to the 1s complement.

$$
\begin{array}{rcccccccc}
\text{Zero in 2s complement} = & 0 & 0 & 0 & 0 & 0 & 0 & 0 & 0 \\
\text{Form the 1s complement} = & 1 & 1 & 1 & 1 & 1 & 1 & 1 & 1 \\
\text{Add 1} = & & & & & & & & 1 \\
\hline
& 0 & 0 & 0 & 0 & 0 & 0 & 0 & 0
\end{array}
$$

Example 1.8   Convert -20_{10} to binary and obtain the 2s complement. Then obtain the 2s complement of $+20_{10}$ using the fast method of keeping the low-order 0s and the first 1 unchanged as the number is scanned from right to left, then inverting all remaining bits.

$$
\underbrace{1110\ 1100}_{-20} \xrightarrow{\ 2s\ } \underbrace{0001\ 0100}_{+20} \xrightarrow{\ 2s\ } \underbrace{1110\ 1100}_{-20}
$$

Example 1.9   Obtain the radix complement (10s complement) of the positive number 08752.43_{10}. Determine the 9s complement as in Example 1.5, then add 1. The 10s complement of 08752.43_{10} is the negative number 91247.57_{10}.

$$\frac{9-0}{9} \quad \frac{9-8}{1} \quad \frac{9-7}{2} \quad \frac{9-5}{4} \quad \frac{9-2}{7} . \frac{9-4}{5} \quad \frac{9-3}{6}$$

$$+) \qquad\qquad\qquad\qquad\qquad\qquad\qquad 1$$

$$\overline{\quad 9 \qquad 1 \qquad 2 \qquad 4 \qquad 7 \quad . \quad 5 \qquad 7 \quad}$$

Adding 1 to the 9s complement in this example is the same as adding 10^{-2} ($.01_{10}$). To verify that the radix complement of 08752.43_{10} is 91247.57_{10}, the sum of the two numbers should equal zero for radix 10. This is indeed the case, as shown below.

$$
\begin{array}{r}
08752.43 \\
+) \quad 91247.57 \\
\hline
00000.00
\end{array}
$$

Example 1.10   Obtain the 10s complement of 0.4572_{10} by adding a power of ten to the 9s complement of the number. The 9s complement of 0.4572_{10} is

$$9-0=9 \quad 9-4=5 \quad 9-5=4 \quad 9-7=2 \quad 9-2=7$$

Therefore, the 10s complement of $0.4572_{10} = 9.5427_{10} + 10^{-4} = 9.5428_{10}$. Adding the positive and negative numbers again produces a zero result.

Example 1.11   Obtain the radix complement of $1111\ 1111_2$ (-1_{10}). The answer can be obtained using two methods: add 1 to the 1s complement or keep the low-order 0s and the first 1 unchanged, then invert all remaining bits. Since there are no low-order 0s, only the rightmost 1 is unchanged. Therefore, the 2s complement of $1111\ 1111_2$ is $0000\ 0001_2$ ($+1_{10}$).

Example 1.12   Obtain the 8s complement of 04360_8. First form the 7s complement, then add 1. The 7s complement of 04360_8 is 73417_8. Therefore, the 8s complement of 04360_8 is $73417_8 + 1$, as shown below, using the rules for octal addition. Adding the positive and negative numbers results in a sum of zero.

$$
\begin{array}{r}
7\ 3\ 4\ 1\ 7 \\
+) \qquad\quad 1 \\
\hline
7\ 3\ 4\ 2\ 0
\end{array}
$$

Example 1.13 Obtain the 16s complement of $F8A5_{16}$. First form the 15s complement, then add 1. The 15s complement of $F8A5_{16}$ is $075A_{16}$. Therefore, the 16s complement of $F8A5_{16} = 075A_{16} + 1 = 075B_{16}$. Adding the positive and negative numbers results in a sum of zero.

Example 1.14 Obtain the 4s complement of 0231_4. The rules for obtaining the radix complement are the same for any radix: generate the diminished-radix complement, then add 1. Therefore, the 4s complement of $0231_4 = 3102_4 + 1 = 3103_4$. To verify the result, add $0231_4 + 3103_4 = 0000_4$, as shown below using the rules for radix 4 addition.

$$
\begin{array}{r}
0\ \ 2\ \ 3\ \ 1 \\
+)\ \ 3\ \ 1\ \ 0\ \ 3 \\
\hline
0\ \ 0\ \ 0\ \ 0
\end{array}
$$

1.2.4 Arithmetic Operations

This section will concentrate on fixed-point binary and binary-coded decimal operations, since these are the dominant number representations in computers — floating-point operations will be discussed in the chapter on computer arithmetic. Examples of addition, subtraction, multiplication, and division will be presented for fixed-point binary using radix complementation and for binary-coded decimal number representations.

Binary Addition Numbers in radix complement representation are designated as signed numbers, specifically as 2s complement numbers in binary. The sign of a binary number can be extended to the left indefinitely without changing the value of the number. For example, the numbers 0000000001010_2 and 00001010_2 both represent a value of $+10_{10}$; the numbers 1111111110110_2 and 11110110_2 both represent a value of -10_{10}.

Thus, when an operand must have its sign extended to the left, the expansion is achieved by setting the extended bits equal to the leftmost (sign) bit. The maximum positive number consists of a 0 followed by a field of all 1s, dependent on the word size of the operand. Similarly, the maximum negative number consists of a 1 followed by a field of all 0s, dependent on the word size of the operand.

The radix (or binary) point can be in any fixed position in the number — thus the radix point is referred to as *fixed-point*. For integers, however, the radix point is positioned to the immediate right of the low-order bit position. There are normally two operands for addition, as shown below.

$$
\begin{array}{rl}
A = & \text{Augend} \\
+)\ \ B = & \underline{\text{Addend}} \\
& \text{Sum}
\end{array}
$$

The rules for binary addition are as follows:

+	0	1
0	0	1
1	1	0 *

* $1 + 1 = 0$ with a carry to the next column on the left. The sum of $1_2 + 1_2$ requires a 2-bit vector 10_2 to represent the value of 2_{10}.

Example 1.15 Add $+51_{10}$ and $+32_{10}$ to yield $+83_{10}$. The sum is also referred to as *true addition*, because the result is the sum of the two numbers, ignoring the sign of the numbers. The carry-out is 0, which can be ignored, since it is the sign extension of the sum.

$$
\begin{array}{rllllcllll l}
A = & 0 & 0 & 1 & 1 & & 0 & 0 & 1 & 1 & (+51) \\
+)\quad B = & 0 & 0 & 1 & 0 & & 0 & 0 & 0 & 0 & (+32) \\
\hline
0 \leftarrow & 0 & 1 & 0 & 1 & & 0 & 1 & 1 & & (+83)
\end{array}
$$

Example 1.16 Add -51_{10} and -32_{10} to yield -83_{10}. The sum is also referred to as *true addition*, because the result is the sum of the two numbers, ignoring the sign of the numbers. The carry-out is 1, which can be ignored, since it is the sign extension of the sum.

$$
\begin{array}{rllllcllll l}
A = & 1 & 1 & 0 & 0 & & 1 & 1 & 0 & 1 & (-51) \\
+)\quad B = & 1 & 1 & 1 & 0 & & 0 & 0 & 0 & 0 & (-32) \\
\hline
1 \leftarrow & 1 & 0 & 1 & 0 & & 1 & 0 & 1 & & (-83)
\end{array}
$$

Example 1.17 Add $+51_{10}$ and -32_{10} to yield $+19_{10}$. The sum is also referred to as *true subtraction*, because the result is the difference of the two numbers, ignoring the sign of the numbers. The carry-out is 1, which can be ignored, since it does not constitute an overflow when adding two numbers of opposite signs.

$$
\begin{array}{rllllcllll l}
A = & 0 & 0 & 1 & 1 & & 0 & 0 & 1 & 1 & (+51) \\
+)\quad B = & 1 & 1 & 1 & 0 & & 0 & 0 & 0 & 0 & (-32) \\
\hline
1 \leftarrow & 0 & 0 & 0 & 1 & & 0 & 1 & 1 & & (+19)
\end{array}
$$

Example 1.18   Add -51_{10} and $+32_{10}$ to yield -19_{10}. The sum is also referred to as *true subtraction*, because the result is the difference of the two numbers, ignoring the sign of the numbers. The carry-out is 0, which can be ignored, since it does not constitute an overflow when adding two numbers of opposite signs.

$$
\begin{array}{rcccccccccl}
& A = & 1 & 1 & 0 & 0 & 1 & 1 & 0 & 1 & (-51) \\
+) & B = & 0 & 0 & 1 & 0 & 0 & 0 & 0 & 0 & (+32) \\
\hline
& 0 \leftarrow & 1 & 1 & 1 & 0 & & 1 & 0 & 1 & (-19)
\end{array}
$$

Overflow   Overflow occurs when the result of an arithmetic operation (usually addition) exceeds the word size of the machine; that is, the sum is not within the representable range of numbers provided by the number representation. For numbers in 2s complement representation, the range is from -2^{n-1} to $+2^{n-1} - 1$. For two n-bit numbers

$$A = a_{n-1} a_{n-2} a_{n-3} \dots a_1 a_0$$

$$B = b_{n-1} b_{n-2} b_{n-3} \dots b_1 b_0$$

a_{n-1} and b_{n-1} are the sign bits of operands A and B, respectively. Overflow can be detected by either of the following two equations:

$$\text{Overflow} = (a_{n-1} \bullet b_{n-1} \bullet s_{n-1}') + (a_{n-1}' \bullet b_{n-1}' \bullet s_{n-1})$$

$$\text{Overflow} = c_{n-1} \oplus c_{n-2} \tag{1.18}$$

where the symbol "$\bullet$" is the logical AND operator, the symbol "$+$" is the logical OR operator, the symbol "$\oplus$" is the exclusive-OR operator as defined below, and c_{n-1} and c_{n-2} are the carry bits out of positions $n-1$ and $n-2$, respectively.

$\oplus$	0	1
0	0	1
1	1	0

Therefore, overflow $= c_{n-1} \oplus c_{n-2}$

$$= (c_{n-1} \bullet c_{n-2}') + (c_{n-1}' \bullet c_{n-2})$$

Thus, overflow produces an erroneous sign reversal and is possible only when both operands have the same sign. An overflow cannot occur when adding two operands of different signs, since adding a positive number to a negative number produces a result that falls within the limit specified by the two numbers.

Example 1.19 Given the following two positive numbers in radix complementation for radix 2, perform an add operation:

$$A = 0100\ 0100\ (+68)$$
$$B = 0101\ 0110\ (+86)$$

$$
\begin{array}{rllll}
A = & 0100 & 0100 & (+68) \\
+)\quad B = & 0101 & 0110 & (+86) \\
\hline
& 1001 & 1010 & (+154) & \text{Overflow. } +154 \text{ takes 9 bits}
\end{array}
$$

If 9 bits are used for the result, then the answer is $+154_{10}$. However, the value of $+154$ cannot be contained in only 8 bits; therefore, an overflow has occurred.

Example 1.20 Given the following two negative numbers in radix complementation for radix 2, perform an add operation:

$$A = 1011\ 1100\ (-68)$$
$$B = 1010\ 1010\ (-86)$$

$$
\begin{array}{rllll}
A = & 1011 & 1100 & (-68) \\
+)\quad B = & 1010 & 1010 & (-86) \\
\hline
& 0110 & 0110 & (-154) & \text{Overflow. } -154 \text{ takes 9 bits}
\end{array}
$$

If all 9 bits are used for the result, then the answer is -154_{10}. However, the value of -154_{10} cannot be contained in only 8 bits; therefore, an overflow has occurred. The range for binary numbers in 2s complement representation is

$$-2^{n-1} \text{ to } +2^{n-1} - 1$$

For $n = 8$: -128 to $+127$

In binary: 1000 0000 to 0111 1111

In the above two examples, Equation 1.18 holds true, indicating that an overflow has occurred. That is, the signs of the two operands are the same, but the sign of the result is different. Also, the carry out of the second high-order bit position and the carry out of the high-order bit position are different.

Binary subtraction   The operands used for subtraction are the minuend and subtrahend, as shown below. The rules for binary subtraction presented in Section 1.1.5, sometimes referred to as the *paper-and-pencil* method, are not easily applicable to computer subtraction. The method used by most processors is to add the 2s complement of the subtrahend to the minuend; that is, change the sign of the minuend and add the resulting two operands. This can be described formally as shown in Equation 1.19.

$$
\begin{array}{rl}
A = & \text{Minuend} \\
-)\ B = & \underline{\text{Subtrahend}} \\
& \text{Difference}
\end{array}
$$

$$
\begin{aligned}
\text{Difference} &= A - B \\
&= A - r^n + r^n - B \\
&= A - r^n + (r^n - B) \\
&= A - r^n + (B' + 1)
\end{aligned}
\tag{1.19}
$$

where B' is the diminished-radix complement (1s complement) of the subtrahend B. An example using Equation 1.19 is shown below.

$$
\begin{array}{rl}
A = & 0111 \\
-)\quad B = & \underline{0011}
\end{array}
\qquad\longrightarrow\qquad
\begin{array}{rl}
A = & 0111 \\
+)\quad B = & \underline{1101} \\
& 0100
\end{array}
$$

Using Equation 1.19 where $n = 4$ and $r = 2$, the following result is observed:

$$
\begin{aligned}
\text{Difference} &= A - r^n + (B' + 1) \\
&= 7 - 2^4 + (12 + 1) \\
&= 7 - 16 + 13 \\
&= 4
\end{aligned}
$$

Example 1.21   Perform the operation $A - B$ for the following two positive operands:

$$A = 0001\ 0000\ (+16)$$
$$B = 0000\ 1010\ (+10)$$

$A - B = A + (B' + 1)$

$$
\begin{array}{rll}
A = & 0001\ 0000 & (+16) \\
+)\ \ B' + 1 = & \underline{1111\ 0110} & (-10)\ \ \text{2s complement of } B \\
A - B = & 0000\ 0110 & (+6)
\end{array}
$$

Example 1.22   Perform the operation $B - A$ for the following two positive operands:

$$A = 0001\ 0000\ (+16)$$
$$B = 0000\ 1010\ (+10)$$

$B - A = B + (A' + 1)$

$$
\begin{array}{rll}
B = & 0000\ 1010 & (+10) \\
+)\ \ A' + 1 = & \underline{1111\ 0000} & (-16)\ \ \text{2s complement of } A \\
B - A = & 1111\ 1010 & (-6)
\end{array}
$$

Example 1.23   Perform the operation $A - B$ for the following two negative operands:

$$A = 1111\ 0000\ (-16)$$
$$B = 1111\ 0110\ (-10)$$

$A - B = A + (B' + 1)$

$$
\begin{array}{rll}
A = & 1111\ 0000 & (-16) \\
+)\ \ B' + 1 = & \underline{0000\ 1010} & (+10)\ \ \text{2s complement of } B \\
A - B = & 1111\ 1010 & (-6)
\end{array}
$$

Example 1.24   Perform the operation $B - A$ for the following two negative operands:

$$A = 1111\ 0000\ (-16)$$
$$B = 1111\ 0110\ (-10)$$

$B - A = B + (A' + 1)$

$$
\begin{array}{rll}
B = & 1111\ 0110 & (-10) \\
+)\quad A' + 1 = & 0001\ 0000 & (+16) \quad \text{2s complement of } A \\
\hline
B - A = & 0000\ 0110 & (+6)
\end{array}
$$

Example 1.25   As stated previously, the diminished-radix complement is rarely used in arithmetic applications because of the dual interpretation of the number zero. This example illustrates another disadvantage of the diminished-radix complement. Given the two radix 2 numbers shown below in 1s complement representation, obtain the difference.

$$
\begin{array}{l}
A = 1111\ 1001\ (-6) \\
B = 1110\ 1101\ (-18) \\
A - B = A + B'\ (\text{1s complement of } B)
\end{array}
$$

$$
\begin{array}{rll}
A = & 1111\ 1001 & (-6) \\
+)\quad B' = & 0001\ 0010 & (+18) \quad \text{1s complement of } B \\
1 \leftarrow & 0000\ 1011 & (+11) \quad \text{Incorrect result} \\
\longrightarrow\ 1 & & \text{End-around carry} \\
\hline
A - B = & 0000\ 1100 & (+12)
\end{array}
$$

When performing a subtract operation using 1s complement operands, an end-around carry will result if at least one operand is negative. As can be seen above, the result will be incorrect $(+11)$ if the carry is not added to the intermediate result. Although 1s complementation may seem easier than 2s complementation, the result that is obtained after an add operation is not always correct. In 1s complement notation, the final carry-out c_{n-1} cannot be ignored. If the carry-out is zero, then the result is correct. Thus, 1s complement subtraction may result in an extra add cycle to obtain the correct result.

Binary multiplication   The multiplication of two n-bit operands results in a product of $2n$ bits, as shown below.

$$
\begin{array}{l}
A = a_{n-1}\, a_{n-2}\, a_{n-3} \cdots a_1\, a_0 \\
B = b_{n-1}\, b_{n-2}\, b_{n-3} \cdots b_1\, b_0
\end{array}
$$

$$
\text{Product of } A \times B = p_{2n-1}\, p_{2n-2}\, p_{2n-3} \cdots p_1\, p_0
$$

$$
\begin{array}{rl}
A = & \text{Multiplicand } (n \text{ bits}) \\
\times)\quad B = & \text{Multiplier } (n \text{ bits}) \\
\hline
P = & \text{Product } (2n \text{ bits})
\end{array}
$$

Multiplication of two fixed-point binary numbers in 2s complement representation is done by a process of successive add and shift operations. The process consists of multiplying the multiplicand by each multiplier bit as the multiplier is scanned right to left. If the multiplier bit is a 1, then the multiplicand is copied as a partial product; otherwise, 0s are inserted as a partial product. The partial products inserted into successive lines are shifted left one bit position from the previous partial product. The partial products are then added to form the product.

The sign of the product is determined by the signs of the operands: If the signs are the same, then the sign of the product is plus; if the signs are different, then the sign of the product is minus. In this sequential add-shift technique, however, the multiplier must be positive. When the multiplier is positive, the bits are treated the same as in the sign-magnitude representation. When the multiplier is negative, the low-order 0s and the first 1 are treated the same as a positive multiplier, but the remaining high-order sign bits are treated as 1s and not the sign bits. Therefore, the algorithm treats the multiplier as unsigned, or positive.

The problem is easily solved by forming the 2s complement of both the multiplier and the multiplicand. An alternative approach is to 2s complement the negative multiplier leaving the multiplicand unchanged. Depending on the signs of the initial operands, it may be necessary to complement the product.

Example 1.26 Multiply the positive operands of 0111_2 ($+7_{10}$) and 0101_2 ($+5_{10}$). Since both operands are positive, the product will be positive ($+35_{10}$).

$$
\begin{array}{cccc|cccc l}
 & & & & 0 & 1 & 1 & 1 & (+7) \\
 & & & \times) & 0 & 1 & 0 & 1 & (+5) \\
\hline
0 & 0 & 0 & 0 & 0 & 1 & 1 & 1 \\
0 & 0 & 0 & 0 & 0 & 0 & 0 \\
0 & 0 & 0 & 1 & 1 & 1 \\
0 & 0 & 0 & 0 & 0 \\
\hline
 & 0 & 1 & 0 & 0 & 0 & 1 & 1 & (+35)
\end{array}
$$

Example 1.27 Multiply a negative multiplicand 1100_2 (-4) by a positive multiplier 0011_2 ($+3$) to obtain a product of $1111\ 0100_2$ (-12). Since the multiplier is positive, the sign of the product will be correct.

$$
\begin{array}{cccc|cccc l}
 & & & & 1 & 1 & 0 & 0 & (-4) \\
 & & & \times) & 0 & 0 & 1 & 1 & (+3) \\
\hline
1 & 1 & 1 & 1 & 1 & 1 & 0 & 0 \\
1 & 1 & 1 & 1 & 1 & 0 & 0 \\
0 & 0 & 0 & 0 & 0 & 0 \\
0 & 0 & 0 & 0 & 0 \\
\hline
 & 1 & 1 & 1 & 0 & 1 & 0 & 0 & (-12)
\end{array}
$$

Example 1.28   In this example, a positive multiplicand 0100_2 ($+4_{10}$) will be multiplied by a negative multiplier 1101_2 (-3_{10}). The product should be -12_{10}; however, the negative multiplier is treated as an unsigned (or positive) integer of $+13_{10}$ by the algorithm. Therefore, both the multiplicand and the multiplier will be 2s complemented (negated) to provide a positive multiplier. This will result in a correctly signed product of -12_{10}.

				0	1	0	0	(+4)
			×)	1	1	0	1	(−3)
0	0	0	0	0	1	0	0	
0	0	0	0	0	0	0		
0	0	0	1	0	0			
0	0	1	0	0				
0	1	1	0	1	0	0	(+52)	

Both the multiplicand and the multiplier will now be 2s complemented so that the multiplier will be positive. Since the multiplicand can be either positive or negative, the product will have the correct sign.

				1	1	0	0	(−4)
			×)	0	0	1	1	(+3)
1	1	1	1	1	1	0	0	
1	1	1	1	1	0	0		
0	0	0	0	0	0			
0	0	0	0	0				
1	1	1	0	1	0	0	(−12)	

Binary division   Division has two operands that produce two results, as shown below. Unlike multiplication, division is not commutative; that is, $A \div B \neq B \div A$, except when $A = B$.

$$\frac{\text{Dividend}}{\text{Divisor}} = \text{Quotient} + \text{Remainder}$$

All operands for a division operation comply with the following equation:

$$\text{Dividend} = (\text{Quotient} \times \text{Divisor}) + \text{Remainder} \qquad (1.20)$$

The remainder has a smaller value than the divisor and has the same sign as the dividend. If the divisor B has n bits, then the dividend A has $2n$ bits and the quotient Q and remainder R both have n bits, as shown below.

$$A = a_{2n-1}\, a_{2n-2} \cdots a_n\, a_{n-1} \cdots a_1\, a_0$$

$$B = b_{n-1}\, b_{n-2} \cdots b_1\, b_0$$

$$Q = q_{n-1}\, q_{n-2} \cdots q_1\, q_0$$

$$R = r_{n-1}\, r_{n-2} \cdots r_1\, r_0$$

The sign of the quotient is q_{n-1} and is determined by the rules of algebra; that is,

$$q_{n-1} = a_{2n-1} \oplus b_{n-1}$$

Multiplication is a shift-add multiplicand operation, whereas division is a shift-subtract divisor operation. The result of a shift-subtract operation determines the next operation in the division sequence. If the partial remainder is negative, then the carry-out is 0 and becomes the low-order quotient bit q_0. The partial remainder thus obtained is restored to the value of the previous partial remainder. If the partial remainder is positive, then the carry-out is 1 and becomes the low-order quotient bit q_0. The partial remainder is not restored. This technique is referred to as *restoring division*.

Example 1.29   An example of restoring division using a hardware algorithm is shown in Figure 1.20, where the dividend is in register-pair $A\,Q$ is $0000\ 0111_2$ ($+7_{10}$) and the divisor is in register B is 0011_2 ($+3_{10}$). The algorithm is implemented with a subtractor and a $2n$-bit shift register.

The first operation in Figure 1.20 is to shift the dividend left 1 bit position, then subtract the divisor, which is accomplished by adding the 2s complement of the divisor. Since the result of the subtraction is negative, the dividend is restored by adding back the divisor, and the low-order quotient bit q_0 is set to 0. This sequence repeats for four cycles — the number of bits in the divisor. If the result of the subtraction was positive, then the partial remainder is placed in register A, the high-order half of the dividend, and q_0 is set to 1.

The division algorithm is slightly more complicated when one or both of the operands are negative. The operands can be preprocessed and/or the results can be post-processed to achieved the desired results. The negative operands are converted to positive numbers by 2s complementation before the division process begins. The resulting quotient is then 2s complemented, if necessary.

Unlike multiplication, overflow can occur in division. This happens when the high-order half of the dividend is greater than or equal to the divisor. Also, division by zero must be avoided. Both of these problems can be detected before the division process begins by subtracting the divisor from the high-order half of the dividend. If the difference is positive, then an overflow or a divide by zero has been detected.

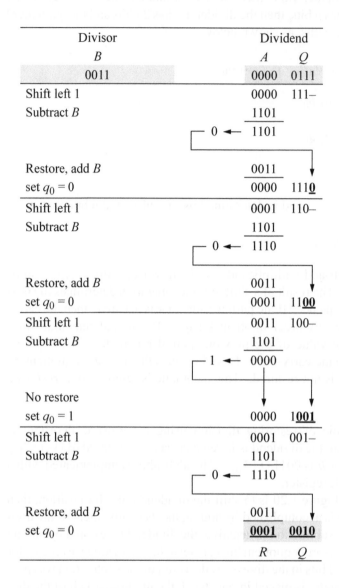

Figure 1.20   Example of binary division using two positive operands.

Figure 1.21 illustrates another example of binary division, this time using a positive dividend $0000\ 0111_2\ (+7_{10})$ and a negative divisor $1101_2\ (-3_{10})$. This will result in a quotient of $1110_2\ (-2_{10})$ and a remainder of $0001_2\ (+1_{10})$. The results can be verified using Equation 1.20. Because the divisor is negative, it is added to the partial remainder, not subtracted from the partial remainder, as was the case in the division example of Figure 1.20.

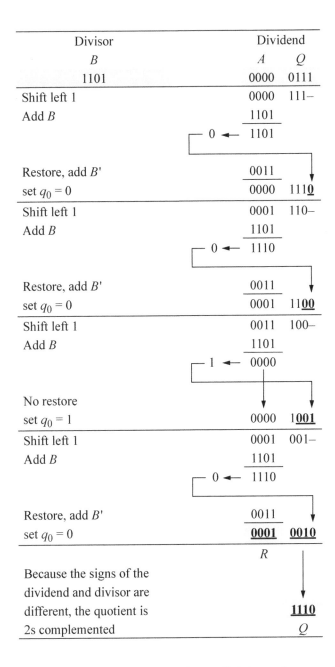

Divisor	Dividend	
B	A	Q
1101	0000	0111

Shift left 1 ... 0000 ... 111–
Add B ... 1101
0 ← 1101

Restore, add B' ... 0011
set $q_0 = 0$ ... 0000 ... 111**0**

Shift left 1 ... 0001 ... 110–
Add B ... 1101
0 ← 1110

Restore, add B' ... 0011
set $q_0 = 0$ ... 0001 ... 11**00**

Shift left 1 ... 0011 ... 100–
Add B ... 1101
1 ← 0000

No restore
set $q_0 = 1$ ... 0000 ... 1**001**

Shift left 1 ... 0001 ... 001–
Add B ... 1101
0 ← 1110

Restore, add B' ... 0011
set $q_0 = 0$ ... **0001**   **0010**

R

Because the signs of the
dividend and divisor are
different, the quotient is
2s complemented ... **1110**
Q

Figure 1.21   Example of binary division using a positive dividend and a negative divisor.

Use Equation 1.20 to verify the results.

$$\text{Dividend} = (\text{Quotient} \times \text{Divisor}) + \text{Remainder}$$
$$7 = (-2 \times -3) + 1$$
$$7 = 6 + 1$$

Binary-coded decimal addition   When two binary-coded decimal (BCD) digits are added, the range is 0 to 18. If the carry-in $c_{in} = 1$, then the range is 0 to 19. If the sum digit is ≥ 10 (1010_2), then it must be adjusted by adding 6 (0110_2). This excess-6 technique generates the correct BCD sum and a carry to the next higher-order digit, as shown below in Figure 1.22 (a). A carry-out of a BCD sum will also cause an adjustment to be made to the sum — called the intermediate sum — even though the intermediate sum is a valid BCD digit, as shown in Figure 1.22 (b).

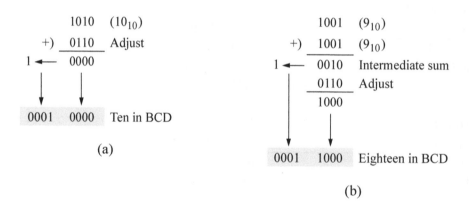

(a)

(b)

Figure 1.22   Example showing adjustment of a BCD sum.

There are three conditions that indicate when the sum of a BCD addition (the intermediate sum) should be adjusted by adding six.

1. Whenever bit positions 8 and 2 are both 1s.

2. Whenever bit positions 8 and 4 are both 1s.

3. Whenever the unadjusted sum produces a carry-out.

The examples of Figure 1.23, Figure 1.24, and Figure 1.25 illustrate these three conditions.

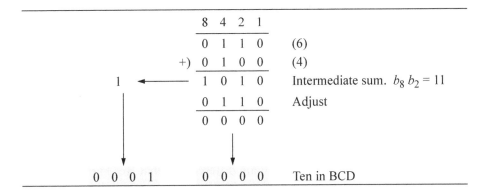

Figure 1.23 Example showing adjustment for BCD addition.

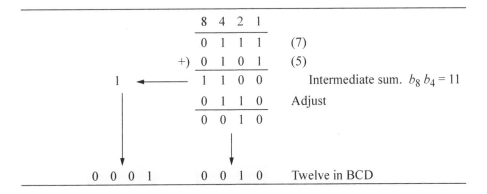

Figure 1.24 Example showing adjustment for BCD addition.

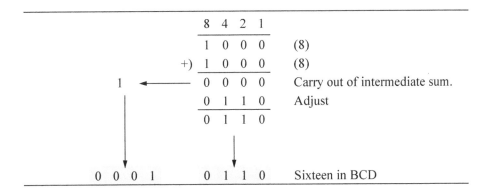

Figure 1.25 Example showing adjustment for BCD addition.

Therefore, all three conditions produce a carry-out to the next higher-order stage; that is,

$$\text{Carry} = c_8 + b_8 b_4 + b_8 b_2 \tag{1.21}$$

where c_8 is the carry-out of the high-order bit; $b_8 b_4$ and $b_8 b_2$ are bits of the unadjusted sum.

The algorithms used for BCD arithmetic are basically the same as those used for fixed-point arithmetic for radix 2. The main difference is that BCD arithmetic treats each digit as four bits, whereas fixed-point arithmetic treats each digit as a bit. Shifting operations are also different — decimal shifting is performed on 4-bit increments.

Binary-coded decimal numbers may also have a sign associated with them. This usually occurs in the *packed* BCD format, as shown below, where the sign digits are the low-order bytes 1100 (+) and 1101 (−). Comparing signs is straightforward. This sign notation is similar to the sign notation used in binary. Note that the low-order bit of the sign digit is 0 (+) or 1 (−).

+53	0 0 0 0	0 1 0 1	0 0 1 1	1 1 0 0	Sign is +

−25	0 0 0 0	0 0 1 0	0 1 0 1	1 1 0 1	Sign is −

Example 1.30 The following decimal numbers will be added using BCD arithmetic 26 and 18, as shown in Figure 1.26.

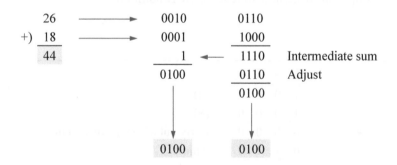

Figure 1.26   Example of BCD addition.

Example 1.31 The following decimal numbers will be added using BCD arithmetic 436 and 825, as shown in Figure 1.27.

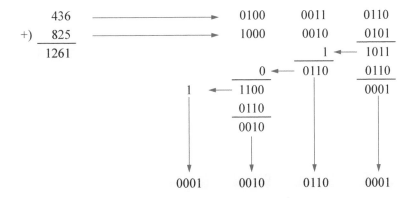

Figure 1.27   Example of BCD addition.

Binary-coded decimal subtraction   Subtraction in BCD is essentially the same as in fixed-point binary: Add the rs complement of the subtrahend to the minuend, which for BCD is the 10s complement. The examples which follow show subtract operations using BCD numbers. Negative results can remain in 10s complement notation or be recomplemented to sign-magnitude notation with a negative sign.

Example 1.32   The following decimal numbers will be subtracted using BCD arithmetic: +30 and +20, as shown in Figure 1.28. This can be considered as true subtraction, because the result is the difference of the two numbers, ignoring the signs. A carry-out of 1 from the high-order decade indicates a positive number. A carry-out of 0 from the high-order decade indicates a negative number in 10s complement notation. The number can be changed to an absolute value by recomplementing the number and changing the sign to indicate a negative value.

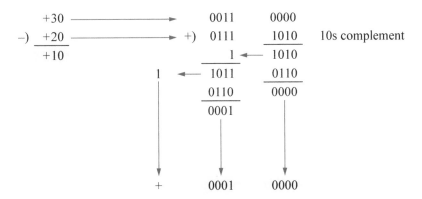

Figure 1.28   Example of BCD subtraction.

Example 1.33   The following decimal numbers will be added using BCD arithmetic: +28 and −20, as shown in Figure 1.29. This can be considered as true subtraction, because the result is the difference of the two numbers, ignoring the signs.

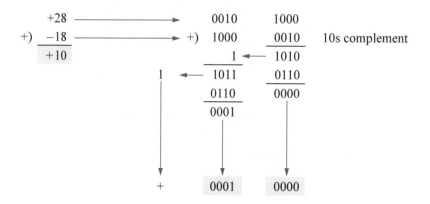

Figure 1.29   Example of BCD subtraction.

Example 1.34   The following decimal numbers will be subtracted using BCD arithmetic: +482 and +627, resulting in a difference of −145, as shown in Figure 1.30.

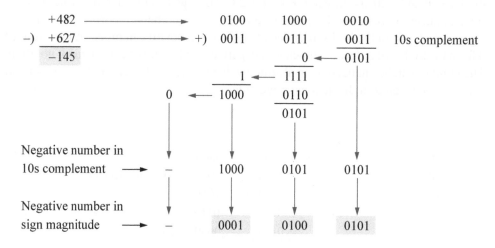

Figure 1.30   Example of BCD subtraction.

Binary-coded decimal multiplication   The multiplication algorithms for decimal multiplication are similar to those for fixed-point binary multiplication except in the way that the partial products are formed. In binary multiplication, the multiplicand is added to the previous partial product if the multiplier bit is a 1. In BCD multiplication, the multiplicand is multiplied by each digit of the multiplier and these subpartial products are aligned and added to form a partial product. When adding digits to obtain a partial product, adjustment may be required to form a valid BCD digit. Two examples of BCD multiplication are shown below.

Example 1.35   The following decimal numbers will be multiplied using BCD arithmetic: $+67$ and $+9$, resulting in a product of $+603$, as shown in Figure 1.31.

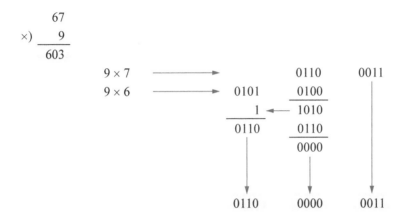

Figure 1.31   Example of BCD multiplication.

Example 1.36   The following decimal numbers will be multiplied using BCD arithmetic: $+6875$ and $+46$, resulting in a product of $+316250$, as shown Figure 1.32 using decimal numbers and in Figure 1.33 using BCD notation.

			6	8	7	5
	$\times)$				4	6
Partial product 1 (pp1)		4	1	2	5	0
Partial product 1 (pp2)	2	7	5	0	0	
Product	3	1	6	2	5	0

Figure 1.32   Example of radix 10 multiplication using decimal numbers.

Each digit in the multiplicand is multiplied by each multiplier digit in turn to produce eight subpartial products, one for each multiplicand digit, as shown in Figure 1.33.

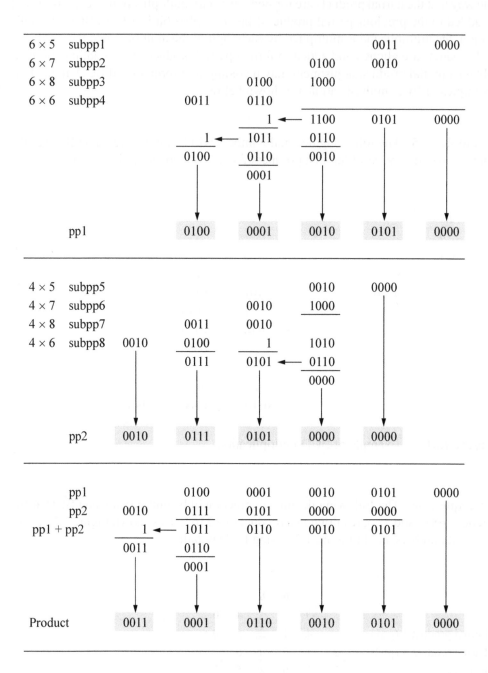

Figure 1.33   Example of decimal multiplication using BCD numbers.

Binary-coded decimal division   The method of decimal division presented here is analogous to the binary search method used in programming, which is a systematic way of searching an ordered database. The method begins by examining the middle of the database. For division, the method adds or subtracts multiples of the divisor or partial remainder. The arithmetic operation is always in the following order:

$$- 8 \times \text{the divisor}$$
$$\pm 4 \times \text{the divisor}$$
$$\pm 2 \times \text{the divisor}$$
$$\pm 1 \times \text{the divisor}$$

This method requires only four cycles for each quotient digit, regardless of the number of quotient digits. The first operation is $- 8 \times$ the divisor. If the result is less than zero, then $4 \times$ the divisor is added to the partial remainder; if the result is greater than or equal to zero, then 8 is added to a quotient counter and $4 \times$ the divisor is subtracted from the partial remainder. The process repeats for $\pm 4 \times$ the divisor, $\pm 2 \times$ the divisor, and $\pm 1 \times$ the divisor. Whenever +8, +4, +2, or +1 is added to the quotient counter, the sum of the corresponding additions is the quotient digit.

Whenever a partial remainder is negative, the next version of the divisor is added to the partial remainder; whenever a partial remainder is positive, the next version of the divisor is subtracted from the partial remainder. The example shown in Figure 1.34 illustrates this technique, where $16 \div 3$. The positive versions of the divisor are $+4 \times$ the divisor and $+1 \times$ the divisor; therefore, the quotient is 5. The remainder is the result of the last cycle, in this case 1.

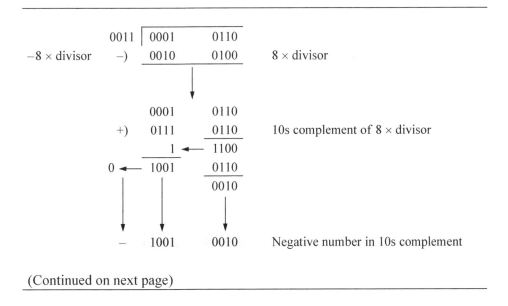

(Continued on next page)

Figure 1.34   Example of BCD division.

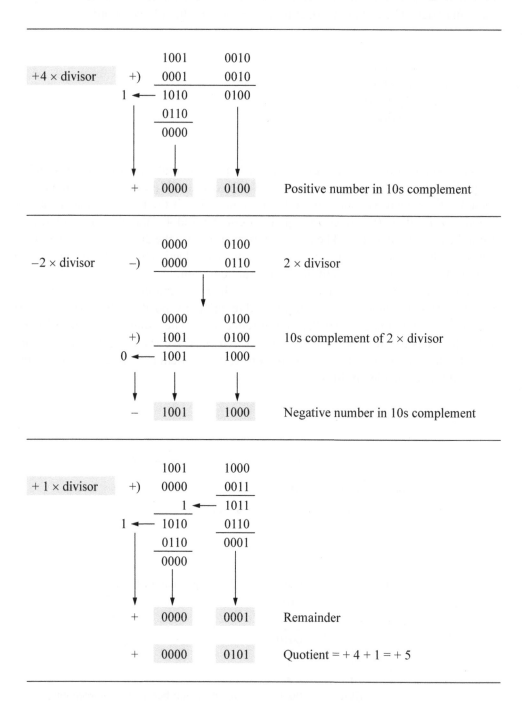

Figure 1.34   (Continued)

1.3 Binary Codes

There are many types of binary codes, many of which will be covered in the following sections. A *code* is simply a set of n-bit vectors that can represent numerical values, alphanumerical values, or some other type of coded information. An unsigned vector of n bits can represent 2^n unique combinations of the n bits and has values that range from 0 to $2^n - 1$. For example, a 3-bit vector has a range of 0 to $2^3 - 1$, or 000 to 111, as shown in Table 1.17 with appropriate weights assigned to each bit position. Although the weights assigned to the individual bits of this code are unique, in general any weight can be assigned.

Table 1.17   Three-Bit Binary Code

Decimal	Three-Bit Code		
	2^2	2^1	2^0
0	0	0	0
1	0	0	1
2	0	1	0
3	0	1	1
4	1	0	0
5	1	0	1
6	1	1	0
7	1	1	1

The bits represented by a fixed-length binary sequence are called a *code word*. This section will discuss additional numerical and nonnumerical codes, weighted and nonweighted codes, and error-detecting codes. If a code word has n bits, this does not necessarily mean that there are 2^n valid code words in the set, as is evident in the binary-coded decimal code, in which six of the sixteen code words are invalid.

1.3.1 Binary Weighted and Nonweighted Codes

The 8421 binary-coded decimal (BCD) code is a *weighted code*, because each bit of the 4-bit code words is assigned a weight. This code was presented in an earlier section and is used to represent the decimal digits 0 through 9. The decimal digit represented by the code word can be obtained by summing the weights associated with the bit positions containing 1s. Table 1.18 shows the 8421 BCD code and other BCD codes in which different weights are assigned to the bit positions. Note that some codes have multiple bit configurations for the same decimal number.

Table 1.18   Examples of Binary Codes for Decimal Digits

Decimal	8421	7421	2421	84–2–1	Excess-3
0	0000	0000	0000	0000	0011
1	0001	0001	0001	0111	0100
2	0010	0010	0010	0110	0101
3	0011	0011	0011	0101	0110
4	0100	0100	0100	0100	0111
5	0101	0101	1011	1011	1000
6	0110	0110	1100	1010	1001
7	0111	1000	1101	1001	1010
8	1000	1001	1110	1000	1011
9	1001	1010	1111	1111	1100

The excess-3 code is a *nonweighted code* and is obtained by adding three to the 8421 BCD code, as shown in Table 1.18. It is interesting to note that some of the codes of Table 1.18 are self-complementing. The excess-3 code is a *self-complementing code* in which the 1s complement of a code word is identical to the 9s complement of the corresponding 8421 BCD code word in excess-3 notation, as shown below for the decimal number 4. The 2421 BCD code is also self-complementing, as shown below for the decimal number 8.

Excess-3 BCD code	8421 BCD code
4	4
↓	↓
0111	0100
↓ 1s complement	↓ 9s complement
1000	0101
	↓ excess-3

2421 BCD code	8421 BCD code
8	8
↓	↓
1110	1000
↓ 1s complement	↓ 9s complement
0001	

An alternative method of explaining the self-complementing characteristic of the excess-3 code is as follows: The 1s complement of the excess-3 number yields the 9s complement of the same excess-3 number. For example, the decimal number 4 in excess-3 notation is 0111 (7_{10}); the 1s complement of 0111 is 1000, which is the excess-3 code for 5, which is the 9s complement of 4, the original decimal number. Two examples are shown below.

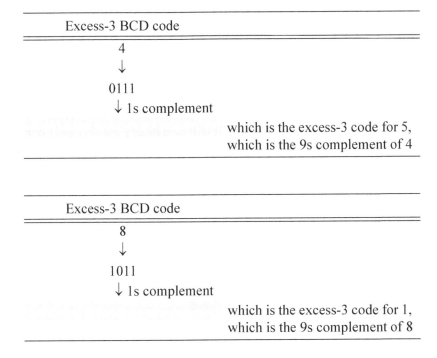

Excess-3 BCD code
4
↓
0111
↓ 1s complement
which is the excess-3 code for 5, which is the 9s complement of 4

Excess-3 BCD code
8
↓
1011
↓ 1s complement
which is the excess-3 code for 1, which is the 9s complement of 8

Addition in excess-3   Adding in excess-3 is relatively easy. Add the excess-3 numbers using the rules for binary addition, then subtract $(n - 1) \times 3$ from the result, where n is the number of decimal digits, as shown below. Alternatively, convert the decimal numbers to BCD, add the numbers using the rules for BCD arithmetic, then convert the result to excess-3, as shown below.

6 =		1001	
8 =		1011	
4 =		0111	
9 =		1100	
27	10	0111	(39 in excess-3)
− 9	00	1001	
		1110	(30 in excess-3, which is 27 in decimal)

```
    54  =  0101      0100
+)  38  =  0011      1000
    92         1  ←  1100
            1001      0110
                      0010
             ↓         ↓     Excess-3
```

Subtraction in excess-3   Subtraction in excess-3 is similar to subtracting in 1s complement. Add the 1s complement of the excess-3 subtrahend to the minuend using the rules for 1s complement arithmetic. The result will be a binary number in 1s complement, as shown below for two cases. Then convert the result to excess-3.

```
     7  =  1010
-)   3  =  0110
     4         ↓
            1010
+)  1001      1s complement of subtrahend
1  ←  0011
        1      End-around carry
      0100     Binary number in 1s complement (+4)
       ↓       Excess-3
               Binary number in excess-3 (7)
```

```
     6  =  1001
-)   9  =  1100
    -3        ↓
           1001
+)  0011      1s complement of subtrahend
    1100      Binary number in 1s complement (−3)
     ↓        Excess-3
              Binary number in excess-3 (0)
```

1.3.2 Binary-to-BCD Conversion

Converting from binary to BCD is accomplished by multiplying by the BCD number by two repeatedly. Multiplying by two is accomplished by a left shift of one bit position followed by an adjustment, if necessary. For example, a left shift of BCD 1001 (9_{10}) results in 1 0010 which is 18 in binary, but only 12 in BCD. Adding six to the low-order BCD digit results in 1 1000, which is the required value of 18.

Instead of adding six after the shift, the same result can be obtained by adding three before the shift since a left shift multiplies any number by two, as shown below. BCD digits in the range 0–4 do not require an adjustment after being shifted left, because the shifted number will be in the range 0–8, which can be contained in a 4-bit BCD digit. However, if the number to be shifted is in the range 5–9, then an adjustment will be required after the left shift, because the shifted number will be in the range 10–18, which requires two BCD digits. Therefore, three is added to the digit prior to being shifted left 1-bit position.

	1001	
+)	0011	
	1100	Shift left 1 yields 0001 1000$_{BCD}$

Figure 1.35 shows the procedure for converting from binary 1 0101 1011$_2$ (347_{10}) to BCD. Since there are 9 bits in the binary number; therefore, nine left-shift operations are required, yielding the resulting BCD number of 0011 0100 0111$_{BCD}$.

	10^2	10^1	10^0	Binary
347_{10}	0000	0000	0000	101011011
Shift left 1	0000	0000	0001	01011011
No addition	0000	0000	0001	
Shift left 1	0000	0000	0010	1011011
No addition	0000	0000	0010	
Shift left 1	0000	0000	0101	011011
Add 3	0000	0000	1000	
Shift left 1	0000	0001	0000	11011
No addition	0000	0001	0000	
Shift left 1	0000	0010	0001	1011
No addition	0000	0010	0001	
(Continued on next page)				

Figure 1.35   Example of binary-to-BCD conversion.

	10^2	10^1	10^0	Binary
Shift left 1	0000	0100	0011	011
No addition	0000	0100	0011	
Shift left 1	0000	1000	0110	11
Add 3	0000	1011	1001	
Shift left 1	0001	0111	0011	1
Add 3	0001	1010	0011	
Shift left 1	0011	0100	0111	$= 347_{10}$

Figure 1.35   (Continued)

1.3.3 BCD-to-Binary Conversion

Conversion from BCD to binary is accomplished by dividing by two repeatedly; that is, by a right shift operation. The low-order bit that is shifted out is the remainder and becomes the low-order binary bit. The BCD digits may require an adjustment after the right shift operation.

   An example of converting 22_{10} to binary is shown in Figure 1.36. After the first shift-right-1 operation, the low-order bit position of the tens decade is a 1, indicating a value of 10_{10}. After the second shift-right-1 operation, the resulting value in the units decade should be 5_{10}; however, the value is 8_{10}. Therefore, a value of 3 must be subtracted to correct the BCD number in the units position. In general, if the high-order bit in any decade is a 1 after the shift-right operation, then a value of 3 must be subtracted from the decade.

	10^1	10^0	Binary
22_{10}	0010	0010	
Shift right 1	0001	0001	0
Shift right 1	0000	1000	10
Subtract 3	0000	0101	
	0000	0101	10_2

Figure 1.36   Example of BCD-to-binary conversion.

A more extensive example is shown in Figure 1.37, which converts 479_{10} to binary. The conversion process is complete when all of the BCD decades contain zeroes. The binary equivalent of 479_{10} is 1110111111_2.

	10^2	10^1	10^0	Binary
479_{10}	0100	0111	1001	
Shift right 1	0010	0011	1100	1
Subtract 3	0010	0011	1001	
Shift right 1	0001	0001	1100	11
Subtract 3	0001	0001	1001	
Shift right 1	0000	1000	1100	111
Subtract 3	0000	0101	1001	
Shift right 1	0000	0010	1100	1111
Subtract 3	0000	0010	1001	
Shift right 1	0000	0001	0100	11111
No subtraction	0000	0001	0100	
Shift right 1	0000	0000	1010	011111
Subtract 3	0000	0000	0111	
Shift right 1	0000	0000	0011	1011111
No subtraction	0000	0000	0011	
Shift right 1	0000	0000	0001	11011111
No subtraction	0000	0000	0001	
Shift right 1	0000	0000	0000	111011111
No subtraction	0000	0000	0000	

Figure 1.37   Example of BCD-to-binary conversion.

1.3.4 Gray Code

The Gray code is an nonweighted code that has the characteristic whereby only one bit changes between adjacent code words. The Gray code belongs to a class of cyclic codes called *reflective codes*, as can be seen in Table 1.19. Notice in the first four rows, that y_4 reflects across the reflecting axis; that is, y_4 in rows 2 and 3 is the mirror image of y_4 in rows 0 and 1. In the same manner, y_3 and y_4 reflect across the reflecting axis drawn under row 3. Thus, rows 4 through 7 reflect the state of rows 0 through 3 for y_3 and y_4. The same is true for y_2, y_3, and y_4 relative to rows 8 through 15 and rows 0 through 7.

Table 1.19 Binary 8421 Code and the Gray Code

	(Binary Code $b_3\,b_2\,b_1\,b_0$)				(Gray Code $g_3\,g_2\,g_1\,g_0$)				
Row	y_1	y_2	y_3	y_4	y_1	y_2	y_3	y_4	
0	0	0	0	0	0	0	0	0	
1	0	0	0	1	0	0	0	1	
2	0	0	1	0	0	0	1	1	← y_4 is reflected
3	0	0	1	1	0	0	1	0	
4	0	1	0	0	0	1	1	0	← y3 and y_4
5	0	1	0	1	0	1	1	1	are reflected
6	0	1	1	0	0	1	0	1	
7	0	1	1	1	0	1	0	0	
8	1	0	0	0	1	1	0	0	← y_2, y_3, and y_4
9	1	0	0	1	1	1	0	1	are reflected
10	1	0	1	0	1	1	1	1	
11	1	0	1	1	1	1	1	0	
12	1	1	0	0	1	0	1	0	
13	1	1	0	1	1	0	1	1	
14	1	1	1	0	1	0	0	1	
15	1	1	1	1	1	0	0	0	

Binary-to-Gray code conversion   A procedure for converting from the binary 8421 code to the Gray code can be formulated. Let an n-bit binary code word be represented as

$$b_{n-1}\,b_{n-2}\,\cdots\,b_1\,b_0$$

and an n-bit Gray code word be represented as

$$g_{n-1}\,g_{n-2}\,\cdots\,g_1\,g_0$$

where b_0 and g_0 are the low-order bits of the binary and Gray codes, respectively. The ith Gray code bit g_i can be obtained from the corresponding binary code word by the following algorithm:

$$g_{n-1} = b_{n-1}$$
$$g_i = b_i \oplus b_{i+1} \tag{1.22}$$

for $0 \le i \le n - 2$, where the symbol $\oplus$ denotes modulo-2 addition defined as:

$$0 \oplus 0 = 0$$
$$0 \oplus 1 = 1$$
$$1 \oplus 0 = 1$$
$$1 \oplus 1 = 0$$

For example, using the algorithm of Equation 1.22, the 4-bit binary code word $b_3 b_2 b_1 b_0 = 1010$ translates to the 4-bit Gray code word $g_3 g_2 g_1 g_0 = 1111$ as follows:

$$g_3 = b_3 \qquad\qquad = 1$$
$$g_2 = b_2 \oplus b_3 = 0 \oplus 1 = 1$$
$$g_1 = b_1 \oplus b_2 = 1 \oplus 0 = 1$$
$$g_0 = b_0 \oplus b_1 = 0 \oplus 1 = 1$$

Gray-to-binary code conversion   The reverse algorithm to convert from the Gray code to the binary 8421 code is shown in Equation 1.23, where an n-bit binary code word is represented as

$$b_{n-1} b_{n-2} \cdots b_1 b_0$$

and an n-bit Gray code word is represented as

$$g_{n-1} g_{n-2} \cdots g_1 g_0$$

where b_0 and g_0 are the low-order bits of the binary and Gray codes, respectively.

$$b_{n-1} = g_{n-1}$$
$$b_i = b_{i+1} \oplus g_i \tag{1.23}$$

For example, using the algorithm of Equation 1.23, the 4-bit Gray code word $g_3 g_2 g_1 g_0 = 1001$ translates to the 4-bit binary code word $b_3 b_2 b_1 b_0 = 1110$ as follows:

$$b_3 = g_3 \qquad\qquad = 1$$

$$b_2 = b_3 \oplus g_2 = 1 \oplus 0 = 1$$

$$b_1 = b_2 \oplus g_1 = 1 \oplus 0 = 1$$

$$b_0 = b_1 \oplus g_0 = 1 \oplus 1 = 0$$

1.4 Error Detection and Correction Codes

Transferring data within a computer or between computers is subject to error, either permanent or transient. Permanent errors can be caused by hardware malfunctions; transient errors can be caused by transmission errors due to noise. In either case, the data error must at least be detected and preferably corrected. The following error detecting codes and error correcting codes will be briefly discussed: parity, Hamming code, cyclic redundancy check (CRC) code, checksum, and two-out-of-five code.

1.4.1 Parity

An extra bit can be added to a message to make the overall parity of the code word either odd or even; that is, the number of 1s in the code word — message bits plus parity bit — will be either odd or even, as shown in Table 1.20 for both odd and even parity.

Table 1.20 Parity Bit Generation

Message	Parity Bit (odd)	Message	Parity Bit (even)
0000	1	0000	0
0001	0	0001	1
0010	0	0010	1
0011	1	0011	0
0100	0	0100	1
0101	1	0101	0
0110	1	0110	0
0111	0	0111	1
1000	0	1000	1

(Continued on next page)

Table 1.20 Parity Bit Generation

Message	Parity Bit (odd)	Message	Parity Bit (even)
1001	1	1001	0
1010	1	1010	0
1011	0	1011	1
1100	1	1100	0
1101	0	1101	1
1110	0	1110	1
1111	1	1111	0

The parity bit to maintain even parity for a 4-bit message $m_3 \, m_2 \, m_1 \, m_0$ can be generated by modulo-2 addition as previously defined and shown in Equation 1.24. The parity bit for odd parity generation is the complement of Equation 1.24.

$$\text{Parity bit (even)} = m_3 \oplus m_2 \oplus m_1 \oplus m_0 \tag{1.24}$$

Parity implementation can detect an odd number of errors, but cannot correct the errors, because the bits in error cannot be determined. If a single error occurred, then an incorrect code word would be generated and the error would be detected. If two errors occurred, then parity would be unchanged and still correct. As can be seen in Table 1.20, every adjacent pair of code words — message plus parity — have a minimum distance of two; that is, they differ in two bit positions. This means that two bits must change to still maintain a correct code word. The distance-2 words shown below are adjacent and have odd parity, because there are an odd number of 1s.

Bit position	7 6 5 4 3 2 1 0 P
Distance 2	1 0 1 0 | 1 0 1 0 | 1
	1 0 1 0 | 1 0 1 1 | 0

When a word is to be transmitted, the parity bit is generated (PG) and the nine bits are transmitted along the 9-bit parallel bus, as shown in Figure 1.38. At the receiving end, the parity of the word is checked (PC) and the parity bit is removed. The parity generator and parity checker are both implemented using modulo-2 addition.

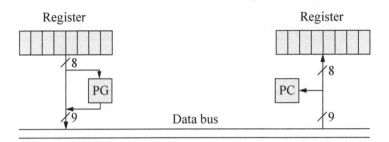

Figure 1.38   General architecture showing parity generation (PG) and parity checking (PC).

1.4.2 Hamming Code

This section provides a brief introduction to the Hamming code error detection and correction technique. An in-depth coverage is presented in a later chapter and includes matrix algebra from which the Hamming code is derived. A common error detection technique is to add a parity bit to the message being transmitted, as shown in the previous section. Thus, single-bit errors — and an odd number of errors — can be detected in the received message. The error bits, however, cannot be corrected because their location in the message is unknown. For example, assume that the message shown below was transmitted using odd parity.

Bit position	7	6	5	4	3	2	1	0	Parity bit
Message	0	1	1	0	0	1	0	0	0

Assume that the message shown below is the received message. The parity of the message is even; therefore, the received message has an error. However, the location of the bit (or bits) in error is unknown. Therefore, in this example, bit 5 cannot be corrected.

Bit position	7	6	5	4	3	2	1	0	Parity bit
Message	0	1	0	0	0	1	0	0	0

Richard W. Hamming developed a code in 1950 that resolves this problem. The *Hamming code* can be considered as an extension of the parity code presented in the previous section, because multiple parity bits provide parity for subsets of the message bits. The subsets overlap such that each message bit is contained in at least two subsets. The basic Hamming code can detect single or double errors and can correct a single error.

Assume a 2-element code as shown below; that is, there are only two code words in the code. If a single error occurred in code word X, then the assumed message is $X = 000$. Similarly, if a single error occurred in code word Y, then the assumed message is $Y = 111$. Therefore, detection and correction is possible, because the code words *differ in three bit positions.*

Code word X = 0 0 0
0 0 1
0 1 0
1 0 0

Code word Y = 1 1 1
0 1 1
1 0 1
1 1 0

A *code word* contains n bits consisting of m message bits plus k parity check bits as shown in Figure 1.39. The m bits represent the information or message part of the code word; the k bits are used for detecting and correcting errors, where $k = n - m$. The *Hamming distance* of two code words X and Y is the number of bits in which the two words differ in their corresponding columns.

For example, the Hamming distance is three for code words X and Y as shown in Figure 1.40. Since the minimum distance is three, single error detection and correction is possible. A later chapter will provide a technique to detect single and double errors and to correct single errors.

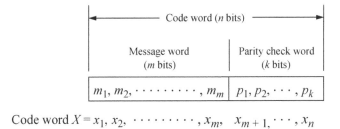

Code word $X = x_1, x_2, \cdots\cdots\cdots, x_m,\ \ x_{m+1}, \cdots, x_n$

Figure 1.39   Code word of n bits containing m message bits and k parity check bits.

X =	1	0	1	1	1	1	0	1
Y =	1	0	1	0	1	1	1	0

Figure 1.40   Two code words to illustrate a Hamming distance of three.

1.4.3 Cyclic Redundancy Check Code

A class of codes has been developed specifically for serial data transfer called cyclic redundancy check (CRC) codes. This section will provide a brief introduction to the CRC codes. A more detailed description of CRC codes will be presented in a later chapter. Cyclic redundancy check codes can detect both single-bit errors and multiple-bit errors and are especially useful for large strings of serial binary data found on single-track storage devices, such as disk drives. They are also used in serial data transmission networks and in 9-track magnetic tape systems, where each track is treated as a serial bit stream.

The generation of a CRC character uses modulo-2 addition, which is a linear operation; therefore, a linear feedback shift register is used in its implementation. The CRC character that is generated is placed at or near the end of the message.

A possible track format for a disk drive is shown in Figure 1.41, which has separate address and data fields. There is a CRC character for each of the two fields; the CRC character in the address field checks the cylinder, head, and sector addresses; the CRC character in the data field checks the data stream.

The address field and data field both have a *preamble* and a *postamble*. The preamble consists of fifteen 0s followed by a single 1 and is used to synchronize the clock to the data and to differentiate between 0s and 1s. The postamble consists of sixteen 0s and is used to separate the address and data fields.

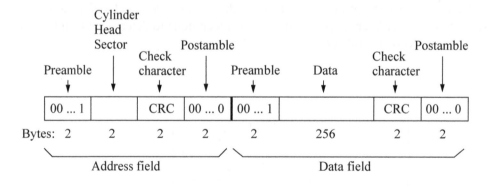

Figure 1.41   Possible track format for a disk drive.

The synchronous data link control (SDLC) — a subset of the high-level data link control (HDLC) — was developed by IBM for their System Network Architecture. It manages synchronous transmission of serial data over a communication channel. The format for the SDLC protocol is illustrated in Figure 1.42, which shows the various fields of a frame.

The first field is the flag field, which is a delimiter indicating the beginning of block. This is followed by an address field, which is decoded by all receivers, then a control field, indicating frame number, last frame, and other control information. The text field contains n bits of serial data followed by a 16-bit CRC character. The final field is a delimiter indicating end of block.

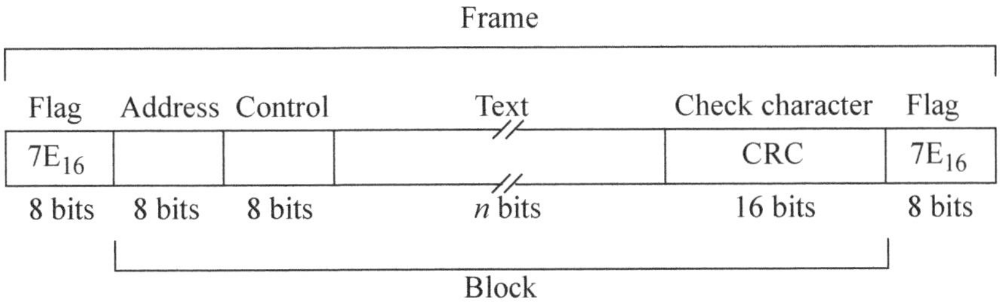

Frame

Flag	Address	Control	Text	Check character	Flag
$7E_{16}$				CRC	$7E_{16}$
8 bits	8 bits	8 bits	n bits	16 bits	8 bits

Block

Figure 1.42   One frame of the synchronous data link control format.

Since there may be flags embedded in the text as data, the transmitter inserts a 0 bit after every five consecutive 1s. The receiver discards the 0 bit after every five consecutive 1s; thus, no flag is transmitted randomly. Figure 1.43 shows two examples of 0s inserted after five consecutive 1s. Note that there is no ambiguity between a flag as data with an inserted 0 or a data bit stream of five 1s followed by a 0.

Text	0 1 1 1 1 1 1 0
Sent as	0 1 1 1 1 1 0 1 0

Text	0 1 1 1 1 1 0
Sent as	0 1 1 1 1 1 0 0

Figure 1.43   Examples of 0s inserted after every five consecutive 1s so that a flag may not be transmitted as data for the SDLC protocol.

1.4.4 Checksum

The *checksum* is the sum derived from the application of an algorithm that is calculated before and after transmission to ensure that the data is free from errors. The checksum character is a numerical value that is based on the number of asserted bits in the message and is appended to the end of the message. The receiving unit then applies the same algorithm to the message and compares the results with the appended checksum character.

There are many versions of checksum algorithms. The parity checking method presented in Section 1.41 is a modulo-2 addition of all bits in the string of data. The modulo-2 sum is 1 if the number of 1s in the 8-bit byte is odd, otherwise, the modulo-2 sum is 0.

If the information consists of n bytes of data, then a simple checksum algorithm is to perform modulo-256 addition on the bytes in the message. The sum thus obtained is the checksum byte and is appended to the last byte creating a message of $n + 1$ bytes. The receiving unit then regenerates the checksum by obtaining the sum of the first n bytes and compares that sum to byte $n + 1$. This method can detect a single byte error.

Alternatively, the sum that is obtained by modulo-256 addition in the transmitting unit is 2s complemented and becomes the checksum character which is appended to the end of the transmitted message. The receiving unit uses the same algorithm and adds the recalculated uncomplemented checksum character to the transmitted checksum character, resulting in a sum of zero if the message had no errors. An example of this algorithm is shown in Figure 1.44 for a 4-byte message.

Message with checksum

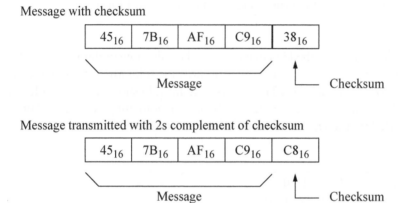

Figure 1.44   Checksum generated for a 4-byte message.

Memories can also include a checksum in their implementation. Consider a small memory as shown in Figure 1.45, containing only eight bytes of data. The checksum is generated by using modulo-2 addition on the columns of each byte.

```
1 1 0 0 1 0 1 0
0 0 1 1 0 1 1 0
0 1 1 1 1 0 0 0
0 1 0 0 1 1 1 1    Data
0 0 1 1 0 1 1 1
0 0 0 0 0 0 1 1
0 1 1 0 1 1 0 0
0 0 0 0 0 0 0 0
1 0 0 1 0 0 1 1    Checksum
```

Figure 1.45   Checksum generation for memories.

1.4.5 Two-Out-Of-Five Code

The two-out-of-five code is 5-bit nonweighted code that is characterized by having exactly two 1s and three 0s in any code word. This code has a minimum distance of two and is relatively easy to provide error detection by counting the number of 1s in a code word.

   An error is detected whenever the number of 1s in a code word is not equal to two. This can result from a change of one or more bits which cause the total number of 1s to differ from two. However, an error will be undetected if there are two simultaneous bit changes which result in a valid code word with two 1s. For example, if code word 01100 (7) were changed during transmission to 01010 (5), then the error would be undetected. The two-out-of-five code is representative of *m*-out-of-*n* codes. Table 1.21 shows a typical two-out-of-five code for the decimal digits 0 through 9.

Table 1.21 Two-out-of-Five Code for the Decimal Digits

Decimal	Two-Out-of-Five Code
0	00011
1	00101
2	01001
3	10001
4	00110
5	01010
6	10010
7	01100
8	10100
9	11000

1.4.6 Horizontal and Vertical Parity Check

For single-error correction and double-error detection, Hamming code is the ideal method for detecting and correcting errors in memory operations. An alternative method to Hamming code for smaller memories — which has less redundancy — is a technique using horizontal and vertical parity. Horizontal parity utilizes an odd parity bit for each word in memory. Vertical parity is the modulo-2 addition of identical bit positions of each word in a block of memory.

   Errors are detected by the horizontal parity, which indicates the word in error. This error is used in conjunction with the vertical parity to detect and correct single-bit errors. Figure 1.46 shows a 10-word block of memory, which will be used to illustrate the technique of single-bit error detection and correction using horizontal and vertical parity.

Word	7	6	5	4	3	2	1	0	Horizontal parity (odd)
0	1	1	1	0	1	0	0	0	1
1	0	1	0	1	0	1	0	1	1
2	0	1	1	0	1	1	1	0	0
3	1	1	0	0	1	1	0	1	0
4	0	1	1	1	1	0	1	0	0
5	1	0	0	1	1	0	0	1	1
6	1	0	1	0	0	0	1	0	0
7	0	0	1	1	0	1	1	0	1
8	0	0	0	1	0	0	1	1	0
9	1	0	1	1	0	1	1	1	1
Vertical parity (odd)	0	0	1	1	0	0	1	0	

Figure 1.46   Ten-word memory to illustrate single-error detection and correction using horizontal and vertical parity.

Assume that word four has an error in bit position four such that word four changes from 0111 1010 0 to 0110 1010 0, which is incorrect for odd parity. To correct the bit in error, the current vertical parity word 0011 0010 is exclusive-ORed with the new vertical parity word obtained by the modulo-2 addition of identical bit positions of each word in a block of memory with the exception of word four. This will generate the original word four, as shown below.

Current vertical parity word		0	0	1	1	0	0	1	0
$\oplus$ New vertical parity word without word four		0	1	0	0	1	0	0	0
Original word four		0	1	1	1	1	0	1	0

As another example, assume that word seven has an error in bit position two such that word seven changes from 0011 0110 1 to 0011 0010 1, which is incorrect for odd parity. Using the procedure outlined above, the correct word seven is obtained, as shown below.

Current vertical parity word		0	0	1	1	0	0	1	0
$\oplus$ New vertical parity word without word seven		0	0	0	0	0	1	0	0
Original word seven		0	0	1	1	0	1	1	0

1.5 Serial Data Transmission

The standard binary code for alphanumeric characters is the American Standard Code for Information Interchange (ASCII). The code uses seven bits to encode 128 characters, which include 26 uppercase letters, 26 lowercase letters, the digits 0 through 9, and 32 special characters, such as &, #, and $. The bits are labeled b_6 through b_0, where b_0 is the low-order bit and is transmitted first. There is also a parity bit which maintains either odd or even parity for the eight bits.

There are two basic types of serial communication: synchronous and asynchronous. In *synchronous* communication, information is transferred using a *self-clocking* scheme; that is, the clock is synchronized to the data, which determines the rate of transfer. Self-clocking is presented in detail in a later chapter. Alternatively, there may be a separate clock signal to determine the bit cell boundaries. Information is normally sent as blocks of data, not as individual bytes, and may contain error checking.

In *asynchronous* communication, the timing occurs for each character. The communications line for asynchronous serial data transfer is in an idle state (a logic 1 level) when characters are not being transmitted. Figure 1.47 shows the format for asynchronous serial data transmission for the ASCII character *W*. The receiver and transmitter must operate at the same clock frequency. The *start bit* causes a transition from a logic 1 level to a logic 0 level and synchronizes the receiver with the transmitter. This synchronization is for one character only — the next character will contain a new start bit. The start bit signals the receiver to start assembling a character.

The start bit is followed by the seven data bits, which in turn are followed by the parity bit (odd parity is shown). The parity bit is followed by two *stop bits* that are asserted high, which ensure that the start bit of the next character will cause a transition from high to low. The stop bits are also used to isolate consecutive characters.

There is a wide range of standard transmission rates, referred to as bauds. A *baud* is a unit of signaling speed, which refers to the number of times the state of a signal changes per second; that is, bits per second.

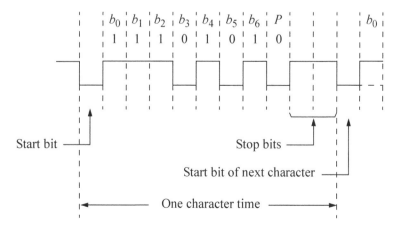

Figure 1.47   Asynchronous serial data transfer for the ASCII character *W*.

1.6 Problems

1.1   Convert the unsigned binary number $0111\ 1101_2$ to radix 10.

1.2   Convert the octal number 5476_8 to radix 10.

1.3   Convert the hexadecimal number $4AF9_{16}$ to radix 10.

1.4   Convert the unsigned binary number 1100.110_2 to radix 10.

1.5   Convert the octal number 173.25_8 to radix 10.

1.6   Convert the hexadecimal number $6BC.5_{16}$ to radix 10.

1.7   Convert the hexadecimal number $4A3CB_{16}$ to binary and octal.

1.8   Represent the following numbers in sign magnitude, diminished-radix complement, and radix complement for radix 2 and radix 10: $+136$ and -136.

1.9   Represent the decimal numbers $+54$, -28, $+127$, and -13 in sign magnitude, diminished-radix complement, and radix complement for radix 2 using 8 bits.

1.10   Convert the following octal numbers to hexadecimal: 6536_8 and 63457_8.

1.11   Convert the binary number $0100\ 1101.1011_2$ to decimal and hexadecimal notation.

1.12   Convert 7654_8 to radix 3.

1.13   Determine the range of positive and negative numbers in radix 2 for sign magnitude, diminished-radix complement, and radix complement.

1.14   Add the unsigned binary numbers $1101\ 0110_2$ and $0111\ 1100_2$.

1.15   Multiply the unsigned binary numbers 1111_2 and 0011_2.

1.16   Divide the following unsigned binary numbers: $101\ 1110_2 \div 1001_2$.

1.17   Convert the following radix -4 number to radix 3 with three fraction digits: 123.13_{-4}.

1.18   Convert 263.2_7 to radix 4 with three fraction digits.

1.19   Obtain the radix complement of $F8B6_{16}$.

1.20   Obtain the radix complement of 54320_6.

1.21   The numbers shown below are in sign-magnitude representation. Convert the numbers to 2s complement representation for radix 2 with the same numerical value using 8 bits.

    1001 1001
    0001 0000
    1011 1111

1.22   The numbers shown below are in 2s complement representation for radix 2. Determine whether the indicated arithmetic operations produce an overflow.

    0011 0110 + 1110 0011
    1001 1000 – 0010 0010
    0011 0110 – 1110 0011

1.23   Convert the following unsigned binary numbers to radix 10:

    (a) 10111_2
    (b) 11111_2
    (c) 1111000.101_2
    (d) 1000001.111_2

1.24   Use direct subtraction to obtain the difference of the following unsigned binary numbers:

    $11010_2 - 10111_2$

1.25   Perform the following binary subtraction using the diminished-radix complement method:

    $101111_2 - 000011_2$

1.26   Convert the following decimal numbers to signed binary numbers in 2s complement representation using 8 bits.

    (a) $+56_{10}$ (b) -27_{10}

1.27   Add the following binary numbers:

    $111111_2 + 111111_2 + 111111_2 + 111111_2$

1.28   Add the following BCD numbers:

    $1001\ 1000_{BCD} + 1001\ 0111_{BCD}$

1.29   Obtain the sum of the following binary numbers:

$$1111\ 1111_2$$
$$1111\ 1111_2$$
$$1111\ 1111_2$$
$$1111\ 1111_2$$
$$1111\ 1111_2$$
$$1111\ 1111_2$$

1.30   Obtain the sum of the following radix 3 numbers:

$$1111_3 + 1111_3 + 1111_3 + 1111_3$$

1.31   Obtain the following hexadecimal numbers:

$$FFFF_{16} + FFFF_{16} + FFFF_{16} + FFFF_{16}$$

1.32   Add the following radix 4 numbers:

$$3213_4$$
$$1010_4$$
$$0213_4$$
$$3010_4$$
$$2101_4$$

1.33   Perform a subtraction on the operands shown below, which are in radix complementation for radix 3.

$$02021_3 - 22100_3$$

1.34   Convert the radix 5 number 2434.1_5 to radix 9 with a precision of one fraction digit.

1.35   Multiply the two binary numbers shown below, which are in radix complementation, using the paper-and-pencil method.

$$11111_2 \times 01011_2$$

1.36   Let A and B be two binary numbers in 2s complement representation as shown below, where A' and B' are the 1s complement of A and B, respectively. Perform the operation listed below.

$$A = 1011\ 0001_2$$
$$B = 1110\ 0100_2$$

$$(A' + 1) - (B' + 1)$$

1.37   The decimal operands shown below are to be added using decimal BCD addition. Obtain the intermediate sum; that is, the sum before adjustment is applied.

$$725_{10} + 536_{10}$$

1.38   Use the paper-and-pencil method for binary restoring division to perform the following divide operation using an 8-bit dividend and a 4-bit divisor:

$$73_{10} \div 5_{10}$$

1.39   Convert the following binary number to BCD using the shift left method:

$$10101110_2$$

1.40   Convert the following decimal number from BCD to binary using the shift right method:

$$363_{10}$$

2

Minimization of Switching Functions

There are three main categories of minimizing Boolean — or switching — functions. *Algebraic minimization* uses the axioms and theorems of Boolean algebra developed by George Boole in 1854 and later put into practice by Claude E. Shannon who showed how to adapt the algebra to describe logic circuits. In 1953, M. Karnaugh developed a method to minimize logic functions using a graphical representation of a Boolean function, hence called a *Karnaugh map*. The *Quine-McCluskey* algorithm is used to minimize a logic function with a large number of variables and is easily converted to a computer program. This chapter will present all three methods of minimization in detail with many examples of each method.

2.1 Boolean Algebra

George Boole introduced a systematic treatment of the logic operations AND, OR, and NOT, which is now called *Boolean algebra*. The symbols (or operators) used for the algebra and the corresponding function definitions are listed in Table 2.1. The table also includes the exclusive-OR function, which is characterized by the three operations of AND, OR, and NOT. Table 2.2 illustrates the truth tables for the Boolean operations AND, OR, NOT, exclusive-OR, and exclusive-NOR, where z_1 is the result of the operation.

A *truth table* is one of several methods utilized to express the relationship between binary variables. A truth table lists all possible combinations of the independent variables and demonstrates the connection between the values of the variables and the logical result of the operation.

Table 2.1 Boolean Operators for Variables x_1 and x_2

Operator	Function	Definition
$\cdot$	AND	$x_1 \cdot x_2$ (Also $x_1 x_2$)
$+$	OR	$x_1 + x_2$
$'$	NOT	x_1'
$\oplus$	Exclusive-OR	$(x_1 x_2') + (x_1' x_2)$

Table 2.2 Truth Table for AND, OR, NOT, Exclusive-OR, and Exclusive-NOR Operations

AND $x_1 x_2$ z_1		OR $x_1 x_2$ z_1		NOT x_1 z_1		Exclusive-OR $x_1 x_2$ z_1		Exclusive-NOR $x_1 x_2$ z_1	
0 0	0	0 0	0	0	1	0 0	0	0 0	1
0 1	0	0 1	1	1	0	0 1	1	0 1	0
1 0	0	1 0	1			1 0	1	1 0	0
1 1	1	1 1	1			1 1	0	1 1	1

The AND operator, which corresponds to the Boolean product, is also indicated by the symbol "$\wedge$" ($x_1 \wedge x_2$) or by no symbol if the operation is unambiguous. Thus, $x_1 x_2, x_1 \cdot x_2$, and $x_1 \wedge x_2$ are all read as "x_1 AND x_2." The OR operator, which corresponds to the Boolean sum, is also specified by the symbol "$\vee$." Thus, $x_1 + x_2$ and $x_1 \vee x_2$ are both read as "x_1 OR x_2." The symbol for the complement operation is usually specified by the prime "$'$" symbol immediately following the variable (x_1'), by a bar over the variable ($\overline{x_1}$), or by the symbol "$\neg$" ($\neg x_1$). This book will use the symbols defined in Table 2.1.

Boolean algebra is a deductive mathematical system which can be defined by a set of variables, a set of operators, and a set of axioms (or postulates). An *axiom* is a statement that is universally accepted as true; that is, the statement needs no proof because its truth is obvious. The axioms of Boolean algebra form the basis from which the theorems and other properties can be derived.

Most axioms and theorems are characterized by two laws. Each law is the dual of the other. The principle of duality specifies that the *dual* of an algebraic expression can be obtained by interchanging the binary operators, $\cdot$ and $+$, and by interchanging the identity elements 0 and 1. Boolean algebra is an algebraic structure consisting of a set of elements B, together with two binary operators, $\cdot$ and $+$, and a unary operator $'$

such that the axioms which follow are true, where the notation $x_1 \in X$ is read as "x_1 is an element of the set X":

Axiom 1: Boolean set definition   The set B contains at least two elements x_1 and x_2, where $x_1 \neq x_2$.

Axiom 2: Closure laws   For every $x_1, x_2 \in B$,

(a) $x_1 + x_2 \in B$
(b) $x_1 \cdot x_2 \in B$

Axiom 3: Identity laws   There exist two unique *identity elements* 0 and 1, where 0 is an identity element with respect to the Boolean sum and 1 is an identity element with respect to the Boolean product. Thus, for every $x_1 \in B$,

(a) $x_1 + 0 = 0 + x_1 = x_1$
(b) $x_1 \cdot 1 = 1 \cdot x_1 = x_1$

Axiom 4: Commutative laws   The commutative laws specify that the order in which the variables appear in a Boolean expression is irrelevant — the result is the same. Thus, for every $x_1, x_2 \in B$,

(a) $x_1 + x_2 = x_2 + x_1$
(b) $x_1 \cdot x_2 = x_2 \cdot x_1$

Axiom 5: Associative laws   The associative laws state that three or more variables can be combined in an expression using Boolean multiplication or addition and that the order of the variables can be altered without changing the result. Thus, for every $x_1, x_2, x_3 \in B$,

(a) $(x_1 + x_2) + x_3 = x_1 + (x_2 + x_3)$
(b) $(x_1 \cdot x_2) \cdot x_3 = x_1 \cdot (x_2 \cdot x_3)$

Axiom 6: Distributive laws   The distributive laws for Boolean algebra are similar, in many respects, to those for college algebra. The interpretation, however, is different and is a function of the Boolean product and the Boolean sum. This is a very useful axiom in minimizing Boolean functions. For every $x_1, x_2, x_3 \in B$,

(a)   The operator + is distributive over the operator • such that
$x_1 + (x_2 \cdot x_3) = (x_1 + x_2) \cdot (x_1 + x_3)$

(b)   The operator • is distributive over the operator + such that
$x_1 \cdot (x_2 + x_3) = (x_1 \cdot x_2) + (x_1 \cdot x_3)$

Axiom 7: Complementation laws   For every $x_1 \in B$, there exists an element x_1' (called the complement of x_1), where $x_1' \in B$, such that

(a) $x_1 + x_1' = 1$
(b) $x_1 \bullet x_1' = 0$

The theorems presented below are derived from the axioms and are listed in pairs, where each relation in the pair is the dual of the other.

Theorem 1: 0 and 1 associated with a variable   Every variable in Boolean algebra can be characterized by the identity elements 0 and 1. Thus, for every $x_1 \in B$,

(a) $x_1 + 1 = 1$
(b) $x_1 \bullet 0 = 0$

Proof of Theorem 1, part (b).

$$
\begin{aligned}
x_1 \bullet 0 &= 0 + x_1 \bullet 0 && \text{Axiom 3 (a): Identity laws} \\
&= (x_1 \bullet x_1') + (x_1 \bullet 0) && \text{Axiom 7 (b): Complementation laws} \\
&= x_1 \bullet (x_1' + 0) && \text{Axiom 6 (b): Distributive laws} \\
&= x_1 \bullet x_1' && \text{Axiom 3 (a): Identity laws} \\
&= 0 && \text{Axiom 7 (b): Complementation laws}
\end{aligned}
$$

Theorem 1 part (a) is true by duality; that is, $x_1 + 1 = 1$. The proof is left as an exercise.

Theorem 2: 0 and 1 complement   The 2-valued Boolean algebra has two distinct identity elements 0 and 1, where $0 \neq 1$. The operations using 0 and 1 are as follows:

$$
\begin{array}{ll}
0 + 0 = 0 & \qquad 0 + 1 = 1 \\
1 \bullet 1 = 1 & \qquad 1 \bullet 0 = 0
\end{array}
$$

Proof: In Axiom 3, x_1 represents any element of B and thus denotes values of 0 and 1. Direct substitution in Axiom 3 provides the above set of relations.

A corollary to Theorem 2 specifies that element 1 satisfies the requirements of the complement of element 0, and vice versa. Thus, each identity element is the complement of the other.

(a) $0' = 1$
(b) $1' = 0$

Theorem 3: Idempotent laws   Idempotency relates to a nonzero mathematical quantity which, when applied to itself for a binary operation, remains unchanged. Thus, if $x_1 = 0$, then $x_1 + x_1 = 0 + 0 = 0$ and if $x_1 = 1$, then $x_1 + x_1 = 1 + 1 = 1$; therefore, one of the elements is redundant and can be discarded. The dual is true for the operator •. The idempotent laws eliminate redundant variables in a Boolean expression and can be extended to any number of identical variables. This law is also referred to as the *law of tautology*, which precludes the needless repetition of the variable. For every $x_1 \in B$,

(a) $x_1 + x_1 = x_1$
(b) $x_1 \bullet x_1 = x_1$

Proof of Theorem 3, part (a).

$$
\begin{aligned}
x_1 + x_1 &= (x_1 + x_1) \bullet 1 & \text{Axiom 3 (b): Identity laws} \\
&= (x_1 + x_1) \bullet (x_1 + x_1') & \text{Axiom 7 (a): Complementation laws} \\
&= x_1 + (x_1 \bullet x_1') & \text{Axiom 6 (a): Distributive laws} \\
&= x_1 + 0 & \text{Axiom 7 (b): Complementation laws} \\
&= x_1 & \text{Axiom 3 (a): Identity laws}
\end{aligned}
$$

Theorem 3, part (b) is true by duality; that is, $x_1 \bullet x_1 = x_1$. The idempotent laws are true for any number of variables. Thus,

$$x_1 + x_1 + x_1 + x_1 + \ldots + x_1 = x_1$$

and

$$x_1 \bullet x_1 \bullet x_1 \bullet x_1 \bullet \ldots \bullet x_1 = x_1$$

Theorem 4: Involution law   The involution law states that the complement of a complemented variable is equal to the variable. There is no dual for the involution law. The law is also called the *law of double complementation*. Thus, for every $x_1 \in B$,

$$x_1'' = x_1$$

The proof is left as an exercise.

Theorem 5: Absorption law 1   This version of the absorption law states that some 2-variable Boolean expressions can be reduced to a single variable without altering the result. Thus, for every $x_1, x_2 \in B$,

(a) $x_1 + (x_1 \bullet x_2) = x_1$ Does not imply $x_1 \bullet x_2 = 0$
(b) $x_1 \bullet (x_1 + x_2) = x_1$

Proof of Theorem 5, part (a).

$$
\begin{aligned}
x_1 + (x_1 \bullet x_2) &= (x_1 \bullet 1) + (x_1 \bullet x_2) && \text{Axiom 3 (b): Identity laws} \\
&= x_1 \bullet (1 + x_2) && \text{Axiom 6 (b): Distributive laws} \\
&= x_1 \bullet (x_2 + 1) && \text{Axiom 4 (a): Commutative laws} \\
&= x_1 \bullet 1 && \text{Theorem 1 (a): 0 and 1 associated with a variable} \\
&= x_1 && \text{Axiom 3 (b): Identity laws}
\end{aligned}
$$

Proof of Theorem 5, part (b) is true by duality; that is, $x_1 \bullet (x_1 + x_2) = x_1$.

Theorem 6: Absorption law 2   This version of the absorption law is used to eliminate redundant variables from certain Boolean expressions. Absorption law 2 eliminates a variable or its complement and is a very useful law for minimizing Boolean expressions.

(a) $x_1 + (x_1' \bullet x_2) = x_1 + x_2$
(b) $x_1 \bullet (x_1' + x_2) = x_1 \bullet x_2$

Proof of Theorem 6, part (a).

$$
\begin{aligned}
x_1 + (x_1' \bullet x_2) &= (x_1 + x_1') \bullet (x_1 + x_2) && \text{Axiom 6 (a): Distributive laws} \\
&= 1 \bullet (x_1 + x_2) && \text{Axiom 7 (a): Complementation laws} \\
&= x_1 + x_2 && \text{Axiom 3 (b): Identity laws}
\end{aligned}
$$

Proof of Theorem 6, part (b) is true by duality; that is, $x_1 \bullet (x_1' + x_2) = x_1 \bullet x_2$.

Theorem 7: DeMorgan's laws   DeMorgan's laws are also useful in minimizing Boolean functions. DeMorgan's laws convert the complement of a sum term or a product term into a corresponding product or sum term, respectively. For every $x_1, x_2 \in B$,

(a) $(x_1 + x_2)' = x_1' \bullet x_2'$
(b) $(x_1 \bullet x_2)' = x_1' + x_2'$

DeMorgan's laws can be generalized for any number of variables such that

(a) $(x_1 + x_2 + \ldots + x_n)' = x_1' \bullet x_2' \bullet \ldots \bullet x_n'$
(b) $(x_1 \bullet x_2 \bullet \ldots \bullet x_n)' = x_1' + x_2' + \ldots + x_n'$

When applying DeMorgan's laws to an expression, the operator • takes precedence over the operator +. For example, use DeMorgan's law to complement the Boolean expression $x_1 + x_2 x_3$.

$$(x_1 + x_2 x_3)' = [x_1 + (x_2 x_3)]'$$
$$= x_1' (x_2' + x_3')$$

Note that: $(x_1 + x_2 x_3)' \neq x_1' \bullet x_2' + x_3'$.

Proof of Theorem 7, part (a). Assume that $x_1' \bullet x_2'$ is the complement of $x_1 + x_2$. Therefore, if it can be shown that

$$(x_1 + x_2) + (x_1' \bullet x_2') = 1 \qquad \text{Axiom 7 (a)}$$

and

$$(x_1 + x_2) \bullet (x_1' \bullet x_2') = 0 \qquad \text{Axiom 7 (b)}$$

then this implies that $(x_1 + x_2)' = x_1' \bullet x_2'$. Since

$(x_1 + x_2) + (x_1' \bullet x_2') =$	$(x_1' \bullet x_2' + x_1) + x_2$	Axiom 5: Associative laws
$=$	$(x_1 + x_1') \bullet (x_1 + x_2') + x_2$	Axiom 6 (a): Distributive laws
$=$	$1 \bullet (x_1 + x_2') + x_2$	Axiom 7 (a): Complementation laws
$=$	$(x_1 + x_2') + x_2$	Axiom 3 (b): Identity laws
$=$	$x_1 + (x_2 + x_2')$	Axiom 5: Associative laws
$=$	$x_1 + 1$	Axiom 7 (a): Complementation laws
$=$	1	Theorem 1 (a): 0 and 1 associated with a variable

and since

$(x_1 + x_2) \bullet (x_1' \bullet x_2') =$	$(x_1 \bullet x_1' \bullet x_2') + (x_2 \bullet x_1' \bullet x_2')$	Axiom 6: Distributive laws
$=$	$0 + 0$	Axiom 7 (b): Complementation laws
$=$	0	Theorem 2: 0 and 1 complement

Therefore, $(x_1 + x_2)' = x_1' \cdot x_2'$. Proof of Theorem 7, part (b) is true by duality; that is, $(x_1 \cdot x_2)' = x_1' + x_2'$.

Minterm   A minterm is the Boolean product of n variables and contains all n variables of the function exactly once, either true or complemented. For example, for the function $z_1(x_1, x_2, x_3)$, $x_1 x_2' x_3$ is a minterm.

Maxterm   A maxterm is the Boolean sum of n variables and contains all n variables of the function exactly once, either true or complemented. For example, for the function $z_1(x_1, x_2, x_3)$, $(x_1 + x_2' + x_3)$ is a maxterm.

Product term   A product term is the Boolean product of variables containing a subset of the possible variables or their complements. For example, for the function $z_1(x_1, x_2, x_3)$, $x_1' x_3$ is a product term because it does not contain all the variables.

Sum term   A sum term is the Boolean sum of variables containing a subset of the possible variables or their complements. For example, for the function $z_1(x_1, x_2, x_3)$, $(x_1' + x_3)$ is a sum term because it does not contain all the variables.

Sum of minterms   A sum of minterms is an expression in which each term contains all the variables, either true or complemented. For example,

$$z_1(x_1, x_2, x_3) = x_1' x_2 x_3 + x_1 x_2' x_3' + x_1 x_2 x_3$$

is a Boolean expression in a sum-of-minterms form. This particular form is also referred to as a *minterm expansion*, a *standard sum of products*, a *canonical sum of products*, or a *disjunctive normal form*. Since each term is a minterm, the expression for z_1 can be written in a more compact sum-of-minterms form as $z_1(x_1, x_2, x_3) = \Sigma_m(3,4,7)$, where each term is converted to its minterm value. For example, the first term in the expression is $x_1' x_2 x_3$, which corresponds to binary 011, representing minterm 3.

Sum of products   A sum of products is an expression in which at least one term does not contain all the variables; that is, at least one term is a proper subset of the possible variables or their complements. For example,

$$z_1(x_1, x_2, x_3) = x_1' x_2 x_3 + x_2' x_3' + x_1 x_2 x_3$$

is a sum of products for the function z_1 because the second term does not contain the variable x_1.

Product of maxterms A product of maxterms is an expression in which each term contains all the variables, either true or complemented. For example,

$$z_1(x_1, x_2, x_3) = (x_1' + x_2 + x_3)(x_1 + x_2' + x_3')(x_1 + x_2 + x_3)$$

is a Boolean expression in a product-of-maxterms form. This particular form is also referred to as a *maxterm expansion*, a *standard product of sums*, a *canonical product of sums*, or a *conjunctive normal form*. Since each term is a maxterm, the expression for z_1 can be written in a more compact product-of-maxterms form as $z_1(x_1, x_2, x_3) = \Pi_M(0,3,4)$, where each term is converted to its maxterm value.

Product of sums A product of sums is an expression in which at least one term does not contain all the variables; that is, at least one term is a proper subset of the possible variables or their complements. For example,

$$z_1(x_1, x_2, x_3) = (x_1' + x_2 + x_3)(x_2' + x_3')(x_1 + x_2 + x_3)$$

is a product of sums for the function z_1 because the second term does not contain the variable x_1.

Summary of Boolean algebra axioms and theorems Table 1.8 provides a summary of the axioms and theorems of Boolean algebra. Each of the laws listed in the table is presented in pairs, where applicable, in which each law in the pair is the dual of the other.

Table 2.3 Summary of Boolean Algebra Axioms and Theorems

Axiom or Theorem	Definition
Axiom 1: Boolean set definition	$x_1, x_2 \in B$
Axiom 2: Closure laws	(a) $x_1 + x_2 \in B$ (b) $x_1 \cdot x_2 \in B$
Axiom 3: Identity laws	(a) $x_1 + 0 = 0 + x_1 = x_1$ (b) $x_1 \cdot 1 = 1 \cdot x_1 = x_1$
Axiom 4: Commutative laws	(a) $x_1 + x_2 = x_2 + x_1$ (b) $x_1 \cdot x_2 = x_2 \cdot x_1$
Axiom 5: Associative laws	(a) $(x_1 + x_2) + x_3 = x_1 + (x_2 + x_3)$ (b) $(x_1 \cdot x_2) \cdot x_3 = x_1 \cdot (x_2 \cdot x_3)$
Axiom 6: Distributive laws	(a) $x_1 + (x_2 \cdot x_3) = (x_1 + x_2) \cdot (x_1 + x_3)$ (b) $x_1 \cdot (x_2 + x_3) = (x_1 \cdot x_2) + (x_1 \cdot x_3)$

(Continued on next page)

Table 2.3 Summary of Boolean Algebra Axioms and Theorems

Axiom or Theorem	Definition
Axiom 7: Complementation laws	(a) $x_1 + x_1' = 1$
	(b) $x_1 \cdot x_1' = 0$
Theorem 1: 0 and 1 associated with a variable	(a) $x_1 + 1 = 1$
	(b) $x_1 \cdot 0 = 0$
Theorem 2: 0 and 1 complement	(a) $0' = 1$
	(b) $1' = 0$
Theorem 3: Idempotent laws	(a) $x_1 + x_1 = x_1$
	(b) $x_1 \cdot x_1 = x_1$
Theorem 4: Involution law	$x_1'' = x_1$
Theorem 5: Absorption laws 1	(a) $x_1 + (x_1 \cdot x_2) = x_1$
	(b) $x_1 \cdot (x_1 + x_2) = x_1$
Theorem 6: Absorption laws 2	(a) $x_1 + (x_1' \cdot x_2) = x_1 + x_2$
	(b) $x_1 \cdot (x_1' + x_2) = x_1 \cdot x_2$
Theorem 7: DeMorgan's laws	(a) $(x_1 + x_2)' = x_1' \cdot x_2'$
	(b) $(x_1 \cdot x_2)' = x_1' + x_2'$

2.2 Algebraic Minimization

The number of terms and variables that are necessary to generate a Boolean function can be minimized by algebraic manipulation. Since there are no specific rules or algorithms to use for minimizing a Boolean function, the procedure is inherently heuristic in nature. The only method available is an empirical procedure utilizing the axioms and theorems, which is based solely on experience and observation without reference to theoretical principles. The examples which follow illustrate the process for minimizing a Boolean function using the axioms and theorems of Boolean algebra.

Example 2.1   The following expression will be minimized using Boolean algebra:
$$x_1 x_2 + x_1' x_3 + x_2 x_3$$

$$
\begin{aligned}
x_1 x_2 + x_1' x_3 + x_2 x_3 &= x_1 x_2 + x_1' x_3 + x_2 x_3 \\
&= x_1 x_2 + x_1' x_3 + x_2 x_3 (x_1 + x_1') && \text{Complementation law} \\
&= x_1 x_2 + x_1' x_3 + x_1 x_2 x_3 + x_1' x_2 x_3 && \text{Distributive law} \\
&= x_1 x_2 + x_1 x_2 x_3 + x_1' x_3 + x_1' x_2 x_3 && \text{Commutative law} \\
&= x_1 x_2 (1 + x_3) + x_1' x_3 (1 + x_2) && \text{Distributive law}
\end{aligned}
$$

(Continued on next page)

$$
\begin{aligned}
&= (x_1 x_2 \cdot 1) + (x_1' x_3 \cdot 1) && \text{Theorem 1} \\
&= x_1 x_2 + x_1' x_3 && \text{Identity law}
\end{aligned}
$$

Example 2.2   The following expression will be minimized using Boolean algebra:
$(x_1 + x_2)(x_1 + x_3)$

$$
\begin{aligned}
(x_1 + x_2)(x_1 + x_3) &= (x_1 + x_2)x_1 + (x_1 + x_2)x_3 && \text{Distributive law} \\
&= x_1 x_1 + x_1 x_2 + x_1 x_3 + x_2 x_3 && \text{Distributive law} \\
&= x_1 + x_1 x_2 + x_1 x_3 + x_2 x_3 && \text{Idempotent law} \\
&= x_1(1 + x_2 + x_3) + x_2 x_3 && \text{Distributive law} \\
&= x_1 + x_2 x_3 && \text{Theorem 1}
\end{aligned}
$$

Example 2.3   The following expression will be minimized using Boolean algebra:
$x_1 x_2 + x_1 x_3 + x_1 x_2' x_3 + x_1' x_2 x_3 + x_1' x_2$

$x_1 x_2 + x_1 x_3 + x_1 x_2' x_3 + x_1' x_2 x_3 + x_1' x_2$

$$
\begin{aligned}
&= x_1 x_2 + x_1 x_3 + x_1 x_2' x_3 + x_1' x_2(x_3 + 1) && \text{Distributive law} \\
&= x_1 x_2 + x_1 x_3 + x_1 x_2' x_3 + x_1' x_2(1) && \text{Theorem 1} \\
&= x_1 x_2 + x_1 x_3 + x_1 x_2' x_3 + x_1' x_2 && \text{Identity law} \\
&= x_1 x_2 + x_1' x_2 + x_1 x_3 + x_1 x_2' x_3 && \text{Commutative law} \\
&= x_2(x_1 + x_1') + x_1 x_3(1 + x_2') && \text{Distributive law} \\
&= x_2(1) + x_1 x_3(1) && \text{Complementation law} \\
& && \text{Theorem 1} \\
&= x_2 + x_1 x_3 && \text{Identity law}
\end{aligned}
$$

Example 2.4   The following sum-of-minterms expression will be minimized using Boolean algebra:
$x_1 x_2 x_3 x_4 + x_1 x_2 x_3' x_4 + x_1 x_2' x_3 x_4 + x_1 x_2' x_3 x_4'$

$x_1 x_2 x_3 x_4 + x_1 x_2 x_3' x_4 + x_1 x_2' x_3 x_4 + x_1 x_2' x_3 x_4'$

$$
\begin{aligned}
&= x_1 x_2 x_4(x_3 + x_3') + x_1 x_2' x_3(x_4 + x_4') && \text{Distributive law} \\
&= x_1 x_2 x_4(1) + x_1 x_2' x_3(1) && \text{Complementation law} \\
&= x_1 x_2 x_4 + x_1 x_2' x_3
\end{aligned}
$$

Example 2.5   The following product-of-sums expression will be minimized using Boolean algebra:

$$(x_3 + x_4)(x_3' + x_4)$$

$$
\begin{aligned}
(x_3 + x_4)(x_3' + x_4) &= (x_3 + x_4)x_3' + (x_3 + x_4)x_4 && \text{Distributive law}\\
&= x_3 x_3' + x_4 x_3' + x_3 x_4 + x_4 x_4 && \text{Distributive law}\\
&= 0 + x_3' x_4 + x_3 x_4 + x_4 && \text{Complementation law}\\
&&& \text{Idempotent law}\\
&= x_3' x_4 + x_4(x_3 + 1) && \text{Distributive law}\\
&= x_3' x_4 + x_4 && \text{Theorem 1}\\
&= x_4(x_3' + 1) && \text{Distributive law}\\
&= x_4 && \text{Theorem 1}
\end{aligned}
$$

Example 2.6   The following equation will be minimized to obtain an exclusive-NOR format using Boolean algebra:

$$z_1 = x_1' x_2' x_3' x_4' + x_1' x_2 x_3' x_4 + x_1 x_2 x_3 x_4 + x_1 x_2' x_3 x_4'$$

$$
\begin{aligned}
z_1 &= x_1' x_3' (x_2' x_4' + x_2 x_4) + x_1 x_3 (x_2 x_4 + x_2' x_4') && \text{Distributive law}\\
&= x_1' x_3' (x_2 \oplus x_4)' + x_1 x_3 (x_2 \oplus x_4)' && \text{Distributive law}\\
&= (x_2 \oplus x_4)' (x_1' x_3' + x_1 x_3) && \text{Distributive law}\\
&= (x_2 \oplus x_4)' (x_1 \oplus x_3)'
\end{aligned}
$$

Example 2.7   The following equation will be minimized to obtain an exclusive-OR format using Boolean algebra:

$$z_1 = x_1' x_2' x_3' x_4 + x_2 x_3' x_4' + x_1' x_2' x_3 x_4 + x_1' x_2 x_3 x_4' + x_1 x_2' x_4 + x_2 x_3 x_4'$$

$$
\begin{aligned}
z_1 &= x_2' x_4(x_1' x_3' + x_1' x_3 + x_1) + x_2 x_4' (x_3' + x_1' x_3 + x_3) && \text{Distributive law}\\
&= x_2' x_4 [x_1' (x_3' + x_3) + x_1] + x_2 x_4' (1) && \text{Distributive law}\\
&= x_2' x_4 + x_2 x_4' && \text{Complementation law}
\end{aligned}
$$

Example 2.8   The following function will be minimized using Boolean algebra:

$$f(x_1, x_2) = x_1 \oplus (x_1 + x_2)$$

$$
\begin{aligned}
x_1 \oplus (x_1 + x_2) &= x_1(x_1 + x_2)' + x_1'(x_1 + x_2) && \text{Exclusive-OR expansion}\\
&= x_1(x_1' x_2') + x_1' x_1 + x_1' x_2 && \text{DeMorgan's law}\\
&= x_1' x_2 && \text{Complementation law}
\end{aligned}
$$

2.3 Karnaugh Maps

Algebraic minimization presented in the previous section becomes tedious when large logical expressions are encountered. The *Karnaugh map* provides a simplified method for minimizing Boolean functions. A Karnaugh map provides a geometrical representation of a Boolean function.

The Karnaugh map is arranged as an array of squares (or cells) in which each square represents a binary value of the input variables. The map is a convenient method of obtaining a minimal number of terms with a minimal number of variables per term for a Boolean function. A Karnaugh map presents a clear indication of function minimization without recourse to Boolean algebra and will generate a minimized expression in either a sum-of-products form or a product-of-sums form.

The Karnaugh map is a diagram consisting of squares, where each square represents a minterm. Since any Boolean function can be expressed as a sum of minterms, a Karnaugh map presents a graphical representation of the Boolean function and is an ideal tool for minimization.

Figure 2.1 illustrates a two-variable Karnaugh map and the corresponding truth table. Since there are four minterms for two variables, the map consists of four squares, one for each minterm. Figure 2.1 (a) shows the truth table for the four minterms that represent the two variables; Figure 2.1 (b) shows the position of the minterms in the map; and Figure 2.1 (c) shows the relationship between the squares and the two variables.

x_1	x_2	Minterm
0	0	m_0
0	1	m_1
1	0	m_2
1	1	m_3

(a)

(b)

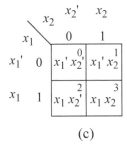

(c)

Figure 2.1   Two-variable Karnaugh map: (a) truth table; (b) minterm placement; and (c) minterms.

Figure 2.2 shows Karnaugh maps for two, three, four, and five variables. Each square in the maps corresponds to a unique minterm. The maps for three or more variables contain column headings that are represented in the *Gray code* format; the maps for four or more variables contain column and row headings that are represented in Gray code. Using the Gray code to designate column and row headings permits physically adjacent squares to be also logically adjacent; that is, to differ by only one variable. Map entries that are adjacent can be combined into a single term. For example, the expression $z_1 = x_1 x_2' x_3 + x_1 x_2 x_3$, which corresponds to minterms 5 and 7 in Figure 2.2 (b), reduces to $z_1 = x_1 x_3 (x_2' + x_2) = x_1 x_3$ using the distributive and complementation laws. Thus, if 1s are entered in minterm locations 5 and 7, then the two minterms can be combined into the single term $x_1 x_3$.

Similarly, in Figure 2.2 (c), if 1s are entered in minterm locations 4, 6, 12, and 14, then the four minterms combine as $x_2 x_4'$. That is, only variables x_2 and x_4' are common to all four squares — variables x_1 and x_3 are discarded by the complementation law. The minimized expression obtained from the Karnaugh map can be verified algebraically by listing the four minterms as a sum-of-minterms expression, then applying the appropriate laws of Boolean algebra as shown below.

$$x_1' x_2 x_3' x_4' + x_1' x_2 x_3 x_4' + x_1 x_2 x_3' x_4' + x_1 x_2 x_3 x_4'$$

$$= x_2 x_4' (x_1' x_3' + x_1' x_3 + x_1 x_3' + x_1 x_3)$$

$$= x_2 x_4'$$

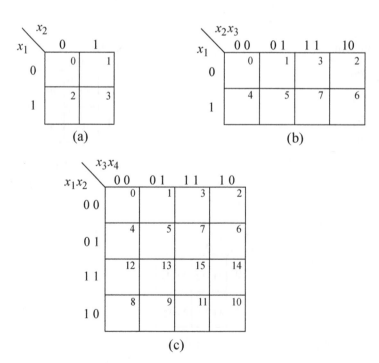

(a)   (b)

(c)

Figure 2.2   Karnaugh maps showing minterm locations: (a) two variables; (b) three variables; (c) four variables; (d) five variables; and (e) alternative map for five variables.

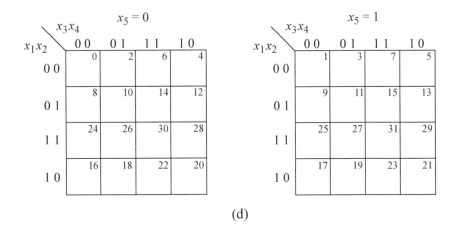

(d)

(e)

Figure 2.2 (Continued)

When minimizing a Boolean expression by grouping the 1s in a Karnaugh map, the result will be in a sum-of-products form; grouping the 0s results in a product-of-sums form. Each product term in a sum-of-products expression is specified as an *implicant* of the function, since the product term implies the function. That is, if the product term is equal to 1, then the function is also equal to 1. Specifically, a *prime implicant* is a unique grouping of 1s (an implicant) that does not imply any other grouping of 1s (other implicants).

Example 2.9   The following function will be minimized using a 4-variable Karnaugh map:

$$z_1(x_1,x_2,x_3,x_4) = x_2x_3' + x_2x_3x_4' + x_1x_2'x_3 + x_1x_3x_4'$$

The minimized result will be obtained in both a sum-of-products form and a prod-uct-of-sums form. To plot the function in the Karnaugh map, 1s are entered in the min-term locations that represent the product terms. For example, the term x_2x_3' is represented by the 1s in minterm locations 4, 5, 12, and 13. Only variables x_2 and x_3' are common to these four minterm locations. The term $x_2x_3x_4'$ is entered in minterm locations 6 and 14. When the function has been plotted, a minimal set of prime im-plicants can be obtained to represent the function. The largest grouping of 1s should always be combined, where the number of 1s in a group is a power of 2. The grouping of 1s is shown in Figure 2.3 and the resulting equation in Equation 2.1 in a sum-of-products notation.

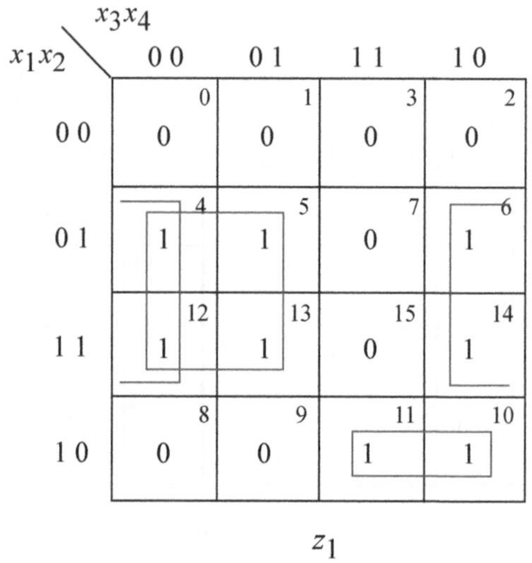

Figure 2.3   Karnaugh map representation of the function $z_1(x_1,x_2,x_3,x_4)=x_2x_3'$ $+ x_2x_3x_4' + x_1x_2'x_3 + x_1x_3x_4'$.

$$z_1(x_1,x_2,x_3,x_4) = x_2x_3' + x_2x_4' + x_1x_2'x_3 \qquad (2.1)$$

The minimal product-of-sums expression can be obtained by combining the 0s in Figure 2.3 to form sum terms in the same manner as the 1s were combined to form product terms. However, since 0s are being combined, each sum term must equal 0. Thus, the four 0s in row $x_1x_2 = 00$ in Figure 1.2 combine to yield the sum term $(x_1 + x_2)$. In a similar manner, the remaining 0s are combined to yield the product-of-sums expression shown in Equation 2.2. When combining 0s to obtain sum terms, treat a variable value of 1 as false and a variable value of 0 as true. Thus, maxterm locations 7 and 15 have variables $x_2x_3x_4 = 111$, providing a sum term of $(x_2' + x_3' + x_4')$.

$$z_1(x_1,x_2,x_3,x_4) = (x_1 + x_2)(x_2 + x_3)(x_2' + x_3' + x_4') \tag{2.2}$$

Equation 2.1 and Equation 2.2 both specify the conditions where z_1 is equal to 1. For example, consider the first term of Equation 2.1. If $x_2 x_3 = 10$, then Equation 2.1 yields $z_1 = 1 + \cdots + 0$, which generates a value of 1 for z_1. Applying $x_2 x_3 = 10$ to Equation 2.2 will cause every term to be equal to 1, such that $z_1 = (1)(1)(1) = 1$.

Figure 2.2 (d) illustrates a 5-variable Karnaugh map. To determine adjacency, the left map is superimposed on the right map. Any cells that are then physically adjacent are also logically adjacent and can be combined. Since x_5 is the low-order variable, the left map contains only even-numbered minterms; the right map is characterized by odd-numbered minterms. If 1s are entered in minterm locations 28, 29, 30, and 31, the four cells combine to yield the term $x_1 x_2 x_3$.

Figure 2.2 (e) illustrates an alternative configuration for a Karnaugh map for five variables. The map hinges along the vertical centerline and folds like a book. Any squares that are then physically adjacent are also logically adjacent. For example, if 1s are entered in minterm locations 24, 25, 28, and 29, then the four squares combine to yield the term $x_1 x_2 x_4'$.

Some minterm locations in a Karnaugh map may contain unspecified entries which can be used as either 1s or 0s when minimizing the function. These "*don't care*" entries are indicated by a dash (–) in the map. A typical situation which includes "don't care" entries is a Karnaugh map used to represent the BCD numbers. This requires a 4-variable map in which minterm locations 10 through 15 contain unspecified entries, since digits 10 through 15 are invalid for BCD.

Example 2.10   A minimized equation will be derived which is asserted whenever a BCD digit is even. All even BCD digits are plotted on a Karnaugh map as shown in Figure 2.4 for function z_1. The unspecified entries in minterm locations 10, 12, and 14 are assigned a value of 1; all remaining unspecified entries are assigned a value of 0. The equation for z_1 is shown in Equation 2.3.

Figure 2.4   Karnaugh map to represent even-numbered BCD digits.

$$z_1 = x_4' \qquad\qquad (2.3)$$

To obtain the equation which specifies BCD digits that are evenly divisible by 3, 1s are entered in minterm locations 0, 3, 6, and 9 to yield Equation 2.4.

$$z_1 = x_1 x_4 + x_2' x_3 x_4 + x_2 x_3 x_4' + x_1' x_2' x_3' x_4' \qquad\qquad (2.4)$$

Example 2.11   The exclusive-OR function and the exclusive-NOR function will be plotted on a 2-variable Karnaugh map. The exclusive-OR function is $f = x_1 x_2' + x_1' x_2$; the exclusive-NOR function is $f = (x_1 x_2' + x_1' x_2)' = x_1' x_2' + x_1 x_2$.

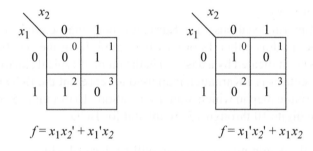

$$f = x_1 x_2' + x_1' x_2 \qquad\qquad\qquad f = x_1' x_2' + x_1 x_2$$

Example 2.12   Given the truth table shown below for the function f, the function will be plotted on a 3-variable Karnaugh map to obtain the minimized expression for f in a sum-of-products notation.

x_1	x_2	x_3	f
0	0	0	0
0	0	1	0
0	1	0	1
0	1	1	1
1	0	0	1
1	0	1	0
1	1	0	1
1	1	1	0

$$f = x_1 x_3' + x_1' x_2$$

Example 2.13 The function shown below will be minimized using a Karnaugh map and the solution verified using Boolean algebra.

$$f = x_1 x_2 x_3' + x_1 x_2' x_3' + x_1 x_2 x_3 + x_1 x_2' x_3$$

The minterms in the function correspond to minterms 6, 4, 7, and 5, respectively. Thus, the function can also be written in disjunctive normal form using decimal notation as

$$f(x_1, x_2, x_3) = \Sigma_m(4, 5, 6, 7), \text{ where the function } f = 1$$

or as

$$f'(x_1, x_2, x_3) = \Sigma_m(0, 1, 2, 3), \text{ where the function } f = 0$$

The minimized expression for the function f as obtained from the Karnaugh map is

$$f = x_1$$

Verify using Boolean algebra.

$$
\begin{aligned}
f = \ &f = x_1 x_2 x_3' + x_1 x_2' x_3' + x_1 x_2 x_3 + x_1 x_2' x_3 \\
= \ &x_1 (x_2 x_3' + x_2' x_3' + x_2 x_3 + x_2' x_3) \\
= \ &x_1 [x_2 (x_3' + x_3) + x_2' (x_3' + x_3)] \\
= \ &x_1 (x_2 + x_2') \\
= \ &x_1
\end{aligned}
$$

Example 2.14 The Karnaugh map of Figure 2.5 shows a redundant grouping of 1s that combines minterm locations 4 and 6. If all the 1s are covered, then any other grouping of 1s is redundant, except for hazards, which are covered in Chapter 3. The minimized equation for the function f is

$$f = x_1 x_2 + x_2' x_3'$$

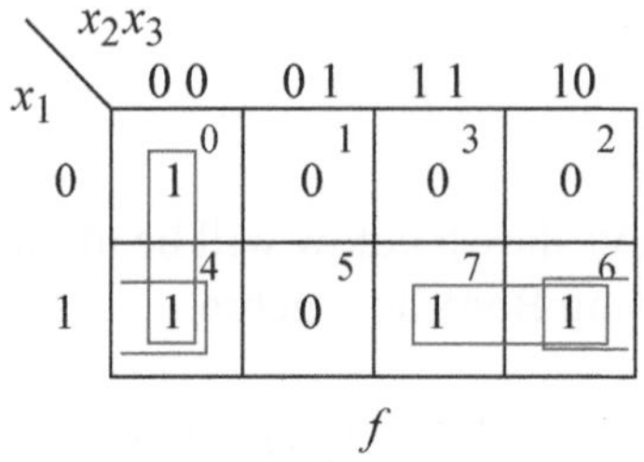

Figure 2.5   Karnaugh map showing a redundant product term of $x_1 x_3'$.

To illustrate that $x_1 x_3'$ is redundant, the term will be included in the equation for f, then minimized using Boolean algebra, as shown below.

$$f = x_1 x_2 + x_2' x_3' + x_1 x_3' \qquad \text{Expand to sum of minterms}$$
$$= x_1 x_2 (x_3 + x_3') + x_2' x_3' (x_1 + x_1') + x_1 x_3' (x_2 + x_2')$$
$$= x_1 x_2 x_3 + x_1 x_2 x_3' + x_1 x_2' x_3' + x_1' x_2' x_3' + x_1 x_2 x_3'$$
$$\quad + x_1 x_2' x_3'$$
$$= x_1 x_2 x_3 + x_1 x_2 x_3' + x_1 x_2' x_3' + x_1' x_2' x_3'$$
$$= x_1 x_2 (x_3 + x_3') + x_2' x_3' (x_1 + x_1')$$
$$= x_1 x_2 + x_2' x_3'$$

Example 2.15 Always group the largest number of 1s; that is, a grouping a 1, 2, 4, 8, or 16 1s, or a any number of 1s that is a power of two. In Figure 2.6 (a), if minterm locations 3 and 7 were combined in a group and minterm locations 2 and 6 were combined in a separate group, then this would result in an incorrect grouping, resulting in an equation for the function f as follows:

$$f = x_2 x_3 + x_2 x_3'$$

The above equation will function correctly, but it is not minimized. The correct grouping of minterm locations should be minterm locations 2, 3, 6, and 7 as a single grouping. This can be verified by Boolean algebra, as follows:

$$f = x_2 x_3 + x_2 x_3'$$
$$= x_2 (x_3 + x_3')$$
$$= x_2$$

Similarly, in Figure 2.6 (b), if minterm locations 2, 3, 6, and 7 are one grouping and minterm locations 1 and 5 are a second grouping, then the second grouping is incorrect to obtain a minimized equation for the function f. The correct groupings are as follows:

- Group minterm locations 2, 3, 6, and 7 to obtain the term x_2
- Group minterm locations 1, 3, 5, and 7 to obtain the term x_3 using minterm locations 3 and 7 a second time.

These groupings provide a minimized equation for the function f as follows:

$$f = x_2 + x_3$$

The non-minimized equation for f can be reduced by Boolean algebra as shown below.

$$f = x_2 + x_2' x_3$$
$$= x_2 + x_3 \qquad \text{Absorption law}$$

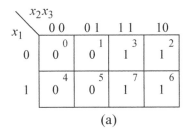

| (a) | (b) |

Figure 2.6   Karnaugh maps for Example 2.15.

Example 2.16 The following equation will be minimized as a sum of products and also as a product of sums:

$$f(x_1, x_2, x_3) = \Sigma_m(0, 1, 2, 5).$$

Figure 2.7 (a) shows the equation plotted on a 3-variable Karnaugh map as a function of x_1, x_2, and x_3, where the 1s are grouped to obtain a minimum sum of products, as shown in Equation 2.5. The value of f in Equation 2.5 is expressed as a logic 1.

$$f = x_1' x_3' + x_2' x_3 \qquad\qquad (2.5)$$

Figure 2.7 (b) shows the same map where the 0s are grouped to obtain a minimum product of sums, as shown in Equation 2.6. The value of f in Equation 2.6 is expressed as a logic 0.

$$f = (x_1' + x_3)(x_2' + x_3') \qquad (2.6)$$

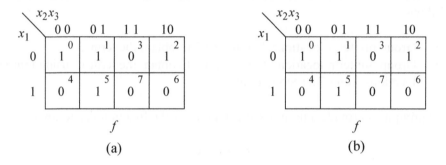

(a)   (b)

Figure 2.7   Karnaugh maps for Example 2.16: (a) combine 1s to obtain a sum of products; and (b) combine 0s to obtain a product of sums.

The complement of f is obtained by combining 0s, as shown in Equation 2.7. The product-of-sums expression can also be obtained by complementing the sum-of-products expression for f', which has a value of 0, as shown below.

$$f' = x_1 x_3' + x_2 x_3 \qquad (2.7)$$

$$
\begin{aligned}
f' &= x_1 x_3' + x_2 x_3 \\
f'' = \quad f &= (x_1 x_3' + x_2 x_3)' \\
&= (x_1' + x_3)(x_2' + x_3')
\end{aligned}
$$

Alternatively, the product of sums can be obtained directly from the map by combining the 0s, where the true value of a variable is treated as a 0 and the false value of a variable is treated as a 1, as shown below.

$$f = (x_1' + x_3)(x_2' + x_3') = 1$$

where $x_1' = 1$, $x_3 = 0$, $x_2' = 1$, and $x_3' = 1$. Thus, the equation for the function f can be expressed in two different ways, resulting in a value of logic 1, as shown in Equation 2.8.

$$f = x_1' x_3' + x_2' x_3 = (x_1' + x_3)(x_2' + x_3') = 1 \qquad (2.8)$$

Example 2.17 Karnaugh maps help to eliminate redundant terms in a minimized expression. The following equation will be minimized using the Karnaugh map of Figure 2.8 to obtain a product of sums:

$$f = x_1 x_2' + x_2 x_3$$

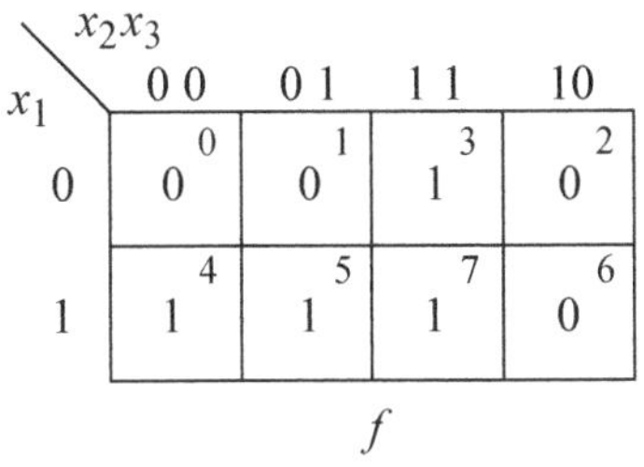

Figure 2.8 Karnaugh map for Example 2.17.

A Karnaugh map provides a clear indication of the function, making it easy to obtain either a sum-of-products expression or a product-of-sums expression. It is readily apparent that the minimized product-of-sums expression will contain two terms by combining two sets of adjacent 0s — maxterm locations 0 and 1; and maxterm locations 2 and 6, as shown in Equation 2.9.

$$f = (x_1 + x_2)(x_2' + x_3) \qquad (2.9)$$

Using Boolean algebra to obtain a minimized product-of-sums expression may introduce redundant terms, since it may not be immediately apparent that a term is redundant. Boolean algebra will now be used to obtain a minimized product-of-sums expression from the original equation $f = x_1 x_2' + x_2 x_3$.

$$
\begin{aligned}
f &= x_1 x_2' + x_2 x_3 \\
&= (x_1 x_2' + x_2)(x_1 x_2' + x_3) && \text{Distributive law} \\
&= (x_1 + x_2)(x_1 + x_3)(x_2' + x_3) && \text{Absorption law, distributive law}
\end{aligned}
$$

The term $(x_1 + x_3)$ is redundant, because all the 0s are covered. This term represents maxterm locations 0 and 2, which are included in the other two terms.

Example 2.18 In the truth table shown in Table 2.4 under the column labeled f, the 1 entries indicate minterms; the 0 entries indicate maxterms. The function f can be expressed as a sum of minterms by combining the 1 entries in a disjunctive normal form.

$$f(x_1, x_2, x_3) = \Sigma_m(1, 3, 4, 6)$$
$$f(x_1, x_2, x_3) = x_1'x_2'x_3 + x_1'x_2x_3 + x_1x_2'x_3' + x_1x_2x_3'$$

or as a product of maxterms by combining the 0s in a conjunctive normal form.

$$f(x_1, x_2, x_3) = \Pi_M(0, 2, 5, 7)$$
$$f(x_1, x_2, x_3) = (x_1 + x_2 + x_3)(x_1 + x_2' + x_3)(x_1' + x_2 + x_3')(x_1' + x_2' + x_3')$$

where the uncomplemented variables represent 0s in Table 2.4 and the complemented variables represent 1s in Table 2.4.

Table 2.4 Truth Table for the Function f of Example 2.18

x_1	x_2	x_3	Function f	
0	0	0	0	Maxterm
0	0	1	1	Minterm
0	1	0	0	Maxterm
0	1	1	1	Minterm
1	0	0	1	Minterm
1	0	1	0	Maxterm
1	1	0	1	Minterm
1	1	1	0	Maxterm

Example 2.19   The function shown in Table 2.5 will be plotted on the 4-variable Karnaugh map shown in Figure 2.9 and then minimized as a sum of products.

Table 2.5 Truth Table for the Function f of Example 2.19

x_1	x_2	x_3	x_4	Function f	x_1	x_2	x_3	x_4	Function f
0	0	0	0	1	1	0	0	0	0
0	0	0	1	1	1	0	0	1	0
0	0	1	0	1	1	0	1	0	0
0	0	1	1	1	1	0	1	1	0
0	1	0	0	0	1	1	0	0	1
0	1	0	1	0	1	1	0	1	1
0	1	1	0	0	1	1	1	0	1
0	1	1	1	0	1	1	1	1	1

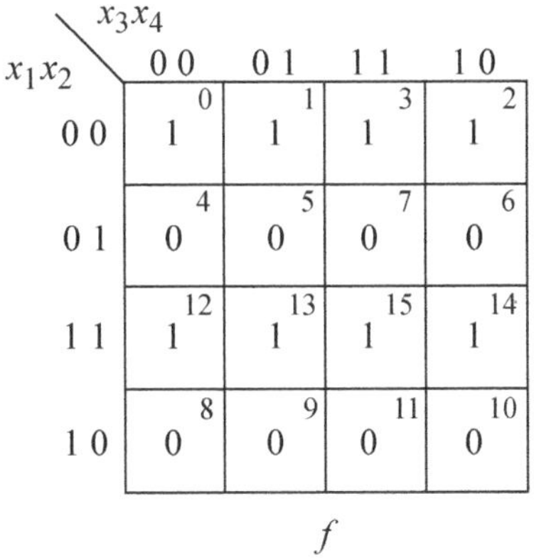

Figure 2.9   Karnaugh map for Example 2.19.

The function $f = x_1'x_2' + x_1x_2$, which is the exclusive-NOR function in which the function is a logic 1 only when the variables are equal. The exclusive-NOR function is also known as the *equality function*.

Example 2.20   The following equation will be minimized utilizing the Karnaugh map of Figure 2.10 and the result will be verified using Boolean algebra:

$$f = x_1'x_3'x_4' + x_1'x_3x_4' + x_1x_3'x_4' + x_2x_3x_4 + x_1x_3x_4'$$

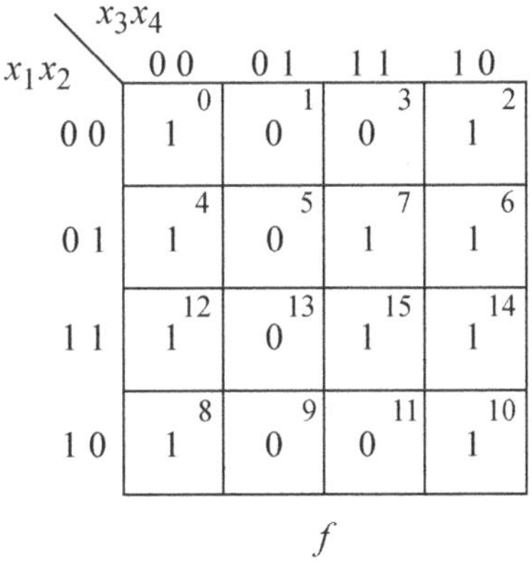

Figure 2.10   Karnaugh map for Example 2.20.

The minimized function is $f = x_4' + x_2x_3$.

Verify using Boolean algebra.

$$
\begin{aligned}
f &= x_1'x_3'x_4' + x_1'x_3x_4' + x_1x_3'x_4' + x_2x_3x_4 + x_1x_3x_4' \\
&= x_3'x_4' + x_3x_4' + x_2x_3x_4 \\
&= x_4' + (x_2x_3)x_4 \\
&= x_4' + x_2x_3
\end{aligned}
$$

Example 2.21 Given the Karnaugh map in Figure 2.11, the function f will be minimized as a sum-of-products expression and also as a product-of-sums expression.

x_1x_2 \ x_3x_4	0 0	0 1	1 1	1 0
0 0	0 (0)	0 (1)	1 (3)	0 (2)
0 1	0 (4)	1 (5)	1 (7)	0 (6)
1 1	0 (12)	1 (13)	1 (15)	0 (14)
1 0	1 (8)	1 (9)	1 (11)	1 (10)

f

Figure 2.11 Karnaugh map for Example 2.21.

Minimum sum of products: $f = x_1x_2' + x_2x_4 + x_3x_4$

Minimum product of sums: $f = (x_2' + x_4)(x_1 + x_4)(x_1 + x_2 + x_3)$

Example 2.22 This example will obtain an equation which generates an output of logic 1 whenever a 4-bit unsigned binary number $f = x_1x_2x_3x_4$ is greater than five, but less than ten, where x_4 is the low-order bit. The equation will be in a minimum sum-of-products form.

The Karnaugh map of Figure 2.12 is used to portray the specifications of Example 2.22. Thus, entries of 1 are placed in minterm locations 6, 7, 8, and 9, which satisfies the requirement of the number being greater than five, but less than ten.

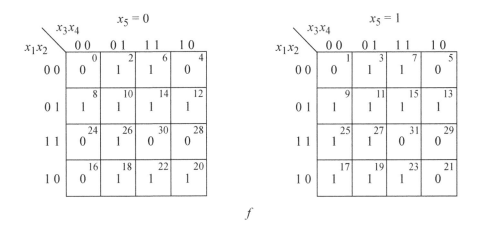

Figure 2.12   Karnaugh map for Example 2.22 to indicate when a number is greater than five, but less than ten.

The following function meets the requirements of Example 2.22:

$$f = x_1' x_2 x_3 + x_1 x_2' x_3'$$

Example 2.23   This example will obtain the minimum product-of-sums expression for the function f represented by the Karnaugh map of Figure 2.13.

Figure 2.13   Karnaugh map for Example 2.23.

Maxterm locations 30, 28, 31, and 29 are all adjacent and combine to yield the sum term of $(x_1' + x_2' + x_3')$ as follows:

$(x_1' + x_2' + x_3' + x_4' + x_5)(x_1' + x_2' + x_3' + x_4 + x_5)(x_1' + x_2' + x_3' + x_4' + x_5')$
$(x_1' + x_2' + x_3' + x_4 + x_5')$

Using the first two maxterms:
$(x_1' + x_2' + x_3' + x_4' + x_5)(x_1' + x_2' + x_3' + x_4 + x_5)$
$= (x_1' + x_2' + x_3') + [(x_4' + x_5)(x_4 + x_5)]$
$= (x_1' + x_2' + x_3') + [(x_4' + x_5)x_4 + (x_4' + x_5)x_5]$
$= (x_1' + x_2' + x_3') + [x_4 x_5 + x_4' x_5 + x_5]$
$= (x_1' + x_2' + x_3') + x_5 + x_5$
$= (x_1' + x_2' + x_3') + x_5$

Using the last two maxterms:
$(x_1' + x_2' + x_3' + x_4' + x_5')(x_1' + x_2' + x_3' + x_4 + x_5')$
$= (x_1' + x_2' + x_3') + [(x_4' + x_5')(x_4 + x_5')]$
$= (x_1' + x_2' + x_3') + [(x_4' + x_5')x_4 + (x_4' + x_5')x_5']$
$= (x_1' + x_2' + x_3') + (x_4 x_5' + x_4' x_5' + x_5')$
$= (x_1' + x_2' + x_3') + x_5' + x_5'$
$= (x_1' + x_2' + x_3') + x_5'$

Combine all four maxterms:
$(x_1' + x_2' + x_3' + x_4' + x_5)(x_1' + x_2' + x_3' + x_4 + x_5)(x_1' + x_2' + x_3' + x_4' + x_5')$
$(x_1' + x_2' + x_3' + x_4 + x_5')$
$= [(x_1' + x_2' + x_3') + x_5][(x_1' + x_2' + x_3') + x_5']$
$= [(x_1' + x_2' + x_3') + x_5][(x_1' + x_2' + x_3')] + [(x_1' + x_2' + x_3') + x_5]x_5'$
$= (x_1' + x_2' + x_3') + (x_1' + x_2' + x_3')x_5 + (x_1' + x_2' + x_3')x_5'$
$= [(x_1' + x_2' + x_3')](1 + x_5 + x_5')$
$= (x_1' + x_2' + x_3')$

In a similar manner, maxterms 0, 4, 1, and 5 combine to yield the sum term of $(x_1 + x_2 + x_4)$; maxterms 24 and 16 combine to yield $(x_1' + x_3 + x_4 + x_5)$; and maxterms 29 and 21 combine to yield $(x_1' + x_3' + x_4 + x_5')$. Therefore, the minimized function f as a product of sums is

$$f = (x_1' + x_2' + x_3')(x_1 + x_2 + x_4)(x_1' + x_3 + x_4 + x_5)(x_1' + x_3' + x_4 + x_5')$$

An equivalent product of sums can be obtained by combining maxterms 5 and 21 for the last term to yield

$$f = (x_1' + x_2' + x_3')(x_1 + x_2 + x_4)(x_1' + x_3 + x_4 + x_5)(x_2 + x_3' + x_4 + x_5')$$

Example 2.24 The following function will be plotted on the 4-variable Karnaugh map of Figure 2.14 and then minimized to obtain a sum-of-products expression:

$$f(x_1, x_2, x_3, x_4) = \Sigma_m(1, 3, 5, 7, 9) + \Sigma_d(6, 12, 13)$$

where $\Sigma_d(6, 12, 13)$ represents "don't care" entries for minterm locations 6, 12, and 13.

x_1x_2 \ x_3x_4	0 0	0 1	1 1	1 0
0 0	(0) 0	(1) 1	(3) 1	(2) 0
0 1	(4) 0	(5) 1	(7) 1	(6) –
1 1	(12) –	(13) –	(15) 0	(14) 0
1 0	(8) 0	(9) 1	(11) 0	(10) 0

f

Figure 2.14 Karnaugh map for Example 2.24.

Minterm locations 1, 5, 13, and 9 combine to yield the product term $x_3'x_4$. Minterm locations 1, 3, 5, and 7 combine to yield the product term $x_1'x_4$. The minimized sum-of-products expression is

$$f = x_3'x_4 + x_1'x_4$$

This can be further minimized by using the distributive law of Boolean algebra to obtain the following expression, which is in a product-of-sums form:

$$f = x_4(x_3' + x_1')$$

The product-of-sums form can also be obtained by combining the 0s and "don't cares" in the Karnaugh map of Figure 2.14.

Example 2.25 The Karnaugh map of Figure 2.15 depicts a function with unspecified entries that will be minimized as a sum of products and as a product of sums. Minterm locations 12, 8, 14, and 10 combine to yield the product term x_1x_4'; minterm locations 7, 6, 15, and 14 combine to yield x_2x_3; minterm locations 2, 6, 14, and 10 combine to yield x_3x_4'; and minterm location 1 yields $x_1'x_2'x_3'x_4$.

Using the same map, 0s are combined with unspecified entries to form sum terms. Maxterm locations 13, 15, 9, and 11 combine to yield the sum term $(x_1' + x_4')$; maxterm locations 4, 5, 12, and 13 combine to yield $(x_2' + x_3)$; maxterm locations 3 and 11 combine to yield $(x_2 + x_3' + x_4')$; and maxterm locations 0 and 4 combine to yield $(x_1 + x_3 + x_4)$.

x_1x_2 \ x_3x_4	0 0	0 1	1 1	1 0
0 0	^{0} 0	^{1} 1	^{3} 0	^{2} 1
0 1	^{4} 0	^{5} 0	^{7} 1	^{6} 1
1 1	^{12} –	^{13} –	^{15} –	^{14} –
1 0	^{8} 1	^{9} 0	^{11} –	^{10} –

f

Figure 2.15 Karnaugh map for Example 2.25.

Sum of products: $f = x_1x_4' + x_2x_3 + x_3x_4' + x_1'x_2'x_3'x_4$
Product of sums: $f = (x_1' + x_4')(x_2' + x_3)(x_2 + x_3' + x_4')(x_1 + x_3 + x_4)$

Example 2.26 Given the four variables $x_1x_2x_3x_4$, an expression will be obtained to satisfy the following requirement: a logic 1 will be generated whenever $x_1x_2 \geq x_3x_4$. This can be solved by means of a truth table; however, a faster approach is to use a 4-variable Karnaugh map, as shown in Figure 2.16, where x_1x_2 and x_3x_4 are two binary numbers with four values for each number — 00, 01, 10, and 11.

A value of 1 is entered in the map whenever the condition $x_1x_2 \geq x_3x_4$ is satisfied. In row $x_1x_2 = 00$, $x_1x_2 = x_3x_4$ in minterm 0; therefore, a 1 is entered in minterm 0 location. In row $x_1x_2 = 01$, $x_1x_2 \geq x_3x_4$ in minterm locations 4 and 5. In a similar manner, 1s are entered in appropriate locations for rows $x_1x_2 = 11$ and $x_1x_2 = 10$. The equation to satisfy the requirement of $x_1x_2 \geq x_3x_4$ is shown in Equation 2.10.

x_1x_2 \ x_3x_4	0 0	0 1	1 1	1 0
0 0	^{0} 1	^{1} 0	^{3} 0	^{2} 0
0 1	^{4} 1	^{5} 1	^{7} 0	^{6} 0
1 1	^{12} 1	^{13} 1	^{15} 1	^{14} 1
1 0	^{8} 1	^{9} 1	^{11} 0	^{10} 1

f

Figure 2.16 Karnaugh map to indicate when $x_1x_2 \geq x_3x_4$.

$$f = x_3'x_4' + x_2x_3' + x_1x_3' + x_1x_2 + x_1x_4' \quad (2.10)$$

2.3.1 Map-Entered Variables

Variables may also be entered in a Karnaugh map as map-entered variables, together with 1s and 0s. A map of this type is more compact than a standard Karnaugh map, but contains the same information. A map containing map-entered variables is particularly useful in analyzing and synthesizing synchronous sequential machines. When variables are entered in a Karnaugh map, two or more squares can be combined only if the squares are adjacent and contain the same variable(s).

Example 2.27 The following Boolean equation will be minimized using a 3-variable Karnaugh map with x_4 as a map-entered variable:

$$z_1(x_1, x_2, x_3, x_4) = x_1x_2'x_3x_4' + x_1x_2 + x_1'x_2'x_3'x_4' + x_1'x_2'x_3'x_4$$

Note that instead of $2^4 = 16$ squares, the map of Figure 2.17 contains only $2^3 = 8$ squares, since only three variables are used in constructing the map. To facilitate plotting the equation in the map, the variable that is to be entered is shown in parentheses as follows:

$$z_1(x_1, x_2, x_3, x_4) = x_1x_2'x_3(x_4') + x_1x_2 + x_1'x_2'x_3'(x_4') + x_1'x_2'x_3'(x_4)$$

x_1 \ x_2x_3	0 0	0 1	1 1	1 0
0	(0) $x_4' + x_4$	(1) 0	(3) 0	(2) 0
1	(4) 0	(5) x_4'	(7) 1	(6) 1

z_1

Figure 2.17 Karnaugh map using x_4 as a map-entered variable for Example 2.27.

The first term in the equation for z_1 is $x_1x_2'x_3$ (x_4') and indicates that the variable x_4' is entered in minterm location 5 $(x_1x_2'x_3)$. The second term x_1x_2 is plotted in the usual manner: 1s are entered in minterm locations 6 and 7. The third term specifies that the variable x_4' is entered in minterm location 0 $(x_1'x_2'x_3')$. The fourth term also applies to minterm 0, where x_4 is entered. The expression in minterm location 0, therefore, is $x_4' + x_4$.

To obtain the minimized equation for z_1 in a sum-of-products form, 1s are combined in the usual manner; variables are combined only if the minterm locations containing the variables are adjacent and the variables are identical. Consider the expression $x_4' + x_4$ in minterm location 0. Since $x_4' + x_4 = 1$, minterm 0 equates to $x_1'x_2'x_3'$. The entry of 1 in minterm location 7 can be restated as $1 + x_4'$ without changing the value of the entry (Theorem 1). This allows minterm locations 5 and 7 to be combined as $x_1 x_3 x_4'$. Finally, minterms 6 and 7 combine to yield the term $x_1 x_2$. The minimized equation for z_1 is shown in Equation 2.11.

$$z_1 = x_1'x_2'x_3' + x_1 x_3 x_4' + x_1 x_2 \tag{2.11}$$

Example 2.28 A Karnaugh map will be used to minimize the following Boolean function where x_2 is a map-entered variable:

$$f = x_1 x_2' + x_1 x_2' x_3' x_4 + x_3 x_4 + x_2 x_3' x_4 + x_1 x_2 x_3 x_4$$

The equation is rewritten as Equation 2.12 with variable x_2 in parentheses for ease of use. The map will have $2^3 = 8$ squares to cover variables x_1, x_3, and x_4 as shown in Figure 2.18.

$$f = x_1(x_2') + x_1(x_2')x_3'x_4 + x_3 x_4 + (x_2)x_3'x_4 + x_1(x_2)x_3 x_4 \tag{2.12}$$

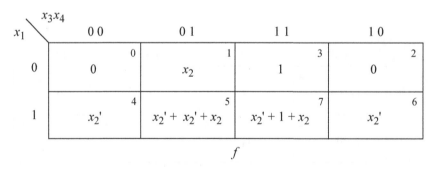

Figure 2.18   Karnaugh map using variable x_2 as a map-entered variable.

The first term in Equation 2.12 specifies that the entire row of x_1 contains the variable x_2'. The second term inserts x_2' in minterm 5 location. The third term does not contain a version of x_2; therefore, 1s are entered in minterm locations 3 and 7. The fourth term places x_2 in minterm locations 1 and 5. The fifth term causes x_2 to be inserted in minterm location 7.

The equation, obtained from the map, is shown in Equation 2.13. The first term in Equation 2.13 is obtained from row x_1 in the Karnaugh map in which each square contains x_2'. Minterm location 5 contains $x_2' + x_2' + x_2 = 1$; minterm location 7 contains $x_2' + 1 + x_2 = 1$. Minterm location 3 contains an entry of 1, which can be restated as $1 + x_2 = 1$. Therefore, the second term in Equation 2.13 combines the x_2 variable in minterm locations 1, 3, 5, and 7 with common variable x_4 — the x_3 variable is not a contributing factor to the $x_2 x_4$ term due to the distributive law and the complementation law. The third term combines the 1 entries in minterm locations 3 and 7.

$$f = x_1 x_2' + x_2 x_4 + x_3 x_4 \qquad (2.13)$$

Example 2.29 The following Boolean equation will be minimized using x_4 and x_5 as map-entered variables:

$$z_1 = x_1' x_2' x_3' (x_4 x_5') + x_1' x_2 + x_1' x_2' x_3' (x_4 x_5) + x_1 x_2' x_3' (x_4 x_5)$$
$$+ x_1 x_2' x_3 + x_1 x_2' x_3' (x_4') + x_1 x_2' x_3' (x_5')$$

Figure 2.19 shows the map entries for Example 2.29. The expression $x_4 x_5' + x_4 x_5$ in minterm location 0 reduces to x_4; the 1 entry in minterm location 2 can be expanded to $1 + x_4$ without changing the value in location 2. Therefore, locations 0 and 2 combine as $x_1' x_3' x_4$. The expression $x_4 x_5 + x_4' + x_5'$ in minterm location 4 reduces to 1. Thus, the 1 entries in the map combine in the usual manner to yield Equation 2.14.

x_1 \ $x_2 x_3$	0 0	0 1	1 1	1 0
0	$x_4 x_5' + x_4 x_5$ 0	0 1	1 3	1 2
1	$x_4 x_5 + x_4' + x_5'$ 4	1 5	0 7	0 6

z_1

Figure 2.19 Karnaugh map for Example 2.29 using x_4 and x_5 as map-entered variables.

$$z_1 = x_1' x_3' x_4 + x_1' x_2 + x_1 x_2' \qquad (2.14)$$

Example 2.30 The Karnaugh map of Figure 2.20 contains two map-entered variables p and q. The entry in minterm locations 2 and 13 is $p + p' = 1$. Since minterm location 2 is equivalent to 1, then that entry can be changed to $p + p' + q = 1$ without changing the value of the location (Theorem 1). Likewise, minterm location 13 can be changed to $p + p' + q' = 1$. These changes will be used to minimize the function. Minterm locations 0, 2, 8, and 10 combine to yield $x_2' x_4' q$ because every location can contain the variable q.

> Minterm location 0 can be changed to $1 + q$
> Minterm location 2 can be changed to $p + p' + q$

In a similar manner, the following minterm locations combine:

> 0 and 2 combine to yield $x_1' x_2' x_4'$
> 5 and 13 combine to yield $x_2 x_3' x_4 p$
> 13 and 15 combine to yield $x_1 x_2 x_4 p'$
> 13 and 9 combine to yield $x_1 x_3' x_4 q'$

$x_3 x_4$

$x_1 x_2$	0 0	0 1	1 1	1 0
0 0	1 ⁰	0 ¹	0 ³	$p + p'$ ²
0 1	0 ⁴	p ⁵	0 ⁷	0 ⁶
1 1	0 ¹²	$p + p'$ ¹³	p' ¹⁵	0 ¹⁴
1 0	q ⁸	q' ⁹	0 ¹¹	q ¹⁰

z_1

Figure 2.20   Karnaugh map for Example 2.30 using two map-entered variables.

Equation 2.15 shows the result of minimizing the Karnaugh map of Figure 2.19.

$$f = x_2' x_4' q + x_1' x_2' x_4' + x_2 x_3' x_4 p + x_1 x_2 x_4 p' + x_1 x_3' x_4 q' \qquad (2.15)$$

Example 2.31   Given the Karnaugh map shown in Figure 2.21 for five variables where a and b are map-entered variables, the function f will be minimized in a sum-of-products form.

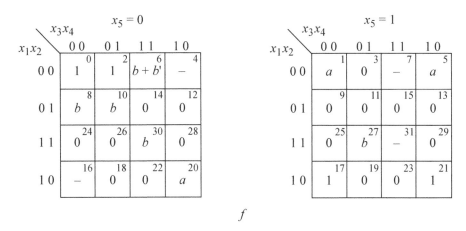

Figure 2.21   Five-variable Karnaugh map with a and b as map-entered variables.

The following minterm locations can be combined to generate the sum-of-minterm expression of Equation 2.16:

> 0, 2, 8, 10 combine to yield $x_1'x_3'x_5'b$;
>> 0 and 2 are equivalent to $1 + b$
> 0, 2, 6, 4 combine to yield $x_1'x_2'x_5'$
> 30 and 31 combine to yield $x_1x_2x_3x_4b$
> 27 and 31 combine to yield $x_1x_2x_4x_5b$
> 17 and 21 combine to yield $x_1x_2'x_4'x_5$
> 0, 4, 16, 20, 1, 5, 17, and 21 combine to yield $x_2'x_4'a$;
>> 1, 17, and 21 are equivalent to $1 + a$

$$f = x_1'x_3'x_5'b + x_1'x_2'x_5' + x_1x_2x_3x_4b + x_1x_2x_4x_5b \qquad (2.16)$$
$$+ x_1x_2'x_4'x_5 + x_2'x_4'a$$

Example 2.32   As a final example of map-entered variables, consider the Karnaugh map of Figure 2.22 in which x_4 is a map-entered variable. The equation for the function f will be obtained in a minimum sum-of-products form. The following minterm locations combine to yield the minimized equation of Equation 2.17:

0 and 2 combine to yield $x_1'x_3'x_4$
2 yields $x_1'x_2x_3'$
5 and 7 combine to yield $x_1x_3x_4'$
4 and 5 combine to yield $x_1x_2'x_4'$
5 yields $x_1x_2'x_3$

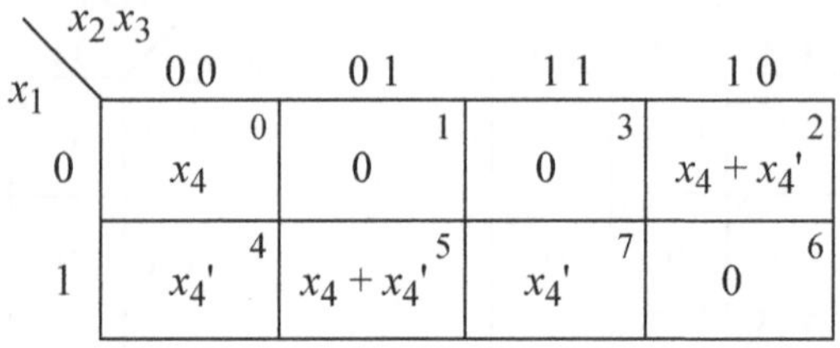

Figure 2.22   Karnaugh map for Example 2.32 using x_4 as a map-entered variable.

$$f = x_1'x_3'x_4 + x_1'x_2x_3' + x_1x_3x_4' + x_1x_2'x_4' + x_1x_2'x_3 \qquad (2.17)$$

2.4 Quine-McCluskey Algorithm

If the number of variables in a function is greater than seven, then the number of squares in the Karnaugh map becomes excess, which makes the selection of adjacent squares tedious. The Quine-McCluskey algorithm is a tabular method of obtaining a minimal set of prime implicants that represents the Boolean function. Because the process is inherently algorithmic, the technique is easily implemented with a computer program. The method consists of two steps: first obtain a set of prime implicants for the function; then obtain a minimal set of prime implicants that represents the function.

The rationale for the Quine-McCluskey method relies on the repeated application of the distributive and complementation laws. For example, for a 4-variable function, minterms $x_1x_2x_3'x_4$ and $x_1x_2x_3'x_4'$ are adjacent because they differ by only one variable. The two minterms can be combined, therefore, into a single product term as follows:

$$x_1x_2x_3'x_4 + x_1x_2x_3'x_4'$$
$$= x_1x_2x_3'(x_4 + x_4')$$
$$= x_1x_2x_3'$$

The resulting product term is specified as $x_1 x_2 x_3 x_4 = 110-$, where the dash (–) represents the variable that has been removed. The process repeats for all minterms in the function. Two product terms with dashes in the same position can be further combined into a single term if they differ by only one variable. Thus, the terms $x_1 x_2 x_3 x_4 = 110-$ and $x_1 x_2 x_3 x_4 = 100-$ combine to yield the term $x_1 x_2 x_3 x_4 = 1-0-$, which corresponds to $x_1 x_3'$.

The minterms are initially grouped according to the number of 1s in the binary representation of the minterm number. Comparison of minterms then occurs only between adjacent groups of minterms in which the number of 1s in each group differs by one. Minterms in adjacent groups that differ by only one variable can then be combined.

Example 2.33   The following function will be minimized using the Quine-McCluskey method: $f(x_1, x_2, x_3, x_4) = \Sigma_m(0,1,3,6,7,8,9,14)$. The first step is to list the minterms according to the number of 1s in the binary representation of the minterm number. Table 2.6 shows the listing of the various groups. Minterms that combine cannot be prime implicants; therefore, a check (✔) symbol is placed beside each minterm that combines with another minterm. When all lists in the table have been processed, the terms that have no check marks are prime implicants.

Consider List 1 in Table 2.6. Minterm 0 differs by only one variable with each minterm in Group 1. Therefore, minterms 0 and 1 combine as 000–, as indicated in the first entry in List 2 and minterms 0 and 8 combine to yield –000, as shown in the second row of List 2. Next, compare minterms in List 1, Group 1 with those in List 1, Group 2. It is apparent that the following pairs of minterms combine because they differ by only one variable: (1,3), (1,9), and (8,9) as shown in List 2, Group 1. Minterms 1 and 3 are in adjacent groups and can combine because they differ by only one variable. The resulting term is 00–1. Minterms 1 and 6 cannot combine because they differ by more than one variable. Minterms 1 and 9 combine as –001 and minterms 8 and 9 combine to yield 100–.

Table 2.6 Minterms Listed in Groups for Example 2.33

List 1			List 2			List 3		
Group	Minterms	$x_1 x_2 x_3 x_4$	Group	Minterms	$x_1 x_2 x_3 x_4$	Group	Minterms	$x_1 x_2 x_3 x_4$
0	0	0 0 0 0 ✔	0	0,1	0 0 0 – ✔	0	0,1,8,9	– 0 0 –
				0,8	– 0 0 0 ✔			
1	1	0 0 0 1 ✔	1	1,3	0 0 – 1			
	8	1 0 0 0 ✔		1,9	– 0 0 1 ✔			
				8,9	1 0 0 – ✔			
2	3	0 0 1 1 ✔	2	3,7	0 – 1 1			
	6	0 1 1 0 ✔		6,7	0 1 1 –			
	9	1 0 0 1 ✔		6,14	– 1 1 0			
3	7	0 1 1 1 ✔						
	14	1 1 1 0 ✔						

In a similar manner, minterms in the remaining groups are compared for possible adjacency. Note that those minterms that combine differ by a power of 2 in the decimal value of their minterm number. For example, minterms 6 and 14 combine as -110, because they differ by a power of 2 ($2^3 = 8$). Note also that the variable x_1 which is removed is located in column 2^3, where the binary weights of the four variables are $x_1 x_2 x_3 x_4 = 2^3\ 2^2\ 2^1\ 2^0$.

List 3 is derived in a similar manner to that of List 2. However, only those terms that are in adjacent groups and have dashes in the same column can be compared. For example, the terms 0,1 (000–) and 8,9 (100–) both contain dashes in column x_4 and differ by only one variable. Thus, the two terms can combine into a single product term as $x_1 x_2 x_3 x_4 = -00- (x_2' x_3')$. If the dashes are in different columns, then the two terms do not represent product terms of the same variables and thus, cannot combine into a single product term.

When all comparisons have been completed, some terms will not combine with any other term. These terms are indicated by the absence of a check symbol and are designated as prime implicants. For example, the term $x_1 x_2 x_3 x_4 = 00-1$ ($x_1' x_2' x_4$) in List 2 cannot combine with any term in either the previous group or the following group. Thus, $x_1' x_2' x_4$ is a prime implicant. The following terms represent prime implicants: $x_1' x_2' x_4$, $x_1' x_3 x_4$, $x_1' x_2 x_3$, $x_2 x_3 x_4'$ and $x_2' x_3'$.

Some of the prime implicants may be redundant, since the minterms covered by a prime implicant may also be covered by one or more other prime implicants. Therefore, the second step in the algorithm is to obtain a minimal set of prime implicants that covers the function. This is accomplished by means of a *prime implicant chart* as shown in Figure 2.23 (a). Each column of the chart represents a minterm; each row of the chart represents a prime implicant. The first row of Figure 2.23 (a) is specified by the minterm grouping of (1, 3), which corresponds to prime implicant $x_1' x_2' x_4$ (00–1). Since prime implicant $x_1' x_2' x_4$ covers minterms 1 and 3, an $\times$ is placed in columns 1 and 3 in the corresponding prime implicant row. The remaining rows are completed in a similar manner. Consider the last row which corresponds to prime implicant $x_2' x_3'$. Since prime implicant $x_2' x_3'$ covers minterms 0, 1, 8, and 9, an $\times$ is placed in the minterm columns 0, 1, 8, and 9.

A single $\times$ appearing in a column indicates that only one prime implicant covers the minterm. The prime implicant, therefore, is an *essential prime implicant*. In Figure 2.23 (a), there are two essential prime implicants: $x_2 x_3 x_4'$ and $x_2' x_3'$. A horizontal line is drawn through all $\times$s in each essential prime implicant row. Since prime implicant $x_2 x_3 x_4'$ covers minterm 6, there is no need to have prime implicant $x_1' x_2 x_3$ also cover minterm 6. Therefore, a vertical line is drawn through all $\times$s in column 6, as shown in Figure 2.23 (a). For the same reason, a vertical line is drawn through all $\times$s in column 1 for the second essential prime implicant $x_2' x_3'$.

The only remaining minterms not covered by a prime implicant are minterms 3 and 7. Minterm 3 is covered by prime implicants $x_1' x_2' x_4$ and $x_1' x_3 x_4$; minterm 7 is covered by prime implicants $x_1' x_3 x_4$ and $x_1' x_2 x_3$, as shown in Figure 2.23 (b). Therefore, a minimal cover for minterms 3 and 7 consists of the *secondary essential prime implicant* $x_1' x_3 x_4$. The complete minimal set of prime implicants for the function z_1 is shown in Equation 2.18. The minimized expression for z_1 can be verified by plotting the function on a Karnaugh map, as shown in Figure 2.24.

Prime implicants		Minterms							
		0	1	3	6	7	8	9	14
1,3	$(x_1'x_2'x_4)$		⨉	×					
3,7	$(x_1'x_3x_4)$			×		×			
6,7	$(x_1'x_2x_3)$				⨉	×			
* 6,14	$(x_2x_3x_4')$				⨉				⨉
* 0,1,8,9	$(x_2'x_3')$	⊗	⨉				⨉	⨉	

(a)

Prime implicants		Minterms							
		0	1	3	6	7	8	9	14
1,3	$(x_1'x_2'x_4)$			×					
3,7	$(x_1'x_3x_4)$			×		×			
6,7	$(x_1'x_2x_3)$					×			

(b)

Figure 2.23   Prime implicant chart for Example 2.33: (a) essential and nonessential prime implicants; and (b) secondary essential prime implicants with minimal cover for remaining prime implicants.

$$f(x_1,x_2,x_3,x_4) = x_2x_3x_4' + x_2'x_3' + x_1'x_3x_4 \qquad (2.18)$$

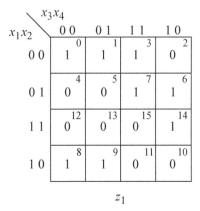

z_1

Figure 2.24   Karnaugh map for Example 2.33.

Example 2.34 Functions which include unspecified entries ("don't cares") are handled in a similar manner. The tabular representation of step 1 lists all the minterms, including "don't cares." The "don't care" conditions are then utilized when comparing minterms in adjacent groups. In step 2 of the algorithm, only the minterms containing specified entries are listed — the "don't care" minterms are not used. Then the minimal set of prime implicants is found as described in Example 2.33. The following function will be minimized using the Quine-McCluskey algorithm, as shown in Table 2.7:

$$f(x_1, x_2, x_3, x_4) = \Sigma_m(0, 1, 2, 7, 8, 9) + \Sigma_d(5, 6)$$

where minterm locations 5 and 6 are "don't cares." The prime implicant chart is shown in Figure 2.25. Notice that there are four solutions to this example, all of which are shown in Equation 2.19. The Karnaugh map of Figure 2.26 verifies the minimized sum-of-products solutions.

Table 2.7 Minterms Listed in Groups for Example 2.34

	List 1			List 2			List 3	
Group	Minterms	$x_1 x_2 x_3 x_4$	Group	Minterms	$x_1 x_2 x_3 x_4$	Group	Minterms	$x_1 x_2 x_3 x_4$
0	0	0 0 0 0 ✓	0	0,1	0 0 0 – ✓	0	0,1,8,9	– 0 0 –
				0,2	0 0 – 0			
				0,8	– 0 0 0 ✓			
1	1	0 0 0 1 ✓	1	1,5	0 – 0 1			
	2	0 0 1 0 ✓		1,9	– 0 0 1 ✓			
	8	1 0 0 0 ✓		2,6	0 – 1 0			
				8,9	1 0 0 – ✓			
2	5 (–)	0 1 0 1 ✓	2	5,7	0 1 – 1			
	6 (–)	0 1 1 0 ✓		6,7	0 1 1 –			
	9	1 0 0 1 ✓						
3	7	0 1 1 1 ✓						

Prime implicants		Minterms					
		0	1	2	7	8	9
0,2	$(x_1'x_2'x_4')$	✶		×			
1,5	$(x_1'x_3'x_4)$		✶				
2,6	$(x_1'x_3x_4')$			×			
5,7	$(x_1'x_2x_4)$				×		
6,7	$(x_1'x_2x_3)$				×		
* 0,1,8,9	$(x_2'x_3')$	✶	✶			⊗	⊗

Figure 2.25   Prime implicant chart for Example 2.34.

$$f = x_2'x_3' + x_1'x_2'x_4' + x_1'x_2x_4$$

$$f = x_2'x_3' + x_1'x_2'x_4' + x_1'x_2x_3$$

$$f = x_2'x_3' + x_1'x_3x_4' + x_1'x_2x_4$$

$$f = x_2'x_3' + x_1'x_3x_4' + x_1'x_2x_3 \tag{2.19}$$

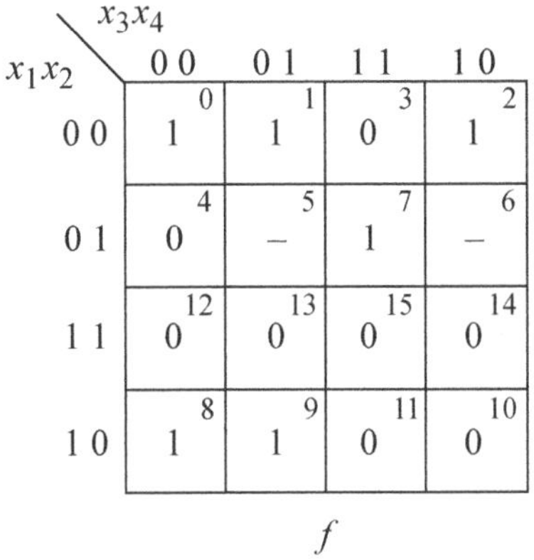

Figure 2.26   Karnaugh map for Example 2.34.

2.4.1 Petrick Algorithm

The function may not always contain an essential prime implicant, or the secondary essential prime implicants may not be intuitively obvious, as they were in previous examples. The technique for obtaining a minimal cover of secondary prime implicants is called the *Petrick algorithm* and can best be illustrated by an example.

Example 2.35   Given the prime implicant chart of Figure 2.27 for function z_1, it is obvious that there are no essential prime implicants, since no minterm column contains a single ×.

Prime implicants	Minterms				
	m_i	m_j	m_k	m_l	m_m
pi_1	×	×			×
pi_2	×		×	×	
pi_3		×		×	
pi_4			×		×

Figure 2.27   Prime implicant chart for Example 2.35.

It is observed that minterm m_i is covered by prime implicants pi_1 or pi_2; m_j is covered by pi_1 or pi_3; m_k is covered by pi_2 or pi_4; m_l is covered by pi_2 or pi_3; and m_m is covered by pi_1 or pi_4. Since the function is covered only if all minterms are covered, Equation 2.20 represents this requirement.

$$\text{Function is covered} = (pi_1 + pi_2)\,(pi_1 + pi_3)\,(pi_2 + pi_4)$$
$$(pi_2 + pi_3)\,(pi_1 + pi_4) \qquad (2.20)$$

Equation 2.20 can be reduced by Boolean algebra or by a Karnaugh map to obtain a minimal set of prime implicants that represents the function. Figure 2.28 illustrates the Karnaugh map in which the sum terms of Equation 2.20 are plotted. The map is then used to obtain a minimized expression that represents the different combinations of prime implicants in which all minterms are covered. Equation 2.21 lists the product terms specified as prime implicants in a sum-of-products notation.

$pi_1 pi_2$ \ $pi_3 pi_4$	0 0	0 1	1 1	1 0
0 0	0	0	0	0
0 1	0	0	1	0
1 1	1	1	1	1
1 0	0	0	1	0

Figure 2.28   Karnaugh map in which the sum terms of Equation 2.20 are entered as 0s.

$$\text{All minterms are covered} = pi_1\,pi_2 + pi_2\,pi_3\,pi_4 + pi_1\,pi_3\,pi_4 \qquad (2.21)$$

The first term of Equation 2.21 represents the fewest number of prime implicants to cover the function. Thus, function z_1 will be completely specified by the expression $z_1 = pi_1\,pi_2$. From any covering equation, the term with the fewest number of variables is chosen to provide a minimal set of prime implicants. Assume, for example, that prime implicant $pi_1 = x_i x_j' x_k$ and that $pi_2 = x_l' x_m x_n$. Thus, the sum-of-products expression is $z_1 = x_i x_j' x_k + x_l' x_m x_n$.

Example 2.36 Using the prime implicant chart of Figure 2.29 for the function f, the minimal sum-of-products expression will be obtained. Any "don't care" conditions are not shown in the chart. There are no essential prime implicants because there is no column with a single $\times$. Prime implicants are selected such that all columns (minterms) are covered by the least number of prime implicants.

	Minterms			
Prime implicants	1	2	3	4
A	$\times$			$\times$
B			$\times$	$\times$
C	$\times$			
D		$\times$	$\times$	
E		$\times$		

Figure 2.29   Prime implicant chart for the function f of Example 2.36.

The following minterm coverage exists:

Minterm 1 is covered by prime implicants $(A + C)$
Minterm 2 is covered by prime implicants $(D + E)$
Minterm 3 is covered by prime implicants $(B + D)$
Minterm 4 is covered by prime implicants $(A + B)$

Therefore, all minterms are covered by Equation 2.22.

$$\text{All minterms are covered} = (A + C)\,(D + E)\,(B + D)\,(A + B) \qquad (2.22)$$

Equation 2.22 is plotted on the Karnaugh map of Figure 2.30 as sum terms, then minimized as a sum of products. This provides a list of all covers for the function f, as shown in Equation 2.23. The product term with the fewest number of variables is chosen to be the minimal cover. The product term AD requires two prime implicants; all others require three prime implicants. Therefore, the minimal cover for the function f is the sum-of-products implementation of $A + D$. If there are two or more prime implicants with the fewest number of variables, then any one can be chosen.

E = 0

C D				
A B	0 0	0 1	1 1	1 0
0 0	0 [0]	0 [2]	0 [6]	0 [4]
0 1	0 [8]	0 [10]	1 [14]	0 [12]
1 1	0 [24]	1 [26]	1 [30]	0 [28]
1 0	0 [16]	1 [18]	1 [22]	0 [20]

E = 1

C D				
A B	0 0	0 1	1 1	1 0
0 0	0 [1]	0 [3]	0 [7]	0 [5]
0 1	0 [9]	0 [11]	1 [15]	1 [13]
1 1	1 [25]	1 [27]	1 [31]	1 [29]
1 0	0 [17]	1 [19]	1 [23]	0 [21]

Figure 2.30   Karnaugh map for the Petrick algorithm of Example 2.36.

$$f = AD + ABE + BCE + BCD \tag{2.23}$$

Example 2.37 As a final example to illustrate the application of the Petrick algorithm when using the Quine-McCluskey minimization technique, Equation 2.24 will be minimized. Table 2.8 shows the minterms partitioned into groups containing identical number of 1s.

$$f(x_1, x_2, x_3, x_4) = \Sigma_m\,(1, 4, 5, 6, 13, 14, 15) + \Sigma_d\,(8, 9) \tag{2.24}$$

Table 2.8 Minterms Listed in Groups for Example 2.37

	List 1			List 2			List 3	
Group	Minterms	$x_1x_2x_3x_4$	Group	Minterms	$x_1x_2x_3x_4$	Group	Minterms	$x_1x_2x_3x_4$
1	1	0 0 0 1 ✓	1	1,5	0 – 0 1 ✓	1	1,5,9,13	– – 0 1
	4	0 1 0 0 ✓		1,9	– 0 0 1 ✓			
	8 (–)	1 0 0 0 ✓		4,5	0 1 0 –			
				4,6	0 1 – 0			
				8,9	1 0 0 –			
2	5	0 1 0 1 ✓	2	5,13	– 1 0 1 ✓			
	6	0 1 1 0 ✓		6,14	– 1 1 0			
	9 (–)	1 0 0 1 ✓		9,13	1 – 0 1 ✓			
3	13	1 1 0 1 ✓	3	13,15	1 1 – 1			
	14	1 1 1 0 ✓		14,15	1 1 1 –			
4	15	1 1 1 1 ✓						

Figure 2.31 shows the prime implicant chart for Example 2.37 in which there is only one essential prime implicant [1, 5, 9, 13 $(x_3'x_4)$] — the remaining prime implicants are secondary essential prime implicants. Since minterms 8 and 9 are "don't care" conditions, there are no columns labeled 8 and 9. Figure 2.32 shows the chart for the secondary essential prime implicants.

Prime implicants		Minterms						
		1	4	5	6	13	14	15
4,5	$(x_1'x_2x_3')$		×	×				
4,6	$(x_1'x_2x_4')$		×		×			
6,14	$(x_2x_3x_4')$				×		×	
13,15	$(x_1x_2x_4)$					×		×
14,15	$(x_1x_2x_3)$						×	×
* 1,5,9,13	$(x_3'x_4)$	⊗		×		×		

Figure 2.31   Prime implicant chart for Example 2.37.

	Prime implicants		Minterms			
			4	6	14	15
A	4,5	$(x_1'x_2x_3')$	×			
B	4,6	$(x_1'x_2x_4')$	×	×		
C	6,14	$(x_2x_3x_4')$		×	×	
D	13,15	$(x_1x_2x_4)$				×
E	14,15	$(x_1x_2x_3)$			×	×

Figure 2.32   Chart for secondary essential prime implicants.

The following minterm coverage exists:

> Minterm 4 is covered by prime implicants $(A + B)$
> Minterm 6 is covered by prime implicants $(B + C)$
> Minterm 14 is covered by prime implicants $(C + E)$
> Minterm 15 is covered by prime implicants $(D + E)$

Therefore, all minterms are covered by Equation 2.25.

$$\text{All minterms are covered} = (A + B)(B + C)(C + E)(D + E) \qquad (2.25)$$

Equation 2.25 is plotted on the Karnaugh map of Figure 2.33 as sum terms, then minimized as a sum of products. This provides a list of all covers for the function f, as shown in Equation 2.26. The product term with the fewest number of variables is chosen to be the minimal cover. The product term BE requires two prime implicants; all others require three prime implicants. Therefore, the minimal cover for the function f is the sum-of-products implementation of $B + E$ plus the essential prime implicant, as shown in Equation 2.27.

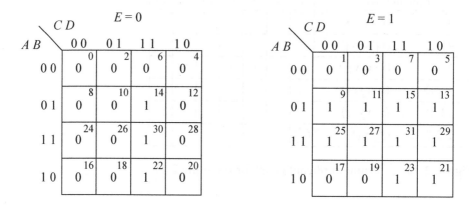

Figure 2.33   Karnaugh map for the Petrick algorithm of Example 2.37.

$$f = BE + ACE + ACD + BCD \tag{2.26}$$

$$f = x_3'x_4 + x_1'x_2x_4' + x_1x_2x_3 \tag{2.27}$$

2.5 Problems

2.1   Minimize the following function using Boolean algebra:

$$f = x_1'x_3'x_1 + x_1x_2x_3x_3' + x_1x_2x_3x_3'$$

2.2   Write the dual for the following statement:

$$z_1 = x_1x_2' + x_1'x_2$$

2.3   Write a Boolean equation using three terms for $f(x_1, x_2, x_3, x_4)$ in

    (a)   Disjunctive normal form.
    (b)   Canonical sum-of-products form.

2.4   Indicate which of the equations shown below will always generate a logic 1.

    (a) $z_1 = x_1 + x_2 x_3 x_4' + x_1' x_3 + x_3'$
    (b) $z_1 = x_1 x_2 + x_1 x_2' + x_2' x_4' + x_1' x_2 x_4$
    (c) $z_1 = x_1 x_2' x_4 + x_1' x_2 x_3' + x_1' x_2 x_4 + x_1' x_2' x_3 + x_1' x_2 x_4'$
    (d) $z_1 = x_4' + x_1' x_4 + x_1 x_4 + x_2' x_4'$

2.5   Minimize the following Boolean expression:

$$f = x_1' x_2 (x_3' x_4' + x_3' x_4) + x_1 x_2 (x_3' x_4' + x_3' x_4) + x_1 x_2' x_3' x_4$$

2.6   Indicate whether the following statement is true or false:

$$x_1' x_2' x_3' + x_1 x_2 x_3' = x_3'$$

2.7   Minimize the following equation using Boolean algebra:

$$z_1 = x_1' x_3' x_4' + x_1' x_3 x_4' + x_1 x_3' x_4' + x_2 x_3 x_4 + x_1 x_3 x_4'$$

2.8   Obtain the canonical product-of-sums form for the following function using Boolean algebra:

$$f(x_1, x_2, x_3, x_4) = x_3$$

2.9   Minimize the following equation to obtain a sum-of-products expression using Boolean algebra:

$$f = (x_1 x_2' + x_1' x_2) x_3' + (x_1 x_2' + x_1' x_2)' x_3 + x_1' x_3 + x_2' x_3$$

2.10   Determine if the following Boolean equation is valid using the axioms and theorems of Boolean algebra:

$$x_1 x_2 = (x_1 + x_3')(x_1' + x_2')(x_1' + x_2)$$

2.11   Prove that $x_1 + 1 = 1$.

2.12   Prove that $x_1'' = x_1$.

2.13   Use DeMorgan's theorem to minimize the following Boolean expression:
$(x_1 + x_2' + x_3 + x_4')' + (x_1 x_2 x_3 x_4')'$

2.14   Given the equation shown below, use x_4 as a map-entered variable in a Karnaugh map and obtain the minimum sum-of-products expression. Then use the original equation using x_2 as a map-entered variable and compare the results. If possible, further minimize both answers using Boolean algebra.

$$f = x_1'x_3x_4 + x_1x_2x_3 + x_1x_2'x_3x_4 + x_1x_2'x_3x_4'$$

2.15   Given the Karnaugh map shown below, obtain the minimum sum-of-products expression and the minimum product-of-sums expression for the function f.

x_1x_2 \ x_3x_4	0 0	0 1	1 1	1 0
0 0	0	0	1	0
0 1	0	1	1	0
1 1	0	1	1	0
1 0	1	1	1	1

f

2.16   Plot the following Boolean expression on a Karnaugh map:

$$f = x_1'x_2(x_3'x_4' + x_3'x_4) + x_1x_2(x_3'x_4' + x_3'x_4) + x_1x_2'x_3'x_4$$

2.17   Obtain the minimized sum-of-products expression for the function z_1 represented by the Karnaugh map shown below.

$x_5 = 0$

x_1x_2 \ x_3x_4	0 0	0 1	1 1	1 0
0 0	0 [0]	1 [2]	1 [6]	0 [4]
0 1	1 [8]	1 [10]	1 [14]	1 [12]
1 1	0 [24]	1 [26]	0 [30]	0 [28]
1 0	0 [16]	1 [18]	1 [22]	1 [20]

$x_5 = 1$

x_1x_2 \ x_3x_4	0 0	0 1	1 1	1 0
0 0	0 [1]	1 [3]	1 [7]	0 [5]
0 1	1 [9]	1 [11]	1 [15]	1 [13]
1 1	1 [25]	1 [27]	0 [31]	0 [29]
1 0	1 [17]	1 [19]	1 [23]	0 [21]

z_1

2.18   Obtain the minimized expression for the function f in a sum-of-products form from the Karnaugh map shown below.

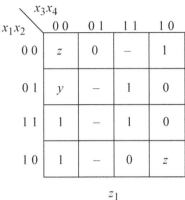

2.19   Write the equation that will generate a logic 1 whenever a 4-bit unsigned binary number $z_1 = x_1 x_2 x_3 x_4$ is greater than six. The equation is to be in a minimum sum-of-products form.

2.20   Given the following Karnaugh map, obtain the minimized expression for z_1 in a sum-of-products form and a product-of-sums form.

<table>
<tr><td rowspan="2">$x_1 x_2$</td><td colspan="4">$x_3 x_4$</td></tr>
<tr><td>0 0</td><td>0 1</td><td>1 1</td><td>1 0</td></tr>
<tr><td>0 0</td><td>0 ₀</td><td>0 ₁</td><td>1 ₃</td><td>0 ₂</td></tr>
<tr><td>0 1</td><td>0 ₄</td><td>1 ₅</td><td>0 ₇</td><td>0 ₆</td></tr>
<tr><td>1 1</td><td>0 ₁₂</td><td>1 ₁₃</td><td>0 ₁₅</td><td>0 ₁₄</td></tr>
<tr><td>1 0</td><td>0 ₈</td><td>1 ₉</td><td>1 ₁₁</td><td>0 ₁₀</td></tr>
</table>

z_1

2.21   Given the Karnaugh map shown below, obtain the minimized expression for z_1 in a sum-of-products form.

x_1 \ x_2x_3	0 0	0 1	1 1	1 0
0	$A + B'$ [0]	$-$ [1]	0 [3]	A [2]
1	B' [4]	1 [5]	$AB + A' + B'$ [7]	0 [6]

$$z_1$$

2.22   Minimize the following expression using a Karnaugh map with x_3 and x_4 as map-entered variables:

$$z_1 = x_1'x_2x_3'x_4' + x_1'x_2x_3'x_4 + x_1'x_2x_3x_4' + x_1x_2'x_3x_4 + x_1x_2x_3'x_4$$

2.23   Plot the following expression on a Karnaugh map using x_4 and x_5 as map-entered variables, then obtain the minimized sum of products:

$$f(x_1, x_2, x_3, x_4, x_5) = x_1'x_2x_3'x_4x_5 + x_1x_2x_3' + x_1x_2'x_3'x_4x_5'$$

2.24   Plot the following function on a Karnaugh map, then obtain the minimum sum-of-products expression and the minimum product-of-sums expression:

$$f(x_1, x_2, x_3, x_4) = \Pi_M \, (0, 1, 2, 8, 9, 12)$$

2.25   Plot the following function on a Karnaugh map, then obtain the minimum sum-of-products expression and the minimum product-of-sums expression:

$$f(x_1, x_2, x_3, x_4) = \Sigma_m \, (0, 1, 2, 3, 5, 7, 10, 12, 15)$$

2.26   Indicate whether the following equation is true or false:

$$x_1'x_2 + x_1'x_2'x_3x_4' + x_1x_2x_3'x_4 = x_1'x_2x_3' + x_1'x_3x_4' + x_1'x_2x_3 + x_2x_3'x_4$$

2.27   Plot the following Boolean expression on a Karnaugh map, then obtain the minimum product of sums:

$$f(x_1, x_2, x_3, x_4, x_5) = x_1x_3x_4(x_2 + x_4x_5') + (x_2' + x_4)(x_1x_3' + x_5)$$

2.28   Plot the following equation on a Karnaugh map, then obtain the minimum sum of products:

$$f = \{[x_1' + (x_1x_2)''] \, [x_2' + (x_1x_2)'']\}'$$

2.29   Obtain the minimum sum-of-products equation that will generate a logic 1 whenever the binary number N shown below satisfies the following criteria:

$N = x_1 x_2 x_3 x_4$, where x_4 is the low-order bit

$2 < N \leq 6$ and $11 \leq N < 14$

2.30   Obtain the minimum sum-of-products expression for the Quine-McCluskey prime implicant table shown below, where $f(x_1, x_2, x_3, x_4, x_5)$.

Prime implicants		Minterms									
		0	1	3	7	15	16	18	19	23	31
0 0 0 0 −	$(x_1'x_2'x_3'x_4')$	×	×								
0 0 0 − 1	$(x_1'x_2'x_3'x_5)$		×	×							
− 0 − 1 1	$(x_2'x_4x_5)$			×	×				×	×	
− − 1 1 1	$(x_3x_4x_5)$				×	×				×	×
1 0 0 1 −	$(x_1x_2'x_3'x_4)$							×	×		
1 0 0 − 0	$(x_1x_2'x_3'x_5')$						×	×			
− 0 0 0 0	$(x_2'x_3'x_4'x_5')$	×					×				

2.31   Minimize the following equation using the Quine-McCluskey algorithm, then verify the result by a Karnaugh map:

$f(x_1, x_2, x_3) = \Sigma_m (1, 2, 5, 6) + \Sigma_d (0, 3)$

2.32   Use the Quine-McCluskey algorithm to obtain the minimum sum-of-products form for the following equation:

$f(x_1, x_2, x_3, x_4) = \Sigma_m (0, 1, 3, 7, 9, 11, 14, 15)$

2.33   Obtain the minimum sum-of-products expression for the Quine-McCluskey prime implicant table shown below, where $f(x_1, x_2, x_3, x_4, x_5)$.

Prime implicants		Minterms									
		0	1	3	7	15	16	18	19	23	31
0,1	$(x_1'x_2'x_3'x_4')$	×	×								
1,3	$(x_1'x_2'x_3'x_5)$		×	×							
3,7,19,23	$(x_2'x_4x_5)$			×	×				×	×	
7,15,23,31	$(x_3x_4x_5)$				×	×				×	×
18,19	$(x_1x_2'x_3'x_4)$							×	×		
16,18	$(x_1x_2'x_3'x_5')$						×	×			
0,16	$(x_2'x_3'x_4'x_5')$	×					×				

2.34   Using any method, list all of the prime implicants (both essential and nonessential) for the following expression:

$$z_1(x_1, x_2, x_3, x_4) = \Sigma_m (0, 2, 3, 5, 7, 8, 10, 11, 13, 15)$$

2.35   Given the prime implicant chart shown below, use the Petrick algorithm to find a minimal set of prime implicants for the function.

Prime implicants	Minterms						
	m1	m2	m3	m4	m5	m6	m7
A	×	×	×			×	×
B	×		×	×	×		
C		×				×	
D				×	×		×

2.36   Use the Quine-McCluskey algorithm to find the minimum sum-of-products expression for the following function, then verify the result by means of a Karnaugh map:

$$f(x_1, x_2, x_3, x_4) = \Sigma_m (0, 1, 7, 8, 10, 12, 14, 15) + \Sigma_d (2, 5)$$

2.37   Use the Quine-McCluskey algorithm and the Petrick algorithm to obtain the minimum sum-of-products expression for the function shown below. Verify the result by means of a Karnaugh map.

$$f(x_1, x_2, x_3, x_4) = \Sigma_m (3, 4, 6, 7, 11, 12, 13, 15)$$

3

Combinational Logic

A combinational logic circuit is one in which the outputs are a function of the present inputs only. The operation of combinational logic circuits can be expressed in terms of fundamental logical operations such as AND, OR, and NOT (Invert). The output f of an exclusive-OR operation with inputs x_1 and x_2 is described as $f = x_1 x_2' + x_1' x_2$, which is read as "$(x_1$ AND NOT $x_2)$ OR (NOT x_1 and $x_2)$." The output f of an exclusive-NOR operation with inputs x_1 and x_2 is described as $f = x_1 x_2 + x_1' x_2'$, which is read as "$(x_1$ AND $x_2)$ OR (NOT x_1 AND NOT $x_2)$."

This chapter will present the analysis and synthesis of combinational logic. During *analysis*, the operation of a given logic circuit is described in terms of a logical expression or a truth table. During *synthesis* (or design), a word description and/or a timing diagram is converted to a logic diagram.

The logic gates presented in this chapter are the basic building blocks of digital systems that can range from simple circuits to more complex circuits such as multiplexers, encoders, and binary adders. A *logic gate* is an electronic switch that allows or prevents signals to be propagated through the device. Logic gates can be of two types: unilateral or bilateral. A *unilateral* device transfers signals in one direction only; a *bilateral* device transfers signals in two directions, depending on the state (or value) of a direction control input.

A combinational logic circuit performs an operation that is specified by a set of Boolean functions and consists of n input variables x, p logic gates, and m output variables z, as shown in Figure 3.1. The logic gates accept signals from the input variables and generate signals to the output variables based on the interconnection of the logic gates.

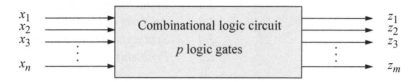

Figure 3.1   Block diagram of a combinational logic circuit.

3.1 Logic Primitive Gates

Figure 3.2 shows the logic gate symbols used for both the ANSI/IEEE Std. 91-1984 uniform-shape symbols and the corresponding distinctive-shape symbols. The symbols for logic macros such as, multiplexers, decoders, and encoders will be covered in Section 3.2 of this chapter. The *polarity symbol* "⌐" in Figure 3.2 indicates an active-low assertion on either an input or an output of a logic symbol. The polarity symbol points in the direction of signal flow.

By convention, input signals enter the logic symbol on the left and exit on the right. Since logic gates are identical in shape, a qualifying symbol is inserted within the rectangle to specify the function of the gate. The *qualifying function symbols* are defined as follows:

Symbol	Description
&	The AND function of two or more inputs. All inputs must be at their active voltage level to assert the output at its active voltage level.
≥1	The OR function of two or more inputs. One or more inputs must be at an active voltage level to assert the output at its active voltage level.
1	The NOT (invert) function. Only one input enters the logic symbol. The input voltage is inverted from high to low or from low to high.
=1	The exclusive-OR function. Only one of the two inputs is active to assert the output at its active voltage level. Thus, if x_1 and x_2 are the inputs and z_1 is the output, then $z_1 = x_1 x_2' + x_1' x_2$. If the output has an active-low polarity symbol, then this represents the exclusive-NOR function. The exclusive-NOR function is also referred to as the equality function, where $z_1 = x_1 x_2 + x_1' x_2'$.

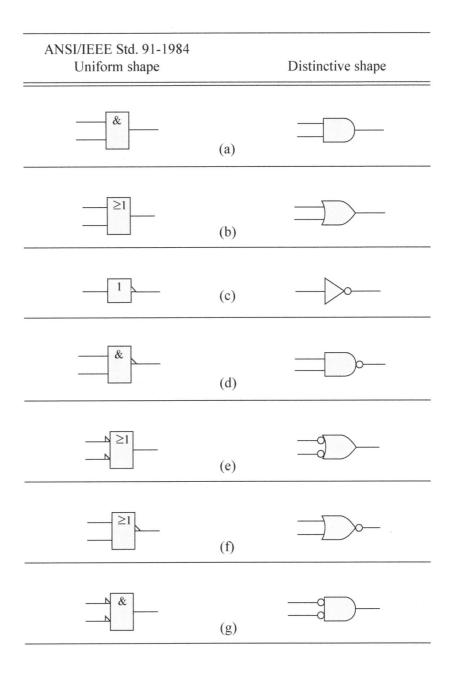

Figure 3.2   ANSI/IEEE Std. 91-1984 uniform-shape logic symbols and the corresponding distinctive-shape logic symbols: (a) AND gate; (b) OR gate; (c) NOT (inverter); (d) NAND gate for the AND function; (e) NAND gate for the OR function; (f) NOR gate for the OR function; (g) NOR gate for the AND function; (h) exclusive-OR function; and (i) exclusive-NOR function.

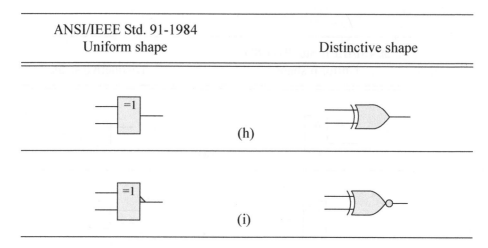

ANSI/IEEE Std. 91-1984
Uniform shape   Distinctive shape

(h)

(i)

Figure 3.2   (Continued)

As can be seen in Figure 3.3, the AND gate can be drawn three ways. Figure 3.3(b) shows the AND function utilizing NAND logic, where the output is inverted. Figure 3.3(c) shows the AND function utilizing NOR logic, where the AND function requires active low inputs. Although only two inputs are shown, both AND and OR gates can have three or more inputs, and depending on the logic family being used, can have two outputs, as shown in Figure 3.3(c).

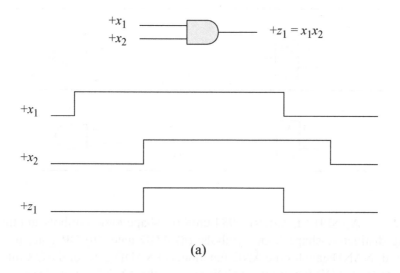

(a)

Figure 3.3   Logic symbols and waveforms for the AND function: (a) AND gate, (b) NAND gate, and (c) NOR gate.

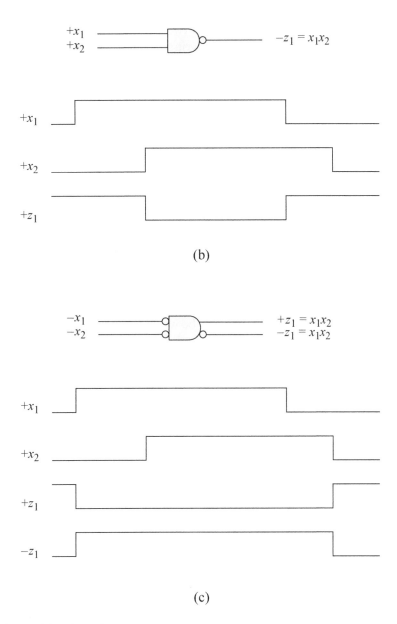

(b)

(c)

Figure 3.3   (Continued)

The plus (+) and minus (−) symbols that are placed to the left of the variables indicate a high or low voltage level, respectively. This indicates the asserted (or active) voltage level for the variables; that is, the *logical 1* (or true) state, in contrast to the *logical 0* (or false) state. Thus, a signal can be asserted either plus or minus, depending upon the active condition of the signal at that point. For example, Figure 3.3(a) specifies that the AND function will be realized when both input x_1 and input

x_2 are at their more positive potential, thus generating an output at its more positive potential. The word *positive* as used here does not necessarily mean a positive voltage level, but merely the more positive of two voltage levels. Therefore, the output of the AND gate of Figure 3.3(a) can be written as $+(x_1 x_2)$.

To illustrate that a plus level does not necessarily mean a positive voltage level, consider two logic families: transistor-transistor logic (TTL) and emitter-coupled logic (ECL). The TTL family uses a +5 volt power supply. A plus level is approximately +3.5 volts; a minus level is approximately +0.2 volts. The ECL family uses a −5.2 volt power supply. A plus level is approximately −0.95 volts; a minus level is approximately −1.7 volts. Although −0.95 volts is a negative voltage, it is the more positive of the two ECL voltages.

The logic symbol of Figure 3.3(b) is a NAND gate in which inputs x_1 and x_2 must both be at their more positive potential for the output to be at its more negative potential. A small circle (or wedge symbol for IEEE Std. 91-1984 logic functions) at the input or output of a logic gate indicates a more negative potential. The output of the NAND gate can be written as $-(x_1 x_2)$.

Figure 3.3(c) illustrates a NOR gate used for the AND function. In this case, inputs x_1 and x_2 must be active (or asserted) at their more negative potential in order for the high output to be at its more positive potential and the low output to be at its more negative potential. The outputs can be written as $+(x_1 x_2)$ and $-(x_1 x_2)$. A variable can be active (or asserted) at a high and a low level at the same time, as shown in Figure 3.3(c).

The OR gate can be drawn three ways, as shown in Figure 3.4. The output of Figure 3.4(a) is at its indicated polarity when one or more of the inputs is at its indicated polarity. Thus, the output of the OR gate will be plus if either x_1 or x_2 is plus or if both x_1 and x_2 are plus. Figure 3.4(b) shows the OR function utilizing NAND logic, where the inputs are active low and the output is inverted. Figure 3.3(c) shows the OR function utilizing NOR logic, where the inputs are asserted as active-high voltage levels. Figure 3.3(c) also illustrates the OR gate with two complementary outputs.

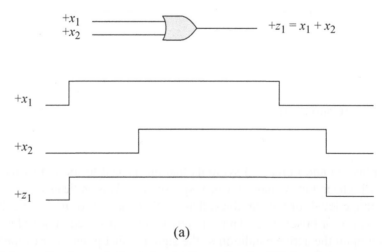

(a)

Figure 3.4   Logic symbols and waveforms for the OR function.

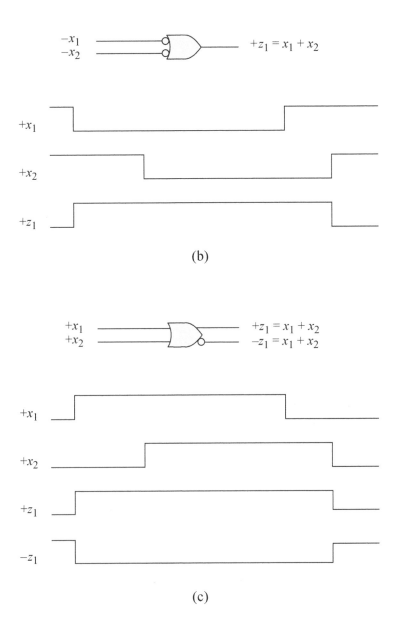

Figure 3.4 (Continued)

Figure 3.5(a) shows the exclusive-OR circuit using discrete logic gates and accompanying waveforms, where the inputs x_1 and x_2 are available in both high and low assertions; output z_1 is asserted high. Figure 3.5(b) shows the same circuit using the exclusive-OR logic symbol. The inputs can be either both active high or both active low.

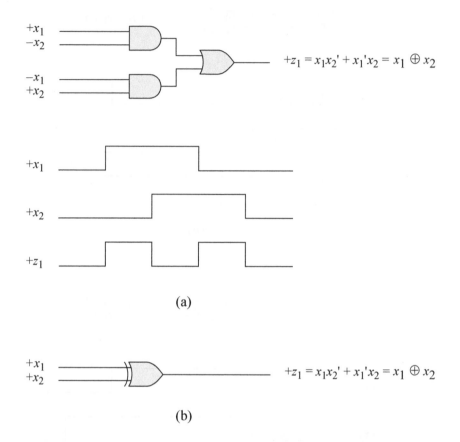

Figure 3.5   Exclusive-OR function: (a) logic diagram and waveforms, and (b) symbol for the exclusive-OR logic function.

Figure 3.6(a) shows the exclusive-NOR circuit using discrete logic gates and accompanying waveforms, where the inputs x_1 and x_2 are available in both high and low assertions; output z_1 is asserted high if inputs are the same logic level. Figure 3.6(b) shows the same circuit using the exclusive-NOR logic symbol. The inputs can be either both active high or both active low.

The truth tables for the 2-input AND, NAND, OR, NOR, exclusive-OR, and exclusive-NOR logic primitives are shown in Table 3.1, Table 3.2, Table 3.3, Table 3.4, Table 3.5, and Table 3.6, respectively, where a logical 1 indicates an asserted (or active) value and a logical 0 indicates a deasserted (or inactive) value. The asserted values can be either a positive or negative voltage, depending on the logic family, as described in Section 3.1.

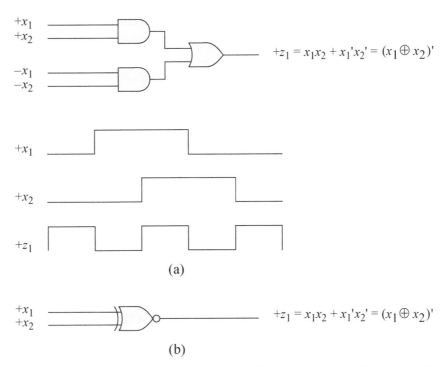

$$+z_1 = x_1 x_2 + x_1' x_2' = (x_1 \oplus x_2)'$$

(a)

$$+z_1 = x_1 x_2 + x_1' x_2' = (x_1 \oplus x_2)'$$

(b)

Figure 3.6   Exclusive-NOR function: (a) logic diagram and waveforms and (b) symbol for the exclusive-NOR function.

Table 3.1 Truth Table for the AND Gate

x_1	x_2	z_1
0	0	0
0	1	0
1	0	0
1	1	1

Table 3.2 Truth Table for the NAND Gate

x_1	x_2	z_1
0	0	1
0	1	1
1	0	1
1	1	0

Table 3.3 Truth Table for the OR Gate

x_1	x_2	z_1
0	0	0
0	1	1
1	0	1
1	1	1

Table 3.4 Truth Table for the NOR Gate

x_1	x_2	z_1
0	0	1
0	1	0
1	0	0
1	1	0

Table 3.5 Truth Table for the Exclusive-OR Function

x_1	x_2	z_1
0	0	0
0	1	1
1	0	1
1	1	0

Table 3.6 Truth Table for the Exclusive-NOR Function

x_1	x_2	z_1
0	0	1
0	1	0
1	0	0
1	1	1

Fan-In Logic gates for the AND and OR functions can be extended to accommodate more than two variables; that is, more than two inputs. The number of inputs available at a logic gate is called the *fan-in*. Current technology allows for gates with a large number of inputs. However, if the fan-in must be further extended, then this can be easily achieved by adding another level of logic, as shown in Figure 3.7, which transforms a 3-input AND gate into a 5-input AND gate, with each gate having a fan-in of three. If NAND gates are being used, then an additional level of delay must be inserted, as shown in Figure 3.8. If output z_1 is to be asserted high, then an additional NAND gate must be utilized to invert the output.

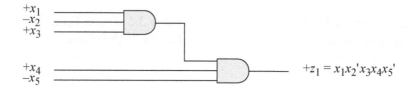

Figure 3.7   Increasing the fan-in capability of an AND gate.

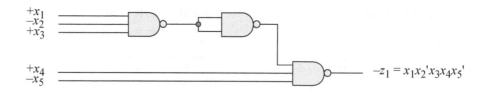

Figure 3.8   Increasing the fan-in capability of a NAND gate.

Fan-Out   The *fan-out* of a logic gate is the maximum number of inputs that the gate can drive and still maintain acceptable voltage and current levels. That is, the fan-out defines the maximum load that the gate can handle. If the fan-out of the AND gate shown in Figure 3.9 is 10, then the gate can drive an additional seven inputs in the same logic family.

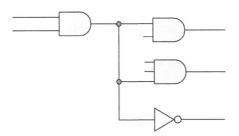

Figure 3.9   Example of fan-out.

Propagation Delay   This is the time associated with a logic circuit that is defined as the time interval between an input change and the resulting output change, either from a logic 1 to a logic 0 or vice versa. The propagation delay for a rising edge input and a falling edge input may be different, as shown for the inverter in Figure 3.10.

Also associated with propagation delay is the rise time and fall time of a signal. The *rise time* of a signal is the time required to go from a logic 0 to a logic 1 voltage level. The *fall time* of a signal is the time required to go from a logic 1 to a logic 0 voltage level.

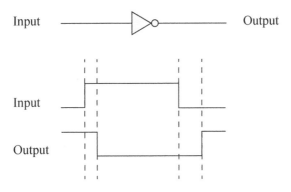

Figure 3.10   Example of propagation delay.

3.1.1 Wired-AND and Wired-OR Operations

Additional logic functions can be realized by wiring together the outputs of certain types of logic gates. The wired-logic function is not a physical gate, but only a symbol that represents the logical function obtained by the wired connection.

For example, open-collector transistor-transistor logic (TTL) NAND gates can be used to add an additional level of logic without adding an additional gate. The output resistor is removed from the collector of all gates and placed external to the circuit, as shown in Figure 3.11 for a *wired-OR* circuit. This circuit will generate a low voltage level for output z_1 if the AND functions for gate 1 or gate 2 are satisfied; that is, z_1 is a logic 0 (active) if $x_1 x_2 = 10$ or if $x_3 x_4 = 11$. Since a low output for TTL circuits is near ground (approximately 0.2 volts), either NAND gate whose AND function is realized will pull the entire output net toward the ground potential. Open-collector devices are often used to generate a higher output voltage.

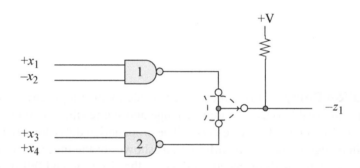

Figure 3.11   Wired-OR circuit for the function $z_1 = x_1 x_2' + x_3 x_4$ using TTL NAND gates.

The same circuit can be used to implement a *wired-AND* function if output z_1 is to be asserted high. In this case, $z_1 = (x_1 x_2' + x_3 x_4)' = (x_1' + x_2) (x_3' + x_4')$ using DeMorgan's theorem. The outputs of both gates will be at a high voltage level (approximately +5 volts).

Emitter-coupled logic (ECL) gates are designed to implement the wired-logic functions without using an external resistor. The power supply for ECL logic is approximately –5.2 volts. A high voltage level is approximately –0.95 volts; a low voltage level is approximately –1.7 volts. The basic gate in ECL is a NOR gate, as shown in Figure 3.12 for the wired-OR function. The AND function using NOR gates has active-low inputs. If the input voltage levels fulfill the requirements of the AND function for either gate, then the output will be at a high voltage level (although still negative), and will pull the entire output net toward –0.95 volts.

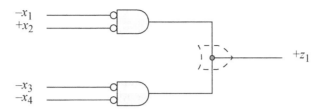

Figure 3.12   Wired-OR circuit for the function $z_1 = x_1 x_2' + x_3 x_4$ using ECL NOR gates.

Inverters can be used to emulate a 4-input OR gate as shown in Figure 3.13. These are open-collector TTL inverters and require an external pull-up resistor. If output z_1 is to be active low, then at least one input must be at a high voltage level, in order to pull the entire output net to a low voltage level and assert z_1, as shown in Figure 3.13(a). This provides the following equation for z_1:

$$z_1 = x_1 + x_2 + x_3 + x_4$$

If the inputs are all active low, then this emulates a 4-input AND gate where z_1 is asserted high, as shown in Figure 3.13(b), providing the following equation for z_1:

$$z_1 = x_1 x_2 x_3 x_4$$

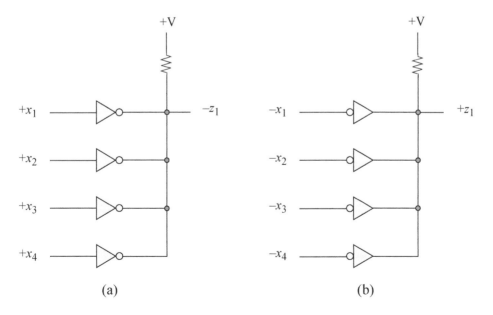

Figure 3.13   Wired-inverters: (a) wired-OR logic and (b) wired AND logic.

3.1.2 Three-State Logic

Three-state logic is used primarily to connect logical devices to a common bus structure. A three-state device can be a gate, a buffer, or a logic macro function. A three-state circuit is one in which the output exhibits three states under control of an *enable* input: (1) a logic 0 state if the input is a logic 0 and the enable input is asserted, (2) a logic 1 state if the input is a logic 1 and the enable input is asserted, and (3) a high-impedance state if the enable input is deasserted, which effectively removes the device from the bus.

A three-state bus is constructed by wiring together the outputs of several three-state devices. When the enable input is disabled, the high-impedance state acts like an open circuit giving the device no logical significance. High-impedance (Hi-Z) is also referred to as a floating state.

Figure 3.14 illustrates a three-state buffer (or driver) and a three-state inverter. The enable inputs can be active high or low. Figure 3.14(a) depicts a noninverting buffer where the output z_1 assumes the state of input x_1 if the enable input is asserted at a high voltage level. Figure 3.14(b) shows an inverter where output z_1 inverts input x_1 if the enable input is asserted at a low voltage level. In both cases, if the enable is deasserted, then the output is in a high-impedance state, regardless of the value of the input.

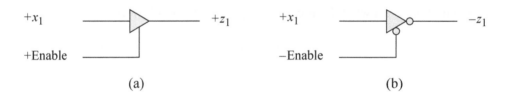

$+x_1$ $+z_1$ $+x_1$ $-z_1$

$+$Enable $-$Enable

(a) (b)

Figure 3.14   Three-state devices: (a) buffer and (b) inverter.

3.1.3 Functionally Complete Gates

Both NAND and NOR gates have the unique characteristic that they can express any Boolean function; that is, they can represent the functions AND, OR, and NOT. Thus, NAND and NOR gates are classified as *functionally complete* gates or *universal* gates. Since every switching function can be expressed in a disjunctive normal form using the set AND, OR, and NOT, it is only necessary to show that NAND and NOR gates can generate the three primitive functions of AND, OR, and NOT to be functionally complete.

Figure 3.15 shows NAND and NOR gates used to implement the NOT function. Using Boolean algebra, it can be shown that the NAND gate can generate the NOT function, as follows:

$$(x_1 x_1)' = x_1' + x_1' \qquad \text{DeMorgan's law}$$
$$= x_1' \qquad \text{Idempotent law}$$

Using Boolean algebra, it can be shown that the NOR gate can generate the NOT function, as follows:

$$(x_1 + x_1)' = x_1' x_1' \qquad \text{DeMorgan's law}$$
$$= x_1' \qquad \text{Idempotent law}$$

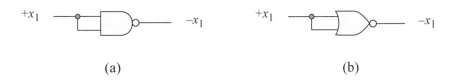

(a)   (b)

Figure 3.15   NAND and NOR gates to implement the NOT (invert) function: (a) NAND gate and (b) NOR gate.

The NAND gate used to implement the AND function is shown in Figure 3.16 in two variations: (a) with an active-low output, and (b) with an active-high output. The Boolean justification for Figure 3.16(b) is as follows:

$$z_1 = [(x_1 x_2)']' \qquad \text{Involution law}$$
$$= (x_1' + x_2')' \qquad \text{DeMorgan's law}$$
$$= x_1 x_2 \qquad \text{DeMorgan's law}$$

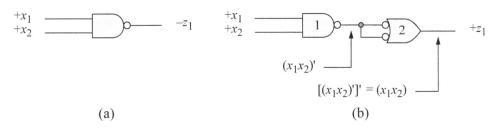

(a)   (b)

Figure 3.16   NAND gate to implement the AND function: (a) active-low output and (b) active-high output.

In Figure 3.16(a), the AND function is realized when the input signals are active high and the output is active low. Thus, if input x_1 is active low and input x_2 is active high, then the equation for output z_1 would be

$$z_1 = x_1' x_2$$

which states that if input x_1 is deasserted (inactive), then the input would be at a logic 1 (high) level, which satisfies the input requirement for a NAND gate with active-high inputs. In Figure 3.16(b), the output of gate 1 is connected to the inputs of gate 2, which is used as an inverter. The truth table for the NAND gate is shown in Table 3.7. The output is a logic 0 only if all inputs are a logic 1.

Table 3.7 Truth Table for the NAND Gate

x_1	x_2	z_1
0	0	1
0	1	1
1	0	1
1	1	0

The NOR gate used to implement the AND function is shown in Figure 3.17 in two variations: (a) with active-low inputs and an active-high output, and (b) with active-high inputs and an active-high output. The Boolean justification for Figure 3.16(b) is as follows:

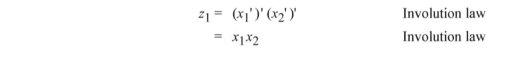

$$z_1 = (x_1')'(x_2')' \qquad \text{Involution law}$$
$$= x_1 x_2 \qquad \text{Involution law}$$

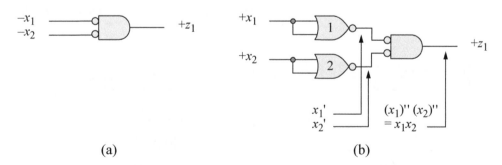

(a)   (b)

Figure 3.17   NOR gate to implement the AND function: (a) active-low inputs with active-high output and (b) active-high inputs with active-high output.

In Figure 3.17(a), the AND function is realized when the input signals are active low and the output is active high. Thus, if input x_1 is active high and input x_2 is active low, then the equation for output z_1 would be

$$z_1 = x_1'x_2$$

which states that if input x_1 was deasserted (inactive), then the input would be at a logic 0 (low) level, which satisfies the input requirement for a NOR gate with active-low inputs. In Figure 3.17(b), the outputs of gate 1 and gate 2 are connected to the inputs of another NOR gate, which is used as an AND function. The truth table for the NOR gate is shown in Table 3.8. The output is a logic 1 only if all inputs are a logic 0.

Table 3.8 Truth Table for the NOR Gate

x_1	x_2	z_1
0	0	1
0	1	0
1	0	0
1	1	0

The NAND gate to implement the OR function is shown in Figure 3.18 in two variations: (a) with active-low inputs and an active-high output, and (b) with active-high inputs and an active-high output. The Boolean justification for Figure 3.18(b) is as follows:

$$\begin{aligned} z_1 &= (x_1')' + (x_2')' \\ &= x_1 + x_2 \qquad \text{Involution law} \end{aligned}$$

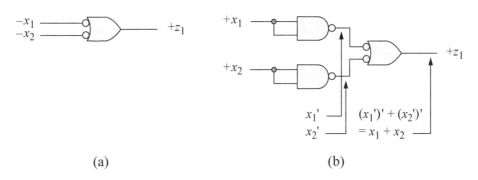

(a)   (b)

Figure 3.18   NAND gate to implement the OR function: (a) active-low inputs with active-high output and (b) active-high inputs with active-high output.

The NOR gate used to implement the OR function is shown in Figure 3.19 in two variations: (a) with active-high inputs and an active-low output, and (b) with active-high inputs and an active-high output. The Boolean justification for Figure 3.19(b) is as follows:

$$
\begin{aligned}
z_1 &= [(x_1 + x_2)'\,(x_1 + x_2)']' & \\
&= [(x_1'x_2')\,(x_1'x_2')]' & \text{DeMorgan's law} \\
&= x_1 + x_2 + x_1 + x_2 & \text{DeMorgan's law} \\
&= x_1 + x_2 & \text{Idempotent law}
\end{aligned}
$$

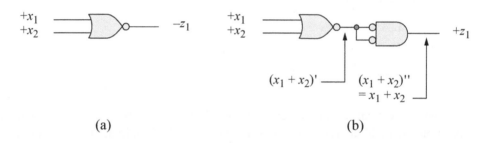

(a)   (b)

Figure 3.19   NOR gate to implement the OR function: (a) active-high inputs with active-low output and (b) active-high inputs with active-high output.

3.2 Logic Macro Functions

Logic macro functions are those circuits that consist of several logic primitives to form larger more complex functions. Combinational logic macros include circuits such as multiplexers, decoders, encoders, comparators, adders, subtractors, array multipliers, array dividers, and error detection and correction circuits. Sequential logic macros include circuits such as: *SR* latches; *D* and *JK* flip-flops; counters of various moduli, including count-up and count-down counters; registers, including shift registers; and sequential multipliers and dividers.

This section will present the functional operation of multiplexers, decoders, encoders, priority encoders, and comparators. The remaining combinational and sequential macro functions will be introduced in their respective chapters.

3.2.1 Multiplexers

A multiplexer is a logic macro device that allows digital information from two or more data inputs to be directed to a single output. Data input selection is controlled by a set of select inputs that determine which data input is gated to the output. The select inputs are labeled $s_0, s_1, s_2, \cdots, s_i, \cdots, s_{n-1}$, where s_0 is the low-order select input with a binary weight of 2^0 and s_{n-1} is the high-order select input with a binary weight of 2^{n-1}. The data inputs are labeled $d_0, d_1, d_2, \cdots, d_j, \cdots, d_2{}^n{}_{-1}$. Thus, if a multiplexer has n select inputs, then the number of data inputs will be 2^n and will be labeled d_0 through $d_2{}^n{}_{-1}$. For example, if $n = 2$, then the multiplexer has two select inputs s_0 and s_1 and four data inputs d_0, d_1, d_2, and d_3.

The logic diagram for a 4:1 multiplexer is shown in Figure 3.20. There can also be an *enable* input which gates the selected data input to the output. Each of the four data inputs x_0, x_1, x_2, and x_3 is connected to a separate 3-input AND gate. The select inputs s_0 and s_1 are decoded to select a particular AND gate. The output of each AND gate is applied to a 4-input OR gate that provides the single output z_1. The truth table for the 4:1 multiplexer is shown in Table 3.9. Input lines that are not selected cannot be transferred to the output and are listed as "don't cares."

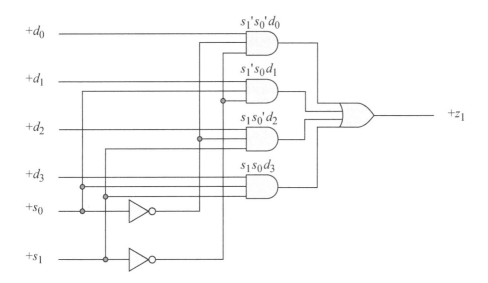

Figure 3.20   Logic diagram for a 4:1 multiplexer.

The equation for output z_1 can be obtained directly from Table 3.9 or from Figure 3.20 and is shown in Equation 3.1.

Table 3.9 Truth Table for the 4:1 Multiplexer of Figure 3.20

s_1	s_0	d_3	d_2	d_1	d_0	z_1
0	0	–	–	–	0	0
0	0	–	–	–	1	1
0	1	–	–	0	–	0
0	1	–	–	1	–	1
1	0	–	0	–	–	0
1	0	–	1	–	–	1
1	1	0	–	–	–	0
1	1	1	–	–	–	1

$$z_1 = s_1's_0'd_0 + s_1's_0d_1 + s_1s_0'd_2 + s_1s_0d_3 \tag{3.1}$$

Figure 3.21 shows four typical multiplexers drawn in the ANSI/IEEE Std. 91-1984 format. The truth tables for the 4:1 (four-to-one) and 8:1 multiplexers are shown in Table 3.10 and Table 3.11, respectively. The truth tables for the 2:1 and 16:1 multiplexers are derived in a similar manner. Consider the 4:1 multiplexer in Table 3.10. If $s_1 s_0 = 00$, then data input d_0 is selected and its value is propagated to the multiplexer output z_1. Similarly, if $s_1 s_0 = 01$, then data input d_1 is selected and its value is directed to the multiplexer output.

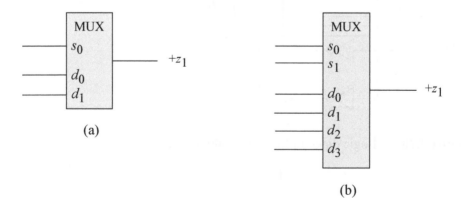

Figure 3.21   ANSI/IEEE Std. 91-1984 symbols for multiplexers: (a) 2:1 multiplexer, (b) 4:1 multiplexer, (c) 8:1 multiplexer, and (d) 16:1 multiplexer.

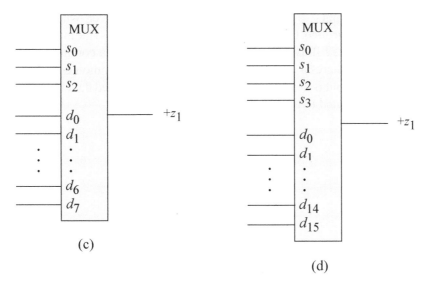

(c)

(d)

Figure 3.21 (Continued)

The equation that represents output z_1 in the 4:1 multiplexer of Figure 3.21 (b) is shown in Equation 3.2. Output z_1 assumes the value of d_0 if $s_1 s_0 = 00$, as indicated by the term $s_1's_0'd_0$. Likewise, z_1 assumes the value of d_1 when $s_1 s_0 = 01$, as indicated by the term $s_1's_0d_1$.

$$z_1 = s_1's_0'd_0 + s_1's_0d_1 + s_1s_0'd_2 + s_1s_0d_3 \tag{3.2}$$

Table 3.10 Truth Table for the 4:1 Multiplexer of Figure 3.20(b)

Select Inputs s_1s_0	Data Input Selected
0 0	d_0
0 1	d_1
1 0	d_2
1 1	d_3

Table 3.11 Truth Table for the 8:1 Multiplexer of Figure 3.20(c)

Select Inputs $s_2s_1s_0$	Data Input Selected
0 0 0	d_0
0 0 1	d_1
0 1 0	d_2
0 1 1	d_3
1 0 0	d_4
1 0 1	d_5
1 1 0	d_6
1 1 1	d_7

The symbols shown in Figure 3.21 represent single multiplexers controlled by dedicated select inputs. The symbology is different, however, when more than one multiplexer is controlled by the same set of select inputs. The multiplexer symbol shown in Figure 3.22 illustrates four 2:1 multiplexers with a common control block. Each multiplexer shares a common select input s_0 and a common enable. Only when the enable (EN) input is at a low voltage level will the selected data input of each multiplexer be propagated to the corresponding output.

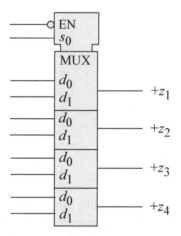

Figure 3.22   ANSI/IEEE Std. 91-1984 logic symbol for four 2:1 multiplexers with a common control block.

Example 3.1   There is a one-to-one correspondence between the data input numbers d_i of a multiplexer and the minterm locations in a Karnaugh map. For example, Figure 3.23 shows a Karnaugh map and a 4:1 multiplexer. Minterm location 0 corresponds to data input d_0 of the multiplexer; minterm location 1 corresponds to data input d_1; minterm location 2 corresponds to data input d_2; and minterm location 3 corresponds to data input d_3. The Karnaugh map and the multiplexer implement Equation 3.3, where x_2 is the low-order variable.

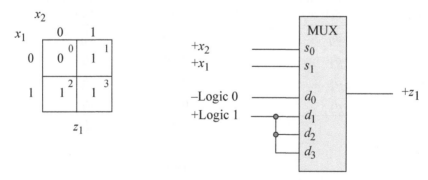

Figure 3.23   One-to-one correspondence between a Karnaugh map and a multiplexer.

$$z_1 = x_1'x_2 + x_1x_2' + x_1x_2$$
$$= x_1 + x_2 \qquad (3.3)$$

Example 3.2 Multiplexers can also be used with Karnaugh maps containing map-entered variables. Equation 3.4 is plotted on the Karnaugh map shown in Figure 3.24(a) using x_3 as a map-entered variable. Figure 3.24(b) shows the implementation using a 4:1 multiplexer.

$$z_1 = x_1x_2(x_3') + x_1x_2'(x_3) + x_1'x_2 \qquad (3.4)$$

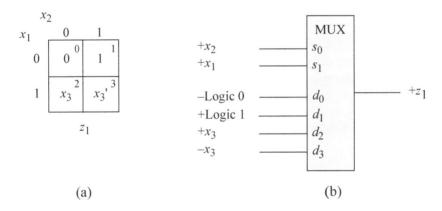

(a)   (b)

Figure 3.24 Multiplexer using a map-entered variable: (a) Karnaugh map and (b) a multiplexer.

Linear-select multiplexers The multiplexer examples described thus far have been classified as *linear-select multiplexers*, because all of the variables of the Karnaugh map coordinates have been utilized as the select inputs for the multiplexer. Since there is a one-to-one correspondence between the minterms of a Karnaugh map and the data inputs of a multiplexer, designing the input logic is relatively straightforward. Simply assign the values of the minterms in the Karnaugh map to the corresponding multiplexer data inputs with the same subscript.

Example 3.3 The Karnaugh map for function z_1 is shown in Figure 3.25(a) using x_3 as a map-entered variable. Assigning a value of logic 0 to the unused state in minterm location 2 allows multiplexer inputs d_0, d_2, and d_3 to be connected to a logic 0. Figure 3.25(b) shows the multiplexer implementation.

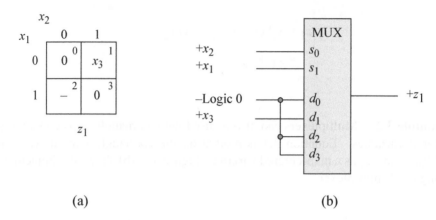

(a)   (b)

Figure 3.25   Linear-select multiplexer using x_3 as a map-entered variable: (a) Karnaugh map and (b) a multiplexer.

Example 3.4   A multiplexer is a fast and simple way to implement Boolean equations. Consider Equation 3.5 using x_4 as a map-entered variable. The five minterms for variables $x_1 x_2 x_3$ correspond to minterms m_4, m_6, m_5, m_2, and m_4, respectively. The equation is plotted on the Karnaugh map shown in Figure 3.26(a) and implemented by the 8:1 linear-select multiplexer shown in Figure 3.26(b), where x_3 is the low-order variable.

$$z_1 = x_1 x_2' x_3' (x_4) + x_1 x_2 x_3' (x_4) + x_1 x_2' x_3 (x_4') +$$
$$x_1' x_2 x_3' (x_4') + x_1 x_2' x_3' (x_4') \qquad (3.5)$$

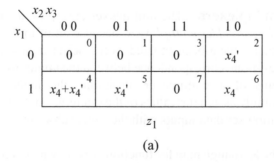

(a)

Figure 3.26   Linear-select 8:1 multiplexer to implement Equation 3.5: (a) Karnaugh map and (b) 8:1 multiplexer.

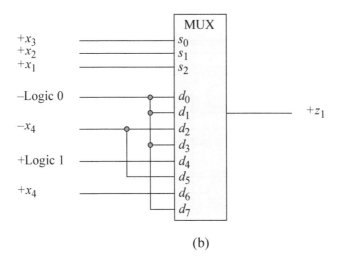

(b)

Figure 3.26 (Continued)

Example 3.5 The function shown in Equation 3.6 will be implemented using an 8:1 linear-select multiplexer using x_4 and x_5 as map-entered variables. The seven minterms for variables $x_1x_2x_3$ correspond to minterms m_5, m_6, m_5, m_0, m_2, m_2, and m_7, respectively. The equation is plotted on the Karnaugh map shown in Figure 3.27(a) and implemented by the 8:1 linear-select multiplexer shown in Figure 3.27(b), where x_3 is the low-order variable.

$$z_1 = x_1 x_2' x_3 (x_4' x_5) + x_1 x_2 x_3' (x_4' x_5') + x_1 x_2' x_3 (x_4' x_5') +$$
$$x_1' x_2' x_3' (x_4 x_5) + x_1' x_2 x_3' (x_4' x_5') + x_1' x_2 x_3' (x_4 x_5') +$$
$$x_1 x_2 x_3 (x_4 x_5) \qquad (3.6)$$

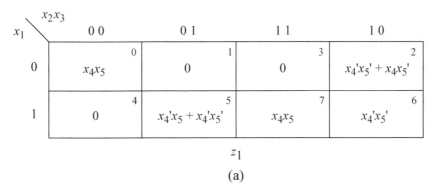

z_1

(a)

Figure 3.27 Linear-select 8:1 multiplexer to implement Equation 3.6: (a) Karnaugh map and (b) 8:1 multiplexer.

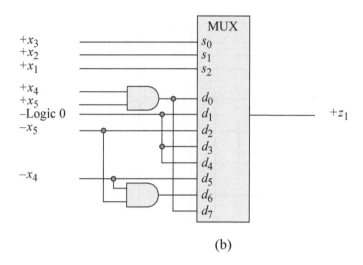

(b)

Figure 3.27   (Continued)

Nonlinear-select multiplexers   In the previous subsection, the logic for the function was implemented with linear-select multiplexers. Although the logic functioned correctly according to the equations, the designs illustrated an inefficient use of the 2^p:1 multiplexers. Smaller multiplexers with fewer data inputs could be effectively utilized with a corresponding reduction in machine cost.

If the number of unique entries in a Karnaugh map satisfies the expression of Equation 3.7, where u is the number of unique entries and p is the number of select inputs, then at most a $(2^p \div 2)$:1 multiplexer will satisfy the requirements. This is referred to as a *nonlinear-select multiplexer*.

$$1 < u \geq (2^p \div 2) \tag{3.7}$$

If, however, $u > 2^p \div 2$, then a 2^p:1 multiplexer is necessary. The largest multiplexer with which to economically implement the logic is a 16:1 multiplexer, and then only if the number of distinct entries in the Karnaugh map warrants a multiplexer of this size. Other techniques, such as a programmable logic device (PLD) implementation, would make more efficient use of current technology.

If a multiplexer has unused data inputs — corresponding to unused states in the input map — then these unused inputs can be connected to logically adjacent multiplexer inputs. The resulting linked set of inputs can be addressed by a common select variable. Thus, in a 4:1 multiplexer, if data input $d_2 = 1$ and $d_3 = $ "don't care," then d_2 and d_3 can both be connected to a logic 1. The two inputs can now be selected by $s_1 s_0 = 10$ or 11; that is, $s_1 s_0 = 1-$. Also, multiple multiplexers containing the same number of data inputs should be addressed by the same select input variables, if possible. This permits the utilization of noncustom technology, where multiplexers in the same integrated circuit share common select inputs.

Example 3.6   The Karnaugh map shown in Figure 3.28 has only two unique entries plus "don't care" entries. Since there are only two distinct entries, a 2:1 Karnaugh map can be used to implement the function. Table 3.12 tabulates the entries in the map and indicates how variables x_2 and x_3 can be assigned to the select inputs of the multiplexer, where the "don't care" entries in minterm locations 1 and 5 are assigned a value of 0. The logic diagram is shown in Figure 3.29 using a 2:1 multiplexer and an AND gate.

$$x_2 x_3$$

x_1	0 0	0 1	1 1	1 0
0	0 ₀	– ₁	1 ₃	1 ₂
1	0 ₄	– ₅	0 ₇	0 ₆

z_1

Figure 3.28   Karnaugh map for Example 3.6.

Table 3.12 Illustrating the Use of a Nonlinear-Select Multiplexer for Figure 3.28

x_1	x_2	x_3	z_1
0	0	0	0
0	0	1	– (0)
0	1	0	1
0	1	1	1
1	0	0	0
1	0	1	– (0)
1	1	0	0
1	1	1	0

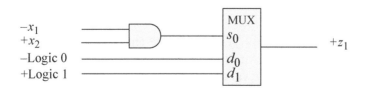

Figure 3.29   A 2:1 nonlinear-select multiplexer to implement the logic of Figure 3.28.

Example 3.7   The Karnaugh map of Figure 3.30 can be implemented with a 4:1 nonlinear-select multiplexer for the function z_1. Variables x_2 and x_3 will connect to select inputs s_1 and s_0, respectively. When select inputs $s_1 s_0 = x_2 x_3 = 00$, data input d_0 is selected; therefore, $d_0 = 0$. When select inputs $s_1 s_0 = x_2 x_3 = 01$, data input d_1 is selected and d_1 contains the complement of x_1; therefore, $d_1 = x_1'$. When select inputs $s_1 s_0 = x_2 x_3 = 10$, data input d_2 is selected; therefore, $d_2 = 1$. When $s_1 s_0 = x_2 x_3 = 11$, data input d_3 is selected and contains the same value as x_1; therefore, $d_3 = x_1$. The logic diagram is shown in Figure 3.31.

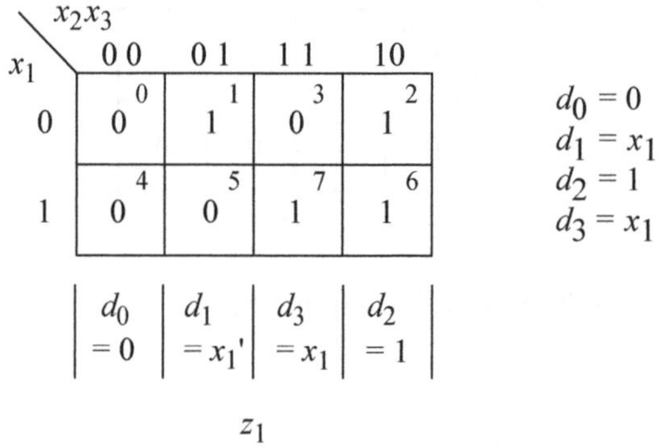

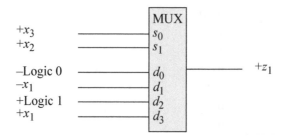

Figure 3.30   Karnaugh map for Example 3.7 which will be implemented by a 4:1 nonlinear-select multiplexer.

Figure 3.31   A 4:1 nonlinear-select multiplexer to implement the Karnaugh map of Figure 3.30.

The multiplexer of Figure 3.31 will now be checked to verify that it operates according to the Karnaugh map of Figure 3.30; that is, for every value of $x_1 x_2 x_3$, output z_1 should generate the same value as in the corresponding minterm location.

Minterm location 1 has a value of 1 if $x_1 x_2 x_3 = 001$, providing multiplexer select inputs of $x_2 x_3 = 01$. Thus, input d_1 is selected, which gates the value of x_1' to the output. Since x_1 is inactive (a logic 0), output z_1 will be equal to a logic 1.

Minterm location 5 has a value of 0 if $x_1 x_2 x_3 = 101$, providing multiplexer select inputs of $x_2 x_3 = 01$. Thus, input d_1 is selected, which gates the value of x_1' to the output. Since x_1 is active (a logic 1), output z_1 will be equal to a logic 0.

Minterm location 2 has a value of 1 if $x_1 x_2 x_3 = 010$, providing multiplexer select inputs of $x_2 x_3 = 10$. Thus, input d_2 is selected, which gates a value of logic 1 to the output. Similarly, minterm location 6 has a value of 1 if $x_1 x_2 x_3 = 110$, providing multiplexer select inputs of $x_2 x_3 = 10$. Thus, input d_2 is selected, which gates a value of logic 1 to the output. In these two cases, output z_1 is independent of the value of x_1, as shown below.

$$\text{Minterm 2: } x_1' x_2 x_3' = 1$$
$$\text{Minterm 6: } x_1 x_2 x_3' = 1$$
$$\text{Therefore, } x_1' x_2 x_3' + x_1 x_2 x_3' = x_2 x_3'(x_1' + x_1) = x_2 x_3' = 1$$

Since there are two unique entries in the Karnaugh map of Figure 3.30, any permutation should produce similar results for output z_1; that is, no additional logic. Figure 3.32 shows one permutation in which the minterm locations are physically moved, but remain logically the same. The multiplexer configuration is shown in Figure 3.33.

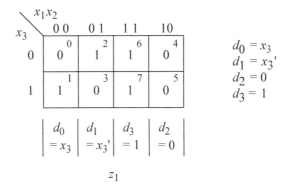

Figure 3.32   A permutation of the Karnaugh map of Figure 3.30.

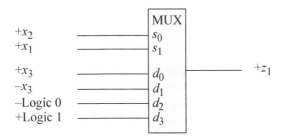

Figure 3.33   A 4:1 nonlinear-select multiplexer to implement the Karnaugh map of Figure 3.32.

The multiplexer of Figure 3.33 will now be checked to verify that it operates according to the Karnaugh map of Figure 3.32; that is, for every value of $x_1 x_2 x_3$, output z_1 should generate the same value as in the corresponding minterm location. The results should be the same as obtained for the multiplexer of Figure 3.31.

Minterm location 2 has a value of 1 if $x_1 x_2 x_3 = 010$, providing multiplexer select inputs of $x_1 x_2 = 01$. Thus, input d_1 is selected, which gates the value of x_3' to the output. Since x_3 is inactive (a logic 0), output z_1 will be equal to a logic 1.

Minterm location 3 has a value of 0 if $x_1 x_2 x_3 = 011$, providing multiplexer select inputs of $x_1 x_2 = 01$. Thus, input d_1 is selected, which gates the value of x_3' to the output. Since x_3 is active (a logic 1), output z_1 will be equal to a logic 0.

Minterm location 6 has a value of 1 if $x_1 x_2 x_3 = 110$, providing multiplexer select inputs of $x_1 x_2 = 11$. Thus, input d_3 is selected, which gates a value of logic 1 to the output. Similarly, minterm location 7 has a value of 1 if $x_1 x_2 x_3 = 111$, providing multiplexer select inputs of $x_1 x_2 = 11$. Thus, input d_3 is selected, which gates a value of logic 1 to the output. In these two cases, output z_1 is independent of the value of x_3, as shown below.

$$\text{Minterm 6: } x_1 x_2 x_3' = 1$$
$$\text{Minterm 7: } x_1 x_2 x_3 = 1$$
$$\text{Therefore, } x_1 x_2 x_3' + x_1 x_2 x_3 = x_1 x_2 (x_3' + x_3) = x_1 x_2 = 1$$

As a further consideration, input x_3 can be used as a map-entered variable and still generate the same Karnaugh map as shown in Figure 3.30, which is reproduced in Figure 3.34(a) for convenience. Figure 3.34(b) illustrates the Karnaugh map using x_3 as a map-entered variable. Figure 3.35 shows the 4:1 linear-select multiplexer, which yields the same results as the permuted map of Figure 3.32.

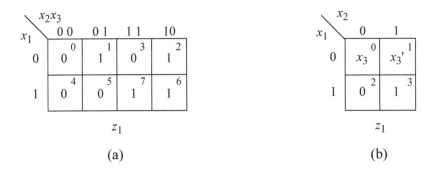

Figure 3.34   Karnaugh maps: (a) the map from Figure 3.30 and (b) using x_3 as a map-entered variable.

Note that when $x_1 x_2 = 00$ in Figure 3.34(a), minterm locations 0 and 1 are the same value as x_3; therefore, x_3 is placed in minterm location 0 in the map of Figure

3.34(b). Likewise, when $x_1 x_2 = 01$, minterm locations 3 and 2 contain the same value as x_3'; therefore, x_3' is placed in minterm location 1 in the map of Figure 3.34(b). In a similar manner, when $x_1 x_2 = 10$ in Figure 3.34(a), minterm locations 4 and 5 contain values of 0, resulting in a 0 being placed in minterm location 2 of the map in Figure 3.34(b); and when $x_1 x_2 = 11$, minterm locations 7 and 6 contain values of 1, which places a 1 in minterm location 3 of the map in Figures 3.34(b).

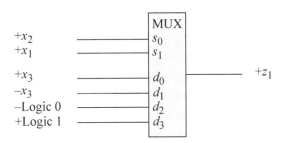

Figure 3.35   Linear-select multiplexer for Figure 3.34(b) using x_3 as a map-entered variable.

Before leaving this example, a final technique will be presented to illustrate another method to implement a nonlinear-select multiplexer for this design. Since the Karnaugh map of Example 3.7 — reproduced in Figure 3.36 — contains only 0s and 1s, the logic could conceivably be implemented with a 2:1 multiplexer plus additional logic, if necessary.

Let variable x_1 connect to the single select input s_0 of a 2:1 multiplexer. If $x_1 = 0$, then $z_1 = 1$ for the following conditions as shown in the Karnaugh map of Figure 3.36: $x_2' x_3 + x_2 x_3' = x_2 \oplus x_3$. If $x_1 = 1$, then $z_1 = 1$ only if $x_2 = 1$. The resulting logic is shown in Figure 3.37 using a 2:1 nonlinear-select multiplexer.

x_1 $\backslash$ $\begin{matrix}x_2 x_3\\ \end{matrix}$	0 0	0 1	1 1	10
0	0 [0]	1 [1]	0 [3]	1 [2]
1	0 [4]	0 [5]	1 [7]	1 [6]

z_1

Figure 3.36   Karnaugh map to be implemented with a 2:1 multiplexer.

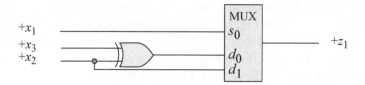

Figure 3.37   A 2:1 nonlinear-select multiplexer to implement the logic for Example 3.7.

Example 3.8   The Karnaugh map shown in Figure 3.38(a) has three unique entries 1, 0, and a map-entered variable A' plus "don't care" entries and will be implemented using a 4:1 nonlinear-select multiplexer. Three variations of the map will be generated: (1) the original map where x_2 and x_3 connect to select inputs s_1 and s_0, respectively; (2) a permutation in which x_1 and x_2 connect to select inputs s_1 and s_0, respectively; and (3) a permutation in which x_1 and x_3 connect to select inputs s_1 and s_0, respectively. The resulting three Karnaugh maps will then be analyzed and the map that provides the least amount of additional logic will be utilized in the implementation.

Figure 3.38(a) requires one AND gate for $x_1'A'$; Figure 3.38(b) requires one OR gate for $x_3' + A'$; and Figure 3.38(c) requires no additional logic. Therefore, the Karnaugh map of Figure 3.38(c) will be utilized to implement the 4:1 nonlinear-select multiplexer.

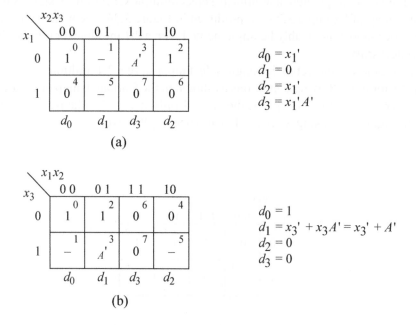

Figure 3.38   Karnaugh maps for Example 3.8 depicting three permutations: (a) original Karnaugh map, (b) first permutation and (c) second permutation.

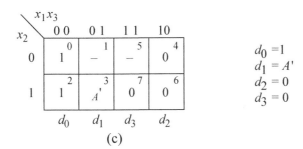

$d_0 = 1$
$d_1 = A'$
$d_2 = 0$
$d_3 = 0$

(c)

Figure 3.38 (Continued)

Refer to the equation for d_1 in Figure 3.38(b). The value of 1 in minterm location 2 equates to x_3' AND 1; the value of A' in minterm location 3 equates to x_3 AND A'. Therefore, the entry for data input d_1 is $x_3' + x_3A' = x_3' + A'$ by the Absorption Law 2. The 4:1 nonlinear-select multiplexer for Figure 3.38(c) is shown in Figure 3.39.

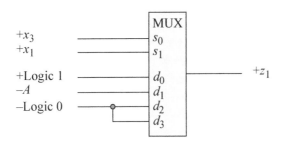

Figure 3.39 The 4:1 nonlinear-select multiplexer for Example 3.8.

Example 3.9 The Karnaugh map shown in Figure 3.40(a) has four unique entries 1, 0, and two map-entered variables A and B. The map will be implemented using a 4:1 nonlinear-select multiplexer. Three variations of the map will be generated: (1) the original map where x_2 and x_3 connect to select inputs s_1 and s_0, respectively; (2) a permutation in which x_1 and x_2 connect to select inputs s_1 and s_0, respectively; and (3) a permutation in which x_1 and x_3 connect to select inputs s_1 and s_0, respectively. The resulting three Karnaugh maps will then be analyzed and the map that provides the least amount of additional logic will be utilized in the implementation.

Figure 3.40(a) requires one OR gate for $x_1' + A$. Figure 3.40(b) requires an OR gate for $x_3' + A$ plus an OR gate for $x_3 + A$. Figure 3.40(c) requires one AND gate for $x_2 A$, plus an OR gate for $x_2' + A$, plus an OR gate for $x_2 + A$. Therefore, the Karnaugh map of Figure 3.40(a) will be utilized to implement the 4:1 nonlinear-select multiplexer, which is shown in Figure 3.41.

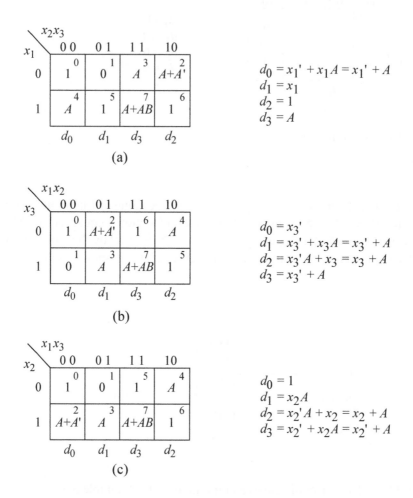

$$d_0 = x_1' + x_1 A = x_1' + A$$
$$d_1 = x_1$$
$$d_2 = 1$$
$$d_3 = A$$

(a)

$$d_0 = x_3'$$
$$d_1 = x_3' + x_3 A = x_3' + A$$
$$d_2 = x_3'A + x_3 = x_3 + A$$
$$d_3 = x_3' + A$$

(b)

$$d_0 = 1$$
$$d_1 = x_2 A$$
$$d_2 = x_2'A + x_2 = x_2 + A$$
$$d_3 = x_2' + x_2 A = x_2' + A$$

(c)

Figure 3.40   Karnaugh maps for Example 3.9 depicting three permutations: (a) original Karnaugh map, (b) first permutation and (c) second permutation.

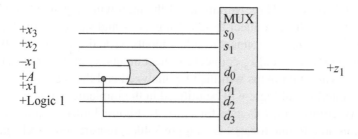

Figure 3.41   The 4:1 nonlinear-select multiplexer for Example 3.9.

Example 3.10 As a final example of nonlinear-select multiplexers, consider the Karnaugh map of Figure 3.42 with E as a map-entered variable. Variables $x_1 x_2$ are connected to select inputs $s_1 s_0$, respectively, of a 4:1 nonlinear-select multiplexer.

$x_1 x_2$ \ $x_3 x_4$	0 0	0 1	1 1	1 0
0 0	0 0	1 1	E 3	– 2
0 1	1 4	0 5	1 7	1 6
1 1	E 12	1 13	1 15	– 14
1 0	1 8	0 9	0 11	1 10

z_1

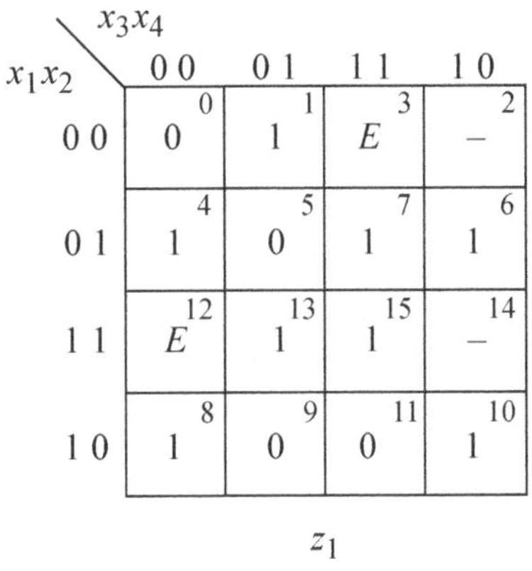

Figure 3.42 Karnaugh map for Example 3.10 to be implemented with a nonlinear-select 4:1 multiplexer.

When $x_1 x_2 = 00$, input d_0 is selected. The entry of 1 in minterm location 1 provides the term $x_3' x_4$. The entry of E in minterm location 3 combines with the "don't care" in minterm location 2 to yield the term $x_3 E$. Therefore, the expression for d_0 is as follows:

$$d_0 = x_3' x_4 + x_3 E$$

When $x_1 x_2 = 01$, input d_1 is selected. The entry of 1 in minterm locations 6 and 7 combine to yield the term x_3. Minterm locations 4 and 6 combine to yield the term x_4'. Therefore, the expression for d_1 is as follows:

$$d_1 = x_3 + x_4'$$

When $x_1 x_2 = 10$, input d_2 is selected. The entry of 1 in minterm locations 8 and 10 combine to yield the term x_4'. Therefore, the expression for d_2 is as follows:

$$d_2 = x_4'$$

When $x_1 x_2 = 11$, input d_3 is selected. The entry of 1 in minterm locations 13 and 15 combine to yield the term x_4. The entry of 1 in minterm locations 13 and 15 can be changed to $1 + E$ without affecting the values. Therefore, the entry of E in minterm location 12 can combine with the entry of 1 in minterm locations 13 and 15 together with the "don't care" in minterm location 14 to yield the term E. Therefore, the expression for d_3 is as follows:

$$d_3 = x_4 + E$$

The above equations for d_0, d_1, d_2, and d_3 can be verified by using a 5-variable Karnaugh with E as the fifth variable, as shown in Figure 3.43. Using $x_1 x_2$ connected to select inputs $s_1 s_0$, the equations can be readily obtained for the data inputs and will corroborate the above equations for the multiplexer data inputs. The logic diagram is shown in Figure 3.44.

$E = 0$

$x_1 x_2$ \ $x_3 x_4$	0 0	0 1	1 1	1 0
0 0	0 [0]	1 [2]	0 [6]	— [4]
0 1	1 [8]	0 [10]	1 [14]	1 [12]
1 1	0 [24]	1 [26]	1 [30]	— [28]
1 0	1 [16]	0 [18]	0 [22]	1 [20]

$E = 1$

$x_1 x_2$ \ $x_3 x_4$	0 0	0 1	1 1	1 0
0 0	0 [1]	1 [3]	1 [7]	— [5]
0 1	1 [9]	0 [11]	1 [15]	1 [13]
1 1	1 [25]	1 [27]	1 [31]	— [29]
1 0	1 [17]	0 [19]	0 [23]	1 [21]

z_1

Figure 3.43 Karnaugh map for Example 3.10 to verify the data input equations of d_0, d_1, d_2, and d_3.

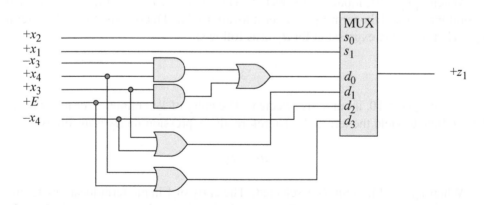

Figure 3.44 Logic diagram for Example 3.10.

3.2.2 Decoders

A decoder is a combinational logic macro that is characterized by the following property: For every valid combination of inputs, a unique output is generated. In general, a decoder has n binary inputs and m mutually exclusive outputs, where $2^n \geq m$. An $n{:}m$ (n-to-m) decoder is shown in Figure 3.45, where the label DX specifies a demultiplexer. Each output represents a minterm that corresponds to the binary representation of the input vector. Thus, $z_i = m_i$, where m_i is the ith minterm of the n input variables. For example, if $n = 3$ and $x_1 x_2 x_3 = 101$, then output z_5 is asserted. A decoder with n inputs, therefore, has a maximum of 2^n outputs. Because the outputs are mutually exclusive, only one output is active for each different combination of the inputs. The decoder outputs may be asserted high or low. Decoders have many applications in digital engineering, ranging from instruction decoding to memory addressing to code conversion.

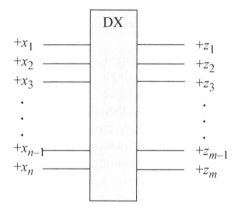

Figure 3.45   An $n{:}m$ decoder.

Figure 3.46 illustrates the logic symbol for a 2:4 decoder, where x_1 and x_2 are the binary input variables and z_0, z_1, z_2, and z_3 are the output variables. Input x_2 is the low-order variable. Since there are two inputs, each output corresponds to a different minterm of two variables as shown in the truth table of Table 3.13.

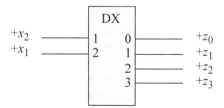

Figure 3.46   Logic symbol for a 2:4 decoder.

Table 3.13 Truth Table for the 2:4 Decoder of Figure 3.46

Inputs $x_1 x_2$	Outputs $z_0 z_1 z_2 z_3$	Minterm Decoding Function
0 0	1 0 0 0	$x_1' x_2'$
0 1	0 1 0 0	$x_1' x_2$
1 0	0 0 1 0	$x_1 x_2'$
1 1	0 0 0 1	$x_1 x_2$

A 3:8 decoder is shown in Figure 3.47 which decodes a binary number into the corresponding octal number. The three inputs are x_1, x_2, and x_3 with binary weights of 2^2, 2^1, and 2^0, respectively. The decoder generates an output that corresponds to the decimal value of the binary inputs. For example, if $x_1 x_2 x_3 = 110$, then output z_6 is asserted high.

A decoder may also have an enable function which allows the selected output to be asserted. The enable function may be a single input or an AND gate with two or more inputs. Figure 3.47 illustrates an enable input consisting of an AND gate with three inputs. If the enable function is deasserted, then the decoder outputs are deasserted. The 3:8 decoder generates all eight minterms z_0 through z_7 of three binary variables x_1, x_2, and x_3. The truth table for the decoder is shown in Table 3.14 and indicates the asserted output that represents the corresponding minterm.

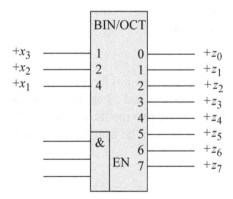

Figure 3.47   A binary-to-octal decoder.

The internal logic for the binary-to-octal decoder of Figure 3.47 is shown in Figure 3.48. The *Enable* gate allows for additional logic functions to control the assertion of the active-high outputs.

Table 3.14 Truth Table for the 3:8 Decoder of Figure 3.47

$x_1 x_2 x_3$	z_0	z_1	z_2	z_3	z_4	z_5	z_6	z_7
0 0 0	1	0	0	0	0	0	0	0
0 0 1	0	1	0	0	0	0	0	0
0 1 0	0	0	1	0	0	0	0	0
0 1 1	0	0	0	1	0	0	0	0
1 0 0	0	0	0	0	1	0	0	0
1 0 1	0	0	0	0	0	1	0	0
1 1 0	0	0	0	0	0	0	1	0
1 1 1	0	0	0	0	0	0	0	1

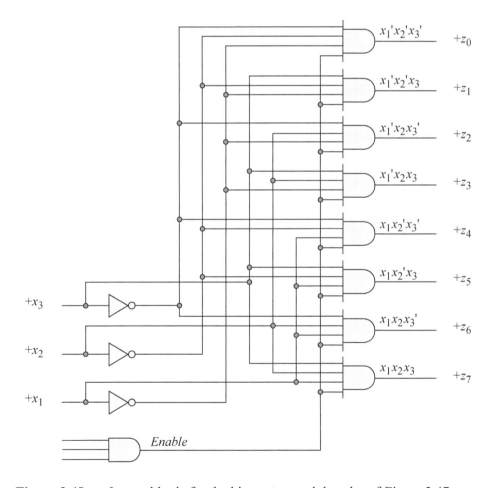

Figure 3.48   Internal logic for the binary-to-octal decoder of Figure 3.47.

Example 3.11   One use for a decoder is to implement a Boolean function. The disjunctive normal form equation shown in Equation 3.8 can be synthesized with a 3:8 decoder and one OR gate as shown in Figure 3.49. The terms in Equation 3.8 represent minterms m_4, m_6, m_5, and m_2, respectively. The equation can also be represented as the following sum-of-minterms expression: $\Sigma_m(2, 4, 5, 6)$. The outputs of the decoder correspond to the eight minterms associated with the three variables $x_1 x_2 x_3$. Therefore, Equation 3.8 is implemented by ORing decoder outputs 2, 4, 5, and 6.

$$z_1(x_1, x_2, x_3) = x_1 x_2' x_3' + x_1 x_2 x_3' + x_1 x_2' x_3 + x_1' x_2 x_3' \qquad (3.8)$$

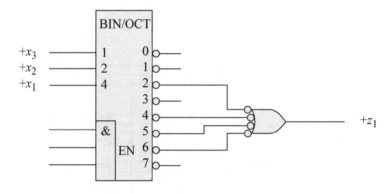

Figure 3.49   Implementation of Equation 3.8 using a 3:8 decoder.

Example 3.12   Equation 3.9 will be implemented using a 4:16 decoder with active-high outputs. Equation 3.9 is also represented in a sum-of-minterms decimal form in Equation 3.10. Minterms 4, 8, 9, 10, and 13 are ORed together to generate the function z_1. All inputs are decoded, but not all outputs are used. The logic diagram is shown in Figure 3.50.

$$\begin{aligned} z_1 = {}& x_1 x_2' x_3' x_4 + x_1 x_2 x_3' x_4 + x_1 x_2' x_3 x_4' + \\ & x_1' x_2 x_3 x_4' + x_1 x_2' x_3' x_4' \end{aligned} \qquad (3.9)$$

$$z_1(x_1, x_2, x_3, x_4) = \Sigma_m(4, 8, 9, 10, 13) \qquad (3.10)$$

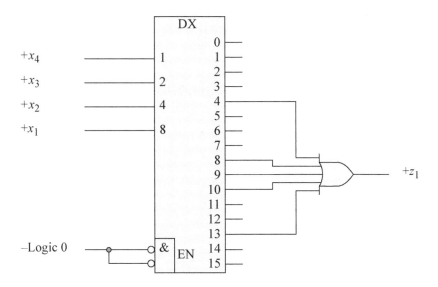

Figure 3.50   Logic diagram to implement Equation 3.9 and Equation 3.10.

Example 3.13   The Karnaugh map of Figure 3.51 will be implemented using NAND gates and also by a 3:8 decoder plus an OR gate. The equations that represent the Karnaugh map are shown in Equation 3.11 and Equation 3.12. The product term $x_2' x_3$ equates to $x_1' x_2' x_3 + x_1 x_2' x_3$ by the distributive law and the complementation law, which corresponds to minterms 1 and 5, respectively. Thus, variable x_1 is not a contributing factor.

Similarly, the product term $x_1 x_2'$ equates to $x_1 x_2' x_3' + x_1 x_2' x_3$, which represents minterms 4 and 5, respectively; thus, variable x_3 is not a contributing factor. In a similar manner, the product term $x_1 x_3$ represents minterms 5 and 7. The term $x_1' x_2 x_3'$ corresponds to minterm 2.

The logic diagram using NAND logic is shown in Figure 3.52 and using a decoder in Figure 3.53.

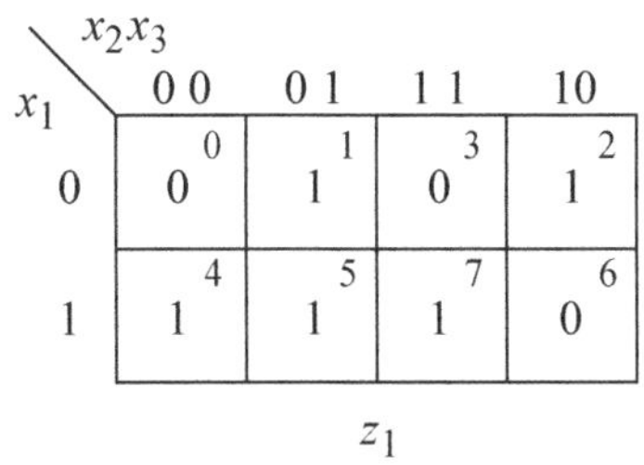

Figure 3.51   Karnaugh map for Example 3.13.

$$z_1 = x_2'x_3 + x_1x_2' + x_1x_3 + x_1'x_2x_3' \qquad (3.11)$$

$$z_1(x_1, x_2, x_3) = \Sigma_m(1, 2, 4, 5, 7) \qquad (3.12)$$

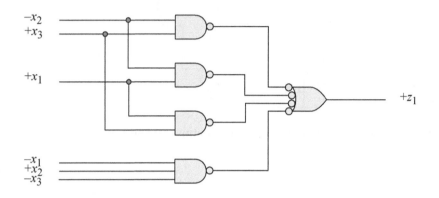

Figure 3.52   Logic diagram using NAND gates for Example 3.13.

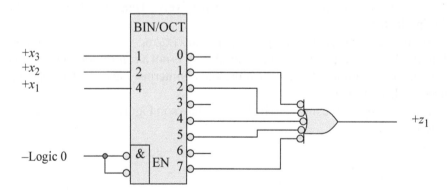

Figure 3.53   Logic diagram using a decoder and an OR gate for Example 3.13.

Example 3.14 A decoder can also be used as a code converter. Table 3.15 shows the binary code and the corresponding Gray code for each binary code word. The Gray code is an unweighted code and belongs to a class of cyclic codes called reflective codes. The Gray code has the unique characteristic whereby each adjacent code word differs in only one bit position. Before presenting the decoder method of converting from binary code to Gray code, a general procedure for the conversion process will be presented.

Table 3.15 Table Showing the Gray Code for each Binary Code Word

Row	Binary Code				Gray Code			
	b_3	b_2	b_1	b_0	g_3	g_2	g_1	g_0
0	0	0	0	0	0	0	0	0
1	0	0	0	1	0	0	0	1
2	0	0	1	0	0	0	1	1
3	0	0	1	1	0	0	1	0
4	0	1	0	0	0	1	1	0
5	0	1	0	1	0	1	1	1
6	0	1	1	0	0	1	0	1
7	0	1	1	1	0	1	0	0
8	1	0	0	0	1	1	0	0
9	1	0	0	1	1	1	0	1
10	1	0	1	0	1	1	1	1
11	1	0	1	1	1	1	1	0
12	1	1	0	0	1	0	1	0
13	1	1	0	1	1	0	1	1
14	1	1	1	0	1	0	0	1
15	1	1	1	1	1	0	0	0

By analyzing Table 3.15, a procedure for converting from the binary 8421 code to the Gray code can be formulated. Let an n-bit binary code word be represented as

$$b_{n-1} \, b_{n-2} \, \cdots \, b_1 \, b_0$$

and an n-bit Gray code word be represented as

$$g_{n-1} \, g_{n-2} \, \cdots \, g_1 \, g_0$$

where b_0 and g_0 are the low-order bits of the binary and Gray codes, respectively. The ith Gray code bit g_i can be obtained from the corresponding binary code word by the following algorithm:

$$g_{n-1} = b_{n-1}$$

$$g_i = b_i \oplus b_{i+1}$$

for $0 \leq i \leq n - 2$, where the symbol $\oplus$ denotes modulo-2 addition defined as:

$$0 \oplus 0 = 0$$

$$0 \oplus 1 = 1$$

$$1 \oplus 0 = 1$$

$$1 \oplus 1 = 0$$

For example, using the algorithm, the 4-bit binary code word $b_3\, b_2\, b_1\, b_0 = 1010$ translates to the 4-bit Gray code word $g_3\, g_2\, g_1\, g_0 = 1111$ as follows:

$$g_3 = b_3 \qquad\qquad\qquad = 1$$

$$g_2 = b_2 \oplus b_3 = 0 \oplus 1 = 1$$

$$g_1 = b_1 \oplus b_2 = 1 \oplus 0 = 1$$

$$g_0 = b_0 \oplus b_1 = 0 \oplus 1 = 1$$

Using Table 3.15, Gray code bit g_0 can be obtained by ORing together the following outputs of a 4:16 decoder: $z_1, z_2, z_5, z_6, z_9, z_{10}, z_{13}$, and z_{14}. Gray code bit g_1 can be obtained by ORing together the following decoder outputs: $z_2, z_3, z_4, z_5, z_{10}, z_{11}, z_{12}$, and z_{13}. Gray bit g_2 can be obtained by ORing together the following decoder outputs: $z_4, z_5, z_6, z_7, z_8, z_9, z_{10}$, and z_{11}. Gray bit g_3 can be obtained by ORing together the following decoder outputs: $z_8, z_9, z_{10}, z_{11}, z_{12}, z_{13}, z_{14}$, and z_{15}. Figure 3.54 shows the logic diagram for the binary-to-Gray code converter using a 4:16 decoder.

Although this section discusses decoders, it is readily apparent that less logic would be required if exclusive-OR circuits were used in the implementation of the binary-to-Gray code converter rather than a 4:16 decoder. This alternative approach is shown in Figure 3.55 using three exclusive-OR circuits and one buffer.

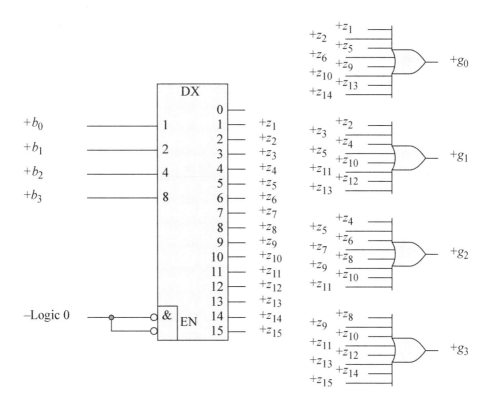

Figure 3.54   Decoder used to convert from binary-to-Gray code.

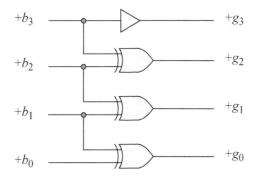

Figure 3.55   Alternative approach for generating a binary-to-Gray code converter.

Example 3.15 A special type of decoder is used to activate the segments of a 7-segment display. The decoder inputs are in binary-coded decimal (BCD) notation and generate multiple simultaneous decoder outputs for the seven segments of the display.

Figure 3.56 illustrates the different segments of a seven-segment display and the segments that are activated for each decimal digit. The decoder is shown in Figure 3.57 with BCD inputs b_3, b_2, b_1, and b_0, where b_0 is the low-order bit. There are seven outputs a, b, c, d, e, f, and g, one for each segment. The decimal values 10 through 15 are invalid for BCD.

Digit	Segments Activated
0	a b c d e f
1	b c
2	a b d e g
3	a b c d g
4	b c f g
5	a c d f g
6	a c d e f g
7	a b c
8	a b c d e f g
9	a b c d f g

Figure 3.56   Seven-segment display and segment activation table.

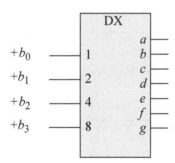

Figure 3.57   BCD-to-seven-segment decoder.

The internal logic to activate segment a is generated by BCD variables $b_3 b_2 b_1 b_0$ = 0000, 0010, 0011, 0101, 0110, 0111, 1000, and 1001. The Karnaugh map representing segment a is shown in Figure 3.58 and the minimized equation to activate segment a is shown in Equation 3.13. In a similar manner, the logic for the other segments is obtained.

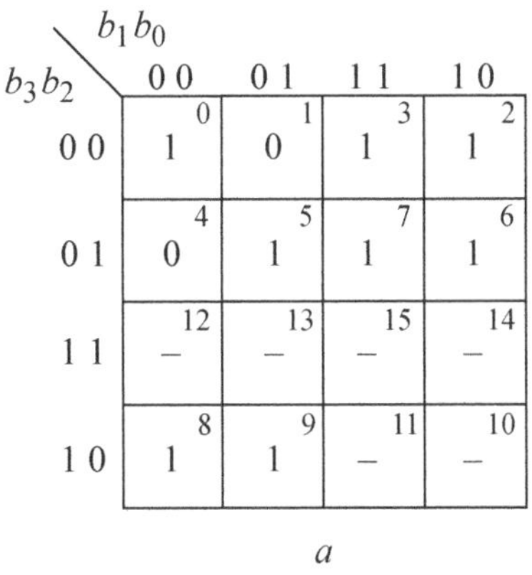

Figure 3.58   Karnaugh map for segment a of a seven-segment display.

$$a = b_1 + b_3 + b_2'b_0' + b_2b_0 \qquad (3.13)$$

Decoder expansion   A 6:64 decoder can be realized by utilizing four 4:16 decoders. Table 3.16 lists the decoder inputs x_1, x_2, x_3, x_4, x_5, and x_6 — where x_6 is the low-order input — together with the numbered outputs. Decoder inputs x_1 and x_2 are connected to the enable inputs of the 4:16 decoders so that each decoder is selected independently, as shown in the logic diagram of Figure 3.59.

Table 3.16 Decoder Inputs to Construct a 6:64 Decoder from Four 4:16 Decoders

32	16	8	4	2	1	
x_1	x_2	x_3	x_4	x_5	x_6	Decoder Outputs
0	0	0	0	0	0	0
			$\cdots$			
0	0	1	1	1	1	15
0	1	0	0	0	0	16
			$\cdots$			
0	1	1	1	1	1	31
1	0	0	0	0	0	32
			$\cdots$			
1	0	1	1	1	1	47
1	1	0	0	0	0	48
			$\cdots$			
1	1	1	1	1	1	63

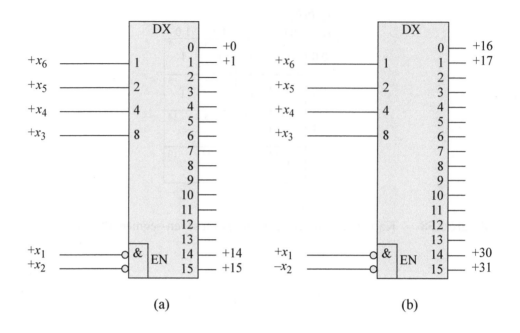

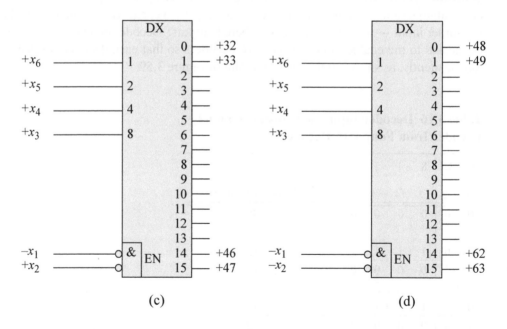

Figure 3.59   Decoder organization to implement a 6:64 decoder using four 4:16 decoders: (a) decode 0 through 15, (b) decode 16 through 31, (c) decode 32 through 47, and (d) decode 48 through 63.

3.2.3 Encoders

An encoder is a macro logic circuit with n mutually exclusive inputs and m binary outputs, where $n \leq 2^m$. The inputs are mutually exclusive to prevent errors from appearing on the outputs. The outputs generate a binary code that corresponds to the active input value. The function of an encoder can be considered to be the inverse of a decoder; that is, the mutually exclusive inputs are encoded into a corresponding binary number.

A general block diagram for an $n{:}m$ encoder is shown in Figure 3.60. An encoder is also referred to as a code converter. In the label of Figure 3.60, X corresponds to the input code and Y corresponds to the output code. The general qualifying label X/Y is replaced by the input and output codes, respectively, such as, OCT/BIN for an octal-to-binary code converter. Only one input x_i is asserted at a time. The decimal value of x_i is encoded as a binary number which is specified by the m outputs.

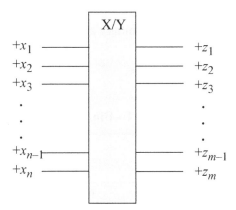

Figure 3.60   An $n{:}m$ encoder or code converter.

An 8:3 octal-to-binary encoder is shown in Figure 3.61(a). Although there are 2^8 possible input combinations of eight variables, only eight combinations are valid. The eight inputs each generate a unique octal code word in binary. If the outputs are to be enabled, then the gating can occur at the output gates.

The truth table for an 8:3 encoder is shown in Table 3.17. The encoder can be implemented with OR gates whose inputs are established from the truth table, as shown in Equation 3.14 and Figure 3.62. The low-order output z_3 is asserted when one of the following inputs are active: $x_1, x_3, x_5,$ or x_7. Output z_2 is asserted when one of the following inputs are active: $x_2, x_3, x_6,$ or x_7. Output z_1 is asserted when one of the following inputs are active: $x_4, x_5, x_6,$ or x_7.

A 10:4 BCD-to-binary encoder is shown in Figure 3.61(b). If input 7 is asserted, then the output values are $z_1z_2z_3z_4 = 0111$, where the binary weight of the outputs is $z_1z_2z_3z_4 = 2^3 2^2 2^1 2^0$.

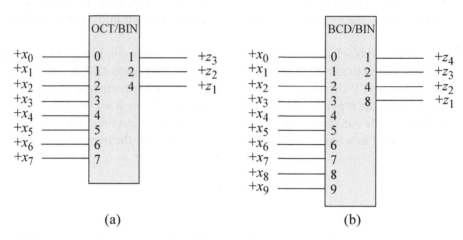

(a)   (b)

Figure 3.61   Encoders: (a) octal-to-binary and (b) BCD-to-binary.

Table 3.17 Truth Table for an Octal-To-Binary Encoder

			Inputs						Outputs	
x_0	x_1	x_2	x_3	x_4	x_5	x_6	x_7	z_1	z_2	z_3
1	0	0	0	0	0	0	0	0	0	0
0	1	0	0	0	0	0	0	0	0	1
0	0	1	0	0	0	0	0	0	1	0
0	0	0	1	0	0	0	0	0	1	1
0	0	0	0	1	0	0	0	1	0	0
0	0	0	0	0	1	0	0	1	0	1
0	0	0	0	0	0	1	0	1	1	0
0	0	0	0	0	0	0	1	1	1	1

$$z_3 = x_1 + x_3 + x_5 + x_7$$

$$z_2 = x_2 + x_3 + x_6 + x_7$$

$$z_1 = x_4 + x_5 + x_6 + x_7 \tag{3.14}$$

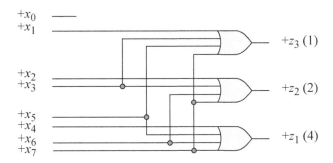

Figure 3.62   Logic diagram for an 8:3 encoder.

Priority encoder   It was stated previously that encoder inputs are mutually exclusive. There may be situations, however, where more than one input can be active at a time. Then a priority must be established to select and encode a particular input. This is referred to as a *priority encoder*.

Usually the input with the highest valued subscript is selected as highest priority for encoding. Thus, if x_i and x_j are active simultaneously and $i < j$, then x_j has priority over x_i. For example, assume that the octal-to-binary encoder of Figure 3.61(a) is a priority encoder. If inputs x_1, x_5, and x_7 are asserted simultaneously, then the outputs will indicate the binary equivalent of decimal 7 such that $z_1 z_2 z_3 = 111$.

The truth table for an octal-to-binary priority encoder is shown in Table 3.18. The outputs $z_1 z_2 z_3$ generate a binary number that is equivalent to the highest priority input. If $x_3 = 1$, the state of x_0, x_1, and x_2 is irrelevant ("don't care") and the output is the binary number 011. The equations for z_1, z_2, and z_3 are shown in Equation 3.15, Equation 3.16, and Equation 3.17, respectively, and are generated from Table 3.18. The implementation of the octal-to-binary priority encoder is obtained directly from Equation 3.15, Equation 3.16, and Equation 3.17.

Table 3.18 Octal-To-Binary Priority Encoder

Inputs								Outputs		
x_0	x_1	x_2	x_3	x_4	x_5	x_6	x_7	z_1	z_2	z_3
1	0	0	0	0	0	0	0	0	0	0
–	1	0	0	0	0	0	0	0	0	1
–	–	1	0	0	0	0	0	0	1	0
–	–	–	1	0	0	0	0	0	1	1
–	–	–	–	1	0	0	0	1	0	0
–	–	–	–	–	1	0	0	1	0	1
–	–	–	–	–	–	1	0	1	1	0
–	–	–	–	–	–	–	1	1	1	1

$$z_1 = x_4 x_5' x_6' x_7' + x_5 x_6' x_7' + x_6 x_7' + x_7$$

$= x_6' x_7' (x_4 x_5' + x_5) + x_6 + x_7$ Distributive law

$= x_6' x_7' (x_4 + x_5) + (x_6 + x_7)$ Absorption law 2
DeMorgan's law

$= x_4 + x_5 + x_6 + x_7$ Absorption law 2   (3.15)

$$z_2 = x_2 x_3' x_4' x_5' x_6' x_7' + x_3 x_4' x_5' x_6' x_7' + x_6 x_7' + x_7$$

$= x_4' x_5' x_6' x_7' (x_2 x_3' + x_3) + x_6 + x_7$ Distributive law

$= x_4' x_5' x_6' x_7' (x_2 + x_3) + x_6 + x_7$ Absorption law 2

$= x_6' x_7' [(x_4' x_5' (x_2 + x_3)] + (x_6 + x_7)$ Associative law

$= x_2 x_4' x_5' + x_3 x_4' x_5' + x_6 + x_7$ Absorption law 2   (3.16)

$$z_3 = x_1 x_2' x_3' x_4' x_5' x_6' x_7' + x_3 x_4' x_5' x_6' x_7' + x_5 x_6' x_7'$$
$$+ x_7$$

$= x_4' x_5' x_6' x_7' (x_1 x_2' x_3' + x_3) + x_5 x_6' x_7' + x_7$ Distributive law

$= x_4' x_5' x_6' x_7' (x_1 x_2' + x_3) + x_5 x_6' + x_7$ Absorption law 2

$= x_1 x_2' x_4' x_5' x_6' x_7' + x_3 x_4' x_5' x_6' x_7' + x_5 x_6' + x_7$ Distributive law

$= (x_1 x_2' x_4' x_5' x_6') x_7' + x_7 + x_6' (x_3 x_4' x_5' x_7' + x_5)$ Associative law
Distributive law

$= x_1 x_2' x_4' x_5' x_6' + x_7 + x_6' (x_3 x_4' x_7' + x_5)$ Absorption law 2

$= x_1 x_2' x_4' x_5' x_6' + x_7 + x_3 x_4' x_6' x_7' + x_5 x_6'$ Distributive law

$= x_1 x_2' x_4' x_5' x_6' + (x_3 x_4' x_6') x_7' + x_7 + x_5 x_6'$ Associative law

$= x_1 x_2' x_4' x_5' x_6' + x_3 x_4' x_6' + x_5 x_6' + x_7$ Absorption law 2

$= x_6' (x_1 x_2' x_4' x_5' + x_5) + x_3 x_4' x_6' + x_7$ Distributive law

$= x_6' (x_5 + x_1 x_2' x_4') + x_3 x_4' x_6' + x_7$ Absorption law 2

$= x_1 x_2' x_4' x_6' + x_3 x_4' x_6' + x_5 x_6' + x_7$ Distributive law
Associative law   (3.17)

3.2.4 Comparators

A comparator is a logic macro circuit that compares the magnitude of two n-bit binary numbers X_1 and X_2. Therefore, there are $2n$ inputs and three outputs that indicate the relative magnitude of the two numbers. The outputs are mutually exclusive, specifying $X_1 < X_2, X_1 = X_2,$ or $X_1 > X_2$. Figure 3.63 shows a general block diagram of a comparator.

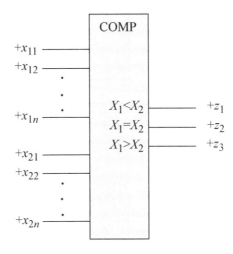

Figure 3.63   General block diagram of a comparator.

If two or more comparators are connected in cascade, then three additional inputs are required for each comparator. These additional inputs indicate the relative magnitude of the previous lower-order comparator inputs and specify $X_1 < X_2, X_1 = X_2,$ or $X_1 > X_2$ for the previous stage. Cascading comparators usually apply only to commercially available comparator integrated circuits.

For example, two 8-bit numbers can be compared by using two 4-bit comparators as shown in Figure 3.64. The magnitude inputs for the low-order comparator are assigned the following values: $(X_1 < X_2) = 0, (X_1 = X_2) = 1,$ and $(X_1 > X_2) = 0$. This assignment initializes the compare operation by specifying that the two operands are equal at that point in the comparison. Any other assignment would have a negative effect on the comparison operation.

If the high-order four bits are equal, then the outputs from the low-order comparator are used to determine the relative magnitude. Figure 3.65 illustrates examples of cascaded comparators to compare two 8-bit operands.

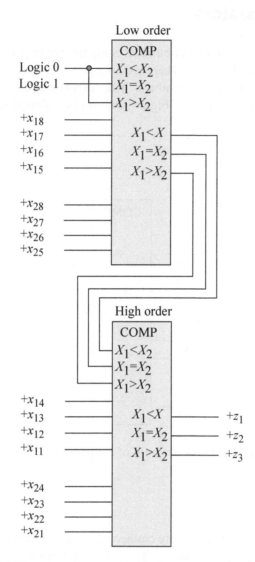

Figure 3.64   Two 4-bit comparators cascaded to compare two 8-bit operands.

Result	High Order					Low Order			
	$X_1 =$ 1 0 1 0					0 1 0 1			
$X_1 < X_2$	$X_1 = X_2$					$X_1 < X_2$			
	$X_2 =$ 1 0 1 0					1 1 0 1			
	$X_1 =$ 1 1 1 0					0 0 1 1			
$X_1 > X_2$	$X_1 > X_2$					$X_1 < X_2$			
	$X_2 =$ 0 1 1 0					1 0 0 1			

Figure 3.65   Cascaded comparators to compare two 8-bit operands.

The design of a comparator is relatively straightforward. Consider two 3-bit unsigned operands $X_1 = x_{11} x_{12} x_{13}$ and $X_2 = x_{21} x_{22} x_{23}$, where x_{13} and x_{23} are the low-order bits of X_1 and X_2, respectively. Three equations will now be derived to represent the three outputs; one equation each for $X_1 < X_2$, $X_1 = X_2$, and $X_1 > X_2$. Comparison occurs in a left-to-right manner beginning at the high-order bits of the two operands.

Operand X_1 will be less than X_2 if $x_{11} x_{21} = 01$. Thus, X_1 cannot be more than 011 while X_2 cannot be less than 100, indicating that $X_1 < X_2$. Therefore, the first term of the equation for $X_1 < X_2$ is $x_{11}' x_{21}$. If, however, $x_{11} = x_{21}$, then the relative magnitude depends on the values of x_{12} and x_{22}. The equality of two bits is represented by the exclusive-NOR symbol, also called the *equality function*. Thus, the second term in the equation for $X_1 < X_2$ is $(x_{11} \oplus x_{21})' x_{12}' x_{22}$. The analysis continues in a similar manner for the remaining bits of the two operands.

The equation for $X_1 < X_2$ is shown in Equation 3.18. The equality of X_1 and X_2 is true if and only if each bit-pair is equal, where $x_{11} = x_{21}$, $x_{12} = x_{22}$, and $x_{13} = x_{23}$; that is, $X_1 = X_2$ if and only if $x_{1i} = x_{2i}$ for $i = 1, 2, 3$. This is indicated by the Boolean product of three equality functions as shown in Equation 3.18 for $X_1 = X_2$.

The final equation, which specifies $X_1 > X_2$, is obtained in a manner analogous to that for $X_1 < X_2$. If the high-order bits are $x_{11} x_{21} = 10$, then it is immediately apparent that $X_1 > X_2$. Using the equality function with the remaining bits yields the equation for $X_1 > X_2$ as shown in Equation 3.18. The design process is modular and can be extended to accommodate any size operands in a well-defined regularity. Two n-bit operands will contain column subscripts of 11 through $1n$ and 21 through $2n$, where n specifies the low-order bits.

An alternative approach which may be used to minimize the amount of hardware is to eliminate the equation for $X_1 = X_2$ and replace it with the following:

$$(X_1 = X_2) \text{ if } (X_1 < X_2)' \text{ AND } (X_1 > X_2)'$$

That is, if X_1 is neither less nor greater than X_2, then X_1 must equal X_2. Designing the hardware for a comparator is relatively straightforward — it consists of AND gates, OR gates, and exclusive-NOR circuits as shown in Equation 3.18.

$$(X_1 < X_2) = x_{11}' x_{21} + (x_{11} \oplus x_{21})' x_{12}' x_{22} + (x_{11} \oplus x_{21})' (x_{12} \oplus x_{22})' x_{13}' x_{23}$$

$$(X_1 = X_2) = (x_{11} \oplus x_{21})' (x_{12} \oplus x_{22})' (x_{13} \oplus x_{23})'$$

$$(X_1 > X_2) = x_{11} x_{21}' + (x_{11} \oplus x_{21})' x_{12} x_{22}' + (x_{11} \oplus x_{21})' (x_{12} \oplus x_{22})' x_{13} x_{23}' \quad (3.18)$$

Example 3.16 Comparators can also be used to determine if a binary word is within a certain range. An output is then generated indicating that the word is either in range or out-of-range of predetermined values. Let 8-bit words represent both the upper and lower limit range of the 8-bit word under test, as shown below.

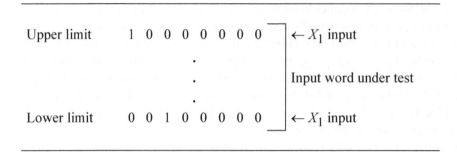

Figure 3.66 illustrates a method to determine whether a test word is within a certain range by utilizing two 8-bit comparators and an OR gate. For the upper comparator, if output $X_1 < X_2$ is asserted, then this means that the input word is greater than the upper limit. Likewise, for the lower comparator, if output $X_1 > X_2$ is asserted, then this means that the input word is less than the lower limit. The result of ORing these two outputs indicates that the word under test is out-of-range.

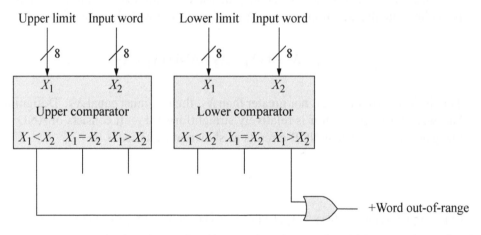

Figure 3.66   Method to determine if a word under test is within a predetermined range.

Example 3.17   This example will determine if a 4-bit binary number $N = n_3 n_2 n_1 n_0$ is within the following range:

$$5 < N \leq 12$$

where n_0 is the low-order bit. The truth table of Table 3.19 shows the possible values of N and the range in which N must be valid.

Figure 3.67 illustrates the method of determining whether the 4-bit word N is within the prescribed range by using two 4-bit comparators and one AND gate. If output $X_1 < X_2$ is asserted in the lower comparator, then this means that N is greater than 5. If output $X_1 > N$ in the higher comparator, then this means that N is less than 13. When these two outputs are ANDed together, the result indicates that $5 < N \le 12$.

Table 3.19 Table Showing the Valid Range of N

n_3	n_2	n_1	n_0	
0	0	0	0	
0	0	0	1	
0	0	1	0	
0	0	1	1	
0	1	0	0	
0	1	0	1	
0	1	1	0	
0	1	1	1	
1	0	0	0	Range for the 4-bit word N
1	0	0	1	
1	0	1	0	$5 < N \le 12$
1	0	1	1	
1	1	0	0	
1	1	0	1	
1	1	1	0	
1	1	1	1	

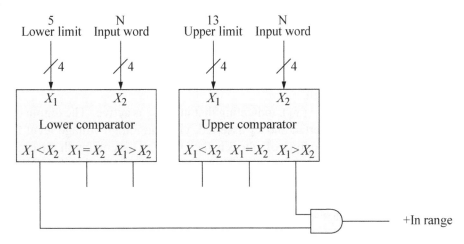

Figure 3.67   Logic to determine if the 4-bit word N is within the range $5 < N \le 12$.

3.3 Analysis of Combinational Logic

The analysis of combinational logic requires determining the function of a given logic circuit by obtaining one or more of the following entities: the equation, truth table, Karnaugh map, or waveforms. The analysis procedure involves obtaining the equation for the output of each gate or logic macro function as an expression of the inputs, then applying those equations to the following gate in the network until the output equation is obtained for the logic circuit. Several examples will now be presented to elaborate the procedure for analyzing combinational logic.

The $+/-$ symbols that precede an input or output variable indicate the active voltage level of the variable. Thus, $+x_1$ or $-x_1$ indicates that variable x_1 is active at a high or low voltage level, respectively. The NAND gates in Figure 3.68 portray two different Boolean equations. In Figure 3.68(a), if both inputs are asserted at a high voltage level, then output z_1 is asserted at a low voltage level. Thus, the equation for an active-low output is $z_1 = x_1 x_2$.

In Figure 3.68(b), however, if x_1 is asserted (producing a high level at the gate input — which requires a high level input) and x_2 is deasserted (producing a high level at the gate input), then z_1 is asserted at a low voltage level. Thus, the equation for an active-low output is $z_1 = x_1 x_2'$. That is, z_1 is asserted (active) if x_1 is asserted (active) and x_2 is deasserted (inactive).

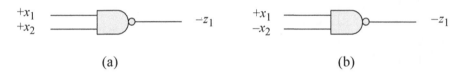

(a)   (b)

Figure 3.68   NAND gates representing two different equations: (a) $z_1 = x_1 x_2$ and $z_1 = x_1 x_2'$.

Example 3.18 The logic diagram of Figure 3.69 will be analyzed to obtain the equation for output z_1. The equation for z_1 will then be converted to a sum-of-products notation and plotted on a Karnaugh map from which a product-of-sums form will be obtained. The expression at the output of each gate indicates the equation for the gate up to that level of the circuit.

The output of gate 1 is $+(x_4)$, indicating that input x_4 is active. The output of gate 2 is the AND of x_3 deasserted and x_4 asserted, yielding a high voltage level for the expression $+(x_3' x_4)$. The output of gate 3 is the OR of x_2 asserted with the output of gate 2, yielding the expression $+(x_2 + x_3' x_4)$. The output of gate 4 is the AND of x_1 with the output of gate 3, which yields the expression $+[x_1(x_2 + x_3' x_4)]$. Gate 5 provides the AND of x_3 and x_4, which yields $+(x_3 x_4)$. The equation for output z_1 is the OR of the output of gates 4 and 5, which generates the final equation for z_1:

$$z_1 = x_1(x_2 + x_3' x_4) + x_3 x_4 \qquad (3.19)$$

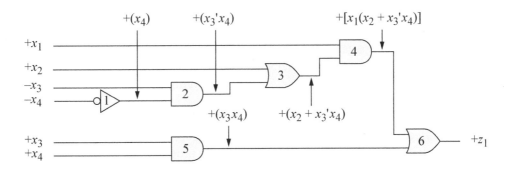

Figure 3.69   Logic diagram for Example 3.18.

Expanding Equation 3.19 provides a sum-of-products expression, as shown in Equation 3.20, which is plotted on the Karnaugh map of Figure 3.70. The equation for z_1 is then obtained as a product of sums, as shown in Equation 3.21. Note that Equation 3.20 and Equation 3.21 both require the same number of gates. Equation 3.20 requires three AND gates and one OR gate. Equation 3.21 requires three OR gates and one AND gate. Equation 3.21, however, uses only 2-input OR gates.

$$z_1 = x_1 x_2 + x_1 x_3' x_4 + x_3 x_4 \tag{3.20}$$

$$
\begin{array}{c|cccc}
\ & x_3 x_4 & & & \\
x_1 x_2 & 0\,0 & 0\,1 & 1\,1 & 1\,0 \\
\hline
0\,0 & 0 & 0 & 1 & 0 \\
0\,1 & 0 & 0 & 1 & 0 \\
1\,1 & 1 & 1 & 1 & 1 \\
1\,0 & 0 & 1 & 1 & 0 \\
\end{array}
$$

cells: 0, 1, 3, 2 / 4, 5, 7, 6 / 12, 13, 15, 14 / 8, 9, 11, 10

z_1

Figure 3.70   Karnaugh map for Equation 3.20.

$$z_1 = (x_1 + x_3)(x_1 + x_4)(x_2 + x_4) \tag{3.21}$$

Example 3.19   Figure 3.71 illustrates a 2-level, multigate network using only NAND gates. The output of gate 1 will be low if the expression $x_1'x_2'$ is true. Since gate 3 requires that at least one input be at a low voltage level to assert the output, the low output from gate 1 propagates through gate 3 to assert output z_1. The polarity symbol " ○ " may be considered as complementation; therefore, the path through gates 1 and 3 is equivalent to the expression $(x_1'x_2')''$, which reduces to $x_1'x_2'$ by the involution law.

Similarly, the output of gate 2 is active low if the expression $x_1x_2x_3'$ is true. Thus, z_1 is asserted high for $(x_1x_2x_3')'' = x_1x_2x_3'$. The fourth input variable is $+x_4$, which specifies that x_4 is active at a high voltage level. In order for the low input requirement to be met for gate 3, input x_4 must be inactive. Thus, if x_4' is true, then z_1 is asserted. The complete equation for z_1, therefore, is $z_1 = x_1'x_2' + x_1x_2x_3' + x_4'$.

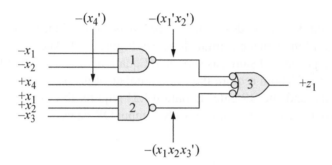

Figure 3.71   Multigate logic diagram using NAND gates for the $z_1 = x_1'x_2' + x_1x_2x_3' + x_4'$.

Example 3.20   Given the logic diagram shown in Figure 3.72, the equation for output z_1 will be obtained. The circuit uses only NOR gates in its implementation. The output level is shown for each gate with the corresponding expression that represents the functionality of the circuit at that point. For example, the output of gate 3 is at a high voltage level if the expression $(x_1 + x_3')x_2'x_4$ is true. Likewise, the output of gate 5 is at a high level if the expression $x_2(x_3' + x_4')$ is true. The complete equation for z_1 is shown in Equation 3.22 in a minimal sum-of-products form.

Example 3.21   The logic diagram shown in Figure 3.73 uses a mixture of NAND and NOR logic plus an exclusive-NOR circuit. The equation for output z_1 is generated in a sum-of-products form directly from the logic diagram, as shown in Equation 3.23. The equation is then plotted on the 4-variable Karnaugh map of Figure 3.74 and regenerated in a minimum sum-of-products form shown in Equation 3.24. The circuit is then redrawn using only NOR logic with z_1 active high, as shown in Figure 3.75.

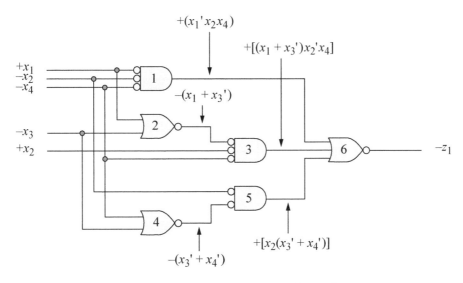

Figure 3.72   Logic diagram for Example 3.20 using only NOR gates. Output $z_1 = x_1'x_2x_4 + x_2'x_4(x_1 + x_3') + x_2(x_3' + x_4')$.

$$z_1 = x_1'x_2 + x_3'x_4 + x_2x_4' + x_1x_2'x_4 \qquad (3.22)$$

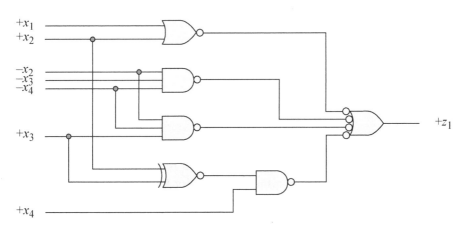

Figure 3.73   Logic diagram for Example 3.21.

$$z_1 = x_1 + x_2 + x_2'x_3'x_4' + x_2'x_3x_4' + (x_2 \oplus x_3)'x_4$$
$$= x_1 + x_2 + x_2'x_3'x_4' + x_2'x_3x_4' + x_2x_3x_4 + x_2'x_3'x_4 \qquad (3.23)$$

	x_3x_4			
x_1x_2	0 0	0 1	1 1	1 0
0 0	1 ⁰	1 ¹	0 ³	1 ²
0 1	1 ⁴	1 ⁵	1 ⁷	1 ⁶
1 1	1 ¹²	1 ¹³	1 ¹⁵	1 ¹⁴
1 0	1 ⁸	1 ⁹	1 ¹¹	1 ¹⁰

z_1

Figure 3.74   Karnaugh map for Example 3.21.

$$z_1 = x_1 + x_2 + x_3' + x_4' \tag{3.24}$$

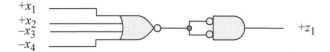

Figure 3.75   Logic diagram redrawn for Example 3.21 in minimum form.

Example 3.22   The mixed-logic diagram shown in Figure 3.76 will be analyzed by obtaining the output expressions that represent the function of each logic gate, then combining the expressions to form the equations for the two outputs z_1 and z_2. It is readily apparent that there are redundant gates for z_1, which will be evident when the equation for z_1 is obtained and then minimized using Boolean algebra.

The resulting equations for z_1 and z_2 will be used to generate a truth table for the two functions. The output of gate 2 will be at a high voltage level if either x_1, x_2, or x_3 is deasserted. Therefore, by DeMorgan's theorem, the output will be at a low voltage level if x_1, x_2, and x_3 are all asserted. That is, the equation for the output of gate 2 will be $-(x_1 x_2 x_3)$. Note that gate 3 is an OR gate that is drawn as an AND gate with active-low inputs and output. The remaining gates in Figure 3.76 are drawn in the standard manner as NAND gates. The equations for output z_1 and z_2 are shown in Equation 3.25 and Equation 3.26, respectively, and generate the truth table of Table 3.20.

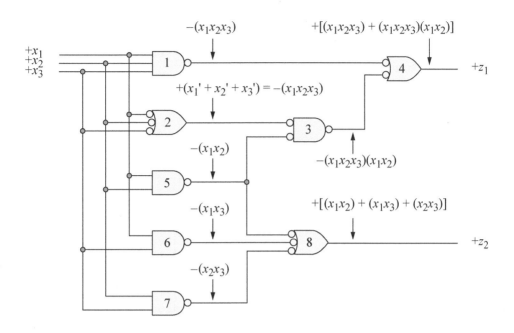

Figure 3.76 Logic diagram for Example 3.22.

$$z_1 = [x_1 x_2 x_3 + x_1 x_2 x_3 (x_1 x_2)]$$

$$= x_1 x_2 x_3 (1 + x_1 x_2)$$

$$= x_1 x_2 x_3 \tag{3.25}$$

$$z_2 = x_1 x_2 + x_1 x_3 + x_2 x_3 \tag{3.26}$$

Table 3.20 Truth Table for the Logic Diagram of Figure 3.76

x_1	x_2	x_3	z_1	z_2	x_1	x_2	x_3	z_1	z_2
0	0	0	0	0	1	0	0	0	0
0	0	1	0	0	1	0	1	0	1
0	1	0	0	0	1	1	0	0	1
0	1	1	0	1	1	1	1	1	1

Example 3.23   The logic diagram shown in Figure 3.77 will be analyzed by obtaining the equation for output z_1 directly from the logic.   Note that z_1 is active low.   The diagram contains a mixture of AND, NAND, and NOR gates, plus inverters.   There are redundant gates in the logic diagram which will become apparent when the equation is obtained then minimized.

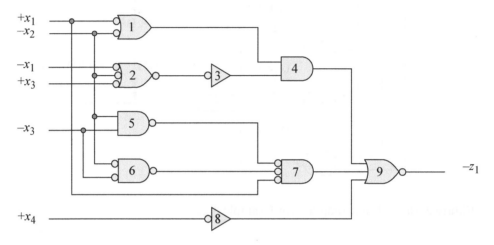

Figure 3.77   Logic diagram for Example 3.23.

Gate 2 is a positive-input AND gate drawn as an OR gate with active-low inputs and output.   Gate 6 is an OR gate drawn as an AND gate with active-low inputs and output.   The output expressions for all logic functions are listed below.

Gate 1:   High level if $x_1' + x_2$ is true.

Gate 2:   Low level if $x_1 + x_2 + x_3'$ is true.

Gate 3:   High level if $x_1 + x_2 + x_3'$ is true.

Gate 4:   High level if $(x_1' + x_2)(x_1 + x_2 + x_3')$ is true.

Gate 5:   Low level if $x_2' x_3'$ is true.

Gate 6:   Low level if $x_2 x_3$ is true.

Gate 7:   High level if $(x_2' x_3')(x_2 x_3)x_1'$ is true.

Gate 8:   High level if x_4' is true.

Gate 9:   Low level if $(x_1' + x_2)(x_1 + x_2 + x_3') + (x_2' x_3')(x_2 x_3)x_1' + x_4'$ is true.

The equation for z_1 is the output of gate 9 and is minimized as follows:

$$
\begin{aligned}
z_1 &= (x_1' + x_2)(x_1 + x_2 + x_3') + (x_2'x_3')(x_2x_3)x_1' + x_4' \\
&= (x_1' + x_2)x_1 + (x_1' + x_2)x_2 + (x_1' + x_2)x_3' + x_4' \\
&= x_1x_2 + x_1'x_2 + x_2 + x_1'x_3' + x_2x_3' + x_4' \\
&= x_2(x_1 + x_1' + 1) + x_1'x_3' + x_2x_3' + x_4' \\
&= x_2 + x_1'x_3' + x_2x_3' + x_4' \\
&= x_2(1 + x_3') + x_1'x_3' + x_4' \\
&= x_2 + x_1'x_3' + x_4'
\end{aligned}
\tag{3.27}
$$

Therefore, the logic to generate output z_1 can be implemented with one 2-input AND gate and one 3-input NOR gate to assert z_1 at a low voltage level. The equation for z_1 will now be plotted on a Karnaugh map and changed from a sum-of-products expression to a product-of-sums expression to determine if there is any reduction in the number of logic gates required. The Karnaugh map is shown in Figure 3.78. The product-of-sums expression is shown in Equation 3.28, which requires two 3-input OR gates and one 2-input NAND gate to assert z_1 at a low voltage level. Therefore, the sum-of-products expression requires the least amount of logic.

x_1x_2 \ x_3x_4	0 0	0 1	1 1	1 0
0 0	1 [0]	1 [1]	0 [3]	1 [2]
0 1	1 [4]	1 [5]	1 [7]	1 [6]
1 1	1 [12]	1 [13]	1 [15]	1 [14]
1 0	1 [8]	0 [9]	0 [11]	1 [10]

z_1

Figure 3.78   Karnaugh map for Example 3.23.

$$
z_1 = (x_1' + x_2 + x_4')(x_2 + x_3' + x_4')
\tag{3.28}
$$

Example 3.24 A majority circuit is a logic circuit whose output is a logic 1 (high) if the majority of the inputs are a logic 1 (high); otherwise, the output is a logic 0 (low). This example will analyze the logic diagram of Figure 3.79 consisting of two majority circuits in a cascade configuration. The equation for Maj_1 will first be obtained, as shown in Equation 3.29. Then the output of Maj_2 will be determined, using the equation of Maj_1 as an input. The Boolean equation for output z_1 (Maj_2) will be obtained in a sum-of-products form, and is shown in Equation 3.30.

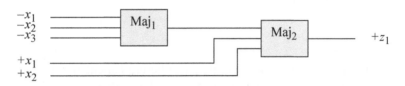

Figure 3.79   Logic diagram for Example 3.24.

$$Maj_1 = x_1'x_2' + x_1'x_3' + x_2'x_3' + x_1'x_2'x_3'$$
$$= x_1'x_2'(1 + x_3') + x_1'x_3' + x_2'x_3'$$
$$= x_1'x_2' + x_1'x_3' + x_2'x_3' \tag{3.29}$$

$$z_1\,(Maj_2) = Maj_1\,x_1 + Maj_1 x_2 + x_1x_2 + Maj_1\,x_1x_2$$
$$= Maj_1\,x_1(1 + x_1) + Maj_1 x_2 + x_1 x_2$$
$$= (x_1'x_2' + x_1'x_3' + x_2'x_3')x_1 +$$
$$\quad (x_1'x_2' + x_1'x_3' + x_2'x_3')x_2 + x_1x_2$$
$$= x_1x_2'x_3' + x_1'x_2x_3' + x_1x_2 \tag{3.30}$$

Example 3.25 The logic diagram in Figure 3.80 will be analyzed by obtaining the equation for z_1 in a product-of-sums form. The logic consists of a mixture of AND, NAND, OR, and NOR gates. Note that gate 2 is an OR gate drawn as an AND gate with active-low inputs and an active-low output. The equation for output z_1 will be obtained directly from the logic diagram, not from the expressions formed by each gate and is shown in Equation 3.31.

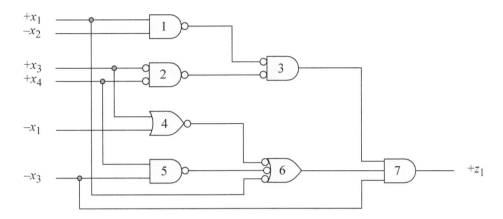

Figure 3.80 Logic diagram for Example 3.25.

$$z_1 = (x_1 x_2' x_3' x_4')(x_1' + x_3 + x_3' x_4 + x_1')(x_3') \qquad (3.31)$$

Example 3.26 The logic diagram shown in Figure 3.81 will be analyzed by means of a truth table from which the equation for output z_1 will be obtained. The truth table is shown in Table 3.21 and the equation for z_1 is shown in Equation 3.32.

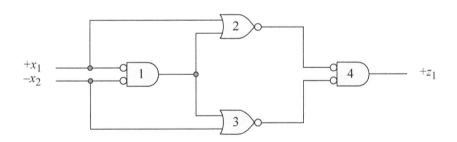

Figure 3.81 Logic diagram for Example 3.26.

Output z_1 will be at a high level when input x_1 is deasserted and x_2 is asserted, providing a high output for gate 1. This will provide a low level at both inputs to gate 4. Output z_1 will also be asserted when x_1 is asserted and x_2 is deasserted, providing high inputs to gates 2 and 3, which in turn will provide low inputs to gate 4.

**Table 3.21 Truth Table
for the Logic Diagram of
Figure 3.81**

x_1	x_2	z_1
0	0	0
0	1	1
1	0	1
1	1	0

$$z_1 = x_1 \oplus x_2 \tag{3.32}$$

Example 3.27 As a final example in the analysis of combinational logic, consider the logic diagram of Figure 3.82. This circuit consists of an OR gate, NAND gates, a NOR gate, an exclusive-OR circuit, and an exclusive-NOR circuit. The equation for output z_1 will be obtained from the logic diagram and then reduced to a minimum sum-of-products form by Boolean algebra, as shown in Equation 3.33.

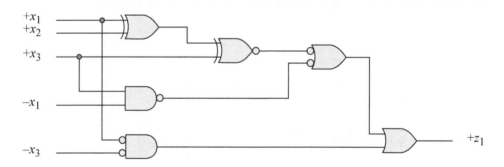

Figure 3.82   Logic diagram for Example 3.27.

$$
\begin{aligned}
z_1 &= \{[(x_1 \oplus x_2) \oplus x_3] + x_1'x_3\} + x_1'x_3 \\
&= [(x_1x_2' + x_1'x_2) \oplus x_3] + x_1'x_3 \\
&= (x_1x_2' + x_1'x_2)x_3' + (x_1x_2' + x_1'x_2)'x_3 + x_1'x_3 \\
&= x_1x_2'x_3' + x_1'x_2x_3' + [(x_1' + x_2)(x_1 + x_2')]x_3 + x_1'x_3
\end{aligned}
$$

$$= x_1 x_2' x_3' + x_1' x_2 x_3' + [(x_1' + x_2)x_1 + (x_1' + x_2)x_2']x_3 + x_1' x_3$$

$$= x_1 x_2' x_3' + x_1' x_2 x_3' + (x_1' + x_2)x_1 x_3 + (x_1' + x_2)x_2' x_3 + x_1' x_3$$

$$= x_1 x_2' x_3' + x_1' x_2 x_3' + x_1 x_2 x_3 + x_1' x_2' x_3 + x_1' x_3$$

$$= x_1 x_2' x_3' + x_1' x_2 x_3' + x_3(x_1 x_2 + x_1' x_2' + x_1')$$

$$= x_1 x_2' x_3' + x_1' x_2 x_3' + x_3(x_1' + x_2 + x_1' x_2')$$

$$= x_1 x_2' x_3' + x_1' x_2 x_3' + x_3(x_1' + x_2 + x_1')$$

$$= x_1 x_2' x_3' + x_1' x_2 x_3' + x_1' x_3 + x_2 x_3 \tag{3.33}$$

3.4 Synthesis of Combinational Logic

Synthesis of combinational logic consists of translating a set of network specifications into minimized Boolean equations and then generating a logic diagram from the equations using the logic primitives. The equations are independent of any logic family and portray the functional operation of the network. The logic primitives can be realized by either AND gates, OR gates, and inverters, or by *functionally complete gates*, such as NAND or NOR gates.

Truth tables are also important techniques that are used to generate equations. They tabulate the functional operation of the logic by utilizing all combinations of the input variables and define the output state for each combination. The equations can be minimized by Boolean algebra, the Quine-McCluskey algorithm, or by Karnaugh maps. Karnaugh maps are the most convenient and easy-to-use method and are key techniques to derive minimized equations in either a sum-of-products or a product-of-sums form.

The synthesis procedure is relatively straightforward. First, the equation is implemented with AND/OR/INVERT symbols without regard for logical complementation. Then, the negative assertion levels are assigned to the gates, where applicable, using the polarity symbol " ○ ." This step establishes the logic family that is used for the implementation. The inputs are usually available in both high and low assertions. Several examples will now be presented to delineate the synthesis procedure.

Example 3.28 Equation 3.34 will be synthesized using NAND gates only. The first term $x_1 x_2' x_3' x_4$ is represented by gate 1 in Figure 3.83. The inputs are shown at their active voltage levels such that all gate inputs are high when the term $x_1 x_2' x_3' x_4$ is true. That is, gate 1 will generate a low output when x_1 and x_4 are asserted and x_2 and x_3 are deasserted. The second expression $x_1'(x_2 x_3 x_4' + x_2' x_3')$ is implemented by gates 2, 3, 4, and 5.

Since only NAND gates are used in the implementation, the exclusive-NOR function is decomposed into a sum-of-products form $x_5 x_6 + x_5' x_6'$ and implemented by gates 6, 7, and 8. If only NOR gates were used in the implementation of Example 3.28, then the active level of all inputs would be complemented and output z_1 would be active at a low voltage level.

$$z_1 = x_1 x_2' x_3' x_4 + x_1'(x_2 x_3 x_4' + x_2' x_3') + x_1 x_3 (x_5 \oplus x_6)' \qquad (3.34)$$

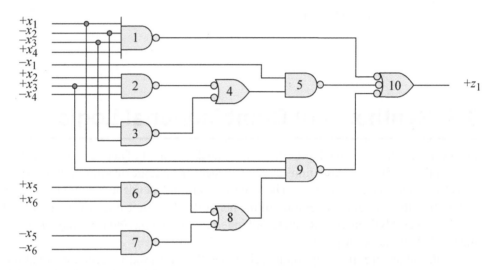

Figure 3.83   NAND gate implementation for Equation 3.34 of Example 3.28.

Example 3.29 Equation 3.35 will be synthesized using NOR gates only. The equation will first be represented as a sum of products, then converted to a minimal product-of-sums form for implementation.

$$z_1 = x_1 x_3' x_4 + [(x_1 + x_2)' + x_3]' + (x_2 \oplus x_4')' \qquad (3.35)$$

The equation can be represented as a sum of products as shown in Equation 3.36.

$$\begin{aligned}
z_1 &= x_1 x_3' x_4 + [(x_1 + x_2)' + x_3]' + (x_2 \oplus x_4')' \\
&= x_1 x_3' x_4 + (x_1' x_2' + x_3)' + x_2 x_4' + x_2' x_4 \\
&= x_1 x_3' x_4 + (x_1 + x_2) x_3' + x_2 x_4' + x_2' x_4 \\
&= x_1 x_3' x_4 + x_1 x_3' + x_2 x_3' + x_2 x_4' + x_2' x_4 \qquad (3.36)
\end{aligned}$$

The equation can now be plotted on a Karnaugh map, as shown in Figure 3.84, then converted to a product-of-sums form, as shown in Equation 3.37. The logic diagram is shown in Figure 3.85 using only NOR gates. The inputs are available in both high and low assertion.

$x_3 x_4$

$x_1 x_2$	0 0	0 1	1 1	1 0
0 0	0 [0]	1 [1]	1 [3]	0 [2]
0 1	1 [4]	1 [5]	0 [7]	1 [6]
1 1	1 [12]	1 [13]	0 [15]	1 [14]
1 0	1 [8]	1 [9]	1 [11]	0 [10]

z_1

Figure 3.84   Karnaugh map for Equation 3.35 of Example 3.29.

$$z_1 = (x_1 + x_2 + x_4)(x_2 + x_3' + x_4)(x_2' + x_3' + x_4') \qquad (3.37)$$

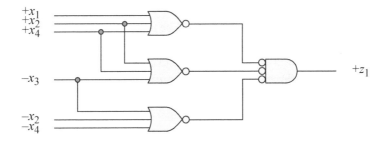

Figure 3.85   NOR gate implementation for Equation 3.35 of Example 3.29.

Example 3.30 A logic circuit will be designed that generates an active-high output whenever a 4-bit binary word contains exactly two 1s. The minterms that contain exactly two 1s are as follows:

$$\Sigma_m(3, 5, 6, 9, 10, 12)$$

The circuit can be implemented using discrete logic gates; however, a simpler solution is to use a 4:16 decoder and OR together with the appropriate minterms, as shown in Figure 3.86.

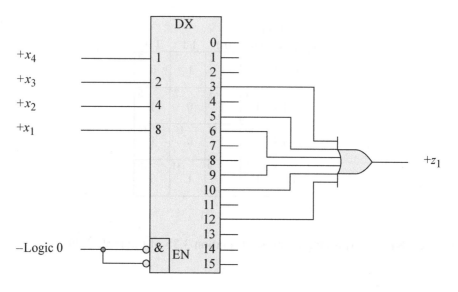

Figure 3.86   Logic diagram for Example 3.30.

Example 3.31   A 5-input majority circuit will be designed that produces a high output on z_1 whenever the majority of inputs x_1, x_2, x_3, x_4, x_5 are at a logic 1, where x_5 is the low-order bit; otherwise, output z_1 will be at a logic 0. In order for there to be a majority, there must be an odd number of inputs. The circuit can be designed by plotting the five variables on a Karnaugh map and inserting 1s in minterm locations in which there are at least three 1s. Then the groups of 1s are combined to form a minimized sum-of-products expression. This implementation would require several logic gates.

An alternative approach is to use a 5:32 decoder and OR the appropriate minterm outputs of the decoder to generate the majority function. This results in a very large decoder and may not be applicable if using standard integrated circuit technology. A third approach — which will be used in this example — is to use two 4:16 decoders with enabling logic. Minterms that contain at least three 1s can be easily portrayed by tabulating all $2^5 = 32$ combinations in a truth table, then selecting the appropriate minterms to be ORed together. Minterms that contain at least three 1s are shown below in a sum-of-products decimal notation.

$$\Sigma_m(7, 11, 13, 14, 15, 19, 21, 22, 23, 25, 26, 27, 28, 29, 30, 31)$$

If a 16-input OR gate is not feasible, then the outputs of two 8-input OR gates can be ORed together to produce the required majority function. The logic diagram is shown in Figure 3.87.

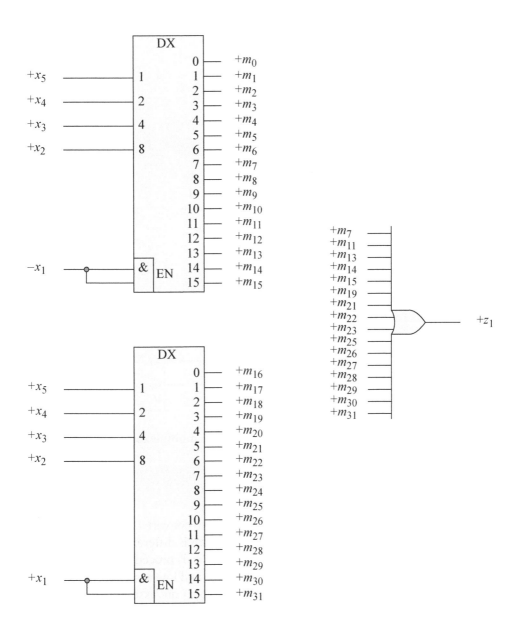

Figure 3.87   Five-input majority circuit for Example 3.31.

Example 3.32   A 4:1 multiplexer can be utilized as a 3:1 multiplexer by using only three of the four combinations of the select inputs; that is, $s_1 s_0 = 00, 01, 10$ and not utilizing select inputs $s_1 s_0 = 11$. As an exercise, this example will use a different approach to designing a 3:1 multiplexer by using two 2:1 multiplexers. Table 3.22 lists the data inputs that are selected by all combinations of the select inputs. The logic diagram is shown in Figure 3.88.

Table 3.22   Truth Table for Example 3.32

x_1	x_2	Input Selected
0	0	In_0
0	1	In_1
1	0	In_2
1	1	In_1

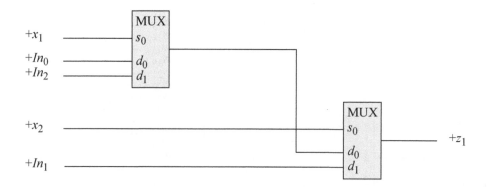

Figure 3.88   Logic diagram to implement a 3:1 multiplexer using two 2:1 multiplexers.

Example 3.33   A code converter will be designed to convert a 4-bit binary number to the corresponding Gray code number. This approach is different than the one used in Example 3.14 and uses Karnaugh maps in the synthesis procedure. The inputs of the binary number $x_1 x_2 x_3 x_4$ are available in both high and low assertion, where x_4 is the low-order bit. The outputs for the Gray code $z_1 z_2 z_3 z_4$ are asserted high, where z_4 is the low-order bit. The binary-to-Gray code conversion table is shown in Table 3.23. There are four Karnaugh maps shown in Figure 3.89, one map for each of the Gray code outputs. The equations obtained from the Karnaugh maps are shown in Equation 3.38. The logic diagram is shown in Figure 3.90.

Table 3.23 Binary-To-Gray Code Conversion

Binary				Gray			
x_1	x_2	x_3	x_4	z_1	z_2	z_3	z_4
0	0	0	0	0	0	0	0
0	0	0	1	0	0	0	1
0	0	1	0	0	0	1	1
0	0	1	1	0	0	1	0
0	1	0	0	0	1	1	0
0	1	0	1	0	1	1	1
0	1	1	0	0	1	0	1
0	1	1	1	0	1	0	0
1	0	0	0	1	1	0	0
1	0	0	1	1	1	0	1
1	0	1	0	1	1	1	1
1	0	1	1	1	1	1	0
1	1	0	0	1	0	1	0
1	1	0	1	1	0	1	1
1	1	1	0	1	0	0	1
1	1	1	1	1	0	0	0

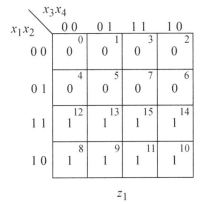 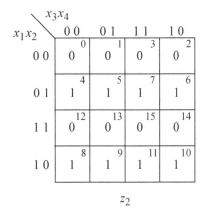

Figure 3.89 Karnaugh maps for the binary-to-Gray code conversion for Example 3.33.

x_3x_4

x_1x_2	0 0	0 1	1 1	1 0
0 0	0	0	1	1
0 1	1	1	0	0
1 1	1	1	0	0
1 0	0	0	1	1

z_3

x_3x_4

x_1x_2	0 0	0 1	1 1	1 0
0 0	0	1	0	1
0 1	0	1	0	1
1 1	0	1	0	1
1 0	0	1	0	1

z_4

Figure 3.89   (Continued)

$$z_1 = x_1$$
$$z_2 = x_1'x_2 + x_1x_2' = x_1 \oplus x_2$$
$$z_3 = x_2x_3' + x_2'x_3 = x_2 \oplus x_3$$
$$z_4 = x_3'x_4 + x_3x_4' = x_3 \oplus x_4 \qquad (3.38)$$

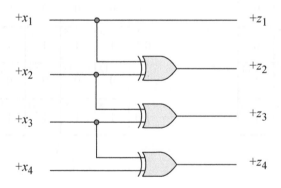

Figure 3.90   Logic diagram for the binary-to-Gray code conversion.

Example 3.34   A binary-coded decimal (BCD)-to-excess-3 code converter will be designed using only NAND gates. The excess-3 code is obtained from the binary 8421 BCD code by adding 3 (0011) to each binary code word. The truth table for converting from BCD to excess-3 is shown in Table 3.24. The BCD code word is $x_1 x_2 x_3 x_4$, where x_4 is the low-order bit; the excess-3 code word is $z_1 z_2 z_3 z_4$, where z_4 is the low-order bit. There are four Karnaugh maps, one for each excess-3 digit, as shown in Figure 3.91. The equations for the excess-3 bits are shown in Equation 3.39. The logic diagram is shown in Figure 3.92.

Table 3.24 BCD-To-Excess-3 Code Conversion

BCD Code				Excess-3 Code			
x_1	x_2	x_3	x_4	z_1	z_2	z_3	z_4
0	0	0	0	0	0	1	1
0	0	0	1	0	1	0	0
0	0	1	0	0	1	0	1
0	0	1	1	0	1	1	0
0	1	0	0	0	1	1	1
0	1	0	1	1	0	0	0
0	1	1	0	1	0	0	1
0	1	1	1	1	0	1	0
1	0	0	0	1	0	1	1
1	0	0	1	1	1	0	0

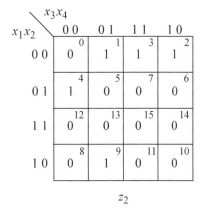

Figure 3.91   Karnaugh maps for the BCD-to-excess-3 code conversion.

x_3x_4 / x_1x_2 map (z_3):

x_1x_2 \ x_3x_4	0 0	0 1	1 1	1 0
0 0	1 (0)	0 (1)	1 (3)	0 (2)
0 1	1 (4)	0 (5)	1 (7)	0 (6)
1 1	0 (12)	0 (13)	0 (15)	0 (14)
1 0	1 (8)	0 (9)	0 (11)	0 (10)

z_3

x_3x_4 / x_1x_2 map (z_4):

x_1x_2 \ x_3x_4	0 0	0 1	1 1	1 0
0 0	1 (0)	0 (1)	0 (3)	1 (2)
0 1	1 (4)	0 (5)	0 (7)	1 (6)
1 1	0 (12)	0 (13)	0 (15)	0 (14)
1 0	1 (8)	0 (9)	0 (11)	0 (10)

z_4

Figure 3.91   (Continued)

$$z_1 = x_1'x_2x_4 + x_1'x_2x_3 + x_1x_2'x_3'$$

$$z_2 = x_1'x_2'x_3 + x_2'x_3'x_4 + x_1'x_2x_3'x_4'$$

$$z_3 = x_1'x_3'x_4' + x_1'x_3x_4 + x_2'x_3'x_4'$$

$$z_4 = x_1'x_4' + x_2'x_3'x_4' \qquad (3.39)$$

Example 3.35 A *half adder* is a combinational circuit that adds two operand bits and produces two outputs: sum and carry-out. A *full adder* is a combinational circuit that adds two operand bits plus a carry-in bit. The carry-in bit represents the carry-out of the previous lower-order stage. A full adder produces two outputs: sum and carry-out. This example will design a full adder consisting of two half adders plus additional logic.

The truth tables for a half adder and full adder are shown in Table 3.25 and Table 3.26, respectively. The corresponding equations for the sum and carry-out are listed in Equation 3.40 and Equation 3.41 and are obtained directly from the truth tables. The synthesis of the full adder can be implemented from Equation 3.41. The logic diagram for the half adder is shown in Figure 3.93 and for the full adder in Figure 3.94. A higher speed full adder can be realized if the sum-of-products expression of Equation 3.41 is utilized, providing only two gate delays.

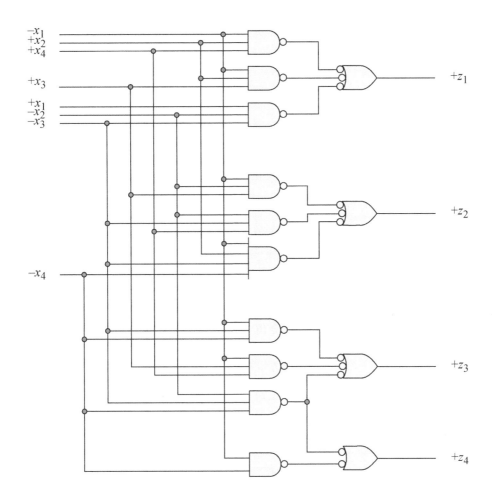

Figure 3.92   Logic diagram for the BCD-to-excess-3 code conversion.

Table 3.25 Truth Table for a Half Adder

a_i	b_i	$cout_i$	sum_i
0	0	0	0
0	1	0	1
1	0	0	1
1	1	1	0

Table 3.26 Truth Table for a Full Adder

a_i	b_i	cin_{i-1}	$cout_i$	sum_i
0	0	0	0	0
0	0	1	0	1
0	1	0	0	1
0	1	1	1	0
1	0	0	0	1
1	0	1	1	0
1	1	0	1	0
1	1	1	1	1

$$sum_i = a_i'b_i + a_ib_i'$$
$$= a_i \oplus b_i$$
$$cout_i = a_ib_i \tag{3.40}$$

$$sum_i = a_i'b_i'cin_{i-1} + a_i'b_icin_{i-1}' + a_ib_i'cin_{i-1}' + a_ib_icin_{i-1}$$
$$= a_i \oplus b_i \oplus cin_{i-1}$$
$$cout_i = a_i'b_icin_{i-1} + a_ib_i'cin_{i-1} + a_ib_icin_{i-1}' + a_ib_icin_{i-1}$$
$$= a_ib_i + a_icin_{i-1} + b_icin_{i-1} \tag{3.41}$$

From Table 3.26, the carry-out for the full adder, $cout_i$, can also be written as

$$cout_i = (a_i \oplus b_i)cin_{i-1} + a_ib_i$$

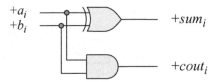

Figure 3.93   Logic diagram for a half adder.

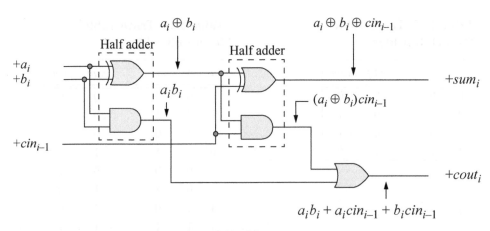

Figure 3.94   Logic diagram for a full adder.

Example 3.36 In Example 3.34, the BCD code was converted to the excess-3 code using NAND gates. In this example, a different approach will be used to convert from the binary code to the corresponding excess-3 code. The full adder of Example 3.35 will be used in which the binary number will be entered into the adder as a 4-bit number representing operand A; the B operand will contain the binary number 3 (0011) which will be added to operand A to produce the excess-3 code as the sum.

The logic macro for the full adder is shown in Figure 3.95. The logic diagram for this method is shown in Figure 3.96, where operand A consists of four bits; $a_3a_2a_1a_0$, where a_0 is the low-order bit. There are four outputs that represent the excess-3 code: sum_3 sum_2 sum_1 sum_0, where sum_0 is the low-order bit.

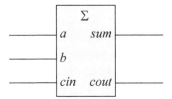

Figure 3.95 Logic macro for a full adder.

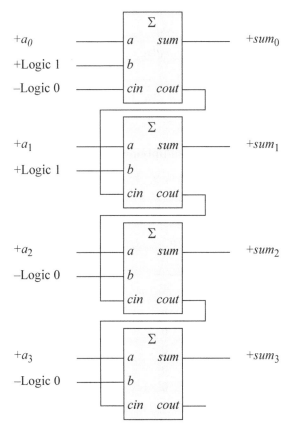

Figure 3.96 Logic for a binary-to-excess-3 code converter using an adder.

Example 3.37   A multiply operation involves two operands. One operand is an n-bit multiplicand; the other operand is an n-bit multiplier. The result is a $2n$-bit product. A multiply circuit will be designed to multiply two 2-bit operands and produce a 4-bit result. The multiplicand is $A = a_1 a_0$, where a_0 is the low-order bit; the multiplier is $B = b_1 b_0$, where b_0 is the low-order bit. The product is $P = p_3 p_2 p_1 p_0$, where p_0 is the low-order bit.

The general procedure for multiplying two 2-bit operands is shown in Figure 3.97. The multiplicand is multiplied by the low-order multiplier bit b_0 to form partial product 0. The multiplicand is then multiplied by the next higher-order multiplier bit b_1 to form partial product 2, shifted left one bit position. The product is obtained by adding all of the partial products.

The technique used in this example, however, uses a truth table to tabulate the multiplicand, multiplier, and product for all combinations of the multiplicand and multiplier, as shown in Table 3.27. The table entries for the product are plotted on four Karnaugh maps, one for each product bit, as shown in Figure 3.98. The equations for the product bits are shown in Equation 3.42 as derived from the Karnaugh maps. The multiplier logic is designed from Equation 3.42 and is shown in Figure 3.99. The multiplicand and multiplier bits are available in both high and low assertion.

Multiplicand			a_1	a_0
Multiplier		$\times$)	b_1	b_0
Partial product 0			$a_1 b_0$	$a_0 b_0$
Partial product 1		$a_1 b_1$	$a_0 b_1$	
Product	p_3	p_2	p_1	p_0
	2^3	2^2	2^1	2^0

Figure 3.97   General procedure for multiplying two 2-bit operands.

Table 3.27 Truth Table to Multiply Two 2-bit Operands

a_1	a_0	b_1	b_0	p_3	p_2	p_1	p_0
0	0	0	0	0	0	0	0
0	0	0	1	0	0	0	0
0	0	1	0	0	0	0	0
0	0	1	1	0	0	0	0
0	1	0	0	0	0	0	0
0	1	0	1	0	0	0	1
0	1	1	0	0	0	1	0

(Continued on next page)

Table 3.27 Truth Table to Multiply Two 2-bit Operands

a_1	a_0	b_1	b_0	p_3	p_2	p_1	p_0
0	1	1	1	0	0	1	1
1	0	0	0	0	0	0	0
1	0	0	1	0	0	1	0
1	0	1	0	0	1	0	0
1	0	1	1	0	1	1	0
1	1	0	0	0	0	0	0
1	1	0	1	0	0	1	1
1	1	1	0	0	1	1	0
1	1	1	1	1	0	0	1

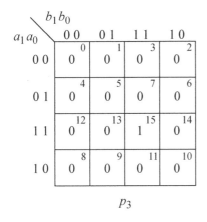

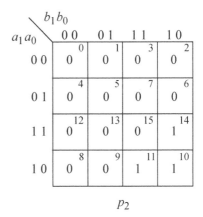

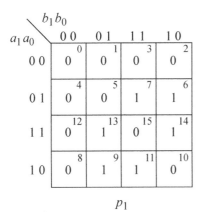

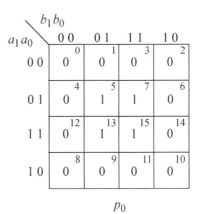

Figure 3.98 Karnaugh maps for the multiplier of Example 3.37.

$$p_3 = a_1 a_0 b_1 b_0$$

$$p_2 = a_1 a_0' b_1 + a_1 b_1 b_0'$$

$$p_1 = a_1' a_0 b_1 + a_0 b_1 b_0' + a_1 b_1' b_0 + a_1 a_0' b_0$$

$$p_0 = a_0 b_0 \tag{3.42}$$

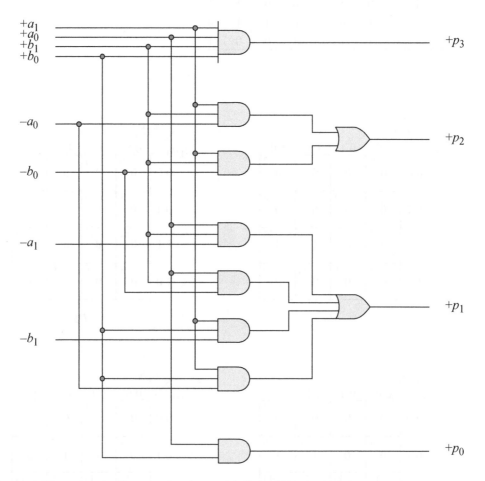

Figure 3.99   Logic diagram for the multiplier of Example 3.37.

Example 3.38   As a final example in the synthesis of combinational logic, a high-speed shifter will be designed to shift an unsigned operand right or left a specified number of bits. Since the operand is unsigned, this is classified as a *logical shifter*. The shifter is implemented with eight 4:1 multiplexers and four OR gates.   Four multiplexers are used for shifting right, and four are used for shifting left. The input data to be shifted consists of a 4-bit vector $b_3b_2b_1b_0$ — shown in Figure 3.100 — that will be shifted right or left under control of a *shft_rt* input or a *shft_lf* input, respectively. The *shft_rt* and *shft_lf* inputs connect to the *enable* input of the appropriate multiplexers.

b_3	b_2	b_1	b_0

Figure 3.100   Data to be shifted by the high-speed shifter.

Since shifting is a logical operation, during a shift-right operation zeroes are shifted into the vacated high-order bit position; during a shift-left operation, zeroes are shifted into the vacated low-order bit positions. The shift amount is determined by the multiplexer select inputs that connect to two shift control inputs $shft\_amt_1$ and $shft\_amt_0$, as shown in Table 3.28.

Table 3.28 Shift Direction and Shift Amounts for the High-Speed Shifter

		$shft\_amt_1$	$shft\_amt_0$	
shft_rt	*shft_lf*	Select 1	Select 0	Shift Amount
1	0	0	0	Shift right 0
1	0	0	1	Shift right 1
1	0	1	0	Shift right 2
1	0	1	1	Shift right 3
0	1	0	0	Shift left 0
0	1	0	1	Shift left 1
0	1	1	0	Shift left 2
0	1	1	1	Shift left 3

The high speed is achieved by pre-wiring the data to the appropriate shift inputs of the multiplexers, then selecting the shift amount by means of the select inputs. The only delay is through the multiplexers and the OR gates. There is really no synthesis procedure for this design — it is simply a matter of understanding how a multiplexer operates and connecting the input data to the desired multiplexer data inputs. The shift amount is then determined by the select inputs.

The logic diagram for the high-speed shifter is shown in Figure 3.101. The multiplexer macros are drawn differently than in previous examples due to space limitations and are positioned horizontally for ease of visualizing the shifting operation. Although only a 4-bit operand is utilized in this design, the implementation can be easily expanded to any size operand. While not explicitly drawn, the select inputs connect to all multiplexers.

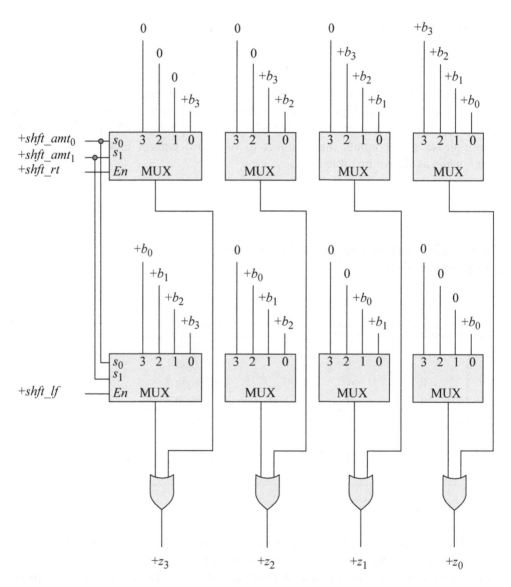

Figure 3.101   High-speed shifter using multiplexers.

3.5 Problems

3.1 Given the logic diagram shown below using NAND gates, obtain the Karnaugh map, the equation in a sum-of-products form, and the equation in a sum-of-products decimal notation.

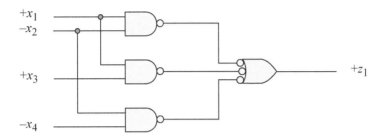

3.2 Analyze the logic diagram shown below by obtaining the equation for output z_1. Then verify the answer by means of a truth table.

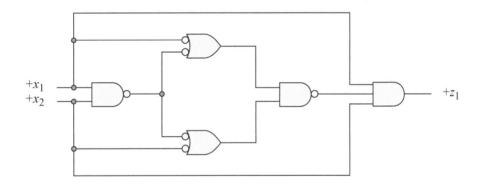

3.3 Analyze the logic diagram shown below by obtaining the equation for output z_1. Then verify the answer by means of a truth table.

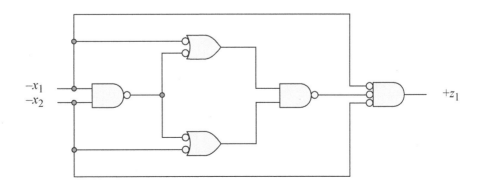

3.4   A majority circuit is a logic circuit whose output is a logic 1 (high) if the majority of the inputs are a logic 1 (high); otherwise, the output is a logic 0 (low). Obtain the Boolean equation for $z_1(x_1, x_2, x_3)$ that is implemented by the majority circuits shown below. The equation for z_1 is to be in a minimum sum-of-products form.

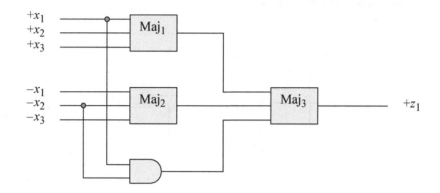

3.5   Obtain the equation for output z_1 in a sum-of-products notation for the logic diagram shown below.

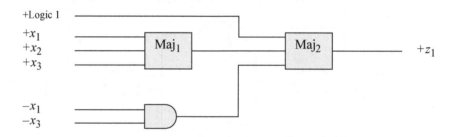

3.6   Given the logic diagram shown below, obtain the equation for output z_1 in a product-of-sums form.

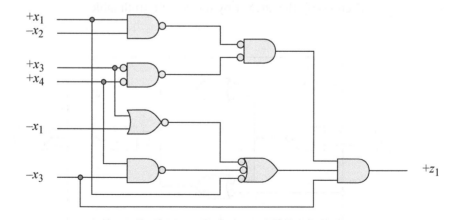

3.7   Obtain the minimum Boolean expression for the following logic circuit in a sum-of-products form.

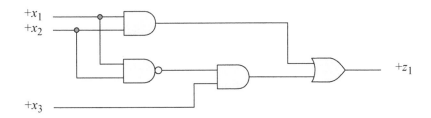

3.8   Given the two multiplexers and inputs as shown below, complete the following truth table for output z_1 as a function of the inputs In_0, In_1, In_2, and In_3.

x_1	x_2	z_1
0	0	
0	1	
1	0	
1	1	

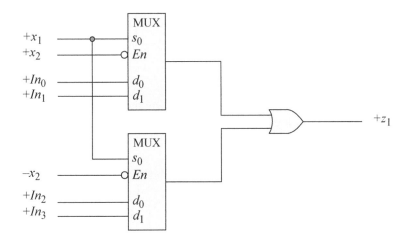

3.9   Given the following equation for z_1, design the logic circuit using AND gates and OR gates.

$$z_1 = [x_1 x_2' + (x_1' + x_2 + x_3)] [(x_1' + x_2 + x_3) + x_4 x_5']$$

3.10   Minimize the following equation using Boolean algebra to obtain a sum-of-products expression, then implement the expression using NAND gates. Assume that only active-high inputs are available. Output z_1 is asserted high.

$$z_1 = x_1'x_2 + x_3x_4 + (x_1 + x_2)' \, [x_1x_3x_4 + (x_2x_5)']$$

3.11   Implement the logic functions shown below using any type of logic gates. The inputs are available in both high and low assertions; the outputs are asserted low. Minimize the equations, if possible, to use the fewest number of gates.

$$z_1 = x_2'x_4'$$
$$z_2 = x_1'x_2 + x_1'x_2'x_4'$$
$$z_3 = x_1x_2'x_3'x_4' + x_1x_2x_3'x_4 + x_1'x_2x_3x_4 + x_1'x_2'x_3x_4'$$

3.12   Design a logic circuit that is represented by the Karnaugh map shown below. Use only NOR logic. The inputs are available in both high and low assertion.

x_1 \ x_2x_3	0 0	0 1	1 1	1 0
0	0 [0]	0 [1]	0 [3]	ab [2]
1	bc [4]	bc [5]	– [7]	1 [6]

z_1

3.13   Design a logic circuit that is represented by the Karnaugh map shown below. Use only NAND logic. The inputs are available in both high and low assertion.

x_1x_2 \ x_3x_4	0 0	0 1	1 1	1 0
0 0	a [0]	1 [1]	0 [3]	0 [2]
0 1	– [4]	1 [5]	0 [7]	0 [6]
1 1	1 [12]	1 [13]	1 [15]	1 [14]
1 0	0 [8]	b [9]	c [11]	– [10]

z_1

3.14   Synthesize a logic circuit that will control the interior lighting of a building. The building contains four rooms separated by removable partitions. There is one switch in each room which, in conjunction with the other switches, provides the following methods of control:

(a)   All partitions are closed forming four separate rooms. Each switch controls only the lights in its respective room.

(b)   All partitions are open forming one large room. Each switch controls all the lights in the building; that is, when the lights are on, they can be turned off by any switch. Conversely, when the lights are off, they can be turned on by any switch.

(c)   Two of the partitions are closed forming three rooms. The middle partition is open, so that the middle room is larger than the other two rooms. Each switch controls the lights in its room only.

(d)   The middle partition is closed forming two rooms. Each switch controls the lights in its room only.

The switch method outlined above is referred to as *four-way switching*. This technique provides control of one set of lights by one or more switches. Implement the circuit using 4:1 multiplexers and additional logic functions. The output of the multiplexers connect to the lights. The partitions are labeled P1, P2, and P3; the lights are labeled L1, L2, L3, and L4; the switches are labeled S1, S2, S3, and S4.

3.15   Implement the Karnaugh map shown below using a linear-select multiplexer and additional logic gates. Use the least amount of logic.

y_1 \ $y_2 y_3$	0 0	0 1	1 1	1 0
0	0	0	1	1
1	$x_1 x_2$	x_3	$x_1 + x_2$	x_1

z_1

3.16   Given the 5-variable Karnaugh map shown on the next page, obtain the equation for z_1. Then plot the equation on a three-variable map using x_4 and x_5 as map-entered variables. Then implement the Karnaugh map using a linear-select multiplexer and additional logic gates.

$x_5 = 0$

x_1x_2 \ x_3x_4	0 0	0 1	1 1	1 0
0 0	0 [0]	0 [2]	0 [6]	1 [4]
0 1	0 [8]	0 [10]	1 [14]	0 [12]
1 1	1 [24]	0 [26]	1 [30]	0 [28]
1 0	1 [16]	1 [18]	1 [22]	1 [20]

$x_5 = 1$

x_1x_2 \ x_3x_4	0 0	0 1	1 1	1 0
0 0	0 [1]	0 [3]	0 [7]	0 [5]
0 1	0 [9]	1 [11]	1 [15]	0 [13]
1 1	1 [25]	0 [27]	1 [31]	0 [29]
1 0	1 [17]	1 [19]	1 [23]	1 [21]

$$z_1$$

3.17 Given the Karnaugh map shown below, obtain the minimized data input equations for a nonlinear-select multiplexer. Show the data input equations for each permutation of the Karnaugh map, then implement the design using the equations with the least amount of logic.

y_1 \ y_2y_3	0 0	0 1	1 1	1 0
0	$(x_1x_2)'$ [0]	$x_1'+x_2'$ [1]	x_1x_2 [3]	– [2]
1	– [4]	1 [5]	1 [7]	1 [6]

$$z_1$$

3.18 The truth table shown below represents the logic for output z_1 of a combinational logic circuit. Implement the logic using a nonlinear-select multiplexer and additional logic gates, if necessary. Use the least amount of logic.

y_1	y_2	y_3	y_4	z_1
0	0	0	0	x_1'
0	0	0	1	1
0	0	1	0	x_1'
0	0	1	1	1
0	1	0	0	–
0	1	0	1	–
0	1	1	0	x_1
0	1	1	1	0

(Continued on next page)

y_1	y_2	y_3	y_4	z_1
1	0	0	0	0
1	0	0	1	0
1	0	1	0	–
1	0	1	1	0
1	1	0	0	0
1	1	0	1	0
1	1	1	0	–
1	1	1	1	–

3.19 Given the following Karnaugh map, implement the function for z_1 using a 4:1 multiplexer and additional logic gates, if necessary.

x_1x_2 \ x_3x_4	0 0	0 1	1 1	1 0
0 0	[0] 1	[1] 1	[3] 0	[2] 1
0 1	[4] 1	[5] 1	[7] 1	[6] 0
1 1	[12] 0	[13] 0	[15] 1	[14] 0
1 0	[8] 0	[9] 0	[11] 1	[10] 0

z_1

3.20 Design an 8-bit odd parity generator. The parity bit is appended to the 8-bit byte of data such that the number of 1s in the nine bits (eight data bits plus the parity bit) is odd. A parity generator can be implemented easily by modulo-2 addition. The truth table for modulo-2 addition (exclusive-OR) is shown below. The output z_1 is a logic 1 if there is an odd number of 1s on the inputs. Therefore, the output can be inverted to obtain odd parity for the byte of data.

x_1	x_2	z_1
0	0	0
0	1	1
1	0	1
1	1	0

3.21   Design an 8-bit odd parity checker.

3.22   A logic circuit has two control inputs c_1 and c_0, two data inputs x_1 and x_2, and one output z_1. The circuit operates as follows:

> If $c_1 c_0 = 00$, then $z_1 = 0$
> If $c_1 c_0 = 01$, then $z_1 = x_2$
> If $c_1 c_0 = 10$, then $z_1 = x_1$
> If $c_1 c_0 = 11$, then $z_1 = 1$

Derive a truth table for output z_1. Then use a Karnaugh map to obtain the Boolean expression for z_1 in a sum-of-products form and implement the logic using NAND gates.

3.23   Synthesize a 4-bit comparator using any logic primitives, including exclusive-NOR functions. There are two 4-bit unsigned binary operands, $A = a_3 a_2 a_1 a_0$ and $B = b_3 b_2 b_1 b_0$, where a_0 and b_0 are the low-order bits of A and B, respectively. There are three outputs:

$$A < B, \ A = B, \ A > B$$

3.24   Design a high-speed shift unit that will shift an 8-bit operand right or left zero, one, two, or three bit positions. Use 4:1 multiplexers as the shifting elements and any additional logic functions. The operand is unsigned; therefore, this will be a logical shift. Zeroes are shifted into the vacated high-order bit positions during a right shift operation. Zeroes are shifted into the vacated low-order bit positions during a left shift operation. Assume that a logic 0 is equivalent to a *ground* potential.

7	6	5	4	3	2	1	0

3.25   Use two 4-bit comparators to determine if a modulo-16 number meets the following requirements: $4 > N > 11$. Use NAND logic only. If the number meets the requirements, then the output of the logic circuit is a high logic level.

4

Combinational Logic Design Using Verilog HDL

This chapter introduces the Verilog hardware description language (HDL) which will be used to design combinational logic. Verilog HDL is the state-of-the-art method for designing digital and computer systems and is able to describe both combinational and sequential logic, including level-sensitive and edge-triggered storage devices. Verilog provides a clear relationship between the language syntax and the physical hardware.

The Verilog simulator used in this book is easy to learn and use, yet powerful enough for any application. It is a logic simulator — called SILOS — developed by Silvaco International for use in the design and verification of digital systems. The SILOS simulation environment is a method to quickly prototype and debug any application-specific integrated circuit (ASIC), field-programmable gate array (FPGA), or complex programmable logic device (CPLD) design. It is an intuitive environment that displays every variable and port from a module to a logic gate. SILOS allows single stepping through the Verilog source code, as well as drag-and-drop ability from the source code to a data analyzer for waveform generation and analysis.

Verilog HDL was developed by Phillip Moorby in 1984 as a proprietary HDL for Gateway Design Automation. Gateway was later acquired by Cadence Design Systems, which placed the language in the public domain in 1990. The Open Verilog International was then formed to promote the Verilog HDL language. In 1995, Verilog was made an IEEE standard HDL (IEEE Standard 1364-1995) and is described in the Verilog Hardware Description Language Reference Manual.

Designs can be modeled in three different modeling constructs: (1) dataflow, (2) behavioral, and (3) structural. Module design can also be done in a mixed-design

231

style, which incorporates the above constructs as well as built-in and user-defined primitives. Structural modeling can be described for any number of module instantiations.

4.1 Built-In Primitives

Logic primitives such as **and**, **nand**, **or**, **nor**, and **not** gates, as well as **xor** (exclusive-OR), and **xnor** (exclusive_NOR) functions are part of the Verilog language and are classified as multiple-input gates. These are built-in primitives that can be instantiated into a module. *Instantiation* means to use one or more lower-level modules in the construction of a higher-level structural module. A module can be a logic gate, an adder, a multiplexer, a counter, or some other logical function.

A module consists of declarative text which specifies the function of the module using Verilog constructs; that is, a Verilog module is a software representation of the physical hardware structure and behavior. The declaration of a module is indicated by the keyword **module** and is always terminated by the keyword **endmodule**.

Verilog's profuse set of built-in primitive gates are used to model nets. The single output of each gate is declared as type **wire**. The inputs are declared as type **wire** or as type **reg** depending on whether they were generated by a structural or behavioral module. This section presents a design methodology that is characterized by a low level of abstraction, where the logic hardware is described in terms of gates. Designing logic at this level is similar to designing logic by drawing gate symbols — there is a close correlation between the logic gate symbols and the Verilog built-in primitive gates. Each predefined primitive is declared by a keyword such as **and**, **or**, **nand**, **nor**, etc.

The primitive gates are used to describe a net and have one or more scalar inputs, but only one scalar output. The output signal is listed first, followed by the inputs in any order. The outputs are declared as **wire**; the inputs can be declared as either **wire** or **reg**. The gates represent combinational logic functions and can be instantiated into a module, as follows, where the instance name is optional:

> **gate_type** inst1 (output, input_1, input_2, . . . , input_n);

Two or more instances of the same type of gate can be specified in the same construct, as shown below. Note that only the last instantiation has a semicolon terminating the line. All previous lines are terminated by a comma.

> **gate_type** inst1 (output_1, input_11, input_12, . . . , input_1n),
>   inst2 (output_2, input_21, input_22, . . . , input_2n),
>
>   .
>   .
>   .
>
>   instm (output_m, input_m1, input_m2, . . . , input_mn);

The best way to learn design methodologies using built-in primitives is by examples. Therefore, several examples will be presented ranging from very simple to moderately complex. When necessary, the theory for the examples will be presented prior to the Verilog design. All examples are carried through to completion at the gate level. Nothing is left unfinished or partially designed.

Example 4.1   The logic diagram of Figure 4.1 will be designed using built-in primitives for the logic gates which consist of NAND gates and one OR gate to generate the two outputs z_1 and z_2. The output of the gate labeled *inst2* (instantiation 2) will be at a high voltage level if either x_1, x_2, or x_3 is deasserted. Therefore, by DeMorgan's theorem, the output will be at a low voltage level if x_1, x_2, and x_3 are all asserted. Note that the gate labeled *inst3* is an OR gate that is drawn as an AND gate with active-low inputs and an active-low output. The output of each gate is assigned a net name, where a *net* is one or more interconnecting wires that connect the output of one logic element to the input of one or more logic elements. The remaining gates in Figure 4.1 are drawn in the standard manner as NAND gates.

The design module is shown in Figure 4.2. The first line is usually reserved for a comment (//) and specifies the function of the module. Comments can also be placed at the end of a line to indicate the function of that line of code. Line 2 is the beginning of the Verilog code and is indicated by the keyword **module** followed by the module name *log_eqn_sop15*. This is followed by the list of input and output ports placed within parentheses followed in turn by a semicolon.

Verilog must know which ports are used for input and which ports are used for output; therefore, lines 4 and 5 list the input and output ports indicated by the keywords **input** and **output**, respectively. Line 7 begins the instantiation of the built-in primitives. The instantiation names and net names in the module correlate directly to the corresponding names in the logic diagram. Thus, line 7 in the module, which is

nand   inst1   (net1, x1, x2, x3);

represents NAND gate *inst1* with inputs x_1, x_2, and x_3 and output *net1* in the logic diagram. Line 14 in the module corresponds to OR function of instantiation *inst8* of the logic diagram whose inputs are *net5*, *net6*, and *net7* and whose active-high output is z_2. The end of the module is indicated by the keyword **endmodule** as shown in line 16. In this simple example, Figure 4.2 correctly describes the hardware that is represented by the logic diagram of Figure 4.1.

However, in order to verify that the module operates correctly according to the functional specifications represented by the logic diagram — especially for more complex designs — the module should be tested. This is accomplished by means of a test bench. Test benches are used to apply input vectors to the module in order to test the functional operation of the module in a simulation environment.

The functionality of the module can be tested by applying stimulus to the inputs and checking the outputs. The test bench can display the inputs and outputs in the following radices: binary (b), octal (o), hexadecimal (h), or decimal (d). Waveforms can also be displayed. It is good practice to keep the design module and test bench module separate.

The test bench contains an instantiation of the unit under test and Verilog code to generate input stimulus and to monitor and display the response to the stimulus. Figure 4.3 shows a test bench to test the correctness of Figure 4.2. Line 1 is a comment indicating that the module is a test bench for the *log_eqn_sop15* module. Line 2 contains the keyword **module** followed by the module name, which includes *tb* indicating a test bench module. The name of the module and the name of the module under test are the same for ease of cross-referencing. The keyword **endmodule** terminates the module.

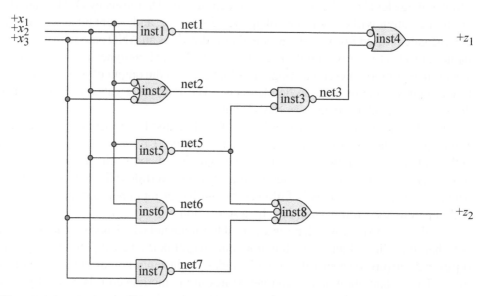

Figure 4.1   Logic diagram to be designed using built-in primitives.

```
 1    //logic diagram using built-in primitives
      module log_eqn_sop15 (x1, x2, x3, z1, z2);

      input x1, x2, x3;
 5    output z1, z2;

      nand    inst1 (net1, x1, x2, x3);
      nand    inst2 (net2, x1, x2, x3);
      or      inst3 (net3, net2, net5);
10    nand    inst4 (z1, net1, net3);
      nand    inst5 (net5, x1, x2);
      nand    inst6 (net6, x1, x3);
      nand    inst7 (net7, x2, x3);
      nand    inst8 (z2, net5, net6, net7);
15
      endmodule
```

Figure 4.2   Design module for Figure 4.1 using built-in primitives.

```
1    //test bench for log_eqn_sop_15
     module log_eqn_sop15_tb;

     reg x1, x2, x3;
5    wire z1, z2;

     //display variables
     initial
     $monitor ("x1=%b, x2=%b, x3=%b, z1=%b, z2=%b",
10             x1, x2, x3, z1, z2);

     //apply input vectors
     initial
     begin
15   #0   x1 = 1'b0;
          x2 = 1'b0;
          x3 = 1'b0;

     #10  x1 = 1'b0;
20        x2 = 1'b0;
          x3 = 1'b1;

     #10  x1 = 1'b0;
          x2 = 1'b1;
25        x3 = 1'b0;

     #10  x1 = 1'b0;
          x2 = 1'b1;
          x3 = 1'b1;
30
     #10  x1 = 1'b1;
          x2 = 1'b0;
          x3 = 1'b0;

35   #10  x1 = 1'b1;
          x2 = 1'b0;
          x3 = 1'b1;

     #10  x1 = 1'b1;
40        x2 = 1'b1;
          x3 = 1'b0;

     #10  x1 = 1'b1;
          x2 = 1'b1;
45        x3 = 1'b1;
     #10  $stop;
     end                 //Continued on next page
```

Figure 4.3   Test bench for the module of Figure 4.2.

```
      //instantiate the module into the test bench
      log_eqn_sop15 inst1 (
50    .x1(x1),
      .x2(x2),
      .x3(x3),
      .z1(z1),
      .z2(z2)
55    );

      endmodule
```

Figure 4.3   (Continued)

Values are assigned to the variables by the notation *1'b0* for example, where the number *1* specifies the width of the variable (1 bit), *b* specifies the radix (binary), and *0* specifies the value (zero). The system task **$stop** causes simulation to stop.

Line 4 specifies that the inputs are **reg** type variables; that is, they contain their values until they are assigned new values. Outputs are assigned as type **wire** in test benches. Output nets are driven by the output ports of the module under test. Line 8 contains an **initial** statement, which executes only once.

Verilog provides a means to monitor a signal when its value changes. This is accomplished by the **$monitor** system task. The **$monitor** continuously monitors the values of the variables indicated in the parameter list that is enclosed in parentheses. It will display the value of the variables whenever a variable changes state. The quoted string within the task is printed and specifies that the variables are to be shown in binary (%b). The **$monitor** is invoked only once. Line 13 is a second **initial** statement that allows the procedural code between the **begin** . . . **end** block statements to be executed only once. Every ten time units (#10) the input variables change state and are displayed by the system task **$monitor**.

Lines 49 through 55 instantiate the design module into the test bench module. The instantiation name is *inst1* followed by a left parenthesis. The port names of the design module are preceded by a period, which is followed by the corresponding port name in the test bench enclosed in parentheses; the port names in the module and the test bench do not necessarily have to be the same. A comma terminates each line of the port instantiation except the line containing the last port name. This is followed by a right parenthesis followed by a semicolon. The keyword **endmodule** terminates the module.

This logic diagram in this example is identical to the one presented in Example 3.22, Figure 3.76. Recall that there were redundant gates in Example 3.22. Analysis of the logic yielded the equations shown in Equation 4.1 and Equation 4.2 for outputs z_1 and z_2, respectively. The outputs obtained from the test bench are shown in Figure 4.4 and correspond to the equations for z_1 and z_2. There is a single active-high output for z_1 and four active-high outputs for z_2, including the case where the outputs overlap. The waveforms are shown in Figure 4.5. Refer to Appendix B in this book for the procedure to create a Verilog project and obtain the outputs and waveforms.

$$z_1 = [x_1 x_2 x_3 + x_1 x_2 x_3 (x_1 x_2)]$$

$$= x_1 x_2 x_3 (1 + x_1 x_2)$$

$$= x_1 x_2 x_3 \qquad (4.1)$$

$$z_2 = x_1 x_2 + x_1 x_3 + x_2 x_3 \qquad (4.2)$$

```
x1=0,  x2=0,  x3=0,  z1=0,  z2=0
x1=0,  x2=0,  x3=1,  z1=0,  z2=0
x1=0,  x2=1,  x3=0,  z1=0,  z2=0
x1=0,  x2=1,  x3=1,  z1=0,  z2=1
x1=1,  x2=0,  x3=0,  z1=0,  z2=0
x1=1,  x2=0,  x3=1,  z1=0,  z2=1
x1=1,  x2=1,  x3=0,  z1=0,  z2=1
x1=1,  x2=1,  x3=1,  z1=1,  z2=1
```

Figure 4.4   Outputs for the logic diagram of Example 4.1 as obtained from the test bench of Figure 4.3.

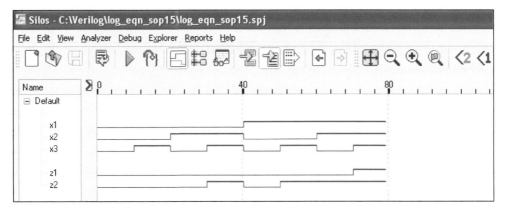

Figure 4.5   Waveforms for the logic diagram of Example 4.1 as obtained from the test bench of Figure 4.3.

Example 4.2   Equation 4.3 will be synthesized using NAND gates and the exclusive-NOR function. The first term $x_1 x_2' x_3' x_4$ is represented by the gate labeled *inst1* in Figure 4.6. The inputs are shown at their active voltage levels such that all gate inputs are high when the term $x_1 x_2' x_3' x_4$ is true. That is, gate *inst1* will generate a low output when x_1 and x_4 are asserted and x_2 and x_3 are deasserted. The second expression $x_1' (x_2 x_3 x_4' + x_2' x_3')$ is implemented by gates labeled *inst2*, *inst3*, *inst4*, and *inst5*.

The design module is shown in Figure 4.7 in which the gate instantiations correspond to those in the logic diagram and the output of each gate is represented by a net name. Input variable names that must be deasserted to generate a logic 1 at the input to a logic gate are prefixed by the tilde (~) symbol; for example, $-x_2$ in the logic diagram is written as ~x2, as shown in *inst1* of Figure 4.7.

$$z_1 = x_1 x_2' x_3' x_4 + x_1'(x_2 x_3 x_4' + x_2' x_3') + x_1 x_3 (x_1 \oplus x_2)' \qquad (4.3)$$

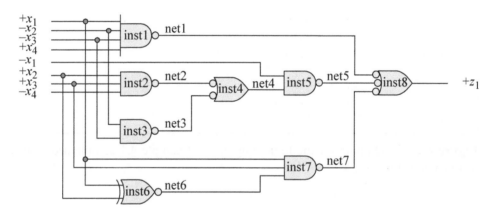

Figure 4.6   Logic diagram to implement Equation 4.3 for Example 4.2.

```
//logic diagram using built-in primitives
module log_eqn_sop16 (x1, x2, x3, x4, z1);

input x1, x2, x3, x4;
output z1;

//instantiate the built-in primitives
nand   inst1 (net1, x1, ~x2, ~x3, x4),
       inst2 (net2, x2, x3, ~x4),
       inst3 (net3, ~x2, ~x3),
       inst4 (net4, net2, net3),
       inst5 (net5, ~x1, net4);        //continued on next page
```

Figure 4.7   Module to implement the logic diagram of Figure 4.6.

```
xnor   inst6 (net6, x1, x2);

nand   inst7 (net7, x1, x3, net6),
       inst8 (z1, net1, net5, net7);

endmodule
```

Figure 4.7 (Continued)

Equation 4.3 is expanded to a sum-of-products notation as shown in Equation 4.4 and plotted on the Karnaugh map of Figure 4.8. The test bench is shown in Figure 4.9 in which all 16 combinations of the four variables x_1, x_2, x_3, and x_4 are applied to the inputs. The outputs are shown in Figure 4.10 which lists all 16 combinations of the inputs. Note that the minterms containing 1s in the Karnaugh map directly correspond to the minterms in the outputs of Figure 4.10. The waveforms are shown in Figure 4.11.

$$z_1 = x_1 x_2' x_3' x_4 + x_1'(x_2 x_3 x_4' + x_2' x_3') + x_1 x_3 (x_1 \oplus x_2)'$$

$$= x_1 x_2' x_3' x_4 + x_1' x_2 x_3 x_4' + x_1' x_2' x_3' + x_1 x_3 (x_1 x_2 + x_1' x_2')$$

$$= x_1 x_2' x_3' x_4 + x_1' x_2 x_3 x_4' + x_1' x_2' x_3' + x_1 x_2 x_3 \tag{4.4}$$

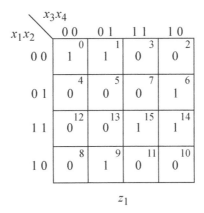

Figure 4.8 Karnaugh map that represents Equation 4.4.

```
//test bench for log_eqn_sop16
module log_eqn_sop_tb;

reg x1, x2, x3, x4;
wire z1;

//display variables
//the brace ({) symbol specifies concatenation
initial
$monitor ("x1x2x3x4 = %b, z1 = %b",
          {x1, x2, x3, x4}, z1);

//apply input vectors
initial
begin
   #0    x1=1'b0; x2=1'b0; x3=1'b0; x4=1'b0; //00
   #10   x1=1'b0; x2=1'b0; x3=1'b0; x4=1'b1; //01
   #10   x1=1'b0; x2=1'b0; x3=1'b1; x4=1'b0; //02
   #10   x1=1'b0; x2=1'b0; x3=1'b1; x4=1'b1; //03
   #10   x1=1'b0; x2=1'b1; x3=1'b0; x4=1'b0; //04
   #10   x1=1'b0; x2=1'b1; x3=1'b0; x4=1'b1; //05
   #10   x1=1'b0; x2=1'b1; x3=1'b1; x4=1'b0; //06
   #10   x1=1'b0; x2=1'b1; x3=1'b1; x4=1'b1; //07
   #10   x1=1'b1; x2=1'b0; x3=1'b0; x4=1'b0; //08
   #10   x1=1'b1; x2=1'b0; x3=1'b0; x4=1'b1; //09
   #10   x1=1'b1; x2=1'b0; x3=1'b1; x4=1'b0; //10
   #10   x1=1'b1; x2=1'b0; x3=1'b1; x4=1'b1; //11
   #10   x1=1'b1; x2=1'b1; x3=1'b0; x4=1'b0; //12
   #10   x1=1'b1; x2=1'b1; x3=1'b0; x4=1'b1; //13
   #10   x1=1'b1; x2=1'b1; x3=1'b1; x4=1'b0; //14
   #10   x1=1'b1; x2=1'b1; x3=1'b1; x4=1'b1; //15

   #10   $stop;
end

//instantiate the module into the test bench
log_eqn_sop16 inst1 (
   .x1(x1),
   .x2(x2),
   .x3(x3),
   .x4(x4),
   .z1(z1)
   );

endmodule
```

Figure 4.9   Test bench for the module of Figure 4.7.

```
x1x2x3x4 = 0000, z1 = 1
x1x2x3x4 = 0001, z1 = 1
x1x2x3x4 = 0010, z1 = 0
x1x2x3x4 = 0011, z1 = 0
x1x2x3x4 = 0100, z1 = 0
x1x2x3x4 = 0101, z1 = 0
x1x2x3x4 = 0110, z1 = 1
x1x2x3x4 = 0111, z1 = 0
x1x2x3x4 = 1000, z1 = 0
x1x2x3x4 = 1001, z1 = 1
x1x2x3x4 = 1010, z1 = 0
x1x2x3x4 = 1011, z1 = 0
x1x2x3x4 = 1100, z1 = 0
x1x2x3x4 = 1101, z1 = 0
x1x2x3x4 = 1110, z1 = 1
x1x2x3x4 = 1111, z1 = 1
```

Figure 4.10   Outputs generated by the test bench of Figure 4.9 for Example 4.2.

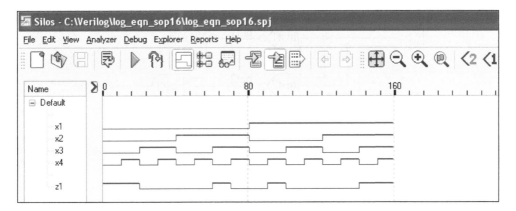

Figure 4.11   Waveforms generated by the test bench of Figure 4.9.

Example 4.3   The Karnaugh map of Figure 4.12 will be implemented using only NOR gates in a product-of-sums format. Equation 4.5 shown the product-of-sums expression obtained from the Karnaugh map. The logic diagram is shown in Figure 4.13 which indicates the instantiation names and net names.

The design module is shown in Figure 4.14 using NOR gate built-in primitives. The test bench is shown in Figure 4.15 using a different approach to generate all 16 combinations of the four inputs. Several new modeling constructs are shown in the test bench. Since there are four inputs to the circuit, all 16 combinations of four variables must be applied to the circuit. This is accomplished by a **for**-loop statement, which is similar in construction to a **for** loop in the C programming language.

x_1x_2 \ x_3x_4	0 0	0 1	1 1	1 0
0 0	0 `0`	1 `1`	3 `1`	2 `0`
0 1	4 `1`	5 `1`	7 `0`	6 `1`
1 1	12 `1`	13 `1`	15 `0`	14 `1`
1 0	8 `1`	9 `1`	11 `1`	10 `0`

z_1

Figure 4.12   Karnaugh map for Example 4.3.

$$z_1 = (x_1 + x_2 + x_4)\,(x_2 + x_3' + x_4)\,(x_2' + x_3' + x_4') \qquad (4.5)$$

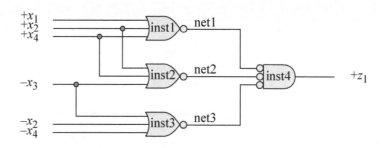

Figure 4.13   Logic diagram for Example 4.3.

```
//logic diagram using built-in primitives
module log_eqn_pos5 (x1, x2, x3, x4, z1);

input x1, x2, x3, x4;
output z1;

//instantiate the nor built-in primitives
nor     inst1 (net1, x1, x2, x4);
nor     inst2 (net2, x2, x4, ~x3);
nor     inst3 (net3, ~x3, ~x2, ~x4);
nor     inst4 (z1, net1, net2, net3);
endmodule
```

Figure 4.14   Module for the product-of-sums logic diagram of Figure 4.13.

Following the keyword **begin** is the name of the block: *apply_stimulus*. In this block, a 5-bit **reg** variable is declared called *invect*. This guarantees that all combinations of the four inputs will be tested by the **for** loop, which applies input vectors of $x_1 x_2 x_3 x_4 = 0000, 0001, 0010, 0011 \ldots 1111$ to the circuit. The **for** loop stops when the pattern 10000 is detected by the test segment (*invect* < 16). If only a 4-bit vector were applied, then the expression (*invect* < 16) would always be true and the loop would never terminate. The increment segment of the **for** loop does not support an increment designated as *invect*++; therefore, the long notation must be used: *invect* = *invect* + 1.

The target of the first assignment within the **for** loop ($\{x_1, x_2, x_3, x_4\}$ = *invect* [4:0]) represents a concatenated target. The concatenation of inputs x_1, x_2, x_3, and x_4 is performed by positioning them within braces: $\{x_1, x_2, x_3, x_4\}$. A vector of five bits ([4:0]) is then assigned to the inputs. This will apply inputs of 0000, 0001, 0010, 0011, $\ldots$ 1111 and stop when the vector is 10000.

The **initial** statement also contains a system task (**$display**) which prints the argument values — within the quotation marks — in binary. The concatenated variables x_1, x_2, x_3, and x_4 are listed first; therefore, their values are obtained from the first argument to the right of the quotation marks: $\{x_1, x_2, x_3, x_4\}$. The value for the second variable z_1 is obtained from the second argument to the right of the quotation marks. The variables to the right of the quotation marks are listed in the same order as the variables within the quotation marks. The delay time (#10) in the system task specifies that the task is to be executed after 10 time units; that is, the delay between the application of a vector and the response of the module. This delay represents the propagation delay of the logic. The simulation results are shown in binary format in Figure 4.16 and the waveforms in Figure 4.17.

```
//test bench for log_eqn_pos5
module log_eqn_pos5_tb;

reg x1, x2, x3, x4;
wire z1;

//apply input vectors
initial
begin: apply_stimulus
   reg [4:0] invect;      //invect[4] terminates the loop
   for (invect = 0; invect < 16; invect = invect + 1)
      begin
         {x1, x2, x3, x4} = invect[4:0];
         #10 $display ("x1x2x3x4 = %b, z1 = %b",
                           {x1, x2, x3, x4}, z1);
      end
end
//continued on next page
```

Figure 4.15   Test bench for the module of Figure 4.14.

```
//instantiate the module into the test bench
log_eqn_pos5 inst1 (
   .x1(x1),
   .x2(x2),
   .x3(x3),
   .x4(x4),
   .z1(z1)
   );
endmodule
```

Figure 4.15   (Continued)

```
x1x2x3x4 = 0000, z1 = 0
x1x2x3x4 = 0001, z1 = 1
x1x2x3x4 = 0010, z1 = 0
x1x2x3x4 = 0011, z1 = 1
x1x2x3x4 = 0100, z1 = 1
x1x2x3x4 = 0101, z1 = 1
x1x2x3x4 = 0110, z1 = 1
x1x2x3x4 = 0111, z1 = 0
x1x2x3x4 = 1000, z1 = 1
x1x2x3x4 = 1001, z1 = 1
x1x2x3x4 = 1010, z1 = 0
x1x2x3x4 = 1011, z1 = 1
x1x2x3x4 = 1100, z1 = 1
x1x2x3x4 = 1101, z1 = 1
x1x2x3x4 = 1110, z1 = 1
x1x2x3x4 = 1111, z1 = 0
```

Figure 4.16   Outputs generated by the test bench of Figure 4.15.

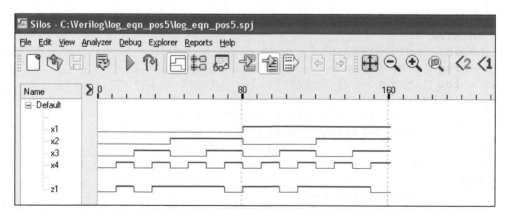

Figure 4.17   Waveforms for the logic diagram of Figure 4.13.

Example 4.4   Equation 4.6 will be minimized as a sum-of-products form and then implemented using built-in primitives of AND and OR with x_4 and x_5 as map-entered variables. The Karnaugh map is shown in Figure 4.18 in which the following minterm locations combine:

Minterm location $0 = x_4 x_5' + x_4 x_5 = x_4$
Minterm location $2 = 1 + x_4$
Combine minterm locations 0 and 2 to yield the sum term $x_1' x_3' x_4$

Combine minterm locations 2 and 3 to yield $x_1' x_2$

Minterm location $4 = x_4 x_5 + x_4' + x_5' = 1$
Minterm location $5 = 1$
Combine minterm locations 4 and 5 to yield $x_1 x_2'$

The minimized sum-of-products equation from the Karnaugh map is shown in Equation 4.7. The logic diagram is shown in Figure 4.19. The design module is shown in Figure 4.20 and the test bench is shown in Figure 4.21. Figure 4.22 lists the outputs obtained from the test bench.

$$z_1 = x_1' x_2' x_3' x_4 x_5' + x_1' x_2 + x_1' x_2' x_3' x_4 x_5 + x_1 x_2' x_3' x_4 x_5$$

$$+ x_1 x_2' x_3 + x_1 x_2' x_3' x_4 + x_1 x_2' x_3' x_5' \qquad (4.6)$$

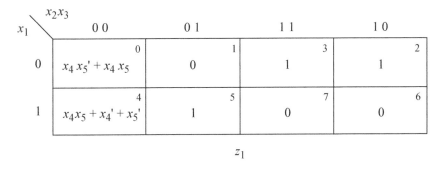

Figure 4.18   Karnaugh map for Example 4.4.

$$z_1 = x_1' x_3' x_4 + x_1' x_2 + x_1 x_2' \qquad (4.7)$$

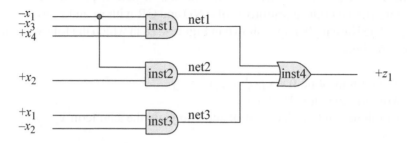

Figure 4.19   Logic diagram for Equation 4.7.

```
//logic equation using map-entered variables
module mev (x1, x2, x3, x4, z1);

input x1, x2, x3, x4;
output z1;

and    inst1 (net1, ~x1, ~x3, x4);
and    inst2 (net2, ~x1, x2);
and    inst3 (net3, x1, ~x2);
or     inst4 (z1, net1, net2, net3);
endmodule
```

Figure 4.20   Module to implement Equation 4.7 using built-in primitives.

```
//test bench for logic equation using map-entered variables
module mev_tb;

reg x1, x2, x3, x4;
wire z1;

//apply input vectors
initial
begin: apply_stimulus
   reg [4:0] invect;
   for (invect=0; invect<16; invect=invect+1)
      begin
         {x1, x2, x3, x4} = invect [4:0];
         #10 $display ("x1x2x3x4 = %b, z1 = %b",
                         {x1, x2, x3, x4}, z1);
      end
end
end                     //continued on next page
```

Figure 4.21   Test bench for the module of Figure 4.20.

```
//instantiate the module into the test bench
mev inst1 (
   .x1(x1),
   .x2(x2),
   .x3(x3),
   .x4(x4),
   .z1(z1)
   );
endmodule
```

Figure 4.21   (Continued)

```
x1x2x3x4 = 0000, z1 = 0
x1x2x3x4 = 0001, z1 = 1
x1x2x3x4 = 0010, z1 = 0
x1x2x3x4 = 0011, z1 = 0
x1x2x3x4 = 0100, z1 = 1
x1x2x3x4 = 0101, z1 = 1
x1x2x3x4 = 0110, z1 = 1
x1x2x3x4 = 0111, z1 = 1
x1x2x3x4 = 1000, z1 = 1
x1x2x3x4 = 1001, z1 = 1
x1x2x3x4 = 1010, z1 = 1
x1x2x3x4 = 1011, z1 = 1
x1x2x3x4 = 1100, z1 = 0
x1x2x3x4 = 1101, z1 = 0
x1x2x3x4 = 1110, z1 = 0
x1x2x3x4 = 1111, z1 = 0
```

Figure 4.22   Outputs for the test bench of Figure 4.21.

Example 4.5   A 4:1 multiplexer will be designed using built-in logic primitives. The 4:1 multiplexer of Figure 4.23 will be designed using built-in primitives of AND, OR, and NOT. The design is simpler and takes less code if a continuous assignment statement is used, but this section presents gate-level modeling only — continuous assignment statements are used in dataflow modeling.

The multiplexer has four data inputs: d_0, d_1, d_2, and d_3, which are specified as a 4-bit vector $d[3:0]$, two select inputs: s_0 and s_1, specified as a 2-bit vector $s[1:0]$, one scalar input *Enable*, and one scalar output z_1. Also, the system function **$time** will be used in the test bench to return the current simulation time measured in nanoseconds (ns). The design module is shown in Figure 4.24, the test bench in Figure 4.25, the outputs in Figure 4.26, and the waveforms in Figure 4.27. The vector waveforms are shown in both vector form with hexadecimal values and as individual bits.

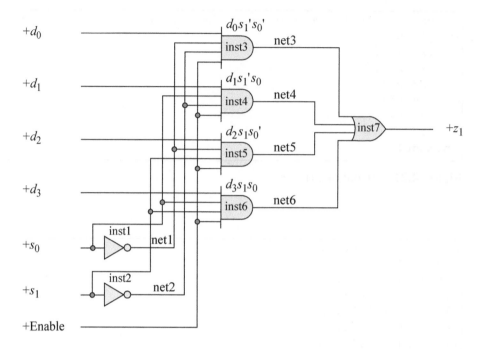

Figure 4.23   Logic diagram of a 4:1 multiplexer to be designed using built-in primitives.

```
//a 4:1 multiplexer using built-in primitives
module mux_4to1 (d, s, enbl, z1);

input [3:0] d;
input [1:0] s;
input enbl;
output z1;

not     inst1 (net1, s[0]),
        inst2 (net2, s[1]);

and     inst3 (net3, d[0], net1, net2, enbl),
        inst4 (net4, d[1], s[0], net2, enbl),
        inst5 (net5, d[2], net1, s[1], enbl),
        inst6 (net6, d[3], s[0], s[1], enbl);

or      inst7 (z1, net3, net4, net5, net6);

endmodule
```

Figure 4.24   Module for a 4:1 multiplexer with *Enable* using built-in primitives.

```
//test bench for 4:1 multiplexer
module mux_4to1_tb;

reg [3:0] d;
reg [1:0] s;
reg enbl;
wire z1;

initial
$monitor ($time,"ns, select:s=%b, inputs:d=%b, output:z1=%b",
         s, d, z1);
initial
begin
   #0    s[0]=1'b0;  s[1]=1'b0;
         d[0]=1'b0;  d[1]=1'b1;  d[2]=1'b0;  d[3]=1'b1;
         enbl=1'b1;  //d[0]=0; z1=0

   #10   s[0]=1'b0;  s[1]=1'b0;
         d[0]=1'b1;  d[1]=1'b1;  d[2]=1'b0;  d[3]=1'b1;
         enbl=1'b1;  //d[0]=1; z1=1

   #10   s[0]=1'b1;  s[1]=1'b0;
         d[0]=1'b1;  d[1]=1'b1;  d[2]=1'b0;  d[3]=1'b1;
         enbl=1'b1;  //d[1]=1; z1=1

   #10   s[0]=1'b0;  s[1]=1'b1;
         d[0]=1'b1;  d[1]=1'b1;  d[2]=1'b0;  d[3]=1'b1;
         enbl=1'b1;  //d[2]=0; z1=0

   #10   s[0]=1'b1;  s[1]=1'b0;
         d[0]=1'b1;  d[1]=1'b0;  d[2]=1'b0;  d[3]=1'b1;
         enbl=1'b1;  //d[1]=1; z1=0

   #10   s[0]=1'b1;  s[1]=1'b1;
         d[0]=1'b1;  d[1]=1'b1;  d[2]=1'b0;  d[3]=1'b1;
         enbl=1'b1;  //d[3]=1; z1=1

   #10   s[0]=1'b1;  s[1]=1'b1;
         d[0]=1'b1;  d[1]=1'b1;  d[2]=1'b0;  d[3]=1'b0;
         enbl=1'b1;  //d[3]=0; z1=0

   #10   s[0]=1'b1;  s[1]=1'b1;
         d[0]=1'b1;  d[1]=1'b1;  d[2]=1'b0;  d[3]=1'b0;
         enbl=1'b0;  //d[3]=0; z1=0

   #10   $stop;
end                      //continued on next page
```

Figure 4.25   Test bench for the 4:1 multiplexer of Figure 4.24.

```
//instantiate the module into the test bench
mux_4to1 inst1 (
    .d(d),
    .s(s),
    .z1(z1),
    .enbl(enbl)
    );

endmodule
```

Figure 4.25   (Continued)

```
0   ns, select:s=00, inputs:d=1010, output:z1=0
10 ns, select:s=00, inputs:d=1011, output:z1=1
20 ns, select:s=01, inputs:d=1011, output:z1=1
30 ns, select:s=10, inputs:d=1011, output:z1=0
40 ns, select:s=01, inputs:d=1001, output:z1=0
50 ns, select:s=11, inputs:d=1011, output:z1=1
60 ns, select:s=11, inputs:d=0011, output:z1=0
```

Figure 4.26   Outputs for the 4:1 multiplexer test bench of Figure 4.25.

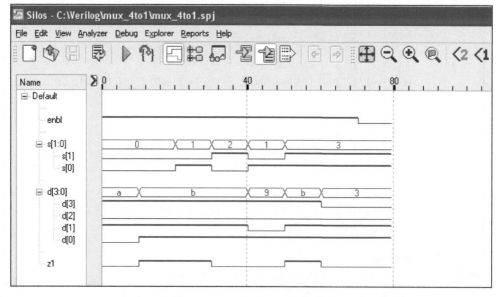

Figure 4.27   Waveforms for the 4:1 multiplexer module of Figure 4.24.

Example 4.6 This example illustrates the design of a majority circuit using built-in primitives. The output of a majority circuit is a logic 1 if the majority of the inputs is a logic 1; otherwise, the output is a logic 0. Therefore, a majority circuit must have an odd number of inputs in order to have a majority of the inputs at the same logic level. A 5-input majority circuit will be designed using the Karnaugh map of Figure 4.28, where a 1 entry indicates that the majority of the inputs is a logic 1.

Equation 4.8 represents the logic for output z_1 in a sum-of-products form. The module is shown in Figure 4.29, which is designed directly from Equation 4.8 without the use of a logic diagram. The test bench is shown in Figure 4.30, the outputs are shown in Figure 4.31, and the waveforms are shown in Figure 4.32.

$x_5 = 0$

x_1x_2 \ x_3x_4	0 0	0 1	1 1	1 0
0 0	0 (0)	0 (2)	0 (6)	0 (4)
0 1	0 (8)	0 (10)	1 (14)	0 (12)
1 1	0 (24)	1 (26)	1 (30)	1 (28)
1 0	0 (16)	0 (18)	1 (22)	0 (20)

$x_5 = 1$

x_1x_2 \ x_3x_4	0 0	0 1	1 1	1 0
0 0	0 (1)	0 (3)	1 (7)	0 (5)
0 1	0 (9)	1 (11)	1 (15)	1 (13)
1 1	0 (25)	1 (27)	1 (31)	1 (29)
1 0	0 (17)	1 (19)	1 (23)	1 (21)

z_1

Figure 4.28 Karnaugh map for the majority circuit of Example 4.6.

$$z_1 = x_3x_4x_5 + x_2x_3x_5 + x_1x_3x_5 + x_2x_4x_5 + x_1x_4x_5 + x_1x_2x_3 + x_1x_2x_4 + x_2x_3x_4 + x_1x_3x_4 \quad (4.8)$$

```
//5-input majority circuit
module majority (x1, x2, x3, x4, x5, z1);
input x1, x2, x3, x4, x5;
output z1;

and    inst1   (net1, x3, x4, x5),
       inst2   (net2, x2, x3, x5),
       inst3   (net3, x1, x3, x5),
       inst4   (net4, x2, x4, x5),    //continued on next page
```

Figure 4.29 Module for the majority circuit of Figure 4.28.

```
        inst5   (net5, x1, x4, x5),
        inst6   (net6, x1, x2, x5),
        inst7   (net7, x1, x2, x4),
        inst8   (net8, x2, x3, x4),
        inst9   (net9, x1, x3, x4);
or      inst10 (z1, net1, net2, net3, net4, net5,
                    net6, net7, net8, net9);
endmodule
```

Figure 4.29   (Continued)

```
//test bench for 5-input majority circuit
module majority_tb;
reg x1, x2, x3, x4, x5;
wire z1;

//apply input vectors
initial
begin: apply_stimulus
        reg [6:0] invect;
        for (invect=0; invect<32; invect=invect+1)
                begin
                        {x1, x2, x3, x4, x5} = invect [6:0];
                        #10 $display ("x1x2x3x4x5 = %b, z1 = %b",
                                        {x1, x2, x3, x4, x5}, z1);
                end
end

//instantiate the module into the test bench
majority inst1 (
        .x1(x1),
        .x2(x2),
        .x3(x3),
        .x4(x4),
        .x5(x5),
        .z1(z1)
        );

endmodule
```

Figure 4.30   Test bench for the majority circuit module of Figure 4.29.

```
x1x2x3x4x5 = 00000, z1 = 0        x1x2x3x4x5 = 10000, z1 = 0
x1x2x3x4x5 = 00001, z1 = 0        x1x2x3x4x5 = 10001, z1 = 0
x1x2x3x4x5 = 00010, z1 = 0        x1x2x3x4x5 = 10010, z1 = 0
x1x2x3x4x5 = 00011, z1 = 0        x1x2x3x4x5 = 10011, z1 = 1
x1x2x3x4x5 = 00100, z1 = 0        x1x2x3x4x5 = 10100, z1 = 0
x1x2x3x4x5 = 00101, z1 = 0        x1x2x3x4x5 = 10101, z1 = 1
x1x2x3x4x5 = 00110, z1 = 0        x1x2x3x4x5 = 10110, z1 = 1
x1x2x3x4x5 = 00111, z1 = 1        x1x2x3x4x5 = 10111, z1 = 1
x1x2x3x4x5 = 01000, z1 = 0        x1x2x3x4x5 = 11000, z1 = 0
x1x2x3x4x5 = 01001, z1 = 0        x1x2x3x4x5 = 11001, z1 = 1
x1x2x3x4x5 = 01010, z1 = 0        x1x2x3x4x5 = 11010, z1 = 1
x1x2x3x4x5 = 01011, z1 = 1        x1x2x3x4x5 = 11011, z1 = 1
x1x2x3x4x5 = 01100, z1 = 0        x1x2x3x4x5 = 11100, z1 = 1
x1x2x3x4x5 = 01101, z1 = 1        x1x2x3x4x5 = 11101, z1 = 1
x1x2x3x4x5 = 01110, z1 = 1        x1x2x3x4x5 = 11110, z1 = 1
x1x2x3x4x5 = 01111, z1 = 1        x1x2x3x4x5 = 11111, z1 = 1
```

Figure 4.31   Outputs for the majority circuit of Figure 4.29.

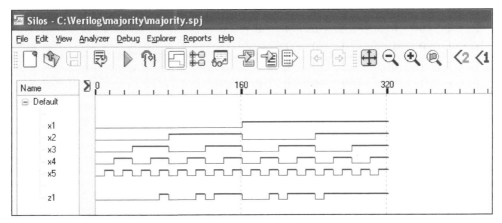

Figure 4.32   Waveforms for the majority circuit of Figure 4.29.

Example 4.7   A code converter will be designed to convert a 4-bit binary number to the corresponding Gray code number. The inputs of the binary number $x_1 x_2 x_3 x_4$ are available in both high and low assertion, where x_4 is the low-order bit. The outputs for the Gray code $z_1 z_2 z_3 z_4$ are asserted high, where z_4 is the low-order bit. The binary-to-Gray code conversion table is shown in Table 4.1.

There are four Karnaugh maps shown in Figure 4.33, one map for each of the Gray code outputs. The equations obtained from the Karnaugh maps are shown in Equation 4.9. The logic diagram is shown in Figure 4.34. The design module, test bench module, outputs, and waveforms are shown in Figure 4.35, Figure 4.36, Figure 4.37, and Figure 4.38, respectively.

Table 4.1 Binary-to-Gray Code Conversion

Binary Code				Gray Code			
x_1	x_2	x_3	x_4	z_1	z_2	z_3	z_4
0	0	0	0	0	0	0	0
0	0	0	1	0	0	0	1
0	0	1	0	0	0	1	1
0	0	1	1	0	0	1	0
0	1	0	0	0	1	1	0
0	1	0	1	0	1	1	1
0	1	1	0	0	1	0	1
0	1	1	1	0	1	0	0
1	0	0	0	1	1	0	0
1	0	0	1	1	1	0	1
1	0	1	0	1	1	1	1
1	0	1	1	1	1	1	0
1	1	0	0	1	0	1	0
1	1	0	1	1	0	1	1
1	1	1	0	1	0	0	1
1	1	1	1	1	0	0	0

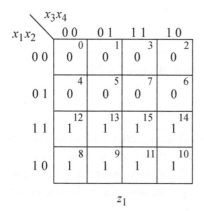
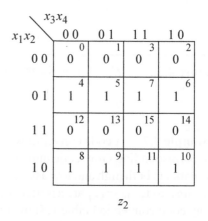

Figure 4.33   Karnaugh maps for the binary-to-Gray code converter.

x_1x_2 \ x_3x_4	0 0	0 1	1 1	1 0
0 0	0 [0]	0 [1]	1 [3]	1 [2]
0 1	1 [4]	1 [5]	0 [7]	0 [6]
1 1	1 [12]	1 [13]	0 [15]	0 [14]
1 0	0 [8]	0 [9]	1 [11]	1 [10]

z_3

x_1x_2 \ x_3x_4	0 0	0 1	1 1	1 0
0 0	0 [0]	1 [1]	0 [3]	1 [2]
0 1	0 [4]	1 [5]	0 [7]	1 [6]
1 1	0 [12]	1 [13]	0 [15]	1 [14]
1 0	0 [8]	1 [9]	0 [11]	1 [10]

z_4

Figure 4.33   (Continued)

$$z_1 = x_1$$
$$z_2 = x_1'x_2 + x_1x_2' = x_1 \oplus x_2$$
$$z_3 = x_2x_3' + x_2'x_3 = x_2 \oplus x_3$$
$$z_4 = x_3'x_4 + x_3x_4' = x_3 \oplus x_4 \qquad\qquad (4.9)$$

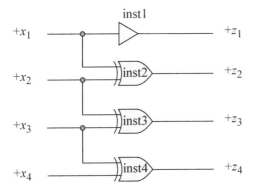

Figure 4.34   Logic diagram for the binary-to-Gray code converter.

```
//binary-to-gray code converter
module bin_to_gray (x1, x2, x3, x4, z1, z2, z3, z4);

input x1, x2, x3, x4;
output z1, z2, z3, z4;

buf inst1 (z1, x1);
xor inst2 (z2, x1, x2);
xor inst3 (z3, x2, x3);
xor inst4 (z4, x3, x4);
endmodule
```

Figure 4.35   Module for the binary-to-Gray code converter.

```
//test bench for binary-to-gray code converter
module bin_to_gray_tb;

reg x1, x2, x3, x4;
wire z1, z2, z3, z4;

//apply input vectors
initial
begin: apply_stimulus
   reg [4:0] invect;
   for (invect=0; invect<16; invect=invect+1)
      begin
         {x1, x2, x3, x4} = invect [4:0];
         #10 $display ("{x1x2x3x4}=%b, {z1z2z3z4}=%b",
                       {x1, x2, x3, x4}, {z1, z2, z3, z4});
      end
end

//instantiate the module into the test bench
bin_to_gray inst1 (
   .x1(x1),
   .x2(x2),
   .x3(x3),
   .x4(x4),
   .z1(z1),
   .z2(z2),
   .z3(z3),
   .z4(z4)
   );

endmodule
```

Figure 4.36   Test bench for the binary-to-Gray code converter.

```
{x1x2x3x4}=0000, {z1z2z3z4}=0000
{x1x2x3x4}=0001, {z1z2z3z4}=0001
{x1x2x3x4}=0010, {z1z2z3z4}=0011
{x1x2x3x4}=0011, {z1z2z3z4}=0010
{x1x2x3x4}=0100, {z1z2z3z4}=0110
{x1x2x3x4}=0101, {z1z2z3z4}=0111
{x1x2x3x4}=0110, {z1z2z3z4}=0101
{x1x2x3x4}=0111, {z1z2z3z4}=0100
{x1x2x3x4}=1000, {z1z2z3z4}=1100
{x1x2x3x4}=1001, {z1z2z3z4}=1101
{x1x2x3x4}=1010, {z1z2z3z4}=1111
{x1x2x3x4}=1011, {z1z2z3z4}=1110
{x1x2x3x4}=1100, {z1z2z3z4}=1010
{x1x2x3x4}=1101, {z1z2z3z4}=1011
{x1x2x3x4}=1110, {z1z2z3z4}=1001
{x1x2x3x4}=1111, {z1z2z3z4}=1000
```

Figure 4.37   Outputs for the binary-to-Gray code converter.

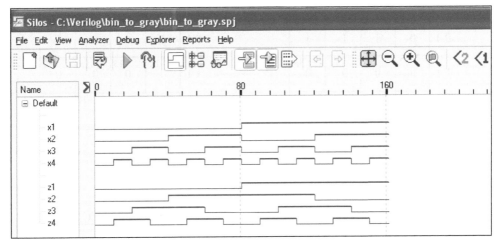

Figure 4.38   Waveforms for the binary-to-Gray code converter.

Example 4.8   A binary-coded decimal (BCD)-to-excess-3 code converter will be designed using only NAND gates. The excess-3 code is obtained from the binary 8421 BCD code by adding 3 (0011) to each binary code word. The truth table for converting from BCD to excess-3 is shown in Table 4.2. The BCD code word is $x_1x_2x_3x_4$, where x_4 is the low-order bit; the excess-3 code word is $z_1z_2z_3z_4$, where z_4 is the low-order bit. There are four Karnaugh maps, one for each excess-3 digit, as shown in Figure 4.39. The equations for the excess-3 bits are shown in Equation 4.10. The logic diagram is shown in Figure 4.40. The design module, test bench module, outputs, and waveforms are shown in Figure 4.41, Figure 4.42, Figure 4.43, and Figure 4.44, respectively.

Table 4.2 BCD-to-Excess-3 Code Conversion

x_1	x_2	x_3	x_4	z_1	z_2	z_3	z_4	x_1	x_2	x_3	x_4	z_1	z_2	z_3	z_4
0	0	0	0	0	0	1	1	0	1	0	1	1	0	0	0
0	0	0	1	0	1	0	0	0	1	1	0	1	0	0	1
0	0	1	0	0	1	0	1	0	1	1	1	1	0	1	0
0	0	1	1	0	1	1	0	1	0	0	0	1	0	1	1
0	1	0	0	0	1	1	1	1	0	0	1	1	1	0	0

(Header for both halves: BCD Code — x_1 x_2 x_3 x_4 ; Excess-3 Code — z_1 z_2 z_3 z_4)

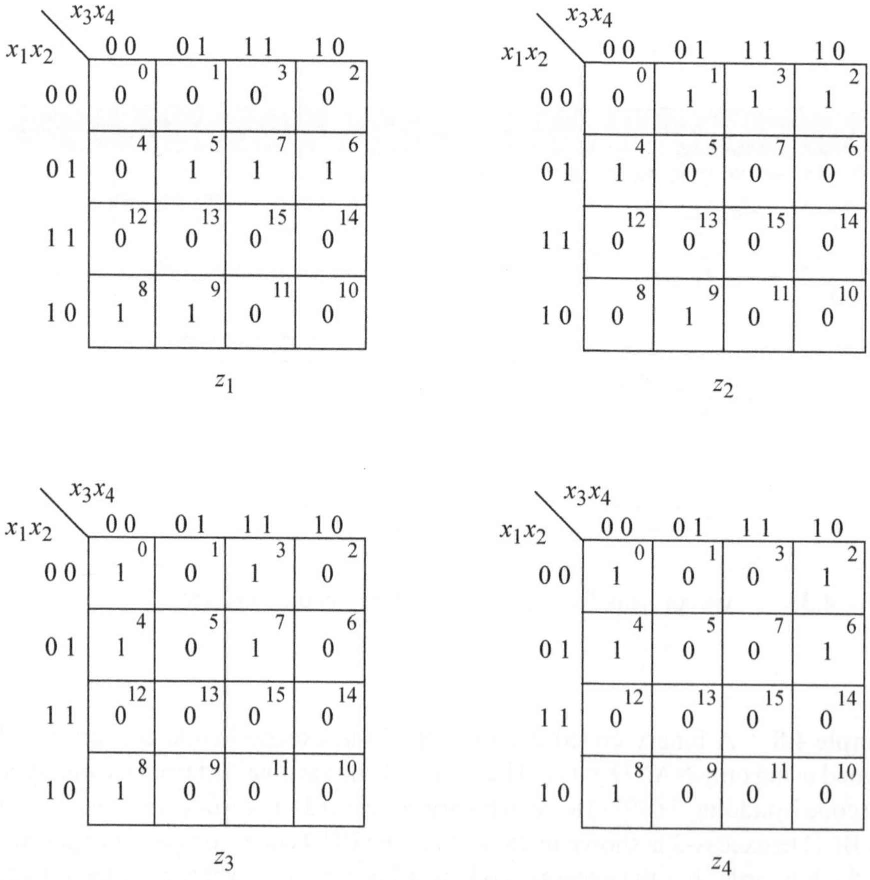

Figure 4.39   Karnaugh maps for BCD-to-excess-3 code conversion.

$$z_1 = x_1'x_2x_4 + x_1'x_2x_3 + x_1x_2'x_3'$$

$$z_2 = x_1'x_2'x_3 + x_2'x_3'x_4 + x_1'x_2x_3'x_4'$$

$$z_3 = x_1'x_3'x_4' + x_1'x_3x_4 + x_2'x_3'x_4'$$

$$z_4 = x_1'x_4' + x_2'x_3'x_4' \qquad\qquad (4.10)$$

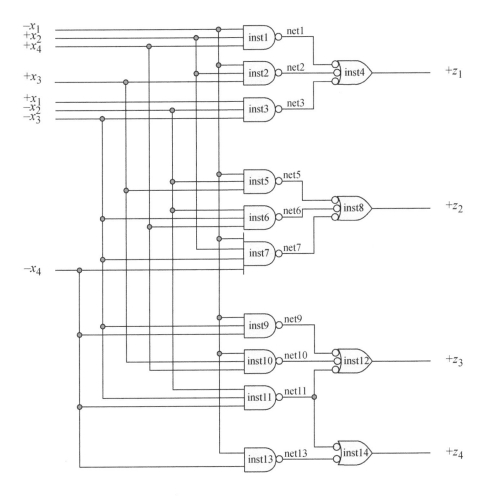

Figure 4.40   Logic diagram for BCD-to-excess-3 code conversion.

```
//bcd-to-excess3 code conversion using built-in primitives
module bcd_to_excess3 (x1, x2, x3, x4, z1, z2, z3, z4);

input x1, x2, x3, x4;
output z1, z2, z3, z4;

//instantiate the nand built-in primitives for z1
nand   inst1      (net1, ~x1, x2, x4),
       inst2      (net2, ~x1, x2, x3),
       inst3      (net3, x1, ~x2, ~x3),
       inst4      (z1, net1, net2, net3);

//instantiate the nand built-in primitives for z2
nand   inst5      (net5, ~x1, ~x2, x3),
       inst6      (net6, ~x2, ~x3, x4),
       inst7      (net7, ~x1, x2, ~x3, ~x4),
       inst8      (z2, net5, net6, net7);

//instantiate the nand built-in primitives for z3
nand   inst9      (net9, ~x1, ~x3, ~x4),
       inst10     (net10, ~x1, x3, x4),
       inst11     (net11, ~x2, ~x3, ~x4),
       inst12     (z3, net9, net10, net11);

//instantiate the nand built-in primitives for z4
nand   inst13     (net13, ~x1, ~x4),
       inst14     (z4, net11, net13);

endmodule
```

Figure 4.41   Module for the BCD-to-excess-3 code conversion using built-in primitives.

```
//test bench for the bcd-to-excess3 module
module bcd_to_excess3_tb;

reg x1, x2, x3, x4;
wire z1, z2, z3, z4;

//apply input vectors
initial
begin: apply_stimulus
   reg [3:0] invect;    //loop stops at 10
   for(invect = 0; invect < 10; invect = invect + 1)
//continued on next page
```

Figure 4.42   Test bench for the BCD-to-excess-3 code conversion.

```
      begin
         {x1, x2, x3, x4} = invect [3:0];
         #10 $display ("x1x2x3x4 = %b, z1z2z3z4 = %b",
                       {x1, x2, x3, x4}, {z1, z2, z3, z4});

      end
end

//instantiate the module into the test bench
bcd_to_excess3 inst1 (
   .x1(x1),
   .x2(x2),
   .x3(x3),
   .x4(x4),
   .z1(z1),
   .z2(z2),
   .z3(z3),
   .z4(z4)
   );
endmodule
```

Figure 4.42   (Continued)

```
x1x2x3x4=0000, z1z2z3z4=0011 | x1x2x3x4=0101, z1z2z3z4=1000
x1x2x3x4=0001, z1z2z3z4=0100 | x1x2x3x4=0110, z1z2z3z4=1001
x1x2x3x4=0010, z1z2z3z4=0101 | x1x2x3x4=0111, z1z2z3z4=1010
x1x2x3x4=0011, z1z2z3z4=0110 | x1x2x3x4=1000, z1z2z3z4=1011
x1x2x3x4=0100, z1z2z3z4=0111 | x1x2x3x4=1001, z1z2z3z4=1100
```

Figure 4.43   Outputs for the BCD-to-excess-3 code conversion.

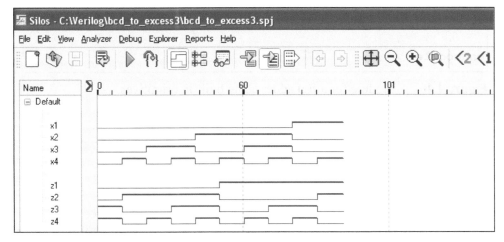

Figure 4.44   Waveforms for the BCD-to-excess-3 code conversion.

Example 4.9   A *full adder* is a combinational circuit that adds two operand bits: *a* and *b* plus a carry-in bit *cin*. The carry-in bit represents the carry-out of the previous lower-order stage. A full adder produces two outputs: a sum bit *sum* and carry-out *cout*. This example will use built-in primitives to design a full adder consisting of two half adders plus additional logic as shown in Figure 4.45.

The design module, test bench module, outputs, and waveforms are shown in Figure 4.46, Figure 4.47, Figure 4.48, and Figure 4.49, respectively.

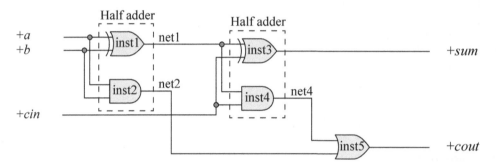

Figure 4.45   Full adder to be designed with built-in primitives.

```
//full adder using built-in primitives
module full_adder_bip (a, b, cin, sum, cout);

input a, b, cin;
output sum, cout;

xor     inst1 (net1, a, b);
and     inst2 (net2, a, b);
xor     inst3 (sum, net1, cin);
and     inst4 (net4, net1, cin);
or      inst5 (cout, net4, net2);

endmodule
```

Figure 4.46   Module for a full adder using built-in primitives.

```
//test bench for full adder using built-in primitives
module full_adder_bip_tb;

reg a, b, cin;
wire sum, cout;                 //continued on next page
```

Figure 4.47   Test bench for the full adder of Figure 4.46.

```
//apply input vectors
initial
begin: apply_stimulus
   reg[3:0] invect;      //invect[3] terminates the for loop
   for (invect = 0; invect < 8; invect = invect + 1)
      begin
         {a, b, cin} = invect [3:0];
         #10 $display ("abcin = %b, cout = %b, sum = %b",
                       {a, b, cin}, cout, sum);
      end
end

//instantiate the module into the test bench
full_adder_bip inst1 (
   .a(a),
   .b(b),
   .cin(cin),
   .sum(sum),
   .cout(cout)
   );
endmodule
```

Figure 4.47   (Continued)

```
abcin=000,  cout=0,  sum=0          abcin=100,  cout=0,  sum=1
abcin=001,  cout=0,  sum=1          abcin=101,  cout=1,  sum=0
abcin=010,  cout=0,  sum=1          abcin=110,  cout=1,  sum=0
abcin=011,  cout=1,  sum=0          abcin=111,  cout=1,  sum=1
```

Figure 4.48   Outputs for the full adder of Figure 4.46.

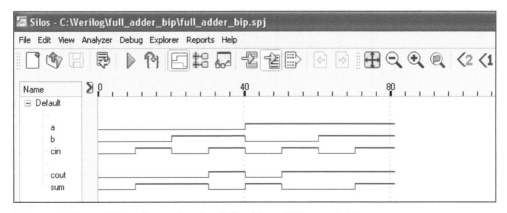

Figure 4.49   Waveforms for the full adder of Figure 4.46.

Example 4.10   This example will design the 2-bit multiplier of Example 3.37 using built-in primitives. A multiply operation involves two operands. One operand is an n-bit multiplicand; the other operand is an n-bit multiplier. The result is a $2n$-bit product. Since the multiplicand and multiplier are two 2-bit operands, a product of 4-bits will result.

The multiplicand is $A = a_1 a_0$, where a_0 is the low-order bit and is specified in the design module as a 2-bit vector $a[1:0]$. The multiplier is $B = b_1 b_0$, where b_0 is the low-order bit and is specified in the design module as a 2-bit vector $b[1:0]$. The product is $P = p_3 p_2 p_1 p_0$, where p_0 is the low-order bit and is specified in the design module as a 4-bit vector $p[3:0]$.

The logic diagram is shown in Figure 4.50. The design module, test bench module, outputs, and waveforms are shown in Figure 4.51, Figure 4.52, Figure 4.53, and Figure 4.54, respectively.

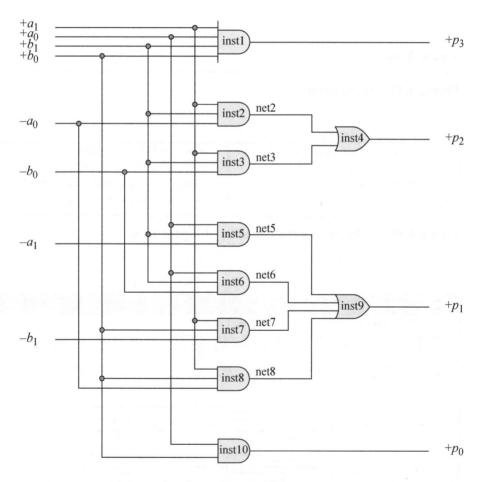

Figure 4.50   A multiplier logic diagram to multiply two 2-bit operands using built-in primitives.

```
//multiplier for 2-bit operands using built-in primitives
module mul_2 (a, b, p);

input[1:0] a, b;
output[3:0] p;

and     inst1     (p[3], a[1], a[0], b[1], b[0]);

and     inst2     (net2, a[1], b[1], ~a[0]),
        inst3     (net3, a[1], b[1], ~b[0]);
or      inst4     (p[2], net2, net3);

and     inst5     (net5, a[0], b[1], ~a[1]),
        inst6     (net6, a[0], b[1], ~b[0]),
        inst7     (net7, a[1], b[0], ~b[1]),
        inst8     (net8, a[1], b[0], ~a[0]);
or      inst9     (p[1], net5, net6, net7, net8);

and     inst10    (p[0], a[0], b[0]);

endmodule
```

Figure 4.51   Module for the 2-bit multiplier of Figure 4.50 using built-in primitives.

```
//test bench for mul_2
module mul_2_tb;

reg [1:0] a, b;
wire [3:0] p;

//apply input vectors
initial
begin: apply_stimulus
   reg [4:0] invect;
   for (invect=0; invect<16; invect=invect+1)
      begin
         {a, b} = invect [4:0];
         #10  $display ("a=%b, b=%b, p=%b", a, b, p);
      end
end

//continued on next page
```

Figure 4.52   Test bench for the 2-bit multiplier of Figure 4.50.

```
//instantiate the module into the test bench
mul_2 inst1 (
    .a(a),
    .b(b),
    .p(p)
    );

endmodule
```

Figure 4.52   (Continued)

```
a=00,  b=00,  p=0000
a=00,  b=01,  p=0000
a=00,  b=10,  p=0000
a=00,  b=11,  p=0000
a=01,  b=00,  p=0000
a=01,  b=01,  p=0001
a=01,  b=10,  p=0010
a=01,  b=11,  p=0011
a=10,  b=00,  p=0000
a=10,  b=01,  p=0010
a=10,  b=10,  p=0100
a=10,  b=11,  p=0110
a=11,  b=00,  p=0000
a=11,  b=01,  p=0011
a=11,  b=10,  p=0110
a=11,  b=11,  p=1001
```

Figure 4.53   Outputs of the test bench for the 2-bit multiplier of Figure 4.51.

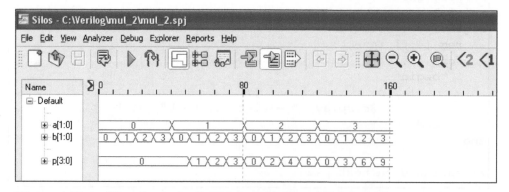

Figure 4.54   Waveforms for the 2-bit multiplier of Figure 4.51.

4.2 User-Defined Primitives

In addition to built-in primitives, Verilog provides the ability to design primitives according to user specifications. These are called *user-defined primitives* (UDPs) and are usually a higher-level logic function than built-in primitives. They are independent primitives and do not instantiate other primitives or modules. UDPs are instantiated into a module the same way as built-in primitives; that is, the syntax for a UDP instantiation is the same as that for a built-in primitive instantiation. A UDP is defined outside the module into which it is instantiated. There are two types of UDPs: combinational and sequential. Sequential primitives include level-sensitive and edge-sensitive circuits.

4.2.1 Defining a User-Defined Primitive

The syntax for a UDP is similar to that for declaring a module. The definition begins with the keyword **primitive** and ends with the keyword **endprimitive**. The UDP contains a name and a list of ports, which are declared as **input** or **output**. For a sequential UDP, the output port is declared as **reg**. UDPs can have one or more scalar inputs, but only one scalar output. The output port is listed first in the terminal list followed by the input ports, in the same way that the terminal list appears in built-in primitives. UDPs do not support **inout** ports.

The UDP table is an essential part of the internal structure and defines the functionality of the circuit. It is a lookup table similar in concept to a truth table. The table begins with the keyword **table** and ends with the keyword **endtable**. The contents of the table define the value of the output with respect to the inputs. The syntax for a UDP is shown below.

```
primitive udp_name (output, input_1, input_2, . . . , input_n);
    output output;
    input input_1, input_2, . . . , input_n;
    reg sequential_output;        //for sequential UDPs

    initial                       //for sequential UDPs

    table
        state table entries
    endtable
endprimitive
```

4.2.2 Combinational User-Defined Primitives

To illustrate the method for defining and using combinational UDPs, several examples will be presented ranging from simple designs to designs with increasing complexity.

UDPs are not compiled separately. They are saved in the same project as the module with a .v extension; for example, *udp_and.v*.

Example 4.11 A 2-input OR gate *udp_or2* will be designed using a UDP. The module is shown in Figure 4.55. The inputs in the state table must be in the same order as in the input list. The table heading is a comment for readability. The inputs and output are separated by a colon and the table entry is terminated by a semicolon. All combinations of the inputs must be entered in the table in order to obtain a correct output; otherwise, the output will be designated as **x** (unknown). To completely specify all combinations of the inputs, a value of **x** should be included in the input values where appropriate.

```
//used-defined primitive for a 2-input OR gate
primitive udp_or2 (z1, x1, x2);//list output first

//input/output declarations
input x1, x2;
output z1;          //must be output (not reg)
                    //...for combinational logic

//state table definition
table
//inputs are in same order as input list
// x1 x2 :  z1;    comment is for readability
   0  0  :  0;
   0  1  :  1;
   1  0  :  1;
   1  1  :  1;
   x  1  :  1;
   1  x  :  1;
endtable
endprimitive
```

Figure 4.55   A user-defined primitive for a 2-input OR gate.

Example 4.12 A 3-input AND gate *udp_and3* will be designed using a UDP. The module is shown in Figure 4.56, which is incompletely specified. For example, if $x_1 x_2 x_3 = \mathbf{x}00$, output z_1 should equal 0. However, z_1 will equal **x** because the corresponding entry was not listed in the state table. Therefore, all combinations of the inputs — including when $x_i = \mathbf{x}$ — should be entered in the state table, together with the appropriate output value. This is true for both combinational and sequential UDPs. However, in order to avoid lengthy state tables for UDP logic gates, this section will provide only the logical functions when all inputs are at a known value. The value **z** (high impedance) is not allowed in a UDP — a **z** value is treated as an **x**.

```
//3-input AND gate as a udp
primitive udp_and3 (z1, x1, x2, x3);//output is listed first

input x1, x2, x3;
output z1;

//state table
table
//inputs are in the same order as the input list
// x1 x2 x3 :  z1;  comment is for readability
   0  0  0  :  0;
   0  0  1  :  0;
   0  1  0  :  0;
   0  1  1  :  0;
   1  0  0  :  0;
   1  0  1  :  0;
   1  1  0  :  0;
   1  1  1  :  1;
endtable
endprimitive
```

Figure 4.56   A user-defined primitive for a 3-input AND gate.

Example 4.13 In this example, a sum-of-products expression will be modeled with UDPs. The equation for the output z_1 is shown in Equation 4.11 and the logic diagram is shown in Figure 4.57.

$$z_1 = x_1 x_2 + x_3 x_4 + x_2' x_3'$$ (4.11)

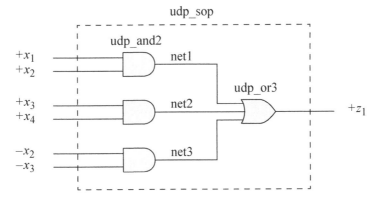

Figure 4.57   A sum-of-products circuit to implemented with a user-defined primitive.

The Karnaugh map for the equation and logic diagram is shown in Figure 4.58, which can be used to verify the outputs obtained from the test bench. UDPs will first be designed for a 2-input AND gate and a 3-input OR gate. These UDPs will then be saved in the project folder *udp_sop* as *udp_and2.v* and *udp_or3.v*. Figure 4.59 shows the Project Properties screen that lists all the source files associated with the *udp_sop* project. The UDPs will then be instantiated into the module *udp_sop*. The Verilog code for the *udp_and2* module is shown in Figure 4.60. The Verilog code for the *udp_or3* module is shown in Figure 4.61. The design module for *udp_sop*, the test bench, the outputs, and the waveforms are shown in Figure 4.62, Figure 4.63, Figure 4.64, and Figure 4.65, respectively.

$x_1 x_2$ \ $x_3 x_4$	0 0	0 1	1 1	1 0
0 0	1	1	1	0
0 1	0	0	1	0
1 1	1	1	1	1
1 0	1	1	1	0

z_1

Figure 4.58   Karnaugh map for the sum-of-products implementation using user-defined primitives.

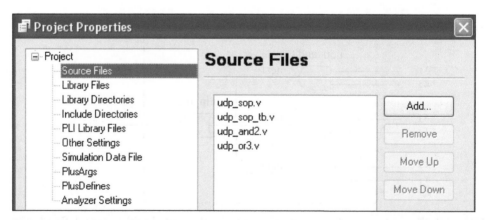

Figure 4.59   The Project Properties screen showing all the source files that are used in the project *udp_sop*.

```
//UDP for a 2-input AND gate
primitive udp_and2 (z1, x1, x2);    //output is listed first

input x1, x2;
output z1;

//define state table
table
//inputs are the same order as the input list
// x1 x2 :  z1;    comment is for readability
   0  0  :  0;
   0  1  :  0;
   1  0  :  0;
   1  1  :  1;
endtable

endprimitive
```

Figure 4.60   User-defined primitive for a 2-input AND gate.

```
//UDP for a 3-input OR gate
primitive udp_or3 (z1, x1, x2, x3);  //output is listed first

input x1, x2, x3;
output z1;

//define state table
table
//inputs are the same order as the input list
// x1 x2 x3 :  z1;    comment is for readability
   0  0  0  :  0;
   0  0  1  :  1;
   0  1  0  :  1;
   0  1  1  :  1;
   1  0  0  :  1;
   1  0  1  :  1;
   1  1  0  :  1;
   1  1  1  :  1;
endtable

endprimitive
```

Figure 4.61   User-defined primitive for a 3-input OR gate.

```
//sum of products using udps for the AND gate and OR gate
module udp_sop (x1, x2, x3, x4, z1);


input x1, x2, x3, x4;
output z1;


//define internal nets
wire net1, net2, net3;


//instantiate the udps
udp_and2 inst1 (net1, x1, x2);
udp_and2 inst2 (net2, x3, x4);
udp_and2 inst3 (net3, ~x2, ~x3);

udp_or3  inst4 (z1, net1, net2, net3);


endmodule
```

Figure 4.62   Module for the sum-of-products circuit of Figure 4.57 using user-defined primitives.

```
//test bench for sum of products using udps
module udp_sop_tb;


reg x1, x2, x3, x4;
wire z1;


//apply input vectors and display variables
initial
begin: apply_stimulus
   reg [4:0] invect;
   for (invect=0; invect<16; invect=invect+1)
      begin
         {x1, x2, x3, x4} = invect [4:0];
         #10 $display ("x1x2x3x4 = %b, z1 = %b",
                       {x1, x2, x3, x4}, z1);
      end
end

//continued on next page
```

Figure 4.63   Test bench for the sum-of-products module of Figure 4.62.

```
//instantiate the module into the test bench
udp_sop inst1 (
   .x1(x1),
   .x2(x2),
   .x3(x3),
   .x4(x4),
   .z1(z1)
   );

endmodule
```

Figure 4.63   (Continued)

```
x1x2x3x4 = 0000, z1 = 1          x1x2x3x4 = 1000, z1 = 1
x1x2x3x4 = 0001, z1 = 1          x1x2x3x4 = 1001, z1 = 1
x1x2x3x4 = 0010, z1 = 0          x1x2x3x4 = 1010, z1 = 0
x1x2x3x4 = 0011, z1 = 1          x1x2x3x4 = 1011, z1 = 1
x1x2x3x4 = 0100, z1 = 0          x1x2x3x4 = 1100, z1 = 1
x1x2x3x4 = 0101, z1 = 0          x1x2x3x4 = 1101, z1 = 1
x1x2x3x4 = 0110, z1 = 0          x1x2x3x4 = 1110, z1 = 1
x1x2x3x4 = 0111, z1 = 1          x1x2x3x4 = 1111, z1 = 1
```

Figure 4.64   Outputs for the test bench of Figure 4.63 for the sum-of-products
module of Figure 4.62.

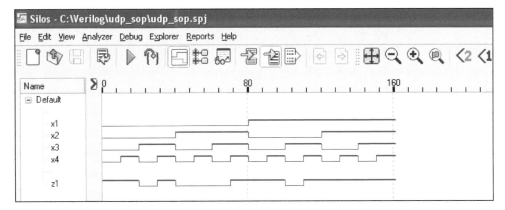

Figure 4.65   Waveforms for the test bench of Figure 4.63 for the sum-of-products
module of Figure 4.62.

Example 4.14   A 5-input majority circuit will be designed using a user-defined primitive. The inputs are x_1, x_2, x_3, x_4, and x_5; the single output is z_1. The design module is shown in Figure 4.66 and the test bench module is shown in Figure 4.67. When instantiating the module into the test bench, the instantiation must be done by position, not by name; that is, in the order listed in the design module — not by the notation .x1(x1). The outputs are shown in Figure 4.68 which lists all combinations of the inputs.

```
//five-input majority circuit as a udp.  Save as a .v file
primitive udp_maj5 (z1, x1, x2, x3, x4, x5);

input x1, x2, x3, x4, x5;
output z1;

table
//inputs are in same order as input list
// x1 x2 x3 x4 x5 :  z1;
   0  0  0  ?  ?  :  0;
   0  0  ?  0  ?  :  0;
   0  0  ?  ?  0  :  0;
   0  ?  0  0  ?  :  0;
   0  ?  ?  0  0  :  0;
   0  ?  0  ?  0  :  0;
   ?  0  0  0  ?  :  0;
   ?  ?  0  0  0  :  0;
   ?  0  0  ?  0  :  0;
   ?  0  ?  0  0  :  0;
   ?  ?  0  0  0  :  0;

   1  1  1  ?  ?  :  1;
   1  1  ?  1  ?  :  1;
   1  1  ?  ?  1  :  1;
   1  ?  1  1  ?  :  1;
   1  ?  ?  1  1  :  1;
   1  ?  1  ?  1  :  1;
   ?  1  1  1  ?  :  1;
   ?  ?  1  1  1  :  1;
   ?  1  1  ?  1  :  1;
   ?  1  ?  1  1  :  1;
   ?  ?  1  1  1  :  1;
endtable

endprimitive
```

Figure 4.66   A user-defined primitive for a 5-input majority circuit.

```
//udp_maj5 test bench
module udp_maj5_tb;

reg x1, x2, x3, x4, x5;      //inputs are reg for test bench
wire z1;                     //outputs are wire for test bench

//Declare a vector that has 1 more bit than the # of inputs.
//This allows the for statement count to go 1 higher than the
//maximum count of the input combinations and prevents
//looping forever.  If only 5 bits were used as the input
//vector, then the count would always be < 32 (the maximum
//count for 5 bits).

initial
begin: name         //a name is required for this method
   reg [5:0] invect;
   for (invect = 0; invect < 32; invect = invect + 1)
      begin
         {x1, x2, x3, x4, x5} = invect [4:0];
         #10 $display ("x1x2x3x4x5 = %b%b%b%b%b, z1=%b",
                          x1, x2, x3, x4, x5, z1);
      end
end

//instantiation must be done by position, not by name.
udp_maj5 inst1 (z1, x1, x2, x3, x4, x5);
endmodule
```

Figure 4.67   Test bench for the 5-input majority circuit of Figure 4.66.

```
x1x2x3x4x5 = 00000, z1=0      x1x2x3x4x5 = 10000, z1=0
x1x2x3x4x5 = 00001, z1=0      x1x2x3x4x5 = 10001, z1=0
x1x2x3x4x5 = 00010, z1=0      x1x2x3x4x5 = 10010, z1=0
x1x2x3x4x5 = 00011, z1=0      x1x2x3x4x5 = 10011, z1=1
x1x2x3x4x5 = 00100, z1=0      x1x2x3x4x5 = 10100, z1=0
x1x2x3x4x5 = 00101, z1=0      x1x2x3x4x5 = 10101, z1=1
x1x2x3x4x5 = 00110, z1=0      x1x2x3x4x5 = 10110, z1=1
x1x2x3x4x5 = 00111, z1=1      x1x2x3x4x5 = 10111, z1=1
x1x2x3x4x5 = 01000, z1=0      x1x2x3x4x5 = 11000, z1=0
x1x2x3x4x5 = 01001, z1=0      x1x2x3x4x5 = 11001, z1=1
x1x2x3x4x5 = 01010, z1=0      x1x2x3x4x5 = 11010, z1=1
x1x2x3x4x5 = 01011, z1=1      x1x2x3x4x5 = 11011, z1=1
x1x2x3x4x5 = 01100, z1=0      x1x2x3x4x5 = 11100, z1=1
x1x2x3x4x5 = 01101, z1=1      x1x2x3x4x5 = 11101, z1=1
x1x2x3x4x5 = 01110, z1=1      x1x2x3x4x5 = 11110, z1=1
x1x2x3x4x5 = 01111, z1=1      x1x2x3x4x5 = 11111, z1=1
```

Figure 4.68   Outputs for the 5-input majority circuit of Figure 4.66.

Example 4.15 The Karnaugh map of Figure 4.69 will be implemented using a 4:1 multiplexer and additional logic. First, the equations for the multiplexer data inputs will be obtained using E as a map-entered variable, where the multiplexer select inputs are $s_1 s_0 = x_1 x_2$. Then the circuit will be designed using UDPs for the multiplexer and associated logic gates.

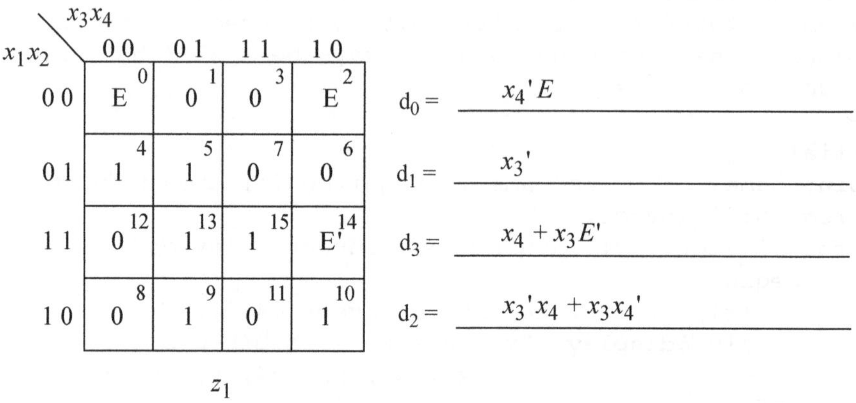

Figure 4.69   Karnaugh map for Example 4.15 using E as a map-entered variable.

To obtain the equation for data input d_0, where $s_1 s_0 = x_1 x_2 = 00$, minterm locations 0 and 2 are adjacent and contain the same variable E; therefore, the term is $x_4' E$. Data input d_1, where $s_1 s_0 = x_1 x_2 = 01$, contains 1s in minterm locations 4 and 5; therefore, $d_1 = x_3'$. To obtain the equation for d_3, where $s_1 s_0 = x_1 x_2 = 11$, minterm locations 13 and 15 combine to yield x_4. Minterm location 15 is equivalent to $1 + E'$; therefore, minterm locations 14 and 15 combine to yield the product term $x_3 E'$. The equation for d_3 is $x_4 + x_3 E'$. Data input d_2 is obtained in a similar manner.

The logic diagram is shown in Figure 4.70 using a 4:1 multiplexer (*udp_mux4*), a 2-input AND gate (*udp_and2*), a 2-input exclusive-OR function (*udp_xor2*), and a 2-input OR gate (*udp_or2*). The 2-input OR gate and the 2-input AND gate have been previously designed. The remaining UDPs will be designed prior to designing the logic of Figure 4.70. The module for *udp_mux4* is shown in Figure 4.71, where the (?) symbol indicates a "don't care" condition. The module for *udp_xor2* is shown in Figure 4.72.

The module for the logic diagram is shown in Figure 4.73 and the test bench is shown in Figure 4.74. The Karnaugh map of Figure 4.69 is expanded to the 5-variable map of Figure 4.75 to better visualize the minterm entries when comparing them with the outputs of Figure 4.76.

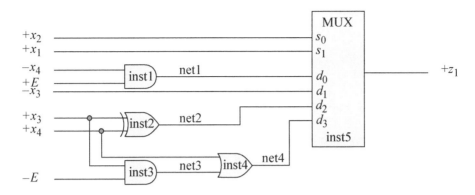

Figure 4.70   Logic diagram for the Karnaugh map of Figure 4.69.

```
//4:1 multiplexer as a user-defined primitive
primitive udp_mux4 (out, s1, s0, d0, d1, d2, d3);


input s1, s0, d0, d1, d2, d3;
output out;


//define state table
table
//inputs are in the same order as the input list
// s1 s0 d0 d1 d2 d3 :   out        comment is for readability
   0  0  1  ?  ?  ?  :   1;         //? is "don't care"
   0  0  0  ?  ?  ?  :   0;

   0  1  ?  1  ?  ?  :   1;
   0  1  ?  0  ?  ?  :   0;

   1  0  ?  ?  1  ?  :   1;
   1  0  ?  ?  0  ?  :   0;

   1  1  ?  ?  ?  1  :   1;
   1  1  ?  ?  ?  0  :   0;
endtable

endprimitive
```

Figure 4.71   A 4:1 multiplexer user-defined primitive module for *udp_mux4*.

```
//user-defined primitive for a 2-input exclusive-OR
primitive udp_xor2 (z1, x1, x2);


input x1, x2;
output z1;


//define state table
table
//inputs are in the same order as the input list
// x1 x2 :  z1;       comment is for readability
   0  0  :  0;
   0  1  :  1;
   1  0  :  1;
   1  1  :  0;
endtable

endprimitive
```

Figure 4.72   User-defined primitive module for a 2-input exclusive-OR function *udp_xor2*.

```
//logic circuit using a multiplexer udp
//together with other logic gate udps
module mux4_mev (x1, x2, x3, x4, E, z1);


input x1, x2, x3, x4, E;
output z1;


//instantiate the udps
udp_and2 inst1 (net1, ~x4, E);
udp_xor2 inst2 (net2, x3, x4);
udp_and2 inst3 (net3, x3, ~E);
udp_or2  inst4 (net4, x4, net3);


//the mux inputs are: s1, s0, d0, d1, d2, d3
udp_mux4 inst5 (z1, x1, x2, net1, ~x3, net2, net4);

endmodule
```

Figure 4.73   Module for the logic diagram of Figure 4.70.

```
//test bench for mux4_mev
module mux4_mev_tb;

reg x1, x2, x3, x4, E;
wire z1;

//apply input vectors
initial
begin: apply_stimulus
   reg [5:0] invect;
   for (invect=0; invect<32; invect=invect+1)
      begin
         {x1, x2, x3, x4, E} = invect [5:0];
         #10 $display ("x1x2x3x4E = %b, z1 = %b",
                        {x1, x2, x3, x4, E}, z1);
      end
end

//instantiate the module into the test bench
mux4_mev inst1 (
   .x1(x1),
   .x2(x2),
   .x3(x3),
   .x4(x4),
   .E(E),
   .z1(z1)
   );
endmodule
```

Figure 4.74   Test bench for Figure 4.73 for the module *mux4_mev.*

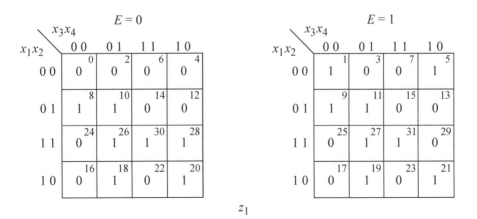

z_1

Figure 4.75   Karnaugh map equivalent to the Karnaugh map of Figure 4.69.

```
x1x2x3x4E = 00000,  z1 = 0
x1x2x3x4E = 00001,  z1 = 1
x1x2x3x4E = 00010,  z1 = 0
x1x2x3x4E = 00011,  z1 = 0
x1x2x3x4E = 00100,  z1 = 0
x1x2x3x4E = 00101,  z1 = 1
x1x2x3x4E = 00110,  z1 = 0
x1x2x3x4E = 00111,  z1 = 0
x1x2x3x4E = 01000,  z1 = 1
x1x2x3x4E = 01001,  z1 = 1
x1x2x3x4E = 01010,  z1 = 1
x1x2x3x4E = 01011,  z1 = 1
x1x2x3x4E = 01100,  z1 = 0
x1x2x3x4E = 01101,  z1 = 0
x1x2x3x4E = 01110,  z1 = 0
x1x2x3x4E = 01111,  z1 = 0
x1x2x3x4E = 10000,  z1 = 0
x1x2x3x4E = 10001,  z1 = 0
x1x2x3x4E = 10010,  z1 = 1
x1x2x3x4E = 10011,  z1 = 1
x1x2x3x4E = 10100,  z1 = 1
x1x2x3x4E = 10101,  z1 = 1
x1x2x3x4E = 10110,  z1 = 0
x1x2x3x4E = 10111,  z1 = 0
x1x2x3x4E = 11000,  z1 = 0
x1x2x3x4E = 11001,  z1 = 0
x1x2x3x4E = 11010,  z1 = 1
x1x2x3x4E = 11011,  z1 = 1
x1x2x3x4E = 11100,  z1 = 1
x1x2x3x4E = 11101,  z1 = 0
x1x2x3x4E = 11110,  z1 = 1
x1x2x3x4E = 11111,  z1 = 1
```

Figure 4.76   Outputs obtained from the test bench of Figure 4.74.

Example 4.16   A binary-coded decimal (BCD)-to-excess-3 code converter will be designed by instantiating four full adders — that were designed using user-defined primitives — into a module labeled *bcd_to_excess3_usg_udp*. The excess-3 code is obtained from the binary 8421 BCD code by adding 3 (0011) to each binary code word. The BCD code word is specified by a 4-bit vector $a[3:0]$, where $a[0]$ is the low-order bit. The excess-3 code word is obtained by adding 0011 to the BCD code $a[3:0]$ and is generated by the vector sum of the adders $sum[3:0]$, where $sum[0]$ is the low-order bit.

The first step is to design a full adder using the user-defined primitives of *udp_xor2*, *udp_and2*, and *udp_or2*, all of which have been previously designed. The logic diagram for a full adder, using two half adders, is shown in Figure 4.77 utilizing the user-defined primitives. The adder is designed and saved as *full_adder_usg_udp2*. The design module for the full adder is shown in Figure 4.78, the test bench is shown in Figure 4.79, and the outputs are shown in Figure 4.80.

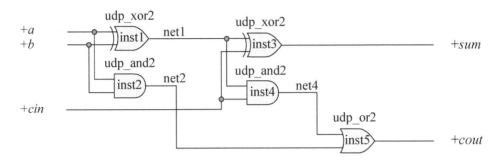

Figure 4.77   A full adder to be used in implementing a BCD-to-excess-3 code converter.

```
//full adder designed with user-defined primitives
module full_adder_usg_udp2 (a, b, cin, sum, cout);

input a, b, cin;
output sum, cout;

//define internal nets
wire net1, net2, net4;

//instantiate the udps
udp_xor2 inst1 (net1, a, b);
udp_and2 inst2 (net2, a, b);
udp_xor2 inst3 (sum, net1, cin);
udp_and2 inst4 (net4, net1, cin);
udp_or2  inst5 (cout, net4, net2);

endmodule
```

Figure 4.78   Module for the full adder that is designed using user-defined primitives.

```
//test bench for the full adder using user-defined primitives
module full_adder_usg_udp2_tb;


reg a, b, cin;
wire sum, cout;


//apply input vectors
initial
begin: apply_stimulus
   reg [3:0] invect;
   for (invect=0; invect<8; invect=invect+1)
      begin
         {a, b, cin} = invect [3:0];
         #10 $display ("a b cin = %b, cout sum = %b",
                       {a, b, cin}, {cout, sum});
      end
end


//instantiate the module into the test bench
full_adder_usg_udp2 inst1 (
   .a(a),
   .b(b),
   .cin(cin),
   .sum(sum),
   .cout(cout)
   );

endmodule
```

Figure 4.79   Test bench for the full adder of Figure 4.78 that is designed using user-defined primitives.

```
a b cin = 000, cout sum = 00
a b cin = 001, cout sum = 01
a b cin = 010, cout sum = 01
a b cin = 011, cout sum = 10
a b cin = 100, cout sum = 01
a b cin = 101, cout sum = 10
a b cin = 110, cout sum = 10
a b cin = 111, cout sum = 11
```

Figure 4.80   Outputs for the full adder of Figure 4.78.

Next, the full adder of Figure 4.77, which is implemented by the module of Figure 4.78, will be utilized in the design of the BCD-to-excess-3 code converter. The logic diagram for the code converter is shown in Figure 4.81. The design module is shown in Figure 4.82 and instantiates the full adder module of Figure 4.78 four times. The test bench is shown in Figure 4.83 and the outputs are shown in Figure 4.84.

Consider the logic diagram of Figure 4.81 and the design module of Figure 4.82 for the discussion which follows. In instantiation *inst0* in both figures, port *a* of the full adder corresponds to *a[0]* in the logic diagram (.*a(a[0])*). Port *b* of the full adder corresponds to a logic 1 in the logic diagram (.*b(1'b1)*). The carry-in *cin* of the full adder connects to a logic 0 in the logic diagram (.*cin(1'b0)*). The sum of the full adder corresponds to *sum[0]* in the logic diagram (.*sum(sum[0])*). The carry-out *cout* of the full adder corresponds to the internal net *cout0* in the logic diagram (.*cout(cout0)*). The other instantiations are similarly defined. The test bench module is shown in Figure 4.83 and the outputs are shown in Figure 4.84.

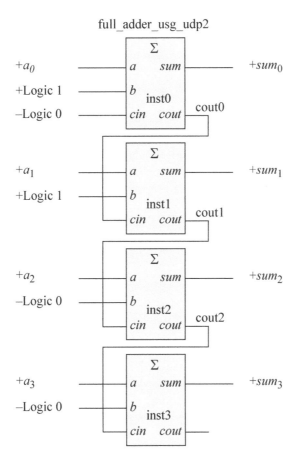

Figure 4.81   BCD-to-excess-3 code converter using the full adder of Figure 4.78.

```
//bcd-to-excess-3 conversion using built-in primitives
module bcd_to_excess3_usg_udp (a, sum);


input [3:0] a;
output [3:0] sum;


//define internal nets
wire cout0, cout1, cout2;


//instantiate the full adder using udps
full_adder_usg_udp2 inst0 (
   .a(a[0]),
   .b(1'b1),
   .cin(1'b0),
   .sum(sum[0]),
   .cout(cout0)
   );


full_adder_usg_udp2 inst1 (
   .a(a[1]),
   .b(1'b1),
   .cin(cout0),
   .sum(sum[1]),
   .cout(cout1)
   );


full_adder_usg_udp2 inst2 (
   .a(a[2]),
   .b(1'b0),
   .cin(cout1),
   .sum(sum[2]),
   .cout(cout2)
   );


full_adder_usg_udp2 inst3 (
   .a(a[3]),
   .b(1'b0),
   .cin(cout2),
   .sum(sum[3]),
   .cout(1'b0)
   );


endmodule
```

Figure 4.82   Module for the BCD-to-excess-3 code converter using four full adders.

```
//test bench for bcd-to-excess-3 using user-defined primitives
module bcd_to_excess3_usg_udp_tb;

reg [3:0] a;
wire [3:0] sum;

//apply input vectors
initial
begin: apply_stimulus
   reg[4:0] invect;
   for (invect=0; invect<10; invect=invect+1)
      begin
         a = invect[4:0];
         #10 $display ("a = %b, sum = %b", a, sum);
      end
end

//instantiate the module into the test bench
bcd_to_excess3_usg_udp inst1 (
   .a(a),
   .sum(sum)
   );

endmodule
```

Figure 4.83   Test bench for the BCD-to-excess-3 code converter of Figure 4.82.

```
a = 0000, sum = 0011
a = 0001, sum = 0100
a = 0010, sum = 0101
a = 0011, sum = 0110
a = 0100, sum = 0111
a = 0101, sum = 1000
a = 0110, sum = 1001
a = 0111, sum = 1010
a = 1000, sum = 1011
a = 1001, sum = 1100
```

Figure 4.84   Outputs for the BCD-to-excess-3 code converter of Figure 4.82.

Example 4.17 As a final example in user-defined primitives, a minimal Boolean expression will be obtained for a logic circuit that generates an output z_1 whenever a 4-bit unsigned binary number N meets the following requirements:

N is an even number or N is evenly divisible by three.

The format for N is: $N = n_3\, n_2\, n_1\, n_0$, where n_0 is the low-order bit. The Karnaugh map that represents output z_1 is shown in Figure 4.85 and the equation is shown in Equation 4.12. The logic diagram is shown in Figure 4.86. The design module uses the user-defined primitives *udp_and3*, which has been previously designed, and *udp_or4*, which is shown in Figure 4.87. The design module, test bench module, and outputs are shown in Figure 4.88, Figure 4.89, and Figure 4.90, respectively.

	$n_1 n_0$			
$n_3 n_2$	0 0	0 1	1 1	1 0
0 0	1 (0)	0 (1)	1 (3)	1 (2)
0 1	1 (4)	0 (5)	0 (7)	1 (6)
1 1	1 (12)	0 (13)	1 (15)	1 (14)
1 0	1 (8)	1 (9)	0 (11)	1 (10)

z_1

Figure 4.85   Karnaugh map for Example 4.17.

$$z_1 = n_0' + n_3' n_2' n_1 + n_3 n_2 n_1 + n_3 n_2' n_1' \qquad (4.12)$$

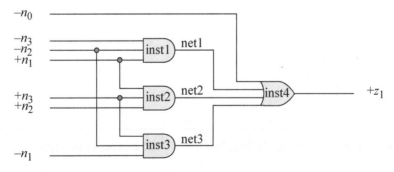

Figure 4.86   Logic diagram for Example 4.17.

```
//a 4-input OR gate as a user-defined primitive
primitive udp_or4 (z1, x1, x2, x3, x4);

input x1, x2, x3, x4;
output z1;

//define state table
table
//inputs are in the same order as the input list
// x1 x2 x3 x4 :  z1;    comment is for readability
    0  0  0  0  :  0;
    0  0  0  1  :  1;
    0  0  1  0  :  1;
    0  0  1  1  :  1;
    0  1  0  0  :  1;
    0  1  0  1  :  1;
    0  1  1  0  :  1;
    0  1  1  1  :  1;
    1  0  0  0  :  1;
    1  0  0  1  :  1;
    1  0  1  0  :  1;
    1  0  1  1  :  1;
    1  1  0  0  :  1;
    1  1  0  1  :  1;
    1  1  1  0  :  1;
    1  1  1  1  :  1;
endtable
endprimitive
```

Figure 4.87   User-defined primitive *udp_or4*.

```
//logic circuit to detect whether a number is
//even or evenly divisible by three
module even_num_udp (n3, n2, n1, n0, z1);

input n3, n2, n1, n0;
output z1;

//instantiate the udps
udp_and3 inst1 (net1, ~n3, ~n2, n1);
udp_and3 inst2 (net2, n3, n2, n1);
udp_and3 inst3 (net3, n3, ~n2, ~n1);
udp_or4  inst4 (z1, net1, net2, net3, ~n0);
endmodule
```

Figure 4.88   Module for Figure 4.86 of Example 4.17.

```
//test bench for even_num_udp
module even_num_udp_tb;

reg n3, n2, n1, n0;
wire z1;

//apply input vectors
initial
begin: apply_stimulus
    reg [4:0] invect;
    for (invect=0; invect<16; invect=invect+1)
        begin
            {n3, n2, n1, n0} = invect [4:0];
            #10 $display ("n3 n2 n1 n0 = %b, z1 = %b",
                          {n3, n2, n1, n0}, z1);
        end
end

//instantiate the module into the test bench
even_num_udp inst1 (
    .n3(n3),
    .n2(n2),
    .n1(n1),
    .n0(n0),
    .z1(z1)
    );

endmodule
```

Figure 4.89   Test bench for the module of Figure 4.88.

```
n3 n2 n1 n0 = 0000, z1 = 1      n3 n2 n1 n0 = 1000, z1 = 1
n3 n2 n1 n0 = 0001, z1 = 0      n3 n2 n1 n0 = 1001, z1 = 1
n3 n2 n1 n0 = 0010, z1 = 1      n3 n2 n1 n0 = 1010, z1 = 1
n3 n2 n1 n0 = 0011, z1 = 1      n3 n2 n1 n0 = 1011, z1 = 0
n3 n2 n1 n0 = 0100, z1 = 1      n3 n2 n1 n0 = 1100, z1 = 1
n3 n2 n1 n0 = 0101, z1 = 0      n3 n2 n1 n0 = 1101, z1 = 0
n3 n2 n1 n0 = 0110, z1 = 1      n3 n2 n1 n0 = 1110, z1 = 1
n3 n2 n1 n0 = 0111, z1 = 0      n3 n2 n1 n0 = 1111, z1 = 1
```

Figure 4.90   Outputs obtained from the test bench of Figure 4.89.

4.3 Dataflow Modeling

Gate-level modeling using built-in primitives is an intuitive approach to digital design because it corresponds one-to-one with traditional digital logic design at the gate level. Dataflow modeling, however, is at a higher level of abstraction than gate-level modeling. Design automation tools are used to create gate-level logic from dataflow modeling by a process called *logic synthesis*. Register transfer level (RTL) is a combination of dataflow modeling and behavioral modeling and characterizes the flow of data through logic circuits.

4.3.1 Continuous Assignment

The *continuous assignment* statement models dataflow behavior and is used to design combinational logic without using gates and interconnecting nets. Continuous assignment statements provide a Boolean correspondence between the right-hand side expression and the left-hand side target. The continuous assignment statement uses the keyword **assign** and has the following syntax with optional drive strength and delay:

assign [drive_strength] [delay] left-hand side target = right-hand side expression

The continuous assignment statement assigns a value to a net (**wire**) that has been previously declared — it cannot be used to assign a value to a register. Therefore, the left-hand target must be a scalar or vector net or a concatenation of scalar and vector nets. The operands on the right-hand side can be registers, nets, or function calls. The registers and nets can be declared as either scalars or vectors.

The following are examples of continuous assignment statements for scalar nets:

$$\textbf{assign } z_1 = x_1 \text{ \& } x_2 \text{ \& } x_3;$$
$$\textbf{assign } z_1 = x_1 \text{ } ^\wedge x_2;$$
$$\textbf{assign } z_1 = (x_1 \text{ \& } x_2) \mid x_3;$$

where the symbol "&" is the AND operation, the symbol "^" is the exclusive-OR operation, and the symbol "|" is the OR operation.

The following are examples of continuous assignment statements for vector and scalar nets, where *sum* is a 9-bit vector to accommodate the *sum* and carry-out, *a* and *b* are 8-bit vectors, and *cin* is a scalar:

$$\textbf{assign } sum = a + b + cin$$
$$\textbf{assign } sum = a \text{ } ^\wedge b \text{ } ^\wedge cin$$

where the symbol "+" is the add operation.

The following is an example of a continuous assignment statement for vector nets and a concatenation of a scalar net and a vector net, where *a* and *b* are 4-bit vectors, and *cin* and *cout* are scalars:

$$\textbf{assign } \{cout, sum\} = a + b + cin;$$

The **assign** statement continuously monitors the right-hand side expression. If a variable changes value, then the expression is evaluated and the result is assigned to the target after any specified delay. If no delay is specified, then the default delay is zero. The drive strength defaults to **strong0** and **strong1**. The continuous assignment statement can be considered to be a form of behavioral modeling because the behavior of the circuit is specified, not the implementation.

Example 4.18   Before designing a 4-input AND gate, the AND function will be further delineated in terms of 2-inputs. The AND function of two variables x_1 and x_2 is also called the *conjunction* of x_1 and x_2 and is stated as x_1 *and* x_2. In general, the AND operator, which corresponds to the Boolean product, is indicated by the symbol "•"($x_1 \bullet x_2$), "∧" ($x_1 \wedge x_2$), or by no symbol $x_1 x_2$ if the operation is unambiguous. Thus, $x_1 x_2, x_1 \bullet x_2$, and $x_1 \wedge x_2$ are all read as "x_1 AND x_2." In Verilog, however, the caret symbol "∧" indicates the exclusive-OR operation and the "&" symbol indicates the AND operation, as previously indicated.

The truth table for a 4-input AND gate is shown in Table 4.3 for inputs x_1, x_2, x_3, x_4 and output z_1. The design module for the 4-input AND gate using the continuous **assign** statement for dataflow modeling is shown in Figure 4.91 and the test bench module is shown in Figure 4.92. The outputs and waveforms are shown in Figure 4.93 and Figure 4.94, respectively.

Table 4.3 Truth Table for a 4-Input AND Gate

x_1	x_2	x_3	x_4	z_1
0	0	0	0	0
0	0	0	1	0
0	0	1	0	0
0	0	1	1	0
0	1	0	0	0
0	1	0	1	0
0	1	1	0	0
0	1	1	1	0
1	0	0	0	0
1	0	0	1	0
1	0	1	0	0
1	0	1	1	0
1	1	0	0	0
1	1	0	1	0
1	1	1	0	0
1	1	1	1	1

```
//dataflow 4-input and gate
module and4_df (x1, x2, x3, x4, z1);

//list all inputs and outputs
input x1, x2, x3, x4;
output z1;

//define signals as wire
wire x1, x2, x3, x4;
wire z1;

//continuous assign used for dataflow
assign z1 = (x1 & x2 & x3 & x4);
endmodule
```

Figure 4.91   Dataflow module for a 4-input AND gate using the continuous **assign** statement.

```
//4-input and gate test bench
module and4_df_tb;

reg x1, x2, x3, x4;
wire z1;

//apply input vectors
initial
begin: apply_stimulus
   reg [4:0] invect;
   for (invect = 0; invect < 16; invect = invect + 1)
      begin
         {x1, x2, x3, x4} = invect [4:0];
         #10 $display ("{x1x2x3x4} = %b, z1 = %b",
                       {x1, x2, x3, x4}, z1);
      end
end

//instantiate the module into the test bench
and4_df inst1 (
   .x1(x1),
   .x2(x2),
   .x3(x3),
   .x4(x4),
   .z1(z1)
   );
endmodule
```

Figure 4.92   Test bench for the 4-input AND gate of Figure 4.91.

```
{x1x2x3x4} = 0000, z1 = 0        {x1x2x3x4} = 1000, z1 = 0
{x1x2x3x4} = 0001, z1 = 0        {x1x2x3x4} = 1001, z1 = 0
{x1x2x3x4} = 0010, z1 = 0        {x1x2x3x4} = 1010, z1 = 0
{x1x2x3x4} = 0011, z1 = 0        {x1x2x3x4} = 1011, z1 = 0
{x1x2x3x4} = 0100, z1 = 0        {x1x2x3x4} = 1100, z1 = 0
{x1x2x3x4} = 0101, z1 = 0        {x1x2x3x4} = 1101, z1 = 0
{x1x2x3x4} = 0110, z1 = 0        {x1x2x3x4} = 1110, z1 = 0
{x1x2x3x4} = 0111, z1 = 0        {x1x2x3x4} = 1111, z1 = 1
```

Figure 4.93   Outputs for the 4-input AND gate of Figure 4.91.

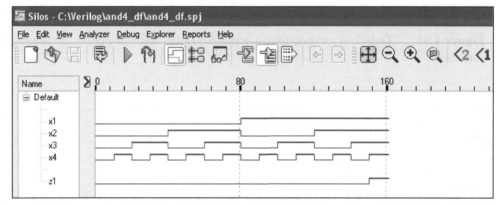

Figure 4.94   Waveforms for the 4-input AND gate of Figure 4.91.

Example 4.19   The logic diagram shown in Figure 4.95 will be implemented using dataflow modeling. The Karnaugh map that represents the logic is shown in Figure 4.96. The equation for output z_1 is shown in Equation 4.13. The design module is shown in Figure 4.97 and the test bench module is shown in Figure 4.98. The outputs and waveforms are shown in Figure 4.99 and Figure 4.100, respectively.

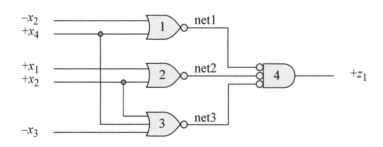

Figure 4.95   Logic diagram for Example 4.19.

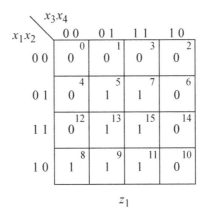

Figure 4.96 Karnaugh map for Example 4.19.

$$z_1 = (x_2' + x_4)(x_1 + x_2)(x_2 + x_3' + x_4) \qquad (4.13)$$

```
//dataflow for product of sums using nor logic
module log_diag_pos_nor (x1, x2, x3, x4, z1);

//define inputs and output
input x1, x2, x3, x4;
output z1;

//define inputs and output as wire
wire x1, x2, x3, x4;
wire z1;

//define internal nets
wire net1, net2, net3;

//define z1 using continuous assignment
assign   net1 = ~(~x2 | x4),
         net2 = ~(x1 | x2),
         net3 = ~(x2 | ~x3 | x4);

assign z1 = (~net1 & ~net2 & ~net3);

endmodule
```

Figure 4.97 Module for the logic diagram of Figure 4.95.

Refer to Figure 4.95 and Figure 4.97 for the discussion which follows. Since gate 1 is a NOR gate, the output must be inverted to realize the expression $x_2' + x_4$. Thus, net 1 is equal to

$$net1 = \sim(\sim x2 \mid x4)$$

where the tilde symbol indicates inversion. Gates 2 and 3 are similarly defined. Gate 4 is a NOR gate drawn as an AND function with active-low inputs. Therefore, each input must be inverted, providing the following expression for z_1:

$$z1 = (\sim net1 \, \& \, \sim net2 \, \& \, \sim net3)$$

```verilog
//test bench for product-of-sums logic diagram
module log_diag_pos_nor_tb;


reg x1, x2, x3, x4;
wire z1;


//apply input vectors and display variables
initial
begin: apply_stimulus
    reg [4:0] invect;
    for (invect = 0; invect < 16; invect = invect + 1)
        begin
            {x1, x2, x3, x4} = invect [4:0];
            #10 $display ("x1 x2 x3 x4 = %b, z1 = %b",
                        {x1, x2, x3, x4}, z1);
        end
end


//instantiate the module into the test bench
log_diag_pos_nor inst1 (
    .x1(x1),
    .x2(x2),
    .x3(x3),
    .x4(x4),
    .z1(z1)
    );


endmodule
```

Figure 4.98   Test bench for the logic diagram of Figure 4.95.

```
x1 x2 x3 x4 = 0000,  z1 = 0        x1 x2 x3 x4 = 1000,  z1 = 1
x1 x2 x3 x4 = 0001,  z1 = 0        x1 x2 x3 x4 = 1001,  z1 = 1
x1 x2 x3 x4 = 0010,  z1 = 0        x1 x2 x3 x4 = 1010,  z1 = 0
x1 x2 x3 x4 = 0011,  z1 = 0        x1 x2 x3 x4 = 1011,  z1 = 1
x1 x2 x3 x4 = 0100,  z1 = 0        x1 x2 x3 x4 = 1100,  z1 = 0
x1 x2 x3 x4 = 0101,  z1 = 1        x1 x2 x3 x4 = 1101,  z1 = 1
x1 x2 x3 x4 = 0110,  z1 = 0        x1 x2 x3 x4 = 1110,  z1 = 0
x1 x2 x3 x4 = 0111,  z1 = 1        x1 x2 x3 x4 = 1111,  z1 = 1
```

Figure 4.99   Outputs for the logic diagram of Figure 4.95. Compare with the Karnaugh map of Figure 4.96.

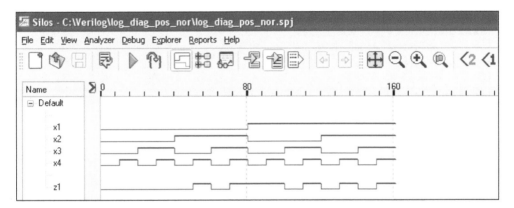

Figure 4.100   Waveforms for the logic diagram of Figure 4.95.

Example 4.20   This example will repeat Example 4.19 using only NAND gates. The inputs are inverted because a NAND gate drawn as an OR function has active-low inputs. The product-of-sums logic will generate a low output for z_1; therefore, the output must be inverted through a NAND gate. The Karnaugh map is the same as shown in Figure 4.96 and the equation for z_1 is the same as Equation 4.13.

The logic diagram is shown in Figure 4.101. The design module and test bench module are shown in Figure 4.102 and Figure 4.103, respectively. The outputs are shown in Figure 4.104 and are identical to the outputs shown in Figure 4.99.

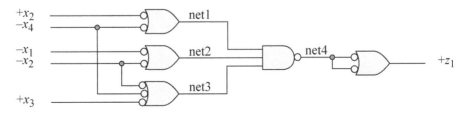

Figure 4.101   Logic diagram for Example 4.20.

```
//dataflow for product of sums using nand logic
module log_diag_pos_nand (x1, x2, x3, x4, z1);

//define inputs and outputs
input x1, x2, x3, x4;
output z1;

//define inputs and output as wire
wire x1, x2, x3, x4;
wire z1;

//define internal nets
wire net1, net2, net3, net4;

//define z1 using continuous assignment
assign    net1 = ~x2 | x4,
          net2 = x1 | x2,
          net3 = x2 | ~x3 | x4;

assign    net4 = ~(net1 & net2 & net3);
assign    z1 = ~net4 | ~net4;

endmodule
```

Figure 4.102   Module for the logic diagram of Figure 4.101.

```
//test bench for product-of-sums logic diagram
module log_diag_pos_nand_tb;

reg x1, x2, x3, x4;
wire z1;

//apply input vectors and display variables
initial
begin: apply_stimulus
   reg [4:0] invect;
   for (invect = 0; invect < 16; invect = invect + 1)
      begin
         {x1, x2, x3, x4} = invect [4:0];
         #10 $display ("x1 x2 x3 x4 = %b, z1 = %b",
                       {x1, x2, x3, x4}, z1);
      end
end
//continued on next page
```

Figure 4.103   Test bench for the logic diagram of Figure 4.101.

```
//instantiate the module into the test bench
log_diag_pos_nand inst1 (
    .x1(x1),
    .x2(x2),
    .x3(x3),
    .x4(x4),
    .z1(z1)
    );
endmodule
```

Figure 4.103   (Continued)

```
x1 x2 x3 x4 = 0000, z1 = 0        x1 x2 x3 x4 = 1000, z1 = 1
x1 x2 x3 x4 = 0001, z1 = 0        x1 x2 x3 x4 = 1001, z1 = 1
x1 x2 x3 x4 = 0010, z1 = 0        x1 x2 x3 x4 = 1010, z1 = 0
x1 x2 x3 x4 = 0011, z1 = 0        x1 x2 x3 x4 = 1011, z1 = 1
x1 x2 x3 x4 = 0100, z1 = 0        x1 x2 x3 x4 = 1100, z1 = 0
x1 x2 x3 x4 = 0101, z1 = 1        x1 x2 x3 x4 = 1101, z1 = 1
x1 x2 x3 x4 = 0110, z1 = 0        x1 x2 x3 x4 = 1110, z1 = 0
x1 x2 x3 x4 = 0111, z1 = 1        x1 x2 x3 x4 = 1111, z1 = 1
```

Figure 4.104   Outputs for the logic diagram of Figure 4.101.

Example 4.21   A comparator will be designed that compares two 2-bit binary operands $x_1 x_2$ and $x_3 x_4$ and generates a high output for z_1 whenever $x_1 x_2 \geq x_3 x_4$. The truth table that defines the comparator is shown in Table 4.4. The comparator will be designed as a sum of products using NAND logic and also as a product of sums using NOR logic.

Table 4.4 Truth Table for Example 4.21

x_1	x_2	x_3	x_4	z_1
0	0	0	0	1
0	0	0	1	0
0	0	1	0	0
0	0	1	1	0
0	1	0	0	1
0	1	0	1	1
0	1	1	0	0
0	1	1	1	0
(Continued on next page)				

Table 4.4 Truth Table for Example 4.21

x_1	x_2	x_3	x_4	z_1
1	0	0	0	1
1	0	0	1	1
1	0	1	0	1
1	0	1	1	0
1	1	0	0	1
1	1	0	1	1
1	1	1	0	1
1	1	1	1	1

The Karnaugh map that represents Table 4.4 is shown in Figure 4.105. The sum-of-products equation is shown in Equation 4.14 and the product-of-sums equation is shown in Equation 4.15. The logic diagram in a sum-of-products notation is shown in Figure 4.106 using NAND logic. The design module, the test bench, and the outputs are shown in Figure 4.107, Figure 4.108, and Figure 4.109, respectively.

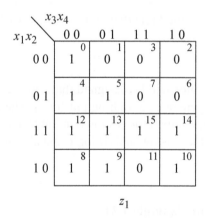

Figure 4.105   Karnaugh map for the 2-bit comparator of Example 4.21.

$$z_1 = x_1 x_2 + x_3' x_4' + x_2 x_3' + x_1 x_3' + x_1 x_4' \qquad (4.14)$$

$$z_1 = (x_1 + x_3')(x_2 + x_3' + x_4')(x_1 + x_2 + x_4') \qquad (4.15)$$

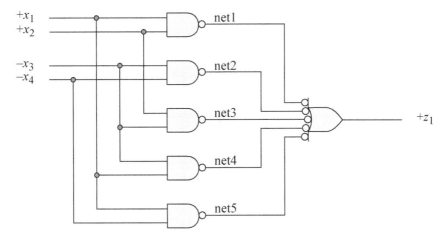

Figure 4.106 Sum-of-products logic diagram for the 2-bit comparator of Example 4.21 using NAND logic.

```
//dataflow for 2-bit comparator using nand logic
module comparator2_nand (x1, x2, x3, x4, z1);


//define inputs and outputs
input x1, x2, x3, x4;
output z1;


//define inputs and output as wire
wire x1, x2, x3, x4;
wire z1;


//define internal nets
wire net1, net2, net3, net4, net5;


//define z1 using continuous assignment
assign    net1 = ~(x1 & x2),
          net2 = ~(~x3 & ~x4),
          net3 = ~(x2 & ~x3),
          net4 = ~(x1 & ~x3),
          net5 = ~(x1 & ~x4);

assign z1 = (~net1 | ~net2 | ~net3 | ~net4 | ~net5);

endmodule
```

Figure 4.107 Module for the comparator of Example 4.21 using NAND logic.

```
//test bench for comparator2 using nand logic
module comparator2_nand_tb;


reg x1, x2, x3, x4;
wire z1;


//apply input vectors and display variables
initial
begin: apply_stimulus
   reg [4:0] invect;
   for (invect = 0; invect < 16; invect = invect + 1)
      begin
         {x1, x2, x3, x4} = invect [4:0];
         #10 $display ("x1 x2 x3 x4 = %b, z1 = %b",
                       {x1, x2, x3, x4}, z1);
      end
end


//instantiate the module into the test bench
comparator2_nand inst1 (
   .x1(x1),
   .x2(x2),
   .x3(x3),
   .x4(x4),
   .z1(z1)
   );

endmodule
```

Figure 4.108   Test bench for the comparator of Figure 4.107.

```
x1 x2 x3 x4 = 0000, z1 = 1      x1 x2 x3 x4 = 1000, z1 = 1
x1 x2 x3 x4 = 0001, z1 = 0      x1 x2 x3 x4 = 1001, z1 = 1
x1 x2 x3 x4 = 0010, z1 = 0      x1 x2 x3 x4 = 1010, z1 = 1
x1 x2 x3 x4 = 0011, z1 = 0      x1 x2 x3 x4 = 1011, z1 = 0
x1 x2 x3 x4 = 0100, z1 = 1      x1 x2 x3 x4 = 1100, z1 = 1
x1 x2 x3 x4 = 0101, z1 = 1      x1 x2 x3 x4 = 1101, z1 = 1
x1 x2 x3 x4 = 0110, z1 = 0      x1 x2 x3 x4 = 1110, z1 = 1
x1 x2 x3 x4 = 0111, z1 = 0      x1 x2 x3 x4 = 1111, z1 = 1
```

Figure 4.109   Outputs for the comparator of Figure 4.107.

The same problem will now be designed as a product-of-sums circuit using NOR logic. The logic diagram is shown in Figure 4.110 as generated from Equation 4.15. The design module, test bench module, and outputs are shown in Figure 4.111, Figure 4.112, and Figure 4.113, respectively. The outputs are identical to those of Figure 4.109. Note that there are fewer logic gates required for the product-of-sums implementation.

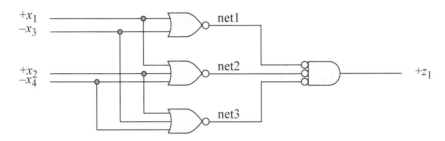

Figure 4.110   Logic diagram for the comparator of Example 4.21 that is implemented as a product of sums.

```
//dataflow for 2-bit comparator using nor logic
module comparator2_nor (x1, x2, x3, x4, z1);

//define inputs and outputs
input x1, x2, x3, x4;
output z1;

//define inputs and output as wire
wire x1, x2, x3, x4;
wire z1;

//define internal nets
wire net1, net2, net3;

//define z1 using continuous assignment
assign    net1 = ~(x1 | ~x3),
          net2 = ~(x1 | x2 | ~x4),
          net3 = ~(x2 | ~x3 | ~x4);

assign    z1 = ~net1 & ~net2 & ~net3;

endmodule
```

Figure 4.111   Module for the comparator of Example 4.21 using NOR logic.

```
//test bench for comparator2 using nor logic
module comparator2_nor_tb;

reg x1, x2, x3, x4;
wire z1;

//apply input vectors and display variables
initial
begin: apply_stimulus
   reg [4:0] invect;
   for (invect = 0; invect < 16; invect = invect + 1)
      begin
         {x1, x2, x3, x4} = invect [4:0];
         #10 $display ("x1 x2 x3 x4 = %b, z1 = %b",
                       {x1, x2, x3, x4}, z1);
      end
end

//instantiate the module into the test bench
comparator2_nor inst1 (
   .x1(x1),
   .x2(x2),
   .x3(x3),
   .x4(x4),
   .z1(z1)
   );

endmodule
```

Figure 4.112   Test bench for the comparator of Example 4.21 using NOR logic.

```
x1 x2 x3 x4 = 0000, z1 = 1        x1 x2 x3 x4 = 1000, z1 = 1
x1 x2 x3 x4 = 0001, z1 = 0        x1 x2 x3 x4 = 1001, z1 = 1
x1 x2 x3 x4 = 0010, z1 = 0        x1 x2 x3 x4 = 1010, z1 = 1
x1 x2 x3 x4 = 0011, z1 = 0        x1 x2 x3 x4 = 1011, z1 = 0
x1 x2 x3 x4 = 0100, z1 = 1        x1 x2 x3 x4 = 1100, z1 = 1
x1 x2 x3 x4 = 0101, z1 = 1        x1 x2 x3 x4 = 1101, z1 = 1
x1 x2 x3 x4 = 0110, z1 = 0        x1 x2 x3 x4 = 1110, z1 = 1
x1 x2 x3 x4 = 0111, z1 = 0        x1 x2 x3 x4 = 1111, z1 = 1
```

Figure 4.113   Outputs for the comparator of Example 4.21 using NOR logic.

Example 4.22 A 4:1 nonlinear-select multiplexer will be designed using dataflow modeling to implement the logic specified by the Karnaugh map of Figure 4.114(a), where y is a map-entered variable. All permutations of the Karnaugh map will be examined, then the map requiring the least amount of hardware will be designed using dataflow modeling in conjunction with the 4:1 nonlinear-select multiplexer.

The two permutations of the map are shown in Figure 4.114(b) and Figure 4.114(c). If the exclusive-OR circuit for input d_0 in Figure 4.114(c) is considered to be three gates $(x_2'y + x_2y')$, then the permutation with the least amount of hardware is the permutation of Figure 4.114(b). The multiplexer block diagram is shown in Figure 4.115 with an active-high enable input. The dataflow module for the multiplexer is shown in Figure 4.116.

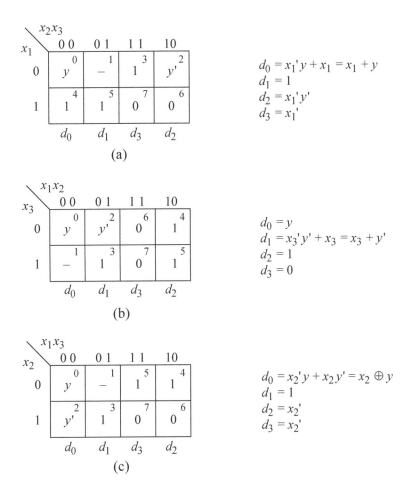

$$d_0 = x_1'y + x_1 = x_1 + y$$
$$d_1 = 1$$
$$d_2 = x_1'y'$$
$$d_3 = x_1'$$

(a)

$$d_0 = y$$
$$d_1 = x_3'y' + x_3 = x_3 + y'$$
$$d_2 = 1$$
$$d_3 = 0$$

(b)

$$d_0 = x_2'y + x_2y' = x_2 \oplus y$$
$$d_1 = 1$$
$$d_2 = x_2'$$
$$d_3 = x_2'$$

(c)

Figure 4.114 Karnaugh map (a) with two permutations (b) and (c) for Example 4.22 to be implemented using a nonlinear-select multiplexer.

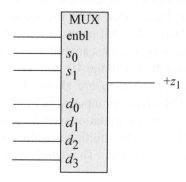

Figure 4.115   Block diagram of a 4:1 multiplexer.

```
module mux4_df (s, d, enbl, z1); //dataflow 4:1 multiplexer
input [1:0] s;
input [3:0] d;
input enbl;
output z1;

wire [1:0] s;
wire [3:0] d;
wire enbl;
wire z1;

assign z1 =(~s[1] & ~s[0] & d[0] & enbl) |
           (~s[1] &  s[0] & d[1] & enbl) |
           ( s[1] & ~s[0] & d[2] & enbl) |
           ( s[1] &  s[0] & d[3] & enbl);
endmodule
```

Figure 4.116   Dataflow module for a 4:1 multiplexer.

The logic diagram for Example 4.22 is shown in Figure 4.117, which is designed from the Karnaugh map of Figure 4.114(b). The input variables $x_1 x_2$ connect to the select inputs of the multiplexer as follows: $x_1 x_2 = s_1 s_0$. The enable input is connected to a logic 1 so that the multiplexer is always enabled.

The design module is shown in Figure 4.118. The test bench module and the outputs are shown in Figure 4.119 and Figure 4.120, respectively. The outputs are identical to the values in the corresponding minterm locations in the Karnaugh map of Figure 4.114(b).

For example, in the Karnaugh map of Figure 4.114(b), if $x_1 x_2 x_3 = 000$, then minterm location 0 contains the variable y. In the logic diagram of Figure 4.117, $x_1 x_2 x_3 = 000$, then input d_0 is selected and output z_1 contains the value of y. Similarly, if $x_1 x_2 x_3$

= 010, then minterm location 2 contains the variable y'. In the logic diagram, if $x_1 x_2 x_3$ = 010, then input d_1 is selected. Since $x_3 = 0$, output z_1 contains the value of y'. All of the outputs in Figure 4.120 can be verified in a similar manner.

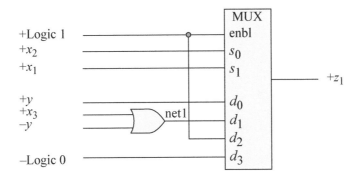

Figure 4.117   Logic diagram for Example 4.22 using a nonlinear-select multiplexer.

```
//dataflow nonlinear-select multiplexer circuit
module mux_nonlinear3 (x1, x2, x3, y, z1);

//define inputs and output
input x1, x2, x3, y;
output z1;

//define inputs and output as wire
wire x1, x2, x3, y;
wire z1;

//define internal net
wire net1;

//instantiate the dataflow multiplexer
mux4_df inst1 (
   .s({x1, x2}),
   .d({1'b0, 1'b1, net1, y}),
   .enbl(1'b1),
   .z1(z1)
   );

//define the or gate
assign net1 = (x3 | ~y);

endmodule
```

Figure 4.118   Module for the logic diagram of Figure 4.117.

```
//test bench for the nonlinear-select multiplexer circuit
module mux_nonlinear3_tb;


reg x1, x2, x3, y;
wire z1;


//apply input vectors and display variables
initial
begin: apply_stimulus
   reg [4:0] invect;
   for (invect = 0; invect < 16; invect = invect + 1)
      begin
         {x1, x2, x3, y} = invect [4:0];
         #10 $display ("x1 x2 x3 y = %b, z1 = %b",
                         {x1, x2, x3, y}, z1);
      end
end


//instantiate the module into the test bench
mux_nonlinear3 inst1 (
   .x1(x1),
   .x2(x2),
   .x3(x3),
   .y(y),
   .z1(z1)
   );

endmodule
```

Figure 4.119   Test bench for the module of Figure 4.118.

```
x1 x2 x3 y = 0000, z1 = 0        x1 x2 x3 y = 1000, z1 = 1
x1 x2 x3 y = 0001, z1 = 1        x1 x2 x3 y = 1001, z1 = 1
x1 x2 x3 y = 0010, z1 = 0        x1 x2 x3 y = 1010, z1 = 1
x1 x2 x3 y = 0011, z1 = 1        x1 x2 x3 y = 1011, z1 = 1
x1 x2 x3 y = 0100, z1 = 1        x1 x2 x3 y = 1100, z1 = 0
x1 x2 x3 y = 0101, z1 = 0        x1 x2 x3 y = 1101, z1 = 0
x1 x2 x3 y = 0110, z1 = 1        x1 x2 x3 y = 1110, z1 = 0
x1 x2 x3 y = 0111, z1 = 1        x1 x2 x3 y = 1111, z1 = 0
```

Figure 4.120   Outputs for the nonlinear-select multiplexer circuit of Figure 4.117.

Example 4.23 The equation of Equation 4.16 will be plotted on the 5-variable Karnaugh map of Figure 4.121. Then the sum-of-products expression and the product-of-sums expression will be obtained, as shown in Equation 4.17 and Equation 4.18, respectively.

The sum-of-products logic diagram is shown in Figure 4.122 and will be implemented in a dataflow module using NAND gates as shown in Figure 4.123. The test bench and outputs are shown in Figure 4.124 and Figure 4.125, respectively.

$$
\begin{aligned}
z_1 &= x_1 x_3 x_4 (x_2 + x_4 x_5') + (x_2' + x_4)(x_1 x_3' + x_5) \\
&= x_1 x_2 x_3 x_4 + x_1 x_3 x_4 x_5' + (x_2' + x_4) x_1 x_3' + (x_2' + x_4) x_5 \\
&= x_1 x_2 x_3 x_4 + x_1 x_3 x_4 x_5' + x_1 x_2' x_3' + x_1 x_3' x_4 + x_2' x_5 + x_4 x_5 \quad (4.16)
\end{aligned}
$$

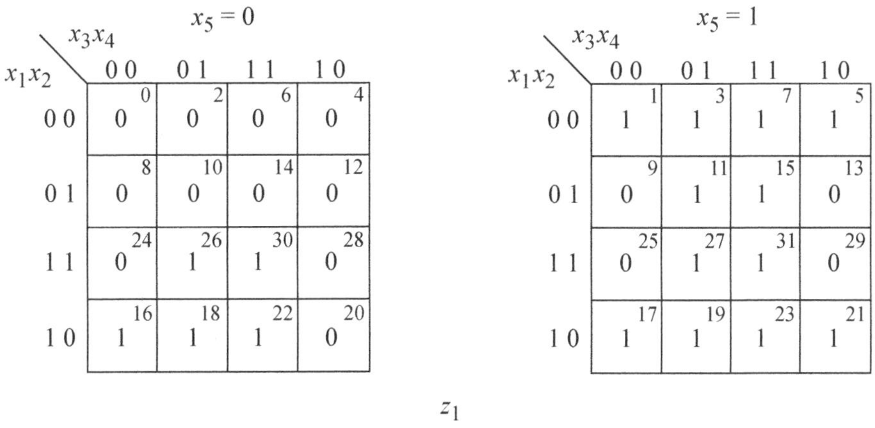

Figure 4.121 Karnaugh map for Equation 4.16.

$$
z_1 = x_1 x_4 + x_1 x_2' x_3' + x_2' x_5 + x_4 x_5 \quad (4.17)
$$

$$
z_1 = (x_1 + x_5)(x_2' + x_4)(x_3' + x_4 + x_5) \quad (4.18)
$$

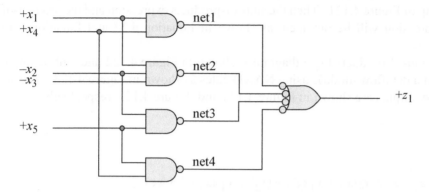

Figure 4.122   Product-of-sums logic diagram for Equation 4.17 using NAND logic.

```
//dataflow module for a sum-of-products
//logic diagram using nand logic
module log_diag_sop1 (x1, x2, x3, x4, x5, z1);


//define inputs and output
input x1, x2, x3, x4, x5;
output z1;


//define inputs and output as wire
wire x1, x2, x3, x4, x5;
wire z1;


//define internal nets
wire net1, net2, net3, net4;


//define z1 using continuous assignment
assign     net1 = ~(x1 & x4),
           net2 = ~(x1 & ~x2 & ~x3),
           net3 = ~(~x2 & x5),
           net4 = ~(x4 & x5);


assign     z1 = (~net1 | ~net2 | ~net3 | ~net4);

endmodule
```

Figure 4.123   Sum of products dataflow module for Equation 4.17.

```
//test bench for logic diagram using nand logic
module log_diag_sop1_tb;


reg x1, x2, x3, x4, x5;
wire z1;


//apply input vectors and display variables
initial
begin: apply_stimulus
   reg [5:0] invect;
   for (invect = 0; invect < 32; invect = invect + 1)
      begin
         {x1, x2, x3, x4, x5} = invect [5:0];
         #10 $display ("x1 x2 x3 x4 x5 = %b, z1 = %b",
                         {x1, x2, x3, x4, x5}, z1);
      end
end


//instantiate the module into the test bench
log_diag_sop1 inst1 (
   .x1(x1),
   .x2(x2),
   .x3(x3),
   .x4(x4),
   .x5(x5),
   .z1(z1)
   );

endmodule
```

Figure 4.124   Test bench for the sum-of-products dataflow module of Figure 4.123.

```
x1 x2 x3 x4 x5 = 00000, z1 = 0   x1 x2 x3 x4 x5 = 01000, z1 = 0
x1 x2 x3 x4 x5 = 00001, z1 = 1   x1 x2 x3 x4 x5 = 01001, z1 = 0
x1 x2 x3 x4 x5 = 00010, z1 = 0   x1 x2 x3 x4 x5 = 01010, z1 = 0
x1 x2 x3 x4 x5 = 00011, z1 = 1   x1 x2 x3 x4 x5 = 01011, z1 = 1
x1 x2 x3 x4 x5 = 00100, z1 = 0   x1 x2 x3 x4 x5 = 01100, z1 = 0
x1 x2 x3 x4 x5 = 00101, z1 = 1   x1 x2 x3 x4 x5 = 01101, z1 = 0
x1 x2 x3 x4 x5 = 00110, z1 = 0   x1 x2 x3 x4 x5 = 01110, z1 = 0
x1 x2 x3 x4 x5 = 00111, z1 = 1   x1 x2 x3 x4 x5 = 01111, z1 = 1

                                 //continued on next page
```

Figure 4.125   Outputs for the sum-of-products dataflow module of Figure 4.123.

```
x1 x2 x3 x4 x5 = 10000, z1 = 1   x1 x2 x3 x4 x5 = 11000, z1 = 0
x1 x2 x3 x4 x5 = 10001, z1 = 1   x1 x2 x3 x4 x5 = 11001, z1 = 0
x1 x2 x3 x4 x5 = 10010, z1 = 1   x1 x2 x3 x4 x5 = 11010, z1 = 1
x1 x2 x3 x4 x5 = 10011, z1 = 1   x1 x2 x3 x4 x5 = 11011, z1 = 1
x1 x2 x3 x4 x5 = 10100, z1 = 0   x1 x2 x3 x4 x5 = 11100, z1 = 0
x1 x2 x3 x4 x5 = 10101, z1 = 1   x1 x2 x3 x4 x5 = 11101, z1 = 0
x1 x2 x3 x4 x5 = 10110, z1 = 1   x1 x2 x3 x4 x5 = 11110, z1 = 1
x1 x2 x3 x4 x5 = 10111, z1 = 1   x1 x2 x3 x4 x5 = 11111, z1 = 1
```

Figure 4.125   (Continued)

The product-of-sums logic diagram is shown in Figure 4.126 and will be implemented in a dataflow module using NOR gates as shown in Figure 4.127. The test bench and outputs are shown in Figure 4.128 and Figure 4.129, respectively. Both designs produce the identical outputs.

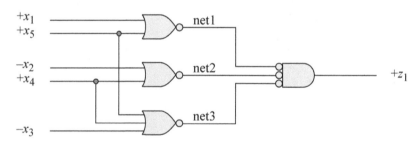

Figure 4.126   Product-of-sums logic diagram for Equation 4.18 using NOR logic.

```
//dataflow module for a product-of-sums
//logic diagram using nor logic
module log_diag_pos1 (x1, x2, x3, x4, x5, z1);

//define inputs and output
input x1, x2, x3, x4, x5;
output z1;

//define inputs and output as wire
wire x1, x2, x3, x4, x5;
wire z1;

//continued on next page
```

Figure 4.127   Product-of-sums dataflow module for Equation 4.18.

```
//define internal nets
wire net1, net2, net3;

//define z1 using continuous assignment
assign   net1 = ~(x1 | x5),
         net2 = ~(~x2 | x4),
         net3 = ~(~x3 | x4 | x5);

assign   z1 = (~net1 & ~net2 & ~net3);

endmodule
```

Figure 4.127   (Continued)

```
//test bench for logic diagram using nor logic
module log_diag_pos1_tb;

reg x1, x2, x3, x4, x5;
wire z1;

//apply input vectors and display variables
initial
begin: apply_stimulus
   reg [5:0] invect;
   for (invect = 0; invect < 32; invect = invect + 1)
      begin
         {x1, x2, x3, x4, x5} = invect [5:0];
         #10 $display ("x1 x2 x3 x4 x5 = %b, z1 = %b",
                       {x1, x2, x3, x4, x5}, z1);
      end
end

//instantiate the module into the test bench
log_diag_pos1 inst1 (
   .x1(x1),
   .x2(x2),
   .x3(x3),
   .x4(x4),
   .x5(x5),
   .z1(z1)
   );

endmodule
```

Figure 4.128   Test bench for the product-of-sums dataflow module of Figure 4.127.

```
x1 x2 x3 x4 x5 = 00000, z1 = 0    x1 x2 x3 x4 x5 = 10000, z1 = 1
x1 x2 x3 x4 x5 = 00001, z1 = 1    x1 x2 x3 x4 x5 = 10001, z1 = 1
x1 x2 x3 x4 x5 = 00010, z1 = 0    x1 x2 x3 x4 x5 = 10010, z1 = 1
x1 x2 x3 x4 x5 = 00011, z1 = 1    x1 x2 x3 x4 x5 = 10011, z1 = 1
x1 x2 x3 x4 x5 = 00100, z1 = 0    x1 x2 x3 x4 x5 = 10100, z1 = 0
x1 x2 x3 x4 x5 = 00101, z1 = 1    x1 x2 x3 x4 x5 = 10101, z1 = 1
x1 x2 x3 x4 x5 = 00110, z1 = 0    x1 x2 x3 x4 x5 = 10110, z1 = 1
x1 x2 x3 x4 x5 = 00111, z1 = 1    x1 x2 x3 x4 x5 = 10111, z1 = 1
x1 x2 x3 x4 x5 = 01000, z1 = 0    x1 x2 x3 x4 x5 = 11000, z1 = 0
x1 x2 x3 x4 x5 = 01001, z1 = 0    x1 x2 x3 x4 x5 = 11001, z1 = 0
x1 x2 x3 x4 x5 = 01010, z1 = 0    x1 x2 x3 x4 x5 = 11010, z1 = 1
x1 x2 x3 x4 x5 = 01011, z1 = 1    x1 x2 x3 x4 x5 = 11011, z1 = 1
x1 x2 x3 x4 x5 = 01100, z1 = 0    x1 x2 x3 x4 x5 = 11100, z1 = 0
x1 x2 x3 x4 x5 = 01101, z1 = 0    x1 x2 x3 x4 x5 = 11101, z1 = 0
x1 x2 x3 x4 x5 = 01110, z1 = 0    x1 x2 x3 x4 x5 = 11110, z1 = 1
x1 x2 x3 x4 x5 = 01111, z1 = 1    x1 x2 x3 x4 x5 = 11111, z1 = 1
```

Figure 4.129   Outputs for the product-of-sums dataflow module of Figure 4.127.

4.3.2 Reduction Operators

The reduction operators are: AND (&), NAND (~&), OR (|), NOR (~|), exclusive-OR ($\wedge$), and exclusive-NOR ($\wedge\sim$ or $\sim\wedge$). Reduction operators are unary operators; that is, they operate on a single vector and produce a single-bit result. Reduction operators perform their respective operations on a bit-by-bit basis from right to left. If any bit in the operand is an **x** or a **z**, then the result of the operation is an **x**.

The reduction operators are defined as follows:

Reduction Operator	Description
& (Reduction AND)	If any bit is a 0, then the result is 0, otherwise the result is 1.
~& (Reduction NAND)	This is the complement of the reduction AND operation.
| (Reduction OR)	If any bit is a 1, then the result is 1, otherwise the result is 0.
~| (Reduction NOR)	This is the complement of the reduction OR operation.
$\wedge$ (Reduction exclusive-OR)	If there are an even number of 1s in the operand, then the result is 0, otherwise the result is 1.
$\sim\wedge$ (Reduction exclusive-NOR)	This is the complement of the reduction exclusive-OR operation.

Example 4.24 This example illustrates the continuous assignment statement to demonstrate the functionality of the reduction operators. Figure 4.130 contains the design module to illustrate the operation of the six reduction operators using an 8-bit operand *a[7:0]*. If no delays are specified for the continuous assignment statement, then only one **assign** keyword is required. Only the final statement is terminated by a semicolon; all other statements are terminated by a comma. The test bench and outputs are shown in Figure 4.131 and Figure 4.132, respectively.

```
//module to illustrate the use of reduction operators
module reduction2 (a, red_and, red_nand, red_or, red_nor,
                   red_xor, red_xnor);

input [7:0] a;
output red_and, red_nand, red_or, red_nor, red_xor, red_xnor;

wire [7:0] a;
wire red_and, red_nand, red_or, red_nor, red_xor, red_xnor;


assign    red_and  = &a,       //reduction AND
          red_nand = ~&a,      //reduction NAND
          red_or   = |a,       //reduction OR
          red_nor  = ~|a,      //reduction NOR
          red_xor  = ^a,       //reduction exclusive-OR
          red_xnor = ^~a;      //reduction exclusive-NOR

endmodule
```

Figure 4.130 Module to illustrate the functionality of the reduction operators.

```
//test bench for reduction2 module
module reduction2_tb;

reg [7:0] a;
wire red_and, red_nand, red_or, red_nor, red_xor, red_xnor;

initial
$monitor ("a=%b, red_and=%b, red_nand=%b, red_or=%b,
         red_nor=%b, red_xor=%b, red_xnor=%b",
         a, red_and, red_nand, red_or, red_nor, red_xor,
         red_xnor);

//continued on next page
```

Figure 4.131 Test bench for the reduction operator module of Figure 4.130.

```
//apply input vectors
initial
begin
   #0    a = 8'b0011_0011;
   #10   a = 8'b1101_0011;
   #10   a = 8'b0000_0000;
   #10   a = 8'b0100_1111;
   #10   a = 8'b1111_1111;
   #10   a = 8'b0111_1111;
   #10   $stop;
end

//instantiate the module into the test bench
reduction2 inst1 (
   .a(a),
   .red_and(red_and),
   .red_nand(red_nand),
   .red_or(red_or),
   .red_nor(red_nor),
   .red_xor(red_xor),
   .red_xnor(red_xnor)
   );
endmodule
```

Figure 4.131   (Continued)

```
a=00110011,
red_and=0, red_nand=1, red_or=1,
red_nor=0, red_xor=0, red_xnor=1

a=11010011,
red_and=0, red_nand=1, red_or=1,
red_nor=0, red_xor=1, red_xnor=0

a=00000000,
red_and=0, red_nand=1, red_or=0,
red_nor=1, red_xor=0, red_xnor=1

a=01001111,
red_and=0, red_nand=1, red_or=1,
red_nor=0, red_xor=1, red_xnor=0

//continued on next page
```

Figure 4.132   Outputs for the test bench of Figure 4.131.

```
a=11111111,
red_and=1, red_nand=0, red_or=1,
red_nor=0, red_xor=0, red_xnor=1

a=01111111,
red_and=0, red_nand=1, red_or=1,
red_nor=0, red_xor=1, red_xnor=0
```

Figure 4.132   (Continued)

4.3.3 Conditional Operator

The conditional operator (**?** **:**) has three operands, as shown in the syntax below. The *conditional_expression* is evaluated. If the result is true (1), then the *true_expression* is evaluated; if the result is false (0), then the *false_expression* is evaluated.

conditional_expression **?** true_expression **:** false_expression;

The conditional operator can be used when one of two expressions is to be selected. For example, in the statement below, if x_1 is greater than or equal to x_2, then z_1 is assigned the value of x_3; if x_1 is less than x_2, then z_1 is assigned the value of x_4.

z1 = (x1 >= x2) **?** x3 **:** x4;

Conditional operators can be nested; that is, each true_expression and false_expression can be a conditional operation, as shown below. This is useful for modeling a 4:1 multiplexer.

conditional_expression **?** (cond_expr1 **?** true_expr1 **:** false_expr1)
                          : (cond_expr2 **?** true_expr2 **:** false_expr2);

Example 4.25 A 4:1 multiplexer will be designed using a conditional operator. This design will declare the multiplexer inputs as scalars instead of vectors. The select inputs are: s_0 and s_1; the data inputs are: in_0, in_1, in_2, and in_3. The design module is shown in Figure 4.133. The test bench and outputs are shown in Figure 4.134 and Figure 4.135, respectively.

```
//dataflow 4:1 mux using conditional operator
module mux4to1_cond (out, in0, in1, in2, in3, s0, s1);

input s0, s1;
input in0, in1, in2, in3;
output out;

//use nested conditional operator
assign out = s1 ? (s0 ? in3 : in2) : (s0 ? in1 : in0);

endmodule

//s1          s0              out
//1(true)   1(true)         in3
//0(false)  1(true)         in1
//1(true)   0(false)        in2
//0(false)  0(false)        in0
```

Figure 4.133   Module for a 4:1 multiplexer using the conditional operator and a continuous assignment statement.

```
//mux4to1_cond test bench
module mux4to1_cond_tb;

reg in0, in1, in2, in3, s0, s1;   //inputs are reg
wire out;                          //outputs are wire

//display signals
initial
$monitor ("s1s0 = %b, in0in1in2in3 = %b, out = %b",
         {s1, s0}, {in0, in1, in2, in3}, out);

//apply stimulus
initial
begin
    #0    s1  = 1'b0;
          s0  = 1'b0;
          in0 = 1'b0;    //out = 0
          in1 = 1'b1;
          in2 = 1'b1;
          in3 = 1'b1;

//continued on next page
```

Figure 4.134   Test bench for the conditional operator multiplexer of Figure 4.133.

```
   #10    s1  = 1'b0;
          s0  = 1'b1;
          in0 = 1'b0;
          in1 = 1'b1;      //out = 1
          in2 = 1'b1;
          in3 = 1'b1;

   #10    s1  = 1'b1;
          s0  = 1'b0;
          in0 = 1'b0;
          in1 = 1'b1;
          in2 = 1'b0;      //out = 0
          in3 = 1'b1;

   #10    s1  = 1'b1;
          s0  = 1'b1;
          in0 = 1'b0;
          in1 = 1'b1;
          in2 = 1'b0;
          in3 = 1'b1;      //out = 1

   #10    $stop;
end

//instantiate the module into the test bench
mux4to1_cond inst1 (
   .s0(s0),
   .s1(s1),
   .in0(in0),
   .in1(in1),
   .in2(in2),
   .in3(in3),
   .out(out)
   );

endmodule
```

Figure 4.134   (Continued)

```
s1s0 = 00, in0in1in2in3 = 0111, out = 0
s1s0 = 01, in0in1in2in3 = 0111, out = 1
s1s0 = 10, in0in1in2in3 = 0101, out = 0
s1s0 = 11, in0in1in2in3 = 0101, out = 1
```

Figure 4.135   Outputs for the conditional operator multiplexer of Figure 4.133.

4.3.4 Relational Operators

Relational operators compare operands and return a Boolean result, either 1 (true) or 0 (false) indicating the relationship between the two operands. There are four relational operators: greater than (>), less than (<), greater than or equal (>=), and less than or equal (<=). These operators function the same as identical operators in the C programming language.

If the relationship is true, then the result is 1; if the relationship is false, then the result is 0. Net or register operands are treated as unsigned values; real or integer operands are treated as signed values. An **x** or **z** in any operand returns a result of **x**. When the operands are of unequal size, the smaller operand is zero-extended to the left. Figure 4.136 shows examples of relational operators using dataflow modeling, where the identifier *gt* means greater than, *lt* means less than, *gte* means greater than or equal, and *lte* means less than or equal. The test bench, which applies several different values to the two operands, is shown in Figure 4.137. The outputs are shown in Figure 4.138.

```verilog
//examples of relational operators
module relational_ops1 (a, b, gt, lt, gte, lte);
input [3:0] a, b;
output gt, lt, gte, lte;

assign gt = a > b;
assign lt = a < b;
assign gte = a >= b;
assign lte = a <= b;

endmodule
```

Figure 4.136   Dataflow module to illustrate the relational operators.

```verilog
//test bench for relational operators
module relational_ops1_tb;

reg [3:0] a, b;
wire gt, lt, gte, lte;

initial
$monitor ("a=%b, b=%b, gt=%d, lt=%d, gte=%d, lte=%d",
          a, b, gt, lt, gte, lte);

//continued on next page
```

Figure 4.137   Test bench for the relational operators module of Figure 4.136.

```
//apply input vectors
initial
begin
     #0    a = 4'b0110;
           b = 4'b1100;


     #5    a = 4'b0101;
           b = 4'b0000;


     #5    a = 4'b1000;
           b = 4'b1001;


     #5    a = 4'b0000;
           b = 4'b0000;


     #5    a = 4'b1111;
           b = 4'b1111;


     #5    $stop;
end

//instantiate the module into the test bench
relational_ops1 inst1 (
   .a(a),
   .b(b),
   .gt(gt),
   .lt(lt),
   .gte(gte),
   .lte(lte)
   );

endmodule
```

Figure 4.137   (Continued)

```
a=0110, b=1100, gt=0, lt=1, gte=0, lte=1
a=0101, b=0000, gt=1, lt=0, gte=1, lte=0
a=1000, b=1001, gt=0, lt=1, gte=0, lte=1
a=0000, b=0000, gt=0, lt=0, gte=1, lte=1
a=1111, b=1111, gt=0, lt=0, gte=1, lte=1
```

Figure 4.138   Outputs for the relational operators using the test bench of Figure 4.137.

4.3.5 Logical Operators

There are three logical operators: the binary logical AND operator (&&), the binary logical OR operator ($||$), and the unary logical negation operator (!). Logical operators evaluate to a logical 1 (true), a logical 0 (false), or an **x** (ambiguous). If a logical operation returns a nonzero value, then it is treated as a logical 1 (true); if a bit in an operand is **x** or **z**, then it is ambiguous and is normally treated as a false condition. For vector operands, a nonzero vector is treated as a 1. Figure 4.139 shows examples of the logical operators using dataflow modeling. Figure 4.140 and Figure 4.141 show the test bench and outputs, respectively.

```
//examples of logical operators
module log_ops1 (a, b, z1, z2, z3);

input [3:0] a, b;
output z1, z2, z3;

assign z1 = a && b;
assign z2 = a || b;
assign z3 = !a;

endmodule
```

Figure 4.139   Examples of logical operators.

```
//test bench for logical operators
module log_ops1_tb;

reg [3:0] a, b;
wire z1, z2, z3;

initial
$monitor ("z1 = %d, z2 = %d, z3 = %d", z1, z2, z3);

//apply input vectors
initial
begin
    #0 a = 4'b0110;          //nonzero vector
       b = 4'b1100;          //nonzero vector
//continued on next page
```

Figure 4.140   Test bench for the logical operators module of Figure 4.139.

```
    #5  a = 4'b0101;              //nonzero vector
        b = 4'b0000;              //zero vector


    #5  a = 4'b1000;
        b = 4'b1001;


    #5  a = 4'b0000;
        b = 4'b0000;


    #5  a = 4'b1111;
        b = 4'b1111;


    #5 $stop;
end


//instantiate the module into the test bench
log_ops1 inst1 (
    .a(a),
    .b(b),
    .z1(z1),
    .z2(z2),
    .z3(z3)
    );

endmodule
```

Figure 4.140   (Continued)

```
 z1 = 1, z2 = 1, z3 = 0        //z1 is logical AND
 z1 = 0, z2 = 1, z3 = 0        //z2 is logical OR
 z1 = 1, z2 = 1, z3 = 0        //z3 is logical negation
 z1 = 0, z2 = 0, z3 = 1
 z1 = 1, z2 = 1, z3 = 0
```

Figure 4.141   Outputs for the logical operators obtained from the test bench of Figure 4.140.

4.3.6 Bitwise Operators

The bitwise operators are: AND (&), OR (|), negation (~), exclusive-OR (^), and exclusive-NOR (^~ or ~^). The bitwise operators perform logical operations on the operands on a bit-by-bit basis and produce a vector result. Except for negation, each bit in one operand is associated with the corresponding bit in the other operand. If one operand is shorter, then it is zero-extended to the left to match the length of the longer operand.

The *bitwise AND* operator performs the AND function on two operands on a bit-by-bit basis. An example of the bitwise AND operator is shown below.

$$
\begin{array}{r}
0\ \ 0\ \ 1\ \ 1\ \ 0\ \ 1\ \ 1\ \ 0 \\
\&)\ \underline{1\ \ 1\ \ 1\ \ 1\ \ 0\ \ 1\ \ 0\ \ 1} \\
0\ \ 0\ \ 1\ \ 1\ \ 0\ \ 1\ \ 0\ \ 0
\end{array}
$$

The *bitwise OR* operator performs the OR function on the two operands on a bit-by-bit basis. An example of the bitwise OR operator is shown below.

$$
\begin{array}{r}
1\ \ 0\ \ 0\ \ 1\ \ 0\ \ 1\ \ 1\ \ 0 \\
|)\ \underline{0\ \ 1\ \ 0\ \ 1\ \ 0\ \ 1\ \ 0\ \ 1} \\
1\ \ 1\ \ 0\ \ 1\ \ 0\ \ 1\ \ 1\ \ 1
\end{array}
$$

The *bitwise negation* operator performs the negation function on one operand on a bit-by-bit basis. Each bit in the operand is inverted. An example of the bitwise negation operator is shown below.

$$
\begin{array}{r}
~)\ \underline{0\ \ 1\ \ 0\ \ 1\ \ 0\ \ 1\ \ 0\ \ 1} \\
1\ \ 0\ \ 1\ \ 0\ \ 1\ \ 0\ \ 1\ \ 0
\end{array}
$$

The *bitwise exclusive-OR* operator performs the exclusive-OR function on two operands on a bit-by-bit basis. An example of the bitwise exclusive-OR operator is shown below.

$$
\begin{array}{r}
0\ \ 0\ \ 1\ \ 1\ \ 0\ \ 1\ \ 1\ \ 0 \\
\text{^})\ \underline{1\ \ 1\ \ 0\ \ 1\ \ 0\ \ 1\ \ 0\ \ 1} \\
1\ \ 1\ \ 1\ \ 0\ \ 0\ \ 0\ \ 1\ \ 1
\end{array}
$$

The *bitwise exclusive-NOR* operator performs the exclusive-NOR function on two operands on a bit-by-bit basis. An example of the bitwise exclusive-NOR operator is shown below.

```
    1  0  1  1  0  1  0  0
^~ ) 1  1  0  1  0  1  0  1
    1  0  0  1  1  1  1  0
```

Bitwise operators perform operations on operands on a bit-by-bit basis and produce a vector result. This is in contrast to logical operators, which perform operations on operands in such a way that the truth or falsity of the result is determined by the truth or falsity of the operands. That is, the logical AND operator returns a value of 1 (true) only if both operands are nonzero (true); otherwise, it returns a value of 0 (false). If the result is ambiguous, it returns a value of **x**. Figure 4.142 shows a coding example to illustrate the use of the five bitwise operators. The test bench and outputs are in Figure 4.143 and Figure 4.144, respectively.

```
//dataflow example of bitwise operators
module bitwise2 (a, b, and_rslt, or_rslt, neg_rslt, xor_rslt,
                 xnor_rslt);

//define inputs and outputs
input [7:0] a, b;
output [7:0] and_rslt, or_rslt, neg_rslt, xor_rslt, xnor_rslt;

wire [7:0] a, b;
wire [7:0] and_rslt, or_rslt, neg_rslt, xor_rslt, xnor_rslt;

//define outputs using continuous assignment
assign    and_rslt = a & b,          //bitwise AND
          or_rslt = a | b,           //bitwise OR
          neg_rslt = ~a,             //bitwise negation
          xor_rslt = a ^ b,          //bitwise exclusive-OR
          xnor_rslt = a ^~ b;        //bitwise exclusive-NOR
endmodule
```

Figure 4.142   Module to illustrate bitwise operators.

```
//test bench for bitwise2 module
module bitwise2_tb;

reg [7:0] a, b;
wire [7:0] and_rslt, or_rslt, neg_rslt, xor_rslt, xnor_rslt;

//continued on next page
```

Figure 4.143   Test bench for the bitwise operator module of Figure 4.142.

```verilog
initial
$monitor ("a=%b, b=%b, and_rslt=%b, or_rslt=%b, neg_rslt=%b,
           xor_rslt=%b, xnor_rslt=%b",
           a, b, and_rslt, or_rslt, neg_rslt,
           xor_rslt, xnor_rslt);


//apply input vectors
initial
begin
   #0    a = 8'b1100_0011;
         b = 8'b1001_1001;


   #10   a = 8'b1001_0011;
         b = 8'b1101_1001;


   #10   a = 8'b0000_1111;
         b = 8'b1101_1001;


   #10   a = 8'b0100_1111;
         b = 8'b1101_1001;


   #10   a = 8'b1100_1111;
         b = 8'b1101_1001;


   #10   $stop;
end


//instantiate the module into the test bench
bitwise2 inst1 (
   .a(a),
   .b(b),
   .and_rslt(and_rslt),
   .or_rslt(or_rslt),
   .neg_rslt(neg_rslt),
   .xor_rslt(xor_rslt),
   .xnor_rslt(xnor_rslt)
   );

endmodule
```

Figure 4.143   (Continued)

```
a = 11000011,
b = 10011001,

and_rslt  = 10000001,
or_rslt   = 11011011,
neg_rslt  = 00111100,
xor_rslt  = 01011010,
xnor_rslt = 10100101

a = 10010011,
b = 11011001,

and_rslt  = 10010001,
or_rslt   = 11011011,
neg_rslt  = 01101100,
xor_rslt  = 01001010,
xnor_rslt = 10110101

a = 00001111,
b = 11011001,

and_rslt  = 00001001,
or_rslt   = 11011111,
neg_rslt  = 11110000,
xor_rslt  = 11010110,
xnor_rslt = 00101001

a = 01001111,
b = 11011001,

and_rslt  = 01001001,
or_rslt   = 11011111,
neg_rslt  = 10110000,
xor_rslt  = 10010110,
xnor_rslt = 01101001

a = 11001111,
b = 11011001,

and_rslt  = 11001001,
or_rslt   = 11011111,
neg_rslt  = 00110000,
xor_rslt  = 00010110,
xnor_rslt = 11101001
```

Figure 4.144   Outputs for the bitwise module of Figure 4.142.

4.3.7 Shift Operators

The shift operators shift a single vector operand left or right a specified number of bit positions. These are logical shift operations, not algebraic; that is, as bits are shifted left or right, zeroes fill in the vacated bit positions. The bits shifted out of the operand are lost; they do not rotate to the high-order or low-order bit positions of the shifted operand. If the shift amount evaluates to **x** or **z**, then the result of the operation is **x**. There are two shift operators, as shown below. The value in parentheses is the number of bits that the operand is shifted.

<< (Left-shift amount)
>> (Right-shift amount)

When an operand is shifted left, this is equivalent to a multiply-by-two operation for each bit position shifted. When an operand is shifted right, this is equivalent to a divide-by-two operation for each bit position shifted. The shift operators are useful to model the sequential add-shift multiplication algorithm and the sequential shift-subtract division algorithm. Figure 4.145 shows examples of the shift-left and shift-right operators using dataflow modeling. The test bench is shown in Figure 4.146 and the outputs in Figure 4.147.

```
//dataflow module to illustrate the shift operators
module shift2 (a, b, a_rslt, b_rslt);


//define inputs and outputs
input [7:0] a, b;
output [7:0] a_rslt, b_rslt;


//define inputs and outputs as wire
wire a, b;
wire a_rslt, b_rslt;


//define outputs using continuous assignment
assign   a_rslt = a << 2,      //multiply by 4
         b_rslt = b >> 3;      //divide by 8


endmodule
```

Figure 4.145   Module to demonstrate shift-left and shift-right operators.

```
//test bench for shift operators module
module shift2_tb;


reg [7:0] a, b;
wire [7:0] a_rslt, b_rslt;


//display variables
initial
$monitor ("a = %b, b = %b, a_rslt = %b, b_rslt = %b",
            a, b, a_rslt, b_rslt);


//apply input vectors
initial
begin
   #0    a = 8'b0000_0010;    //2;      a_rslt = 8
         b = 8'b0000_1000;    //8;      b_rslt = 1


   #10   a = 8'b0000_0110;    //6;      a_rslt = 24
         b = 8'b0001_1000;    //24;     b_rslt = 3


   #10   a = 8'b0000_1111;    //15;     a_rslt = 60
         b = 8'b0011_1000;    //56;     b_rslt = 7


   #10   a = 8'b1110_0000;    //-32;    a_rslt = -128
         b = 8'b0000_0011;    //3;      b_rslt = 0


   #10   $stop;
end


//instantiate the module into the test bench
shift2 inst1 (
   .a(a),
   .b(b),
   .a_rslt(a_rslt),
   .b_rslt(b_rslt)
   );


endmodule
```

Figure 4.146   Test bench for the shift operators module of Figure 4.145.

```
a = 00000010,            //a = 2
a_rslt = 00001000,       //multiply a by 4; a_rslt = 8

b = 00001000,            //b = 8
b_rslt = 00000001        //divide b by 8; b_rslt = 1

a = 00000110,            //a = 6
a_rslt = 00011000,       //multiply a by 4; a_rslt = 24

b = 00011000,            //b = 24
b_rslt = 00000011        //divide b by 8; b_rslt = 3

a = 00001111,            //a = 15
a_rslt = 00111100,       //multiply a by 4; a_rslt = 60

b = 00111000,            //b = 56
b_rslt = 00000111        //divide b by 8; b_rslt = 7

a = 11100000,            //a = -32
a_rslt = 10000000,       //multiply a by 4; a_rslt = -128

b = 00000011,            //b = 3
b_rslt = 00000000        //divide b by 8; b_rslt = 0
```

Figure 4.147   Outputs for the test bench of Figure 4.146.

4.4 Behavioral Modeling

Describing a module in *behavioral* modeling is an abstraction of the functional operation of the design. It does not describe the implementation of the design at the gate level. The outputs of the module are characterized by their relationship to the inputs. The behavior of the design is described using procedural constructs. These constructs are the **initial** statement and the **always** statement.

The **initial** statement is executed only once during a simulation — beginning at time 0 — and then suspends forever. The **always** statement also begins at time 0 and executes the statements in the **always** block repeatedly in a looping manner. Both statements use only register data types. Objects of the **reg** data types resemble the behavior of a hardware storage device because the data retains its value until a new value is assigned.

Behavioral modeling is an algorithmic approach to hardware implementation and represents a higher level of abstraction than previous modeling methods. A Verilog module may contain a mixture of built-in primitives, UDPs, dataflow constructs, and behavioral constructs.

A *procedure* is a series of operations taken to design a module. A Verilog module that is designed using behavioral modeling contains no internal structural details, it simply defines the behavior of the hardware in an abstract, algorithmic description. A behavior may consist of a single statement or a block of statements delimited by the keywords **begin** ... **end**. A module may contain multiple **initial** and **always** statements. These statements are the basic statements used in behavioral modeling and execute concurrently starting at time zero in which the order of execution is not important. All other behavioral statements are contained inside these structured procedure statements.

4.4.1 Initial Statement

All statements within an **initial** statement comprise an **initial** block. An **initial** statement executes only once beginning at time zero, then suspends execution. An **initial** statement provides a method to initialize and monitor variables before the variables are used in a module; it is also used to generate waveforms. For a given time unit, all statements within the **initial** block execute sequentially. Execution or assignment is controlled by the # symbol. The syntax for an **initial** statement is shown below.

>   **initial** [optional timing control] procedural statement or
>   block of procedural statements

Each **initial** block executes concurrently at time zero and each block ends execution independently. If there is only one procedural statement, then the statement does not require the keywords **begin** ... **end**. However, if there are two or more procedural statements, they are delimited by the keywords **begin** ... **end**.

Example 4.26   A module showing the use of the **initial** statement is shown in Figure 4.148, where the variables x_1, x_2, x_3, x_4, and x_5 are initialized to specific values. Multiple **initial** statements are used for both a single procedural statement and a block of procedural statements. The outputs and waveforms are shown in Figure 4.149 and Figure 4.150, respectively.

```
//module showing use of the initial keyword
module initial_ex (x1, x2, x3, x4, x5);

output x1, x2, x3, x4, x5;

reg x1, x2, x3, x4, x5;
//continued on next page
```

Figure 4.148   Module to illustrate the use of the **initial** statement.

```verilog
initial      //display variables
$monitor ($time, " x1x2x3x4x5 = %b", {x1, x2, x3, x4, x5});

//initialize variables to 0
//multiple statements require begin . . . end
initial
begin
    #0      x1 = 1'b0
            x2 = 1'b0;
            x3 = 1'b0;
            x4 = 1'b0;
            x5 = 1'b0;
end

//set x1
//single statement requires no begin . . . end
initial
    #10     x1 = 1'b1;

//set x2 and x3
initial
begin
    #10     x2 = 1'b1;
    #10     x3 = 1'b1;
end

//set x4 and x5
initial
begin
    #10     x4 = 1'b1;
    #10     x5 = 1'b1;
end

//reset variables
initial
begin
    #20     x1 = 1'b0;
    #10     x2 = 1'b0;
    #10     x3 = 1'b0;
    #10     x4 = 1'b0;
    #10     x5 = 1'b0;
end

//determine length of simulation
initial
    #70     $finish;
endmodule
```

Figure 4.148   (Continued)

```
0   x1x2x3x4x5 = 00000
10  x1x2x3x4x5 = 11010
20  x1x2x3x4x5 = 01111
30  x1x2x3x4x5 = 00111
40  x1x2x3x4x5 = 00011
50  x1x2x3x4x5 = 00001
60  x1x2x3x4x5 = 00000
```

Figure 4.149   Outputs for the module of Figure 4.148.

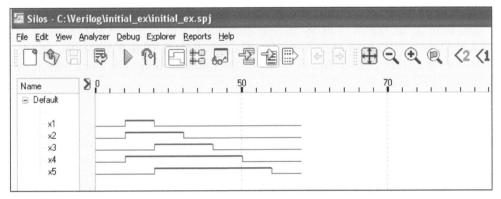

Figure 4.150   Waveforms for the module of Figure 4.148.

Refer to the module of Figure 4.148 and the waveforms of Figure 4.150 for the discussion that follows. Figure 4.148 contains seven **initial** statements. The first **initial** statement invokes the system task **$monitor**, which causes the specified string (enclosed in quotation marks) to be printed whenever a variable changes in the argument list (enclosed in braces). The **$time** system function returns the simulation time as a 64-bit number.

The second **initial** statement initializes all variables to zero. The third **initial** statement sets x_1 at 10 time units. Since all **initial** statements begin execution at time zero, the fourth **initial** statement sets x_2 at 10 time units also, and sets x_3 at time 20 time units (#10 plus #10). This can be seen in the waveforms of Figure 4.150. Variable x_4 is set at 10 time units by the fifth **initial** statement, which also sets x_5 at 20 time units. The sixth **initial** statement resets all variables. The seventh **initial** statement invokes the system task **$finish** which causes the simulator to exit the module and return control to the operating system.

4.4.2 Always Statement

The **always** statement executes the behavioral statements within the **always** block repeatedly in a looping manner and begins execution at time zero. Execution of the statements continues indefinitely until the simulation is terminated. The keywords **initial** and **always** specify a behavior and the statements within a behavior are classified as *behavioral* or *procedural*. The syntax for the **always** statement is shown below.

> **always** [optional timing control] procedural statement or
> block of procedural statements

A typical application of the **always** statement is shown in Figure 4.151 to generate a series of clock pulses as used in test bench constructs. There are two **initial** statements and one **always** statement. The clock is first initialized to zero, then the **always** statement cycles the clock every 10 time units for a clock period of 20 time units. The clock stops cycling after 100 time units. The system task **$finish** causes the simulator to exit the module and return control to the operating system. The clock waveform is shown in Figure 4.152.

```verilog
//clock generation using initial and always statements
module clk_gen2 (clk);


output clk;
reg clk;


//initialize clock to 0
initial
    clk = 1'b0;


//toggle clock every 10 time units
always
    #10   clk = ~clk;


//determine length of simulation
initial
    #100   $finish;

endmodule
```

Figure 4.151   Clock waveform generation.

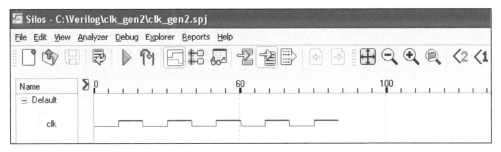

Figure 4.152 Waveform for the clock generation module of Figure 4.151.

Example 4.27 Figure 4.153 shows a 3-input OR gate, which will be designed using behavioral modeling. The behavioral module is shown in Figure 4.154 using an **always** statement. The expression within the parentheses is called an *event control* or *sensitivity list*. Whenever a variable in the event control list changes value, the statements in the **begin** ... **end** block will be executed; that is, if either x_1 or x_2 or x_3 changes value, the following statement will be executed:

$$z_1 = x_1 \mid x_2 \mid x_3;$$

where the symbol (|) signifies the logical OR operation.

If only a single statement appears after the **always** statement, then the keywords **begin** and **end** are not required. It is often useful, however, to include the **begin** and **end** keywords because additional statements may have to be added later. The **always** statement has a sequential block (**begin** ... **end**) associated with an event control. The statements within a **begin** ... **end** block execute sequentially and execution suspends when the last statement has been executed. When the sequential block completes execution, the **always** statement checks for another change of variables in the event control list.

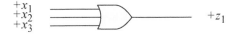

Figure 4.153 Three-input OR gate to be implemented using behavioral modeling.

The test bench for the module is shown in Figure 4.155. As stated previously, the inputs for a test bench are of type **reg** because they retain their value until changed, and the outputs are of type **wire**. To ensure that all eight combinations of the inputs are tested, a register vector called *invect* is specified as four bits: 3 through 0, where bit 0 is the low-order bit. This guarantees that a vector of $x_1 x_2 x_3 = 111$ will be applied to the OR gate inputs. The inputs are applied in sequence, $x_1 x_2 x_3 = 000$ through 111. When an input vector of $x_1 x_2 x_3 = 1000$ is reached, the test in the **for** loop returns a value of false and the simulator exits the statements in the **begin** . . . **end** sequence. The binary outputs of the simulator are shown in Figure 4.156 listing the output value for z_1 for all combinations of inputs. The waveforms are shown in Figure 4.157.

```
//behavioral 3-input or gate
module or3 (x1, x2, x3, z1);

//define inputs and output
input x1, x2, x3;
output z1;

//declare inputs as type wire and output as type reg
wire x1, x2, x3;
reg z1;

//define output z1
always @ (x1 or x2 or x3)   //the sensitivity list is x1, x2, x3
begin
   z1 = x1 | x2 | x3;
end
endmodule
```

Figure 4.154   Behavioral module for the 3-input OR gate of Figure 4.153.

```
//test bench for the or3 module
module or3_tb;

reg x1, x2, x3;
wire z1;

//apply input vectors
initial
begin: apply_stimulus
   reg [3:0] invect;
   for (invect = 0; invect < 8; invect = invect + 1)
      begin
         {x1, x2, x3} = invect [3:0];
         #10 $display ("{x1x2x3} = %b, z1 = %b",
               {x1, x2, x3}, z1);
      end
end          //continued on next page
```

Figure 4.155   Test bench for the 3-input OR gate module of Figure 4.154.

```
//instantiate the module into the test bench
or3 inst1 (
   .x1(x1),
   .x2(x2),
   .x3(x3),
   .z1(z1)
   );
endmodule
```

Figure 4.155   (Continued)

```
{x1x2x3} = 000, z1 = 0
{x1x2x3} = 001, z1 = 1
{x1x2x3} = 010, z1 = 1
{x1x2x3} = 011, z1 = 1
{x1x2x3} = 100, z1 = 1
{x1x2x3} = 101, z1 = 1
{x1x2x3} = 110, z1 = 1
{x1x2x3} = 111, z1 = 1
```

Figure 4.156   Outputs for the test bench of Figure 4.155.

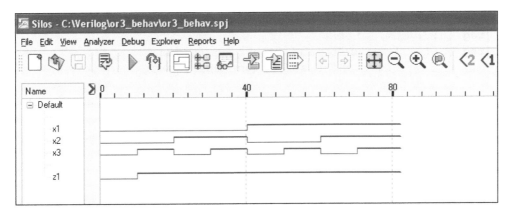

Figure 4.157   Waveforms for the 3-input OR gate module of Figure 4.154.

Example 4.28   This example presents the design of an add-subtract-shift unit that illustrates behavioral modeling for an 8-bit adder/subtractor together with shift left and shift right capabilities. The sensitivity list in the **always** statement contains the augend/minuend a and the addend/subtrahend b. When either a or b changes value, the sum and difference are obtained. The sum is then shifted left four bit positions for an equivalent multiply-by-16 operation; the sum is also shifted right three bit positions for an equivalent divide-by-8 operation. The difference is shifted left three bit positions for an equivalent multiply-by-8 operation; the difference is also shifted right two bit positions for an equivalent divide-by-4 operation.

The add operation produces a 9-bit result to accommodate the sum plus carry-out. The *left_shift_add*, *right_shift_add*, *left_shift_sub*, and *right_shift_sub* are 16-bit registers to allow for larger shift amounts. The module provides no information on the detailed design or architecture of the unit — it simply states the behavior.

The design module is shown in Figure 4.158 and the test bench is shown in Figure 4.159, providing six different input vectors. The outputs are shown in Figure 4.160 and the waveforms are shown in Figure 4.161.

```
//behavioral add-sub-shift unit
module add_sub_shift (a, b, sum, diff,
                        left_shift_add, right_shift_add,
                        left_shift_sub, right_shift_sub);

//define inputs and outputs
input [7:0] a, b;
output [8:0] sum, diff;
output [15:0] left_shift_add, right_shift_add,
              left_shift_sub, right_shift_sub;

//declare inputs as wire and outputs as reg
wire [7:0] a, b;
reg [8:0] sum, diff;
reg [15:0] left_shift_add, right_shift_add,
           left_shift_sub, right_shift_sub;

//use always with a sensitivity list and a begin ... end block
always @ (a or b)
begin
   sum = a + b;
   diff = a - b;

   left_shift_add = sum << 4;      //multiply by 16
   right_shift_add = sum >> 3;     //divide by 8

   left_shift_sub = diff << 3;     //multiply by 8
   right_shift_sub = diff >> 2;    //divide by 4
end

endmodule
```

Figure 4.158   Module for the add-subtract-shift unit of Example 4.28.

```verilog
//test bench for the add-subtract-shift unit
module add_sub_shift_tb;

reg [7:0] a, b;
wire [8:0] sum, diff;
wire [15:0] left_shift_add, right_shift_add,
            left_shift_sub, right_shift_sub;

//display variables
initial
$monitor ("a=%b, b=%b, sum=%b, diff=%b,
            left_shift_add=%b, right_shift_add=%b,
            left_shift_sub=%b, right_shift_sub=%b",
            a, b, sum, diff,
            left_shift_add, right_shift_add,
            left_shift_sub, right_shift_sub);

//apply input vectors
initial
begin
    #0     a = 8'b0011_1111;     b = 8'b0011_1111;

    #10    a = 8'b1100_1100;     b = 8'b0011_1111;

    #10    a = 8'b0100_0011;     b = 8'b0011_1001;

    #10    a = 8'b0011_1100;     b = 8'b0110_1111;

    #10    a = 8'b0111_0001;     b = 8'b0011_1000;

    #10    a = 8'b1111_1111;     b = 8'b1111_1111;

    #10    $stop;
end

//instantiate the module into the test book
add_sub_shift inst1 (
    .a(a),
    .b(b),
    .sum(sum),
    .diff(diff),
    .left_shift_add(left_shift_add),
    .right_shift_add(right_shift_add),
    .left_shift_sub(left_shift_sub),
    .right_shift_sub(right_shift_sub)
    );
endmodule
```

Figure 4.159   Test bench for the add-subtract-shift unit.

```
a = 00111111,  b = 00111111,
sum = 001111110, diff = 000000000,

left_shift_add  = 0000011111100000, <<4
right_shift_add = 0000000000001111, >>3
left_shift_sub  = 0000000000000000, <<3
right_shift_sub = 0000000000000000  >>2

a = 11001100,  b = 00111111,
sum = 100001011, diff = 010001101,

left_shift_add  = 0001000010110000, <<4
right_shift_add = 0000000000100001, >>3
left_shift_sub  = 0000010001101000, <<3
right_shift_sub = 0000000000100011  >>2

a = 01000011,  b = 00111001,
sum = 001111100, diff = 000001010,

left_shift_add  = 0000011111000000, <<4
right_shift_add = 0000000000001111, >>3
left_shift_sub  = 0000000001010000, <<3
right_shift_sub = 0000000000000010  >>2

a = 00111100,  b = 01101111,
sum = 010101011, diff = 111001101,

left_shift_add  = 0000101010110000, <<4
right_shift_add = 0000000000010101, >>3
left_shift_sub  = 0000111001101000, <<3
right_shift_sub = 0000000001110011  >>2

a = 01110001,  b = 00111000,
sum = 010101001, diff = 000111001,

left_shift_add  = 0000101010010000, <<4
right_shift_add = 0000000000010101, >>3
left_shift_sub  = 0000000111001000, <<3
right_shift_sub = 0000000000001110  >>2

a = 11111111,  b = 11111111,
sum = 111111110, diff = 000000000,

left_shift_add  = 0001111111100000, <<4
right_shift_add = 0000000000111111, >>3
left_shift_sub  = 0000000000000000, <<3
right_shift_sub = 0000000000000000  >>2
```

Figure 4.160   Outputs for the add-subtract-shift unit.

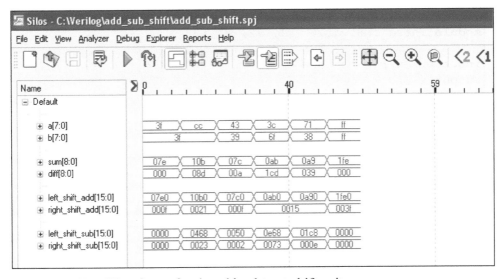

Figure 4.161 Waveforms for the add-subtract-shift unit.

4.4.3 Intrastatement Delay

An *intrastatement* delay is a delay on the right-hand side of the statement and indicates that the right-hand side is to be evaluated, wait the specified number of time units, and then assign the value to the left-hand side. This can be used to simulate logic gate delays. Equation 4.19 is an example of an intrastatement delay.

$$z_1 = \#5 \ x_1 \ \& \ x_2 \tag{4.19}$$

The statement evaluates the logical function x_1 AND x_2, waits five time units, then assigns the result to z_1. If no delay is specified in a procedural assignment, then zero delay is the default delay and the assignment occurs instantaneously.

The module of Figure 4.162 illustrates the intrastatement delay of five time units for Equation 4.19 using an **always** statement. Figure 4.163 shows the test bench in which all combinations of the two inputs are applied to x_1 and x_2. The waveforms are shown in Figure 4.164 that clearly display the intrastatement delay that generates z_1 for Equation 4.19. When x_1 and x_2 both become asserted, a delay of five time units takes place, then output z_1 is asserted.

```
//behavioral model to demonstrate intrastatement delay
module intra_stmt_dly4 (x1, x2, z1);

input x1, x2;
output z1;

reg z1;

always @ (x1 or x2)
begin
   z1 = #5 (x1 & x2);
end
endmodule
```

Figure 4.162   Module to illustrate intrastatement delay.

```
//test bench for intrastatement delay
module intra_stmt_dly4_tb;

reg x1, x2;
wire z1;

//display variables
initial
$monitor ("x1 x2 = %b, z1 = %b", {x1, x2}, z1);

//apply input vectors and display variables
initial
begin
   #0    x1 = 1'b0;   x2 = 1'b0;
   #10   x1 = 1'b0;   x2 = 1'b1;
   #10   x1 = 1'b1;   x2 = 1'b0;
   #10   x1 = 1'b1;   x2 = 1'b1;
   #10   x1 = 1'b0;   x2 = 1'b0;
end

//instantiate the module into the test bench
intra_stmt_dly4 inst1 (
   .x1(x1),
   .x2(x2),
   .z1(z1)
   );
endmodule
```

Figure 4.163   Test bench for the intrastatement delay module of Figure 4.162.

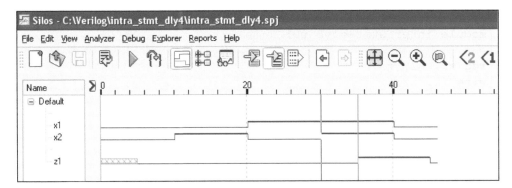

Figure 4.164   Waveforms for the intrastatement delay module of Figure 4.162.

4.4.4 Interstatement Delay

An *interstatement* delay is the delay by which a statement's execution is delayed; that is, it is the delay between statements. The code segment of Equation 4.20 is an example of an interstatement delay.

$$z_1 = x_1 \mid x_2$$
$$\#5 \; z_2 = x_1 \; \& \; x_2 \tag{4.20}$$

When the first statement has completed execution, a delay of five time units will be taken before the second statement is executed. If no delays are specified in a procedural assignment, then there is zero delay in the assignment. The behavioral module of Figure 4.165 illustrates the use of an interstatement delay. The test bench is shown in Figure 4.166 and the waveforms are shown in Figure 4.167.

```
//behavioral module to illustrate interstatement delay
module inter_stmt_dly2 (x1, x2, z1, z2);

input x1, x2;
output z1, z2;

reg z1, z2;

always @ (x1 or x2)
begin
        z1 = (x1 | x2);
    #5 z2 = (x1 & x2);
end
endmodule
```

Figure 4.165   Module to illustrate interstatement delay.

```
//test bench for interstatement delay
module inter_stmt_dly2_tb;

reg x1, x2, x3;
wire z1, z2;

//display variables
initial
$monitor ("x1 x2 = %b, z1 = %b, z2 = %b", {x1, x2}, z1, z2);

//apply input vectors and display variables
initial
begin
    #0      x1 = 1'b0;   x2 = 1'b0;

    #10     x1 = 1'b0;   x2 = 1'b1;
    #10     x1 = 1'b0;   x2 = 1'b0;

    #10     x1 = 1'b1;   x2 = 1'b1;
    #10     x1 = 1'b0;   x2 = 1'b0;
end

//instantiate the module into the test bench
inter_stmt_dly2 inst1 (
    .x1(x1),
    .x2(x2),
    .z1(z1),
    .z2(z2)
    );
endmodule
```

Figure 4.166   Test bench for the interstatement delay module of Figure 4.165.

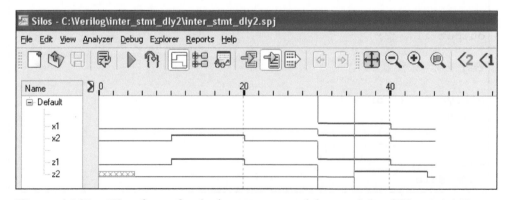

Figure 4.167   Waveforms for the interstatement delay module of Figure 4.165.

4.4.5 Blocking Assignments

A blocking procedural assignment completes execution before the next statement executes. The assignment operator ($=$) is used for blocking assignments. The right-hand expression is evaluated, then the assignment is placed in an internal temporary register called the *event queue* and scheduled for assignment. If no time units are specified, then the scheduling takes place immediately. The event queue is covered in Appendix A in this book.

In the code segment below, an interstatement delay of one time unit is specified for the assignment to z_2. The evaluation of z_2 is delayed by the timing control; that is, the expression for z_2 will not be evaluated until the expression for z_1 has been executed, plus one time unit. The execution of any following statements is blocked until the assignment occurs.

```
initial
   begin
            z1 = x1 & x2;
      #1  z2 = x2 | x3;
   end
```

Example 4.29 The module of Figure 4.168 shows delayed blocking assignments for a block of three statements, each with an interstatement delay of one time unit. The blocking statement for z_1 is assigned to be executed one time unit later than the current simulation time t at $t + 1$. The right-hand side expression is evaluated at time $t + 1$ and assigned to z_1 at time $t + 1$. The statement for z_2 is evaluated at time $t + 2$, then assigned to z_2. The statement for z_3 is evaluated at time $t + 3$, then assigned to z_3.

The test bench is shown in Figure 4.169 and the waveforms are shown in Figure 4.170, which show the delay for each blocking statement. Observe the waveforms of Figure 4.170 for the discussion which follows.

At 60 time units, x_1 and x_2 are both asserted; therefore, at $t + 1$ time units (61), the following equation for output z_1 is true

$$\#1 \; z_1 = x_1 \; \& \; x_2;$$

In blocking assignments, the time units are cumulative; therefore, the equation for z_2 does not execute until $t + 2$ time units. This is clearly observed in the waveforms. At 10 time units, the equation for z_2 becomes true

$$\#1 \; z_2 = (x_1 \; \& \; x_2) \,|\, x_3;$$

because $x_1 x_2 x_3 = 001$, where x_3 is asserted and two time units later at $t + 2$ (12) output z_2 is asserted.

Output z_3 is asserted three time units later at $t + 3$ if the input conditions are established. The equation for z_3

$$\#1 \; z_3 = (x_1 \; {}^\wedge x_2) \; \& \; x_3;$$

becomes true at 30 time units, where $x_1 x_2 x_3 = 011$. Therefore, three time units later at $t + 3$ (33) output z_3 is asserted.

```
//example of blocking assignment
module blocking5 (x1, x2, x3, z1, z2, z3);

input x1, x2, x3;
output z1, z2, z3;

reg z1, z2, z3;

always @ (x1 or x2 or x3)
begin
    #1 z1 = x1 & x2;
    #1 z2 = (x1 & x2) | x3;
    #1 z3 = (x1 ^ x2) & x3;
end

endmodule
```

Figure 4.168   Module to illustrate interstatement delays.

```
//test bench for blocking assignment
module blocking5_tb;

reg x1, x2, x3;
wire z1, z2, z3;

//apply input vectors and display variables
initial
begin: apply_stimulus
    reg [3:0] invect;
    for (invect = 0; invect < 8; invect = invect + 1)
        begin
            {x1, x2, x3} = invect [3:0];
            #10 $display ("x1 x2 x3 = %b, z1=%b, z2=%b, z3=%b",
                        {x1, x2, x3}, z1, z2, z3);
        end
end

//continued on next page
```

Figure 4.169   Test bench for the interstate delay module of Figure 4.168.

```
//instantiate the module into the test bench
blocking5 inst1 (
    .x1(x1),
    .x2(x2),
    .x3(x3),
    .z1(z1),
    .z2(z2),
    .z3(z3)
    );
endmodule
```

Figure 4.169 (Continued)

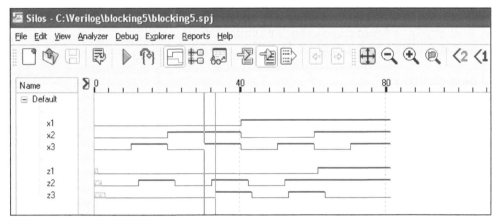

Figure 4.170 Waveforms for the interstatement delay module of Figure 4.168.

4.4.6 Nonblocking Assignments

The assignment symbol ($<=$) is used to represent a nonblocking procedural assignment. Nonblocking assignments allow the scheduling of assignments without blocking execution of the following statements in a sequential procedural block. A nonblocking assignment is used to synchronize assignment statements so that they appear to execute at the same time.

The Verilog simulator schedules a nonblocking assignment statement to execute, then proceeds to the next statement in the block without waiting for the previous nonblocking statement to complete execution. That is, the right-hand expression is evaluated and the value is stored in the event queue and is *scheduled* to be assigned to the left-hand target. The assignment is made at the end of the current time step if there are no intrastatement delays specified.

Nonblocking assignments are typically used to model several concurrent assignments that are caused by a common event such as the low-to-high transition of a clock pulse or a change to any variable in a sensitivity list (event control). The order of the assignments is irrelevant because the right-hand side evaluations are stored in the event queue before any assignments are made.

Example 4.30   This example uses nonblocking assignments and intrastatement delays to create a waveform. Because nonblocking assignments are used, the delays are not cumulative. All the statements in Figure 4.171 begin execution at time zero. The execution of the first statement results in a value of 0 scheduled to be assigned to *clk* at time unit 5. The second statement schedules a value of 1 to be assigned to *clk* at time unit 10. The third statement schedules *clk* to receive a value of 0 at time unit 15. The assignments continue relative to time zero. The waveform is shown in Figure 4.172.

```verilog
//example of nonblocking statements
module nonblock (clk);

output clk;
reg clk;

initial
begin
  clk <= #5  1'b0;
  clk <= #10 1'b1;
  clk <= #15 1'b0;
  clk <= #20 1'b1;
  clk <= #30 1'b0;
end
endmodule
```

Figure 4.171   Module to illustrate the use of nonblocking assignments with intrastatement delays.

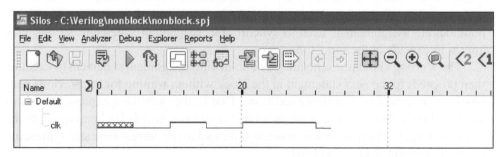

Figure 4.172   Waveform for Figure 4.171.

Example 4.31 A register is a logic macro device that stores data; the data is retained until new data is entered. This example will model register assignments using blocking and nonblocking constructs with intrastatement delays. The first three statements in the **initial** block of the module shown in Figure 4.173 use blocking assignments and execute sequentially at time 0. Because of the nonblocking behavioral construct in the **initial** block, the next three statements are processed at the same simulation time, but are scheduled to execute at different times due to the intrastatement delays.

```
//example of nonblocking assignment
module nonblock3 (data_reg_a, data_reg_b, index);

output [7:0] data_reg_a, data_reg_b;
output [3:0] index;

reg [7:0] data_reg_a, data_reg_b;
reg [3:0] index;

initial
begin
   data_reg_a = 8'h84;
   data_reg_b = 8'h0f;
   index = 4'b0;

   data_reg_a [1:0] <= #10 2'b11;
   data_reg_b [7:0] <= #5 {data_reg_a [3:0], 4'b0011};
   index <= index +1;
end

endmodule
```

Figure 4.173 Module to model blocking and nonblocking assignments.

The statement *data_reg_a[1:0] <= #10 2'b11;* is scheduled to execute at time unit 10, at which time the binary number 11 (3) replaces the two low-order bits of data register *a*, as shown below to yield 1000_0111. This is clearly seen in the waveforms of Figure 4.174, where the new contents of data register *a* are shown at time unit 10.

$$
\begin{array}{llr}
\text{Data register } a = & 1000\ 0100 & 84_{16} \\
+) & \underline{\hspace{1.4em}11} & \\
& 1000\ 0111 & 87_{16}
\end{array}
$$

The statement *data_reg_b[7:0] <= #5 {data_reg_a[3:0], 4'b0011};* is scheduled to execute at time unit 5, at which time the contents of data register *b* are replaced by the low-order four bits of data register *a* right-concatenated with the binary number 0011 (3) to yield 0100_0011. Since the nonblocking assignments are processed at the same simulation time, the low-order four bits of data register *a* have not yet changed and are still equal to *data_reg_a[3:0]* = 0100. This is shown in the waveforms at time unit five.

The last statement *index <= index + 1;* executes at time unit 0 because there is no intrastatement delay. The waveforms of Figure 4.174 show the assignments to the registers based on their scheduling in the event queue.

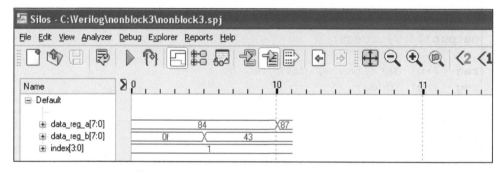

Figure 4.174   Waveforms for Figure 4.173.

Example 4.32 A behavioral module will be used to design a full adder using non-blocking statements with intrastatement delays of 5 time units. There are three scalar inputs: the augend *a*, the addend *b*, and the carry-in *cin*. There are two outputs: the sum of the adder *sum* and the carry-out *cout*. The truth table for a full adder is shown in Table 4.5.

Table 4.5 Truth Table for a Full Adder

a	b	cin	cout	sum
0	0	0	0	0
0	0	1	0	1
0	1	0	0	1
0	1	1	1	0
1	0	0	0	1
1	0	1	1	0
1	1	0	1	0
1	1	1	1	1

The Karnaugh map that represents the truth table of Table 4.5 is show in Figure 4.175. The equations for a full adder are shown in Equation 4.21 as obtained from the Karnaugh map. The design module is shown in Figure 4.176 and the test bench module is shown in Figure 4.177. The outputs and waveforms are presented in Figure 4.178 and Figure 4.179, respectively. The waveforms show that when an input changes value, the outputs are delayed by the intrastatement delay of five time units, then displayed.

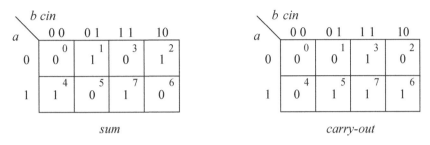

Figure 4.175   Karnaugh maps for the sum and carry-out of a full adder.

$$sum = a \oplus b \oplus cin$$

$$cout = ab + acin + bcin \tag{4.21}$$

```
//behavioral full adder using nonblocking assignments
module full_adder_nonblock (a, b, cin, sum, cout);
input a, b, cin;
output sum, cout;

wire a, b, cin;        //inputs are wire in behavioral
reg sum, cout;

initial                //initialize sum and cout to avoid Xs
begin
   sum = 1'b0; cout = 1'b0;
end

always @ (a or b or cin)
begin
   sum <= #5 (a ^ b ^ cin);    //nonblocking
   cout <= #5 ((a & b) | (a & cin) | (b & cin));
end
endmodule
```

Figure 4.176   Design module for a full adder using nonblocking assignments.

```
//test bench for full adder using nonblocking statements
module full_adder_nonblock_tb;

reg a, b, cin;
wire sum, cout;

//apply stimulus and display variables
initial
begin: apply_stimulus
   reg [3:0] invect;
   for (invect = 0; invect < 8; invect = invect + 1)
   begin
      {a, b, cin} = invect [3:0];
      #10 $display ("a b cin = %b, cout = %b, sum = %b",
                    {a, b, cin}, cout, sum);
   end
end

//instantiate the module into the test bench
full_adder_nonblock inst1 (
   .a(a),
   .b(b),
   .cin(cin),
   .sum(sum),
   .cout(cout)
   );

endmodule
```

Figure 4.177   Test bench for a full adder using intrastatement delays.

```
a b cin = 000, cout = 0, sum = 0
a b cin = 001, cout = 0, sum = 1
a b cin = 010, cout = 0, sum = 1
a b cin = 011, cout = 1, sum = 0
a b cin = 100, cout = 0, sum = 1
a b cin = 101, cout = 1, sum = 0
a b cin = 110, cout = 1, sum = 0
a b cin = 111, cout = 1, sum = 1
```

Figure 4.178   Outputs for a full adder using intrastatement delays.

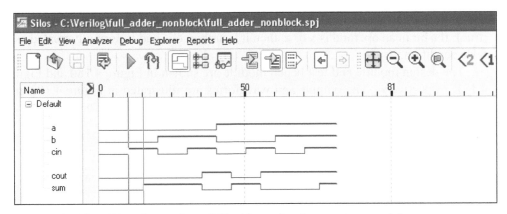

Figure 4.179   Waveforms for a full adder using intrastatement delays.

4.4.7 Conditional Statement

Conditional statements alter the flow within a behavior based upon certain conditions. The choice among alternative statements depends on the Boolean value of an expression. The alternative statements can be a single statement or a block of statements delimited by the keywords **begin . . . end**. The keywords **if** and **else** are used in conditional statements. There are three categories of the conditional statement as shown below. A true value is 1 or any nonzero value; a false value is 0, **x**, or **z**. If the evaluation is false, then the next expression in the activity flow is evaluated.

//no **else** statement
if (expression) statement1;   //if expression is true, then statement1 is executed.

//one **else** statement   //choice of two statements. Only one is executed.
if (expression) statement1;   //if expression is true, then statement1 is executed.
else statement2;   //if expression is false, then statement2 is executed.

//nested **if-else if**   //choice of multiple statements. Only one is executed.
if (expression1) statement1;   //if expression1 is true, then statement1 is executed.
else if (expression2) statement2; //if expression2 is true, then statement2 is executed.
else if (expression3) statement3; //if expression3 is true, then statement3 is executed.
else default statement;   //else the default statement is executed.

Example 4.33   This example uses scalar variables $x_1 x_2 x_3$ to illustrates the use of an **if . . . else** conditional statement to implement the expression $((x_1 \char94 x_2) || (x_2 \&\& x_3))$, where the symbol $||$ represents the logical OR operation and the symbol $\&\&$ represents the logical AND operation. Recall that the logical OR and logical AND operators are binary operations that evaluate to a logical 1 (true), a logical 0 (false), or an **x** (ambiguous). If a logical operation returns a nonzero value, then it is treated as a logical 1 (true); if a bit in an operand is **x** or **z**, then it is ambiguous and is normally treated as a false condition.

The design module and test bench module are shown in Figure 4.180 and Figure 4.181, respectively. The outputs and waveforms are illustrated in Figure 4.182 and Figure 4.183, respectively, both of which display the correct value for output z_1 for all combinations of the input values.

```verilog
//behavioral conditional statement using if ... else
module cond_if_else (x1, x2, x3, z1);

input x1, x2, x3;
output z1;
reg z1;

always @ (x1 or x2 or x3)
begin
   if ((x1 ^ x2) || (x2 && x3))//|| logical or; && logical and
      z1 = 1'b1;
   else
      z1 = 1'b0;
end
endmodule
```

Figure 4.180   Module to illustrate the conditional statement **if . . . else**.

```verilog
//test bench for cond_if_else module
module cond_if_else_tb;

reg x1, x2, x3;
wire z1;

//apply input vectors and display variables
initial
begin: apply_stimulus
   reg [3:0] invect;
   for (invect = 0; invect < 8; invect = invect + 1)
      begin
         {x1, x2, x3} = invect [3:0];
         #10 $display ("x1 x2 x3 =%b, z1 = %b",
                        {x1, x2, x3}, z1);
      end
end            //continued on next page
```

Figure 4.181   Test bench for the conditional statement module of Figure 4.180.

```
//instantiate the module into the test bench
cond_if_else inst1 (
   .x1(x1),
   .x2(x2),
   .x3(x3),
   .z1(z1)
   );

endmodule
```

Figure 4.181   (Continued)

```
x1 x2 x3 =000,  z1 = 0
x1 x2 x3 =001,  z1 = 0
x1 x2 x3 =010,  z1 = 1
x1 x2 x3 =011,  z1 = 1
x1 x2 x3 =100,  z1 = 1
x1 x2 x3 =101,  z1 = 1
x1 x2 x3 =110,  z1 = 0
x1 x2 x3 =111,  z1 = 1
```

Figure 4.182   Outputs for the conditional statement module of Figure 4.180.

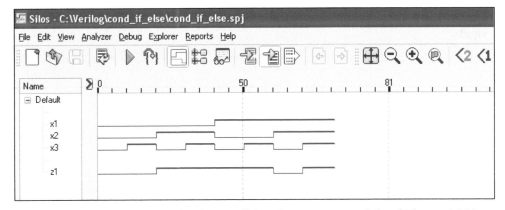

Figure 4.183   Waveforms for the conditional statement module of Figure 4.180.

Example 4.34   Now assume that the inputs are vector variables of eight bits for the same expression as utilized in Example 4.33; that is

$$z_1 = ((x_1 \,^\wedge x_2) \,||\, (x_2 \,\&\&\, x_3))$$

The same rules for conditional statements apply to vector variables as well as to scalar variables; that is, if a logical operation returns a nonzero value, then it is treated as a logical 1 (true); otherwise, it is treated as a logical 0 (false).

The design module and test bench module are shown in Figure 4.184 and Figure 4.185, respectively. The outputs and waveforms are shown in Figure 4.186 and Figure 4.187, respectively. Note that for a logical AND operation, if any vector is all zeroes, then the result is a logical 0 (false). For the logical OR operation, if at least one vector is nonzero, then the result is a logical 1 (true).

```
//behavioral conditional statement using if ... else
module cond_if_else_vect (x1, x2, x3, z1);

input [7:0] x1, x2, x3;
output z1;

reg z1;

always @ (x1 or x2 or x3)
begin
   if ((x1 ^ x2) || (x2 && x3))//|| logical or; && logical and
      z1 = 1'b1;
   else
      z1 = 1'b0;
end
endmodule
```

Figure 4.184   Module to illustrate the conditional statement **if . . . else** for vector inputs.

```
//test bench for conditional if ... else vectors
module cond_if_else_vect_tb;

reg [7:0] x1, x2, x3;
wire z1;

//display variables
initial
$monitor ("x1 = %b, x2 = %b, x3 = %b, z1 = %b",
          x1, x2, x3, z1);      //continued on next page
```

Figure 4.185   Test bench for Figure 4.184.

```
initial   //apply input vectors
begin
//x1 ^ x2 = 1; x2 && x3 = 1; z1 = 1
   #0     x1 = 8'b1111_0000;
          x2 = 8'b0000_1111;
          x3 = 8'b0101_0101;

//x1 ^ x2 = 1; x2 && x3 = 1; z1 = 1
   #10    x1 = 8'b1010_0110;
          x2 = 8'b0110_1010;
          x3 = 8'b0111_0100;

//x1 ^ x2 = 1; x2 && x3 = 1; z1 = 1
   #10    x1 = 8'b0000_0110;
          x2 = 8'b0110_1010;
          x3 = 8'b0111_0100;

//x1 ^ x2 = 0; x2 && x3 = 0; z1 = 0
   #10    x1 = 8'b1111_1111;
          x2 = 8'b1111_1111;
          x3 = 8'b0000_0000;

//x1 ^ x2 = 1; x2 && x3 = 1; z1 = 1
   #10    x1 = 8'b1010_0110;
          x2 = 8'b0110_1010;
          x3 = 8'b0111_0100;

//x1 ^ x2 = 0; x2 && x3 = 0; z1 = 0
   #10    x1 = 8'b0000_0000;
          x2 = 8'b0000_0000;
          x3 = 8'b0111_0100;

//x1 ^ x2 = 1; x2 && x3 = 1; z1 = 1
   #10    x1 = 8'b1010_0110;
          x2 = 8'b0110_1010;
          x3 = 8'b0111_0100;
   #10    $stop;
end

//instantiate the module into the test bench
cond_if_else_vect inst1 (
   .x1(x1),
   .x2(x2),
   .x3(x3),
   .z1(z1)
   );
endmodule
```

Figure 4.185   (Continued)

```
z1 = ((x1 ^ x2) || (x2 && x3))

x1 = 11110000, x2 = 00001111, x3 = 01010101, z1 = 1
x1 = 10100110, x2 = 01101010, x3 = 01110100, z1 = 1
x1 = 00000110, x2 = 01101010, x3 = 01110100, z1 = 1
x1 = 11111111, x2 = 11111111, x3 = 00000000, z1 = 0
x1 = 10100110, x2 = 01101010, x3 = 01110100, z1 = 1
x1 = 00000000, x2 = 00000000, x3 = 01110100, z1 = 0
x1 = 10100110, x2 = 01101010, x3 = 01110100, z1 = 1
```

Figure 4.186   Outputs for Figure 4.184.

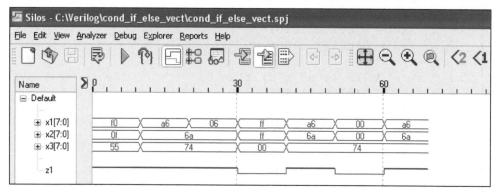

Figure 4.187   Waveforms for Figure 4.184.

4.4.8 Case Statement

The **case** statement is an alternative to the **if** . . . **else if** construct and may simplify the readability of the Verilog code. The **case** statement is a multiple-way conditional branch. It executes one of several different procedural statements depending on the comparison of an expression with a case item. The expression and the case item are compared bit-by-bit and must match exactly. The statement that is associated with a case item may be a single procedural statement or a block of statements delimited by the keywords **begin** . . . **end.** The **case** statement has the following syntax:

```
case (expression)
    case_item1 : procedural_statement1;
    case_item2 : procedural_statement2;
    case_item3 : procedural_statement3;
                .
                .
    case_itemn : procedural_statementn;
    default : default_statement;
endcase
```

The case expression may be an expression or a constant. The case items are evaluated in the order in which they are listed. If a match occurs between the case expression and the case item, then the corresponding procedural statement, or block of statements, is executed. If no match occurs, then the optional default statement is executed.

There are two other variations of the **case** statement that handle "don't cares": **casex** and **casez**. In the **casex** construct, the values of **x** and **z** that appear in either the case expression or in the case item are treated as "don't cares." In the **casez** construct, a value of **z** that appears in either the case expression or the case item is treated as a "don't care." The **casex** and **casez** statements are useful for comparing only certain bits of the case expression and case item. Wherever an **x** or **z** appears, that bit is ignored. Bit positions that correspond to **z** can be replaced by a **?** in those bit positions. An example of the use of "don't cares" is shown below in a **casex** construct. The **casez** construct can be used in a similar situation.

```
        casex (opcode)
            4'b1xxx : a + b;
            4'bx1xx : a − b;
            4'bxx1x : a * b;
            4'bxxx1 : a / b;
        endcase
```

Examples will now be presented that demonstrate the use of the **case** statement using behavioral modeling.

Example 4.35 An 8:1 multiplexer will be designed using the **case** statement. There is a vector select input *sel[2:0]*, a vector data input *data[7:0]*, and a scalar enable input *enbl*. There is one scalar output *out*. The design module is shown in Figure 4.188. The test bench module is shown in Figure 4.189 in which each data input is selected in turn. The outputs and waveforms are shown in Figure 4.190 and Figure 4.191, respectively.

```
//behavioral 8:1 multiplexer using the case statement
module mux_8to1_case (sel, data, enbl, out);

input [2:0] sel;
input [7:0] data;
input enbl;
output out;

reg out;

//continued on next page
```

Figure 4.188 Module to implement an 8:1 multiplexer using the **case** statement.

```
always @ (sel or data)
begin
   case (sel)
      (0): out = data[0];
      (1): out = data[1];
      (2): out = data[2];
      (3): out = data[3];
      (4): out = data[4];
      (5): out = data[5];
      (6): out = data[6];
      (7): out = data[7];
      default : out = 1'b0;
   endcase
end

endmodule
```

Figure 4.188   (Continued)

```
//test bench for 8:1 mux using case
module mux_8to1_case_tb;

reg [2:0] sel;
reg [7:0] data;
reg enbl;

wire out;

initial
$monitor ("sel = %b, data = %b, out = %b",
            sel, data, out);

//apply stimulus
initial
begin
   #0    sel = 3'b000;
         data = 8'b0000_0001; //out = 1
         enbl = 1'b0;

   #10   sel = 3'b001;
         data = 8'b0100_1001; //out = 0
         enbl = 1'b0;

//continued on next page
```

Figure 4.189   Test bench for the 8:1 multiplexer of Figure 4.188.

```
   #10    sel = 3'b010;
          data = 8'b0001_1010; //out = 0
          enbl = 1'b0;

   #10    sel = 3'b011;
          data = 8'b0010_1101; //out = 1
          enbl = 1'b0;

   #10    sel = 3'b100;
          data = 8'b0001_1001; //out = 1
          enbl = 1'b0;

   #10    sel = 3'b101;
          data = 8'b0001_1001; //out = 0
          enbl = 1'b0;

   #10    sel = 3'b110;
          data = 8'b1001_1001; //out = 0
          enbl = 1'b0;

   #10    sel = 3'b111;
          data = 8'b1001_1001; //out = 1
          enbl = 1'b0;

   #10    $stop;
end

//instantiate the module into the test bench
mux_8to1_case inst1 (
   .sel(sel),
   .data(data),
   .enbl(enbl),
   .out(out)
   );
endmodule
```

Figure 4.189   (Continued)

```
sel = 000, data = 00000001, out = 1
sel = 001, data = 01001001, out = 0
sel = 010, data = 00011010, out = 0
sel = 011, data = 00101101, out = 1
sel = 100, data = 00011001, out = 1
sel = 101, data = 00011001, out = 0
sel = 110, data = 10011001, out = 0
sel = 111, data = 10011001, out = 1
```

Figure 4.190   Outputs for the 8:1 multiplexer of Figure 4.188.

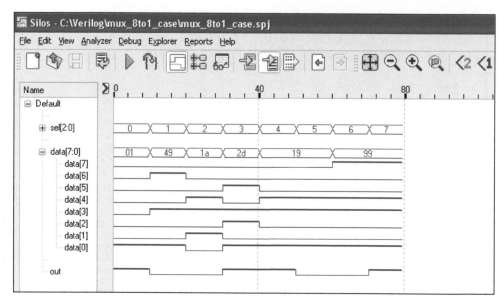

Figure 4.191   Waveforms for the 8:1 multiplexer of Figure 4.188.

Example 4.36   An 8-function arithmetic and logic unit (ALU) will be designed in this example using the **case** construct. There are two 4-bit inputs: operands *a[3:0]* and *b[3:0]* and one 3-bit input *opcode[2:0]*. There is one 8-bit output *z[7:0]* which contains the result of the operations and is declared to be 8 bits to accommodate the $2n$-bit product. The **parameter** keyword will declare and assign values to the operation codes. The operation codes are shown in Table 4.6.

Table 4.6 Operation Codes for the 8-Function ALU

Operation Code	Operation	Mnemonic
000	Add	add_op
001	Subtract	sub_op
010	Multiply	mul_op
011	Divide	div_op
100	AND	and_op
101	OR	or_op
110	Exclusive-OR	xor_op
111	Exclusive-NOR	xnor_op

A block diagram is shown in Figure 4.192. The behavioral module is shown in Figure 4.193. The test bench is shown in Figure 4.194 in which input vectors are applied for each operation code. Notice that the results for the add and subtract operations are correctly represented in 2s complement representation; thus, $4 - 8 = -4$ (11111100) and $6 - 15 = -9$ (11110111). The multiply and divide operations assume that the operands are positive. This will be explained more fully in Chapter 5 on computer arithmetic. The outputs are shown in Figure 4.195. The outputs indicate the result of each operation. If the previous result is not affected by an operation, then the previous result is unchanged. The waveforms are shown in Figure 4.196.

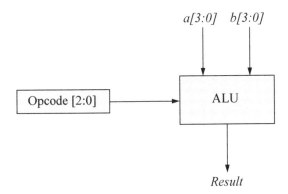

Figure 4.192   Block diagram for the 8-function ALU of Example 4.36.

```
//behavioral 8-function ALU
module alu_8fctn (a, b, opcode, rslt, rslt_mul);

input [3:0] a, b;
input [2:0] opcode;
output [3:0] rslt;
output [7:0] rslt_mul;

wire [3:0] a, b;        //inputs are wire
wire [2:0] opcode;
reg [3:0] rslt;         //outputs are reg
reg [7:0] rslt_mul;

//define operation codes
parameter   add_op  = 3'b000,
            sub_op  = 3'b001,
            mul_op  = 3'b010,
            div_op  = 3'b011,
            and_op  = 3'b100,
            or_op   = 3'b101,
            xor_op  = 3'b110,
            xnor_op = 3'b111;
//continued on next  page
```

Figure 4.193   Module for 8-function ALU.

```verilog
//perform the operations
always @(a or b or opcode)
begin
   case (opcode)
      add_op :  rslt = a + b;
      sub_op :  rslt = a - b;
      mul_op :  rslt_mul = a * b;
      div_op :  rslt = a / b;
      and_op :  rslt = a & b;
      or_op  :  rslt = a | b;
      xor_op :  rslt = a ^ b;
      xnor_op:  rslt = a ^~ b;
      default:  rslt = 8'b0000_0000;
   endcase
end

endmodule
```

Figure 4.193   (Continued)

```verilog
//test bench for 8-function ALU
module alu_8fctn_tb;

reg [3:0] a, b;
reg [2:0] opcode;
wire [3:0] rslt;
wire [7:0] rslt_mul;

//display variables
initial
$monitor ("a=%b, b=%b, opcode=%b, result=%b, result_mul=%d",
            a, b, opcode, rslt, rslt_mul);

//apply input vectors
initial
begin
//add operation
   #0  a=4'b0001; b=4'b0001; opcode=3'b000; //sum=2
   #10 a=4'b0010; b=4'b1101; opcode=3'b000; //sum=15(f)
   #10 a=4'b1111; b=4'b1111; opcode=3'b000; //sum=30(1e)

//continued on next page
```

Figure 4.194   Test bench for the 8-function ALU.

```
//subtract operation
   #10 a=4'b1000; b=4'b0100; opcode=3'b001;//diff=4
   #10 a=4'b1111; b=4'b0101; opcode=3'b001;//diff=10(a)
   #10 a=4'b1110; b=4'b0011; opcode=3'b001;//diff=11(b)
   #10 a=4'b0100; b=4'b1000; opcode=3'b001;//diff=-4(fc)
   #10 a=4'b0110; b=4'b1111; opcode=3'b001;//diff=-9(f7)

//multiply operation
   #10 a=4'b0100; b=4'b0111; opcode=3'b010;//product=28(1c)
   #10 a=4'b0101; b=4'b0011; opcode=3'b010;//product=15(f)
   #10 a=4'b1111; b=4'b1111; opcode=3'b010;//product=225(e1)

//divide operation
   #10 a=4'b1111; b=4'b0101; opcode=3'b011;//quotient=3
   #10 a=4'b1100; b=4'b0011; opcode=3'b011;//quotient=4
   #10 a=4'b1110; b=4'b0010; opcode=3'b011;//quotient=7
   #10 a=4'b0011; b=4'b1100; opcode=3'b011;//quotient=0

//and operation
   #10 a=4'b1100; b=4'b0100; opcode=3'b100;//result=0100
   #10 a=4'b1111; b=4'b0011; opcode=3'b100;//result=0011
   #10 a=4'b1010; b=4'b0101; opcode=3'b100;//result=0000

//or operation
   #10 a=4'b0101; b=4'b1010; opcode=3'b101;//result=1111
   #10 a=4'b0011; b=4'b1001; opcode=3'b101;//result=1011
   #10 a=4'b1000; b=4'b0000; opcode=3'b101;//result=1000

//exclusive-OR operation
   #10 a=4'b1010; b=4'b0101; opcode=3'b110;//result=1111
   #10 a=4'b0000; b=4'b0101; opcode=3'b110;//result=0101
   #10 a=4'b1111; b=4'b1010; opcode=3'b110;//result=0101

//exclusive-NOR operation
   #10 a=4'b1010; b=4'b1010; opcode=3'b111;//result=1111
   #10 a=4'b1111; b=4'b0000; opcode=3'b111;//result=0000
   #10 a=4'b0110; b=4'b0011; opcode=3'b111;//result=1010

   #10    $stop;

end

//continued on next page
```

Figure 4.194   (Continued)

```
//instantiate the module into the test bench
alu_8fctn inst1 (
    .a(a),
    .b(b),
    .opcode(opcode),
    .rslt(rslt),
    .rslt_mul(rslt_mul)
    );

endmodule
```

Figure 4.194   (Continued)

```
a=0001, b=0001, opcode=000, result=0010, result_mul=x    add
a=0010, b=1101, opcode=000, result=1111, result_mul=x
a=1111, b=1111, opcode=000, result=1110, result_mul=x
a=1000, b=0100, opcode=001, result=0100, result_mul=x    sub
a=1111, b=0101, opcode=001, result=1010, result_mul=x
a=1110, b=0011, opcode=001, result=1011, result_mul=x
a=0100, b=1000, opcode=001, result=1100, result_mul=x
a=0110, b=1111, opcode=001, result=0111, result_mul=x
a=0100, b=0111, opcode=010, result=0111, result_mul= 28 mul
a=0101, b=0011, opcode=010, result=0111, result_mul= 15
a=1111, b=1111, opcode=010, result=0111, result_mul=225
a=1111, b=0101, opcode=011, result=0011, result_mul=225 div
a=1100, b=0011, opcode=011, result=0100, result_mul=225
a=1110, b=0010, opcode=011, result=0111, result_mul=225
a=0011, b=1100, opcode=011, result=0000, result_mul=225
a=1100, b=0100, opcode=100, result=0100, result_mul=225 and
a=1111, b=0011, opcode=100, result=0011, result_mul=225
a=1010, b=0101, opcode=100, result=0000, result_mul=225
a=0101, b=1010, opcode=101, result=1111, result_mul=225 or
a=0011, b=1001, opcode=101, result=1011, result_mul=225
a=1000, b=0000, opcode=101, result=1000, result_mul=225
a=1010, b=0101, opcode=110, result=1111, result_mul=225 xor
a=0000, b=0101, opcode=110, result=0101, result_mul=225
a=1111, b=1010, opcode=110, result=0101, result_mul=225
a=1010, b=1010, opcode=111, result=1111, result_mul=225 xnor
a=1111, b=0000, opcode=111, result=0000, result_mul=225
a=0110, b=0011, opcode=111, result=1010, result_mul=225
```

Figure 4.195   Outputs for the 8-function ALU.

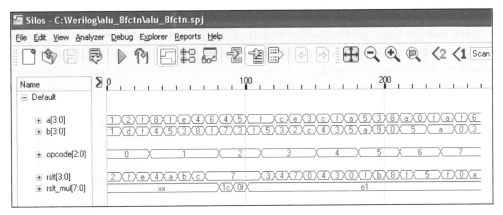

Figure 4.196   Waveforms for the 8-function ALU.

4.4.9 Loop Statements

There are four types of loop statements in Verilog: **for**, **while**, **repeat**, and **forever**. Loop statements must be placed within an **initial** or an **always** block and may contain delay controls. The loop constructs allow for repeated execution of procedural statements within an **initial** or an **always** block.

For loop   The **for** loop contains three parts:

1. An *initial* condition to assign a value to a register control variable. This is executed once at the beginning of the loop to initialize a register variable that controls the loop.

2. A *test* condition to determine when the loop terminates. This is an expression that is executed before the procedural statements of the loop to determine if the loop should execute. The loop is repeated as long as the expression is true. If the expression is false, the loop terminates and the activity flow proceeds to the next statement in the module.

3. An *assignment* to modify the control variable, usually an increment or a decrement. This assignment is executed after each execution of the loop and before the next test to terminate the loop.

The syntax of a **for** loop is shown below. The body of the loop can be a single procedural statement or a block of procedural statements.

for (initial control variable assignment; test expression; control variable assignment)
    procedural statement or block of procedural statements

The **for** loop is generally used when there is a known beginning and an end to a loop. The **for** loop is similar in function to the **for** loop in the C programming language and has been used in the test bench of several previous examples. A code segment is shown in Figure 4.197 using a **for** loop which provides all 16 combinations of the four bits x_1, x_2, x_3, and x_4.

```
//apply input vectors and display variables
initial
begin: apply_stimulus
    reg [4:0] invect;
    for (invect = 0; invect < 16; invect = invect + 1)
        begin
            {x1, x2, x3, x4} = invect [4:0];
            #10 $display ("{x1x2x3x4} = %b, z1 = %b",
                            {x1, x2, x3, x4}, z1);
        end
end
```

Figure 4.197   Code segment using a **for** loop.

While loop   The **while** loop executes a procedural statement or a block of procedural statements as long as a Boolean expression returns a value of true. When the procedural statements are executed, the Boolean expression is reevaluated. The loop is executed until the expression returns a value of false. If the evaluation of the expression is false, then the **while** loop is terminated and control is passed to the next statement in the module. If the expression is false before the loop is initially entered, then the **while** loop is never executed.

The Boolean expression may contain any of the following types: arithmetic, logical, relational, equality, bitwise, reduction, shift, concatenation, replication, or conditional. If the **while** loop contains multiple procedural statements, then they are contained within the **begin** . . . **end** keywords. The syntax for a **while** statement is as follows:

> **while** (expression)
>   procedural statement or block of procedural statements

Example 4.37   This example demonstrates the use of the **while** construct to count the number of 1s in a 16-bit register *reg_a*. The module is shown in Figure 4.198. The variable *count* is declared as type **integer** and is used to obtain the cumulative count of the number of 1s. The first **begin** keyword must have a name associated with the keyword because this declaration is allowed only with named blocks.

The register is initialized to contain ten 1s (*16'hd37c*). Alternatively, the register can be loaded from any other register. If *reg_a* contains a 1 bit in any bit position, then the **while** loop is executed. If *reg_a* contains all zeroes, then the **while** loop is terminated.

The low-order bit position (*reg_a[0]*) is tested for a 1 bit. If a value of 1 (true) is returned, *count* is incremented by one and the register is shifted right one bit position. There is only one procedural statement following the **if** statement; therefore, if a value of 0 (false) is returned, then *count* is not incremented and the register is shifted right one bit position.

The **$display** system task then displays the number of 1s that were contained in the register *reg_a* as shown in Figure 4.199. Notice that the count changes value only when there is a 1 bit in the low-order bit position of *reg_a*. If *reg_a[0]* = 0, then the count is not incremented, but the total count is still displayed.

```verilog
//example of a while loop
//count the number of 1s in a 16-bit register
module while_loop3;

integer count;

initial
begin: number_of_1s
   reg [16:0] reg_a;

   count = 0;

   reg_a = 16'hd37c;              //set reg_a to a known value

   while (reg_a)                  //do while reg_a contains 1s
      begin
         if (reg_a[0])            //check low-order bit
            count = count + 1;//if true, add one to count
         reg_a = reg_a >> 1;   //shift right 1 bit position

         $display ("count = %d", count);
      end
end

endmodule
```

Figure 4.198   Module to illustrate the use of the **while** construct.

count = 0	count = 5	count = 8
count = 0	count = 5	count = 8
count = 1	count = 6	count = 9
count = 2	count = 7	count = 10
count = 3	count = 7	
count = 4	count = 7	

Figure 4.199 Outputs for the module of Figure 4.198 that illustrates the use of the **while** loop.

Repeat loop The **repeat** loop executes a procedural statement or a block of procedural statements a specified number of times. The **repeat** construct can contain a constant, an expression, a variable, or a signed value. The syntax for the **repeat** loop is as follows:

> **repeat** (loop count expression)
> procedural statement or block of procedural statements

If the loop count is **x** or **z**, then the loop count is treated as zero. The value of the loop count expression is evaluated once at the beginning of the loop.

Example 4.38 An example of the **repeat** loop is shown in Figure 4.200. This module increments an integer *count* by two; the process is repeated 16 times. Since the integer *count* is initialized to zero, the count will stop when *count* is equal to 30. The outputs are shown in Figure 4.201 in decimal notation.

```
//example of the repeat keyword
module repeat_example2;

integer count;

initial
begin
   count = 0;
   repeat (16)
   begin
      $display ("count = %d", count);
      count = count + 2;
   end
end
endmodule
```

Figure 4.200 Example of the **repeat** construct.

```
count = 0          count = 16
count = 2          count = 18
count = 4          count = 20
count = 6          count = 22
count = 8          count = 24
count = 10         count = 26
count = 12         count = 28
count = 14         count = 30
```

Figure 4.201   Outputs for the **repeat** module of Figure 4.200.

Forever loop   The **forever** loop executes the procedural statement continuously until the system tasks **$finish** or **$stop** are encountered. It can also be terminated by the **disable** statement. The **disable** statement is a procedural statement; therefore, it must be used within an **initial** or an **always** block. It is used to prematurely terminate a block of procedural statements or a system task. When a **disable** statement is executed, control is transferred to the statement immediately following the procedural block or task. The **forever** loop is similar to a **while** loop in which the expression always evaluates to true (1). A timing control must be used with the **forever** loop; otherwise, the simulator would execute the procedural statement continuously without advancing the simulation time. The syntax of the **forever** loop is as follows:

> **forever**
> > procedural statement

The **forever** statement is typically used for clock generation as shown in Figure 4.202 together with the system task **$finish**. The variable *clk* will toggle every 10 time units for a period of 20 time units. The length of simulation is 100 time units.

```
//define clock
initial
begin
   clk = 1'b0;
   forever
      #10  clk = ~clk;
end

//define length of simulation
initial
   #100  $finish;
```

Figure 4.202   Clock generation using the **forever** statement.

4.4.10 Tasks

Verilog provides tasks and functions that are similar to procedures or subroutines found in other programming languages. These constructs allow a behavioral module to be partitioned into smaller segments. Tasks and functions permit modules to execute common code segments that are written once then called when required, thus reducing the amount of code needed. They enhance the readability and maintainability of the Verilog modules.

Tasks can be invoked only from within a behavior in the module. That is, they are called from an **always** block, an **initial** block, or from other tasks or functions. A task cannot be invoked from a continuous assignment statement and does not return values to an expression, but places the values on the **output** or **inout** ports. Tasks can contain delays, timing, or event control statements and can execute in nonzero simulation time when event control is applied. A task can invoke other tasks and functions and can have arguments of type **input**, **output**, or **inout**.

Task declaration   A task is delimited by the keywords **task** and **endtask**. The syntax for a task declaration is as follows:

```
task task_name
    input arguments
    output arguments
    inout  arguments
    task declarations
    local variable declarations
    begin
        statements
    end
endtask
```

Arguments (or parameters) that are of type **input** or **inout** are processed by the task statements; arguments that are of type **output** or **inout**, resulting from the task construct, are passed back to the task invocation statement. The keywords **input**, **output**, and **inout** are not ports of the module, they are ports used to pass values between the task invocation statement and the task construct. Additional local variables can be declared within a task, if necessary. Since tasks cannot be synthesized, they are used only in test benches. When a task completes execution, control is passed to the next statement in the module.

Task invocation   A task can be invoked (or called) from a procedural statement; therefore, it must appear within an **always** or an **initial** block. A task can call itself or be invoked by tasks that it has called. The syntax for a task invocation is as follows, where the expressions are parameters passed to the task:

```
task_name (expression1, expression2, . . . , expressionN);
```

Values for arguments of type **output** and **inout** are passed back to the variables in the task invocation statement upon completion of the task. The list of arguments in the task invocation must match the order of **input**, **output**, and **inout** variables in the task declaration. The **output** and **inout** arguments must be of type **reg** because a task invocation is a procedural statement.

Example 4.39   This example illustrates a module that contains a task to perform logical operations on two 8-bit vectors *a[7:0]* and *b[7:0]*. The logical operations are: AND, NAND, OR, NOR, exclusive-OR, and exclusive-NOR. Variables *a* and *b* are passed to the task *logical*, which returns the results as *a_and_b*, *a_nand_b*, *a_or_b*, *a_nor_b*, *a_xor_b*, and *a_xnor_b*. Figure 4.203 shows a block diagram of the task *logical* embedded in the task module *task_logical*. The task module is shown in Figure 4.204; the outputs are shown in Figure 4.205.

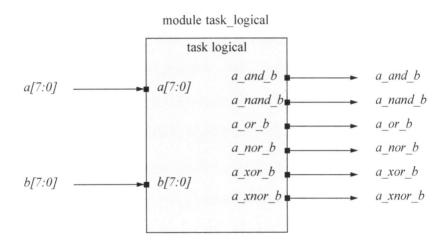

Figure 4.203   Block diagram for the task of Example 4.39.

```
//module to illustrate a task for logical operations
module task_logical;


reg [7:0] a, b;
reg [7:0] a_and_b, a_nand_b, a_or_b, a_nor_b,
          a_xor_b, a_xnor_b;


//continued on next page
```

Figure 4.204   Module for the logical task of Example 4.39.

```
initial
begin
      a=8'b1010_1010;    b=8'b1100_1100;
   logical (a, b, a_and_b, a_nand_b, a_or_b, a_nor_b,
          a_xor_b, a_xnor_b);      //invoke the task

     a=8'b1110_0111;    b=8'b1110_0111;
   logical (a, b, a_and_b, a_nand_b, a_or_b, a_nor_b,
          a_xor_b, a_xnor_b);      //invoke the task

     a=8'b0000_0111;    b=8'b0000_0111;
   logical (a, b, a_and_b, a_nand_b, a_or_b, a_nor_b,
          a_xor_b, a_xnor_b);      //invoke the task

     a=8'b0101_0101;    b=8'b1010_1010;
   logical (a, b, a_and_b, a_nand_b, a_or_b, a_nor_b,
          a_xor_b, a_xnor_b);      //invoke the task
end


task logical;
   input [7:0] a, b;
   output [7:0] a_and_b, a_nand_b, a_or_b, a_nor_b,
                a_xor_b, a_xnor_b;
   begin
      a_and_b = a & b;
      a_nand_b = ~(a & b);
      a_or_b = a | b;
      a_nor_b = ~(a | b);
      a_xor_b = a ^ b;
      a_xnor_b = ~(a ^ b);

      $display ("a=%b, b=%b, a_and_b=%b, a_nand_b=%b,
          a_or_b=%b, a_nor_b=%b, a_xor_b=%b, a_xnor_b=%b",
          a, b, a_and_b, a_nand_b, a_or_b, a_nor_b,
          a_xor_b, a_xnor_b);
   end
endtask

endmodule
```

Figure 4.204   (Continued)

```
a=10101010, b=11001100,
a_and_b=10001000, a_nand_b=01110111,
a_or_b=11101110, a_nor_b=00010001,
a_xor_b=01100110, a_xnor_b=10011001

a=11100111, b=11100111,
a_and_b=11100111, a_nand_b=00011000,
a_or_b=11100111, a_nor_b=00011000,
a_xor_b=00000000, a_xnor_b=11111111

a=00000111, b=00000111,
a_and_b=00000111, a_nand_b=11111000,
a_or_b=00000111, a_nor_b=11111000,
a_xor_b=00000000, a_xnor_b=11111111

a=01010101, b=10101010,
a_and_b=00000000, a_nand_b=11111111,
a_or_b=11111111, a_nor_b=00000000,
a_xor_b=11111111, a_xnor_b=00000000
```

Figure 4.205   Outputs for the logical task of Example 4.39.

Example 4.40   A module will be designed that contains a task to add two 8-bit operands $a[7:0]$, $b[7:0]$, and a carry-in *cin*. The module is shown in Figure 4.206 and the outputs are shown in Figure 4.207. Note that the last set of inputs generates a sum that produces an overflow.

```
//module to illustrate the use of a task
module task2_adder8;

reg [7:0] a, b;
reg cin;
reg [7:0] sum;

initial
begin
     a=2; b=5; cin=0;
   add (a, b, cin, sum);   //call the add task

     a=3; b=2; cin=1;
   add (a, b, cin, sum);
//continued on next page
```

Figure 4.206   A task module to add two 8-bit operands.

```
         a=4;  b=6;  cin=1;
     add (a, b, cin, sum);

         a=14; b=63; cin=1;
     add (a, b, cin, sum);

         a=150; b=225; cin=0;
     add (a, b, cin, sum);
end

task add;
    input [7:0] a, b;
    input cin;
    output [7:0] sum;

    begin
       sum = a + b + cin;
       $display ("a=%b, b=%b, cin=%b, sum=%b", a, b, cin, sum);
    end
endtask
endmodule
```

Figure 4.206   (Continued)

```
a=00000010, b=00000101, cin=0, sum=00000111
a=00000011, b=00000010, cin=1, sum=00000110
a=00000100, b=00000110, cin=1, sum=00001011
a=00001110, b=00111111, cin=1, sum=01001110
a=10010110, b=11100001, cin=0, sum=01110111
```

Figure 4.207   Outputs for the add task of Figure 4.206.

4.4.11 Functions

Functions are similar to tasks, except that functions return only a single value to the expression from which they are called. Like tasks, functions provide the ability to execute common procedures from within a module. A function can be invoked from a continuous assignment statement or from within a procedural statement and is represented by an operand in an expression.

Functions cannot contain delays, timing, or event control statements and execute in zero simulation time. Although functions can invoke other functions, they are not recursive. Functions cannot invoke a task. Functions must have at least one **input** argument, but cannot have **output** or **inout** arguments. The syntax for a function declaration is as follows:

function [range or type] function name
    input declaration
    other declarations
    begin
        statements
    end
endfunction

Function invocation   A function is invoked from an expression. The function is invoked by specifying the function name together with the input parameters. The syntax is shown below.

function name (expression1, expression2, . . . , expressionN);

All local registers that are declared within a function are static; that is, they retain their values between invocations of the function. When the function execution is finished, the return value is positioned at the location where the function was invoked.

Example 4.41   A function will be designed for a half adder that adds operands *a* and *b* and produces a result *sum*. The **case** statement will be used in the function. A block diagram of the module is shown in Figure 4.208. The module is shown in Figure 4.209 and the outputs are shown in Figure 4.210.

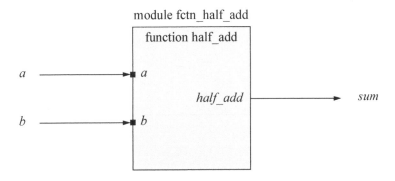

Figure 4.208   Block diagram of the half adder module.

```
//module for a half adder using a function
module fctn_half_add;

reg a, b;
reg [1:0] sum;

//continued on next page
```

Figure 4.209   Function for a half adder using the **case** statement.

```
initial
begin
   sum = half_add (1'b0, 1'b0);
      $display ("a=0, b=0, cout, sum = %b", sum);

   sum = half_add (1'b0, 1'b1);
      $display ("a=0, b=1, cout, sum = %b", sum);

   sum = half_add (1'b1, 1'b0);
      $display ("a=1, b=0, cout, sum = %b", sum);

   sum = half_add (1'b1, 1'b1);
      $display ("a=1, b=1, cout, sum = %b", sum);
end

function [1:0] half_add;
input a, b;
reg [1:0] sum;

begin
   case ({a,b})
      2'b00:  sum = 2'b00;
      2'b01:  sum = 2'b01;
      2'b10:  sum = 2'b01;
      2'b11:  sum = 2'b10;
      default:sum = 2'bxx;
   endcase

      half_add = sum;
end
endfunction
endmodule
```

Figure 4.209   (Continued)

```
a=0, b=0, cout, sum = 00
a=0, b=1, cout, sum = 01
a=1, b=0, cout, sum = 01
a=1, b=1, cout, sum = 10
```

Figure 4.210   Outputs for the half adder function.

Example 4.42 A module will be designed that contains a function to perform the logical operations of AND, NAND, OR, NOR, exclusive-OR, and exclusive-NOR on two 8-bit operands *a[7:0]* and *b[7:0]*. There is also an 8-bit result *rslt[7:0]*. A 3-bit mode control *mode[2:0]* is passed to the function, together with the two operands, to specify the function to be performed. The module is shown in Figure 4.211 and the outputs are shown in Figure 4.212.

```verilog
//function to perform logical operations on two 8-bit operands
module fctn_logical;
reg [7:0] a, b, rslt;
reg [2:0] mode;

initial
begin
  rslt = logic (8'b0000_1111, 8'b1111_1010, 3'b000);
  $display ("and, a=0000_1111, b=1111_1010, result=%b",rslt);

  rslt = logic (8'b0000_1111, 8'b1111_1010, 3'b001);
  $display ("nand, a=0000_1111, b=1111_1010, result=%b",rslt);

  rslt = logic (8'b0000_1111, 8'b1111_1010, 3'b010);
  $display ("or, a=0000_1111, b=1111_1010, result=%b",rslt);

  rslt = logic (8'b0000_1111, 8'b1111_1010, 3'b011);
  $display ("nor, a=0000_1111, b=1111_1010, result=%b",rslt);

  rslt = logic (8'b0000_1111, 8'b1111_1010, 3'b100);
  $display ("xor, a=0000_1111, b=1111_1010, result=%b",rslt);

  rslt = logic (8'b0000_1111, 8'b1111_1010, 3'b101);
  $display ("xnor, a=0000_1111, b=1111_1010, result=%b",rslt);
end

function [7:0] logic;
input [7:0] a, b;
input [2:0] mode;
reg [7:0] rslt;

begin
  case (mode)
    3'b000:  rslt = a & b;
    3'b001:  rslt = ~(a & b);
    3'b010:  rslt = a | b;
    3'b011:  rslt = ~(a | b);
    3'b100:  rslt = a ^ b;
    3'b101:  rslt = a ~^ b;
    default: rslt = 8'bxxxx_xxxx;
  endcase
    logic = rslt;
end
endfunction
endmodule
```

Figure 4.211   Module for a function that performs logical operations.

```
and,   a=0000_1111, b=1111_1010, result = 00001010
nand,  a=0000_1111, b=1111_1010, result = 11110101
or,    a=0000_1111, b=1111_1010, result = 11111111
nor,   a=0000_1111, b=1111_1010, result = 00000000
xor,   a=0000_1111, b=1111_1010, result = 11110101
xnor,  a=0000_1111, b=1111_1010, result = 00001010
```

Figure 4.212   Outputs for the logical function of Figure 4.211.

4.5 Structural Modeling

Structural modeling consists of instantiation of one or more of the following design objects:

- Built-in primitives
- User-defined primitives (UDPs)
- Design modules

Instantiation means to use one or more lower-level modules — including logic primitives — that are interconnected in the construction of a higher-level structural module. A module can be a logic gate, an adder, a multiplexer, a counter, or some other logical function. The objects that are instantiated are called *instances*. Structural modeling is described by the interconnection of these lower-level logic primitives or modules. The interconnections are made by wires that connect primitive terminals or module ports.

4.5.1 Module Instantiation

Design modules were instantiated into every test bench module in previous chapters. The ports of the design module were instantiated by name and connected to the corresponding net names of the test bench. Each named instantiation was of the form

.design_module_port_name (test_bench_module_net_name)

Design module ports can be instantiated by name explicitly or by position. Instantiation by position is not recommended when a large number of ports are involved. Instantiation by name precludes the possibility of making errors in the instantiation process. Modules cannot be nested, but they can be instantiated into other modules. Structural modeling is analogous to placing the instances on a logic diagram and then connecting them by wires. When instantiating built-in primitives, an instance name is optional; however, when instantiating a module, an instance name must be used. Instances that are instantiated into a structural module are connected by nets of type **wire**.

A structural module may contain behavioral statements (**always**), continuous assignment statements (**assign**), built-in primitives (**and**, **or**, **nand**, **nor**, etc.), UDPs (*mux4*, *half_adder*, *adder4*, etc.), design modules, or any combination of these objects. Design modules can be instantiated into a higher-level structural module in order to achieve a hierarchical design.

Each module in Verilog is either a top-level (higher-level) module or an instantiated module. There is only one top-level module and it is not instantiated anywhere else in the design project. Instantiated primitives or modules, however, can be instantiated many times into a top-level module and each instance of a module is unique.

4.5.2 Ports

Ports provide a means for the module to communicate with its external environment. Ports, also referred to as terminals, can be declared as **input**, **output**, or **inout**. A port is a net by default; however, it can be declared explicitly as a net. A module contains an optional list of ports, as shown below for a full adder.

module full_adder (a, b, cin, sum, cout);

Ports *a*, *b*, and *cin* are input ports; ports *sum* and *cout* are output ports. The test bench for the full adder contains no ports as shown below because it does not communicate with the external environment.

module full_adder_tb;

There are two methods of associating ports in the module being instantiated and the module doing the instantiation: instantiation by position and instantiation by name (the preferred method). The two methods cannot be mixed. Instantiation by position must have the ports in the module instantiation listed in the same order as in the module definition. Instantiation by name does not require the ports to be listed in the same order. Figure 4.213 shows a module for a full adder with ports *a*, *b*, *cin*, *sum*, and *cout*. Figure 4.214 shows the two methods of instantiation used by a test bench.

```
//dataflow full adder
module full_adder (a, b, cin, sum, cout);
input a, b, cin;               //list all inputs and outputs
output sum, cout;
wire a, b, cin;                //define wires
wire sum, cout;

assign sum = (a ^ b) ^ cin;    //continuous assignment
assign cout = cin & (a ^ b) | (a & b);
endmodule
```

Figure 4.213   Module for a full adder showing the port list.

```
//instantiate by name          //instantiate by position

full_adder inst1 (             full_adder inst1 (
     .b(b),                         a,
     .a(a),                         b,
     .sum(sum),                     cin,
     .cin(cin),                     sum,
     .cout(cout)                    cout
     );                             );
```

Figure 4.214   Two methods of instantiation.

Input ports   Input ports are those that allow signals to enter the module from external sources. The width of the input port is declared within the module. The size of the input port can be declared as either a scalar such as a, b, cin or as a vector such as *[3:0] a, b*, where a and b are the augend and addend inputs, respectively, to a 4-bit adder. The format of the declarations shown below is the same for both behavioral and structural modeling.

      input [3:0] a, b;   //declared as 4-bit vectors
      input cin;   //declared as a scalar

Output ports   Output ports are those that allow signals to exit the module to external destinations. The width of the output port is declared within the module. For behavioral modeling, the output is declared as type **reg** with a specified width, either scalar or vector. For structural modeling, the output port is declared as type **wire** with a specified width. The format for output ports is shown below.

      output [3:0] sum;   //declared as a 4-bit vector
      output cout;   //declared as a scalar

      reg [3:0] sum;   //for behavioral modeling
      reg cout;   //for behavioral modeling

      wire [3:0] sum;   //for structural modeling
      wire cout;   //for structural modeling

Inout ports   An **inout** port is bidirectional — it transfers signals to and from the module depending on the value of a direction control signal. Ports of type **inout** are declared internally as type **wire**; externally, they connect to nets of type **wire**. Since port declarations are implicitly declared as type **wire**, it is not necessary to explicitly declare a port as **wire**. However, an output can also be redeclared as a **reg** type variable if it is used within an **always** statement or an **initial** statement.

Unconnected ports   Ports can be left unconnected in an instantiation by leaving the port name blank as shown in the following for both instantiation by name and instantiation by position. If instantiating by name, the port corresponding to x_2 is left blank, indicating no connection. If instantiating by position, no port name is specified for x_3, indicating no connection. Input ports that are unconnected are assigned a value of **z**; output ports that are unconnected are unused.

```
//instantiate by name          //instantiate by position

xor_xnor inst1 (               xor_xnor inst1 (
    .x1(x1),                         x1,
    .x2(),                           x2,
    .x3(x3),                         ,
    .x4(x4),                         x4,
    .xor_out(xor_out),               xor_out,
    .xnor_out(xnor_out)              xnor_out
    );                               );
```

Port connection rules   A port is an entry into a module from an external source. It connects the external unit to the internal logic of the module. When a module is instantiated within another module, certain rules apply. An error message is indicated if the port connection rules are not followed. Figure 4.215 illustrates the rules for port connections.

Input ports must always be of type **wire** (net) internally except for test benches; externally, input ports can be **reg** or **wire**. Output ports can be of type **reg** or **wire** internally; externally, output ports must always be connected to a **wire**. The input port names can be different, but the net (**wire**) names connecting the input ports must be the same, as shown in Figure 4.215.

When making intermodule port connections, it is permissible to connect ports of different widths. Port width matching occurs by right justification or truncation. Figure 4.216 shows an example of connecting ports of different widths. The bit positions that are not connected are assigned a value of **z**.

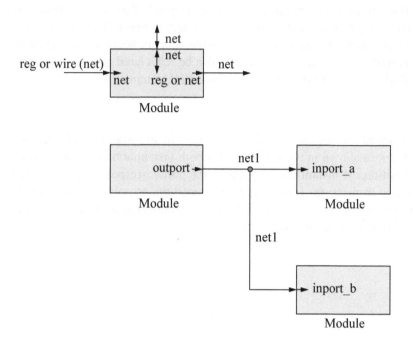

Figure 4.215   Diagram illustrating port connection rules.

```
module name (x1, z1);              module top;
                                   wire [1:0] x2;
input [3:0] x1;                    wire [4:0] z2;
output [1:0] z1;                      .
   .                                  .
   .                                  .
   .                               name inst1 (
endmodule                             .x1(x2),
                                      .z1(z2)
                                      );
                                   endmodule
```

	z	z		
x1	3	2	1	0

| z1 | | | | | | 1 | 0 |

| x2 | | | | | | 1 | 0 |

			z	z	z		
z2			4	3	2	1	0

Figure 4.216   Figure to illustrate connecting ports of different widths.

4.5.3 Design Examples

Several examples will be presented that illustrate the structural modeling technique for combinational logic. These examples include logic equations, majority circuits, non-linear-select multiplexers, a comparator, an adder and high-speed shifter, and an iterative network array. Each example will be completely designed in detail and will include appropriate theory where applicable.

Example 4.43   A combinational logic circuit will be designed using structural modeling that will implement the following function:

$$z_1 = (x_1 x_2 x_3 + x_2' x_3' x_4 + x_1 x_2' x_4)(x_2 + x_3 x_4')$$

This example will be a two-part problem. First, the logic will be designed directly from the equation — with no minimization — by instantiating NAND gates that were designed using dataflow modeling. The logic diagram is shown in Figure 4.217. The design module is shown in Figure 4.218. The test bench and outputs are shown in Figure 4.219 and Figure 4.220, respectively. The function will then be minimized using Boolean algebra and the results compared with the outputs of Figure 4.220.

   For the second part, the function for z_1 will be designed directly from the equation by instantiating NOR gates that were designed using dataflow modeling. The logic diagram is shown in Figure 4.221. The design module is shown in Figure 4.222 and uses a test bench that is equivalent to that shown in Figure 4.219. The outputs are shown in Figure 4.223.

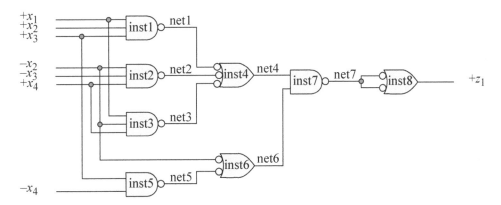

Figure 4.217   Logic diagram for the product-of-sums expression of Example 4.43 using NAND gates.

```
//structural logic equation       nand3_df inst4 (
//as a product of sums               .x1(net1),
//using only NAND logic              .x2(net2),
module log_eqn_pos_nand             .x3(net3),
      (x1, x2, x3, x4, z1);         .z1(net4)
                                    );
input x1, x2, x3, x4;
output z1;                        nand2_df inst5 (
                                    .x1(x3),
//define internal nets              .x2(~x4),
wire net1, net2, net3, net4,        .z1(net5)
      net3, net6, net7;             );

//instantiate the logic gates     nand2_df inst6 (
nand3_df inst1 (                    .x1(~x2),
   .x1(x1),                         .x2(net5),
   .x2(x2),                         .z1(net6)
   .x3(x3),                         );
   .z1(net1)
   );                             nand2_df inst7 (
                                    .x1(net4),
nand3_df inst2 (                    .x2(net6),
   .x1(~x2),                        .z1(net7)
   .x2(~x3),                        );
   .x3(x4),
   .z1(net2)                      nand2_df inst8 (
   );                               .x1(net7),
                                    .x2(net7),
nand3_df inst3 (                    .z1(z1)
   .x1(x1),                         );
   .x2(~x2),
   .x3(x4),                       endmodule
   .z1(net3)
   );
```

Figure 4.218   Module for the product-of-sums expression using NAND gates for Example 4.43.

```
//test bench for the product of sums using NAND gates
module log_eqn_pos_nand_tb;

reg x1, x2, x3, x4;
wire z1;
//continued on next page
```

Figure 4.219   Test bench for the module of Figure 4.218.

```
//apply input vectors and display variables
initial
begin: apply_stimulus
   reg [4:0] invect;
   for (invect = 0; invect < 16; invect = invect + 1)
      begin
         {x1, x2, x3, x4} = invect [4:0];
         #10 $display ("x1 x2 x3 x4 = %b, z1 = %b",
                       {x1, x2, x3, x4}, z1);
      end
end

//instantiate the module into the test bench
log_eqn_pos_nand inst1 (
   .x1(x1),
   .x2(x2),
   .x3(x3),
   .x4(x4),
   .z1(z1)
   );

endmodule
```

Figure 4.219   (Continued)

```
x1 x2 x3 x4 = 0000, z1 = 0
x1 x2 x3 x4 = 0001, z1 = 0
x1 x2 x3 x4 = 0010, z1 = 0
x1 x2 x3 x4 = 0011, z1 = 0
x1 x2 x3 x4 = 0100, z1 = 0
x1 x2 x3 x4 = 0101, z1 = 0
x1 x2 x3 x4 = 0110, z1 = 0
x1 x2 x3 x4 = 0111, z1 = 0
x1 x2 x3 x4 = 1000, z1 = 0
x1 x2 x3 x4 = 1001, z1 = 0
x1 x2 x3 x4 = 1010, z1 = 0
x1 x2 x3 x4 = 1011, z1 = 0
x1 x2 x3 x4 = 1100, z1 = 0
x1 x2 x3 x4 = 1101, z1 = 0
x1 x2 x3 x4 = 1110, z1 = 1
x1 x2 x3 x4 = 1111, z1 = 1
```

Figure 4.220   Outputs for the module of Figure 4.218.

The function $z_1 = (x_1 x_2 x_3 + x_2' x_3' x_4 + x_1 x_2' x_4)(x_2 + x_3 x_4')$ will now be minimized using Boolean algebra.

$$
\begin{aligned}
z_1 &= (x_1 x_2 x_3 + x_2' x_3' x_4 + x_1 x_2' x_4)(x_2 + x_3 x_4') \\
&= (x_1 x_2 x_3 x_2 + x_1 x_2 x_3 x_3 x_4') + (x_2' x_3' x_4 x_2 + x_2' x_3' x_4 x_3 x_4') + \\
&\quad (x_1 x_2' x_4 x_2 + x_1 x_2' x_4 x_3 x_4') \\
&= x_1 x_2 x_3 + x_1 x_2 x_3 x_4' \\
&= x_1 x_2 x_3 (1 + x_4') \\
&= x_1 x_2 x_3
\end{aligned}
$$

The equation for z_1 exactly replicates the outputs as shown in Figure 4.220. The logic diagram using NOR gates is shown below in Figure 4.221. The module and outputs are shown in Figure 4.222 and Figure 4.223, respectively. The Karnaugh map that represents the function z_1 is shown in Figure 4.224 and clearly displays the values of the minterm locations for z_1 for the sum-of-products form and the product-of-sums form.

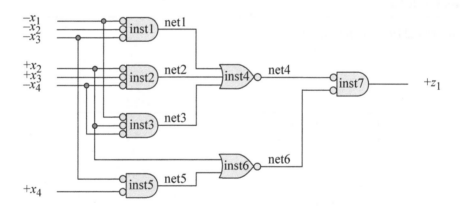

Figure 4.221   Logic diagram for the product-of-sums expression of Example 4.43 using NOR gates.

```
//structural logic equation as a product of sums
//using only NOR logic
module log_eqn_pos_nor (x1, x2, x3, x4, z1);
input x1, x2, x3, x4;
output z1;
//continued on next page
```

Figure 4.222   Module for the product-of-sums expression using NOR gates.

```verilog
wire net1, net2, net3, net4, net5, net6;//define internal nets
nor3_df inst1 (    //instantiate the logic gates
   .x1(~x1),
   .x2(~x2),
   .x3(~x3),
   .z1(net1)
   );

nor3_df inst2 (
   .x1(x2),
   .x2(x3),
   .x3(~x4),
   .z1(net2)
   );

nor3_df inst3 (
   .x1(~x1),
   .x2(x2),
   .x3(~x4),
   .z1(net3)
   );

nor3_df inst4 (
   .x1(net1),
   .x2(net2),
   .x3(net3),
   .z1(net4)
   );

nor2_df inst5 (
   .x1(~x3),
   .x2(x4),
   .z1(net5)
   );

nor2_df inst6 (
   .x1(x2),
   .x2(net5),
   .z1(net6)
   );

nor2_df inst7 (
   .x1(net4),
   .x2(net6),
   .z1(z1)
   );
endmodule
```

Figure 4.222   (Continued)

```
x1 x2 x3 x4 = 0000,  z1 = 0
x1 x2 x3 x4 = 0001,  z1 = 0
x1 x2 x3 x4 = 0010,  z1 = 0
x1 x2 x3 x4 = 0011,  z1 = 0
x1 x2 x3 x4 = 0100,  z1 = 0
x1 x2 x3 x4 = 0101,  z1 = 0
x1 x2 x3 x4 = 0110,  z1 = 0
x1 x2 x3 x4 = 0111,  z1 = 0
x1 x2 x3 x4 = 1000,  z1 = 0
x1 x2 x3 x4 = 1001,  z1 = 0
x1 x2 x3 x4 = 1010,  z1 = 0
x1 x2 x3 x4 = 1011,  z1 = 0
x1 x2 x3 x4 = 1100,  z1 = 0
x1 x2 x3 x4 = 1101,  z1 = 0
x1 x2 x3 x4 = 1110,  z1 = 1
x1 x2 x3 x4 = 1111,  z1 = 1
```

Figure 4.223   Outputs for the module of Figure 4.222.

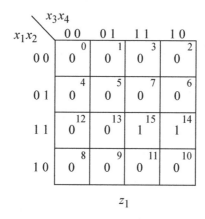

Figure 4.224   Karnaugh map for z_1 for Example 4.43.

From the Karnaugh map, the equation for z_1 in a product-of-sums notation is $z_1 = (x_1)(x_2)(x_3)$ which is the same as the sum-of-products notation.

Observe *inst1* of the logic diagram of Figure 4.221 and *inst1* of the design module of Figure 4.222 as shown in Figure 4.225. In Figure 4.225(a), the NOR gate is drawn as an AND function. The Verilog code may seem contradictory when compared to the logic — the logic specifies *net1* to be the AND of x_1, x_2, and x_3, all asserted low (not deasserted). However, the Verilog code specifies that the inputs be deasserted ; that is, $(\sim x_1 \sim x_2 \sim x_3)$. Figure 4.225(b) shows the same gate drawn as a NOR gate for the OR function with the inputs again asserted low. Using DeMorgan's law, *net1* is specified as $(x_1' + x_2' + x_3')'$, which becomes $x_1 x_2 x_3$ as indicated by the first term in the equation for z_1.

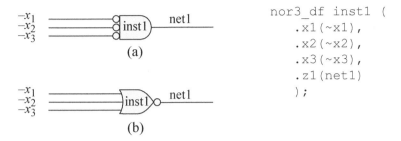

Figure 4.225 shows on the left (a) a gate with inputs $-x_1$, $-x_2$, $-x_3$ into inst1 producing net1, and (b) a gate with inputs $-x_1$, $-x_2$, $-x_3$ into inst1 producing net1.

On the right:

```
nor3_df inst1 (
    .x1(~x1),
    .x2(~x2),
    .x3(~x3),
    .z1(net1)
    );
```

Figure 4.225   Gate configuration for *inst1* of the logic diagram of Figure 4.221.

The same rationale applies to *inst2* in both the logic diagram and the Verilog code. The logic specifies that *net2* is the AND of x_2', x_3', and x_4. However, the Verilog code specifies that *net2* is the NOR of x_2, x_3, and x_4'. This is also correct if DeMorgan's law is applied; that is, $net2 = (x_2 + x_3 + x_4')' = x_2' x_3' x_4$, which is indicated by the second term of the equation for z_1.

Example 4.44   A logic diagram is shown in Figure 4.226 consisting of two majority circuits and one AND gate. The output of a majority circuit is a logic 1 if the majority of the inputs are a logic 1 as shown in Equation 4.22 for majority circuit Maj_1. The term $x_1 x_2 x_3$ is redundant.

A majority circuit will be designed using structural modeling and then instantiated twice into a structural module for the majority network of Figure 4.226, where z_1 is represented by Equation 4.23. The majority circuit is shown in Figure 4.227 using structural modeling. The design module for the majority network is shown in Figure 4.228. The test bench and outputs are shown in Figure 4.229 and Figure 4.230, respectively.

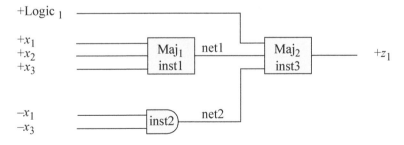

Figure 4.226   A majority network for Example 4.44.

$$Maj_1 = x_1 x_2 + x_1 x_3 + x_2 x_3 \qquad (4.22)$$

$$\text{Maj}_1 = x_1 x_2 + x_1 x_3 + x_2 x_3$$

$$z_1 = 1\text{Maj}_1 + 1 x_1' x_3' + \text{Maj}_1 x_1' x_3'$$

$$= x_1 x_2 + x_1 x_3 + x_2 x_3 + x_1' x_3' + 0$$

$$= x_2 + x_1 x_3 + x_1' x_3' \qquad\qquad (4.23)$$

```
//structural 3-input majority
module majority3 (x1, x2, x3, z1);

input x1, x2, x3;
output z1;

//define internal nets
wire net1, net2, net3;

//instantiate the logic gates
and2_df inst1 (
   .x1(x1),
   .x2(x2),
   .z1(net1)
   );

and2_df inst2 (
   .x1(x1),
   .x2(x3),
   .z1(net2)
   );

and2_df inst3 (
   .x1(x2),
   .x2(x3),
   .z1(net3)
   );

or3_df inst4 (
   .x1(net1),
   .x2(net2),
   .x3(net3),
   .z1(z1)
   );

endmodule
```

Figure 4.227   Structural module for a 3-input majority circuit.

```
//structural module for a majority network
module majority_network (x1, x2, x3, z1);
input x1, x2, x3;
output z1;

//define internal nets
wire net1, net2;

//instantiate the majority circuits and gate
majority3 inst1 (
    .x1(x1),
    .x2(x2),
    .x3(x3),
    .z1(net1)
    );

and2_df inst2 (
    .x1(~x1),
    .x2(~x3),
    .z1(net2)
    );

majority3 inst3 (
    .x1(1'b1),
    .x2(net1),
    .x3(net2),
    .z1(z1)
    );

endmodule
```

Figure 4.228   Structural module for the majority network of Figure 4.226.

```
//test bench for the majority network
module majority_network_tb;

reg x1, x2, x3;
wire z1;

//continued on next page
```

Figure 4.229   Test bench for the majority network of Figure 4.228.

```
//apply input vectors and display variables
initial
begin: apply_stimulus
   reg [3:0] invect;
   for (invect = 0; invect < 8; invect = invect + 1)
      begin
         {x1, x2, x3} = invect [3:0];
         #10 $display ("x1 x2 x3 = %b, z1 = %b",
                        {x1, x2, x3}, z1);
      end
end


//instantiate the number into the test bench
majority_network inst1 (
   .x1(x1),
   .x2(x2),
   .x3(x3),
   .z1(z1)
   );

endmodule
```

Figure 4.229   (Continued)

```
x1 x2 x3 = 000, z1 = 1
x1 x2 x3 = 001, z1 = 0
x1 x2 x3 = 010, z1 = 1
x1 x2 x3 = 011, z1 = 1
x1 x2 x3 = 100, z1 = 0
x1 x2 x3 = 101, z1 = 1
x1 x2 x3 = 110, z1 = 1
x1 x2 x3 = 111, z1 = 1
```

Figure 4.230   Outputs for the majority network of Figure 4.228.

Example 4.45 Structural modeling will be used to implement the truth table of Table 4.7 with a 4:1 multiplexer and additional logic gates. Variables x_1 and x_2 will be used as select inputs to the multiplexer, where $x_1 x_2 = s_1 s_0$. The data input equations are shown in Equation 4.24 as obtained from the truth table. The logic diagram is shown in Figure 4.231. The design module is shown in Figure 4.232 using dataflow modeling for the logic gates and multiplexer. The test bench and outputs are shown in Figure 4.233 and Figure 4.234, respectively.

Table 4.7 Truth Table for Example 4.45

x_1	x_2	x_3	x_4	z_1
0	0	0	0	y'
0	0	0	1	1
0	0	1	0	y'
0	0	1	1	1
0	1	0	0	–
0	1	0	1	–
0	1	1	0	y
0	1	1	1	0
1	0	0	0	0
1	0	0	1	0
1	0	1	0	–
1	0	1	1	0
1	1	0	0	0
1	1	0	1	0
1	1	1	0	–
1	1	1	1	–

$$\text{Data input } 0 = x_4'y' + x_4 = x_4 + y'$$

$$\text{Data input } 1 = x_4'y$$

$$\text{Data input } 2 = 0$$

$$\text{Data input } 3 = 0 \tag{4.24}$$

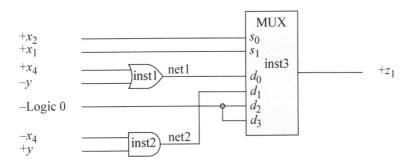

Figure 4.231   Logic diagram for Example 4.45.

```
//structural nonlinear multiplexer
module mux_nonlinear4 (x1, x2, x3, x4, y, z1);

input x1, x2, x3, x4, y;
output z1;

//define internal nets
wire net1, net2;

//instantiate the multiplexer and gates
or2_df inst1 (
   .x1(x4),
   .x2(~y),
   .z1(net1)
   );

and2_df inst2 (
   .x1(~x4),
   .x2(y),
   .z1(net2)
   );

mux4a_df inst3 (
   .s({x1, x2}),
   .d({1'b0, 1'b0, net2, net1}),
   .z1(z1)
   );
endmodule
```

Figure 4.232   Structural module for Figure 4.231.

```
//test bench for 4:1 nonlinear-select multiplexer
module mux_nonlinear4_tb;
reg x1, x2, x3, x4, y;
wire z1;

initial  //apply input vectors and display variables
begin: apply_stimulus
   reg [5:0] invect;
   for (invect = 0; invect < 32; invect = invect + 1)
      begin
         {x1, x2, x3, x4, y} = invect [5:0];
         #10 $display ("x1 x2 = %b, x3 x4 = %b, y = %b, z1 = %b",
                       {x1, x2}, {x3, x4}, y, z1);
      end
end          //continued on next page
```

Figure 4.233   Test bench for the nonlinear multiplexer of Figure 4.232.

```
//instantiate the module into the test bench
mux_nonlinear4 inst1 (
    .x1(x1),
    .x2(x2),
    .x3(x3),
    .x4(x4),
    .y(y),
    .z1(z1)
    );

endmodule
```

Figure 4.233   (Continued)

```
x1 x2 = 00, x3 x4 = 00, y = 0, z1 = 1
x1 x2 = 00, x3 x4 = 00, y = 1, z1 = 0
x1 x2 = 00, x3 x4 = 01, y = 0, z1 = 1
x1 x2 = 00, x3 x4 = 01, y = 1, z1 = 1
x1 x2 = 00, x3 x4 = 10, y = 0, z1 = 1
x1 x2 = 00, x3 x4 = 10, y = 1, z1 = 0
x1 x2 = 00, x3 x4 = 11, y = 0, z1 = 1
x1 x2 = 00, x3 x4 = 11, y = 1, z1 = 1
x1 x2 = 01, x3 x4 = 00, y = 0, z1 = 0   don't care = y
x1 x2 = 01, x3 x4 = 00, y = 1, z1 = 1   don't care = y
x1 x2 = 01, x3 x4 = 01, y = 0, z1 = 0   don't care = 0
x1 x2 = 01, x3 x4 = 01, y = 1, z1 = 0   don't care = 0
x1 x2 = 01, x3 x4 = 10, y = 0, z1 = 0
x1 x2 = 01, x3 x4 = 10, y = 1, z1 = 1
x1 x2 = 01, x3 x4 = 11, y = 0, z1 = 0
x1 x2 = 01, x3 x4 = 11, y = 1, z1 = 0
x1 x2 = 10, x3 x4 = 00, y = 0, z1 = 0
x1 x2 = 10, x3 x4 = 00, y = 1, z1 = 0
x1 x2 = 10, x3 x4 = 01, y = 0, z1 = 0
x1 x2 = 10, x3 x4 = 01, y = 1, z1 = 0
x1 x2 = 10, x3 x4 = 10, y = 0, z1 = 0   don't care = 0
x1 x2 = 10, x3 x4 = 10, y = 1, z1 = 0   don't care = 0
x1 x2 = 10, x3 x4 = 11, y = 0, z1 = 0
x1 x2 = 10, x3 x4 = 11, y = 1, z1 = 0
x1 x2 = 11, x3 x4 = 00, y = 0, z1 = 0
x1 x2 = 11, x3 x4 = 00, y = 1, z1 = 0
x1 x2 = 11, x3 x4 = 01, y = 0, z1 = 0
x1 x2 = 11, x3 x4 = 01, y = 1, z1 = 0
x1 x2 = 11, x3 x4 = 10, y = 0, z1 = 0   don't care = 0
x1 x2 = 11, x3 x4 = 10, y = 1, z1 = 0   don't care = 0
x1 x2 = 11, x3 x4 = 11, y = 0, z1 = 0   don't care = 0
x1 x2 = 11, x3 x4 = 11, y = 1, z1 = 0   don't care = 0
```

Figure 4.234   Outputs for the structural module of Figure 4.232.

Example 4.46   Structural modeling will be used to design a 3-bit comparator for the following operands:

$$A = a_2 a_1 a_0$$
$$B = b_2 b_1 b_0$$

where a_0 and b_0 are the low-order bits of A and B, respectively.   The following outputs will be used:

$$a\_lt\_b \text{ indicating } A < B$$
$$a\_eq\_b \text{ indicating } A = B$$
$$a\_gt\_b \text{ indicating } A > B$$

The equations for the comparator are shown below.

$$(A < B) = a_2' \, b_2 + (a_2 \oplus b_2)' \, a_1' \, b_1 + (a_2 \oplus b_2)' \, (a_1 \oplus b_1)' \, a_0' \, b_0$$
$$(A = B) = (a_2 \oplus b_2)' \, (a_1 \oplus b_1)' \, (a_0 \oplus b_0)'$$
$$(A > B) = \; a_2 \, b_2' + (a_2 \oplus b_2)' \, a_1 \, b_1' + (a_2 \oplus b_2)' \, (a_1 \oplus b_1)' \, a_0 \, b_0'$$

The structural module instantiates the following dataflow modules: *and2_df*, *xnor2_df*, *and3_df*, *and4_df*, and *or3_df*. The design module, test bench module, outputs, and waveforms are shown in Figure 4.235, Figure 4.236, Figure 4.237, and Figure 4.238, respectively. The test bench applies eight sets of inputs to demonstrate the relative magnitude of the two operands.

```
//structural 3-bit comparator
module comp3_struc (a, b, a_lt_b, a_eq_b, a_gt_b);

input [2:0] a, b;
output a_lt_b, a_eq_b, a_gt_b;

wire [2:0] a, b;
wire a_lt_b, a_eq_b, a_gt_b;
wire net1, net2, net3, net4, net5, net7, net9, net10, net11;

//instantiate the logic for a_lt_b
and2_df inst1 (
   .x1(~a[2]),
   .x2(b[2]),
   .z1(net1)
   );

//continued on next page
```

Figure 4.235   Structural module for a 3-bit comparator.

```
xnor2_df inst2 (
   .x1(a[2]),
   .x2(b[2]),
   .z1(net2)
   );

xnor2_df inst3 (
   .x1(a[1]),
   .x2(b[1]),
   .z1(net3)
   );

and3_df inst4 (
   .x1(net2),
   .x2(~a[1]),
   .x3(b[1]),
   .z1(net4)
   );

and4_df inst5 (
   .x1(net2),
   .x2(net3),
   .x3(~a[0]),
   .x4(b[0]),
   .z1(net5)
   );

or3_df inst6 (
   .x1(net1),
   .x2(net4),
   .x3(net5),
   .z1(a_lt_b)
   );

//instantiate the logic for a_eq_b
xnor2_df inst7 (
   .x1(a[0]),
   .x2(b[0]),
   .z1(net7)
   );

and3_df inst8 (
   .x1(net2),
   .x2(net3),
   .x3(net7),
   .z1(a_eq_b)
   );                //continued on next page
```

Figure 4.235 (Continued)

```
//instantiate the logic for a_gt_b
and2_df inst9 (
   .x1(a[2]),
   .x2(~b[2]),
   .z1(net9)
   );

and3_df inst10 (
   .x1(net2),
   .x2(a[1]),
   .x3(~b[1]),
   .z1(net10)
   );

and4_df inst11 (
   .x1(net2),
   .x2(net3),
   .x3(a[0]),
   .x4(~b[0]),
   .z1(net11)
   );

or3_df inst12 (
   .x1(net9),
   .x2(net10),
   .x3(net11),
   .z1(a_gt_b)
   );

endmodule
```

Figure 4.235   (Continued)

```
//test bench for structural 3-bit comparator
module comp3_struc_tb;

reg [2:0] a, b;
wire a_lt_b, a_eq_b, a_gt_b;

//display inputs and outputs
initial
$monitor ("a=%b, b=%b, a_lt_b=%b, a_eq_b=%b, a_gt_b=%b",
          a, b, a_lt_b, a_eq_b, a_gt_b);
//continued on next page
```

Figure 4.236   Test bench for the 3-bit comparator of Figure 4.235.

```
//apply input vectors
initial
begin
   #0      a=3'b000;    b=3'b000;
   #10     a=3'b001;    b=3'b010;
   #10     a=3'b100;    b=3'b100;
   #10     a=3'b110;    b=3'b111;
   #10     a=3'b111;    b=3'b101;
   #10     a=3'b110;    b=3'b011;
   #10     a=3'b111;    b=3'b111;
   #10     a=3'b011;    b=3'b111;

   #10     $stop;
end


//instantiate the module into the test bench
comp3_struc inst1 (
   .a(a),
   .b(b),
   .a_lt_b(a_lt_b),
   .a_eq_b(a_eq_b),
   .a_gt_b(a_gt_b)
   );

endmodule
```

Figure 4.236   (Continued)

```
a=000, b=000, a_lt_b=0, a_eq_b=1, a_gt_b=0
a=001, b=010, a_lt_b=1, a_eq_b=0, a_gt_b=0
a=100, b=100, a_lt_b=0, a_eq_b=1, a_gt_b=0
a=110, b=111, a_lt_b=1, a_eq_b=0, a_gt_b=0
a=111, b=101, a_lt_b=0, a_eq_b=0, a_gt_b=1
a=110, b=011, a_lt_b=0, a_eq_b=0, a_gt_b=1
a=111, b=111, a_lt_b=0, a_eq_b=1, a_gt_b=0
a=011, b=111, a_lt_b=1, a_eq_b=0, a_gt_b=0
```

Figure 4.237   Outputs for the 3-bit comparator of Figure 4.235.

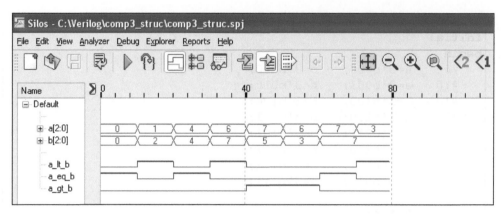

Figure 4.238   Waveforms for the 3-bit comparator of Figure 4.235.

Example 4.47   In this example, an adder and high-speed shifter will be designed by instantiating a 4-bit adder and eight 4:1 multiplexers into a structural module. The multiplexers will be used as a combinational shifter. Designing a shifter in this manner results in a shifting unit that is faster than a sequential shift register that shifts one bit per clock cycle because all the shift amounts are prewired. In order to shift the operand a specified number of bits, the shift amount is simply selected and the operand is shifted the requisite number of bits — the speed is a function only of the inertial delay of the multiplexer gates.

Four multiplexers are used for shifting left and four multiplexers are used for shifting right. Shifting a binary number left one bit position corresponds to multiplying the number by two. Shifting a binary number right one bit position corresponds to dividing the number by two.

This is a *logical* shifter; that is, zeroes are shifted into the vacated low-order bit positions for a shift-left operation and zeroes are shifted into the vacated high-order bit positions for a shift-right operation. The carry-out of the adder is not used in this design, although it could easily be gated into the low- or high-order positions of the multiplexers. Alternatively, a larger multiplexer could be used to accommodate the carry-out.

There are two operands: the augend $A[3:0] = a_3 a_2 a_1 a_0$ and the addend $B[3:0] = b_3 b_2 b_1 b_0$, where a_0 and b_0 are the low-order bits of A and B, respectively. There are also two shift direction inputs: *shiftleft* and *shiftright* and one input vector *shiftcount[1:0]* to determine the number of bits that the sum is to be shifted. In addition, there are two output vectors: *slmux[3:0]* and *srmux[3:0]* that are the outputs of the shift-left and shift-right multiplexers, respectively. Table 4.8 lists the control variables for shifting the sum.

Figure 4.239 pictorially depicts the shift-left and shift-right operations. The logic diagram is shown in Figure 4.240 displaying the instantiation names and the net names.

Table 4.8 Shift Control Variables

Shift Direction	Shiftcount [1]	Shiftcount [0]	Shift Amount
Shiftleft	0	0	0
	0	1	1
	1	0	2
	1	1	3
Shiftright	0	0	0
	0	1	1
	1	0	2
	1	1	3

Shift Left				
Adder output	Σ_3	Σ_2	Σ_1	Σ_0
No shift	Σ_3	Σ_2	Σ_1	Σ_0
Shift left 1	Σ_2	Σ_1	Σ_0	0
Shift left 2	Σ_1	Σ_0	0	0
Shift left 3	Σ_0	0	0	0

Shift Right				
Adder output	Σ_3	Σ_2	Σ_1	Σ_0
No shift	Σ_3	Σ_2	Σ_1	Σ_0
Shift right 1	0	Σ_3	Σ_2	Σ_1
Shift right 2	0	0	Σ_3	Σ_2
Shift right 3	0	0	0	Σ_3

Figure 4.239   Pictorial representation of the shift-left and shift-right operations.

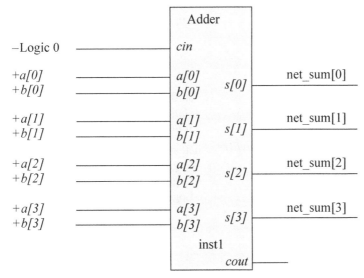

(Continued on next page)

Figure 4.240   Logic diagram for the adder and high-speed shifter.

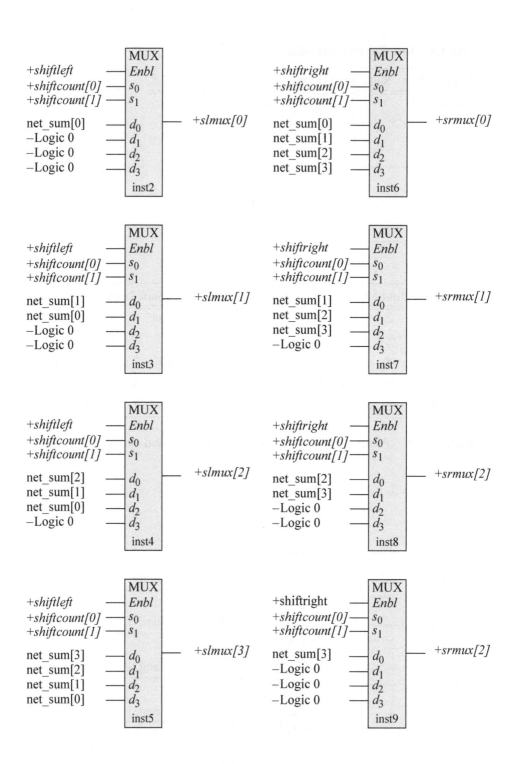

Figure 4.240   (Continued)

The structural module is shown in Figure 4.241. The instantiated multiplexer was designed as a dataflow module with the following declarations: *input [1:0] s; input [3:0] d; input enbl; and output z1.* Therefore, when the multiplexer is instantiated, the data inputs must be in the order as specified; that is, *d[3]*, *d[2]*, *d[1]*, and *d[0]*. The test bench, outputs, and waveforms are shown in Figure 4.242, Figure 4.243, and Figure 4.244, respectively.

```
//structural adder and high-speed shifter
module adder_shifter (a, b, shiftcount, shiftleft,
                            shiftright, slmux, srmux);

input [3:0] a, b;
input [1:0] shiftcount;
input shiftleft, shiftright;

output [3:0] slmux, srmux;


//define internal nets
wire [3:0] net_sum;

//instantiate the adder
adder4_df inst1 (
   .a(a),
   .b(b),
   .cin(1'b0),
   .sum(net_sum)
   );


//instantiate the multiplexers for shifting left
mux4_df inst2 (
   .s(shiftcount),
   .d({1'b0, 1'b0, 1'b0, net_sum[0]}),
   .enbl(shiftleft),
   .z1(slmux[0])
   );

mux4_df inst3 (
   .s(shiftcount),
   .d({1'b0, 1'b0, net_sum[0], net_sum[1]}),
   .enbl(shiftleft),
   .z1(slmux[1])
   );
//continued on next page
```

Figure 4.241   Structural module for the adder and high-speed shifter.

```
mux4_df inst4 (
   .s(shiftcount),
   .d({1'b0, net_sum[0], net_sum[1], net_sum[2]}),
   .enbl(shiftleft),
   .z1(slmux[2])
   );

mux4_df inst5 (
   .s(shiftcount),
   .d({net_sum[0], net_sum[1], net_sum[2], net_sum[3]}),
   .enbl(shiftleft),
   .z1(slmux[3])
   );


//instantiate the multiplexers for shifting right
mux4_df inst6 (
   .s(shiftcount),
   .d({net_sum[3], net_sum[2], net_sum[1], net_sum[0]}),
   .enbl(shiftright),
   .z1(srmux[0])
   );


mux4_df inst7 (
   .s(shiftcount),
   .d({1'b0, net_sum[3], net_sum[2], net_sum[1]}),
   .enbl(shiftright),
   .z1(srmux[1])
   );


mux4_df inst8 (
   .s(shiftcount),
   .d({1'b0, 1'b0, net_sum[3], net_sum[2]}),
   .enbl(shiftright),
   .z1(srmux[2])
   );


mux4_df inst9 (
   .s(shiftcount),
   .d({1'b0, 1'b0, 1'b0, net_sum[3]}),
   .enbl(shiftright),
   .z1(srmux[3])
   );
endmodule
```

Figure 4.241   (Continued)

```
//test bench for the adder and high-speed shifter
module adder_shifter_tb;

reg [3:0] a, b;
reg [1:0] shiftcount;
reg enbl;
reg shiftleft, shiftright;

wire [3:0] slmux, srmux;

initial
$monitor ("a=%b, b=%b, shiftcount=%b, shiftleft=%b,
          shiftright=%b, slmux=%b, srmux=%b",
      a, b, shiftcount, shiftleft, shiftright, slmux, srmux);


//apply input vectors
initial
begin
//no shift
   #0    a=4'b0011;
         b=4'b0001;        //sum=0100
         shiftcount=2'b00; //no shift
         shiftleft=1'b1;
         shiftright=1'b0;


//shift left
   #10   a=4'b0111;
         b=4'b0011;        //sum=1010
         shiftcount=2'b01; //shift one
         shiftleft=1'b1;   //shift left
         shiftright=1'b0;

   #10   a=4'b1100;
         b=4'b0011;        //sum=1111
         shiftcount=2'b10; //shift two
         shiftleft=1'b1;   //shift left
         shiftright=1'b0;

   #10   a=4'b1100;
         b=4'b0011;        //sum=1111
         shiftcount=2'b11; //shift three
         shiftleft=1'b1;   //shift left
         shiftright=1'b0;

//continued on  next page
```

Figure 4.242   Test bench for the adder and high-speed shifter.

```
//shift right
   #10    a=4'b1100;
          b=4'b0011;          //sum=1111
          shiftcount=2'b01; //shift one
          shiftleft=1'b0;
          shiftright=1'b1;   //shift right

   #10    a=4'b0110;
          b=4'b0111;          //sum=1101
          shiftcount=2'b10; //shift two
          shiftleft=1'b0;
          shiftright=1'b1;   //shift right

   #10    a=4'b0110;
          b=4'b1001;          //sum=1111
          shiftcount=2'b11; //shift three
          shiftleft=1'b0;
          shiftright=1'b1;   //shift right
   #10    $stop;
end

//instantiate the module into the test bench
adder_shifter inst1 (
   .a(a),
   .b(b),
   .shiftcount(shiftcount),
   .shiftleft(shiftleft),
   .shiftright(shiftright),
   .slmux(slmux),
   .srmux(srmux)
   );
endmodule
```

Figure 4.242   (Continued)

```
a=0011, b=0001, shiftcount=00, shiftleft=1, shiftright=0,
                slmux=0100, srmux=0000

a=0111, b=0011, shiftcount=01, shiftleft=1, shiftright=0,
                slmux=0100, srmux=0000

a=1100, b=0011, shiftcount=10, shiftleft=1, shiftright=0,
                slmux=1100, srmux=0000
//continued on next page
```

Figure 4.243   Outputs for the adder and high-speed shifter.

```
a=1100, b=0011, shiftcount=11, shiftleft=1, shiftright=0,
               slmux=1000, srmux=0000

a=1100, b=0011, shiftcount=01, shiftleft=0, shiftright=1,
               slmux=0000, srmux=0111

a=0110, b=0111, shiftcount=10, shiftleft=0, shiftright=1,
               slmux=0000, srmux=0011

a=0110, b=1001, shiftcount=11, shiftleft=0, shiftright=1,
               slmux=0000, srmux=0001
```

Figure 4.243   (Continued)

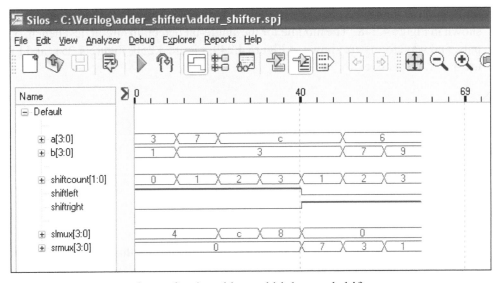

Figure 4.244   Waveforms for the adder and high-speed shifter.

Example 4.48   An *iterative network* is a logical structure composed of an array of identical cells. It is a cascade of identical combinational or sequential circuits (cells) in which the first or last cells may be different than the other cells in the network. Since an iterative network consists of identical cells, it is only necessary to design a typical cell, and then to replicate that cell for the entire network.

A circuit will be designed to determine if an input vector (x_1, x_2, x_3, x_4) contains only a single 1 bit. The output will be a logical 1 if there is only a single 1 bit; the output will be a logical 0 if there are no 1 bits or two or more 1 bits. First, a typical cell will be designed, then instantiated four times into a higher-level structural module.

The structural module will detect a single bit in the 4-bit input vector. The block diagram for a typical $cell_i$ is shown in Figure 4.245, where the input and output lines are defined as follows:

- y_{1_in} is an active-high input line indicating that a single 1 bit was detected up to that cell.

- y_{0_in} is an active-high input line indicating that no 1 bits were detected up to that cell.

- y_{1_out} is an active-high output line indicating that a single 1 bit was detected up to and including that cell.

- y_{0_out} is an active-high output line indicating that no 1 bits were detected up to and including that cell.

Figure 4.246 shows the internal logic of the cell, which will be instantiated four times into the higher-level circuit of Figure 4.247. The module for the typical cell is shown in Figure 4.248. The module for the single-bit detection circuit, test bench, outputs, and waveforms are shown in Figure 4.249, Figure 4.250, Figure 4.251, and Figure 4.252, respectively.

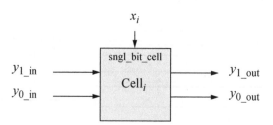

Figure 4.245   Typical cell for a single-bit detection circuit that will be instantiated four times into a higher-level module.

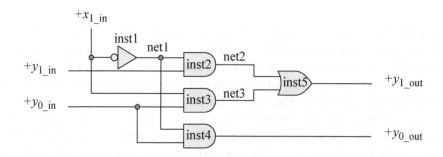

Figure 4.246   Internal logic for a typical cell for the single-bit detection circuit.

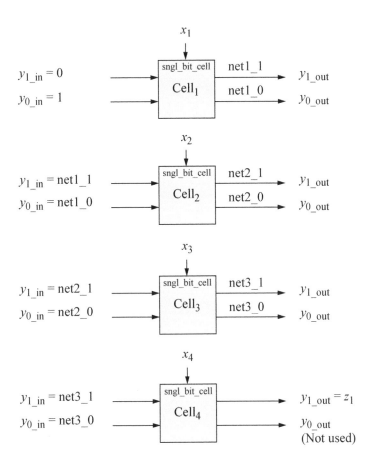

Figure 4.247   Block diagram to detect a single 1 bit in a 4-bit input vector.

```
//typical cell for single-bit detection
module sngl_bit_cell (x1_in, y1_in, y0_in, y1_out, y0_out);

input x1_in, y1_in, y0_in;
output y1_out, y0_out;

not inst1 (net1, x1_in);
and inst2 (net2, net1, y1_in);
and inst3 (net3, x1_in, y0_in);
and inst4 (y0_out, net1, y0_in);
or  inst5 (y1_out, net2, net3);

endmodule
```

Figure 4.248   Typical cell that is instantiated four times to detect a single bit in an input vector $x[1:4]$.

```
//structural single-bit detection module
//instantiate a typical cell four times
module sngl_bit_detect2 (x1, x2, x3, x4, z1);

input x1, x2, x3, x4;
output z1;

//instantiate the single-bit cell modules
//cell 1 *********************************************
sngl_bit_cell inst1(
   .x1_in(x1),
   .y1_in(1'b0),
   .y0_in(1'b1),
   .y1_out(net1_1),
   .y0_out(net1_0)
   );

//cell 2 *********************************************
sngl_bit_cell inst2(
   .x1_in(x2),
   .y1_in(net1_1),
   .y0_in(net1_0),
   .y1_out(net2_1),
   .y0_out(net2_0)
   );

//cell 3 *********************************************
sngl_bit_cell inst3(
   .x1_in(x3),
   .y1_in(net2_1),
   .y0_in(net2_0),
   .y1_out(net3_1),
   .y0_out(net3_0)
   );

//cell 4 *********************************************
sngl_bit_cell inst4(
   .x1_in(x4),
   .y1_in(net3_1),
   .y0_in(net3_0),
   .y1_out(z1)
   );

endmodule
```

Figure 4.249   Module to detect a single bit in a 4-bit input vector in which the typical cell of Figure 4.248 is instantiated four times.

```
//test bench for the single-bit detection
//using a typical cell instantiation
module sngl_bit_detect2_tb;

reg x1, x2, x3, x4;
wire z1;


//apply input vectors
initial
begin: apply_stimulus
   reg [4:0] invect;
   for (invect=0; invect<16; invect=invect+1)
      begin
         {x1, x2, x3, x4} = invect [4:0];
         #10 $display ("x1x2x3x4 = %b, z1 = %b",
                       {x1, x2, x3, x4}, z1);
      end
end


//instantiate the module into the test bench
sngl_bit_detect2 inst1 (
   .x1(x1),
   .x2(x2),
   .x3(x3),
   .x4(x4),
   .z1(z1)
   );

endmodule
```

Figure 4.250   Test bench for the single-bit detection module.

```
x1x2x3x4 = 0000, z1 = 0        x1x2x3x4 = 1000, z1 = 1
x1x2x3x4 = 0001, z1 = 1        x1x2x3x4 = 1001, z1 = 0
x1x2x3x4 = 0010, z1 = 1        x1x2x3x4 = 1010, z1 = 0
x1x2x3x4 = 0011, z1 = 0        x1x2x3x4 = 1011, z1 = 0
x1x2x3x4 = 0100, z1 = 1        x1x2x3x4 = 1100, z1 = 0
x1x2x3x4 = 0101, z1 = 0        x1x2x3x4 = 1101, z1 = 0
x1x2x3x4 = 0110, z1 = 0        x1x2x3x4 = 1110, z1 = 0
x1x2x3x4 = 0111, z1 = 0        x1x2x3x4 = 1111, z1 = 0
```

Figure 4.251   Outputs for the single-bit detector module.

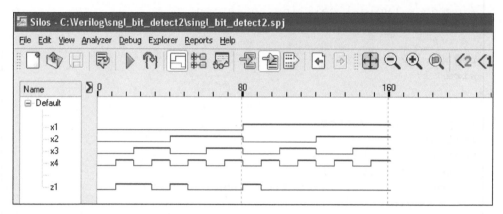

Figure 4.252   Waveforms for the single-bit detector module.

4.6 Problems

4.1   Use AND gate and OR gate built-in primitives to implement a circuit in a sum-of-products form that will generate an output z_1 if an input is greater than or equal to 2 and less than 5; and also greater than or equal to 14 and less than 13. Then obtain the design module, test bench module, and outputs.

4.2   Use only NOR gate built-in primitives to implement the following function:

$$z_1 = x_1 x_3 + (x_2' + x_3')x_4 + (x_1 x_3' x_4')'$$

The inputs are available in both high and low assertion; the output is asserted high. First, minimize the equation to a sum-of-products notation using Boolean algebra, then convert the equation to a product-of-sums notation using any method. Obtain the design module, the test bench module, and the outputs.

4.3   Obtain the minimum sum-of-products equation for the logic diagram shown below using Boolean algebra. Then redesign the logic using NAND gate built-in primitives and generate the design module, test bench module, and outputs.

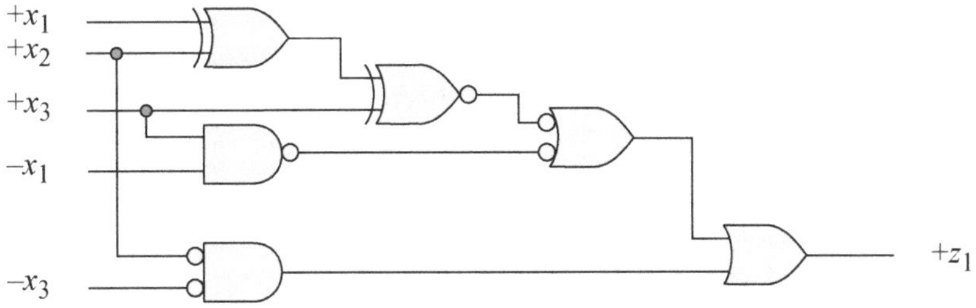

4.4 Use the full adder user-defined primitive that was designed in Chapter 4 to design an excess-3-to-binary code converter. Obtain the design module, the test bench module, and the outputs.

4.5 Given the Karnaugh map shown below for the function z_1 with a as a map-entered variable, obtain the minimum expression for z_1 in a sum-of-products form. Then use NAND logic as user-defined primitives to implement the function. Obtain the design module, the test bench module, and the outputs.

x_1 \ $x_2 x_3$	0 0	0 1	1 1	1 0
0	1 (0)	a' (1)	0 (3)	$a + a'$ (2)
1	a (4)	1 (5)	0 (7)	a (6)

z_1

4.6 Obtain a minimal product-of-sums form for the following expression:

$$z_1(x_1, x_2, x_3, x_4) = \Sigma_{\mathrm{m}}(1, 2, 3, 5, 9, 12, 13) + \Sigma_{\mathrm{d}}(4, 7, 8, 15)$$

Then use NOR user-defined primitives to implement the logic. Obtain the design module, the test bench module, and the outputs.

4.7 Design a 4:1 multiplexer as a user-defined primitive. The select inputs are $s_1 s_0$, where s_0 is the low-order input. The data inputs are $d_3 d_2 d_1 d_0 = 0110$, where d_0 is the low-order input. The output of the multiplexer is *out*. Apply the waveforms for the select inputs as shown below and determine the waveform for output *out*.

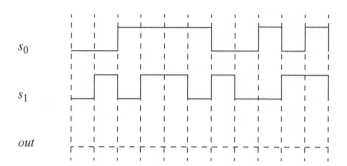

4.8   Obtain the equation for a logic circuit that will generate a logic 1 on output z_1 if a 4-bit unsigned binary number $N = x_1 x_2 x_3 x_4$ satisfies the following criteria, where x_4 is the low-order bit:

$$2 < N \le 6 \text{ or } 11 \le N < 14$$

Use NOR user-defined primitives. Obtain the design module, the test bench module, and outputs.

4.9   Given the Karnaugh map shown below with x_1 and x_2 as map-entered variables, obtain the data input equations for a nonlinear-select multiplexer with z_1 as the output. Obtain the multiplexer input equations in a sum-of-products form for all permutations of the Karnaugh map. Use the permutation that produces the fewest number of primitive gates and implement the circuit using a user-defined primitive for the multiplexer and built-in primitive gates for any additional logic functions.

Obtain the design module, the test bench module, and the outputs. Verify that the outputs are consistent with the values in the corresponding minterm locations of the Karnaugh map.

y_1 \ $y_2 y_3$	0 0	0 1	1 1	1 0
0	1 $\quad^0$	1 $\quad^1$	$x_1 x_2$ $\quad^3$	$-$ $\quad^2$
1	$-$ $\quad^4$	$(x_1 x_2)' + x_1 x_2$ $\quad^5$	$x_1' + x_2'$ $\quad^7$	$(x_1 x_2)'$ $\quad^6$

z_1

4.10   The logic block shown below generates an output of a logic 1 when the inputs contain an odd number of 1s. Use only this type of logic block to generate a parity bit for an 8-bit byte of data. The parity bit will be a logic 1 when there are an even number of 1s in the byte of data. All inputs must be used. Using user-defined primitives, obtain the design module, test bench, and outputs for eight combinations of the input variables.

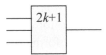

4.11   Write the function shown below in the conjunctive normal form, then implement the function in dataflow modeling using only NAND gates. Output z_1 is asserted as a high voltage level. Obtain the design module, the test bench module, and the outputs.

$$z_1(x_1, x_2, x_3) = \Pi_M(0, 2, 4, 5, 7)$$

4.12   Obtain the minimal product-of-sums expression for the function z_1 represented by the Karnaugh map shown below. Obtain the dataflow module using only NOR logic, the test bench module, and the outputs.

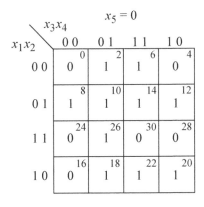
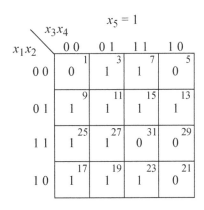

z_1

4.13   Indicate whether the following equation is true or false using dataflow modeling:

$$x_1'x_2 + x_1'x_2'x_3x_4' + x_1x_2x_3'x_4 = x_1'x_2x_3' + x_1'x_3x_4' + x_1'x_2x_3 + x_2x_3'x_4$$

4.14   Design a dataflow module that will generate a logic 1 when two 4-bit unsigned operands are unequal. The operands are: $a[3:0]$ and $b[3:0]$; the output is z_1. Obtain the test bench and outputs for eight of the $2^8 = 256$ combinations of the inputs.

4.15   Use dataflow modeling to design an 8-bit adder. The augend and addend are $a[7:0]$ and $b[7:0]$, respectively, where $a[0]$ and $b[0]$ are the low-order bits of a and b. The sum is $sum[7:0]$, where $sum[0]$ is the low-order bit. Obtain the design module, the test bench module for eight combinations of the augend and addend, the outputs, and the waveforms.

4.16   Use dataflow modeling to indicate which of the expressions shown below will always generate a logic 1. Use only NAND gates.

$$z_1(x_1, x_2) = x_1'x_2' + x_1x_2$$
$$z_2(x_1, x_2, x_3) = x_1' + x_2 + x_3 + x_1'x_2x_3'$$
$$z_3(x_1, x_2, x_3) = x_1'x_3' + x_3 + x_1x_2 + x_1x_2'x_3'$$
$$z_4(x_1, x_2, x_3) = x_3' + x_1'x_3 + x_1x_2'x_3$$
$$z_5(x_1, x_2, x_3) = x_1' + x_1x_2 + x_2x_3' + x_1'x_2'x_3' + x_2'$$

4.17   Use dataflow modeling to implement the following Boolean function using a 3:8 decoder and a minimum amount of additional logic, if necessary:

$$f(x_1, x_2, x_3) = x_1'x_2x_3 + x_1x_3 + x_1x_2' + x_1x_2'x_3'$$

Obtain the design module, the test bench module, and the outputs.

4.18   Using dataflow modeling, design an 8-bit odd parity generator. Obtain the design module, test bench module, and outputs for eight of the $2^8 = 256$ combinations of the inputs. There will be one output labeled z_1.

4.19   Using dataflow modeling, obtain the equation that specifies the function for the logic diagram shown below. The equation is to be in minimum form. Obtain the design module, the test bench module, and the outputs. The equation for output z_1 can be obtained directly from the outputs.

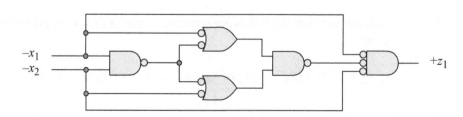

4.20   Use dataflow modeling to design an odd parity generator for an 8-bit byte of data. The parity bit that is generated will be a logic 1 when there are an even number of 1s in the byte of data; that is, the number of 1s in the 9 bits (8 data bits plus 1 parity bit) will be an odd number. Design the logic circuit and assign net names to the outputs of the logic functions. Obtain the design module, the test bench module for eight different combinations of the inputs, and the outputs.

   Then repeat Problem 4.20 and use dataflow modeling to design the circuit directly without the use of a logic diagram by using the exclusive-OR and the exclusive-NOR operators. This is a much simpler approach to designing a parity generator. Obtain the design module, the test bench module for eight combinations of the inputs, and the outputs.

4.21   Use dataflow modeling to implement the function shown below in a sum-of-products form and also in a product-of-sums form. Obtain the design module, the test bench module, and the outputs. Compare the outputs for both forms.

$$z_1(x_1, x_2, x_3, x_4) = \Sigma_m(0, 2, 3, 6, 7) + \Sigma_d(5, 8, 10, 11, 15)$$

4.22   Use dataflow modeling to design an adder/subtractor unit that adds two 8-bit operands $a[7:0]$ and $b[7:0]$, where $a[0]$ and $b[0]$ are the low-order bits of the augend/minuend and the addend/subtrahend, respectively. There are two add/subtract results: $add\_rslt[7:0]$ and $sub\_rslt[7:0]$ and two scalar overflow results: add_ovfl and sub_ovfl. Obtain the design module, the test bench module for several different operands, and the corresponding outputs.

4.23   The adders shown below in parts (a) and (b) perform operations on the 8-bit input operands $a[7:0]$ and $b[7:0]$. The designations a' and b' are the 1s complement of operands a and b, respectively. Use dataflow modeling to determine which operation (i) through (v) is the correct operation for parts (a) and (b). Obtain the design module, the test bench module for several combinations of the inputs, and the outputs.

(a)

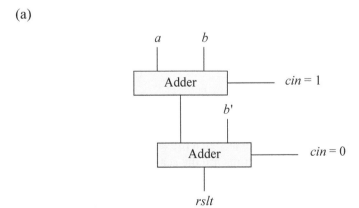

(i) $rslt = a - b$
(ii) $rslt = a + b$
(iii) $rslt$ = Transfer a
(iv) $rslt$ = Increment a
(v) $rslt = b - a$

(b)

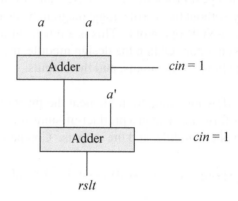

(i) $rslt = a - a$
(ii) $rslt$ = Transfer a
(iii) $rslt$ = Zero
(iv) $rslt$ = 2s complement a
(v) $rslt$ = Increment a

4.24   Using behavioral modeling, design a 2-input exclusive-NOR circuit. Obtain the design module, the test bench module, the outputs, and the waveforms.

4.25   Use behavioral modeling to design the logic circuit shown below. Provide an intrastatement delay of five time units. Obtain the design module, the test bench module for all combinations of the inputs, the outputs, and the waveforms.

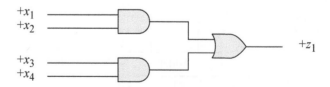

4.26   Use behavioral modeling to design a full adder. The inputs are the scalar variables a, b, and cin; the outputs are the scalar variables sum and $cout$. Obtain the design module, the test bench module for all combinations of the inputs, the outputs, and the waveforms.

4.27   Use behavioral modeling combined with dataflow modeling to design a 4:1 multiplexer. There is a vector data input $d[3:0]$, a vector select input $s[1:0]$, and a scalar enable input *enbl*. There is one scalar output z_1. Obtain the design module, the test bench module for eight combinations of the data inputs and select inputs, and the corresponding outputs.

4.28   Determine the logic function of the design module shown below, then verify the answer by means of a test bench and outputs.

```
//behavioral function
module fctn (x, z1);

input [3:0] x;
output z1;

reg z1;

always @ (x)
begin
   z1 = & x;
end
endmodule
```

4.29   The design module shown below describes a relationship between two 4-bit operands $a[3:0]$ and $b[3:0]$. Specify the relationship between a and b when output z_1 is at a logic 1 level. Verify the answer by means of a test bench for eight different combinations of the inputs, then display the outputs.

```
//function using dataflow and built-in primitives
module fctn1 (a, b, z1);

input [3:0] a, b;
output z1;

wire [3:0] a, b;
wire z1;

//define internal nets
wire net1, net2, net3, net4;

xor    (net1, a[0], b[0]),
       (net2, a[1], b[1]),
       (net3, a[2], b[2]),
       (net4, a[3], b[3]);

assign z1 = (net1 | net2 | net3 | net4);

endmodule
```

4.30   A Karnaugh map is shown below using x_5 as a map-entered variable. Obtain the input equations for a nonlinear-select multiplexer using $x_1 x_2 = s_1 s_0$. Then use behavioral and dataflow modeling with the **case** statement to implement the design module. Provide all combinations of the five variables $x_1 x_2 x_3 x_4 x_5$ in the test bench. Obtain the outputs and verify that they conform to the minterm entries of the Karnaugh map.

$x_1 x_2$ \ $x_3 x_4$	0 0	0 1	1 1	1 0
0 0	x_5 0	0 1	0 3	x_5 2
0 1	1 4	1 5	0 7	0 6
1 1	0 12	1 13	1 15	x_5' 14
1 0	0 8	1 9	0 11	1 10

z_1

4.31   Use behavioral modeling with the **case** statement to design a 5-function arithmetic and logic unit for the following five functions: add, subtract, multiply, divide, and modulus. The operands are 4-bit vectors: $a[3:0]$ and $b[3:0]$. Obtain the behavioral module, test bench module, outputs and waveforms for one input vector for each operation.

4.32   Use the **while** construct to execute a loop if $reg\_a \neq reg\_b$ and increment a variable *count* if the registers are not equal. Register reg_b will be incremented by one until $reg\_a = reg\_b$. The variable *count* will increment by one for each increment of reg_b. Set the two registers to known unequal values. Obtain the module and outputs.

4.33   Design a task module that performs both arithmetic and logical operations. There are three inputs: $a[7:0]$, $b[7:0]$, and $c[7:0]$, where $a[0]$, $b[0]$, and $c[0]$ are the low-order bits of a, b, and c, respectively. There are four outputs: z_1, z_2, z_3, and z_4 that perform the operations shown below. Obtain the outputs for four sets of inputs.

$$z_1 = (a + b) \& (c)$$
$$z_2 = (a + b) | (c)$$
$$z_3 = (a \& b) + (c)$$
$$z_4 = (a | b) + (c)$$

4.34   Design a module that contains a task to count the number of 1s in an 8-bit register *reg_a*. The task returns the number of 1s to a 4-bit register *count*. Obtain the outputs from the module.

4.35   Design a module that contains a function to calculate the parity of a 16-bit address and returns 1 bit indicating whether the parity is even (1) or odd (0). If the parity is even, then *parity is even* is displayed; if parity is odd, then *parity is odd* is displayed. The function module, like tasks, has no ports to communicate with the external environment. The only ports are input ports that receive parameters from the function invocation.

Draw the block diagram of the module *fctn_parity* with the function *calc_parity* embedded in the module. Obtain the design module and show the outputs.

4.36   Design a function to implement a full adder using the **case** statement. Draw the block diagram for the full adder module *fctn_full_adder* which contains the full adder function *full_add*. The only ports are those in the function — there are no ports in the module to the external environment. Operands *a*, *b*, and *cin* are scalar inputs to the function; *sum* is a 3-bit vector of type **reg** that is returned to function invocation *full_add* where it is assigned to the variable *sum*, which is then displayed. Show the outputs for the function.

4.37   A small corporation has 100 shares of stock. Each share entitles the owner to one vote at a stockholders' meeting. The shares are distributed as follows:

10 shares   20 shares   30 shares   40 shares

A two-thirds majority is required in order to pass a measure at a stockholders' meeting. Each of the four persons has a switch that closes to vote "yes" for all of that person's shares or opens to vote "no" for all of that person's shares.

Derive a truth table for an output z_1 which indicates that the two-thirds majority had been met. Write the expression for z_1 in a sum-of-minterms decimal notation. Then minimize the expression for z_1 to a minimum sum-of-products notation using Boolean algebra. Write the expression for z_1 in a product-of-maxterms decimal notation. Then minimize the expression for z_1 to a minimum product-of-sums notation using a Karnaugh map. Design a structural module for both the sum-of-products form and the product-of-sums form. Obtain the test bench and outputs for both forms and compare the results.

4.38   Design a BCD-to-decimal decoder using built-in primitives. The inputs and outputs are asserted high. Then use the decoder to implement the two functions shown on the next page.

$$z_1(x_1, x_2, x_3, x_4) = \Sigma_m(1, 2, 4, 8)$$
$$z_2(x_1, x_2, x_3, x_4) = \Pi_M(0, 1, 2, 3, 6, 8)$$

Obtain the structural module, test bench, and outputs.

4.39   Design a logic circuit using structural modeling that will control the interior lighting of a building. The building contains four rooms separated by removable partitions. There is one switch in each room which, in conjunction with the other switches, provides the following methods of control:

(a)   All partitions are closed forming separate rooms. Each switch controls only the light in its respective room.

(b)   All partitions are open forming one large room. Each switch controls all the lights in the building. That is, when the lights are on, they can be turned off by any switch. Conversely, when the lights are off, they can be turned on by any switch.

(c)   Two of the partitions are closed forming three rooms. The middle partition is open, so that the middle room is larger than the other two rooms. Each switch controls the lights in its room only.

(d)   The middle partition is closed forming two rooms. Each switch controls the lights in its room only.

The switch control method outlined above is referred to as *four-way switching*. This technique provides control of one set of lights by one or more switches. Draw the logic diagram and use 4:1 multiplexers in the design and additional logic functions. Obtain the structural module, the test bench to thoroughly test the circuit, and the outputs.

4.40   Design a comparator for two binary-coded decimal (BCD) operands using structural modeling. The BCD operands are *a[3:0]* and *b[3:0]*, where *a[0]* and *b[0]* are the low-order bits of *a* and *b*, respectively. There are three outputs that indicate the relative magnitude of the two operands: $a < b$, $a = b$, and $a > b$. There are also two outputs that indicate if *a* or *b* is invalid for BCD. If either operand is invalid, then the comparator outputs are disabled.

Obtain the structural module using the following dataflow modules: *xnor2_df, or2_df, or4_df, and2_df, and3_df, and4_df,* and *and5_df*. Obtain the test bench module for several combinations of the operands, which will include the cases where an operand is invalid for BCD, and the outputs.

4.41   Design a structural module that will generate a high output z_1 if a 4-bit binary input *x[3:0]* has a value less than or equal to five or greater than nine.

Generate a Karnaugh map and obtain the equation for z_1 in a sum-of-products form.   Instantiate dataflow modules into the structural module.   Obtain the design module, the test bench module for all combinations of the inputs, and the outputs.

5

Computer Arithmetic

This chapter presents computer arithmetic for the three number representations of fixed-point, decimal, and floating-point. The four basic operations of addition, subtraction, multiplication, and division are described for each number representation. The heart of any computer is the arithmetic processor, which performs the four basic operations for all three number representations.

Emphasis is placed on high-speed operations, although some low-speed methods are discussed because they demonstrate some interesting concepts. The arithmetic algorithms and processor architectures are presented in sufficient detail to permit the concepts to be easily understood.

Fixed-point addition includes a low-speed ripple-carry adder and a high-speed carry lookahead adder that generates the carries between stages in parallel. Subtraction is simply an extension of addition and uses the adders previously designed with appropriate modifications.

Fixed-point multiplication is discussed for the following four methods: sequential add-shift algorithm, bit-pair recoding, and planar array multiplication. Fixed-point division expounds on the restoring and nonrestoring division techniques.

Decimal addition and subtraction is presented in which all bits of both digits are processed in parallel. Subtraction is accomplished by adding the rs complement (10s complement) of the subtrahend to the minuend. Decimal multiplication and division are described using a table lookup method.

Floating-point algorithms for addition, subtraction, multiplication, and division are presented. These operations are based on the Institute of Electrical and Electronics Engineers (IEEE) Standard 754.

5.1 Fixed-Point Addition

In fixed-point arithmetic, the radix point (or binary point) can be fixed in a particular location within the string of bits that represent the number. The radix point is usually located to the immediate right of the number for integers or to the immediate left of the number for fractions. The radix point is implicitly defined to be in a certain location, and therefore does not require a physical location in memory.

Adders are utilized for all four operations of addition, subtraction, multiplication, and division; therefore, high-speed adders are essential for high-performance arithmetic processors. Two operands are used for addition: the *augend* and the *addend*. The addend is added to the augend to form the *sum*.

The adder treats both unsigned and signed numbers the same; that is, it does not differentiate between unsigned and signed numbers. The leftmost digit of a signed radix number is the sign digit. For example, consider the n-digit signed number in radix r in Equation 5.1, where a_{n-1} is the sign digit. The sign of the number is determined by Equation 5.2.

$$A = (a_{n-1}\, a_{n-2}\, a_{n-3} \cdots a_1\, a_0)_r \tag{5.1}$$

$$a_{n-1} = \begin{cases} 0 & \text{if } A \geq 0 \\ r-1 & \text{if } A < 0 \end{cases} \tag{5.2}$$

Thus, for radix 2, the sign bit is 0 for positive numbers (including zero) and 1 for negative numbers. The remaining bits specify either the true magnitude if $a_{n-1} = 0$ or the complemented magnitude if $a_{n-1} = 1$. The complemented magnitude is in either 1s complement notation or 2s complement notation. The rules for binary addition are restated in Table 5.1.

Table 5.1 Rules for Binary Addition

+	0	1
0	0	1
1	1	0*

* Indicates a carry-out to the next higher-order column.

The primary number representation for binary operations is the 2s complement number representation. The maximum positive number consists of a 0 bit followed by an integer field of all 1s. The maximum negative number consists of a 1 bit followed by an integer field of all 0s. Thus, the range for numbers in 2s complement notation is from -2^{n-1} to $+2^{n-1} - 1$. For 8-bit operands, the range is from 0111 1111 (+127) to 1000 0000 (−128).

Overflow Overflow occurs when the result of an arithmetic operation (usually addition) exceeds the word size of the machine; that is, the sum is not within the representable range of numbers provided by the number representation. For two *n*-bit numbers

$$A = a_{n-1} a_{n-2} a_{n-3} \cdots a_1 a_0$$

$$B = b_{n-1} b_{n-2} b_{n-3} \cdots b_1 b_0$$

where a_{n-1} and b_{n-1} are the sign bits of operands A and B, respectively, overflow can be detected by either of the following two equations:

$$\text{Overflow} = (a_{n-1} \bullet b_{n-1} \bullet sum_{n-1}') + (a_{n-1}' \bullet b_{n-1}' \bullet sum_{n-1})$$

$$\text{Overflow} = cout_{n-1} \oplus cout_{n-2} \tag{5.3}$$

where the symbol "$\bullet$" is the logical AND operator, the symbol "$+$" is the logical OR operator, the symbol "$\oplus$" is the exclusive-OR operator, and $cout_{n-1}$ and $cout_{n-2}$ are the carry bits out of positions $n-1$ and $n-2$, respectively.

Equation 5.3 can be restated as follows: Overflow occurs whenever the sign of the sum bit sum_{n-1} is different than the signs of the operands a_{n-1} and b_{n-1} when the signs of the operands are the same; also, overflow occurs if the carry-out of the high-order numeric position $cout_{n-2}$ is different than the carry-out of the sign position $cout_{n-1}$. An overflow cannot occur when adding two numbers with opposite signs, because adding a positive number to a negative number produces a sum — either positive or negative — that resides within the range of the two numbers. Examples of overflow are shown below for two 8-bit numbers in 2s complement representation.

$$
\begin{array}{lcccccccccc}
A = & 0 & 0 & 1 & 1 & & 1 & 1 & 1 & 0 & +62 \\
+) \; B = & 0 & 1 & 1 & 0 & & 0 & 0 & 1 & 1 & +99 \\
\hline
Sum = & 1 & 0 & 1 & 0 & & 0 & 0 & 0 & 1 & +161 \\
\end{array}
$$

$$
\begin{array}{lcccccccccc}
A = & 1 & 0 & 1 & 1 & & 0 & 1 & 0 & 1 & -75 \\
+) \; B = & 1 & 1 & 0 & 0 & & 1 & 0 & 0 & 0 & -56 \\
\hline
Sum = & 0 & 1 & 1 & 1 & & 1 & 1 & 0 & 1 & -131 \\
\end{array}
$$

5.1.1 Ripple-Carry Addition

A ripple adder adds two n-bit operands and requires n full adders. Full adders were presented in a previous chapter; however, the equations for the sum and carry-out are replicated in Equation 5.4 for any stage i. Since the carries propagate — or ripple — through the adder, the final sum bit sum_{i-1} may take an inordinately long time to be generated, especially for large operands.

$$sum_i = a_i \oplus b_i \oplus cin_{i-1}$$

$$cout_i = a_i b_i + a_i cin_{i-1} + b_i cin_{i-1} \tag{5.4}$$

The carry-in c_0 of the adder is usually connected to a logical 0 for addition, and the carry-out $cout_{n-1}$ can be used for overflow detection as indicated in Equation 5.3. Figure 5.1 shows a block diagram of a full adder. Figure 5.2 shows an n-bit ripple adder using n full adders.

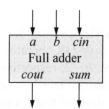

Figure 5.1   Block diagram of a full adder.

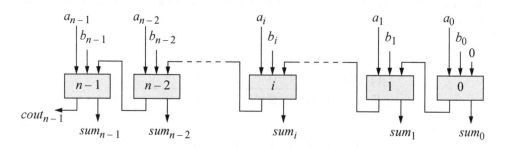

Figure 5.2   Organization of an n-bit ripple adder.

5.1.2 Carry Lookahead Addition

The slow speed of a ripple adder is the result of specifying the carry-in to a stage as a function of the carry-out of the previous lower-order stage. This section presents an adder in which the carry-out of any stage can be represented as a function of the augend and addend of that stage and the low-order carry-in to the adder. Equation 5.4 can be modified as shown in Equation 5.5.

$$cout_i = a_i b_i + a_i cin_{i-1} + b_i cin_{i-1}$$
$$= a_i b_i + (a_i + b_i) cin_{i-1} \tag{5.5}$$

A technique will now be presented that allows the carry-in to all stages of an adder to be generated simultaneously, thereby negating the ripple-carry effect. This results in a constant addition time regardless of the length of the operands. Let A and B be the augend and addend inputs to an n-bit adder as shown below.

$$A = a_{n-1} a_{n-2} a_{n-3} \cdots a_1 a_0$$
$$B = b_{n-1} b_{n-2} b_{n-3} \cdots b_1 b_0$$

The carry-in cin_{i-1} is the carry-in to stage $_i$ from stage $_{i-1}$. The sum and carry-out of stage $_i$ are sum_i and $cout_i$, respectively. Since the low-order stage is 0, the carry-in to stage $_0$ is specified as cin_{-1}. Two auxiliary functions, *generate* and *propagate*, are defined for carry generation as follows for any stage $_i$:

$$\text{Generate:} \quad G_i = a_i b_i$$
$$\text{Propagate:} \quad P_i = a_i + b_i$$

The generate function G_i specifies that, if $a_i b_i = 11$, then a carry-out will be generated for stage $_i$, since $1 + 1$ produces a $cout_i sum_i = 10$. Similarly, if a_i or $b_i = 1$, then a carry-in cin_{i-1} from the previous lower-order stage $_{i-1}$ to stage $_i$ will be propagated through stage $_i$ as $cout_i$ to the next higher stage $_{i+1}$. Note that if a_i and b_i are both 1s, then that constitutes a generate function. Examples of generate and propagate are shown below for two 8-bit operands in 2s complement notation.

	$A =$	1	1	0	0	1	1	0	1	-51
+)	$B =$	0	1	0	1	1	1	0	0	$+92$
		G'	G	G'	G'	G	G	G'	G'	
		P	P'	P'	P	P'	P'	P'	P	
									0	$cin = 0$
$cout = 1$		0	0	1	0	1	0	0	1	$+41$

	$A =$	0	0	1	0	0	1	0	1	$+37$
+)	$B =$	0	1	0	0	1	1	1	1	$+79$
		G'	G'	G'	G'	G'	G	G'	G	
		P'	P	P	P'	P	P'	P	P'	
									1	$cin = 1$
$cout = 0$		0	1	1	1	0	1	0	1	$+117$

Equation 5.5 can now be restated as shown in Equation 5.6 in terms of the generate and propagate functions. Equation 5.6 states that the carry-out from any stage can be obtained independently and concurrently from the operand bits of that stage together with the low-order carry-in to the adder. Equation 5.6 can be applied recursively to a 4-bit adder to yield Equation 5.7.

$$cout_i = a_i b_i + a_i cin_{i-1} + b_i cin_{i-1}$$
$$= a_i b_i + (a_i + b_i) cin_{i-1}$$
$$= G_i + P_i cin_{i-1} \tag{5.6}$$

$$cout_0 = G_0 + P_0 cin_{-1}$$

$$cout_1 = G_1 + P_1 cout_0$$
$$= G_1 + P_1(G_0 + P_0 cin_{-1})$$
$$= G_1 + P_1 G_0 + P_1 P_0 cin_{-1} \tag{5.7}$$

(Continued on next page)

$$cout_2 = G_2 + P_2 cout_1$$

$$= G_2 + P_2(G_1 + P_1 G_0 + P_1 P_0 cin_{-1})$$

$$= G_2 + P_2 G_1 + P_2 P_1 G_0 + P_2 P_1 P_0 cin_{-1}$$

$$cout_3 = G_3 + P_3 cout_2$$

$$= G_3 + P_3(G_2 + P_2 G_1 + P_2 P_1 G_0 + P_2 P_1 P_0 cin_{-1})$$

$$= G_3 + P_3 G_2 + P_3 P_2 G_1 + P_3 P_2 P_1 G_0 + P_3 P_2 P_1 P_0 cin_{-1} \qquad (5.7)$$

Consider the expression for $cout_2$ in Equation 5.7 to further explain the generate and propagate functions to produce a carry-out. Each of the product terms shown below will produce a carry-out of 1 for $cout_2$.

$cout_2 =$ G_2	+	$P_2 G_1$	+	$P_2 P_1 G_0$	+	$P_2 P_1 P_0$
1		0 1		0 1 1		1 0 0
1		1 1		1 0 1		0 1 1
						1 $\leftarrow$ cin_{-1}
1 $\leftarrow$ 0	1 $\leftarrow$ 0 0	1 $\leftarrow$ 0 0 0	1 $\leftarrow$ 0 0 0

It can be seen from Equation 5.7 that each carry is now an expression consisting of only three gate delays: one delay each for the generate and propagate functions, one delay to AND the generate and propagate functions, and one delay to OR all of the product terms. If a high-speed full adder is used in the implementation, then the sum bits can be generated with only two gate delays, providing a maximum of only five delays for an add operation. This technique provides an extremely fast addition of two n-bit operands.

Group generate and propagate   As n becomes large, the number of inputs to the high-order gates also becomes large, which may be a problem for some technologies. The problem can be alleviated to some degree by partitioning the adder stages into 4-bit groups. Additional auxiliary functions can then be defined for *group generate* and *group propagate*, as shown in Equation 5.8 for group$_j$, which consists of individual adder stages $i + 3$ through i. In this method, each group of four adders is considered as a unit with its individual group carry sent to the next higher-order group.

Group generate: $GG_j = G_{i+3} + P_{i+3}G_{i+2} + P_{i+3}P_{i+2}G_{i+1} +$

$$P_{i+3}P_{i+2}P_{i+1}G_i$$

Group propagate: $GP_j = P_{i+3}P_{i+2}P_{i+1}P_i$ (5.8)

The group generate GG_j signifies a carry that is generated out of the high-order $(i+3)$ position that originated from within the group. The group propagate GP_j indicates that a carry was propagated through the group. The group carry can now be written in terms of the group generate and group propagate functions, as shown in Equation 5.9. The term GC_{j-1} is the carry-in to the group from the previous lower-order group. If group$_j$ is the low-order group, then $GC_{j-1} = cin_{-1}$.

Group carry: $GC_j = GG_j + GP_jGC_{j-1}$ (5.9)

Section generate and propagate   If the fan-in limitation is still a problem for very large operands, then the group generate and group propagate concept can be extended to partition four groups into one section. For a 64-bit adder, there would be four sections with four groups per section, with four full adders per group. Two additional auxiliary functions can now be defined as *section generate* and *section propagate* for section$_k$, as shown in Equation 5.10.

Section generate: $SG_k = GG_{j+3} + GP_{j+3}GG_{j+2} +$

$$GP_{j+3}GP_{j+2}GG_{j+1} +$$

$$GP_{j+3}GP_{j+2}GP_{j+1}GG_j$$

Section propagate: $SP_k = GP_{j+3}GP_{j+2}GP_{j+1}GP_j$ (5.10)

The section generate SG_k signifies a carry that is generated out of the high-order $(j+3)$ position that originated from within the section. The section propagate SP_k indicates that a carry was propagated through the section. The section carry can now be written in terms of the section generate and section propagate functions, as shown in Equation 5.11. The term SC_{k-1} is the carry-in to the section from the previous lower-order section. If section$_k$ is the low-order section, then $SC_{k-1} = cin_{-1}$.

$$\text{Section carry:} \quad SC_k = SG_k + SP_k SC_{k-1} \tag{5.11}$$

The carry-out of the high-order section SC_{k+3} is also the carry-out of the adder and can be written as $cout_{n-1}$. This section has presented a method to increase the speed of addition by partitioning the adder into sections and groups and developing carry lookahead logic within individual groups and within individual sections. Figure 5.4 shows a block diagram of a 64-bit adder consisting of four sections, with four groups per section, and four full adders per group.

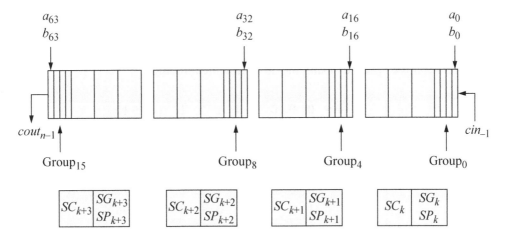

Figure 5.3   Block diagram of a 64-bit carry lookahead adder.

5.2 Fixed-Point Subtraction

Subtraction using the paper-and-pencil method is accomplished by subtracting the subtrahend from the minuend according to the rules shown in Table 5.2. An example is shown in Figure 5.4 in which the subtrahend 0101 0011 (+83) is subtracted from the minuend 0110 0101 (+101), resulting in a difference of 0001 0010 (+18).

Table 5.2 Rules for Binary Subtraction

$-$	0	1
0	0	1*
1	1	0

* Indicates a borrow from the minuend in the next higher-order column.

	2^7	2^6	2^5	2^4	2^3	2^2	2^1	2^0
Minuend (+101)	0	1	1	0	0	1	0	1
−) Subtrahend (+83)	0	1	0	1	0	0	1	1
Difference (+18)	0	0	0	1	0	0	1	0

Figure 5.4   Example of the paper-and-pencil method of binary subtraction.

The paper-and-pencil method is not appropriate for subtraction in a computer. Recall that the rs complement is obtained from the $r - 1$ complement by adding 1. For radix 2, the 2s complement is obtained by adding 1 to the 1s complement. Arithmetic processors use the adder to perform subtraction by adding the 2s complement of the subtrahend to the minuend. Examples of subtraction are shown below for both positive and negative operands using the 2s complement method.

$$
\begin{array}{lllllllll}
A = & 0 & 0 & 0 & 0 & 1 & 1 & 1 & 1 & \quad +15 \\
-) \; B = & 0 & 1 & 1 & 0 & 0 & 0 & 0 & 0 & \quad +96 \\
\end{array}
$$

$$\downarrow$$

$$
\begin{array}{llllllllll}
 & 0 & 0 & 0 & 0 & 1 & 1 & 1 & 1 & \quad +15 \\
+) & 1 & 0 & 1 & 0 & 0 & 0 & 0 & 0 & \quad -96 \\
\hline
 & 1 & 0 & 1 & 0 & 1 & 1 & 1 & 1 & \quad -81 \\
\end{array}
$$

$$
\begin{array}{lllllllll}
A = & 1 & 0 & 1 & 1 & 0 & 0 & 0 & 1 & \quad -79 \\
-) \; B = & 1 & 1 & 1 & 0 & 0 & 1 & 0 & 0 & \quad -28 \\
\end{array}
$$

$$\downarrow$$

$$
\begin{array}{llllllllll}
 & 1 & 0 & 1 & 1 & 0 & 0 & 0 & 1 & \quad -79 \\
+) & 0 & 0 & 0 & 1 & 1 & 1 & 0 & 0 & \quad +28 \\
\hline
 & 1 & 1 & 0 & 0 & 1 & 1 & 0 & 1 & \quad -51 \\
\end{array}
$$

$$
\begin{array}{lllllllll}
A = & 1 & 0 & 0 & 0 & 0 & 1 & 1 & 1 & \quad -121 \\
-) \; B = & 1 & 1 & 1 & 0 & 0 & 1 & 1 & 0 & \quad -26 \\
\end{array}
$$

$$\downarrow$$

$$
\begin{array}{llllllllll}
 & 1 & 0 & 0 & 0 & 0 & 1 & 1 & 1 & \quad -121 \\
+) & 0 & 0 & 0 & 1 & 1 & 0 & 1 & 0 & \quad +26 \\
\hline
 & 1 & 0 & 1 & 0 & 0 & 0 & 0 & 1 & \quad -95 \\
\end{array}
$$

$$
\begin{array}{rccccccccc}
A = & 0 & 0 & 0 & 1 & 0 & 0 & 1 & 1 & +19 \\
-)\ B = & 0 & 1 & 0 & 1 & 1 & 1 & 0 & 0 & +92 \\
\end{array}
$$

$$\downarrow$$

$$
\begin{array}{rccccccccc}
 & 0 & 0 & 0 & 1 & 0 & 0 & 1 & 1 & +19 \\
+) & 1 & 0 & 1 & 0 & 0 & 1 & 0 & 0 & -92 \\
\hline
 & 1 & 0 & 1 & 1 & 0 & 1 & 1 & 1 & -73 \\
\end{array}
$$

The adder of the previous section can be modified slightly so that both subtraction and addition can be performed using the same hardware. The 1s complement must be obtained for the subtrahend plus a carry-in to the low-order adder stage, thus forming the 2s complement of the subtrahend. The logic that is used to invert the subtrahend should also permit the noninverted addend to be applied to the inputs of the adder. This can be accomplished by the exclusive-OR circuit, as shown in Table 5.3.

Table 5.3 Rules for the Exclusive-OR Circuit

		Addend/Subtrahend	
	$\oplus$	0	1
Mode Control (m)*	0	0	1
	1	1	0

* Mode Control (m) = 0 specifies an add operation.
  Mode Control (m) = 1 specifies a subtract operation.

The mode control input is connected to the low-order carry-in of the adder. When the mode control input = 0, the addend is not inverted and the carry-in is 0, executing an add operation. When the mode control input = 1, the subtrahend is 1s complemented and the carry-in is 1, providing the requisite 2s complement of the subtrahend for executing a subtract operation. Figure 5.5 shows the logic diagram of a 4-bit carry lookahead adder/subtractor for $stage_0$ through $stage_3$.

When the subtrahend has been 2s complemented, high-speed addition occurs using the carry lookahead method. The 2s complementation does not affect the operation of the adder — the complementation simply applies a different operand to the adder inputs.

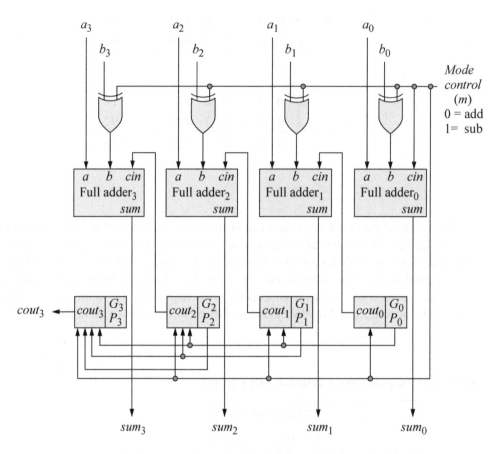

Figure 5.5   Logic diagram of a 4-bit adder/subtractor.

5.3 Fixed-Point Multiplication

Multiplication of two fixed-point binary numbers will be presented for the following methods: sequential add-shift, Booth algorithm, bit-pair recoding, and array multiplier. The operands are normally n-bits and the result is $2n$-bits, as shown below.

$$
\begin{aligned}
\text{Multiplicand:} \quad A &= a_{n-1}\, a_{n-2}\, a_{n-3} \cdots a_1\, a_0 \\
\text{Multiplier:} \quad B &= b_{n-1}\, b_{n-2}\, b_{n-3} \cdots b_1\, b_0 \\
\text{Product:} \quad P &= p_{2n-1}\, p_{2n-2}\, p_{2n-3} \cdots p_1\, p_0
\end{aligned}
$$

The algorithm consists of multiplying the multiplicand by the low-order multiplier bit to obtain a partial product. If the multiplier bit is a 1, then the multiplicand becomes the partial product; if the multiplier bit is a 0, then zeroes become the partial product. The partial product is then shifted left 1 bit position. The multiplicand is then multiplied by the next higher-order multiplier bit to obtain a second partial product. The process repeats for all remaining multiplier bits, at which time the partial products are added to obtain the product. The sign of the product is positive if the operands have the same sign. If the signs of the operands are different, then the sign of the product is negative. Multiplication of two fixed-point binary numbers is a process of repeated add-shift operations and is best illustrated by examples.

Example 5.1   Let the multiplicand and multiplier be two 4-bit operands as shown below. When obtaining the partial products, the sign of the multiplicand is extended left to maintain $2n$ bits in the product. In 2s complement notation, the sign bit is treated the same as any other bit in the operands; that is, the sign bit indicates neither a positive number nor a negative number.

Multiplicand:						0	1	1	1		+7
Multiplier:					×)	0	1	1	0		+6
	0	0	0	0	0	0	0	0			
	0	0	0	0	1	1	1				
	0	0	0	1	1	1					
	0	0	0	0	0						
Product:	0	0	1	0	1	0	1	0			+42

Example 5.2   The only restriction in the add-shift technique is that the multiplier must be positive; the multiplicand can be either positive or negative. An example is shown below with a negative multiplicand and a positive multiplier.

Multiplicand:						1	0	1	0		−6
Multiplier:					×)	0	1	0	1		+5
	1	1	1	1	1	0	1	0			
	0	0	0	0	0	0	0				
	1	1	1	0	1	0					
	0	0	0	0	0						
Product:	1	1	1	0	0	0	1	0			−30

Example 5.3   A problem occurs when the multiplier is negative, because the value of the multiplier is not the absolute value. In this case, the multiplier must first be negated through 2s complementation. Then the multiplication operation is performed and the product is 2s complemented. An example is shown below for a positive

multiplicand (+7) and a negative multiplier (–3). The product is 91, because the multiplier is an unsigned number of 13.

Multiplicand:					0	1	1	1	+7 (7)
Multiplier:			×)	1	1	0	1		–3 (13)
	0	0	0	0	0	1	1	1	
	0	0	0	0	0	0	0		
	0	0	0	1	1	1			
	0	0	1	1	1				
Product:	0	1	0	1	1	0	1	1	91

The same example will now initially 2s complement the multiplier, perform the multiply operation, then will 2s complement the product to obtain the correct result of –21.

Multiplicand:					0	1	1	1	+7
Multiplier:			×)	0	0	1	1		+3
	0	0	0	0	0	1	1	1	
	0	0	0	0	1	1	1		
	0	0	0	0	0	0			
	0	0	0	0	0				
	0	0	0	1	0	1	0	1	+21
Product:	1	1	1	0	1	0	1	1	–21

Example 5.4   When both operands are negative, the correct result can be realized by 2s complementing both operands before the operation begins, since a negative multiplicand multiplied by a negative multiplier yields a positive product. The first example below shows the result of not negating both operands, where the multiplicand is –4 and the multiplier is –5; that is, multiplying –4 by 11 to yield –44. The second example 2s complements both operands to obtain a correct product of +20.

Multiplicand:					1	1	0	0	–4
Multiplier:			×)	1	0	1	1		–5 (11)
	1	1	1	1	1	1	0	0	
	1	1	1	1	1	0	0		
	0	0	0	0	0	0			
	1	1	1	0	0				
Product:	1	1	0	1	0	1	0	0	–44

Multiplicand:					0	1	0	0		+4
Multiplier:				×)	0	1	0	1		+5
	0	0	0	0	0	1	0	0		
	0	0	0	0	0	0	0			
	0	0	0	1	0	0				
	0	0	0	0	0					
Product:	0	0	0	1	0	1	0	0		+20

5.3.1 Sequential Add-Shift

In this method, the multiplier must be positive. If the multiplier is negative, then the purpose of the multiplier bits is not always the same during the generation of the partial products. Any low-order 0s and the first 1 bit are treated the same as for a positive multiplier; however, the remaining higher-order bits are complemented and have an inverse effect. If the multiplier is negative, then it can be 2s complemented as shown previously, leaving the multiplicand either positive or negative. Negative multipliers are presented in a later section.

An alternative approach is to 2s complement both the multiplicand and multiplier if the multiplier is negative. This is equivalent to multiplying both operands by −1, but does not change the sign of the product.

Examples will now be presented to illustrate the hardware add-shift technique for both positive and negative multiplicands. A register is a logic macro device that stores information. The information is retained in the register until changed by setting new information into the register. A shift register is a register that is capable of shifting the contents either left or right a specified number of bits.

The n-bit multiplicand is stored in an n-bit register. The multiplier is placed in the low-order n bits of a $2n$-bit shift register that will ultimately contain the $2n$-bit product. The high-order n bits of this shared register are set to zeroes. The low-order multiplier bit determines the operand to be added to the previous partial product: if the low-order multiplier bit is 0, then no addition takes place and the partial product is shifted right 1 bit position; otherwise, the multiplicand is added to the partial product with sign extended to $2n$ bits and then shifted right 1 bit position.

Example 5.5 The add-shift multiply hardware algorithm is shown in Figure 5.6 for a multiplicand of +5 and a multiplier of +6. There are n cycles for n-bit operands. During the first cycle, if the low-order multiplier bit is 0, then the shift register is shifted right 1 bit position and the high-order bit of the shift register is propagated to the right. If the low-order multiplier bit is 1, then the multiplicand is added to the high-order n bits of the shift register and the register is shifted right 1 bit position with the sign of the multiplicand placed in the high-order bit position of the shift register.

The sign of the partial product is not determined until the first add-shift cycle. Any carry-out during the add operation is shifted into the resulting partial product during the subsequent shift operation of an add-shift cycle.

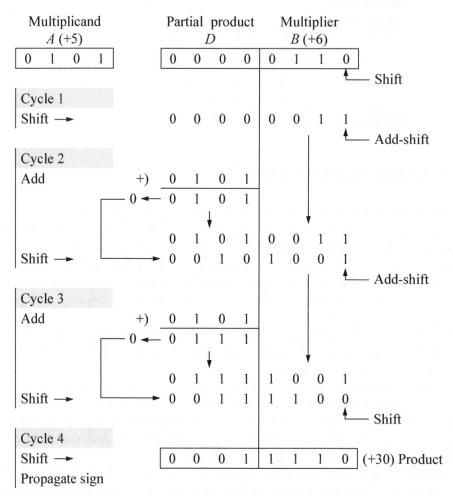

Figure 5.6   Hardware multiply algorithm for a positive multiplicand and a positive multiplier.

Example 5.6   In Figure 5.7, the multiplicand is negative (−5) and the multiplier is positive (+7) resulting in a product of −35. The sign of the product is again determined during the first add-shift cycle in which the sign of the multiplicand becomes the sign of the product. The sign can also be determined by the following expression for two n-bit operands $A = a_{n-1} a_{n-2} a_{n-3} \ldots a_1 a_0$ and $B = b_{n-1} b_{n-2} b_{n-3} \ldots b_1 b_0$:

$$a_{n-1} \oplus b_{n-1}$$

This example has three add-shift cycles. The second and third add-shift cycles shift the carry-out of the add operation into the subsequent partial product during the shift-right operation.

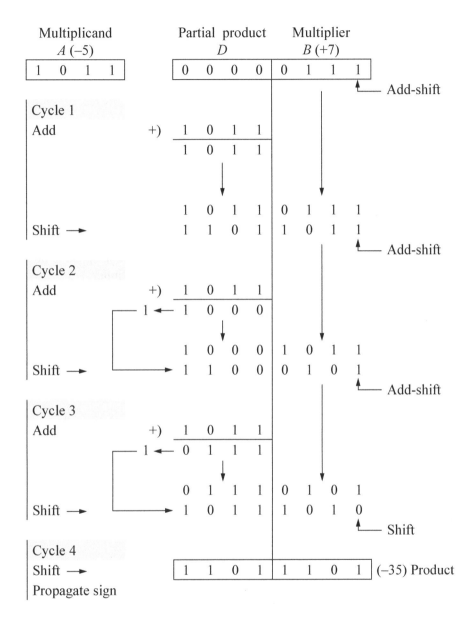

Figure 5.7   Hardware multiply algorithm for a negative multiplicand and a positive multiplier.

During the add operation, a carry lookahead adder of n bits is required since only the high-order n bits are used for the add operation. The shift register, however, must be $2n$ bits. When multiplication is implemented in a computer, a single adder is used to add the multiplicand to the high-order n bits of the shifted partial product and the result is placed in register D, which is part of the $2n$-bit shift register DB.

A count-down counter may be used to control the number of cycles during the multiply operation. The counter is set to the number of bits that is represented by the value of n. When the counter reaches a value of zero, the operation is finished. Alternatively, the operation can be controlled by a microprogram sequencer.

5.3.2 Booth Algorithm

The Booth algorithm is an effective technique for 2s complement multiplication. Unlike the sequential add-shift method, it treats both positive and negative numbers uniformly. In the sequential add-shift method, each multiplier bit generates a rendering of the multiplicand that is added to the partial product. For large operands, the delay to obtain the product can be substantial. The Booth algorithm reduces the number of partial products by shifting over strings of zeroes in a recoded version of the multiplier. This method is referred to as *skipping over zeroes*. The increase in speed is proportional to the number of zeroes in the recoded version of the multiplier.

Consider a string of k consecutive 1s in the multiplier as shown below. The multiplier will be recoded so that the k consecutive 1s will be transformed into $k - 1$ consecutive 0s. The multiplier hardware will then shift over the $k - 1$ consecutive 0s without having to generate partial products.

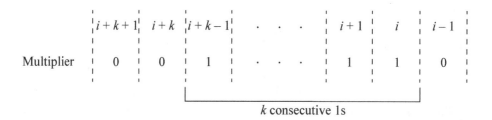

In the sequential add-shift method, the multiplicand would be added k times to the shifted partial product. The number of additions can be reduced by the following property of binary strings:

$$2^{i+k} - 2^i = 2^{i+k-1} + 2^{i+k-2} + \cdots 2^{i+1} + 2^i \qquad (5.8)$$

The right-hand side of the equation is a binary string that can be replaced by the difference of two numbers on the left-hand side of the equation. Thus, the k consecutive 1s can be replaced by the following string:

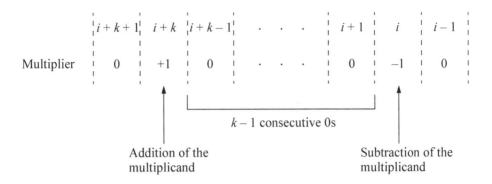

k − 1 consecutive 0s

Addition of the
multiplicand

Subtraction of the
multiplicand

An example will help to clarify the procedure. Let the multiplier be +30 as shown below.

	2^{i+k}						
	2^{i+4}				2^i		
	2^5	2^4	2^3	2^2	2^1	2^0	
Multiplier	0	1	1	1	1	0	+30

$k = 4$

The validity of Equation 5.8 can be verified using the multiplier of +30 shown above.

2^{i+k}	2^i	=	2^{i+k-1}			
2^5	-2^1	=	2^4	$+2^3$	$+2^2$	$+2^1$
32	-2	=	16	$+8$	$+4$	$+2$
	30	=	30			

Thus, the multiplier 011110 (+30) can be regarded as the difference of two numbers: $32 - 2$, as shown below.

$$
\begin{array}{rccccccl}
 & 0 & 1 & 0 & 0 & 0 & 0 & 0 & (32) \\
-) & 0 & 0 & 0 & 0 & 0 & 1 & 0 & (2) \\
\hline
 & 0 & 0 & 1 & 1 & 1 & 1 & 0 & (30)
\end{array}
$$

The product can be generated by one subtraction in column 2^i and one addition in column 2^{i+k}, as shown below; that is, by adding 32 and subtracting 2. In this case, adding 2^5 times the multiplicand and subtracting 2^1 times the multiplicand yields the appropriate result.

	2^6	2^5	2^4	2^3	2^2	2^1	2^0
Standard multiplier	0	0	+1	+1	+1	+1	0
			←	$k=4$		→	
Recoded multiplier	0	+1	0	0	0	−1	0
			←$k-1=3$→				

As the recoded multiplier is scanned from right to left, observe that an entry of −1 occurs at $0 \rightarrow 1$ boundaries, indicating that the left-shifted multiplicand is to be 2s complemented. An entry of +1 in the recoded multiplier occurs at $1 \rightarrow 0$ boundaries. The Booth algorithm can be applied to any number of groups of 1s in a multiplier, including the case where the group consists of a single 1 bit. Table 5.4 shows the multiplier recoding table for two consecutive bits that specify which version of the multiplicand will be added to the shifted partial product.

Table 5.4 Booth Multiplier Recoding Table

Multiplier Bit i	Bit $i-1$	Version of Multiplicand
0	0	$0 \times$ multiplicand
0	1	$+1 \times$ multiplicand
1	0	$-1 \times$ multiplicand
1	1	$0 \times$ multiplicand

The Booth algorithm converts both positive and negative multipliers into a form that generates versions of the multiplicand to be added to the shifted partial products. Since the increase in speed is a function of the bit configuration of the multiplier, the efficiency of the Booth algorithm is data dependent. Several examples will now be presented to illustrate the Booth algorithm.

Example 5.7   An 8-bit positive multiplicand (+53) will be multiplied by an 8-bit positive multiplier (+30) — first using the standard sequential add-shift technique, then using the Booth algorithm to show the reduced number of partial products. In the sequential add-shift technique, there are six additions; using the Booth algorithm, there are only two additions. The second partial product in the Booth algorithm is the 2s complement of the multiplicand shifted left. The third partial product is the result of shifting left over three 0s.

Standard sequential add-shift																	
Multiplicand								0	0	1	1	0	1	0	1		+53
Multiplier ×)								0	0	0	1	1	1	1	0		+30
								0	0	0	0	0	0	0	0		
							0	0	1	1	0	1	0	1			
						0	0	1	1	0	1	0	1				
					0	0	1	1	0	1	0	1					
				0	0	1	1	0	1	0	1						
			0	0	0	0	0	0	0	0							
		0	0	0	0	0	0	0	0								
	0	0	0	0	0	0	0	0									
	0	0	0	0	0	1	1	0	0	0	1	1	0	1	1	0	+1590

Booth algorithm																	
Multiplicand								0	0	1	1	0	1	0	1		+53
Recoded multiplier ×)								0	0	+1	0	0	0	−1	0		
	0	0	0	0	0	0	0	0	0	0	0	0	0	0	0	0	
	1	1	1	1	1	1	1	1	1	0	0	1	0	1	1		
	0	0	0	0	0	1	1	0	1	0	1						
	0	0	0	0	0	1	1	0	0	0	1	1	0	1	1	0	+1590

Example 5.8   This is an example of a 5-bit positive multiplicand (+13) and a 5-bit negative multiplier (−12).

Multiplicand	0	1	1	0	1	+13
Multiplier ×)	1	0	1	0	0	−12
						−156

Booth algorithm											
Multiplicand						0	1	1	0	1	+13
Recoded multiplier ×)						−1	+1	−1	0	0	
	0	0	0	0	0	0	0	0	0	0	
	1	1	1	1	0	0	1	1			
	0	0	0	1	1	0	1				
	1	1	0	0	1	1					
		1	0	1	1	0	0	1	0	0	−156

Example 5.9   If the multiplier has a low-order bit of 1, then an implied 0 is placed to the right of the low-order multiplier bit. This provides a boundary of $0 \rightarrow 1$ (-1 times the multiplicand) for the low-order bit as the multiplier is scanned from right to left. This is shown in the example below, which multiplies a 5-bit positive multiplicand by a 5-bit positive multiplier.

Multiplicand		0	1	1	1	1		+15
Multiplier	×)	0	0	0	1	1	0	+3
								+45

Booth algorithm											
Multiplicand						0	1	1	1	1	+15
Recoded multiplier					×)	0	0	+1	0	−1	
	1	1	1	1	1	1	0	0	0	1	
		0	0	0	0	1	1	1	1		
		0	0	0	1	0	1	1	0	1	+45

Example 5.10   This is an example of a 6-bit negative multiplicand (-19) and a 6-bit positive multiplier ($+14$).

Multiplicand		1	0	1	1	0	1	−19
Multiplier	×)	0	0	1	1	1	0	+14
								−266

Booth algorithm												
Multiplicand						1	0	1	1	0	1	−19
Recoded multiplier					×)	0	+1	0	0	−1	0	
		0	0	0	0	0	0	0	0	0	0	
			0	0	0	0	1	0	0	1	1	
				1	0	1	1	0	1			
			0	1	1	1	1	0	1	1	0	−266

Example 5.11   This is an example of a 5-bit negative multiplicand (−13) and a 5-bit negative multiplier (−7). There is an implied 0 to the right of the low-order multiplier bit to provide a boundary of $0 \to 1$ (−1 times the multiplicand) for the low-order bit.

Multiplicand			1	0	0	1	1			−13
Multiplier	×)	1	1	0	0	1	0			−7
										+91

Booth algorithm										
Multiplicand				1	0	0	1	1		−13
Recoded multiplier		×)	0	−1	0	+1	−1			
0	0	0	0	0	0	1	1	0	1	
1	1	1	1	1	0	0	1	1		
0	0	0	1	1	0	1				
	0	0	1	0	1	1	0	1	1	+91

Example 5.12   A final example shows a multiplier that has very little advantage over the sequential add-shift method because of alternating 1s and 0s. The multiplicand is a positive 6-bit operand (+19) and the multiplier is a negative 6-bit operand (−11). There is an implied 0 to the right of the low-order multiplier bit.

Multiplicand		0	1	0	0	1	1		+19
Multiplier	×)	1	1	0	1	0	1	0	−11
									−209

Booth algorithm											
Multiplicand					0	1	0	0	1	1	+19
Recoded multiplier			×)	0	−1	+1	−1	+1	−1		
1	1	1	1	1	0	1	1	0	1		
0	0	0	0	1	0	0	1	1			
1	1	1	0	1	1	0	1				
0	0	1	0	0	1	1					
1	0	1	1	0	1						
	1	0	0	1	0	1	1	1	1	1	−209

The Booth algorithm accomplishes two objectives: (1) it uniformly transforms both positive and negative multipliers to a configuration that selects appropriate versions of the multiplicand to be added to the shifted partial product and (2) it increases the speed of the multiply operation by shifting over 0s in a recoded version of the multiplier.

5.3.3 Bit-Pair Recoding

The speed increase of the Booth algorithm depended on the bit configuration of the multiplier and is, therefore, data dependent. This section presents a speedup technique that is derived from the Booth algorithm and assures that an n-bit multiplier will have no more than $n/2$ partial products. It also treats both positive and negative multipliers uniformly; that is, there is no need to 2s complement the multiplier before multiplying or to 2s complement the product after multiplying.

As in the Booth algorithm, bit-pair recoding regards a string of 1s as the difference of two numbers. This property of binary strings is restated in Equation 5.12, where k is the number of consecutive 1s and i is the position of the rightmost 1 in the string of 1s.

$$2^{i+k} - 2^i = 2^{i+k-1} + 2^{i+k-2} + \cdots 2^{i+1} + 2^i \tag{5.12}$$

Consider the multiplier example introduced in the Booth algorithm in which there were four consecutive 1s for a value of 30, as shown below. This indicates that the number 0011110 (30) has the same value as $2^5 - 2^1 = 30$.

	2^6	2^5	2^4	2^3	2^2	2^1	2^0	
	0	1	0	0	0	0	0	(32)
$-)$	0	0	0	0	0	1	0	(2)
	0	0	1	1	1	1	0	(30)

The same result can be obtained by examining pairs of bits in the multiplier in conjunction with the bit to the immediate right of the bit-pair under consideration, as shown below. Only the binary weights of 2^1 and 2^0 are required because each pair of bits is examined independently of the other pairs.

	2^6	2^5	2^4	2^3	2^2	2^1	2^0	
	0	0	1	1	1	1	0	(30)

$$\downarrow$$

2^1 2^0	2^1 2^0	2^1 2^0	2^1 2^0

| ▢ 0 | 0 1 | 1 1 | 1 0 | ▢ | (30) |

Bit-pair $2^1 2^0$ is examined in conjunction with the implied 0 to the right of the low-order 0; bit-pair $2^3 2^2$ is examined in conjunction with bit 2^1; bit-pair $2^5 2^4$ is examined in conjunction with bit 2^3; and bit-pair $2^7 2^6$ is examined in conjunction with bit 2^5, where bit 2^7 is the sign extension of the multiplier in order to obtain a bit-pair. Each pair of bits is scanned from right to left with the rightmost bit of each pair used as the column reference for the partial product placement, because it is the center of the three bits under consideration. Using the Booth algorithm technique, the additions and subtractions take place on the multiplier as shown in Figure 5.9.

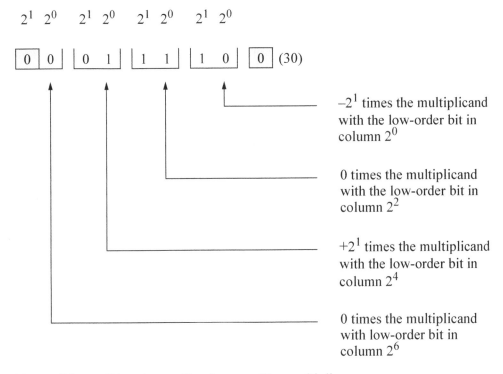

Figure 5.8   Bit-pair recoding for a positive multiplier.

Consider other examples with strings consisting of 010 and 101, as shown in Figure 5.9 and Figure 5.10.

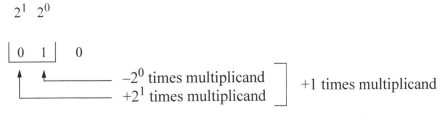

Figure 5.9   Example of a string of a single 1 for the beginning and end of a string.

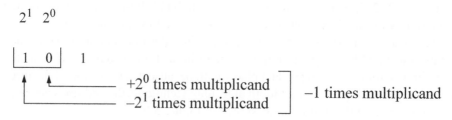

Figure 5.10   Example of a string of a 1 at the beginning and end of a string.

For Figure 5.9 the beginning of the string specifies that -2^0 times the multiplicand is added to the partial product; the end of the string specifies that $+2^1$ is added to the partial product. The net result of this string is that $+1$ times the multiplicand is added to the partial product. The net result of Figure 5.10 is that -1 times the multiplicand is added to the partial product.

Since the bit-pair is examined in conjunction with the bit to the immediate right of the bit-pair under consideration, there are a total of eight versions of the multiplicand to be added or subtracted to the partial product, as shown in Table 5.5. Note that -2 times the multiplicand is the 2s complement of the multiplicand shifted left 1 bit position and $+2$ times the multiplicand is the multiplicand shifted left 1 bit position.

Figure 5.11 shows a positive multiplicand and a positive multiplier; Figure 5.12 shows a negative multiplicand and a positive multiplier; Figure 5.13 shows a positive multiplicand and a negative multiplier; and Figure 5.14 shows a negative multiplicand and a negative multiplier, all using bit-pair recoding.

Table 5.5 Multiplicand Versions for Multiplier Bit-Pair Recoding

Multiplier Bit-Pair		Multiplier Bit on the Right		
2^1	2^0			
$i+1$	i	$i-1$	Multiplicand Versions	Explanation
0	0	0	$0 \times$ multiplicand	No string
0	0	1	$+1 \times$ multiplicand	End of string
0	1	0	$+1 \times$ multiplicand	Single 1 $(+2 -1)$
0	1	1	$+2 \times$ multiplicand	End of string
1	0	0	$-2 \times$ multiplicand	Beginning of string
1	0	1	$-1 \times$ multiplicand	End/beginning of string $(+1 -2)$
1	1	0	$-1 \times$ multiplicand	Beginning of string
1	1	1	$0 \times$ multiplicand	String of 1s

```
            0 0   1 1   0 1   0 1        +53
     ×)     0 1   0 0   1 1   0 0        +76
                           ↓             +4028
            0 0   1 1   0 1   0 1
           [0 1] [0 0] [1 1] [0 0]0
            +1    +1    −1     0

 0 0   0 0   0 0   0 0   0 0   0 0   0 0   0 0
 1 1   1 1   1 1   1 1   0 0   1 0   1 1
 0 0   0 0   0 0   1 1   0 1   0 1
 0 0   0 0   1 1   0 1   0 1
 ─────────────────────────────────────────────
 0 0   0 0    1     1   1 0   1 1     1     0
      4096          256         16         1
         0          3840       176        12    +4028
```

Figure 5.11   Bit-pair recoding using a positive multiplicand and a positive multiplier.

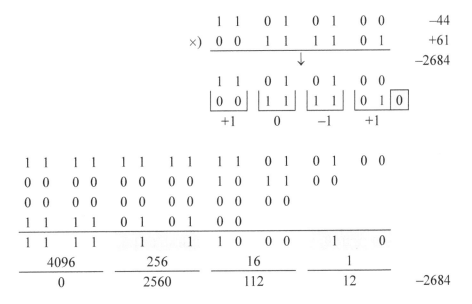

Figure 5.12   Bit-pair recoding using a negative multiplicand and a positive multiplier.

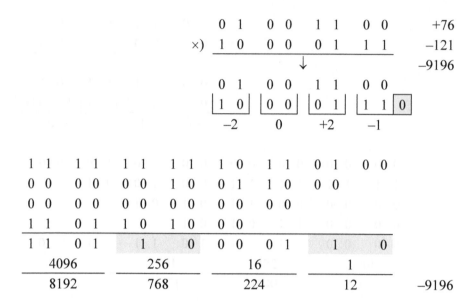

Figure 5.13   Bit-pair recoding using a positive multiplicand and a negative multiplier.

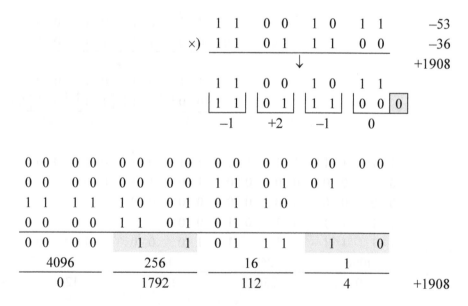

Figure 5.14   Bit-pair recoding using a negative multiplicand and a negative multiplier.

In Figure 5.11 through Figure 5.14, each partial product is shifted left 2 bit positions for correct alignment with the previous partial product. This is in contrast to the standard sequential add-shift method in which the partial products were aligned by shifting left 1 bit position.

The binary weight of each 4-bit segment is shown beneath the product together with the numerical value of each segment, which are added together to provide the product. Note that for a negative product, the binary weight of the 0s is used to determine the value, plus one. For example, the product of Figure 5.13 is shown below for each 4-bit segment together with the corresponding binary weight for the segment.

8	4	2	1		8	4	2	1		8	4	2	1		8	4	2	1	
1	1	0	1			1		0		0	0	0	1			1		0	
	4096					256					16					1			
	8192					768					224					12			−9196

The high-order segment is 1101, where the 0 has a weight of 2; therefore, $2 \times 4096 = 8192$. In a similar manner, the next lower-order segment is 1100 were the two 0s represent a value of three; therefore, $3 \times 256 = 768$. Likewise, the next lower-order segment is 0001, where the three 0s have a weight of 14; therefore, $14 \times 16 = 224$. Finally, the low-order segment is 0100, where the three 0s represent a value of 11. Recall that the rs complement is obtained from the $r - 1$ complement by adding one. Therefore, for radix 2, the low-order segment has a value of $11 + 1 = 12$. Since the sign bit of the product is 1 indicating a negative number, the aggregate sum of all the segments in the product is −9196.

In Figure 5.11, the second partial product was obtained from the operation of −1 times the multiplicand; that is, the 2s complement of the multiplicand. In Figure 5.13, the second partial product was obtained from the operation of +2 times the multiplicand; that is, the multiplicand shifted left 1 bit position. Also, in Figure 5.13, the fourth partial product was obtained from the operation of −2 times the multiplicand; that is, the 2s complement of the multiplicand shifted left 1 bit position.

5.3.4 Array Multiplier

This section presents a method for achieving high-speed multiplication using a planar array. The sequential add-shift technique requires less hardware, but is relatively slow when compared to the array multiplier method. Multiplication of the multiplicand by a 1 bit in the multiplier simply copies the multiplicand. If the multiplier bit is a 1, then the multiplicand is entered in the appropriately shifted position as a partial product to be added to other partial products to form the product. If the multiplier bit is 0, then 0s are entered as a partial product.

Although the array multiplier method is applicable to any size operands, an example will be presented that uses two 3-bit operands as shown in Figure 5.15. The multiplicand is $A [2:0] = a_2 a_1 a_0$ and the multiplier is $B[2:0] = b_2 b_1 b_0$, where a_0 and b_0 are the low-order bits of A and B, respectively. Each bit in the multiplicand is multiplied by the low-order bit b_0 of the multiplier. This is equivalent to the AND function and generates the first of three partial products.

Each bit in the multiplicand is then multiplied by bit b_1 of the multiplier. The resulting partial product is shifted 1 bit position to the left. The process is repeated for bit b_2 of the multiplier. The partial products are then added together to form the product. A carry-out of any column is added to the next higher-order column.

				a_2	a_1	a_0
			$\times$)	b_2	b_1	b_0
Partial product 1				$a_2 b_0$	$a_1 b_0$	$a_0 b_0$
Partial product 2			$a_2 b_1$	$a_1 b_1$	$a_0 b_1$	
Partial product 3		$a_2 b_2$	$a_1 b_2$	$a_0 b_2$		
Product	2^5	2^4	2^3	2^2	2^1	2^0

Figure 5.15   General array multiply algorithm for two 3-bit operands.

An array multiplier assumes that the multiplier is positive. Two examples are shown below using 3-bit operands, one where the multiplicand is positive and one where the multiplicand is negative.

Example 5.13   In this example, the multiplicand and the multiplier are both positive. Since the multiplicand is positive, the partial products are zero-extended to the left to accommodate the $2n$ bits. The multiplicand and multiplier both have a value of $+7$.

			1	1	1		+7
		$\times$)	1	1	1		+7
0	0	0	1	1	1		
	0	1	1	1			
	1	1	1				
1	1	0	0	0	1		+49

Example 5.14   In this example, the multiplicand is negative and the multiplier is positive. Since the multiplicand is negative, the partial products are sign-extended to the left to accommodate the 2n bits. The multiplicand has a value of -1 and multiplier has a value of $+7$.

$$
\begin{array}{ccccccc|c}
 & & & 1 & 1 & 1 & & -1 \\
 & & \times) & 1 & 1 & 1 & & +7 \\
\hline
1 & 1 & 1 & 1 & 1 & 1 & & \\
 & 1 & 1 & 1 & 1 & & & \\
 & & 1 & 1 & 1 & & & \\
\hline
1 & 1 & 1 & 0 & 0 & 1 & & -7 \\
\end{array}
$$

A block diagram of an array multiplier is shown in Figure 5.16 together with a full adder to be used as the planar array elements as shown in the array multiplier of Figure 5.17.

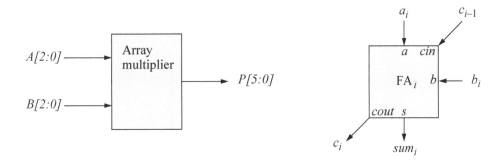

Figure 5.16   Array multiplier block diagram and full adder block diagram.

In Figure 5.17, partial product 1 is shown as $a_2b_0 \ a_1b_0 \ a_0b_0$. Since this is the first partial product, the carry-in to both full adders is 0. Partial product 2 is indicated as $a_2b_1 \ a_1b_1 \ a_0b_1$, which is shifted left and added to partial product 1. Partial product 3 is specified as $a_2b_2 \ a_1b_2 \ a_0b_2$, which is shifted left and added to partial product 2 plus the carry-outs from the previous full adders. The product is $p_5 \ p_4 \ p_3 \ p_2 \ p_1 \ p_0$ with a binary weight of $2^5 \ 2^4 \ 2^3 \ 2^2 \ 2^1 \ 2^0$, respectively. The product is obtained from the sum of all the partial products plus a carry-in of 0. The partial products and the product shown in Figure 5.17 are identical to those shown in the general array multiply algorithm of Figure 5.15.

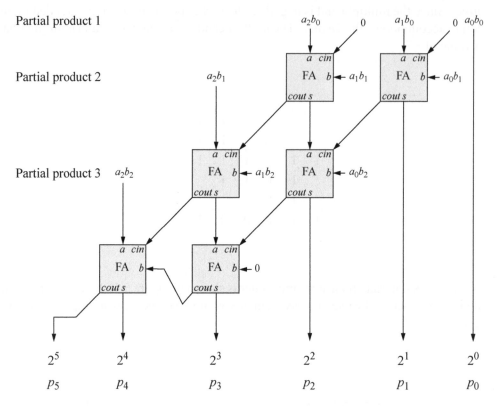

Figure 5.17   Array multiplier for 3-bit operands.

5.4 Fixed-Point Division

There are two operands in division; the dividend A and the divisor B that yield a quotient Q and a remainder R, as shown below. The remainder is smaller than the divisor and has the same sign as the dividend. Unlike multiplication, division is not always commutative; that is, $A/B \neq B/A$, except when $A = B$.

$$\frac{\text{Dividend }(2n\text{ bits})}{\text{Divisor }(n\text{ bits})} = \text{Quotient }(n\text{ bits}), \text{Remainder }(n\text{ bits})$$

Division can be considered as the inverse of multiplication in which the dividend, divisor, and quotient correspond to the product, multiplicand, and multiplier, respectively. Division employs the same general principles as multiplication. Multiplication is an add-shift operation, whereas division is a shift-subtract/add operation. In division, the result of each subtraction determines the next operation in the division sequence. There can be no overflow in multiplication because the product can never exceed $2n$ bits. In division, however, the dividend may be so large compared to the divisor that the value of the quotient exceeds n bits.

In decimal division, the divisor is compared to the current partial remainder — or to the dividend initially — to determine a number that is equal to or greater than the divisor. This is inherently a trial-and-error process. Binary division is simpler, because there are only two possible results: 1 or 0. This section will present the sequential shift-subtract/add restoring division and nonrestoring division. An example will now be given that shows both decimal and binary division using unsigned operands of the same numerical values.

Example 5.15 Figure 5.18 shown two unsigned division examples: Figure 5.18(a) displays the decimal example of dividing 327 by 14; Figure 5.18(b) displays the binary example of dividing 10100011 (327) by 1110 (14). This paper-and-pencil approach involves repeated shifting and subtraction or addition.

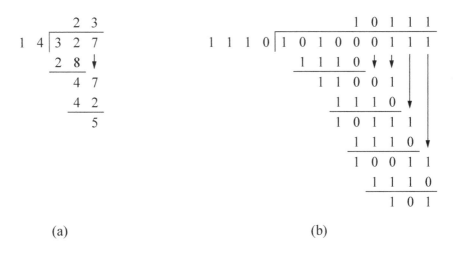

(a)   (b)

Figure 5.18   Examples of division: (a) decimal division and (b) binary division.

Decimal division is a familiar operation; therefore, only binary division will be described. Initially, the bits of the dividend are examined in a left-to-right sequence until the set of bits has a value that is greater than or equal to the divisor. A 1 bit is then placed in the quotient and the divisor is subtracted from the corresponding set of bits in the dividend. The result of this subtraction is called the *partial remainder*. The next lower-order dividend bit is then appended to the right of the partial remainder until the partial remainder is greater than or equal to the divisor. If the partial remainder is less than the divisor, then a 0 is placed in the quotient. The process repeats by subtracting the divisor from the partial remainders.

5.4.1 Restoring Division

Let A and B be the dividend and divisor, respectively, where

$$A = a_{2n-1} \, a_{2n-2} \cdots a_n \, a_{n-1} \cdots a_1 \, a_0$$

$$B = b_{n-1} \, b_{n-2} \cdots b_1 \, b_0$$

The dividend is a $2n$-bit positive integer and the divisor is an n-bit positive integer. The quotient Q and remainder R are n-bit positive integers, where

$$Q = q_{n-1} \, q_{n-2} \cdots q_1 \, q_0$$

$$R = r_{n-1} \, r_{n-2} \cdots r_1 \, r_0$$

When division is implemented in hardware, the process is slightly different than previously described. The dividend is initially shifted left 1 bit position. Then the divisor is subtracted from the dividend. Subtraction is accomplished by adding the 2s complement of the divisor. If the carry-out of the subtract operation is 1, then a 1 is placed in the next lower-order bit position of the quotient; if the carry-out is 0, then a 0 is placed in the next lower-order bit position of the quotient. The concatenated partial remainder and dividend are then shifted left 1 bit position. Fixed-point binary restoring division requires one subtraction for each quotient bit.

Figure 5.19 shows an example of the hardware algorithm for restoring division. Since there are 4 bits in the divisor, there are four cycles. Each cycle begins with a left shift operation. The low-order quotient bit is left blank after each left shift operation —it will be set before the next left shift. Then the divisor is subtracted from the high-order half of the dividend. If the resulting difference is negative — indicated by a carry-out of 0 — then the low-order quotient bit q_0 is set to 0 and the previous partial remainder is restored. If the carry-out is 1, then q_0 is set to 1 and there is no restoration — the partial remainder thus obtained is loaded into the high-order half of the dividend. This sequence repeats for all n bits of the divisor. If the sign of the remainder is different than the sign of the dividend, then the previous partial remainder is restored.

Overflow can occur in division if the value of the dividend and the value of the divisor are disproportionate, yielding a quotient that exceeds the range of the machine's word size. Since the dividend has double the number of bits of the divisor, *overflow* can occur when the high-order half of the dividend has a value that is greater than of equal to that of the divisor.

If the high-order half of the dividend ($a_{2n-1} \ldots a_n$) and the divisor ($b_{n-1} \ldots b_0$) are equal, then the quotient will exceed n bits. If the value of the high-order half of the dividend is greater than the value of the divisor, then value of the quotient will be even greater. Overflow can be detected by subtracting the divisor from the high-order half of the dividend before the first shift-subtract cycle.

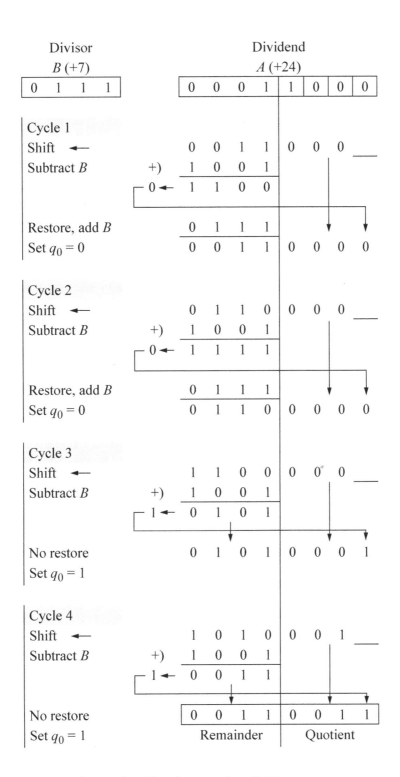

Figure 5.19   Hardware algorithm for restoring division.

If the difference is positive, then an overflow has been detected; if the difference is negative, then the condition for overflow has not been met. Division by zero must also be avoided. This can be detected by the preceding method, since any high-order dividend bits will be greater than or equal to a divisor of all zeroes.

5.4.2 Nonrestoring Division

The speed of the shift-subtract/add restoring division can be increased by redefining the algorithm. Instead of restoring the partial remainder to the previous partial remainder if the result of a subtraction is negative, the algorithm allows both positive and negative partial remainders to be used. That is, a negative partial remainder is used unchanged in the following cycle. Thus, the absolute value of the partial remainder is reduced every cycle by adding or subtracting the divisor from the partial remainder.

In restoring division, if the partial remainder is positive after a subtract operation, then the dividend is shifted left 1 bit position and the divisor is subtracted; that is, the operation is $2A - B$. If the partial remainder is negative after a subtract operation, then the partial remainder is restored by adding the divisor $(A + B)$, then it is shifted left 1 bit position and the divisor B is subtracted. This is equivalent to $2A + B$, as shown below.

$$2(A + B) - B = 2A + B$$

Thus, in nonrestoring division, only two operations are required: $2A - B$ and $2A + B$ for each cycle. The value of q_0 can be determined by the carry-out of the addition or subtraction operation as shown below.

$$q_0 = \begin{cases} 0 \text{ if carry-out is } 0 \\ \\ 1 \text{ if carry-out is } 1 \end{cases}$$

The initial left shift of the dividend must be followed by a subtraction of the divisor in order to establish a starting point for the nonrestoring procedure. Only the final partial remainder is restored to a positive value if it is negative. Thus, if the final value for q_0 is 0, then the previous partial remainder must be restored in order to obtain the correct remainder. Therefore, in nonrestoring division, there are n or $n + 1$ shift-add/subtract cycles.

Figure 5.20 shows an example of the hardware algorithm for nonrestoring division. As in the previous restoring division example, subtraction is done by adding the 2s complement of the divisor. In cycle 2, the operation is $2A + B$, because the previous partial remainder was negative. There is a final restore cycle, because the sign of the remainder is different than the sign of the quotient.

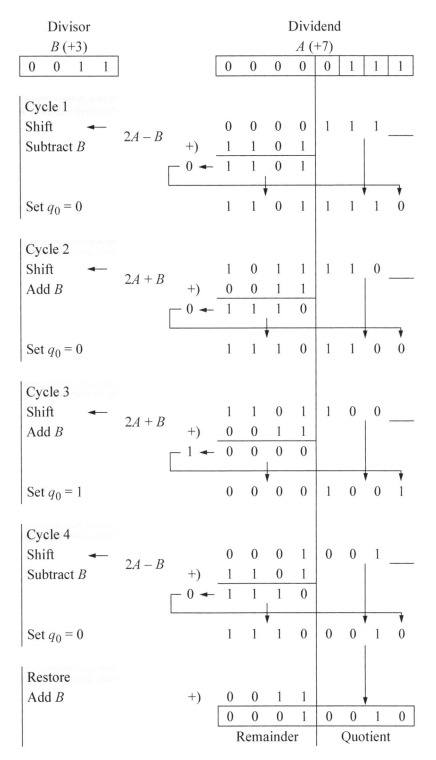

Figure 5.20   Hardware algorithm for nonrestoring division.

5.5 Decimal Addition

If a small amount of decimal arithmetic processing is required, then a processor can convert the decimal numbers into binary, execute the operations in binary, then convert the results to decimal. If a large amount of data is to be processed, then it is more efficient to perform the calculations in decimal.

The single element in decimal arithmetic has nine inputs and five outputs, as shown in Figure 5.21. Each of the two decimal operands is represented by a 4-bit binary-coded decimal (BCD) digit. A carry-in is also provided from the previous lower-order decimal element. The outputs of the decimal element are a 4-bit valid BCD digit and a carry-out.

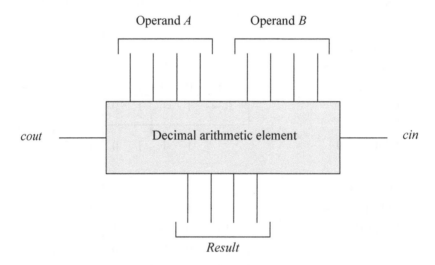

Figure 5.21   A one-digit binary-coded decimal (BCD) arithmetic element.

Fixed-point radix 2 arithmetic treats each bit as a digit; therefore, shifting operations shift only 1 bit at a time. Since decimal arithmetic treats four bits as a digit, shifting operations shift 4 bits at a time. Thus, to perform a logical (unsigned) left shift of two digits positions on the following decimal number requires a shift of 8 bits with zeroes filling the vacated low-order digit positions:

    0111 1101 1111 0110 0101 yields 1111 0110 0101 0000 0000

BCD instructions operate on decimal numbers that are encoded as 4-bit binary numbers in the 8421 code. For example, the decimal number 576 is encoded in BCD as 0101 0111 0110. BCD numbers have a range of 0 to 9; therefore, any number greater than 9 must be adjusted by adding a value of 6 to the number to yield a valid BCD number. For example, if the result of an operation is 1010, then 0110 is added to this intermediate sum. This yields a value of 0000 with a carry of 1, or 0001 0000 in

BCD which is 10 in decimal. The condition for a correction (adjustment) of an inter-mediate sum that also produces a carry-out is shown in Equation 5.13. This specifies that a carry-out will be generated whenever bit position b_8 is a 1 in both decades, when bit positions b_8 and b_4 are both 1s, or when bit positions b_8 and b_2 are both 1s. Decimal (BCD) arithmetic operations were covered in Chapter 2; therefore, this section will concentrate on the hardware required for BCD operations.

$$\text{Carry} = c_8 + b_8 b_4 + b_8 b_2 \tag{5.13}$$

5.5.1 Addition With Sum Correction

A decimal adder adds two BCD digits in parallel and produces a valid decimal digit. The adder must include correction logic to yield a valid decimal digit according to Equation 5.13; that is, to correct an intermediate sum digit that is equal to 1010 through 1111. Two adders are used for each decade of an n-digit decimal adder. These adders are 4-bit fixed-point binary adders of the type previously described. The augend A and the addend B are defined as shown in Equation 5.14 for n-bit operands.

$$A = a[4(n-1)+3 : 4(n-1)]$$

$$\cdots$$

$$a[15 : 12]$$

$$a[11 : 8]$$

$$a[7 : 4]$$

$$a[3 : 0]$$

$$B = b[4(n-1)+3 : 4(n-1)]$$

$$\cdots$$

$$b[15 : 12]$$

$$b[11 : 8]$$

$$b[7 : 4]$$

$$b[3 : 0] \tag{5.14}$$

The low-order decade of an n-digit decimal adder is shown in Figure 5.22. The carry-out of the decade is generated from Equation 5.13. The carry-out of adder(1) — which is the carry-out of the decade — is connected to inputs b_4 and b_2 of adder(2) with $b_8 \, b_1 = 00$. This corrects an invalid decimal digit. The carry-out of adder(2) can be ignored, because it provides no new information.

This decimal adder stage can be used in conjunction with other identical stages to design an n-digit parallel decimal adder. The carry-out of stage$_i$ connects to the carry-in of stage$_{i+1}$; therefore, this is a ripple adder for decimal operands.

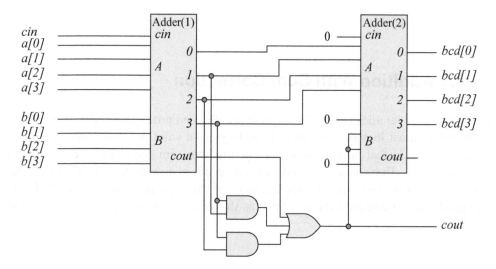

Figure 5.22   Low-order decade of an n-digit decimal adder.

5.5.2 Addition Using Multiplexers for Sum Correction

An alternative approach to determining whether to add six to correct an invalid decimal number is to use a multiplexer. The two operands are added in adder(1) as before; however, this intermediate sum is always added to six in adder(2), as shown in Figure 5.23. The sums from adder(1) and adder(2) are then applied to four 2:1 multiplexers. Selection of the adder(1) sum or the adder(2) sum is determined by ORing together the carry-out of both adders — $cout3$ and $cout8$ — to generate a select input $cout$ to the multiplexers, as shown below. Numerical examples of a decimal adder with multiplexers are shown in Figure 5.24.

$$\text{Multiplexer select} = \begin{cases} \text{Adder(1) sum if } cout \text{ is } 0 \\ \\ \text{Adder(2) sum if } cout \text{ is } 1 \end{cases}$$

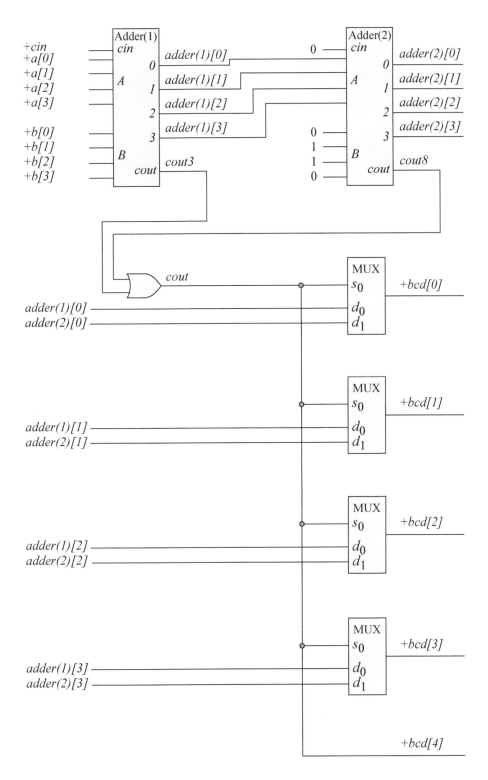

Figure 5.23 Decimal addition with multiplexers to obtain a valid decimal digit.

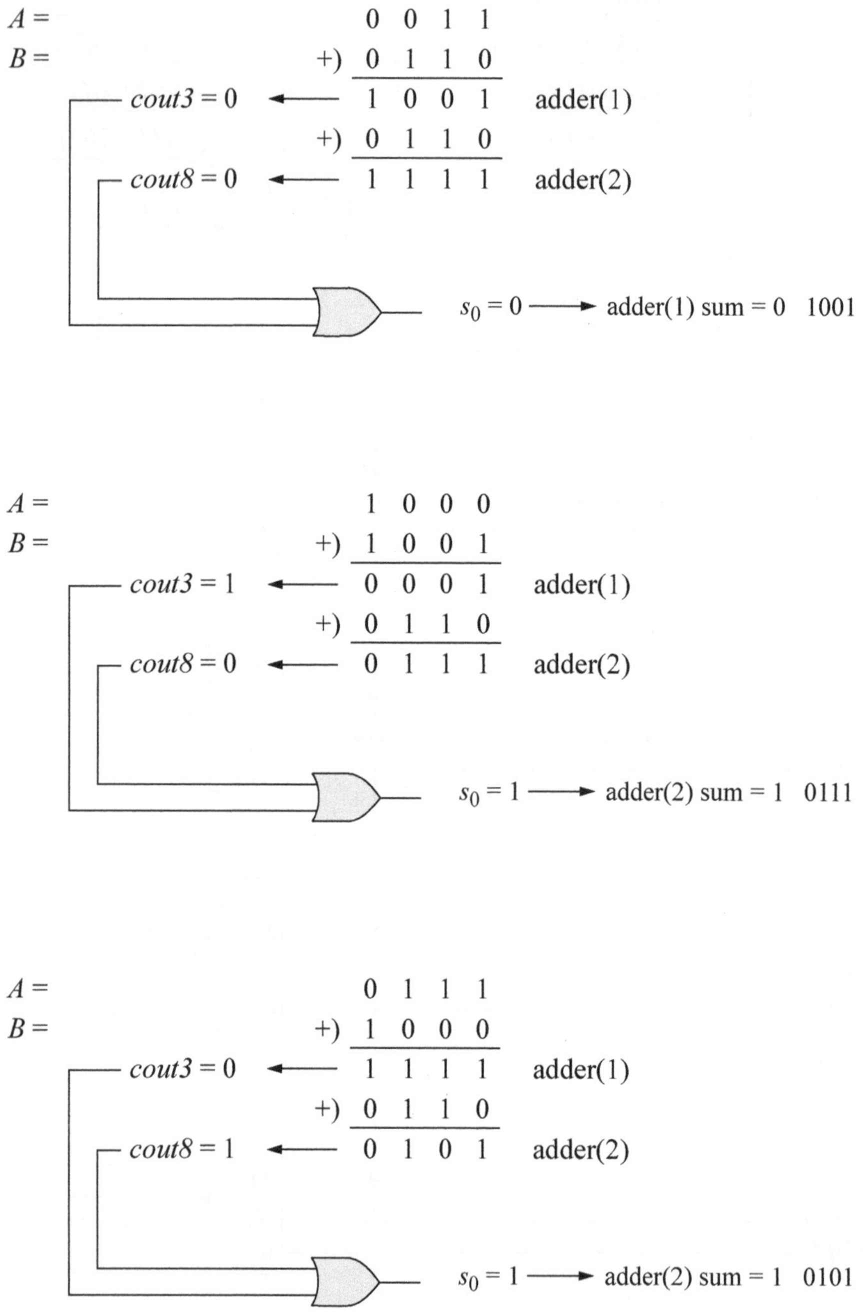

$A =$ 0   0   1   1

$B =$ +)   0   1   1   0

$cout3 = 0$ ←—— 1   0   0   1   adder(1)

             +)   0   1   1   0

$cout8 = 0$ ←—— 1   1   1   1   adder(2)

$s_0 = 0$ ——▶ adder(1) sum = 0   1001

$A =$ 1   0   0   0

$B =$ +)   1   0   0   1

$cout3 = 1$ ←—— 0   0   0   1   adder(1)

             +)   0   1   1   0

$cout8 = 0$ ←—— 0   1   1   1   adder(2)

$s_0 = 1$ ——▶ adder(2) sum = 1   0111

$A =$ 0   1   1   1

$B =$ +)   1   0   0   0

$cout3 = 0$ ←—— 1   1   1   1   adder(1)

             +)   0   1   1   0

$cout8 = 1$ ←—— 0   1   0   1   adder(2)

$s_0 = 1$ ——▶ adder(2) sum = 1   0101

Figure 5.24   Examples using a decimal adder with multiplexers.

5.6 Decimal Subtraction

In fixed-point binary arithmetic, subtraction is performed by adding the rs complement of the subtrahend (B) to the minuend (A); that is, by adding the 2s complement of the subtrahend as shown below.

$$A - B = A + (B' + 1)$$

where B' is the 1s complement ($r - 1$) and 1 is added to the low-order bit position to form the 2s complement. The same rationale is used in decimal arithmetic; however, the rs complement is obtained by adding 1 to the 9s complement ($r - 1$) of the subtrahend to form the 10s complement.

Example 5.16 The number 42_{10} will be subtracted from the number 76_{10}. This yields a result of $+34_{10}$. The 10s complement of the subtrahend is obtained as follows using radix 10 numbers: $9 - 4 = 5$; $9 - 2 = 7 + 1 = 8$, where 5 and 7 are the 9s complement of 4 and 2, respectively. A carry-out of the high-order decade indicates that the result is a positive number in BCD.

$$
\begin{array}{rrrr}
 & 76 & 0111 & 0110 \\
-) & 42 & +)\ 0101 & 1000 \\
\hline
 & 34 & 1 \leftarrow & 1110 \\
 & & 1 \leftarrow 1101 & 0110 \\
 & & \underline{0110} & \underline{0100} \\
 & & 0011 & \\
\hline
 & & +\quad 0011 & 0100 \\
\end{array}
$$

Example 5.17 The number 87_{10} will be subtracted from the number 76_{10}. This yields a difference of -11_{10}, which is 1000 1001 represented as a negative BCD number in 10s complement. The 10s complement of the subtrahend is obtained as follows using radix 10 numbers: $9 - 8 = 1$; $9 - 7 = 2 + 1 = 3$, where 1 and 2 are the 9s complement of 8 and 7, respectively. A carry-out of 0 from the high-order decade indicates that the result is a negative BCD number in 10s complement. To obtain the result in radix 10, form the 10s complement of 89_{10}; this will yield 11_{10}, which is interpreted as a negative number.

$$
\begin{array}{rrrr}
 & 76 & 0111 & 0110 \\
-) & 87 & +)\ 0001 & 0011 \\
\hline
 -11\ (89) & 0 \leftarrow 1000 & 1001 \\
\end{array}
$$

$$
\begin{array}{rrr}
- & 1000 & 1001 \quad \text{Negative number in 10s complement} \\
\end{array}
$$

Before presenting the organization for the BCD adder/subtractor, a 9s complementer will be designed which will be used in the adder/subtractor together with a carry-in to form the 10s complement of the subtrahend. A 9s complementer is required because BCD is not a self-complementing code; that is, the $r-1$ complement cannot be obtained by simply inverting the bits.

The truth table for a 9s complementer is shown in Table 5.6. The f_is are the outputs of the 9s complementer, and are established by a mode control input m, such that

$$Add: \quad m = 0$$
$$Subtract: \quad m = 1$$

The mode control input m will determine whether operand $B[3:0] = b_3 b_2 b_1 b_0$ will be added to operand $A[3:0] = a_3 a_2 a_1 a_0$ or subtracted from operand A, as shown in the block diagram of the 9s complementer of Figure 5.25. The function $F[3:0] = f_3 f_2 f_1 f_0$ is the 9s complement of the subtrahend B.

Table 5.6 Nines Complementer

Subtrahend				9s Complement ($m = 1$)			
b_3	b_2	b_1	b_0	f_3	f_2	f_1	f_0
0	0	0	0	1	0	0	1
0	0	0	1	1	0	0	0
0	0	1	0	0	1	1	1
0	0	1	1	0	1	1	0
0	1	0	0	0	1	0	1
0	1	0	1	0	1	0	0
0	1	1	0	0	0	1	1
0	1	1	1	0	0	1	0
1	0	0	0	0	0	0	1
1	0	0	1	0	0	0	0

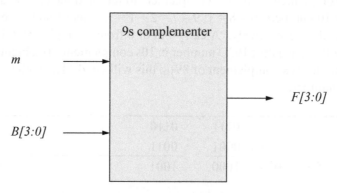

Figure 5.25   Block diagram for a 9s complementer.

From Table 5.6, if the operation is addition, then the B operand is unchanged; that is, $f_i = b_i$. If the operation is subtraction ($m = 1$), then b_0 is inverted; otherwise, b_0 is unchanged. Therefore, the equation for f_0 is

$$f_0 = b_0 \oplus m$$

The b_1 variable is unchanged for both addition and subtraction; therefore, the equation for f_1 is

$$f_1 = b_1$$

The equations for f_2 and f_3 are less obvious. The Karnaugh maps representing f_2 and f_3 for a subtract operation are shown in Figure 5.26 and Figure 5.27, respectively. Since f_2 does not depend upon b_3, input b_3 can be ignored; that is, $f_2 = 1$ whenever b_2 and b_1 are different. The equations for the 9s complement outputs are shown in Equation 5.15. The logic diagram for the 9s complementer is shown in Figure 5.28.

b_2 \ $b_1 b_0$	0 0	0 1	1 1	10
0	0 (0)	0 (1)	1 (3)	1 (2)
1	1 (4)	1 (5)	0 (7)	0 (6)

f_2 ($m = 1$)

Figure 5.26 Karnaugh map for f_2 of the 9s complementer.

$b_3 b_2$ \ $b_1 b_0$	0 0	0 1	1 1	1 0
0 0	1 (0)	1 (1)	0 (3)	0 (2)
0 1	0 (4)	0 (5)	0 (7)	0 (6)
1 1	0 (12)	0 (13)	0 (15)	0 (14)
1 0	0 (8)	0 (9)	0 (11)	0 (10)

f_3 ($m = 1$)

Figure 5.27 Karnaugh map for f_3 of the 9s complementer.

$$f_0 = b_0 \oplus m$$

$$f_1 = b_1$$

$$f_2 = b_2 m' + (b_2 \oplus b_1) m$$

$$f_3 = b_3 m' + b_3' b_2' b_1' m \qquad\qquad (5.15)$$

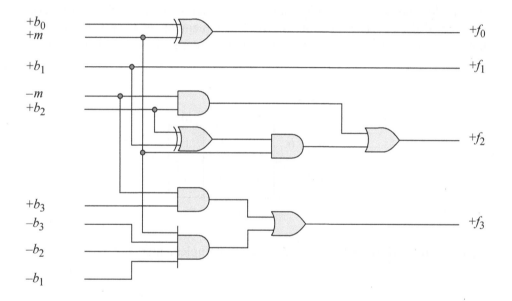

Figure 5.28   Logic diagram for a 9s complementer.

The logic diagram for a two-decade BCD adder/subtractor is shown in Figure 5.29. Operand $A[3:0]$ is connected directly to the A inputs of a fixed-point adder; operand $B[3:0]$ is connected to the inputs of a 9s complementer whose outputs $F[3:0]$ connect to the B inputs of the adder. There is also a mode control input m to the 9s complementer which specifies either an add operation ($m = 0$) or a subtract operation ($m = 1$). The mode control is also connected to the carry-in of the low-order adder.

If the operation is addition, then the mode control adds 0 to the uncomplemented version of operand B. If the operation is subtraction, then the mode control adds a 1 to the 9s complement of operand B to form the 10s complement. The outputs of the adder are a 4-bit intermediate $SUM[3:0]$ and a carry-out labeled $cout3$. The intermediate sum is connected to a second fixed-point adder, which will add 0000 to the intermediate sum if no adjustment is required or add 0110 to the intermediate sum if adjustment is required.

The carry-out of the low-order BCD stage is specified as an auxiliary carry *aux_cy* and is defined by Equation 5.16. The auxiliary carry determines whether 0000 or 0110 is added to the intermediate sum produce a valid BCD digit, as well as providing a carry-in to the high-order stage.

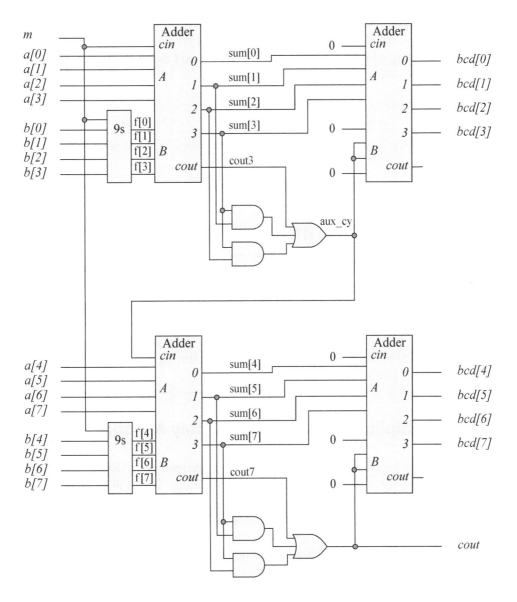

Figure 5.29   Logic diagram for a two-decade BCD adder/subtractor.

$$aux\_cy = cout3 + sum[3] \; sum[1] + sum[3] \; sum[2] \qquad (5.16)$$

The same rationale applies to the high-order BCD stage for operand $A[7:4]$ and operand $B[7:4]$. The 9s complementer produces outputs $F[7:4]$ which are connected to the B inputs of the adder. The mode control input is also connected to the input of the 9s complementer. The fixed-point adder generates an intermediate sum of $sum[7:4]$ and a carry-out $cout7$. The carry-out of the BCD adder/subtractor is labeled $cout$ and determines whether 0000 or 0110 is added to the intermediate sum labeled $sum[7:4]$. The equation for $cout$ is shown in Equation 5.17. The BCD adder/subtractor can be extended to any size operand by simply adding more stages.

$$cout = cout7 + sum[7] \; sum[5] + sum[7] \; sum[6] \qquad (5.17)$$

5.7 Decimal Multiplication

The algorithms for fixed-point multiplication and decimal multiplication are similar. The main difference is how the partial products are formed. Binary multiplication uses digits of 0 and 1; whereas, decimal multiplication uses digits with a range of 0 to 9. In binary multiplication, the multiplicand is added to the previous partial product if the multiplier bit is 1; otherwise, zeroes are added. In decimal multiplication, the multiplicand is first multiplied by the low-order multiplier digit. This result is then added to the previous partial product.

5.7.1 Multiplication Using Read-Only Memory

The advent of high-speed, high-capacity read-only memories (ROMs) makes these devices extremely practical for decimal multiplication. The multiplicand and multiplier are connected to the address inputs of a ROM. The addressed data contains two valid BCD digits that are sent the ROM outputs, as shown in Figure 5.30.

The outputs of each ROM of an n-digit multiplicand are then aligned and added to form a partial product. The partial product is then added to the previous shifted partial product. This process continues until all partial products have been added to form the product. Table 5.7 lists the ROM entries that are used for partial product generation as addressed by the multiplicand and multiplier.

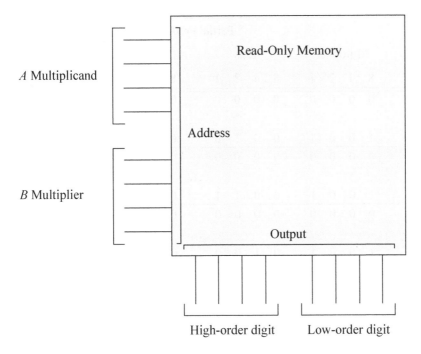

Figure 5.30   Generation of partial products using a ROM.

Table 5.7 ROM Entries for Generating Partial Products

										Partial Product						
Multiplicand					Multiplier				Tens					Units		
8	4	2	1		8	4	2	1	8	4	2	1		8	4	2	1
0	0	0	0		0	0	0	0	0	0	0	0		0	0	0	0
		...									...						
0	0	0	0		1	0	0	1	0	0	0	0		0	0	0	0
0	0	0	1		0	0	0	0	0	0	0	0		0	0	0	0
		...									...						
0	0	0	1		1	0	0	1	0	0	0	0		1	0	0	1
0	0	1	0		0	0	0	0	0	0	0	0		0	0	0	0
		...									...						
0	0	1	0		1	0	0	1	0	0	0	1		1	0	0	0

(Continued on next page)

Table 5.7 ROM Entries for Generating Partial Products

								Partial Product							
Multiplicand				Multiplier				Tens				Units			
8	4	2	1	8	4	2	1	8	4	2	1	8	4	2	1
0	0	1	1	0	0	0	0	0	0	0	0	0	0	0	0
		…								…					
0	0	1	1	1	0	0	1	0	0	1	0	0	1	1	1
0	1	0	0	0	0	0	0	0	0	0	0	0	0	0	0
		…								…					
0	1	0	0	1	0	0	1	0	0	1	1	0	1	1	0
0	1	0	1	0	0	0	0	0	0	0	0	0	0	0	0
		…								…					
0	1	0	1	1	0	0	1	0	1	0	0	0	1	0	1
0	1	1	0	0	0	0	0	0	0	0	0	0	0	0	0
		…								…					
0	1	1	0	1	0	0	1	0	1	0	1	0	1	0	0
0	1	1	1	0	0	0	0	0	0	0	0	0	0	0	0
		…								…					
0	1	1	1	1	0	0	1	0	1	1	0	0	0	1	1
1	0	0	0	0	0	0	0	0	0	0	0	0	0	0	0
		…								…					
1	0	0	0	1	0	0	1	0	1	1	1	0	0	1	0
1	0	0	1	0	0	0	0	0	0	0	0	0	0	0	0
		…								…					
1	0	0	1	1	0	0	1	1	0	0	0	0	0	0	1

Four ROMs are required for a 4-digit multiplicand that is multiplied by a 1-digit multiplier, as shown in Figure 5.31. The partial product is shifted and stored in a register. Then the multiplicand is multiplied by the next higher-order multiplier digit and the new partial product is added to the previous partial product. Figure 5.32 shows two partial products and their alignment. The partial products are then added in decimal adders to form the product. The carry lookahead technique may be applied to the adders to increase the speed of the decimal multiplication operation.

A numerical example is shown in Figure 5.33 in which a 2-digit multiplicand (57) is multiplied by a 2-digit multiplier (57) to generate a 4-digit product (3249). The partial products are shown for each multiplier digit together with their alignment before being added in a decimal adder to obtain the product.

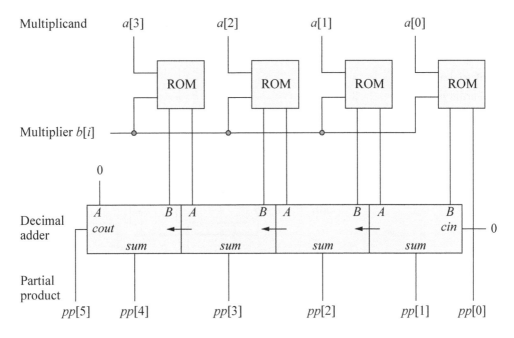

Figure 5.31   Partial product generation using ROMs for decimal multiplication.

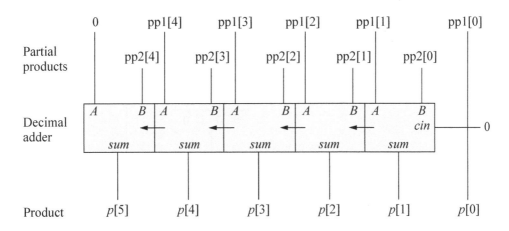

Figure 5.32   Partial product alignment for decimal multiplication.

```
         A =              5   7
     ×)  B =              5   7
Partial product 1     3   9   9
Partial product 2   2   8   5
Product             3   2   4   9
```

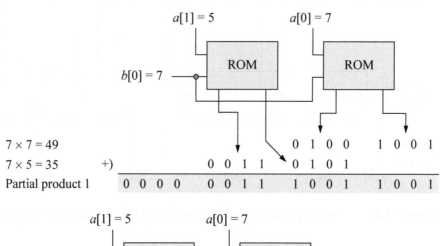

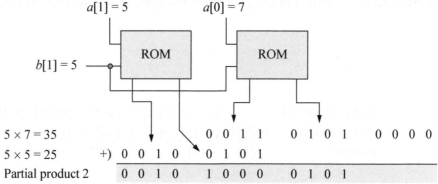

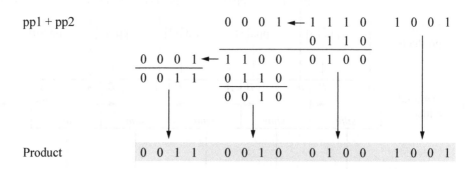

Figure 5.33   Numerical example for decimal multiplication using ROMs.

5.8 Decimal Division

Decimal division is similar to fixed-point division, except for the digits used in the operations — 0 and 1 for fixed-point; 0 through 9 for decimal. Also, correct alignment must be established initially between the dividend and divisor. The divisor is subtracted from the dividend or previous partial remainder in much the same way as the restoring division method in fixed-point division. If the division is unsuccessful — producing a negative result — then the divisor is added to the negative partial remainder.

5.8.1 Division Using Table Lookup

Decimal division using the table lookup technique is similar in many respects to the binary search algorithm used in programming, which is a systematic method of searching through an ordered table. The algorithm examines the entry at the middle of the table — or the middle ± 1 for tables with an even number of entries — and compares the value at that location with a keyword. If the keyword is less than the value, then the bottom half of the table is used for a new binary search. If the keyword is equal to the value, then the entry was found. If the keyword is greater than the value, then the top half of the table is used for a new binary search.

The table lookup method always adds or subtracts the divisor from the dividend or previous partial remainder in the following order:

$$\pm 8 \text{ times the divisor}$$
$$\pm 4 \text{ times the divisor}$$
$$\pm 2 \text{ times the divisor}$$
$$\pm 1 \text{ times the divisor}$$

However, in order to establish an initial digit, the first operation is always a subtraction of eight times the divisor. The table lookup method requires only four cycles for each quotient digit. Figure 5.34 and Figure 5.35 present two numerical examples of decimal division using the table lookup method. Whenever the result of any of the operations shown above is positive, the digits 8, 4, 2, or 1 become part of the quotient and are summed to generate the final quotient digit for that cycle.

5.9 Floating-Point Arithmetic

This section will introduce a method by which very large and very small numbers can be represented. Fixed-point notation assumes that the radix point is in a fixed location within the number, either at the right end of the number for integers or at the left end of the number for fractions. Fixed-point integers in 2s complement representation have the following range: -2^{n-1} to $2^{n-1}-1$.

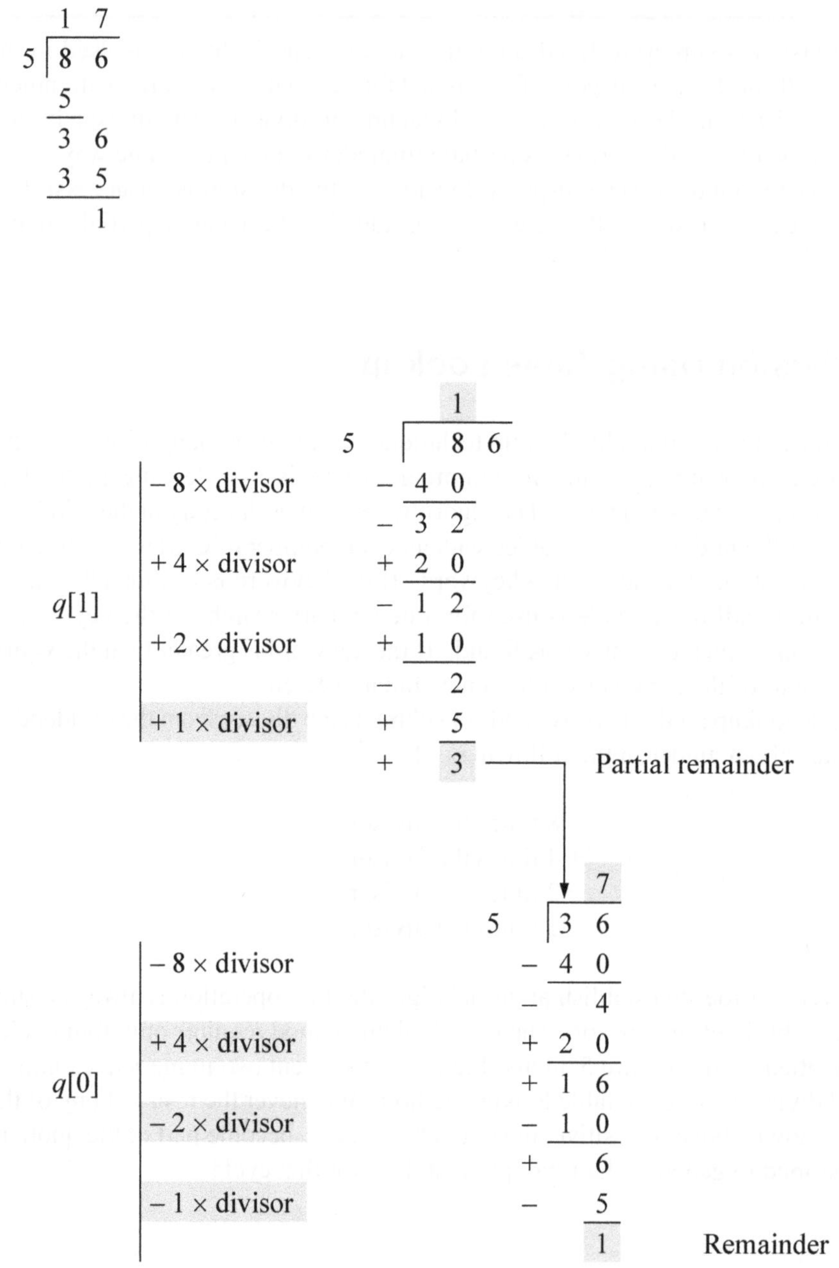

```
      1 7
   5 | 8 6
      5
    ──────
      3 6
      3 5
    ──────
        1
```

$q[1]$

$-8 \times \text{divisor}$

$+4 \times \text{divisor}$

$+2 \times \text{divisor}$

$+1 \times \text{divisor}$

```
                1
        5   |   8  6
          -     4  0
          ─────────
          -     3  2
          +     2  0
          ─────────
          -     1  2
          +     1  0
          ─────────
          -        2
          +        5
          ─────────
          +        3   ──── Partial remainder
```

$q[0]$

$-8 \times \text{divisor}$

$+4 \times \text{divisor}$

$-2 \times \text{divisor}$

$-1 \times \text{divisor}$

```
                       7
        5   |      3  6
          -        4  0
          ─────────────
          -           4
          +        2  0
          ─────────────
          +        1  6
          -        1  0
          ─────────────
          +           6
          -           5
          ─────────────
                      1   Remainder
```

Quotient = 17   Remainder = 1

Figure 5.34   Example of decimal division using the table lookup technique.

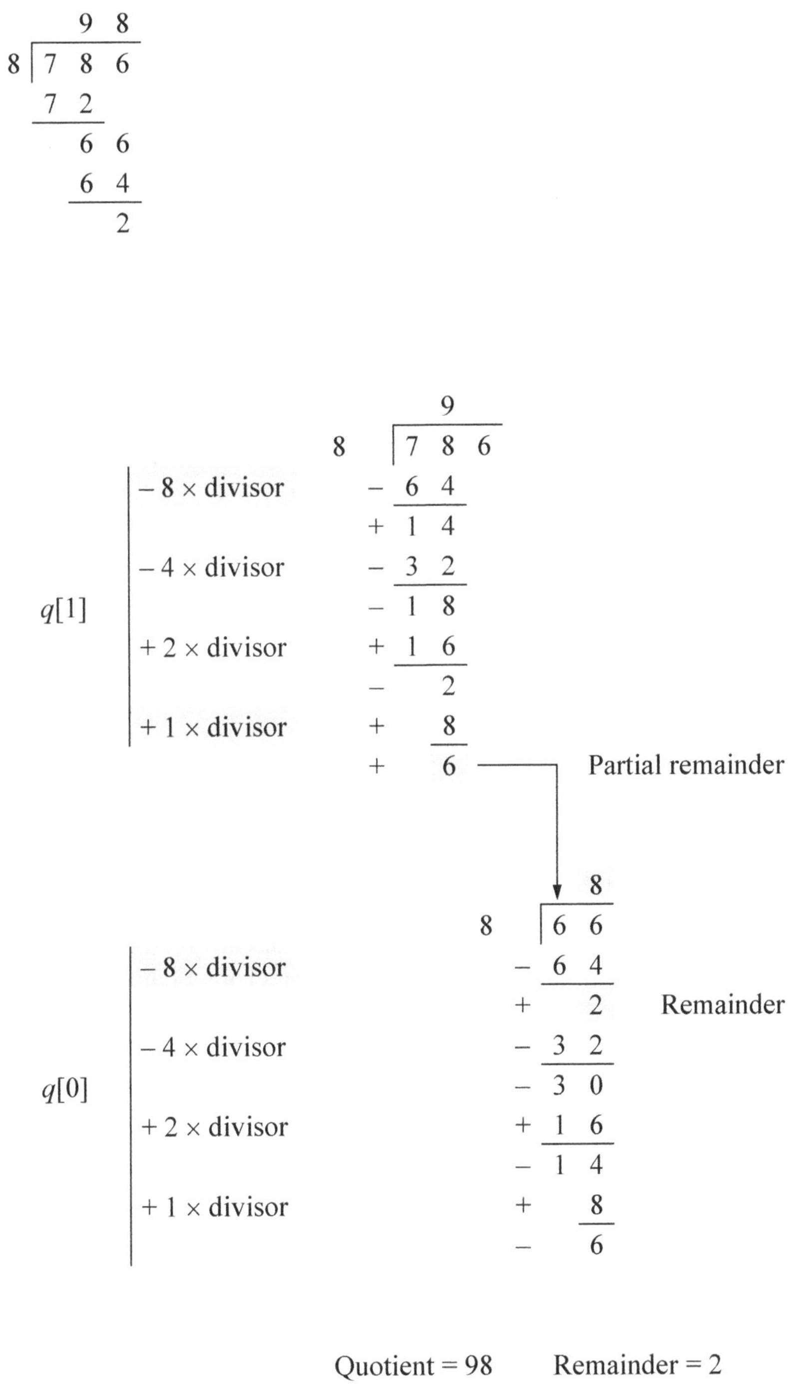

```
            9  8
       8 | 7  8  6
           7  2
          ─────
              6  6
              6  4
             ─────
                 2
```

```
                              9
                     8  | 7   8  6
     − 8 × divisor        −  6  4
                          +  1  4
     − 4 × divisor        −  3  2
q[1]                      −  1  8
     + 2 × divisor        +  1  6
                          −     2
     + 1 × divisor        +     8
                          ────────
                          +     6  ─────────  Partial remainder


                                          8
                              8  |  6   6
     − 8 × divisor               −  6   4
                                 +      2      Remainder
     − 4 × divisor               −  3   2
q[0]                             −  3   0
     + 2 × divisor               +  1   6
                                 −  1   4
     + 1 × divisor               +      8
                                 ────────
                                 −      6
```

Quotient = 98   Remainder = 2

Figure 5.35   Example of decimal division using the table lookup technique.

Neither of these limits is adequate for scientific computations, which may include numbers of the following magnitude:

$$28{,}500{,}000{,}000 \times 0.0000000485$$

The preceding multiplication can also be written in scientific notation as

$$(285 \times 10^8) \times (4.85 \times 10^{-8})$$

Floating-point notation is simply scientific notation. A floating-point number consists of two parts: a fraction f and an exponent e. There is also a sign bit associated with the number. The floating-point number A is obtained by multiplying the fraction f by a radix r that is raised to the power of e, as shown below,

$$A = f \times r^e$$

where the fraction is in sign-magnitude notation. The fraction and exponent are also referred to as the *mantissa* (or *significand*) and *characteristic*, respectively.

By adjusting the magnitude of the exponent e, the radix point can be made to float around the fraction, thus, the notation $A = f \times r^e$ is referred to as *floating-point* notation. Consider an example of a floating-point number in radix 10, as shown below.

$$A = 0.00006758 \times 10^{+4}$$

The number A can also be written as $A = 0.6758 \times 10^0$ or as $A = 67.58 \times 10^{-6}$. When the fraction is shifted k positions to the left, the exponent is decreased by k; when the fraction is shifted k positions to the right, the exponent is increased by k. The Institute of Electrical and Electronics Engineers (IEEE) has developed a standard for representing floating-point numbers, as shown in Figure 5.36(a) for single-precision floating-point numbers and in Figure 5.36(b) for double-precision floating-point numbers.

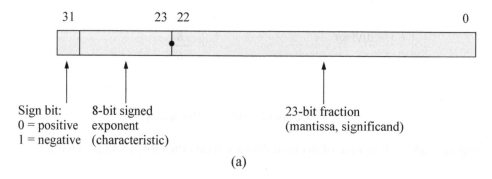

Figure 5.36   IEEE floating-point formats: (a) 32-bit single precision and (b) 64-bit double precision.

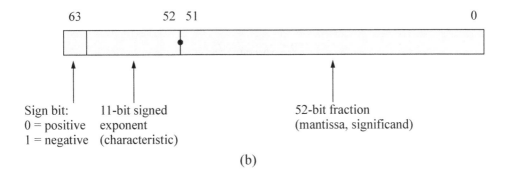

(b)

Figure 5.36   (Continued)

Fractions in the IEEE format are normalized; that is, the leftmost significant bit is a 1. Figure 5.37 shows unnormalized and normalized numbers in the 32-bit format. Since there will always be a 1 to the immediate right of the radix point, the 1 bit is not explicitly shown — it is an implied 1.

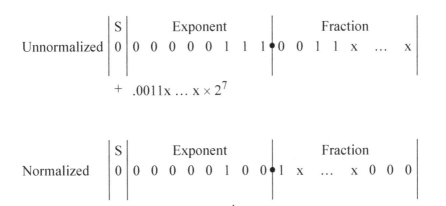

Figure 5.37   Unnormalized and normalized floating-point numbers.

As mentioned earlier, the exponent is a signed number in 2s complement. When adding or subtracting floating-point numbers, the exponents are compared and made equal resulting in a right shift of the fraction with the smaller exponent. The comparison is easier if the exponents are unsigned — a simple comparator can be used for the comparison. As the exponents are being formed, a *bias* constant is added to the exponents such that all exponents are positive internally.

For the single-precision format, the bias constant is +127 — also called excess-127; therefore, the biased exponent has a range of

$$0 \le e_{\text{biased}} \le 255$$

The lower and upper limits of a biased exponent have special meanings. A value of $e_{\text{biased}} = 0$ with a fraction = 0 is used to represent a number with a value of zero. A value of $e_{\text{biased}} = 255$ with a fraction = 0 is used to represent a number with a value of infinity.

5.9.1 Floating-Point Addition/Subtraction

The addition of two fractions is identical to the addition algorithm presented in fixed-point addition. If the signs of the operands are the same ($A_{\text{sign}} \oplus B_{\text{sign}} = 0$), then this is referred to as *true addition* and the fractions are added. True addition corresponds to one of the following conditions:

$$
\begin{array}{ccc}
(+A) & + & (+B) \\
(-A) & + & (-B) \\
(+A) & - & (-B) \\
(-A) & - & (+B)
\end{array}
$$

Floating-point addition is defined in Equation 5.18 for two numbers A and B, where $A = f_A \times r^{eA}$ and $B = f_B \times r^{eB}$:

$$
\begin{aligned}
A + B &= (f_A \times r^{eA}) + (f_B \times r^{eB}) \\
&= [f_A + (f_B \times r^{-(eA - eB)})] \times r^{eA} \text{ for } e_A > e_B \\
&= [(f_A \times r^{-(eB - eA)}) + f_B] \times r^{eB} \text{ for } e_A \le e_B
\end{aligned}
\qquad (5.18)
$$

The terms $r^{-(eA - eB)}$ and $r^{-(eB - eA)}$ are shifting factors to shift right the fraction with the smaller exponent. This is analogous to a divide operation, since $r^{-(eA - eB)}$ is equivalent to $1/r^{(eA - eB)}$. For $e_A > e_B$, fraction f_B is shifted right the number of bit positions specified by the absolute value of $|e_A - e_B|$. An example of using the shifting factor for addition is shown in Figure 5.38 for two operands A and B.

Before alignment

$$A = f_A \times r^5$$

$$A = 0 . 1\ 1\ 0\ 1\ 0\ 1\ 0\ 0 \quad \times 2^5 \qquad +26.5$$

$$B = f_B \times r^3$$

$$B = 0 . 1\ 0\ 0\ 0\ 1\ 1\ 0\ 0 \quad \times 2^3 \qquad +4.375$$

After alignment

$$A = 0 . 1\ 1\ 0\ 1\ 0\ 1\ 0\ 0 \quad \times 2^5 \qquad +26.5$$

$$B = 0 . 0\ 0\ 1\ 0\ 0\ 0\ 1\ 1 \quad \times 2^5 \qquad +4.375$$

$$A + B = 0 . 1\ 1\ 1\ 1\ 0\ 1\ 1\ 1 \quad \times 2^5 \qquad +30.875$$

Figure 5.38   Addition example showing fraction alignment and exponent adjustment.

Operand B is shifted right an amount equal to $|e_A - e_B| = |5 - 3| = 2$; that is,

$$B = f_B \times r^{-2}$$

which is a divide by four operation accomplished by a right shift of 2 bit positions.

Figure 5.39 shows another addition example of fraction alignment and exponent adjustment. Since the fractions must be properly aligned before addition can take place, the fraction with the smaller exponent is shifted right and the exponent is adjusted by increasing the exponent by one for each bit position shifted.

Before alignment

$$A = 0 . 1\ 0\ 1\ 1\ 0\ 0 \quad \times 2^5 \qquad +22$$

$$+)\, B = 0 . 1\ 1\ 1\ 0\ 0\ 0 \quad \times 2^2 \qquad +3.5$$

After alignment

$$A = 0 . 1\ 0\ 1\ 1\ 0\ 0 \quad \times 2^5 \qquad +22$$

$$+)\, B = 0 . 0\ 0\ 0\ 1\ 1\ 1 \quad \times 2^5 \qquad +3.5$$

$$0 . 1\ 1\ 0\ 0\ 1\ 1 \quad \times 2^5 \qquad +25.5$$

Figure 5.39   Addition example showing fraction alignment and exponent adjustment.

The subtraction of two fractions is identical to the subtraction algorithm presented in fixed-point addition. If the signs of the operands are the same ($A_{sign} \oplus B_{sign} = 0$), then this is referred to as *true subtraction* and the fractions are subtracted. True subtraction corresponds to one of the following conditions:

$$
\begin{aligned}
(+A) &\ -\ (+B) \\
(-A) &\ -\ (-B) \\
(+A) &\ +\ (-B) \\
(-A) &\ +\ (+B)
\end{aligned}
$$

Floating-point subtraction is defined in Equation 5.19 for two numbers A and B, where $A = f_A \times r^{eA}$ and $B = f_B \times r^{eB}$:

$$
\begin{aligned}
A - B &= (f_A \times r^{eA}) - (f_B \times r^{eB}) \\
&= [f_A - (f_B \times r^{-(eA - eB)})] \times r^{eA} \text{ for } e_A > e_B \\
&= [(f_A \times r^{-(eB - eA)}) - f_B] \times r^{eB} \text{ for } e_A \le e_B
\end{aligned}
\tag{5.19}
$$

The flowchart of Figure 5.40 illustrates the alignment of fractions and the addition or subtraction of operands. If the exponents are not equal, then the fraction with the smaller exponent is shifted right and its exponent is increased by 1. A right shift of the operand A fraction is specified by

$$
A = 0\ a_{n-1}\ a_{n-2}\ \dots\ a_1
$$

This process repeats until the exponents are equal. The fractions are added or subtracted depending on the operation and on the signs of the operands. If the operation is addition and the signs are the same; that is, $A_{sign} \oplus B_{sign} = 0$, then the fractions are added. If the signs are different; that is, $A_{sign} \oplus B_{sign} = 1$, then the fractions are subtracted by adding the 2s complement of the addend to the augend. If the operation is subtraction and the signs are the same, that is, $A_{sign} \oplus B_{sign} = 0$, then the 2s complement of the subtrahend is added to the minuend. If the signs are different, such that, $A_{sign} \oplus B_{sign} = 1$, then the fractions are added.

Postnormalization   When the result of a subtraction produces a carry-out $= 0$ from the high-order bit position $n - 1$, the result must be 2s complemented and the sign bit inverted. This difference may be positive, negative, or zero. A negative result is in 2s complement notation and occurs when $|A < B|$. Since the floating-point number must be in sign-magnitude notation, the fraction cannot be a signed number — there cannot be two signs associated with the number — therefore, the result is 2s complemented in order to express the result in the correct sign-magnitude representation.

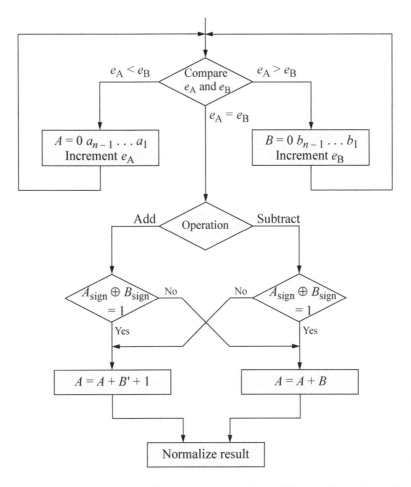

Figure 5.40   Flowchart for aligning fractions for addition/subtraction of two floating-point operands.

If an operation produces a 0 in bit position $n-1$ of the result, then the result is normalized by shifting the fraction left until a 1 bit appears in bit position $n-1$ and the exponent is decreased accordingly. When the sum of a true addition is equal to or greater than 1.000 … 00, the result is shifted right 1 bit position and the exponent is incremented by 1; that is, the carry bit is shifted right into the high-order bit position a_{n-1} of the fraction and the sign is unchanged. Several examples will now be presented to illustrate the addition and subtraction of floating-point numbers and the associated postnormalization.

Example 5.18 The following two decimal numbers will be added as 8-bit floating-point operands: $(+15) + (+27)$ to yield a result of $+42$.

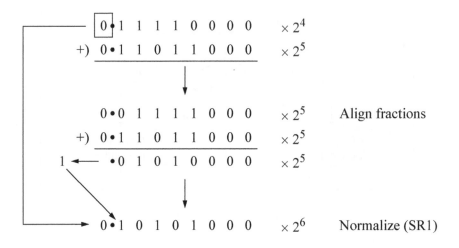

Example 5.19 The following two decimal numbers will be added as 8-bit floating-point operands: $(-15) + (-27)$ to yield a result of -42.

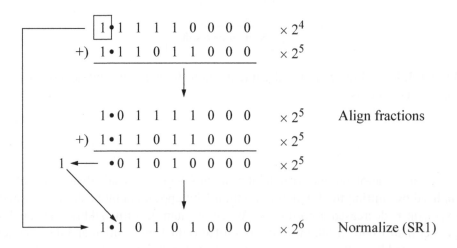

Example 5.20   The following two decimal numbers will be added as 8-bit floating-point operands: $(+15) + (-27)$ to yield a result of -12.

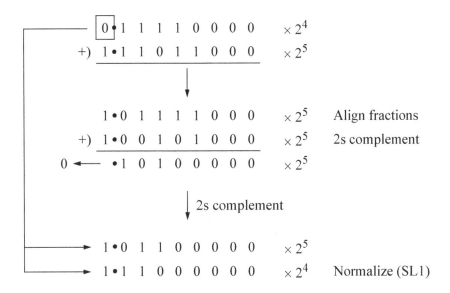

Example 5.21   The following two decimal numbers will be subtracted as 8-bit floating-point operands: $(+27) - (+15)$ to yield a result of $+12$.

```
        ┌──────  ┌0┐•1 1 0 1 1 0 0 0    × 2⁵
        │        └─┘
        │     −)  1•1 1 1 1 0 0 0 0    × 2⁴
        │        ─────────────────────
        │                 │
        │                 ▼
        │
        │         0•1 1 0 1 1 0 0 0    × 2⁵
        │
        │     −)  0•0 1 1 1 1 0 0 0    × 2⁵    Align fractions
        │        ─────────────────────
        │                 │
        │                 ▼
        │
        │         0•1 1 0 1 1 0 0 0    × 2⁵
        │     +)  0•1 0 0 0 1 0 0 0    × 2⁵    2s complement
        │        ─────────────────────
        │    1 ◄───  •0 1 1 0 0 0 0 0    × 2⁵
        ├──────►   0•0 1 1 0 0 0 0 0    × 2⁵
        └──────►   0•1 1 0 0 0 0 0 0    × 2⁴    Normalize (SL1)
```

Example 5.22 The following two decimal numbers will be subtracted as 8-bit floating-point operands: $(+15) - (+27)$ to yield a result of -12.

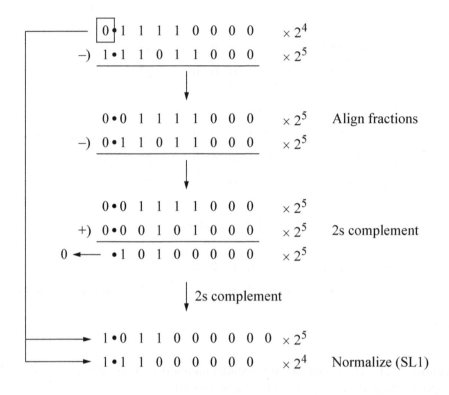

Example 5.23 The following two decimal numbers will be subtracted as 8-bit floating-point operands: $(-15) - (-27)$ to yield a result of $+12$.

5.9.2 Floating-Point Multiplication

In floating-point multiplication, the fractions are multiplied and the exponents are added. The fractions are multiplied by any of the methods described in fixed-point multiplication and operate on two normalized floating-point operands. Floating-point multiplication is simpler than floating-point addition or subtraction because there is no comparison of exponents and no alignment of fractions. Fraction multiplication and exponent addition are two independent operations and can be done in parallel. Floating-point multiplication is defined as shown in Equation 5.20.

$$
\begin{aligned}
A \times B &= (f_A \times r^{eA}) \times (f_B \times r^{eB}) \\
&= (f_A \times f_B) \times r^{(eA + eB)}
\end{aligned}
\tag{5.20}
$$

The sign of the product is determined by the signs of the operands as shown below.

$$
A_{sign} \oplus B_{sign}
$$

If both exponents are positive before biasing, then exponent overflow may occur, signifying that the product exceeds the capacity of the machine. In a similar manner, if both exponents are negative, then exponent underflow may occur, indicating that the product is too small to be represented. Multiplication generates a product of $2n$ bits which, in conjunction with the exponent, may be of sufficient precision without utilizing the low-order half of the product.

Although it will not be apparent in the following paper-and-pencil floating-point multiplication examples, there is a minor problem when adding two biased exponents. Since both exponents are biased, there will a double bias in the resulting exponent, as shown below.

$$
(e_A + bias) + (e_B + bias) = (e_A + e_B) + 2\,bias
$$

The resulting exponent should be restored to a single bias before the multiplication operation begins. This is accomplished by subtracting the bias.

The multiplication algorithm is partitioned into five parts:

1. Check for zero operands. If $A = 0$ or $B = 0$, then the product $= 0$.
2. Determine the sign of the product.
3. Add exponents and subtract the bias.
4. Multiply fractions. Steps 3 and 4 can be done in parallel, but both must be completed before step 5.
5. Normalize the product.

Examples will now be shown that illustrate floating-point multiplication using 8-bit fractions. The add-shift fixed-point algorithm will be used in all cases.

Example 5.24   The decimal numbers +15 and +7 will be converted to floating-point numbers and then multiplied using the fixed-point multiplication algorithm to generate a product of +105. Both operands are normalized prior to the multiplication operation, as shown below.

$$\text{Multiplicand } A = \; 0.11110000 \times 2^4 \qquad (+15)$$
$$\text{Multiplier } B = \; 0.11100000 \times 2^3 \qquad (+7)$$

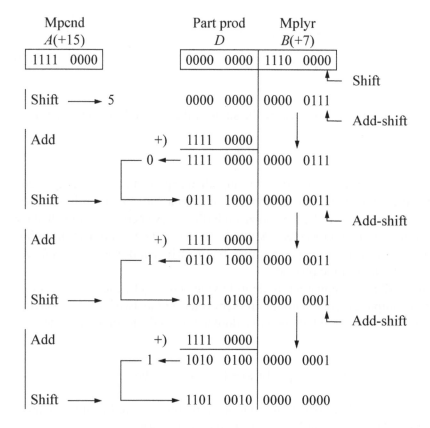

No postnormalization

Product $= 0.11010010 \times 2^{(4+3)\,=\,7} = +105$

Example 5.25   The decimal numbers +25.5 and +9.5 will be converted to floating-point numbers and then multiplied using the fixed-point multiplication algorithm to generate a product of +242.25. Both operands are normalized prior to the multiplication operation, as shown below.

$$\text{Multiplicand } A = \quad 0.11001100 \times 2^5 \quad (+25.5)$$
$$\text{Multiplier } B = \quad 0.10011000 \times 2^4 \quad (+9.5)$$

Since floating-point numbers are in sign-magnitude notation, the fractions are true magnitude; that is, they are unsigned (or positive). Therefore, zeroes are shifted into the partial product during a shift-only operation. During an add-shift operation, the carry-out of the add cycle is shifted into the partial product during the shift cycle.

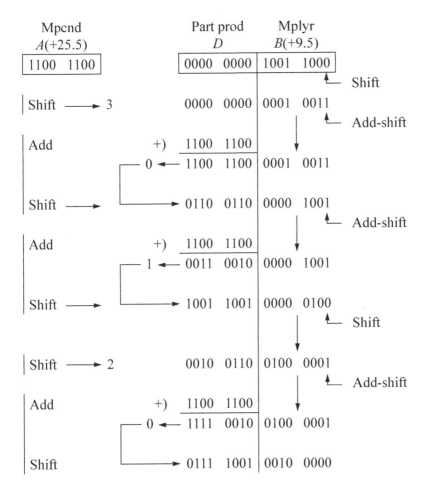

Product = $0.0111100100100000 \times 2^{(5+4)\,=\,9}$

Postnormalize = $0.1111001001000000 \times 2^8$

Normalized product = $0.11110010010 \times 2^8 = +242.25$

Example 5.26   The decimal numbers -13 and $+18$ will be converted to floating-point numbers and then multiplied using the fixed-point multiplication algorithm to generate a product of -234. Both operands are normalized prior to the multiplication operation, as shown below.

$$\text{Multiplicand } A = \quad 1.11010000 \times 2^4 \qquad (-13)$$
$$\text{Multiplier } B = \quad 0.10010000 \times 2^5 \qquad (+18)$$

Since the floating-point numbers are in sign-magnitude notation, the fractions are unsigned and can therefore be multiplied without any modification. The sign of the product, however, is determined by the following equation:

$$\text{Sign of product} = A_{\text{sign}} \oplus B_{\text{sign}} = 1 \oplus 0 = 1$$

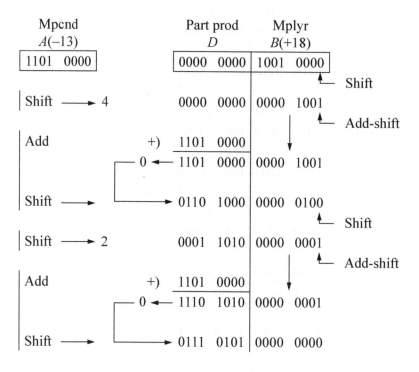

Product = $1.0111010100000000 \times 2^{(5+4) = 9}$

Postnormalize = $1.1110101000000000 \times 2^8$

Normalized product = $1.11101010 \times 2^8 = -234$

5.9.3 Floating-Point Division

The division of two floating-point numbers is accomplished by dividing the fractions and subtracting the exponents. The fractions are divided by any of the methods presented in the section on fixed-point division and overflow is checked in the same manner. Fraction division and exponent subtraction are two independent operations and can be done in parallel. Floating-point division is defined as shown in Equation 5.21.

$$A \ / \ B = (f_A \times r^{eA}) \ / \ (f_B \times r^{eB})$$
$$= (f_A \ / \ f_B) \times r^{(eA - eB)} \qquad\qquad (5.21)$$

Divide overflow occurs when the high-order half of the dividend is greater than or equal to the divisor. This presents no problem with floating-point division — the dividend is shifted right 1 bit position and the exponent is incremented by one. This guarantees that the dividend is smaller than the divisor.

The division algorithm is partitioned into six parts.

1. Check for zero operands.
2. Determine the sign of the quotient.
3. Align the dividend, if necessary.
4. Subtract the exponents and add the bias.
5. Divide the fractions. Steps 4 and 5 can be done in parallel, but both must be properly synchronized before ending the divide operation.
6. Normalize the result, if necessary.

The sign of the quotient is determined by the signs of the operands as shown below.

$$A_{sign} \oplus B_{sign}$$

Like floating-point multiplication, there is also a minor problem with exponent arithmetic. Although it will not be apparent in the following paper-and-pencil floating-point division example, when subtracting two biased exponents the resulting exponent will have zero bias, as shown below.

$$(e_A + \text{bias}) - (e_B + \text{bias}) = (e_A - e_B)$$

The resulting exponent should be restored to a single bias before the division operation begins by adding back the bias. In the example that follows, the restoring division method is used. For divide operations, the dividend is $2n$ bits and the divisor is n bits.

Example 5.27   The decimal numbers +12 and +4 will be converted to floating-point numbers and then divided using the fixed-point restoring algorithm to generate a quotient of +3. Both operands are normalized prior to the division operation, as shown below.

$$\text{Dividend } A.Q = \quad 0.11000000 \times 2^4 \qquad (+12)$$
$$\text{Divisor } B = \quad 0.1000 \qquad \times 2^3 \qquad (+4)$$

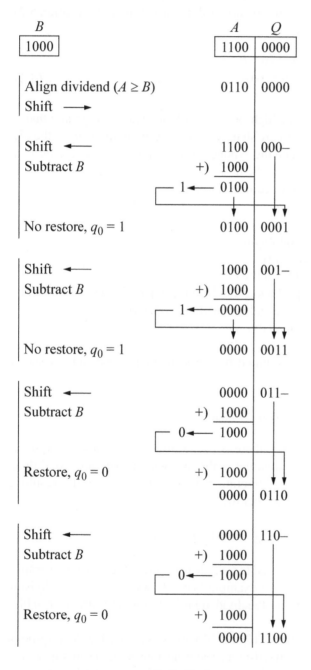

B

| 1000 |

$A \qquad Q$

| 1100 | 0000 |

Align dividend ($A \geq B$)   0110 | 0000
Shift ⟶

Shift ⟵   1100 | 000–
Subtract B +) 1000
                 1 ⟵ 0100

No restore, $q_0 = 1$ 0100 | 0001

Shift ⟵   1000 | 001–
Subtract B +) 1000
                 1 ⟵ 0000

No restore, $q_0 = 1$ 0000 | 0011

Shift ⟵   0000 | 011–
Subtract B +) 1000
                 0 ⟵ 1000

Restore, $q_0 = 0$ +) 1000
                       0000 | 0110

Shift ⟵   0000 | 110–
Subtract B +) 1000
                 0 ⟵ 1000

Restore, $q_0 = 0$ +) 1000
                       0000 | 1100

$$\text{Quotient} = 0.1100 \times 2^{(4+1)-3} = 2$$

5.9.4 Rounding Methods

Rounding deletes one or more of the low-order bits and adjusts the retained bits according to some rule. Rounding reduces the number of bits in an operand in order to permit the operand to be retained within the word size of the machine. Since bits are deleted, this limits the precision of the result. Rounding can occur when adding two n-bit numbers which result in a sum of $n + 1$ bits. The overflow is handled by shifting the fraction right 1 bit position, resulting in the low-order bit being lost unless it is saved. Rounding attempts to dispose of the extra bits and yet preserve a high degree of accuracy. Three common methods for rounding are presented in this section.

Truncation   This method of rounding is also called *chopping*. Truncation deletes extra bits and makes no changes to the retained bits. Aligning fractions during addition or subtraction could result is losing several low-order bits, so there is obviously an error associated with truncation. Assume that the following fraction is to be truncated to four bits:

$$0.b_{-1}\, b_{-2}\, b_{-3}\, b_{-4}\, b_{-5}\, b_{-6}\, b_{-7}\, b_{-8}$$

Then all fractions in the range $0.b_{-1}\, b_{-2}\, b_{-3}\, b_{-4}\, 0000$ to $0.b_{-1}\, b_{-2}\, b_{-3}\, b_{-4}\, 1111$ will be truncated to $0.b_{-1}\, b_{-2}\, b_{-3}\, b_{-4}$. The error ranges from 0 to .00001111. In general, the error ranges from 0 to approximately 1 in the low-order position of the retained bits.

Adder-based rounding   This technique adds a 1 to the low-order bit position of the bits being retained if there is a 1 in the high-order bit position of the bits being removed. For example, if the fraction

$$0.b_{-1}\, b_{-2}\, b_{-3}\, b_{-4}\, 1\, b_{-6}\, b_{-7}\, b_{-8}$$

is to be rounded to four bits using adder-based rounding, the result would be

$$0.b_{-1}\, b_{-2}\, b_{-3}\, b_{-4} + 0.0001$$

This approaches the true value from above. An example is shown below.

```
                     | Delete
      0 . 0 1 1 1 | 1 0 0 1    × 2^8  (+121)
 +)   0 . 0 0 0 1 |
      0 . 1 0 0 0 |            × 2^8  (+128)
```

In a similar manner, if the fraction

$$0.b_{-1}\,b_{-2}\,b_{-3}\,b_{-4}\,0\,b_{-6}\,b_{-7}\,b_{-8}$$

is to be rounded to four bits using adder-based rounding, the result would be

$$0.b_{-1}\,b_{-2}\,b_{-3}\,b_{-4}$$

This approaches the true value from below. An example is shown below.

```
                  | Delete
  0 . 0  1  1  1  | 0  1  1  1    × 2^8  (+119)
        ↓         |
  0 . 0  1  1  1  |               × 2^8  (+112)
```

von Neumann rounding This is also referred to as *jamming*. If the bits to be deleted are all zeroes, then the bits are truncated and there is no change to the retained bits. However, if any bit in the bits to be deleted is a 1, then the low-order bit of the retained bits is set to 1. Examples are shown below.

```
                  | Delete
  0 . 0  1  1  0  | 0  0  0  0
        ↓         |
  0 . 0  1  1  0  |
```

```
                  | Delete
  0 . 0  1  1  0  | 0  1  0  0
        ↓         |
  0 . 0  1  1  1  |
```

This section on floating-point arithmetic has been a short introduction to the topic. A more exhaustive treatise would require an entire chapter in order to present floating-point topics in detail such as, read-only memory rounding, guard digits, exponent overflow and underflow, multiple precision, hexadecimal radix, a denormalized number, and detection of *not a number* (NaN), among others.

5.10 Problems

5.1   For $n = 8$, let A and B be two fixed-point binary numbers in 2s complement representation, where $A = 01101101$ and $B = 11001111$. Determine $A + B' + 1 + B$, where B' is the 1s complement.

5.2   Perform the indicated operations on the fixed-point numbers shown below in 2s complement representation. In each case, indicate if there is an overflow.

(a)

```
      0  1  0  0  0  0  0  0
 +)   0  1  0  0  0  0  0  0
```

(b)

```
      0  0  1  1  0  1  1  0
 +)   1  1  1  0  0  0  1  1
```

(c)

```
      1  0  0  1  1  0  0  0
 -)   0  0  1  0  0  0  1  0
```

(d)

```
      0  0  1  1  0  1  1  0
 -)   1  1  1  0  0  0  1  1
```

5.3   Let A and B be two fixed-point binary numbers in 2s complement representation as shown below, where A' and B' are the 1s complement of A and B, respectively. Perform the indicated operations.

$$A = 10110001 \qquad B = 11100100$$

(a) $A - (B + 1)$
(b) $A' + 1 + B' + 1$
(c) $A' + 1 - (B' + 1)$

5.4   Obtain the sum of the following unsigned hexadecimal numbers:

```
        1  2  3  4  5
        6  7  8  9  A
 +)     B  C  D  E  F
```

5.5   Obtain the sum of the following 2s complement radix 2 numbers:

$$
\begin{array}{r}
1\ 1\ 1\ 1\ 1\ 0 \\
1\ 1\ 1\ 1\ 0\ 0 \\
1\ 1\ 1\ 0\ 0\ 0 \\
+)\quad 1\ 1\ 0\ 0\ 0\ 0 \\
\hline
\end{array}
$$

5.6   Perform the following radix 2 subtraction using the 1s complement method:

$$
\begin{array}{r}
1\ 1\ 0\ 0\ 1\ 1\ 0\ 0 \\
-)\quad 0\ 0\ 1\ 1\ 0\ 0\ 1\ 1 \\
\hline
\end{array}
$$

5.7   Obtain the sum of the following radix 5 numbers:

$$
\begin{array}{r}
1\ 2\ 3\ 4 \\
4\ 3\ 2\ 1 \\
3\ 3\ 3\ 3 \\
\hline
\end{array}
$$

5.8   Perform a subtraction on the operands shown below, which are in radix complementation for radix 3.

$$
\begin{array}{r}
0\ 2\ 0\ 2\ 1 \\
-)\quad 2\ 2\ 1\ 0\ 0 \\
\hline
\end{array}
$$

5.9   Write the equations for two ways to detect overflow for two n-bit operands in fixed-point addition assuming radix complementation for radix 2.

5.10   Indicate whether overflow occurs for the following decimal numbers when they are converted to 8-bit fixed-point binary numbers for radix 2 for the indicated operations.

(a) $(+35) + (+42)$
(b) $(-62) - (+67)$
(c) $(-31) + (-34)$
(d) $(-31) - (+33)$

5.11   Add the following numbers, which are shown in radix complementation. Obtain the sum in radix 4: $0201_{10} + 321_4$.

5.12   Show the radix 3 result of the following subtraction:

$$
\begin{array}{r}
2\ \ 0\ \ 2\ \ 2\ \ 0\ \ 2\ \ 0\ \ 1_3 \\
-)\quad 2\ \ 1\ \ 0\ \ 2\ \ 0\ \ 2\ \ 0\ \ 2_3 \\
\hline
\end{array}
$$

5.13   Use the paper-and-pencil method to multiply the following numbers, which are in 2s complement representation:

> 001110
> 000111

5.14   Use the paper-and-pencil method to multiply the following numbers, which are in 2s complement representation:

> 111111
> 001011

5.15   Use the sequential add-shift method to multiply the following operands, which are in 2s complement representation:

> Multiplicand $A =$ 1111
> Multiplier $B =$ 0111

5.16   Use the sequential add-shift method to multiply the following operands, which are in 2s complement representation:

> Multiplicand $A =$ 0101
> Multiplier $B =$ 0101

5.17   Use the Booth algorithm to multiply the following operands, which are in 2s complement representation:

> Multiplicand $A =$ 010111
> Multiplier $B =$ 001101

5.18   Use the Booth algorithm to multiply the following operands, which are in 2s complement representation:

> Multiplicand $A =$ 010110
> Multiplier $B =$ 110110

5.19   Use the Booth algorithm to multiply the following operands, which are in 2s complement representation:

> Multiplicand $A =$ 110001
> Multiplier $B =$ 100110

5.20   Use bit-pair recoding to determine the multiplicand multiples to be added for the following multipliers, which are in 2s complement representation.

(a) 100111
(b) 110011
(c) 001101

5.21   Use bit-pair recoding to multiply the following operands, which are in 2s complement representation:

$$\text{Multiplicand } A = \quad 110010$$
$$\text{Multiplier } B = \quad 011001$$

5.22   Use bit-pair recoding to multiply the following operands, which are in 2s complement representation:

$$\text{Multiplicand } A = \quad 0100110$$
$$\text{Multiplier } B = \quad 1001101$$

5.23   Multiply the following two unsigned fixed-point operands using an array multiplier:

$$\text{Multiplicand } A = \quad 0111$$
$$\text{Multiplier } B = \quad 1100$$

5.24   Use the sequential restoring method to perform a divide operation on the following operands:

$$\text{Dividend } A = \quad 00110000$$
$$\text{Divisor } B = \quad 0111$$

5.25   Use the sequential nonrestoring method to perform a divide operation on the following operands:

$$\text{Dividend } A = \quad 00000111$$
$$\text{Divisor } B = \quad 0100$$

5.26   Use the sequential nonrestoring method to perform a divide operation on the following operands:

$$\text{Dividend } A = \quad 01100011$$
$$\text{Divisor } B = \quad 0111$$

5.27   The decimal operands shown below are to be added using decimal (BCD) addition. Obtain the answer that correctly represents the intermediate sum; that is, the sum before correction (adjustment) is applied.

$(+725) + (+536)$

5.28   Perform the indicated decimal operation on the operands shown below.

$(+20) + (-32)$

5.29   Perform decimal multiplication using a read-only memory (ROM) on the following operands:

$$\text{Multiplicand } A = \ 736$$
$$\text{Multiplier } B = \ \ \ 48$$

5.30   Perform the following floating-point addition operation for positive operands:

$$A = 0 \ . \ 1 \ 0 \ 0 \ 1 \ 0 \ 0 \ 0 \ 0 \quad \times 2^6$$
$$+) \ \ B = 0 \ . \ \underline{1 \ 1 \ 1 \ 1 \ 0 \ 0 \ 0 \ 0} \quad \times 2^2$$

5.31   Under what conditions can there be an exponent overflow during floating-point arithmetic?

5.32   Perform the operation listed below for normalized floating-point numbers using 8-bit fractions.

$$- \ 1 \ 0$$
$$+) \ \underline{+ \ 3 \ 3}$$

5.33   Multiply the floating-point fractions shown below using the sequential add-shift method.

$$\text{Multiplicand } A = 0 \ . \ 1 \ 1 \ 0 \ 0 \ 1 \ 0 \ 0 \ 0 \quad \times 2^5$$
$$\text{Multiplier } B = 0 \ . \ 1 \ 0 \ 0 \ 1 \ 0 \ 0 \ 0 \ 0 \quad \times 2^4$$

5.34   Multiply the floating-point fractions shown below using the sequential add-shift method.

$$\text{Multiplicand } A = 0 \ . \ 1 \ 1 \ 0 \ 0 \ 1 \ 0 \ 0 \ 0 \quad \times 2^5$$
$$\text{Multiplier } B = 0 \ . \ 1 \ 1 \ 0 \ 1 \ 0 \ 0 \ 0 \ 0 \quad \times 2^4$$

5.35   Multiply the floating-point fractions shown below using the sequential add-shift method.

Multiplicand $A =$ 1 . 1 1 0 0 1 0 0 0 $\times 2^5$

Multiplier $B =$ 0 . 1 1 1 0 0 0 0 0 $\times 2^3$

5.36   Multiply the floating-point fractions shown below using the sequential add-shift method.

Multiplicand $A =$ 1 . 1 1 0 0 0 0 0 0 $\times 2^4$

Multiplier $B =$ 0 . 1 0 0 0 1 0 0 0 $\times 2^5$

5.37   How is quotient overflow determined in floating-point division and how is the problem solved?

5.38   Explain the biasing problem that occurs during floating-point multiplication and division.

6

Computer Arithmetic Design Using Verilog HDL

This chapter augments the topics contained in Chapter 5 by designing select arithmetic circuits using Verilog HDL. In order to keep the page count of the book to a reasonable number, only some of the more common arithmetic circuits will be designed using Verilog HDL. The sections on fixed-point arithmetic consists of the following arithmetic circuits: a high-speed full adder, a 4-bit ripple adder, a 4-bit carry lookahead adder, an adder/subtractor, a Booth algorithm multiplier, and an array multiplier.

The sections on decimal arithmetic consist of the following circuits: a decimal adder and a decimal adder/subtractor. Various modeling styles will be used to design both fixed-point and decimal circuits, including built-in primitives, dataflow modeling, behavioral modeling, and structural modeling.

6.1 Fixed-Point Addition

Three examples are presented in this section utilizing adders. Each example includes the associated theory for the design together with tables, Karnaugh maps, and equations. The first is a high-speed full adder with only two gate delays. This is followed by a 4-bit low-speed ripple adder, then a 4-bit high-speed carry lookahead adder.

6.1.1 High-Speed Full Adder

This section designs a high-speed full adder with only NAND gates using structural modeling. Recall that a full adder has three scalar inputs: augend a, addend b, and a carry-in cin from the previous lower-order stage. There are two scalar outputs: sum and carry-out $cout$. The truth table for a general full adder for stage i is shown in Table 6.1. Note that the sum is 1 for an odd number of 1s; the carry-out $cout$ is a 1 for two or more 1s. The equations for the sum and carry-out are shown in Equation 6.1 and Equation 6.2, respectively, as obtained from the truth table. Equation 6.2 can be further simplified by using the Karnaugh map of Figure 6.1, which yields Equation 6.3. The logic diagram is shown in Figure 6.2.

Table 6.1 Truth Table for a Full Adder

		Carry-in	Sum	Carry-out
a	b	cin	sum	$cout$
0	0	0	0	0
0	0	1	1	0
0	1	0	1	0
0	1	1	0	1
1	0	0	1	0
1	0	1	0	1
1	1	0	0	1
1	1	1	1	1

$$sum = a'b'\,cin + a'\,b\,cin' + ab'cin' + abcin \qquad (6.1)$$

$$cout = a'bcin + ab'cin + abcin' + abcin \qquad (6.2)$$

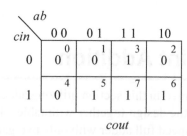

Figure 6.1   Karnaugh map for the carry-out of a full adder.

$$cout_i = ab + acin + bcin \qquad (6.3)$$

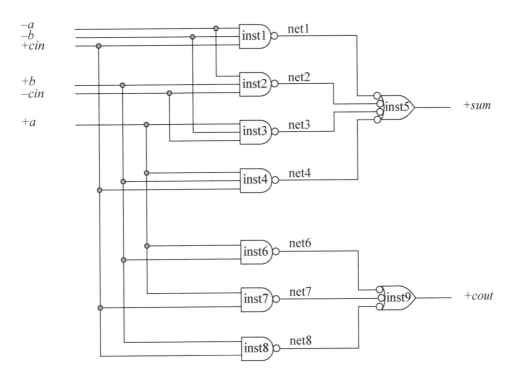

Figure 6.2   Logic diagram for a high-speed full adder.

The design module is shown in Figure 6.3, which correlates the instantiation names and the net names with the logic diagram. The following NAND gates are instantiated into the structural module: *nand2_df*, *nand3_df*, and *nand4_df*. These NAND gates have been previously designed. The test bench module is shown in Figure 6.4. The outputs and waveforms are shown in Figure 6.5 and Figure 6.6, respectively.

```
//structural high-speed full adder
module full_adder_hi_spd_struc (a, b, cin, sum, cout);

input a, b, cin;
output sum, cout;

//continued on next page
```

Figure 6.3   Design module for the high-speed full adder.

```
//define internal nets
wire net1, net2, net3, net4, net6, net7, net8;

//instantiate the nand gates for the sum
nand3_df inst1 (
    .x1(~a),
    .x2(~b),
    .x3(cin),
    .z1(net1)
    );

nand3_df inst2 (
    .x1(~a),
    .x2(b),
    .x3(~cin),
    .z1(net2)
    );

nand3_df inst3 (
    .x1(a),
    .x2(~b),
    .x3(~cin),
    .z1(net3)
    );

nand3_df inst4 (
    .x1(a),
    .x2(b),
    .x3(cin),
    .z1(net4)
    );

nand4_df inst5 (
    .x1(net1),
    .x2(net2),
    .x3(net3),
    .x4(net4),
    .z1(sum)
    );

//instantiate the nand gates for the carry-out
nand2_df inst6 (
    .x1(a),
    .x2(b),
    .z1(net6)
    );
//continued on next page
```

Figure 6.3   (Continued)

```
nand2_df inst7 (
   .x1(a),
   .x2(cin),
   .z1(net7)
   );

nand2_df inst8 (
   .x1(b),
   .x2(cin),
   .z1(net8)
   );

nand3_df inst9 (
   .x1(net6),
   .x2(net7),
   .x3(net8),
   .z1(cout)
   );

endmodule
```

Figure 6.3   (Continued)

```
//test bench for high-speed full adder
module full_adder_hi_spd_struc_tb;

reg a, b, cin;
wire sum, cout;

//apply input vectors and display variables
initial
begin: apply_stimulus
   reg [3:0] invect;
   for (invect = 0; invect < 8; invect = invect + 1)
      begin
         {a, b, cin} = invect [3:0];
         #10 $display ("a b cin = %b, cout sum = %b",
                          {a, b, cin}, {cout, sum});
      end
end

//continued on next page
```

Figure 6.4   Test bench for the high-speed full adder.

```
//instantiate the module into the test bench
full_adder_hi_spd_struc inst1 (
   .a(a),
   .b(b),
   .cin(cin),
   .sum(sum),
   .cout(cout)
   );

endmodule
```

Figure 6.4   (Continued)

```
a b cin = 000, cout sum = 00
a b cin = 001, cout sum = 01
a b cin = 010, cout sum = 01
a b cin = 011, cout sum = 10
a b cin = 100, cout sum = 01
a b cin = 101, cout sum = 10
a b cin = 110, cout sum = 10
a b cin = 111, cout sum = 11
```

Figure 6.5   Outputs for the high-speed full adder.

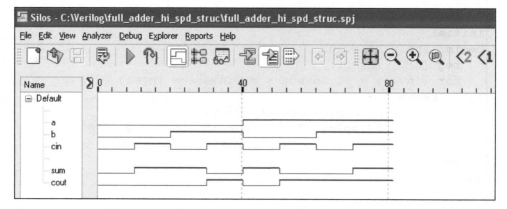

Figure 6.6   Waveforms for the high-speed full adder.

6.1.2 Four-Bit Ripple Adder

A ripple adder is a relatively low-speed adder because there is no carry-lookahead feature. A 4-bit ripple adder consists of four full adders connected serially in which the carry-out of stage $_i$ is carry-in to stage $_{i+1}$. A full adder adds two operand bits — the augend and the addend — plus the carry-in from the previous lower-order stage and produces two outputs: sum and carry-out. Each full adder can be designed using two half adders and one OR gate; therefore, this type of full adder is not considered a high-speed full adder. A half adder adds the two operand bits only, and produces two outputs: sum and carry. Each half adder is designed using one exclusive-OR gate and one AND gate.

   The truth tables for a half adder and full adder are shown in Table 6.2 and Table 6.3, respectively. The corresponding equations are listed in Equation 6.4 and Equation 6.5. The logic diagram for the half adder is shown in Figure 6.7 and for the full adder in Figure 6.8.

Table 6.2 Truth Table for a Half Adder

a	b	sum	cout
0	0	0	0
0	1	1	0
1	0	1	0
1	1	0	1

Table 6.3 Truth Table for a Full Adder

a	b	cin	sum	cout
0	0	0	0	0
0	0	1	1	0
0	1	0	1	0
0	1	1	0	1
1	0	0	1	0
1	0	1	0	1
1	1	0	0	1
1	1	1	1	1

$$
\begin{aligned}
sum &= a'b + ab' \\
&= a \oplus b \\
cout &= ab
\end{aligned}
\tag{6.4}
$$

$$
\begin{aligned}
sum &= a'b'cin + a'bcin' + ab'cin' + abcin \\
&= cin\,(a \oplus b)' + cin'\,(a \oplus b) \\
&= a \oplus b \oplus cin \\
cout &= a'bcin + ab'cin + abcin' + abcin \\
&= cin\,(a \oplus b) + ab
\end{aligned}
\tag{6.5}
$$

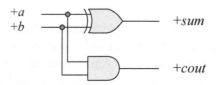

Figure 6.7   Logic diagram for a half adder.

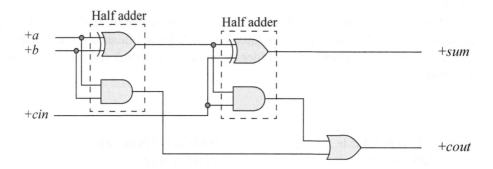

Figure 6.8   Logic diagram for a full adder.

The symbol for a full adder is shown in Figure 6.9 and the logic diagram for a 4-bit ripple adder is shown in Figure 6.10 for augend *A[3:0]* and addend *B[3:0]*, where *a[0]* and *b[0]* are the low-order bits of *A* and *B*, respectively. In order to design the ripple adder, a full adder will be designed using behavioral modeling, then instantiated four times into the structural module of the ripple adder.

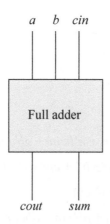

Figure 6.9   Symbol for a full adder.

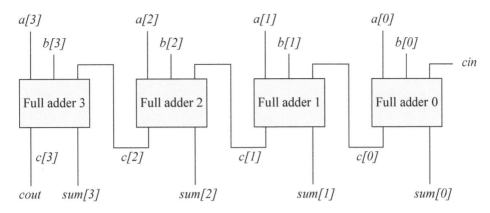

Figure 6.10   Logic diagram for a 4-bit ripple adder.

The behavioral module for a full adder is shown in Figure 6.11 and will be instantiated four times into the 4-bit ripple adder structural module as shown in Figure 6.12. Target variables used in an **always** statement are declared as type **reg**. The test bench module for the ripple adder is shown in Figure 6.13. The outputs and waveforms are shown in Figure 6.14 and Figure 6.15, respectively.

```
//behavioral full adder
module full_adder_bh (a, b, cin, sum, cout);

//define inputs and outputs
input a, b, cin;
output sum, cout;

//inputs are wire by default, but can be declared if so desired
wire a, b, cin;
reg sum, cout;        //target variables in always are type reg

always @ (a or b or cin)
begin
   sum = a ^ b ^ cin;
   cout = (a & b) | (a & cin) | (b & cin);
end

endmodule
```

Figure 6.11   Behavioral module for a full adder.

```
//structural 4-bit ripple adder
module adder_ripple4_struc (a, b, cin, sum, cout);

input [3:0] a, b;
input cin;
output [3:0] sum;
output cout;

wire [3:0] a, b;
wire cin;
wire [3:0] sum;
wire [3:0] c;          //define internal nets for carries
wire cout;

assign cout = c[3];
full_adder_bh inst1 (
   .a(a[0]),
   .b(b[0]),
   .cin(cin),
   .sum(sum[0]),
   .cout(c[0])
   );

full_adder_bh inst2 (
   .a(a[1]),
   .b(b[1]),
   .cin(c[0]),
   .sum(sum[1]),
   .cout(c[1])
   );

full_adder_bh inst3 (
   .a(a[2]),
   .b(b[2]),
   .cin(c[1]),
   .sum(sum[2]),
   .cout(c[2])
   );

full_adder_bh inst4 (
   .a(a[3]),
   .b(b[3]),
   .cin(c[2]),
   .sum(sum[3]),
   .cout(c[3])
   );
endmodule
```

Figure 6.12   Structural design module for a 4-bit ripple adder.

```
//structural adder_ripple4_struc test bench
module adder_ripple4_struc_tb;


reg [3:0]a, b;
reg cin;
wire [3:0] sum;
wire cout;


//display variables
initial
$monitor ("a=%b, b=%b, cin=%b, cout=%b, sum=%b",
          a, b, cin, cout, sum);


initial
begin
   #0    a = 4'b1111;    b = 4'b0000;    cin = 1'b0;
   #10   a = 4'b1111;    b = 4'b0001;    cin = 1'b0;
   #10   a = 4'b0011;    b = 4'b0111;    cin = 1'b0;
   #10   a = 4'b0101;    b = 4'b0101;    cin = 1'b0;
   #10   a = 4'b1001;    b = 4'b1001;    cin = 1'b0;
   #10   a = 4'b1110;    b = 4'b0001;    cin = 1'b0;
   #10   a = 4'b1101;    b = 4'b1101;    cin = 1'b0;
   #10   a = 4'b1111;    b = 4'b1111;    cin = 1'b0;
   #10   a = 4'b1111;    b = 4'b1111;    cin = 1'b0;
   #10   a = 4'b1110;    b = 4'b0001;    cin = 1'b1;
   #10   a = 4'b1101;    b = 4'b1101;    cin = 1'b1;
   #10   a = 4'b1110;    b = 4'b1111;    cin = 1'b1;
   #10   a = 4'b1111;    b = 4'b1111;    cin = 1'b1;

   #10   $stop;
end


//instantiate the module into the test bench
adder_ripple4_struc inst1 (
   .a(a),
   .b(b),
   .cin(cin),
   .sum(sum),
   .cout(cout)
   );

endmodule
```

Figure 6.13   Test bench for the structural ripple adder.

```
a=1111, b=0000, cin=0, cout=0, sum=1111
a=1111, b=0001, cin=0, cout=1, sum=0000
a=0011, b=0111, cin=0, cout=0, sum=1010
a=0101, b=0101, cin=0, cout=0, sum=1010
a=1001, b=1001, cin=0, cout=1, sum=0010
a=1110, b=0001, cin=0, cout=0, sum=1111
a=1101, b=1101, cin=0, cout=1, sum=1010
a=1111, b=1111, cin=0, cout=1, sum=1110
a=1110, b=0001, cin=1, cout=1, sum=0000
a=1101, b=1101, cin=1, cout=1, sum=1011
a=1110, b=1111, cin=1, cout=1, sum=1110
a=1111, b=1111, cin=1, cout=1, sum=1111
```

Figure 6.14   Outputs for the structural full adder.

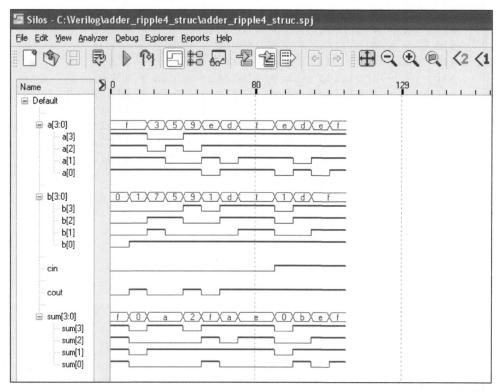

Figure 6.15   Waveforms for the structural ripple adder.

6.1.3 Carry Lookahead Adder

The speed limitation in the ripple adder arises from specifying $cout_i$ as a function of the carry-out from the previous lower-order stage $cout_{i-1}$. A considerable increase in speed can be realized by expressing the carry-out $cout_i$ of any stage i as a function of the two operand bits a_i and b_i and the carry-in cin_{-1} to the low-order stage$_0$ of the adder, where the adder is an n-bit adder $n_{-1} n_{-2} \ldots n_1 n_0$. The Karnaugh map that represents the carry-out from stage$_i$ is shown in Figure 6.16, which yields Equation 6.6.

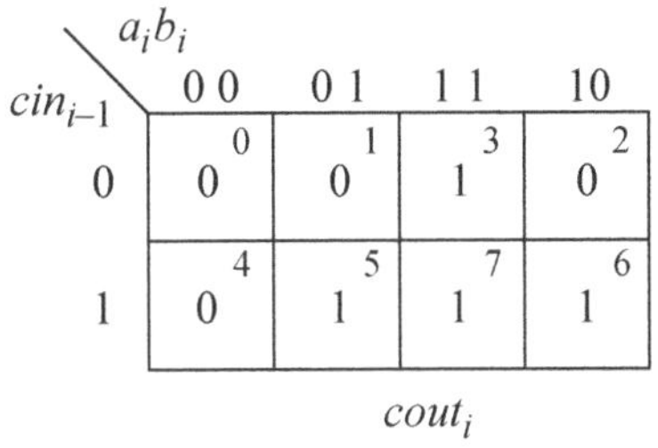

Figure 6.16   Karnaugh map for the carry-out of stage$_i$ of an n-bit adder.

$$cout_i = a_i' b_i cin_{i-1} + a_i b_i' cin_{i-1} + a_i b_i cin_{i-1}' + a_i b_i cin_{i-1}$$

$$= a_i b_i + (a_i \oplus b_i) cin_{i-1} \tag{6.6}$$

Equation 6.6 states that a carry will be generated whenever $a_i = b_i = 1$, or when either $a_i = 1$ or $b_i = 1$ — but not both — with $c_{i-1} = 1$. Note that if $a_i = b_i = 1$, then this represents a generate function, not a propagate function. Verilog requires a propagate function to be the exclusive-OR of a_i and b_i. A technique will now be presented that increases the speed of the carry propagation in a parallel adder. The carries entering all the bit positions of the adder can be generated simultaneously by a *carry lookahead* generator. This results in a constant addition time that is independent of the length of the adder. Two auxiliary functions will be defined as follows:

Generate $G_i = a_i b_i$
Propagate $P_i = a_i \oplus b_i$

The carry *generate* function G_i reflects the condition where a carry is generated at the ith stage. The carry *propagate* function P_i is true when the ith stage will pass through (or propagate) the incoming carry cin_{i-1} to the next higher stage$_{i+1}$. Equation 6.6 can now be restated as Equation 6.7.

$$cout_i = a_i b_i + (a_i \oplus b_i) \, cin_{i-1}$$

$$= G_i + P_i \, cin_{i-1} \tag{6.7}$$

Equation 6.7 indicates that the generate G_i and propagate P_i functions for any carry out $cout_i$ can be obtained independently and in parallel when the operand inputs are applied to the n-bit adder. The equation can be applied recursively to obtain a set of carry-out equations in terms of the variables G_i, P_i, and cin_{-1} for a 4-bit adder, where c_{-1} is the carry-in to the low-order stage$_0$ of the adder. The equations are shown in Equation 6.8.

$$c_0 = G_0 + P_0 \, cin_{-1}$$

$$c_1 = G_1 + P_1 \, c_0$$
$$= G_1 + P_1 \, (G_0 + P_0 \, cin_{-1})$$
$$= G_1 + P_1 G_0 + P_1 P_0 \, cin_{-1}$$

$$c_2 = G_2 + P_2 \, c_1$$
$$= G_2 + P_2 \, (G_1 + P_1 G_0 + P_1 P_0 \, cin_{-1})$$
$$= G_2 + P_2 G_1 + P_2 P_1 G_0 + P_2 P_1 P_0 \, cin_{-1}$$

$$c_3 = G_3 + P_3 \, c_2$$
$$= G_3 + P_3 (G_2 + P_2 G_1 + P_2 P_1 G_0 + P_2 P_1 P_0 \, cin_{-1})$$
$$= G_3 + P_3 G_2 + P_3 P_2 G_1 + P_3 P_2 P_1 G_0 + P_3 P_2 P_1 P_0 \, cin_{-1} \tag{6.8}$$

A 4-bit carry lookahead adder will be designed using dataflow modeling for Equation 6.8. The block diagram of the adder is shown in Figure 6.17, where the augend is $A[3:0]$, the addend is $B[3:0]$, and the sum is $SUM[3:0]$. The low-order bits are $a[0]$, $b[0]$, and $sum[0]$ of A, B, and SUM, respectively. There is also a carry-in cin and a carry-out $cout$.

The dataflow module is shown in Figure 6.18 using the **assign** statement for the generate, propagate, internal carries, and the sum. The carry equations c_i correspond directly with Equation 6.8. The test bench is shown in Figure 6.19 for 10 values of A,

B, and *cin*. The outputs are shown in Figure 6.20 and the waveforms are shown in Figure 6.21 in hexadecimal notation and also as single bits.

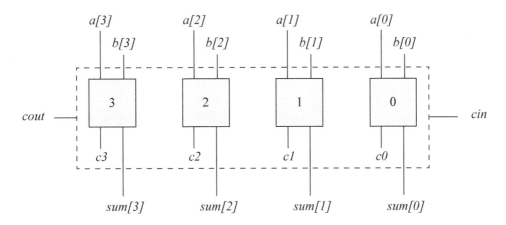

Figure 6.17   Block diagram of a 4-bit carry lookahead adder.

```
//dataflow for a 4-bit carry lookahead adder
module adder4_cla (a, b, cin, sum, cout);

//define inputs and outputs
input [3:0] a, b;
input cin;
output [3:0] sum;
output cout;

//define internal wires
wire g3, g2, g1, g0;      //for generate functions
wire p3, p2, p1, p0;      //for propagate functions
wire c3, c2, c1, c0;      //for carries

//define generate functions
assign
   g0 = a[0] & b[0],      //multiple statements using 1 assign
   g1 = a[1] & b[1],
   g2 = a[2] & b[2],
   g3 = a[3] & b[3];
//continued on next page
```

Figure 6.18   Design module for a 4-bit carry lookahead adder using dataflow modeling.

```
//define propagate functions
assign
   p0 = a[0] ^ b[0],
   p1 = a[1] ^ b[1],
   p2 = a[2] ^ b[2],
   p3 = a[3] ^ b[3];

//obtain the carry equations
assign
   c0 = g0| (p0 & cin),
   c1 = g1| (p1 & g0) | (p1 & p0 & cin),
   c2 = g2| (p2 & g1) | (p2 & p1 & g0) | (p2 & p1 & p0 & cin),
   c3 = g3| (p3 & g2) | (p3 & p2 & g1) | (p3 & p2 & p1 & g0) |
            (p3 & p2 & p1 & p0 & cin);

//obtain the sum equations
assign
   sum[0] = p0 ^ cin,
   sum[1] = p1 ^ c0,
   sum[2] = p2 ^ c1,
   sum[3] = p3 ^ c2;

//obtain cout
assign cout = c3;

endmodule
```

Figure 6.18   (Continued)

```
//test bench for the dataflow 4-bit carry lookahead adder
module adder4_cla_tb;

reg [3:0] a, b;
reg cin;

wire [3:0] sum;
wire cout;

//display variables
initial
$monitor ("a = %b, b = %b, cin = %b, cout = %b, sum = %b",
            a, b, cin, cout, sum);
//continued on next page
```

Figure 6.19   Test bench for the 4-bit carry lookahead adder.

```
//apply input vectors
initial
begin
   #0     a = 4'b0000;   b = 4'b0000;   cin = 1'b0;
          //cout = 0, sum = 0000

   #10    a = 4'b0001;   b = 4'b0010;   cin = 1'b0;
          //cout = 0, sum = 0011

   #10    a = 4'b0010;   b = 4'b0110;   cin = 1'b0;
          //cout = 0, sum = 1000

   #10    a = 4'b0111;   b = 4'b0111;   cin = 1'b0;
          //cout = 0, sum = 1110

   #10    a = 4'b1001;   b = 4'b0110;   cin = 1'b0;
          //cout = 0, sum = 1111

   #10    a = 4'b1100;   b = 4'b1100;   cin = 1'b0;
          //cout = 1, sum = 1000

   #10    a = 4'b1111;   b = 4'b1110;   cin = 1'b0;
          //cout = 1, sum = 1101

   #10    a = 4'b1110;   b = 4'b1110;   cin = 1'b1;
          //cout = 1, sum = 1101

   #10    a = 4'b1111;   b = 4'b1111;   cin = 1'b1;
          //cout = 1, sum = 1111

   #10    $stop;

end

//instantiate the module into the test bench
adder4_cla inst1 (
   .a(a),
   .b(b),
   .cin(cin),
   .sum(sum),
   .cout(cout)
   );

endmodule
```

Figure 6.19   (Continued)

```
a = 0000, b = 0000, cin = 0, cout = 0, sum = 0000
a = 0001, b = 0010, cin = 0, cout = 0, sum = 0011
a = 0010, b = 0110, cin = 0, cout = 0, sum = 1000
a = 0111, b = 0111, cin = 0, cout = 0, sum = 1110
a = 1001, b = 0110, cin = 0, cout = 0, sum = 1111
a = 1100, b = 1100, cin = 0, cout = 1, sum = 1000
a = 1111, b = 1110, cin = 0, cout = 1, sum = 1101
a = 1110, b = 1110, cin = 1, cout = 1, sum = 1101
a = 1111, b = 1111, cin = 1, cout = 1, sum = 1111
```

Figure 6.20   Outputs for the 4-bit carry lookahead adder.

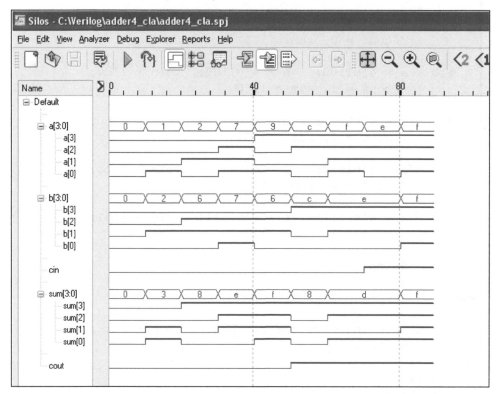

Figure 6.21   Waveforms for the 4-bit carry lookahead adder.

6.2 Fixed-Point Subtraction

This section presents the design of a 4-bit fixed-point ripple adder/subtractor. The design of a carry lookahead adder/subtractor is similar except that the carry logic uses the carry lookahead technique as described in Section 6.1.3. It is desirable to have the adder unit perform both addition and subtraction since there is no advantage to having a separate adder and subtractor. A ripple-carry adder will be modified so that it can perform subtraction while still maintaining the ability to add.

Subtraction is accomplished by adding the 2s complement of the subtrahend to the minuend as shown below for minuend A and subtrahend B, where B' is the 1s complement of B. The 2s complement of a number is obtained by adding 1 to the 1s complement of the number.

$$A - B = A + (B' + 1)$$

An inverter could be used to invert each subtrahend bit, but this would not allow the noninverted addend bits to be used for addition. The logic that inverts the subtrahend bits should also allow for addition. The exclusive-OR operation for two variables is defined as

$$A \oplus B = AB' + A'B$$

Thus, $A \oplus B = 1$ only if $A \neq B$. The truth table for the exclusive-OR function is shown in Table 6.4. Note that when $m = 1$, B is inverted; when $m = 0$, B is noninverted. Therefore, the variable m can be used as a *mode control* input to determine whether the operation is addition or subtraction. If the mode control line is zero, then the operation is addition; if the mode control line is one, then the operation is subtraction.

Table 6.4 Rules for Exclusive-OR Operation

		b	
	$\oplus$	0	1
	0	0	1
m	1	1	0

The logic diagram for a 4-bit adder/subtractor is shown in Figure 6.22 in which four full adders are used together with the requisite exclusive-OR functions. The logic diagram also indicates the instantiation names for the full adders that will be used in the Verilog design. The augend/minuend is a 4-bit vector $A[3:0]$ and the addend/subtrahend is a 4-bit vector $B[3:0]$. Note that the mode control input m is connected to each exclusive-OR function and also to the carry-in of the low-order stage. The result is a 4-bit vector $RSLT[3:0]$; the carry-out of each stage is also a 4-bit vector $COUT[3:0]$.

Examples of subtraction are shown in Figure 6.23 for 4-bit operands in 2s complement representation. The term *true addition* means that the result is the sum of the two operands, ignoring the sign bit; whereas *true subtraction* means that the result is the difference of the two operands, ignoring the sign bit. To obtain the 2s complement (negation) of an operand, the low-order zeroes and the first 1 bit are unchanged; then the remaining higher-order bits are complemented.

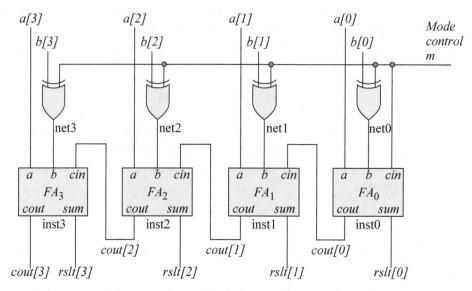

Figure 6.22   Logic diagram for a 4-bit adder/subtractor.

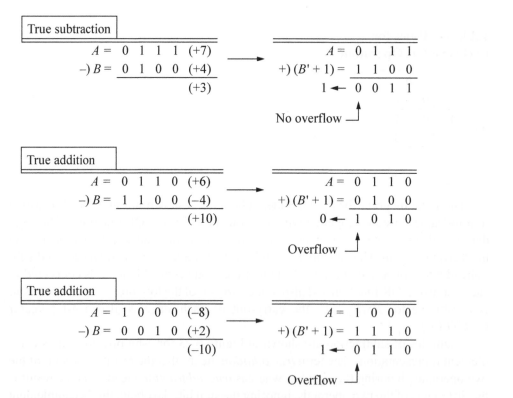

Figure 6.23   Examples of fixed-point subtraction.

In the last two examples of Figure 6.23, an overflow occurred because the result was too large to fit in the word size of the operands. The range for numbers in 2s complement representation is

$$-2^{n-1} \text{ to } +2^{n-1}-1$$

where n is the number of bits in the operands. Thus, for $n = 4$, the range is from -8 to $+7$. The result of the last two examples is $+10$ and -10, both of which exceed the range for four bits. There are two ways to detect overflow for n-bit operands, as shown below, where a_{n-1} and b_{n-1} are the sign bits of operands A and B, respectively.

$$A = a_{n-1}\, a_{n-2} \cdots a_1\, a_0$$

$$B = b_{n-1}\, b_{n-2} \cdots b_1\, b_0$$

$$\text{Overflow} = cout_{n-1} \oplus cout_{n-2}$$
$$\text{Overflow} = a_{n-1}\, b_{n-1}\, rslt_{n-1}{}' + a_{n-1}{}'\, b_{n-1}{}'\, rslt_{n-1}$$

The module for a full adder is shown in Figure 6.24 using dataflow modeling. This will be instantiated four times into the structural module of Figure 6.25 — which also uses built-in primitives — to implement the 4-bit adder/subtractor. The test bench is shown in Figure 6.26. The outputs and waveforms are shown in Figure 6.27 and Figure 6.28, respectively.

```
//dataflow full adder
module full_adder (a, b, cin, sum, cout);

//list all inputs and outputs
input a, b, cin;
output sum, cout;

//define wires
wire a, b, cin;
wire sum, cout;

//continuous assignment
assign sum = (a ^ b) ^ cin;
assign cout = cin & (a ^ b) | (a & b);

endmodule
```

Figure 6.24   Full adder to be instantiated into a structural module to implement a 4-bit adder/subtractor.

```
//structural module for an adder/subtractor
module adder_subtr_struc (a, b, m, rslt, cout, ovfl);

input [3:0] a, b;
input m;
output [3:0] rslt, cout;
output ovfl;

//define internal nets
wire net0, net1, net2, net3;

//define overflow
xor (ovfl, cout[3], cout[2]);


//instantiate the xor and the full adder for FA0
xor (net0, b[0], m);
full_adder inst0 (
   .a(a[0]),
   .b(net0),
   .cin(m),
   .sum(rslt[0]),
   .cout(cout[0])
   );

//instantiate the xor and the full adder for FA1
xor (net1, b[1], m);
full_adder inst1 (
   .a(a[1]),
   .b(net1),
   .cin(cout[0]),
   .sum(rslt[1]),
   .cout(cout[1])
   );


//instantiate the xor and the full adder for FA2
xor (net2, b[2], m);
full_adder inst2 (
   .a(a[2]),
   .b(net2),
   .cin(cout[1]),
   .sum(rslt[2]),
   .cout(cout[2])
   );


//continued on next page
```

Figure 6.25   Structural module for a 4-bit adder/subtractor.

```
//instantiate the xor and the full adder for FA3
xor (net3, b[3], m);
full_adder inst3 (
   .a(a[3]),
   .b(net3),
   .cin(cout[2]),
   .sum(rslt[3]),
   .cout(cout[3])
   );
endmodule
```

Figure 6.25   (Continued)

```
//test bench for structural adder/subtractor
module adder_subtr_struc_tb;

reg [3:0] a, b;
reg m;
wire [3:0] rslt, cout;
wire  ovfl;

initial          //display variables
$monitor ("a=%b, b=%b, m=%b, rslt=%b, cout[3]=%b, cout[2]=%b,
           ovfl=%b", a, b, m, rslt, cout[3], cout[2], ovfl);

initial          //apply input vectors
begin
//addition
   #0    a = 4'b0000;   b = 4'b0001;   m = 1'b0;
   #10   a = 4'b0010;   b = 4'b0101;   m = 1'b0;
   #10   a = 4'b0110;   b = 4'b0001;   m = 1'b0;
   #10   a = 4'b0101;   b = 4'b0001;   m = 1'b0;

//subtraction
   #10   a = 4'b0111;   b = 4'b0101;   m = 1'b1;
   #10   a = 4'b0101;   b = 4'b0100;   m = 1'b1;
   #10   a = 4'b0110;   b = 4'b0011;   m = 1'b1;
   #10   a = 4'b0110;   b = 4'b0010;   m = 1'b1;

//overflow
   #10   a = 4'b0111;   b = 4'b0101;   m = 1'b0;
   #10   a = 4'b1000;   b = 4'b1011;   m = 1'b0;
   #10   a = 4'b0110;   b = 4'b1100;   m = 1'b1;
   #10   a = 4'b1000;   b = 4'b0010;   m = 1'b1;
   #10   $stop;
end                      //continued on next page
```

Figure 6.26   Test bench for the 4-bit adder/subtractor.

```
adder_subtr_struc inst1 (       //instantiate the module
   .a(a),
   .b(b),
   .m(m),
   .rslt(rslt),
   .cout(cout),
   .ovfl(ovfl)
   );
endmodule
```

Figure 6.26   (Continued)

```
a=0000, b=0001, m=0, rslt=0001, cout[3]=0, cout[2]=0, ovfl=0
a=0010, b=0101, m=0, rslt=0111, cout[3]=0, cout[2]=0, ovfl=0
a=0110, b=0001, m=0, rslt=0111, cout[3]=0, cout[2]=0, ovfl=0
a=0101, b=0001, m=0, rslt=0110, cout[3]=0, cout[2]=0, ovfl=0
a=0111, b=0101, m=1, rslt=0010, cout[3]=1, cout[2]=1, ovfl=0
a=0101, b=0100, m=1, rslt=0001, cout[3]=1, cout[2]=1, ovfl=0
a=0110, b=0011, m=1, rslt=0011, cout[3]=1, cout[2]=1, ovfl=0
a=0110, b=0010, m=1, rslt=0100, cout[3]=1, cout[2]=1, ovfl=0
a=0111, b=0101, m=0, rslt=1100, cout[3]=0, cout[2]=1, ovfl=1
a=1000, b=1011, m=0, rslt=0011, cout[3]=1, cout[2]=0, ovfl=1
a=0110, b=1100, m=1, rslt=1010, cout[3]=0, cout[2]=1, ovfl=1
a=1000, b=0010, m=1, rslt=0110, cout[3]=1, cout[2]=0, ovfl=1
```

Figure 6.27   Outputs for the 4-bit adder/subtractor.

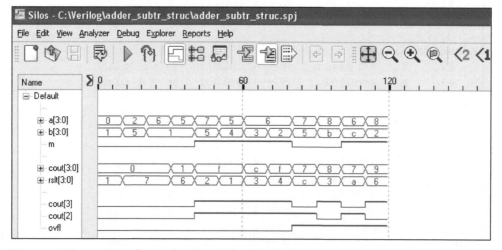

Figure 6.28   Waveforms for the 4-bit adder/subtractor.

6.3 Fixed-Point Multiplication

Multiplication of two fixed-point binary operands — multiplicand and multiplier — will be presented in this section. Each step of the multiplication process generates a partial product. The multiplicand and multiplier and both n-bit operands and the product is $2n$ bits.

6.3.1 Booth Algorithm

In the sequential add-shift technique, the multiplier must be positive. The Booth algorithm treats both positive and negative operands uniformly, including the multiplier. Chapter 5 discussed the Booth algorithm in detail. This chapter will present the design of a Booth algorithm multiplier using Verilog HDL.

The behavioral and dataflow design module for the Booth algorithm is shown in Figure 6.29. The operands are 4-bit vectors $A[3:0]$ and $B[3:0]$; the product is an 8-bit result $RSLT[7:0]$. The following internal wires are defined: $a\_ext\_pos[7:0]$, which is operand A with sign extended, and $a\_ext\_neg[7:0]$, which is the negation (2s complement) of operand A with sign extended.

The following internal registers are defined: $a\_neg[3:0]$, which is the negation of operand A, and $pp1[7:0]$, $pp2[7:0]$, $pp3[7:0]$, and $pp4[7:0]$, which are the partial products to be added together to obtain the product $RSLT[7:0]$. The example below illustrates the use of the internal registers. The right-most 4 bits of partial product 1 ($pp1$) are the negation [a_neg (2s complement)] of operand A, which is generated as a result of the -1 times operand A operation. The entire row of partial product 1 ($pp1$) corresponds to a_neg with the sign bit extended; that is, a_ext_neg.

Assume that the multiplicand is 0111 and the multiplier is 0101 as shown above. In Figure 6.29, each pair of bits is examined according to the Booth algorithm. Bits $b[1:0]$ are examined for all $2^2 = 4$ possibilities. If $b[1:0] = 00$, then partial product 1 ($pp1$) and partial product 2 ($pp2$) are 0000 0000. If $b[1:0] = 01$, then partial product 1 is the 2s complement of operand A with sign extended ($a\_ext\_neg = 1111\ 1001$) and partial product 2 is +1 times A, which is operand A aligned to the left of 1 bit position ($\{\{3\{a[3]\}\}, a[3:0], 1'b0\} = 0000\ 1110$).

If $b[1:0] = 10$, then partial product 1 is 0000 0000 and partial product 2 is the 2s complement of operand A aligned to the left 1 bit position ($\{a\_ext\_neg[6:0],1'b0\} = 1111\ 0010$). If $b[1:0] = 11$, then partial product 1 is the 2s complement of operand A ($a\_ext\_neg = 1111\ 1001$) and partial product 2 is 0000 0000. In a similar manner, multiplier bits $b[2:1]$ and bits $b[3:2]$ are examined.

The test bench is shown in Figure 6.30, which tests all pairs of bits for bits $b[1:0]$, $b[2:1]$, and $b[3:2]$. The outputs and waveforms are shown in Figure 6.31 and Figure 6.32, respectively.

```
//behavioral and dataflow for the Booth multiply algorithm
module booth3 (a, b, rslt);

input [3:0] a, b;
output [7:0] rslt;

wire [3:0] a, b;
wire [7:0] rslt;

wire [3:0] a_bar;


//define internal wires and registers
wire [7:0] a_ext_pos;
wire [7:0] a_ext_neg;
reg [3:0] a_neg;
reg [7:0] pp1, pp2, pp3, pp4;

assign a_bar = ~a;

//the following will cause synthesis of a single adder
//rather than multiple adders in the case statement

always @ (a_bar)
    a_neg = a_bar + 1;

assign a_ext_pos = {{4{a[3]}}, a};
assign a_ext_neg = {{4{a_neg[3]}}, a_neg};
//continued on next page
```

Figure 6.29   Mixed-design module for the Booth algorithm.

```
//test b[1:0] ---------------------------------------
always @ (b, a_ext_neg)
begin
   case (b[1:0])
      2'b00 :
         begin
            pp1 = 8'h00;
            pp2 = 8'h00;
         end

      2'b01 :
         begin
            pp1 = a_ext_neg;
            pp2 = {{3{a[3]}}, a[3:0], 1'b0};
         end

      2'b10 :
         begin
            pp1 = 8'h00;
            pp2 = {a_ext_neg[6:0], 1'b0};
         end

      2'b11 :
         begin
            pp1 = a_ext_neg;
            pp2 = 8'h00;
         end
   endcase
end

//test b[2:1] ---------------------------------------
always @ (b, a_ext_pos, a_ext_neg)

begin
   case (b[2:1])
      2'b00: pp3 = 8'h00;

      2'b01: pp3 = {a_ext_pos[5:0], 2'b0};

      2'b10: pp3 = {a_ext_neg[5:0], 2'b00};

      2'b11: pp3 = 8'h00;
   endcase
end

//continued on next page
```

Figure 6.29   (Continued)

```
//test b[3:2] --------------------------------------
always @ (b, a_ext_pos, a_ext_neg)

begin
   case (b[3:2])
      2'b00: pp4 = 8'h00;

      2'b01: pp4 = {a_ext_pos[4:0], 3'b000};

      2'b10: pp4 = {a_ext_neg[4:0], 3'b000};

      2'b11: pp4 = 8'h00;
   endcase
end
assign rslt = pp1 + pp2 + pp3 + pp4;
endmodule
```

Figure 6.29   (Continued)

```
//test bench for booth algorithm
module booth3_tb;

reg [3:0] a, b;
wire [7:0] rslt;

//display operands a, b, and rslt --------------------
initial
$monitor ("a = %b, b = %b, rslt = %h", a, b, rslt);

//apply input vectors
initial
begin
//test b[1:0] --------------------------------------
   #0    a = 4'b0111;
         b = 4'b0111;

   #10   a = 4'b0111;
         b = 4'b0110;

   #10   a = 4'b1110;
         b = 4'b1100;

   #10   a = 4'b1011;
         b = 4'b1001;        //continued on next page
```

Figure 6.30   Test bench for the Booth algorithm module.

```
//test b[2:1] ----------------------------------------
   #10    a = 4'b0101;
          b = 4'b1000;

   #10    a = 4'b1111;
          b = 4'b0111;

   #10    a = 4'b1011;
          b = 4'b1100;

   #10    a = 4'b0111;
          b = 4'b0011;

//test b[3:2] ----------------------------------------
   #10    a = 4'b0110;
          b = 4'b0110;

   #10    a = 4'b1001;
          b = 4'b1001;

   #10    a = 4'b1001;
          b = 4'b0010;

   #10    a = 4'b1101;
          b = 4'b1111;

   #10    $stop;
end

//instantiate the module into the test bench
booth3 inst1 (
   .a(a),
   .b(b),
   .rslt(rslt)
   );
endmodule
```

Figure 6.30   (Continued)

```
a = 0111, b = 0111, rslt = 31 | a = 1011, b = 1100, rslt = 14
a = 0111, b = 0110, rslt = 2a | a = 0111, b = 0011, rslt = 15
a = 1110, b = 1100, rslt = 08 | a = 0110, b = 0110, rslt = 24
a = 1011, b = 1001, rslt = 23 | a = 1001, b = 1001, rslt = 31
a = 0101, b = 1000, rslt = d8 | a = 1001, b = 0010, rslt = f2
a = 1111, b = 0111, rslt = f9 | a = 1101, b = 1111, rslt = 03
```

Figure 6.31   Outputs for the Booth algorithm module.

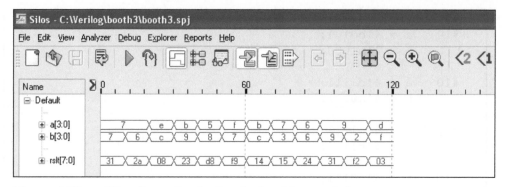

Figure 6.32   Waveforms for the Booth algorithm module.

6.3.2 Array Multiplier

Consider the array multiplier presented in Section 5.3.4 which multiplied two 3-bit operands. The multiplicand is $A[2:0]$ and the multiplier is $B[2:0]$, where $a[0]$ and $b[0]$ are the low-order bits of A and B, respectively. The two operands generate a product of $P[5:0]$.

Each bit in the multiplicand is multiplied by the low-order bit b_0 of the multiplier. This is equivalent to the AND function and generates the first of three partial products. Each bit in the multiplicand is then multiplied by bit b_1 of the multiplier. The resulting partial product is shifted one bit position to the left. The process is repeated for bit b_2 of the multiplier. The partial products are then added together to form the product. A carry-out of any column is added to the next higher-order column. An example of a general array multiply algorithm is shown in Figure 6.33 for two 3-bit operands.

			a_2	a_1	a_0
$\times$)			b_2	b_1	b_0
Partial product 1			a_2b_0	a_1b_0	a_0b_0
Partial product 2		a_2b_1	a_1b_1	a_0b_1	
Partial product 3	a_2b_2	a_1b_2	a_0b_2		
2^5	2^4	2^3	2^2	2^1	2^0

Figure 6.33   General array multiply algorithm for two 3-bit operands.

The logic diagram for the array multiplier is shown in Figure 6.34 utilizing full adders as the array elements and showing the generated partial products that correspond to those shown in Figure 6.33. The third row of full adders add the sum and carry-out of the previous columns. The structural module is shown in Figure 6.35 in which the net names and instantiation names correspond to those shown in Figure 6.34. A full adder is instantiated six times and a 2-input AND gate is instantiated nine times. The test bench is shown in Figure 6.36 using all combinations of the three multiplicand bits and the three multiplier bits. The outputs are shown in Figure 6.37 in decimal notation.

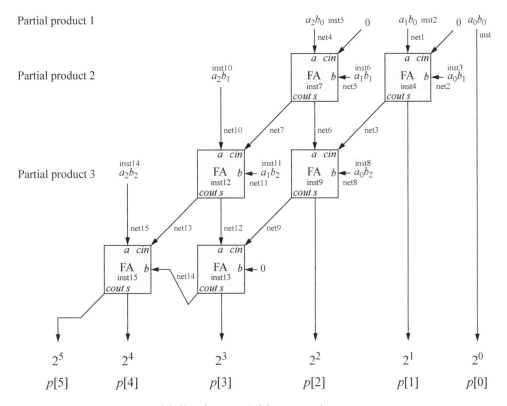

Figure 6.34   Array multiplier for two 3-bit operands.

```
//structural array multiplier
module array_mul3 (a, b, p);

input [2:0] a, b;
output [5:0] p;

wire [2:0] a, b;

//continued on next page
```

Figure 6.35   Structural module for an array multiplier for two 3-bit operands.

```
//declare internal nets
wire net1, net2, net3, net4, net5, net6, net7, net8;
wire net9, net10, net11, net12, net13, net14, net15;

wire [5:0] p; //product is six bits


//instantiate the logic for product p[0]
and2_df inst1 (
    .x1(a[0]),      //AND gate input x1 connected to a[0]
    .x2(b[0]),      //AND gate input x2 connected to b[0]
    .z1(p[0])       //AND gate output z1 connected to p[0]
    );


//instantiate the logic for product p[1]
and2_df inst2 (
    .x1(a[1]),
    .x2(b[0]),
    .z1(net1)
    );


and2_df inst3 (
    .x1(a[0]),
    .x2(b[1]),
    .z1(net2)
    );


full_adder inst4 (
    .a(net1),
    .b(net2),
    .cin(1'b0),
    .sum(p[1]),
    .cout(net3)
    );


//instantiate the logic for product p[2]
and2_df inst5 (
    .x1(a[2]),
    .x2(b[0]),
    .z1(net4)
    );

//continued on next page
```

Figure 6.35   (Continued)

```
and2_df inst6 (
   .x1(a[1]),
   .x2(b[1]),
   .z1(net5)
   );

full_adder inst7 (
   .a(net4),
   .b(net5),
   .cin(1'b0),
   .sum(net6),
   .cout(net7)
   );

and2_df inst8 (
   .x1(a[0]),
   .x2(b[2]),
   .z1(net8)
   );

full_adder inst9 (
   .a(net6),
   .b(net8),
   .cin(net3),
   .sum(p[2]),
   .cout(net9)
   );


//instantiate the logic for product p[3]
and2_df inst10 (
   .x1(a[2]),
   .x2(b[1]),
   .z1(net10)
   );

and2_df inst11 (
   .x1(a[1]),
   .x2(b[2]),
   .z1(net11)
   );

//continued on next page
```

Figure 6.35   (Continued)

```
full_adder inst12 (
   .a(net10),
   .b(net11),
   .cin(net7),
   .sum(net12),
   .cout(net13)
   );

full_adder inst13 (
   .a(net12),
   .b(1'b0),
   .cin(net9),
   .sum(p[3]),
   .cout(net14)
   );


//instantiate the logic for product p[4] and p[5]
and2_df inst14 (
   .x1(a[2]),
   .x2(b[2]),
   .z1(net15)
   );

full_adder inst15 (
   .a(net15),
   .b(net14),
   .cin(net13),
   .sum(p[4]),
   .cout(p[5])
   );
endmodule
```

Figure 6.35   (Continued)

```
//test bench for structural array multiplier
module array_mul3_tb;

reg [2:0] a, b;
wire [5:0] p;

//continued on next page
```

Figure 6.36   Test bench for the array multiplier.

```
//apply stimulus and display variables
initial
begin: apply_stimulus
   reg [6:0] invect;
      for (invect=0; invect<64; invect=invect+1)
      begin
         {a, b} = invect [6:0];
         #10 $display ("a=%d, b=%d, p=%d", a, b, p);
      end
end

//instantiate the module into the test bench
array_mul3 inst1 (
   .a(a),
   .b(b),
   .p(p)
   );

endmodule
```

Figure 6.36   (Continued)

a=0, b=0, p= 0	a=2, b=6, p=12	a=5, b=4, p=20
a=0, b=1, p= 0	a=2, b=7, p=14	a=5, b=5, p=25
a=0, b=2, p= 0	a=3, b=0, p= 0	a=5, b=6, p=30
a=0, b=3, p= 0	a=3, b=1, p= 3	a=5, b=7, p=35
a=0, b=4, p= 0	a=3, b=2, p= 6	a=6, b=0, p= 0
a=0, b=5, p= 0	a=3, b=3, p= 9	a=6, b=1, p= 6
a=0, b=6, p= 0	a=3, b=4, p=12	a=6, b=2, p=12
a=0, b=7, p= 0	a=3, b=5, p=15	a=6, b=3, p=18
a=1, b=0, p= 0	a=3, b=6, p=18	a=6, b=4, p=24
a=1, b=1, p= 1	a=3, b=7, p=21	a=6, b=5, p=30
a=1, b=2, p= 2	a=4, b=0, p= 0	a=6, b=6, p=36
a=1, b=3, p= 3	a=4, b=1, p= 4	a=6, b=7, p=42
a=1, b=4, p= 4	a=4, b=2, p= 8	a=7, b=0, p= 0
a=1, b=5, p= 5	a=4, b=3, p=12	a=7, b=1, p= 7
a=1, b=6, p= 6	a=4, b=4, p=16	a=7, b=2, p=14
a=1, b=7, p= 7	a=4, b=5, p=20	a=7, b=3, p=21
a=2, b=0, p= 0	a=4, b=6, p=24	a=7, b=4, p=28
a=2, b=1, p= 2	a=4, b=7, p=28	a=7, b=5, p=35
a=2, b=2, p= 4	a=5, b=0, p= 0	a=7, b=6, p=42
a=2, b=3, p= 6	a=5, b=1, p= 5	a=7, b=7, p=49
a=2, b=4, p= 8	a=5, b=2, p=10	
a=2, b=5, p=10	a=5, b=3, p=15	

Figure 6.37   Outputs for the array multiplier.

6.4 Decimal Addition

There are three approaches for addition and subtraction of binary-coded decimal (BCD) operands: parallel, in which all digits of both operands are processed in parallel; digit-serial, bit-parallel, in which all four bits of both operands are added in parallel, but the 4-bit digits enter the adder in serially; digit-serial, bit-serial, in which the digits and bits of both operands enter a full adder serially. Only the digit-parallel, bit-parallel method is presented.

A decimal arithmetic element has two 4-bit BCD operands as inputs and one carry-in; there is one 4-bit output that represents a valid BCD sum and one carry-out, as shown in Figure 6.38. The most commonly used code in BCD arithmetic is the 8421 code, which allows binary fixed-point arithmetic to be used. In fixed-point arithmetic, each bit is treated as a digit, whereas in decimal arithmetic a digit consists of four bits encoded in the 8421 code.

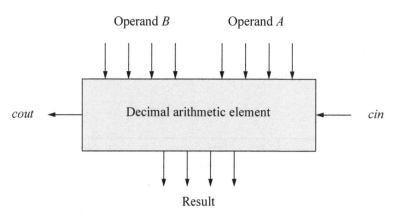

Figure 6.38   Single-stage BCD arithmetic element.

In Chapter 5, BCD addition was presented. Recall that six (0110) was added to the intermediate sum to correct an invalid BCD digit — a digit between 1010 and 1111 — and a carry-out was added to the next higher-order decade, as specified by Equation 6.9.

$$\text{Carry} = c_8 + b_8 b_4 + b_8 b_2 \qquad (6.9)$$

Before proceeding with the Verilog design, BCD addition will be briefly reviewed. Figure 6.39 shows an example of adding the decimal numbers 976 and 943 to yield a sum of 1919. In the rightmost column, the intermediate sum is a valid BCD number (1001); therefore, no adjustment is required. In the middle column, the intermediate sum is an invalid BCD number (1011); therefore, the sum is corrected by adding 6 (0110). In the leftmost column, the intermediate sum is a valid BCD number; however, there is a carry-out of the high-order decade. Whenever the unadjusted sum produces a carry-out, the intermediate sum must be corrected by adding six.

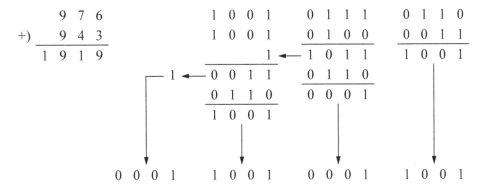

Figure 6.39   Example of BCD addition.

6.4.1 BCD Addition With Sum Correction

Figure 6.40 shows the logic diagram of a single-stage BCD arithmetic element that adds two BCD operands. If the intermediate sum inputs *sum[0]*, *sum[1]*, *sum[2]*, and *sum[3]* to *adder_2* are invalid for BCD, they are adjusted by adding six (0110) to intermediate sum. Instantiations *inst3*, *inst4*, and *inst5* represent the equation expressed in Equation 6.9.

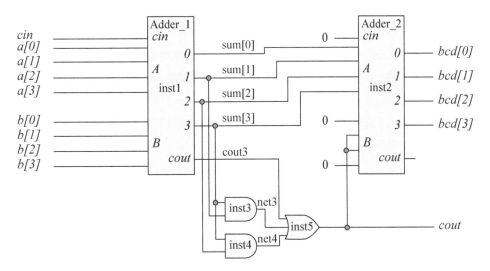

Figure 6.40   Logic diagram for a single-stage BCD adder.

The structural module is shown in Figure 6.41 in which the instantiation names and the net names correspond to those in the logic diagram. The test bench is shown in Figure 6.42, which applies several BCD numbers to the design module. The outputs and waveforms are shown in Figure 6.43 and Figure 6.44, respectively.

```
//structural bcd adder
module add_bcd (a, b, cin, bcd, cout);

//define inputs and outputs
input [3:0] a, b;
input cin;
output [3:0] bcd;
output cout;

//define internal nets
wire [3:0] sum;
wire cout3, net3, net4;

//instantiate the logic for adder_1
adder4 inst1 (
   .a(a[3:0]),
   .b(b[3:0]),
   .cin(cin),
   .sum(sum[3:0]),
   .cout(cout3)
   );

//instantiate the logic for adder_2
adder4 inst2 (
   .a(sum[3:0]),
   .b({1'b0, cout, cout, 1'b0}),
   .cin(1'b0),
   .sum(bcd[3:0])
   );

//instantiate the logic for intermediate sum adjustment
and2_df inst3 (
   .x1(sum[3]),
   .x2(sum[1]),
   .z1(net3)
   );

and2_df inst4 (
   .x1(sum[3]),
   .x2(sum[2]),
   .z1(net4)
   );                   //continued on next page
```

Figure 6.41   Structural module for the BCD single-stage adder.

```
or3_df inst5 (
   .x1(cout3),
   .x2(net3),
   .x3(net4),
   .z1(cout)
   );

endmodule
```

Figure 6.41   (Continued)

```
//test bench for structural add_bcd
module add_bcd_tb;

reg [3:0] a, b;
reg cin;
wire [3:0] bcd;
wire cout;

//display variables
initial
$monitor ("a=%b, b=%b, cin=%b, cout=%b, bcd=%b",
          a, b, cin, cout, bcd);

//apply input vectors
initial
begin
   #0    a = 4'b0011;   b = 4'b0011;   cin = 1'b0;
   #10   a = 4'b0101;   b = 4'b0110;   cin = 1'b0;
   #10   a = 4'b0111;   b = 4'b1000;   cin = 1'b0;
   #10   a = 4'b0111;   b = 4'b0111;   cin = 1'b0;
   #10   a = 4'b1000;   b = 4'b1001;   cin = 1'b0;
   #10   a = 4'b1001;   b = 4'b1001;   cin = 1'b0;
   #10   a = 4'b0101;   b = 4'b0110;   cin = 1'b1;
   #10   a = 4'b0111;   b = 4'b1000;   cin = 1'b1;
   #10   a = 4'b1001;   b = 4'b1001;   cin = 1'b1;

   #10   $stop;
end

//continued on next page
```

Figure 6.42   Test bench for the BCD single-stage adder.

```
//instantiate the module into the test bench
add_bcd inst1 (
    .a(a),
    .b(b),
    .cin(cin),
    .bcd(bcd),
    .cout(cout)
    );

endmodule
```

Figure 6.42   (Continued)

```
a=0011, b=0011, cin=0, cout=0, bcd=0110
a=0101, b=0110, cin=0, cout=1, bcd=0001
a=0111, b=1000, cin=0, cout=1, bcd=0101
a=0111, b=0111, cin=0, cout=1, bcd=0100
a=1000, b=1001, cin=0, cout=1, bcd=0111
a=1001, b=1001, cin=0, cout=1, bcd=1000
a=0101, b=0110, cin=1, cout=1, bcd=0010
a=0111, b=1000, cin=1, cout=1, bcd=0110
a=1001, b=1001, cin=1, cout=1, bcd=1001
```

Figure 6.43   Outputs for the single-stage BCD adder.

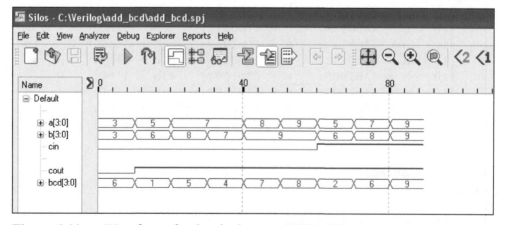

Figure 6.44   Waveforms for the single-stage BCD adder.

6.4.2 BCD Addition Using Multiplexers for Sum Correction

An alternative approach to determining whether to add six to correct an invalid decimal number is to use a multiplexer. The two operands are added in adder_1 as before; however, this intermediate sum is always added to six in adder_2, as shown in Figure 6.45, which replicates the numerical examples from Chapter 5. The sums from adder_1 and adder_2 are then applied to four 2:1 multiplexers. Selection of the adder_1 sum or the adder_2 sum is determined by ORing together the carry-out of both adders — $cout3$ and $cout8$ — to generate a select input $cout$ to the multiplexers, as shown below. The logic diagram is shown in Figure 6.46, which shows the instantiation names and net names that will be used in the structural design module.

$$\text{Multiplexer select} = \begin{cases} \text{Adder\_1 sum if } cout \text{ is } 0 \\ \\ \text{Adder\_2 sum if } cout \text{ is } 1 \end{cases}$$

Figure 6.45   Examples using a decimal adder with multiplexers.

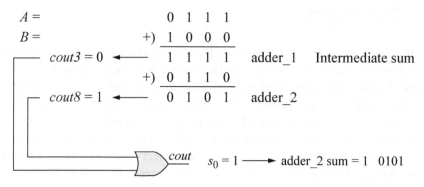

Figure 6.45   (Continued)

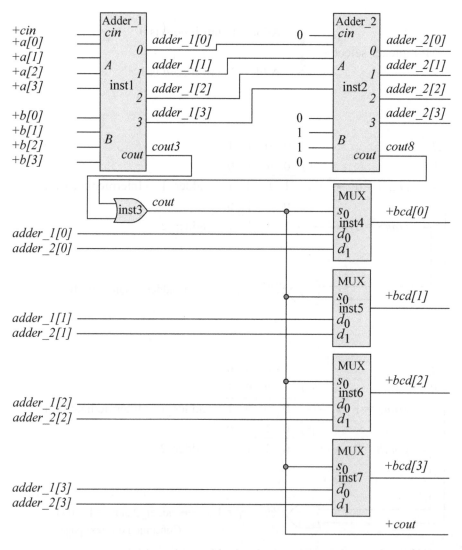

Figure 6.46   Decimal addition with multiplexers to obtain a valid decimal digit.

The behavioral module for a 4-bit adder is shown in Figure 6.47 and the dataflow module for a 2:1 multiplexer is shown in Figure 6.48, both of which will be instantiated into the structural module of the BCD adder using multiplexers of Figure 6.49. The test bench for the BCD adder is shown in Figure 6.50. The outputs and waveforms are shown in Figure 6.51 and Figure 6.52, respectively.

```verilog
//behavioral model for a 4-bit adder
module adder4 (a, b, cin, sum, cout);

input [3:0] a, b;
input cin;
output [3:0] sum;
output cout;

wire [3:0] a, b;
wire cin;
reg [3:0] sum;
reg cout;

always @ (a or b or cin)
begin
   sum  = a + b + cin;
   cout = (a[3] & b[3]) |
          ((a[3] | b[3]) & (a[2] & b[2])) |
          ((a[3] | b[3]) & (a[2] | b[2]) & (a[1] & b[1])) |
          ((a[3] | b[3]) & (a[2] | b[2]) & (a[1] | b[1]) &
             (a[0] & b[0])) |
          ((a[3] | b[3]) & (a[2] | b[2]) & (a[1] | b[1]) &
             (a[0] | b[0]) & cin);
end
endmodule
```

Figure 6.47 Four-bit adder to be instantiated into the structural module of the BCD adder, which uses multiplexers.

```verilog
//dataflow 2:1 multiplexer
module mux2_df (sel, data, z1);

input sel;
input [1:0] data;
output z1;

assign z1 = (~sel & data[0]) | (sel & data[1]);
endmodule
```

Figure 6.48 Two-to-one multiplexer to be instantiated into the structural module of the BCD adder, which uses multiplexers.

```
//structural bcd adder using multiplexers
module add_bcd_mux (a, b, cin, bcd, cout);

//define inputs and outputs
input [3:0] a, b;
input cin;
output [3:0] bcd;
output cout;

//define internal nets
wire [3:0] adder_1, adder_2;
wire cout3, cout8;

//instantiate the adder for adder_1
adder4 inst1 (
   .a(a[3:0]),
   .b(b[3:0]),
   .cin(cin),
   .sum(adder_1),
   .cout(cout3)
   );

//instantiate the adder for adder_2
adder4 inst2 (
   .a(adder_1),
   .b({1'b0, 1'b1, 1'b1, 1'b0}),
   .cin(1'b0),
   .sum(adder_2),
   .cout(cout8)
   );

//instantiate the multiplexer select logic
or2_df inst3 (
   .x1(cout8),
   .x2(cout3),
   .z1(cout)
   );

//instantiate the 2:1 multiplexers
mux2_df inst4 (
   .sel(cout),
   .data({adder_2[0], adder_1[0]}),
   .z1(bcd[0])
   );

//continued on next page
```

Figure 6.49   Structural module for the BCD adder using multiplexers to adjust the intermediate sum.

```
mux2_df inst5 (
   .sel(cout),
   .data({adder_2[1], adder_1[1]}),
   .z1(bcd[1])
   );

mux2_df inst6 (
   .sel(cout),
   .data({adder_2[2], adder_1[2]}),
   .z1(bcd[2])
   );

mux2_df inst7 (
   .sel(cout),
   .data({adder_2[3], adder_1[3]}),
   .z1(bcd[3])
   );
endmodule
```

Figure 6.49 (Continued)

```
//test bench for add_bcd_mux
module add_bcd_mux_tb;

reg [3:0] a, b;
reg cin;
wire [3:0] bcd;
wire cout;

initial      //display variables
$monitor ("a=%b, b=%b, cin=%b, cout=%b, bcd=%b",
          a, b, cin, cout, bcd);

initial      //apply input vectors
begin
   #0    a = 4'b0011;b = 4'b0011;cin = 1'b0;
   #10   a = 4'b0101;b = 4'b0110;cin = 1'b0;
   #10   a = 4'b0111;b = 4'b1000;cin = 1'b0;
   #10   a = 4'b0111;b = 4'b0111;cin = 1'b0;
   #10   a = 4'b1000;b = 4'b1001;cin = 1'b0;
   #10   a = 4'b1001;b = 4'b1001;cin = 1'b0;
   #10   a = 4'b0101;b = 4'b0110;cin = 1'b1;
   #10   a = 4'b0111;b = 4'b1000;cin = 1'b1;
   #10   a = 4'b1001;b = 4'b1001;cin = 1'b1;
   #10   $stop;
end                  //continued on next page
```

Figure 6.50 Test bench for the BCD adder using multiplexers.

```
//instantiate the module into the test bench
add_bcd_mux inst1 (
    .a(a),
    .b(b),
    .cin(cin),
    .bcd(bcd),
    .cout(cout)
    );

endmodule
```

Figure 6.50   (Continued)

```
a=0011, b=0011, cin=0, cout=0, bcd=0110
a=0101, b=0110, cin=0, cout=1, bcd=0001
a=0111, b=1000, cin=0, cout=1, bcd=0101
a=0111, b=0111, cin=0, cout=1, bcd=0100
a=1000, b=1001, cin=0, cout=1, bcd=0111
a=1001, b=1001, cin=0, cout=1, bcd=1000
a=0101, b=0110, cin=1, cout=1, bcd=0010
a=0111, b=1000, cin=1, cout=1, bcd=0110
a=1001, b=1001, cin=1, cout=1, bcd=1001
```

Figure 6.51   Outputs for the BCD adder using multiplexers.

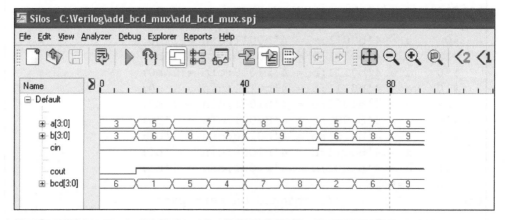

Figure 6.52   Waveforms for the BCD adder using multiplexers.

6.5 Decimal Subtraction

In fixed-point binary arithmetic, subtraction is performed by adding the rs complement of the subtrahend (B) to the minuend (A); that is, by adding the 2s complement of the subtrahend as shown below.

$$A - B = A + (B' + 1)$$

where B' is the 1s complement ($r - 1$) and 1 is added to the low-order bit position to form the 2s complement. The same rationale is used in decimal arithmetic; however, the rs complement is obtained by adding 1 to the 9s complement ($r - 1$) of the subtrahend to form the 10s complement.

Figure 6.53 shows two examples of decimal subtraction. Figure 6.53(a) produces a positive result of +34 when subtracting +42 from +76, where a carry-out of 1 indicates a positive number. The 10s complement of the subtrahend is obtained as follows using radix 10 numbers:

$$9 - 4 = 5;$$
$$9 - 2 = 7 + 1 = 8$$

where 5 and 7 are the 9s complement of 4 and 2, respectively.

Figure 6.53(b) produces a negative result of −11 when subtracting +87 from +76, where a carry-out of 0 indicates a negative number in 10s complement notation. To obtain the result in radix 10, form the 10s complement of 89, which will yield 11. This is interpreted as a negative number.

```
        76              0111      0110
  -)    42         +)   0101      1000
        34                 1  ←   1110
                   1  ←  1101      0110
                        0110      0100
                        0011
                   +    0011      0100

                        (a)

        76              0111      0110
  -)    87         +)   0001      0011
       -11  (89)   0  ←  1000      1001

                   −    0001      0001

                        (b)
```

Figure 6.53 Examples of decimal (BCD) subtraction.

Analogous to the fixed-point adder/subtractor, the decimal subtractor must also perform addition. Before presenting the organization for the BCD adder/subtractor, a 9s complementer will be designed which will be used in the adder/subtractor module together with a carry-in to form the 10s complement of the subtrahend.

A 9s complementer is required because BCD is not a self-complementing code; that is, it cannot form the $r - 1$ complement by inverting the bits. The truth table for a 9s complementer is shown in Table 6.5. A mode control input m will be used to determine whether operand $B[3:0]$ will be added to operand $A[3:0]$ or subtracted from operand $A[3:0]$ as shown in the block diagram of the 9s complementer of Figure 6.54. The function $F[3:0]$ is the 9s complement of the subtrahend $B[3:0]$.

Table 6.5 9s Complementer

Subtrahend				9s Complement			
$b[3]$	$b[2]$	$b[1]$	$b[0]$	$f[3]$	$f[2]$	$f[1]$	$f[0]$
0	0	0	0	1	0	0	1
0	0	0	1	1	0	0	0
0	0	1	0	0	1	1	1
0	0	1	1	0	1	1	0
0	1	0	0	0	1	0	1
0	1	0	1	0	1	0	0
0	1	1	0	0	0	1	1
0	1	1	1	0	0	1	0
1	0	0	0	0	0	0	1
1	0	0	1	0	0	0	0

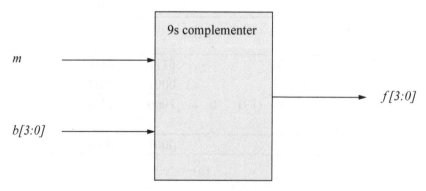

Figure 6.54   Block diagram for a 9s complementer.

The equations for the output $F[3:0]$ of the 9s complementer in conjunction with the mode control input m are shown in Equation 6.10. The equations are obtained directly from Table 6.5. Note that if the operation is add ($m = 0$), then $b[0]$ is passed through the 9s complementer unchanged; if the operation is subtract ($m = 1$), then $b[0]$ is inverted. The $b[1]$ bit is unchanged. The logic diagram for the 9s complementer is shown in Figure 6.55. The mixed-design module is shown in Figure 6.56, which will be instantiated into the structural module of the BCD adder/subtractor.

$$f[0] = b[0] \oplus m$$

$$f[1] = b[1]$$

$$f[2] = m'\, b[2] + m(b[2] \oplus b[1])$$

$$f[3] = m'\, b[3] + m\, b[3]'\, b[2]'\, b[1]' \qquad (6.10)$$

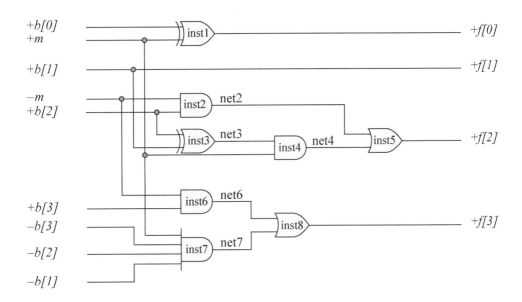

Figure 6.55   Logic diagram for a 9s complementer.

```
//mixed-design module for 9s complementer
module nines_compl (m, b, f);

input m;
input [3:0] b;
output [3:0] f;
//continued on next  page
```

Figure 6.56   Mixed-design module for a 9s complementer.

```
//define internal nets
wire net2, net3, net4, net6, net7;

//instantiate the logic gates for the 9s complementer
xor2_df inst1 (
   .x1(b[0]),
   .x2(m),
   .z1(f[0])
   );

assign f[1] = b[1];

and2_df inst2 (
   .x1(~m),
   .x2(b[2]),
   .z1(net2)
   );

xor2_df inst3 (
   .x1(b[2]),
   .x2(b[1]),
   .z1(net3)
   );

and2_df inst4 (
   .x1(net3),
   .x2(m),
   .z1(net4)
   );

or2_df inst5 (
   .x1(net2),
   .x2(net4),
   .z1(f[2])
   );

and2_df inst6 (
   .x1(~m),
   .x2(b[3]),
   .z1(net6)
   );


//continued on next page
```

Figure 6.56   (Continued)

```
and4_df inst7 (
   .x1(m),
   .x2(~b[3]),
   .x3(~b[2]),
   .x4(~b[1]),
   .z1(net7)
   );

or2_df inst8 (
   .x1(net6),
   .x2(net7),
   .z1(f[3])
   );
endmodule
```

Figure 6.56   (Continued)

The logic diagram for a BCD adder/subtractor is shown in Figure 6.57. Operand *A[3:0]* is connected directly to the *A* inputs of a fixed-point adder; operand *B[3:0]* is connected to the inputs of a 9s complementer whose outputs *F[3:0]* connect to the *B* inputs of the adder. There is also a mode control input *m* to the 9s complementer which specifies either an add operation (*m* = 0) or a subtract operation (*m* = 1). The mode control is also connected to the carry-in of the low-order adder.

If the operation is addition, then the mode control adds 0 to the uncomplemented version of operand *B[3:0]*. If the operation is subtraction, then the mode control adds a 1 to the 9s complement of operand *B[3:0]* to form the 10s complement. The outputs of the adder are a 4-bit intermediate sum *SUM[3:0]* and a carry-out labeled *cout3*. The intermediate sum is connected to a second fixed-point adder, which will add 0000 to the intermediate sum if no adjustment is required or add 0110 to the intermediate sum if adjustment is required.

The carry-out of the low-order BCD stage is specified as an auxiliary carry *aux_cy* and is defined by Equation 6.11. The auxiliary carry determines whether 0000 or 0110 is added to the intermediate sum produce a valid BCD digit, as well as providing a carry-in to the high-order stage.

$$aux\_cy = cout3 + sum[3] \ sum[1] + sum[3] \ sum[2] \qquad (6.11)$$

The same rationale applies to the high-order BCD stage for operand *A[7:4]* and operand *B[7:4]*. The 9s complementer produces outputs *F[7:4]* which are connected to the *B* inputs of the adder. The mode control input is also connected to the input of the 9s complementer. The fixed-point adder generates an intermediate sum of *SUM[7:4]* and a carry-out *cout7*. The carry-out of the BCD adder/subtractor is labeled *cout* and determines whether 0000 or 0110 is added to the intermediate sum. The equation for *cout* is shown in Equation 6.12. The BCD adder/subtractor can be extended to any size operand by simply adding more stages.

$$cout = cout7 + sum[7]\ sum[5] + sum[7]\ sum[6] \qquad (6.12)$$

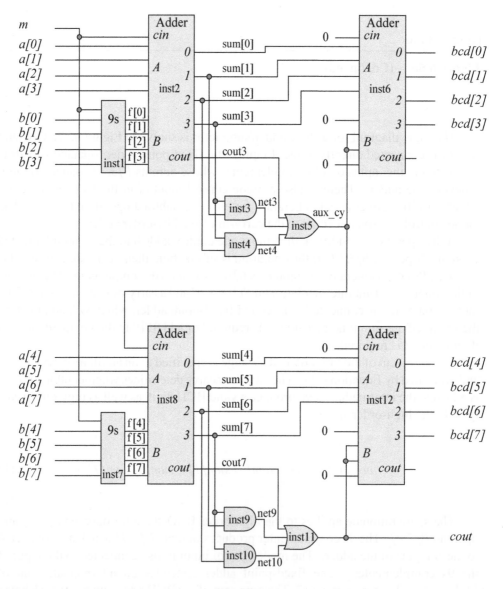

Figure 6.57   Logic diagram for a BCD adder/subtractor.

The module for the BCD adder/subtractor is shown in Figure 6.58 using structural modeling. There are two input operands *A[7:0]* and *B[7:0]* and one input mode control *m*. There are two outputs, *BCD[7:0]*, which represent a valid BCD number, and a carry-out *cout*. The test bench is shown in Figure 6.59 and contains operands for addition and subtraction, including numbers that result in negative differences in BCD. The outputs are shown in Figure 6.60 for both addition and subtraction and the waveforms are shown in Figure 6.61.

```
//structural bcd adder subtractor
module add_sub_bcd (a, b, m, bcd, cout);

input [7:0] a, b;
input m;
output [7:0] bcd;
output cout;

//define internal nets
wire [7:0] f;
wire [7:0] sum;
wire cout3, aux_cy, cout7;
wire net3, net4, net9, net10;


//instantiate the logic for the low-order stage [3:0]
//instantiate the 9s complementer
nines_compl inst1 (
    .m(m),
    .b(b[3:0]),
    .f(f[3:0])
    );


//instantiate the adder for the intermediate sum
adder4 inst2 (
    .a(a[3:0]),
    .b(f[3:0]),
    .cin(m),
    .sum(sum[3:0]),
    .cout(cout3)
    );


//instantiate the logic gates
and2_df inst3 (
    .x1(sum[3]),
    .x2(sum[1]),
    .z1(net3)
    );                  //continued on next page
```

Figure 6.58 Structural module for the BCD adder/subtractor.

```
and2_df inst4 (
   .x1(sum[2]),
   .x2(sum[3]),
   .z1(net4)
   );

or3_df inst5 (
   .x1(cout3),
   .x2(net3),
   .x3(net4),
   .z1(aux_cy)
   );

//instantiate the adder for the bcd sum [3:0]
adder4 inst6 (
   .a(sum[3:0]),
   .b({1'b0, aux_cy, aux_cy, 1'b0}),
   .cin(1'b0),
   .sum(bcd[3:0])
   );

//instantiate the logic for the high-order stage [7:4]
//instantiate the 9s complementer
nines_compl inst7 (
   .m(m),
   .b(b[7:4]),
   .f(f[7:4])
   );

//instantiate the adder for the intermediate sum
adder4 inst8 (
   .a(a[7:4]),
   .b(f[7:4]),
   .cin(aux_cy),
   .sum(sum[7:4]),
   .cout(cout7)
   );

//instantiate the logic gates
and2_df inst9 (
   .x1(sum[7]),
   .x2(sum[5]),
   .z1(net9)
   );

//continued on next page
```

Figure 6.58   (Continued)

```
and2_df inst10 (
   .x1(sum[6]),
   .x2(sum[7]),
   .z1(net10)
   );

or3_df inst11 (
   .x1(cout7),
   .x2(net9),
   .x3(net10),
   .z1(cout)
   );

//instantiate the adder for the bcd sum [7:0]
adder4 inst12 (
   .a(sum[7:4]),
   .b({1'b0, cout, cout, 1'b0}),
   .cin(1'b0),
   .sum(bcd[7:4])
   );

endmodule
```

Figure 6.58   (Continued)

```
//test bench for the bcd adder subtractor
module add_sub_bcd_tb;

reg [7:0] a, b;
reg m;

wire [7:0] bcd;
wire cout;


//display variables
initial
$monitor ("a=%b, b=%b, m=%b, cout=%b, bcd=%b",
          a, b, m, cout, bcd);


//continued on next page
```

Figure 6.59   Test bench for the BCD adder/subtractor.

```
//apply input vectors
initial
begin

//add bcd
   #0     a = 8'b1001_1001;     b = 8'b0110_0110;     m = 1'b0;
   #10    a = 8'b0010_0110;     b = 8'b0101_1001;     m = 1'b0;
   #10    a = 8'b0001_0001;     b = 8'b0011_0011;     m = 1'b0;
   #10    a = 8'b0000_1000;     b = 8'b0000_0101;     m = 1'b0;
   #10    a = 8'b0110_1000;     b = 8'b0011_0101;     m = 1'b0;
   #10    a = 8'b1000_1001;     b = 8'b0101_1001;     m = 1'b0;
   #10    a = 8'b1001_0110;     b = 8'b1001_0011;     m = 1'b0;
   #10    a = 8'b1001_1001;     b = 8'b0000_0001;     m = 1'b0;
   #10    a = 8'b1001_1001;     b = 8'b0110_0110;     m = 1'b0;

//subtract bcd
   #10    a = 8'b1001_1001;     b = 8'b0110_0110;     m = 1'b1;
   #10    a = 8'b1001_1001;     b = 8'b0110_0110;     m = 1'b1;
   #10    a = 8'b0011_0011;     b = 8'b0110_0110;     m = 1'b1;
   #10    a = 8'b0111_0110;     b = 8'b0100_0010;     m = 1'b1;
   #10    a = 8'b0111_0110;     b = 8'b1000_0111;     m = 1'b1;
   #10    a = 8'b0001_0001;     b = 8'b1001_1001;     m = 1'b1;
   #10    a = 8'b0001_1000;     b = 8'b0010_0110;     m = 1'b1;
   #10    a = 8'b0001_1000;     b = 8'b0010_1000;     m = 1'b1;
   #10    a = 8'b1001_0100;     b = 8'b0111_1000;     m = 1'b1;

   #10    $stop;

end

//instantiate the module into the test bench
add_sub_bcd inst1 (
   .a(a),
   .b(b),
   .m(m),
   .bcd(bcd),
   .cout(cout)
   );

endmodule
```

Figure 6.59   (Continued)

```
Addition
a=1001_1001, b=0110_0110, m=0, cout=1, bcd=0110_0101
a=0010_0110, b=0101_1001, m=0, cout=0, bcd=1000_0101
a=0001_0001, b=0011_0011, m=0, cout=0, bcd=0100_0100
a=0000_1000, b=0000_0101, m=0, cout=0, bcd=0001_0011
a=0110_1000, b=0011_0101, m=0, cout=1, bcd=0000_0011
a=1000_1001, b=0101_1001, m=0, cout=1, bcd=0100_1000
a=1001_0110, b=1001_0011, m=0, cout=1, bcd=1000_1001
a=1001_1001, b=0000_0001, m=0, cout=1, bcd=0000_0000
a=1001_1001, b=0110_0110, m=0, cout=1, bcd=0110_0101


Subtraction
a=1001_1001, b=0110_0110, m=1, cout=1, bcd=0011_0011
a=0011_0011, b=0110_0110, m=1, cout=0, bcd=0110_0111
a=0111_0110, b=0100_0010, m=1, cout=1, bcd=0011_0100
a=0111_0110, b=1000_0111, m=1, cout=0, bcd=1000_1001
a=0001_0001, b=1001_1001, m=1, cout=0, bcd=0001_0010
a=0001_1000, b=0010_0110, m=1, cout=0, bcd=1001_0010
a=0001_1000, b=0010_1000, m=1, cout=0, bcd=1001_0000
a=1001_0100, b=0111_1000, m=1, cout=1, bcd=0001_0110
```

Figure 6.60   Outputs for the BCD adder/subtractor.

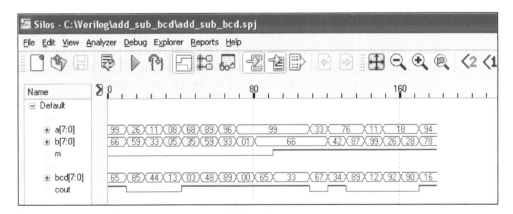

Figure 6.61   Waveforms for the BCD adder/subtractor.

6.6 Problems

6.1   Use dataflow modeling to design a full adder. Obtain the design module, the test bench module, the outputs for all combinations of the inputs, and the waveforms.

6.2   Use the full adder designed in Problem 6.1 to design a 2-bit ripple adder using structural modeling. Obtain the design module, the test bench module, the outputs, and the waveforms.

6.3   Use behavioral modeling to design a 4-function arithmetic and logic unit (ALU) for the following operations:

$$00 = \text{ADD}$$
$$01 = \text{SUB}$$
$$10 = \text{MUL}$$
$$11 = \text{DIV}$$

Use the **parameter** keyword to define the operation codes; use the **case** construct to select the operation that is to be performed. There are two 4-bit operands: $A[3:0]$, $B[3:0]$, and one 2-bit operation code $OPCODE[1:0]$. There is one 8-bit output $RSLT[7:0]$ to accommodate the $2n$-bit product. The results must be correctly represented in 2s complement representation; that is; $4 - 8 = -4$ (11111100). Obtain the design module, the test bench module, the outputs, and the waveforms.

6.4   Use structural modeling to design a 4-function ALU for two 4-bit operands $A[3:0]$ and $B[3:0]$. The operation codes are shown below, where $c[1]$ and $c[0]$ are two control inputs with $c[0]$ being the low-order bit.

$c[1]$	$c[0]$	Operation
0	0	Add
0	1	Subtract
1	0	And
1	1	Or

Draw the logic diagram, then design the following dataflow modules: a 4-bit ripple adder, a 4:1 multiplexer, which will select one of the four operations, a 2-input AND gate, a 2-input OR gate, and an exclusive-OR circuit, all of which will be instantiated into the structural module. Obtain the structural module, the test bench module that applies different input vectors for each operation, the outputs, and the waveforms.

6.5   Use behavioral modeling to design a 5-function ALU for the following five functions: add, subtract, multiply, divide, and modulus. The operands are 4-bit vectors *A[3:0]* and *B[3:0]*. Obtain the behavioral module, the test bench module which applies one input vector for each operation, the outputs, and the waveforms.

6.6   Design an 8-bit carry lookahead adder using dataflow modeling. The adder has two groups of four bits per group. The high-order group has operands *A[7:4]* and *B[7:4]* that produce a sum of *SUM[7:4]*; the low-order group has operands *A[3:0]* and *B[3:0]* that produce a sum of *SUM[3:0]*. The carry-in is *cin*; the carry-out is *cout*, which is the carry-out of bit 7. Obtain the design module, the test bench module for several different operands, the outputs in decimal notation, and the waveforms.

6.7   Use behavioral modeling to design an 8-bit adder together with shift-left and shift-right capabilities. Use the **always** statement in which the sensitivity list contains the augend *A[7:0]* and the addend *B[7:0]*. When either *A* or *B* changes value, the sum is obtained and then shifted left 4 bit positions for an equivalent multiply-by-16 operation; then the original sum is shifted right 4 bit positions for an equivalent divide-by-16 operation. Although the shift count is only 4 bit positions, the design must be able to accommodate larger shift amounts.

Obtain the design module, the test bench module that provides five different input vectors, the outputs, and the waveforms.

6.8   Use structural modeling to design an add-subtract-shift unit. There are two 4-bit operands *A[30]* and *B[3:0]*, where *A* is the augend/minuend and *B* is the addend/subtrahend. Use dataflow modeling to design a full adder, a 4:1 multiplexer, and a 2-input exclusive-OR circuit, all of which will be instantiated four times into the structural module. The 4:1 multiplexer is used to shift the sum or difference right or left 1 bit position. This will be a logical shift; that is, during a shift-left operation, a zero is entered into the low-order bit position; during a shift-right operation, a zero is entered into the high-order bit position.

Draw the logic diagram, then obtain the structural design module, the test bench module for several input vectors for addition and subtraction and for left- and right-shift operations, the outputs, and the waveforms.

There is one mode control input (*m*) to determine whether the operation is addition or subtraction; if *m* = 0, then the operation is addition; if *m* = 1, then the operation is subtraction. There are also two control inputs *c[1]* and *c[0]* defined as shown on the following page.

c[1]	c[0]	Function
0	0	$A + B$. The sum is not shifted
0	1	$A - B$. The difference is not shifted
1	0	Shift the result left 1 bit position
1	1	Shift the result right 1 bit position

6.9   Use structural modeling to design an adder and high-speed shifter. The operands are augend $A[3:0]$ and addend $B[3:0]$. The design is accomplished by instantiating a 4-bit adder and eight 4:1 multiplexers. This design represents a high-speed shifter because the shift amounts are prewired. In order to shift the operand a specified number of bits, the shift amount is simply selected and the operand is shifted the requisite number of bits — the speed is a function only of the inertial delay of the multiplexer gates.

Four multiplexers are used for shifting left and four multiplexers are used for shifting right. Shifting a binary number left 1 bit position corresponds to multiplying the number by 2. Shifting a binary number right 1 bit position corresponds to dividing the number by 2.

This is a *logical* shifter; that is, zeroes are shifted into the vacated low-order bit positions for a shift-left operation and zeroes are shifted into the vacated high-order bit positions for a shift-right operation. The carry-out of the adder is not used in this design. The following table lists the control variables for shifting the sum:

Shift Direction	Shiftcount [1]	Shiftcount [0]	Shift Amount
Shiftleft	0	0	0
	0	1	1
	1	0	2
	1	1	3
Shiftright	0	0	0
	0	1	1
	1	0	2
	1	1	3

Generate the logic diagram and obtain the structural module and the test bench module for various combinations of no shift, shift left, and shift right. Obtain the outputs and waveforms.

6.10   Shift registers are used in various arithmetic applications; for example, the sequential add-shift multiply algorithm and the shift-subtract division algorithm. Design a 4-bit shift register that operates as a parallel-in, parallel-out register and shifts the contents right 1 bit position. There is a scalar input called *fctn* that allows the register to load data (*fctn* = 1) and shift data (*fctn* = 0). Use the **parameter** and **case** keywords to define the load and shift-right operations.

The concept of *clock* will be introduced in this problem and will be covered in more detail in the chapter on sequential logic. Clock pulses are synchronization signals provided by a clock, usually an astable multivibrator, which generates a periodic series of pulses. A clock signal can be generated by the following code segment in which *clk* changes state every 10 time units:

```
//define clock
initial
begin
    clk = 1'b0;
    forever
        #10  clk = ~clk;
end
```

The keywords **posedge** and **negedge** refer to the rising and falling edge of the clock, respectively. Obtain the design module, the test bench module for four different operands, the outputs, and the waveforms.

6.11   Use structural modeling to design a 4-bit array multiplier. Use dataflow modeling to design the logic primitives and structural modeling to design a full adder, all of which will be instantiated into the multiplier module. Draw the logic diagram for the array multiplier showing all instantiations and net names. Obtain the design module and the test bench module for all combinations of the multiplicand and multiplier. Obtain the outputs using decimal notation. The multiplicand is *A[3:0]*, the multiplier is *B[3:0]*, and the product is *PROD[7:0]*.

6.12   Full adders can be used to add multiple rows in a single column; that is, multioperand addition, such as occurs in an array of partial products. These are referred to as *carry-save* adders because they save the carry propagation until all additions are complete. The carry-save adder is ideal for this type of operation because it requires very little hardware.

A carry-save adder is shown below that adds 3 bits in the same column 2^i of three operands. The sum bit represents column 2^i and the carry bit represents column 2^{i+1}.

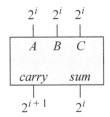

Use carry-save adders to add 4 bits in the same column. When using carry-save adders, the carry bit cannot be added to the sum bit from the same adder, as shown below.

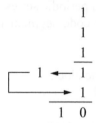

6.13   Use carry-save adders to add 7 bits from the same column. Check the design for correct operation.

7

Sequential Logic

This chapter will analyze and synthesize both synchronous and asynchronous sequential machines. Analysis is achieved by applying certain techniques to existing machines to observe their behavior. Synchronization for synchronous machines is accomplished by *clock* pulses, where a clock is a device that produces pulses at regular intervals.

Before proceeding with the analysis of synchronous sequential machines, the concept of a state will be defined. A *state* is a set of values that is measured at different locations within the machine. The values correspond to the set or reset condition of storage elements. The storage elements may be clocked flip-flops or *SR* latches. A storage element can store one bit of information, and therefore, has two stable states: 0 and 1. A machine is stable in a particular state when no input signals are changing; that is, the input variables and the clock, if applicable, are stable.

If there are p storage elements, then the state of the machine is the p-tuple of the storage element states and consists of 0s and 1s, where each bit is the value of a particular storage element cell. Thus, there are 2^p possible states, some of which may be unused. For example, if there are four storage elements, then possible states (also called state codes) are: 0101, 1001, 1100, etc.

A sequential logic circuit consists of combinational logic and storage elements. The circuit is called sequential because operations are performed in sequence. A sequential circuit can assume a finite number of internal states and can, therefore, be regarded as a *finite-state machine* (or simply a state machine) with a finite number of inputs and a finite number of outputs. In general, there are two types of sequential circuits: synchronous sequential machines (clocked) and asynchronous sequential machines (not clocked).

7.1 Analysis of Synchronous Sequential Machines

A *synchronous sequential machine* is a machine whose present outputs are a function of the present state only or the present state and present inputs. The present states are determined by the sequence of previous inputs (the input history) and the previous states. The sequence of previous inputs is the order in which the inputs occurred.

A requirement of a synchronous sequential machine is that state changes occur only when the machine is clocked, either on the positive or negative transition of the clock. Thus, input changes do not affect the present state of the machine until the occurrence of the next active clock transition. Clock pulses are synchronization signals provided by a clock, usually an astable multivibrator, which generates a periodic series of pulses. The storage elements are affected only at the positive or negative transition of the clock pulses.

7.1.1 Machine Alphabets

A sequential machine is a mathematical model of a sequential logic circuit and consists of the following three alphabets:

Input alphabet   There is a set of external inputs consisting of n binary variables

$$\{x_1, x_2, \cdots, x_n\}$$

where $x_i = 0$ or 1 for $1 \le i \le n$. The inputs generate 2^n input combinations of ordered n-tuples. Each combination of the input variables is referred to as a vector (or symbol) of the input alphabet. Thus, the input alphabet X is the set of 2^n input symbols as shown in Equation 7.1.

$$X = \{X_0, X_1, X_2, \cdots, X_{2^n-2}, X_{2^n-1}\} \tag{7.1}$$

For example, if a machine has three input variables, x_1, x_2, and x_3, then the input alphabet consists of eight symbols X_0 through X_7, as shown below.

$$X = \{000, 001, 010, \cdots, 110, 111\}$$

where

$$X_0 = x_1 x_2 x_3 = 000$$

$$X_1 = x_1 x_2 x_3 = 001$$

$$X_2 = x_1 x_2 x_3 = 010$$

$$\cdots$$

$$X_7 = x_1 x_2 x_3 = 111$$

Some vectors of the input alphabet may not be used.

State alphabet   There is a set of present internal state variables consisting of p synchronous storage elements

$$\{y_1, y_2, \cdots, y_p\}$$

where $y_i = 0$ or 1 for $1 \leq i \leq p$. The storage elements generate 2^p possible states, although the machine may not use all states.

Each combination of the storage element values is referred to as a state (or *state code*) of the state alphabet. Thus, the state alphabet Y is the set of 2^p states, as shown in Equation 7.2.

$$Y = \{Y_0, Y_1, Y_2, \cdots, Y_{2^p-2}, Y_{2^p-1}\} \tag{7.2}$$

For example, if a machine has four storage elements y_1, y_2, y_3, and y_4, then the state alphabet consists of 16 states Y_0 through Y_{15}:

$$Y = \{0000, 0001, 0010, \cdots, 1110, 1111\}$$

where

$$Y_0 = y_1 y_2 y_3 y_4 = 0000$$

$$Y_1 = y_1 y_2 y_3 y_4 = 0001$$

$$Y_2 = y_1 y_2 y_3 y_4 = 0010$$

$$\cdots$$

$$Y_{15} = y_1 y_2 y_3 y_4 = 1111$$

A 0 state indicates a reset storage element and a 1 indicates a set storage element. The state of the machine, therefore, is the p-tuple of the storage element states.

Output alphabet   There is a set of outputs consisting of m binary variables

$$\{z_1, z_2, \cdots, z_m\}$$

where $z_i = 0$ or 1 for $1 \leq i \leq m$. The outputs generate 2^m output combinations of ordered m-tuples. Each combination of the output variables is referred to as a vector (or symbol) of the output alphabet. Thus, the output alphabet Z is the set of 2^m output vectors as shown in Equation 7.3.

$$Z = \{Z_0, Z_1, Z_2, \cdots, Z_2{}^m{}_{-2}, Z_2{}^m{}_{-1}\} \tag{7.3}$$

For example, if a machine has two outputs z_1 and z_2, then the output alphabet consists of four symbols Z_0 through Z_3:

$$Z = \{00, 01, 10, 11\}$$

where

$$Z_0 = z_1 z_2 = 00$$

$$Z_1 = z_1 z_2 = 01$$

$$Z_2 = z_1 z_2 = 10$$

$$Z_3 = z_1 z_2 = 11$$

Some symbols of the output alphabet may not be used.

The present external inputs $x_1, x_2, \cdots, x_n$ and the present values of the state variables $y_1, y_2, \cdots, y_p$, which were obtained at clock (t), combine to produce the present outputs $z_1, z_2, \cdots, z_m$ and also the next state of the machine at the occurrence of the next clock $(t + 1)$.

The following notation will be used to represent the specified vectors and states:

$X_{i(t)}$ is the present input vector, where $X_{i(t)} \in X$.
$Y_{j(t)}$ is the present state, where $Y_{j(t)} \in Y$.
$Y_{k(t+1)}$ is the next state, where $Y_{k(t+1)} \in Y$.
$Z_{r(t)}$ is the present output vector, where $Z_{r(t)} \in Z$.

The *Cartesian product* of two sets is defined as follows: For any two sets S and T, the Cartesian product of S and T is written as $S \times T$ and is the set of all *ordered pairs* of S and T, where the first member of the ordered pair is an element of S and the second member is an element of T. Thus, the general classification of a synchronous sequential machine M can be defined as the 5-tuple shown in Equation 7.4.

$$M = (X, Y, Z, \delta, \lambda) \tag{7.4}$$

where

1. X is a nonempty finite set of inputs.
2. Y is a nonempty finite set of states.
3. Z is a nonempty finite set of outputs.
4. δ is the next-state function which maps the Cartesian product of $X \times Y$ into Y.
5. λ is the output function which maps the Cartesian product of $X \times Y$ into Z.

A synchronous sequential machine is deterministic and can now be defined in terms of the machine alphabets and the next-state function δ. The next state $Y_{k(t+1)}$ is uniquely determined by the present inputs $X_{i(t)}$ and the present state $Y_{j(t)}$. Thus, the next state can be expressed as shown in Equation 7.5 as a function of the δ next-state logic.

$$Y_{k(t+1)} = \delta(X_{i(t)}, Y_{j(t)}) \tag{7.5}$$

7.1.2 Storage Elements

This section will review the operating characteristics of the *SR* latch, *D* flip-flop, *JK* flip-flop, and *T* flip-flop. A latch is a level-sensitive storage element in which a change to an input signal affects the output directly without recourse to a clock input. The set (s) and reset (r) inputs may be active high or active low. The D, JK, and T flip-flops, however, are triggered on the application of a clock signal and are positive- or negative-edge-triggered devices.

SR latch   The *SR* latch is usually implemented using either NAND gates or NOR gates, as shown in Figure 7.1(a) and Figure 7.1(b), respectively. When a negative pulse (or level) is applied to the −*set* input of the NAND gate latch, the output +y_1 becomes active at a high voltage level. This high level is also connected to the input of NAND gate 2. Since the set and reset inputs cannot both be active simultaneously, the reset input is at a high level, providing a low voltage level on the output of gate 2 which is fed back to the input of gate 1. The negative feedback, therefore, provides a second set input to the latch. The original set pulse can now be removed and the latch will remain set. Concurrent set and reset inputs represent an invalid condition, since both outputs will be at the same voltage level; that is, outputs +y_1 and −y_1 will both be at the more positive voltage level — an invalid state for a bistable device with complementary outputs.

If the NAND gate latch is set, then a low voltage level on the *–reset* input will cause the output of gate 2 to change to a high level which is fed back to gate 1. Since both inputs to gate 1 are now at a high level, the $+y_1$ and $-y_1$ outputs will change to a low and high level, respectively, which is the reset state for the latch. The *characteristic table* for an *SR* latch is shown in Table 7.1, where $Y_{j(t)}$ and $Y_{k(t+1)}$ are the present state and next state of the latch, respectively. The excitation equation is shown in Equation 7.6.

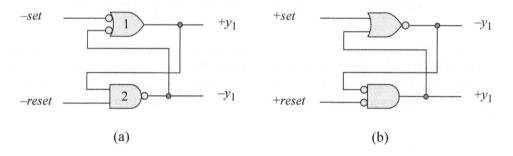

Figure 7.1   *SR* latches: (a) using NAND gates and (b) using NOR gates.

Table 7.1 *SR* Latch Characteristic Table

Data Inputs $S\ R$	Present State $Y_{j(t)}$	Next State $Y_{k(t+1)}$
0 0	0	0
0 0	1	1
0 1	0	0
0 1	1	0
1 0	0	1
1 0	1	1
1 1	0	Invalid
1 1	1	Invalid

$$Y_{k(t+1)} = S + R'\, Y_{j(t)} \qquad\qquad (7.6)$$

***D* flip-flop**   A *D* flip-flop is an edge-triggered device with one data input and one clock input. Figure 7.2 illustrates a positive-edge-triggered *D* flip-flop. The $+y_1$ output will assume the state of the *D* input at the next positive clock transition. After the occurrence of the clock's positive edge, any change to the *D* input will not affect the output until the next active clock transition. The characteristic table for a *D* flip-flop is shown in Table 7.2 and the corresponding excitation equation in Equation 7.7.

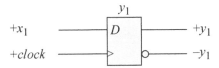

Figure 7.2   A positive-edge-triggered *D* flip-flop.

Table 7.2 *D* Flip-Flop Characteristic Table

Data Input D	Present State $Y_{j(t)}$	Next State $Y_{k(t+1)}$
0	0	0
0	1	0
1	0	1
1	1	1

$$Y_{k(t+1)} = D \qquad (7.7)$$

***JK* flip-flop**   The *JK* flip-flop is also an edge-triggered storage device. The active clock transition can be either the positive or negative edge. Figure 7.3 illustrates a negative-edge-triggered *JK* flip-flop. The functional characteristics of the *JK* data inputs are defined in Table 7.3. The characteristic table of Table 7.4 lists the next state $Y_{k(t+1)}$ for each combination of *J*, *K*, and the present state $Y_{j(t)}$ based on the functional characteristics of *J* and *K*. Table 7.5 shows an excitation table in which a particular state transition predicates a set of values for *J* and *K*. This table is especially useful in the synthesis of synchronous sequential machines.

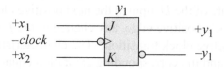

Figure 7.3   A negative-edge-triggered *JK* flip-flop.

Table 7.3 *JK* **Functional Characteristic Table**

JK	Function
0 0	No change
0 1	Reset
1 0	Set
1 1	Toggle

Table 7.4 *JK* **Flip-Flop Characteristic Table**

Data Inputs JK	Present State $Y_{j(t)}$	Next State $Y_{k(t+1)}$
0 0	0	0
0 0	1	1
0 1	0	0
0 1	1	0
1 0	0	1
1 0	1	1
1 1	0	1
1 1	1	0

Table 7.5 Excitation Table for a *JK* Flip-Flop

Present State $Y_{j(t)}$	Next State $Y_{k(t+1)}$	Data Inputs JK	
0	0	0 –	A dash (–) indicates a "don't care" condition
0	1	1 –	
1	0	– 1	
1	1	– 0	

The excitation equation for a *JK* flip-flop is derived from Table 7.5 and is shown in Equation 7.8.

$$Y_{k(t+1)} = Y_{j(t)}{}' J + Y_{j(t)} K' \tag{7.8}$$

***T* flip-flop** The toggle (T) flip-flop is shown in Figure 7.4 as a positive-edge-triggered device. When the T input is at a logic 1 level, the flip-flop will toggle (change state) at the next active clock transition. The characteristic table is shown in Table 7.6 and the excitation table in Table 7.7. The corresponding excitation equation is shown in Equation 7.9.

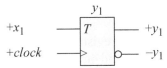

Figure 7.4   A T flip-flop.

Table 7.6 *T* Flip-Flop Characteristic Table		
Data Input T	Present State $Y_{j(t)}$	Next State $Y_{k(t+1)}$
0	0	0
0	1	1
1	0	1
1	1	0

Table 7.7 *T* Flip-Flop Excitation Table		
Present State $Y_{j(t)}$	Next State $Y_{k(t+1)}$	Data Input T
0	0	0
0	1	1
1	0	1
1	1	0

$$Y_{k(t+1)} = Y_{j(t)}' T + Y_{j(t)} T' \qquad (7.9)$$

7.1.3 Classes of Sequential Machines

This section will present classes of sequential machines. With the exception of the last class of machines (asynchronous sequential machines), all other machines are synchronous, where changes to states and outputs occur only in synchronization with clock pulses. The storage elements in synchronous sequential machines are bistable multivibrators such as, D flip-flops and JK flip-flops. The storage elements in asynchronous sequential machines are SR latches. Some of the machines in the following sections contain no data inputs; however, all of the machines produce outputs. Each machine consists of at least one of the following three logical units:

1. Input combinational logic
2. Synchronous storage elements
3. Output combinational logic

If input combinational logic is used, then the δ next-state function maps the present state or the present state and the present inputs into the state alphabet. That is, the δ mapping generates the input equations for the storage elements. If output combinational logic is used, then the λ function maps the present state or the present state and the present inputs into the output alphabet. Thus, the λ mapping generates outputs from the machine.

Registers   In their simplest form, registers contain only storage elements. They may, however, consist of input logic and output logic as shown in Figure 7.5. If input logic is specified in the design, then the next-state function δ is a function of the input alphabet X, and maps X into Y, as shown in Equation 7.10. Also, δ is a function of the present input vector $X_{i(t)}$, and maps $X_{i(t)}$ into the next state $Y_{k(t+1)}$, as shown in Equation 7.11. Equation 7.12 defines the next state.

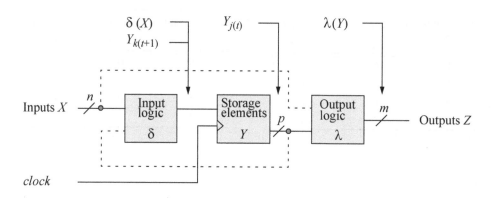

Figure 7.5   Register general block diagram.

$$\delta(X) : X \to Y \tag{7.10}$$
$$\delta(X_{i(t)}) : X_{i(t)} \to Y_{k(t+1)} \tag{7.11}$$
$$Y_{k(t+1)} = \delta(X_{i(t)}) \tag{7.12}$$

When the output function λ is implemented in a register, then λ is a function of the state alphabet Y and maps Y into the output alphabet Z, as shown in Equation 7.13. The present output vector $Z_{r(t)}$ is shown in Equation 7.14.

$$\lambda(Y) : Y \to Z \tag{7.13}$$

$$Z_{r(t)} = \lambda(Y_{j(t)}) \tag{7.14}$$

Registers can be Moore- or Mealy-type machines depending on the configuration of the output logic. Moore and Mealy machines are presented in Section 7.2.4 and Section 7.2.5, respectively. Registers are used to store binary information in a digital

system. The information can consist of a single bit or of several bits which define instructions or operands. An n-bit register consists of n storage elements and can store
n bits of binary information, one bit in each storage element.

Data can be loaded into a register either in parallel or in serial format. Parallel
loading is faster, because all storage elements receive new information during one
clock pulse. During a serial load operation, each storage element receives new data
from the element to its immediate left or right, depending on the direction of loading
(shifting). The first storage element receives its data from an external source.

A typical 8-bit parallel-in, parallel-out register is shown in Figure 7.6, which consists of the next-state function δ, storage elements $Y[7:0] = y_7, y_6, y_5, \ldots, y_0$, and the
output function λ. Both the δ and λ mappings use combinational logic consisting of
eight AND gates for each of the two functions. The clock pulse sets the register to the
contents of the input vector $X[7:0] = x_7, x_6, x_5, \cdots, x_0$.

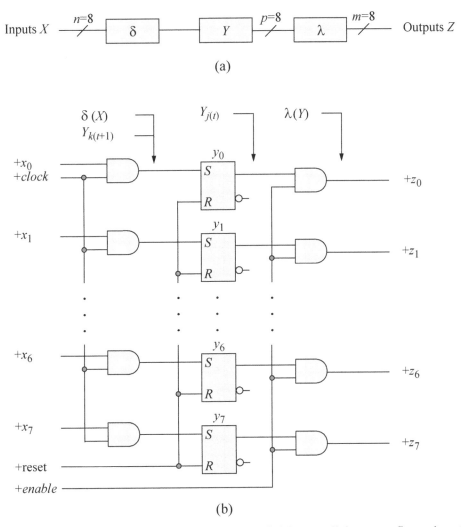

Figure 7.6   Register implemented in a parallel-in, parallel-out configuration: (a)
block diagram and (b) logic diagram using latches for the storage elements.

Because the storage elements are latches, the input data must not change while the *clock* is positive. An *enable* signal is one of the inputs to the λ output function combinational logic and transforms the state of the register to the output vector $Z[7:0] = z_7, z_6, z_5, \cdots, z_0$.

Figure 7.7 illustrates an 8-bit serial-in, serial-out shift register implemented with D flip-flops. There is no logic shown for either the δ next-state function or the λ output function, although the input logic is implied by the D input of the flip-flop. The shift register is reset initially, then information is shifted into stage_0 one bit per clock pulse. The positive transition of the clock signal causes the following shift sequence to occur:

$$x_0 \rightarrow y_0$$
$$y_j \rightarrow y_{j+1}$$
$$y_7 \text{ is lost}$$

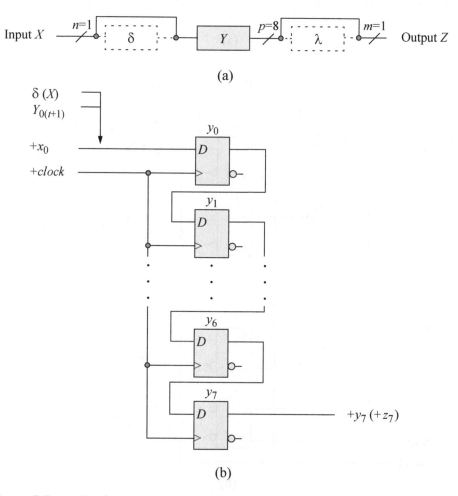

(a)

(b)

Figure 7.7   Register implemented in a serial-in, serial-out configuration: (a) block diagram and (b) logic diagram using D flip-flops for the storage elements.

Counters   This section will present some general comments regarding counters. The details of synthesis will be covered in Section 7.2.3. Counters are one of the simplest types of sequential machines, requiring only one input in most cases. The single input is a clock pulse. Although most counters can be categorized as a type of Moore machine, counters are of sufficient importance to warrant a separate classification. A *counter* is constructed from one or more flip-flops that change state in a prescribed sequence upon the application of a series of clock pulses. The sequence of states in a counter may generate a binary count, a binary-coded decimal (BCD) count, or any other counting sequence. The counting sequence does not have to be sequential.

Counters are used for counting the number of occurrences of an event and for general timing sequences. A block diagram of a synchronous counter is shown in Figure 7.8. The diagram depicts a typical counter consisting of combinational input logic for the δ next-state function, storage elements, and combinational output logic for the λ output function. Input logic is required when an initial count must be loaded into the counter. The input logic then differentiates between a clock pulse that is used for loading and a clock pulse that is used for counting. Not all counters are implemented with input and output logic, however. Some counters contain only storage elements that are connected in cascade.

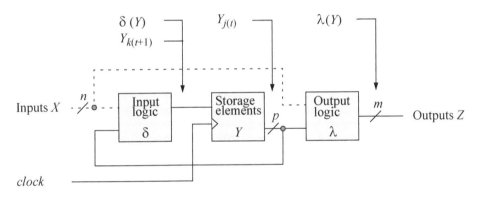

Figure 7.8   Counter block diagram.

Counters can be designed as count-up counters, in which the counting sequence increases numerically, or as count-down counters, in which the counting sequence decreases numerically. A counter may also be designed as both a count-up counter and a count-down counter, the mode of operation being controlled by a separate input.

Figure 7.9 illustrates a synchronous modulo-8 count-up binary counter. Synchronous counters are faster than asynchronous counters because the clock pulse is transmitted to all stages simultaneously. The counter of Figure 7.9 is implemented with JK flip-flops as the storage elements, where each flip-flop is wired in toggle mode; that is, $JK = 11$. The clock inputs to the storage elements are negative edge-triggered inputs; thus, the counter is incremented on each negative transition of the clock pulse, as shown in Figure 7.9(c). The counter is designed as a modulo-8 counter; therefore, after being reset initially, the counting sequence is $000, 001, 010, \ldots 110, 111, 000, \ldots$.

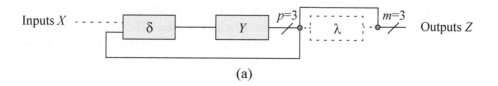

(a)

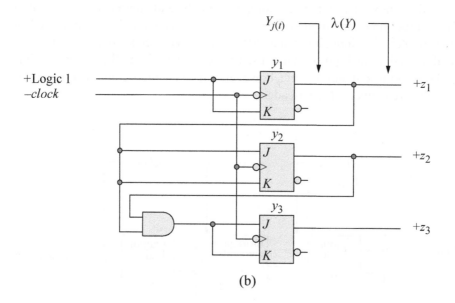

(b)

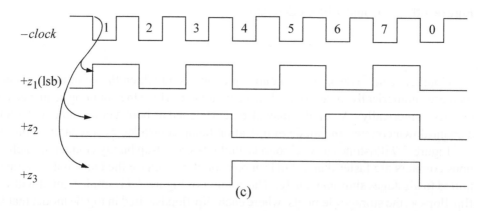

(c)

Figure 7.9   Counter implemented as a modulo-8 count-up binary counter using *JK* flip-flops: (a) block diagram; (b) logic diagram; and (c) timing diagram.

Moore machines   Moore machines are synchronous sequential machines in which the output function λ produces an output vector Z_r which is determined by the present state only, and is not a function of the present inputs. The general configuration of a Moore machine is shown in Figure 7.10. The next-state function δ is an $(n + p)$-input, p-output switching function. The output function λ is a p-input, m-output switching function. If a Moore machine has no data input, then it is referred to as an *autonomous* machine. Autonomous circuits are independent of the inputs. The clock signal is not considered as a data input. An autonomous Moore machine is an important class of synchronous sequential machines, the most common application being a counter, as discussed previously. A Moore machine may be synchronous or asynchronous; however, this section pertains to synchronous organizations only.

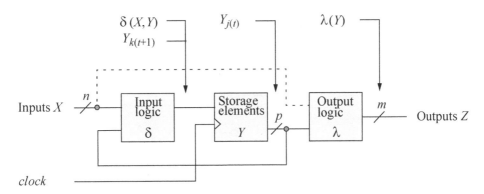

Figure 7.10   Moore synchronous sequential machine in which the outputs are a function of the present state only.

A Moore machine is a 5-tuple and can be defined as shown in Equation 7.15,

$$M = (X, Y, Z, \delta, \lambda) \tag{7.15}$$

where

1. X is a nonempty finite set of inputs such that,
    $$X = \{X_0, X_1, X_2, \cdots, X_{2^n-2}, X_{2^n-1}\}$$

2. Y is a nonempty finite set of states such that,
    $$Y = \{Y_0, Y_1, Y_2, \cdots, Y_{2^p-2}, Y_{2^p-1}\}$$

3. Z is a nonempty finite set of outputs such that,
    $$Z = \{Z_0, Z_1, Z_2, \cdots, Z_{2^m-2}, Z_{2^m-1}\}$$

4. $\delta(X, Y) : X \times Y \rightarrow Y$

5. $\lambda(Y) : Y \rightarrow Z$

A simple Moore machine is shown in Figure 7.11 in which the input alphabet is $X = \{X_0, X_1\}$, the state alphabet is $Y = \{Y_0, Y_1, Y_2, Y_3\}$, and the output alphabet is $Z = \{Z_0, Z_1\}$.

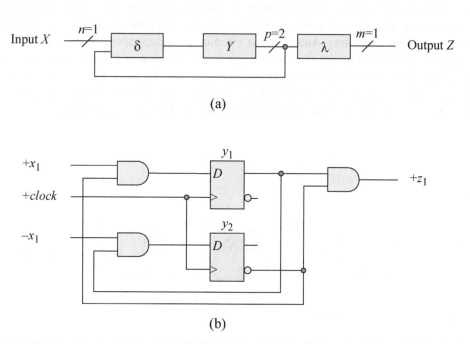

(a)

(b)

Figure 7.11   Moore machine: (a) block diagram and (b) logic diagram.

Mealy machines   Mealy machines are synchronous sequential machines in which the output function λ produces an output vector $Z_{r(t)}$ which is determined by both the present input vector $X_{i(t)}$ and the present state of the machine $Y_{j(t)}$. The general configuration of a Mealy machine is shown in Figure 7.12.

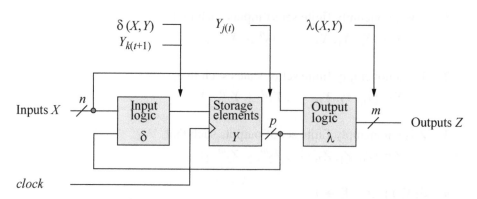

Figure 7.12   Mealy machine in which the outputs are a function of both the present state and the present inputs.

The next-state function δ is an $(n + p)$-input, p-output switching function. The output function λ is an $(n + p)$-input, m-output switching function. A Mealy machine is not an autonomous machine because the outputs are a function of the present state and the input signals. A Mealy machine may be synchronous or asynchronous; however, this section pertains to synchronous organizations only.

A Mealy machine is a 5-tuple and can be formally defined as shown in Equation 7.16,

$$M = (X, Y, Z, \delta, \lambda) \tag{7.16}$$

where

1. X is a nonempty finite set of inputs such that,
   $$X = \{X_0, X_1, X_2, \cdots, X_{2^n-2}, X_{2^n-1}\}$$

2. Y is a nonempty finite set of states such that,
   $$Y = \{Y_0, Y_1, Y_2, \cdots, Y_{2^p-2}, Y_{2^p-1}\}$$

3. Z is a nonempty finite set of outputs such that,
   $$Z = \{Z_0, Z_1, Z_2, \cdots, Z_{2^m-2}, Z_{2^m-1}\}$$

4. $\delta(X, Y) : X \times Y \to Y$

5. $\lambda(X, Y) : Y \to Z$

A simple Mealy machine is shown in Figure 7.13, in which the input alphabet is $X = \{X_0, X_1\}$, the state aphabet is $Y = \{Y_0, Y_1, Y_2, Y_3\}$, and the output alphabet is $Z = \{Z_0, Z_1\}$.

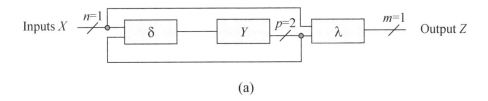

(a)

Figure 7.13   Mealy machine: (a) block diagram and (b) logic diagram.

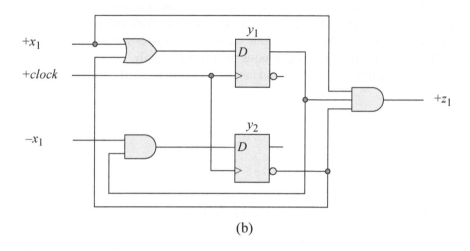

(b)

Figure 7.13   (Continued)

7.1.4 Methods of Analysis

Analysis is the methodical investigation of a problem and the decomposition of the problem into smaller related units for further detailed study. The problem, in this case, is a synchronous sequential machine which will be studied using various analytical techniques. These techniques include a next-state table, a present-state map, next-state maps, input maps and their associated input equations, output maps and equations, a timing diagram, and a state diagram.

Understanding a synchronous sequential machine through analysis is ideal preparation for later synthesizing (or designing) sequential machines. Different machines will be presented in this section and, as each machine is analyzed, the various units of the machine will be identified with the corresponding equations. First, the techniques, or methods, that are used in the analysis procedure will be defined.

Next-state table   The next-state table is a convenient method of describing the operation of a machine in tabular form. The table lists all possible present states and input values, together with the next state and present output. Table 7.8 shows a typical next-state table for a Moore machine with two D flip-flops. All combinations of two variables are listed under the present-state heading. The machine is assumed to be reset initially, which is represented by the first two rows, where $y_1 y_2 = 00$. In this example, each pair of rows in the table corresponds to a state in the machine. For example, the first and second rows represent state $y_1 y_2 = 00$. The state can also be expressed as a state name, such as "a."

Table 7.8 Typical Next-State Table for a Moore Synchronous Sequential Machine Using D Flip-Flops

State Name	Present State y_1y_2	Input x_1	Flip-Flop Inputs $Dy_1\ Dy_2$	Next State y_1y_2	Output z_1
a	0 0	0	0 0	0 0	0
	0 0	1	0 1	0 1	0
b	0 1	0	0 1	0 1	0
	0 1	1	1 0	1 0	0
c	1 0	0	1 0	1 0	1
	1 0	1	1 1	1 1	1
d	1 1	0	1 1	1 1	0
	1 1	1	0 0	0 0	0

The entries in Table 7.8 denote the state transitions and output that correspond to a given sequence of inputs. In the first row, the present state is a ($y_1y_2 = 00$). If $x_1 = 0$, the machine remains in state a and the present output $z_1 = 0$. If, however, $x_1 = 1$ in state a, then the machine moves to state b ($y_1y_2 = 01$) at the next assertion of the clock and state b becomes the new present state. No indication is given in the next-state table as to the active assertion of the clock; the machine may be clocked on either the positive or negative clock transition.

Since D flip-flops are used in the synchronous sequential machine of Table 7.8, the next state is identical to the values of Dy_1 and Dy_2 after the active clock transition. Output z_1 is active when the present state is $y_1y_2 = 10$, regardless of the value of x_1. Since the output is a function of the present state only, the next-state table represents a Moore machine.

Next-state map   The next-state map is simply the next-state table represented in Karnaugh map form as shown in Figure 7.14 for the machine of Table 7.8. The information that is specified in the next-state map can also be obtained from the logic diagram, if one is given. Since there are two flip-flops in the implementation of this machine, there are two next-state maps — one for each flip-flop. The map contains eight squares to accommodate the two flip-flops and input x_1.

For a D flip-flop, the next state corresponds to the present value of the D input. Referring to Table 7.8 and Figure 7.14, if the present state is $y_1y_2 = 00$ and $x_1 = 0$, then $Dy_1 = 0$ and the next state for y_1 will be 0. For a present state of $y_1y_2 = 00$ and $x_1 = 1$, then $Dy_1 = 0$ and the next state for $y_1 = 0$. Similarly, in state c, where $y_1y_2 = 10$, $Dy_1 = 1$ regardless of the value of x_1; therefore, the next state for y_1 is 1. Using this procedure, the information contained in the next-state table is transferred to the next-state maps.

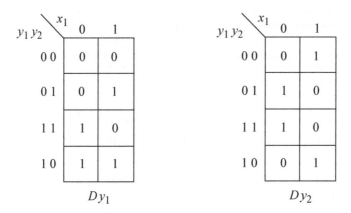

Figure 7.14   Next-state maps for the Moore machine of Table 7.8. These are the same as the input maps when D flip-flops are used.

Input map   The input map represents the δ next-state function from which equations are generated for the data input logic of the flip-flop. Because D flip-flops are used in the implementation of the machine shown in Table 7.8, the next-state maps also specify the input maps; thus, the two types of maps are identical. In Figure 7.14, the input maps for y_1 and y_2 yield the input equations as shown in Equation 7.17, from which the δ next-state logic can be implemented directly.

$$Dy_1 = x_1'y_1 + y_1y_2' + x_1y_1'y_2$$
$$Dy_2 = x_1'y_2 + x_1y_2' \tag{7.17}$$

Using the same next-state table as Table 7.8, but replacing the D flip-flops with JK flip-flops yields the next-state table of Table 7.9. The next-state maps are identical to those shown in Figure 7.14. However, the input maps change due to the characteristics of a JK flip-flop, which are reproduced below.

State Transition From To		Values of J K	
0	→ 0	0	–
0	→ 1	1	–
1	→ 0	–	1
1	→ 1	–	0

Table 7.9 Typical Next-State Table for a Moore Machine Using *JK* Flip-Flops

State Name	Present State y_1y_2	Input x_1	Flip-Flop Inputs $Jy_1\ Ky_1$	$Jy_2\ Ky_2$	Next State y_1y_2	Output z_1
a	0 0	0	0 –	0 –	0 0	0
	0 0	1	0 –	1 –	0 1	0
b	0 1	0	0 –	– 0	0 1	0
	0 1	1	1 –	– 1	1 0	0
c	1 0	0	– 0	0 –	1 0	1
	1 0	1	– 0	1 –	1 1	1
d	1 1	0	– 0	– 0	1 1	0
	1 1	1	– 1	– 1	0 0	0

The input maps for the machine of Table 7.9 are shown in Figure 7.15. There are two maps for each flip-flop, one for the J input and one for the K input. The maps are constructed directly from Table 7.9. Figure 7.15 also specifies the JK input equations, which define the logic for the δ next-state function.

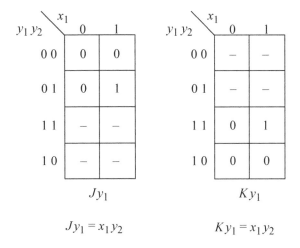

$$Jy_1 = x_1 y_2 \qquad\qquad Ky_1 = x_1 y_2$$

(Continued on next page)

Figure 7.15 Input maps for the Moore machine of Table 7.9.

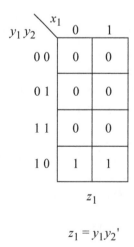

$y_1 y_2$ x_1	0	1
0 0	0	1
0 1	–	–
1 1	–	–
1 0	0	1

$J y_2$

$J y_2 = x_1$

$y_1 y_2$ x_1	0	1
0 0	–	–
0 1	0	1
1 1	0	1
1 0	–	–

$K y_2$

$K y_2 = x_1$

Figure 7.15   (Continued)

Output map   The output map represents the λ output function from which equations are generated for the output logic of the machine. The output map for the machines of Table 7.8 and Table 7.9 is shown in Figure 7.16. Output z_1 is asserted in state $y_1 y_2 = 10$ regardless of the value of x_1, which corresponds to the definition of a Moore machine.

$y_1 y_2$ x_1	0	1
0 0	0	0
0 1	0	0
1 1	0	0
1 0	1	1

z_1

$z_1 = y_1 y_2'$

Figure 7.16   Output map for the Moore machine of Table 7.8 and Table 7.9.

Timing diagram   Another useful tool for analyzing synchronous sequential machines is a timing diagram (or waveforms) which illustrates the voltage levels of the inputs, storage elements, and outputs as the machine progresses through a sequence of states. Using the Moore machine of Table 7.8 and assuming an initial reset state of $y_1 y_2 = 00$, the timing diagram for this machine is shown in Figure 7.17 for an arbitrary input sequence of $x_1 = 1101$.

The D flip-flops are clocked on the positive clock transition. To assure that the flip-flops do not become *metastable*, any changes to input x_1 will occur on the negative clock transition. This guarantees that the D inputs will be stable before the next positive clock transition, thus meeting the setup requirements for the flip-flop. Metastability is a condition of instability on the output of a flip-flop caused by a change to the data input at or near the active clock transition.

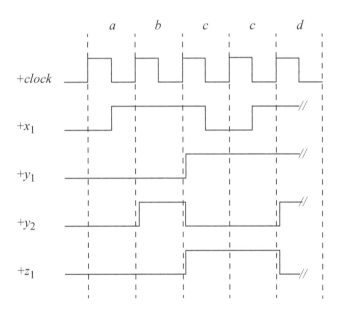

Figure 7.17   Timing diagram for the Moore machine of Table 7.8 for an input sequence of $x_1 = 1101$.

The machine begins in state a, where $y_1 y_2 = 00$. At the positive clock transition at the end of state a, both flip-flops are clocked and the following events occur:

$y_1 = 0$, because all the terms in Equation 7.17 for $D y_1$ were zero
$y_2 = 1$, because the term $x_1 y_2' = 1$ in Equation 7.17 for $D y_2$

Notice the small delay in the assertion of y_2. This is the propagation delay of the internal logic of flip-flop associated with the assertion of y_2. At the time when the positive transition of *clk* occurred at the end of state a, flip-flop y_2 was deasserted and x_1 was asserted; therefore, the term $x_1 y_2'$ was true. The machine then enters state b, where $y_1 y_2 = 01$.

In state b, x_1 remains asserted and the following events occur at the positive clock transition at the end of state b:

$y_1 = 1$, because the term $x_1 y_1' y_2 = 1$ in Equation 7.17 for Dy_1
$y_2 = 0$, because all the terms in Equation 7.17 for Dy_2 were zero

The machine then enters state c the first time where $y_1 y_2 = 10$. Output z_1 is asserted in state c regardless of the value of x_1. As before, the flip-flops are clocked on the positive clock transition and the flip-flop outputs become stable after the appropriate propagation delay of the devices. Input x_1 becomes deasserted at the negative clock transition of the first state c; flip-flop y_2 is already deasserted. Therefore, when the positive clock transition occurs at the end of the first state c, the following conditions exist:

$y_1 = 1$, because the term $x_1' y_1 = 1$ in Equation 7.17 for Dy_1
$y_2 = 0$, because all the terms in Equation 7.17 for Dy_2 were zero

The machine then enters state c the second time and x_1 becomes asserted. At the positive clock transition at the end of the second state c, the machine enters state d, where $y_1 y_2 = 11$, because the following events occur:

$y_1 = 1$, because the term $y_1 y_2' = 1$ in Equation 7.17 for Dy_1
$y_2 = 1$, because the term $x_1 y_2' = 1$ in Equation 7.17 for Dy_2

From Table 7.8, it can be seen that the next state will be either state d if $x_1 = 0$ or state a if $x_1 = 1$. When analyzing a synchronous sequential machine, a timing diagram shows more detail than any other analytical method — the state times are precisely defined and the propagation delays are clearly illustrated. Since the clock is an astable multivibrator, the clock has no stable level. The clock signal in Figure 7.17 is specified as $+clk$, where the plus sign indicates that the positive clock transition is used to clock the flip-flops which are positive-edge-triggered devices. All other signals are active high ($+$).

State diagram   A *state diagram* is a directed graph which is used in conjunction with the state table. The state diagram portrays the same information as the state table, but presents a graphical representation in which the state transitions are more easily followed. The state diagrams that are used in this book are similar to flow chart diagrams in which the transition sequences and thus, the operational characteristics of the machine, are clearly delineated. Two symbols are used: a state symbol and an output symbol.

The *state symbol* is designated by a circle as shown in Figure 7.18. These nodes (or vertices) correspond to the state of the machine; the state name, such as state a, is placed inside the circle. The connecting directed lines between states correspond to the allowable state transitions. There are one or more entry paths and one or more exit paths as indicated by the arrows, unless the vertex is a *terminal state*, in which case there is no exit.

The flip-flop names are positioned alongside the state symbol. In Figure 7.18, the machine is designed using three flip-flops which are designated as $y_1 y_2 y_3$, where y_3 is the low-order flip-flop. Directly beneath the flip-flop names, the *state code* is specified. The state code represents the state of the individual flip-flops. In Figure 7.18, the state code is 101, which corresponds to $y_1 y_2' y_3$. If an input causes a transition from state a to another state, this is indicated by placing the name of the input variable adjacent to the exit arrow as shown in Figure 7.17 for x_1 and x_1'.

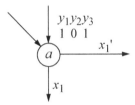

Figure 7.18   State diagram state symbol indicating state a.

The *output symbol* is represented by a rectangle and is placed immediately following the state symbol, as shown in Figure 7.19(a) for a Moore machine, or placed immediately after an input variable that causes the output to become active, as shown in Figure 7.19(b) for a Mealy machine. Figure 7.19(a) specifies a Moore machine in which output z_1 is a function of the present state only; that is, state b, where $y_1 y_2 = 01$. Figure 7.19(b) indicates a Mealy machine in which output z_1 is a function of both the present state b ($y_1 y_2 = 01$) and input x_1.

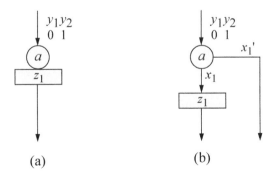

(a)   (b)

Figure 7.19   State diagram output symbol indicating output z_1: (a) Moore machine and (b) Mealy machine.

For a Moore machine, the outputs can be asserted for segments of the clock period rather than for the entire clock period only. This is illustrated in Figure 7.20 where the positive clock transitions define the clock cycles, and hence, the state times. Two clock cycles are shown, one for the present state $Y_{j(t)}$ and one for the next state $Y_{k(t+1)}$.

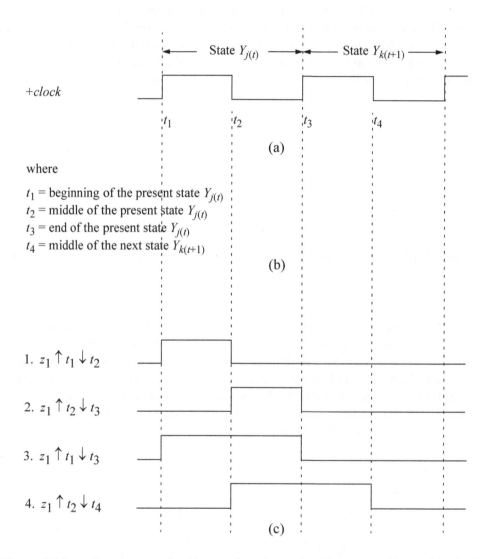

where

t_1 = beginning of the present state $Y_{j(t)}$
t_2 = middle of the present state $Y_{j(t)}$
t_3 = end of the present state $Y_{j(t)}$
t_4 = middle of the next state $Y_{k(t+1)}$

1. $z_1 \uparrow t_1 \downarrow t_2$

2. $z_1 \uparrow t_2 \downarrow t_3$

3. $z_1 \uparrow t_1 \downarrow t_3$

4. $z_1 \uparrow t_2 \downarrow t_4$

Figure 7.20   Output assertion/deassertion times for Moore machines: (a) clock pulses; (b) definition of assertion/deassertion times; and (c) assertion/deassertion statements with corresponding asserted outputs.

The leading edge of the clock pulse, which defines the beginning of the present state, is labeled t_1. The leading edge may be a positive or negative clock transition and is used for clocking positive- or negative-edge-triggered devices, respectively. All assertion/deassertion times are referenced to the present state $Y_{j(t)}$. Time t_2 occurs at the middle of the present state; time t_3 occurs at the end of the present state; and time t_4 occurs at the midpoint of the next state $Y_{k(t+1)}$. The assertion of an output is indicated by an up-arrow ($\uparrow$); deassertion is indicated by a down-arrow ($\downarrow$). The output assertion/deassertion times for a Mealy machine cannot be uniquely specified as for a Moore machine, because the outputs are contingent not only upon a specific state but also upon the input variables, whose assertion times may not be known.

Asserting the output signals at various times and for different durations, as shown in Figure 7.20(c), provides more flexibility in the λ output logic. Waveforms 2 and 4 are especially useful in avoiding glitches, because the assertion of the outputs is delayed from the active clock transition where flip-flops change state — not all flip-flops may change state simultaneously.

7.1.5 Analysis Examples

Five synchronous sequential machines will be analyzed in this section. The first is a Moore implementation with D flip-flops; the second is a Moore machine implemented with JK flip-flops; the third is a Moore machine implemented with D flip-flops and linear-select multiplexers; the fourth is a Mealy machine implemented with one JK flip-flop; the fifth is a Mealy-Moore machine implemented with two JK flip-flops.

Example 7.1   The first synchronous sequential machine to be analyzed is the Moore machine of Figure 7.21 consisting of a single input x_1, two D flip-flops y_1 and y_2, and a single output z_1. Since there is only one input, the input alphabet is quite simple, containing only one scalar input: $x_1 = 0$ or $x_1 = 1$. Two storage elements y_1 and y_2 assume four unique values to specify four states $y_1 y_2 = 00, 01, 10$, and 11.

State names are assigned to the states as shown in the next-state table of Table 7.10 where state a corresponds to $y_1 y_2 = 00$. The state names do not necessarily have to be in ascending sequence when tabulated; they are, however, arranged in an ascending systematic sequence when entered in the state diagram, as will be shown later in this example.

The output alphabet consists of one scalar output: $z_1 = 0$ or $z_1 = 1$. Output z_1 is asserted during the last half of the clock pulse when $-clock$ becomes a high logic level and $y_1 y_2 = 11$. This indicates a Moore machine since the output is a function of the present state only and is not a function of the input. The machine will be analyzed by obtaining the next-state table, the input maps, the output map, the timing diagram, and the state diagram.

The next state of the Moore machine in this example is defined as

$$Y_{k(t+1)} = \begin{cases} \delta y_1 (x_1, y_2) = x_1 y_2 \\ \delta y_2 (x_1, y_1) = x_1{}' + y_1{}' \end{cases}$$

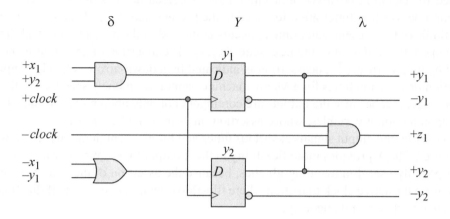

Figure 7.21   Moore machine for Example 7.1.

Table 7.10 Next-State Table for the Moore Machine of Figure 7.21

State Name	Present State $y_1 y_2$	Input x_1	Flip-Flop Inputs $Dy_1\ Dy_2$	Next State $y_1 y_2$	Output z_1
a	0 0	0	0 1	0 1	0
	0 0	1	0 1	0 1	0
b	0 1	0	0 1	0 1	0
	0 1	1	1 1	1 1	0
d	1 0	0	0 1	0 1	0
	1 0	1	0 0	0 0	0
c	1 1	0	0 1	0 1	1
	1 1	1	1 0	1 0	1

The input maps are shown in Figure 7.22 and can be derived from either the next-state table or from the logic diagram. For example, in Table 7.10, states b and c specify that Dy_1 is asserted only when both x_1 and y_2 are asserted. Therefore, using x_1 as a map-entered variable, the input map for Dy_1 contains the entry x_1 in column y_2. In the same manner, the input map for y_2 and the output map for z_1 are derived.

The equation for Dy_2 can be derived by one of two methods:

1. Read directly from the map without changing the minterm entries. This yields $Dy_2 = y_1' + y_1 x_1'$. Then using the absorption law, $Dy_2 = x_1' + y_1'$, or

2. Change the map entries as follows:

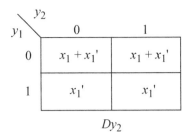

$$Dy_2$$

This does not change the minterm values. Since every square contains x_1', therefore, $Dy_2 = x_1'$. Now reassign $x_1 + x_1'$ as a value of 1, then combine the two 1s in row $y_1 = 0$ as $Dy_2 = y_1'$. Thus, $Dy_2 = x_1' + y_1'$.

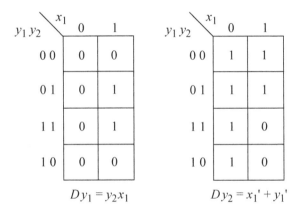

Figure 7.22   Input maps for the Moore machine of Figure 7.21.

The output map is shown in Figure 7.23 and can also be derived from either the next-state table or from the logic diagram. The output map does not show the time when z_1 is asserted, only the logic that causes the assertion.

$y_1 y_2$ \ x_1	0	1
0 0	0	0
0 1	0	0
1 1	1	1
1 0	0	0

$$z_1 = y_1 y_2$$

Figure 7.23   Output map for the Moore machine of Figure 7.21.

The timing diagram is shown in Figure 7.24 using an arbitrary input sequence of $x_1 = 1011$. The machine is reset initially to state a. Using the timing diagram in conjunction with the logic diagram of Figure 7.21 or the next-state table of Table 7.10, the machine proceeds to state b at the next positive clock transition, regardless of the value of x_1. At the positive clock transition at the end of state b, the following conditions exist: $x_1 = 0$, $y_1 = 0$, and $y_2 = 1$, and the machine remains in state b ($y_1 y_2 = 01$), because

$$Dy_1 = x_1 y_2 = 01 = 0$$

$$Dy_2 = x_1' + y_1' = 1 + 1 = 1$$

Using a similar procedure for the remaining states, it can be determined that the state transition sequence for $x_1 = 1011$ is $abbcd \cdots$. Output z_1 is asserted when $y_1 y_2$ $clock = 110$; that is, $y_1 y_2$ $clock'$ causes z_1 to be active during the last half of the clock cycle in which $y_1 y_2 = 11$.

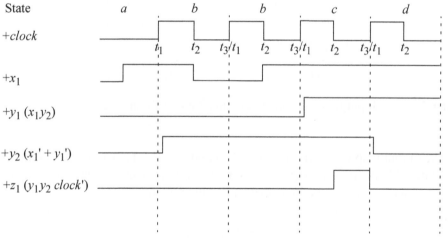

Figure 7.24   Timing diagram for the Moore machine of Figure 7.21 using an arbitrary sequence of $x_1 = 1011$.

The final analysis technique is the state diagram, which is a graphical representation of the functions δ and λ. The state diagram, illustrated in Figure 7.25, is derived directly from the next-state table of Table 7.10. Notice the assertion and deassertion of z_1 in state c. Output z_1 is asserted in state c at the midpoint (t_2) of the cycle and is deasserted at the end (t_3) of the cycle. This prevents a possible erroneous output (or glitch) from occurring when the state transition is from state d to state b.

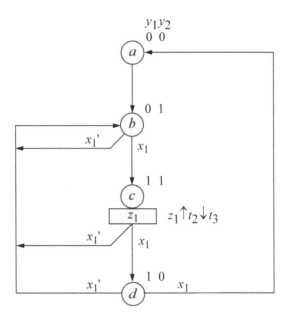

Figure 7.25   State diagram for the Moore machine of Figure 7.21.

Output glitches can occur when two or more flip-flops change state and the output assertion is at time t_1. When the active clock transition triggers the machine to initiate a state transition from state d to state b, both flip-flops change state. The machine then enters a period of instability until the machine stabilizes in state b ($y_1y_2 = 01$). If y_2 is faster at setting than y_1 is at resetting, then the machine will momentarily enter state c ($y_1y_2 = 11$). Because state c contains a Moore-type output, whenever the machine enters state c, output z_1 will be asserted. Thus, an erroneous output will be generated on z_1 as the machine passes through transient state c for a state transition sequence of $d \rightarrow b$.

The state diagram presents a much clearer state transition sequence than either the logic diagram or the timing diagram. The state diagram is hardware-independent; that is, it shows the complete operation of the machine at a glance for all possible input sequences, but does not indicate the type of storage elements or gates that are used in the implementation. The logic diagram shows the physical realization of the machine; however, the sequence of state transitions is obtained only after a tedious examination of inputs and present states. The timing diagram is an illustration of actual waveforms that would be observed on an oscilloscope for a given input sequence.

Example 7.2   Given the Moore machine of Figure 7.26, the machine will be analyzed by deriving next-state table and the state diagram. The next-state table is shown in Table 7.11. The machine is reset to state a ($y_1y_2 = 00$). In state a, the value of $Jy_1 Ky_1 = 10$ regardless of the value of x_1. This represents a set condition for a *JK* flip-flop; therefore, the next state for flip-flop $y_1 = 1$. Similarly, the value of $Jy_2 Ky_2 = 01$, regardless of the value of x_1. This represents a reset condition for a JK flip-flop; therefore, the next state for flip-flop $y_2 = 0$. Thus, at the next negative clock transition the next state is state b ($y_1 y_2 = 10$) and output z_1 is asserted, as shown in the state diagram of Figure 7.27.

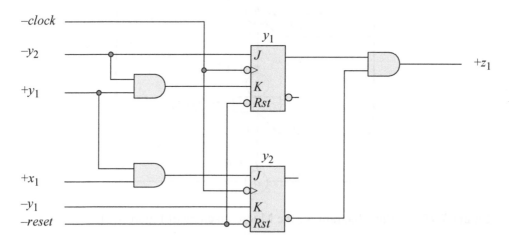

Figure 7.26   Moore machine to be analyzed for Example 7.2.

Table 7.11 Next-State Table for the Moore Machine of Figure 7.26

State Name	Present State $y_1 y_2$	Input x_1	Flip-Flop Inputs $Jy_1 Ky_1$	$Jy_2 Ky_2$	Next State $y_1 y_2$	Output z_1
a	0 0	0	1 0	0 1	1 0	0
	0 0	1	1 0	0 1	1 0	0
b	1 0	0	1 1	0 0	0 0	1
	1 0	1	1 1	1 0	0 1	1
c	0 1	0	0 0	0 1	0 0	0
	0 1	1	0 0	0 1	0 0	0

In state b, if $x_1 = 0$, then $Jy_1 Ky_1 = 11$, which represents a toggle condition for a JK flip-flop and y_1 toggles from 1 to 0 and $Jy_2 Ky_2 = 00$, which represents a no change condition for a JK flip-flop; therefore, the next state is a ($y_1 y_2 = 00$). In state b, if $x_1 = 1$, then $Jy_1 Ky_1 = 11$, which toggles flip-flop y_1 from 1 to 0. In state b, if $x_1 = 1$, then $Jy_2 Ky_2 = 10$, which represents a set condition for a JK flip-flop and $y_2 = 1$; therefore, the next state is $y_1 y_2 = 01$. In a similar manner, state c is obtained. The is no state d ($y_1 y_2 = 11$) for this machine.

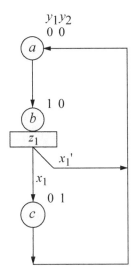

Figure 7.27   State diagram for the Moore machine of Figure 7.26.

Example 7.3   The logic diagram for the Moore machine shown in Figure 7.28 will be analyzed by obtaining the state diagram, the input maps, and the output maps.

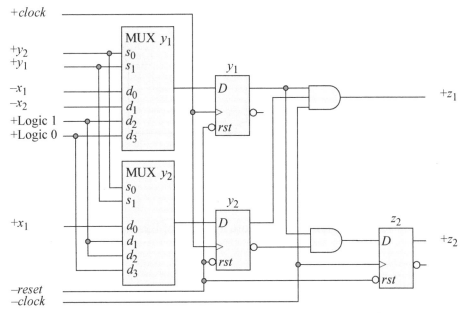

Figure 7.28   Logic diagram for the Moore machine of Example 7.3.

Output z_1 has the following assertion/deassertion times: $z_1 \uparrow t_2 \downarrow t_3$; that is, z_1 is asserted during the last half of the clock cycle. Output z_2 has the following assertion/deassertion times: $z_2 \uparrow t_2 \downarrow t_4$; that is, z_2 is asserted during the last half of the present clock cycle and during the first half of the following clock cycle.

The multiplexers are linear-select multiplexers with select inputs of $s_0 s_1 = y_1 y_2$, where y_2 is the low-order select input. Flip-flops y_1 and y_2 are clocked on the positive transition of the clock. Note that output z_1 is asserted by the term $y_1 y_2 clk'$, which is the last half of the present clock cycle. In order to have output z_2 asserted from t_2 to t_4, a D flip-flop is used, which is clocked on the negative transition of the clock.

The logic diagram will now be analyzed. The machine is reset to state a ($y_1 y_2 = 00$). In state a, if $x_1 = 0$, then the D input for flip-flop $y_1 = 1$ and the D input for flip-flop $y_2 = 0$. Therefore, the next state is d ($y_1 y_2 = 10$) in which the flip-flop for z_2 is clocked on the negative transition of the clock. In state a, if $x_1 = 1$, then the D input for flip-flop $y_1 = 0$ and the D input for flip-flop $y_2 = 1$. Therefore, the next state is b ($y_1 y_2 = 01$).

In state b, if $x_2 = 1$, then flip-flop $y_1 = 0$ at the next positive clock transition and flip-flop $y_2 = 1$. Therefore, the machine remains in state b. In state b, if $x_2 = 0$, then the next state is c ($y_1 y_2 = 11$), in which output z_1 is asserted during the last half of the clock cycle. In a similar manner, the remaining states are derived, producing the state diagram of Figure 7.29.

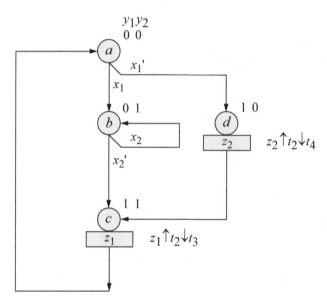

Figure 7.29   State diagram of the Moore machine of Example 7.3.

The input maps for flip-flops y_1 and y_2 are shown in Figure 7.30 using x_1 and x_2 as map-entered variables. The input maps are derived from the logic diagram or from the next-state table. In state a ($y_1 y_2 = 00$), flip-flop $y_1 = 1$ only if x_1 is deasserted and flip-flop $y_2 = 1$ only if x_1 is asserted. Therefore, minterm location 0 in the map for Dy_1 contains the variable x_1' and minterm location 0 in the map for Dy_2 contains the variable x_1. In state b ($y_1y_2 = 01$), flip-flop $y_1 = 1$ only if x_2 is deasserted and flip-flop $y_2 = 1$ regardless of value of x_2. In a similar manner, the remaining minterm locations in both maps are derived. The input equations for Dy_1 and Dy_2 are not necessary, because multiplexers are used for the input logic.

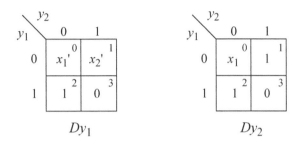

Figure 7.30   Input maps for the Moore machine of Figure 7.28.

The output maps for z_1 and z_2 are shown in Figure 7.31 and are derived directly from the state diagram or from the logic diagram. The equations for z_1 and z_2 are shown in Equation 7.18.

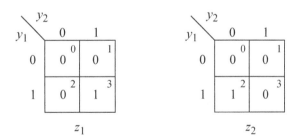

Figure 7.31   Output maps for the Moore machine of Figure 7.28.

$$z_1 = y_1y_2$$

$$z_2 = y_1y_2' \tag{7.18}$$

Example 7.4   The Mealy machine shown in Figure 7.32 will be analyzed by obtaining the next-state table and the state diagram. By applying all combinations of $x_1 x_2$ in both states of the flip-flop, the next-state table shown in Table 7.12 is obtained. Either the next-state table or the logic diagram can be used to obtain the state diagram of Figure 7.33.

The machine is reset initially to state a ($y_1 = 0$). Since the K input of the JK flip-flop is connected to a logic 1, the flip-flop will either reset or toggle upon the application of a positive clock transition. In state a ($y_1 = 0$), the machine will sequence to state b (1) for the following conditions:

$$x_1 + x_1'x_2 = x_1 + x_2.$$

Output z_1 will be asserted in state b ($y_1 = 1$) if input x_1 is asserted.

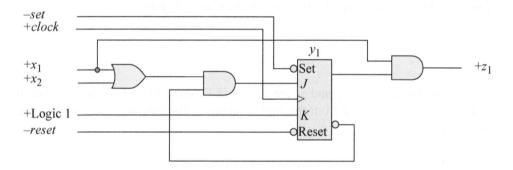

Figure 7.32   Logic diagram for the Mealy machine of Example 7.4.

Table 7.12 Next-State Table for the Mealy Machine of Example 7.4

State Name	Present State y_1	Inputs $x_1 x_2$	Flip-Flop Inputs $Jy_1 Ky_1$	Next State y_1	Output z_1
a	0	0 0	0 1	0	0
	0	0 1	1 1	1	0
	0	1 0	1 1	1	0
	0	1 1	1 1	1	0
b	1	0 0	0 1	0	0
	1	0 1	0 1	0	0
	1	1 0	0 1	0	1
	1	1 1	0 1	0	1

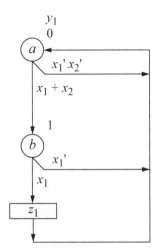

Figure 7.33 State diagram for the Mealy machine of Example 7.4.

Example 7.5 The synchronous sequential machine shown in Figure 7.34 is implemented with JK flip-flops and contains both Moore- and Mealy-type outputs. The machine will be analyzed by obtaining the next-state table for all combinations of the input variables, the state diagram, the input maps for Jy_1Ky_1 and Jy_2Ky_2, and the output maps for z_1 and z_2. All Karnaugh maps will use x_1 and x_2 as map-entered variables. The machine is reset to state a ($y_1y_2 = 00$).

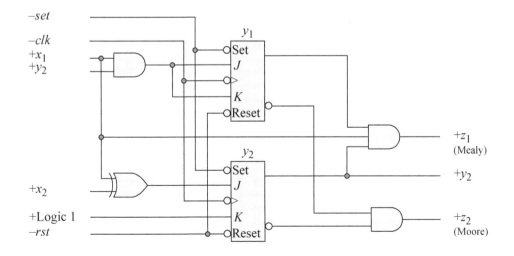

Figure 7.34 Logic diagram for the synchronous sequential machine of Example 7.5.

The δ next-state logic consists of one AND gate and one exclusive-OR function. There are four possible input vectors: $x_1 x_2 = 00, 01, 10,$ and 11. There are four possible states: $y_1 y_2 = 00, 01, 10,$ and 11. The λ output logic consists of two AND gates: one for z_1 and one for z_2. By applying all combinations of the inputs beginning in state a ($y_1 y_2 = 00$), the next-state table of Table 7.13 is obtained.

For example, in state a, $y_1 = 0$; therefore, $Jy_1 Ky_1 = 00$ and there is no change to the present state of flip-flop y_1. For flip-flop y_2, $Jy_2 = 1$ whenever x_1 and x_2 contain different values. Since $Ky_2 = 1$, the JK values are either $JK = 01$ or 11 and the next-state for flip-flop y_2 is either $y_2 = 0$ or $y_2 = 1$. Therefore, in state a, the next state for the machine is either $y_1 y_2 = 00$ or $y_1 y_2 = 01$. In a similar manner, the remaining next states are obtained.

By progressing through Table 7.13 line by line, the next-state table can be easily translated into the state diagram of Figure 7.35, which depicts the same information as the next-state table, but exhibits a graphical representation in which the state transitions are more readily observed.

Table 7.13 Next-State Table for the Synchronous Sequential Machine of Figure 7.34

State Name	Present State $y_1 y_2$	Inputs $x_1 x_2$	Flip-Flop Inputs $Jy_1 Ky_1$	$Jy_2 Ky_2$	Next State $y_1 y_2$	Outputs $z_1 z_2$
a	0 0	0 0	0 0	0 1	0 0	0 1
		0 1	0 0	1 1	0 1	0 1
		1 0	0 0	1 1	0 1	0 1
		1 1	0 0	0 1	0 0	0 1
b	0 1	0 0	0 0	0 1	0 0	0 0
		0 1	0 0	1 1	0 0	0 0
		1 0	1 1	1 1	1 0	0 0
		1 1	1 1	0 1	1 0	0 0
c	1 0	0 0	0 0	0 1	1 0	0 0
		0 1	0 0	1 1	1 1	0 0
		1 0	0 0	1 1	1 1	0 0
		1 1	0 0	0 1	1 0	0 0
d	1 1	0 0	0 0	0 1	1 0	0 0
		0 1	0 0	1 1	1 0	0 0
		1 0	1 1	1 1	0 0	1 0
		1 1	1 1	0 1	0 0	1 0

The input maps for $Jy_1 Ky_1$ and $Jy_2 Ky_2$ are shown in Figure 7.36 using x_1 and x_2 as map-entered variables and yield the equations of Equation 7.19. The output maps are shown in Figure 7.37, also using x_1 and x_2 as map-entered variables and yield the equations of Equation 7.20. All Karnaugh maps are derived from either the logic diagram, the next-state table, or the state diagram.

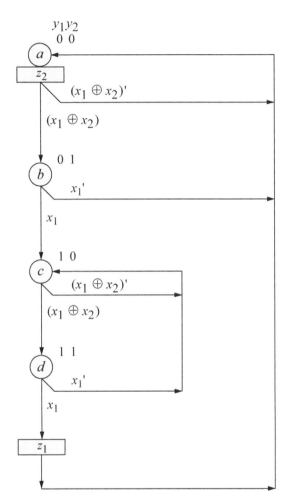

Figure 7.35 State diagram for the synchronous sequential machine of Figure 7.34.

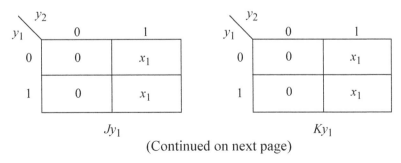

(Continued on next page)

Figure 7.36 Input maps for the synchronous sequential machine of Figure 7.34.

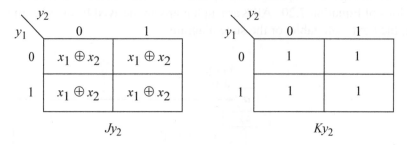

$$Jy_2$$

$$Ky_2$$

Figure 7.36　(Continued)

$$Jy_1 = y_2 x_1$$

$$Ky_1 = y_2 x_1$$

$$Jy_2 = x_1 \oplus x_2$$

$$Ky_2 = 1 \tag{7.19}$$

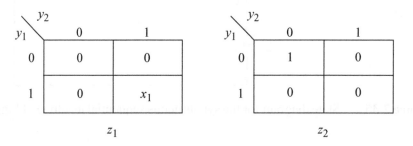

$$z_1$$

$$z_2$$

Figure 7.37　Output maps for the synchronous sequential machine of Figure 7.34.

$$z_1 = y_1 y_2 x_1$$

$$z_2 = y_1' y_2' \tag{7.20}$$

7.2   Synthesis of Synchronous Sequential Machines

Techniques for synthesizing (designing) synchronous sequential machines are introduced. A detailed procedure is presented to synthesize a synchronous sequential machine from a given set of machine specifications. The design process begins with a set of specifications and culminates with a logic diagram or a list of Boolean functions from which the logic can be designed.

Unlike combinational logic, which can be completely specified by a truth table, a synchronous sequential machine requires a state diagram or state table for its precise description. The state diagram depicts the sequence of events that must occur in order for the machine to perform the functions which are defined in the machine specifications. A *synchronous sequential machine* consists of storage elements, usually flip-flops, and δ next-state combinational logic that connects to the flip-flop data inputs. The machine may also contain combinational logic for the λ output function. In some cases, the output logic may require one or more storage elements, depending on the assertion and deassertion of the output signals.

The number of flip-flops is determined by the number of states required by the machine. The combinational logic is derived directly from either the state diagram or from the state table. When the type and quantity of storage elements has been determined, the design process proceeds in a manner analogous to that of combinational logic design.

The requirement for hardware minimization is of paramount importance. Hardware minimization is realized by reducing (or minimizing) the number of states in the machine, thus minimizing the number of storage elements and logic gates, while maintaining the input-output requirements. Reducing the number of states in a machine may not always reduce the number of flip-flops, since the number of eliminated states may not reduce the total state count by a power of two.

For example, if a 16-state machine (requiring $p = 4$ storage elements) is reduced to a 12-state machine, then four storage elements are still required. There will be no reduction in the number of flip-flops until the number of states has been reduced to at least eight ($2^p = 2^3 = 8$ flip-flops). However, the increased number of unused states may result in less combinational logic, because these "don't care" states can be combined with other machine states in a Karnaugh map, resulting in a reduction of combinational logic.

A proper choice of state code assignments may also reduce the number of gates in the δ next-state function logic. Since there are p storage elements, the binary values of these p-tuples can usually be chosen such that the combinational input logic is minimized. A judicious choice of state codes permits more entries in the Karnaugh map to be combined. Since the map entries represent the input logic, combining a greater number of minterm locations results in input equations with fewer terms and fewer variables per term. Thus, an overall reduction in logic gates is realized for the δ next-state logic.

The synthesis procedure utilizes a hierarchical method — also referred to as a *top-down* approach — for machine design. This is a systematic and orderly procedure that commences with the machine specifications and advances down through increasing levels of detail to arrive at a final logic diagram. Thus, the machine is decomposed into modules which are independent of previous and following modules, yet operate together as a cohesive system to satisfy the machine's requirements.

7.2.1 Synthesis Procedure

This section develops a detailed method for designing synchronous sequential machines using various types of storage elements. The hierarchical design algorithm is shown below.

1. Develop a state diagram from the problem definition, which may be either a word description and/or a timing diagram.

2. Check for equivalent states and then eliminate redundant states.

3. Assign state codes for the storage elements in the form of a binary p-tuple.

4. Generate a next-state table.

5. Select the type of storage element to be used, then generate the input maps for the δ next-state function and derive the input equations.

6. Generate the output maps for the λ output function and derive the output equations.

7. Design the logic diagram using the input equations, the storage elements, and the output equations.

A critical step in synthesizing synchronous sequential machines is the derivation of the state diagram. The *state diagram* specifies the machine performance and gives a clear indication of the state transitions and the output assertion, both of which are a function of the input sequence. If the state diagram is correct, then the remaining steps are relatively straightforward and will result in a logic circuit that performs according to the machine specifications. If, however, the state diagram does not reflect the exact performance of the machine, then the remaining steps — although correct in themselves — will not result in a machine that adheres to the prescribed specifications.

Methods will be described to identify equivalent states, after which the redundant states can be eliminated and the state diagram redrawn as a reduced state diagram. State codes are assigned according to rules which will be defined later. If p storage elements are required to implement the machine, then a state code in the form of a binary p-tuple is assigned to each state in the state diagram. The next-state table is then derived from the state diagram. This step is not always necessary and can be eliminated in many cases, especially when D flip-flops are used. A next-state map may also prove useful for completeness.

When the type of storage elements has been determined, the input maps can then be obtained using the next-state table, and from these maps the corresponding input equations. An alternative approach is to derive the input maps from the state diagram directly. The output maps can be derived directly from the state diagram. The λ output logic is usually combinational, but may require storage elements, depending upon the assertion/deassertion specifications. Finally, the logic diagram is designed using the input equations, the storage elements, and the output equations.

Equivalent states   Before exemplifying the steps of the synthesis algorithm, two methods will be presented for determining equivalent states. When equivalent states have been found, all but one are redundant and should be eliminated before implementing the state diagram with hardware. At each node in the state diagram, two events occur: the outputs (if applicable) for the present state are generated as a function of the present state only (Moore) or the present state and inputs (Mealy); the next state is determined as a function of the present state only or the present state and inputs.

When deriving a state diagram from machine specifications, some states might be included which contain no new information that is pertinent to the machine's performance. These are classified as redundant states and should be eliminated since they may increase the amount of logic that is required.

If the state diagram is sufficiently small, then redundant states can be easily recognized as the state diagram is being constructed. For larger state diagrams, redundant states may be inadvertently inserted due to the complexity of the machine specifications. During construction of the state diagram, it is best to obtain a diagram that accurately reflects the machine specifications regardless of the number of states that are included. When a correct state diagram has been established, a simple algorithm can then be utilized to find equivalent states. Redundant states can then be eliminated, yielding a reduced state diagram which still completely characterizes the behavior of the machine.

Two states Y_i and Y_j of a machine are equivalent if, for every input sequence, the output sequence when started in state Y_i is identical to the output sequence when started in state Y_j. Therefore, two states Y_i and Y_j are equivalent if and only if the following two conditions are true:

1.   For every input sequence X_i, the output sequence Z_i is the same whether the machine begins in state Y_i or Y_j; that is, $\lambda(Y_i, X_i) = \lambda(Y_j, X_i)$, where $\lambda(Y_i, X_i)$ is the output from present state Y_i with input vector X_i and $\lambda(Y_j, X_i)$ is the output from present state Y_j with input vector X_i.

2.   Both states Y_i and Y_j have the same or equivalent next state; that is, $\delta(Y_i, X_i) \equiv \delta(Y_j, X_i)$, where $\delta(Y_i, X_i)$ is the next state for a present state of Y_i with input vector X_i and $\delta(Y_j, X_i)$ is the next state for a present state of Y_j with input vector X_i and the symbol $\equiv$ specifies equivalence.

Row matching   Finding equivalent states and then eliminating redundant states will be explained using the state diagram illustrated in Figure 7.38. The technique used in the first method is a *row-matching* procedure employing an approach that is more heuristic than algorithmic. The state diagram in Figure 7.38 depicts a Moore machine with the corresponding next-state table shown in Table 7.14. The machine examines a serial 3-bit word on an serial input line x_1 and generates an output z_1 whenever the 3-bit word is 111 or 000. There is one bit space (one clock period) between contiguous words during which time output z_1 is asserted, as shown below.

$$x_1 = \quad \cdots \ \left| \ b_1\ b_2\ b_3\ \right| \quad \left| \ b_1\ b_2\ b_3 \quad \cdots \ \right|$$

$$\uparrow$$

$$z_1$$

where b_i is 0 or 1.

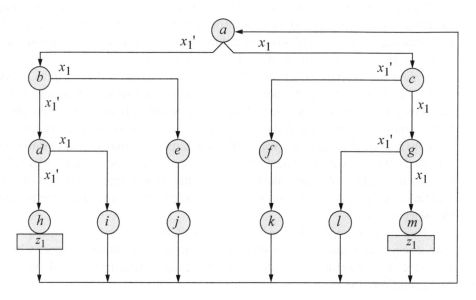

Figure 7.38   Moore machine to detect a sequence of $x_1 = 111$ or 000.

By carefully considering the machine specifications as the state diagram is being generated, it is relatively easy — in this example — to obtain a state diagram that has no redundant states. However, in order to illustrate the techniques used to identify equivalent states and then to eliminate redundant states, superfluous states have been deliberately inserted.

Using the two rules for equivalence, states i, j, k, and l of Table 7.14 are seen to be equivalent, because they all have state a as the next state and all have output $z_1 = 0$. By convention, the lowest number or the lowest ranked letter is retained for the state name and all other equivalent states are given this state name. Therefore, all js, ks, and ls are changed to i wherever they appear in the state diagram. This will eliminate rows j, k, and l from the next-state table. The renamed states are indicated by a slash followed by the new name. Although state m also has a next state of a, output $z_1 = 1$; thus, state m is not equivalent to states i, j, k, and l. It is also apparent that states h and m are equivalent: both have state a as the next state and both assert output z_1. Therefore, all ms are replaced with hs in Table 7.14.

Table 7.14 Next-State Table for the Moore Machine of Figure 7.38

Present State	Input x_1	Next State	Output z_1	
a	0	b	0	
	1	c	0	
b	0	d	0	
	1	e	0	
c	0	f / e	0	
	1	g	0	
d	0	h	0	
	1	i	0	
e	0	j / i	0	Equivalent states
	1	j / i	0	(Eliminate state f)
f / e	0	k / i	0	
	1	k / i	0	
g	0	l / i	0	
	1	m / h	0	
h	0	a	1	Equivalent states
	1	a	1	(Eliminate state m)
i	0	a	0	Equivalent states
	1	a	0	(Eliminate states
j / i	0	a	0	j, k, and l)
	1	a	0	
k / i	0	a	0	
	1	a	0	
l / i	0	a	0	
	1	a	0	
m / h	0	a	1	
	1	a	1	

After the equivalent states have been renamed, a check is made for equivalent states using the modified state names to determine if any further equivalences exist. In states d and g, output $z_1 = 0$. Therefore, states d and g are equivalent if $h \equiv i$. The outputs for states h and i, however, are different; thus, states h and i are not equivalent and consequently states d and g are not equivalent.

Continuing with the examination of Table 7.14, it is observed that states e and f are now equivalent, because in both states the next state is i and output $z_1 = 0$. Therefore, all fs are changed to es wherever fs appear in Table 7.14. Further inspection produces no additional equivalent states. Equivalent states are: $e \equiv f$, $h \equiv m$, $i \equiv j \equiv k \equiv l$. The reduced next-state table is shown in Table 7.15; the reduced state diagram with only eight states is illustrated in Figure 7.39.

Table 7.15 Reduced Next-State Table for the Moore Machine of Figure 7.38

Present state	Input x_1	Next State	Output z_1
a	0	b	0
	1	c	0
b	0	d	0
	1	e	0
c	0	e	0
	1	g	0
d	0	h	0
	1	i	0
e	0	i	0
	1	i	0
g	0	i	0
	1	h	0
h	0	a	1
	1	a	1
i	0	a	0
	1	a	0

Implication table   The second method for determining equivalent states is by use of an *implication table*. This technique is more algorithmic in nature than the previous method, because it follows a set of well-defined rules to find equivalent states in a finite number of steps. The example of Table 7.14 will again be used, this time to illustrate the steps required to find equivalent states using an implication table. For convenience and clarity, Table 7.14 is reproduced in Table 7.16.

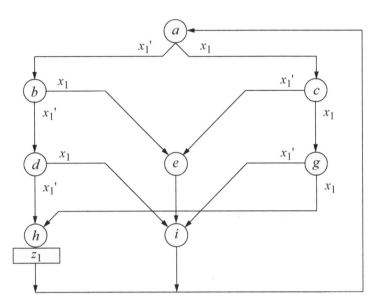

Figure 7.39   Reduced state diagram for the Moore machine of Figure 3.1.

Table 7.16 Next-State Table for the Moore Machine of Figure 7.38

Present State	Input x_1	Next State	Output z_1
a	0	b	0
	1	c	0
b	0	d	0
	1	e	0
c	0	f	0
	1	g	0
d	0	h	0
	1	i	0
e	0	j	0
	1	j	0
f	0	k	0
	1	k	0
g	0	l	0
	1	m	0
h	0	a	1
	1	a	1

(Continued on next page)

Table 7.16 Next-State Table for the Moore Machine of Figure 7.38

Present State	Input x_1	Next State	Output z_1
i	0	a	0
	1	a	0
j	0	a	0
	1	a	0
k	0	a	0
	1	a	0
l	0	a	0
	1	a	0
m	0	a	1
	1	a	1

The implication table is a lower-left triangular matrix whose rows are labeled with the state names in ascending sequence with the exception of the first state. The columns are also labeled with the state names in ascending sequence with the exception of the last state. The reason for omitting the first state name in the upper-left corner of the matrix is because $a \equiv a$, negating the necessity of inserting a square to determine if a is equivalent to a. The same rationale applies to omitting the last state name in the lower-right corner of the matrix. The square at the intersection of a row-column pair is marked with an $\times$ if the corresponding states are not equivalent or marked with the symbol $\equiv$ if the states are equivalent.

The first step is to construct a chart of the form shown in Figure 7.40. The chart contains a square for every possible pair of states. The square in row f, column c, for example, corresponds to state pair (f,c). Thus, the squares in the first column correspond to state pairs (b,a), (c,a), (d,a), etc. Squares above the diagonal are not required, because they represent the symmetric property of equivalence; that is, if $c \equiv d$, then $d \equiv c$. Thus, only one of the state pairs is required. Also, squares such as (a,a), (b,b), etc. are omitted, since the state pair within the parentheses is obviously equivalent.

To fill in the first column of the chart, row a of the next-state table shown in Table 7.16 is compared with each of the remaining rows. Consider rows a and b. Since output $z_1 = 0$ for states a and b, then

$$a \equiv b \text{ if and only if } b \equiv d \ (x_1 = 0) \text{ and } c \equiv e \ (x_1 = 1)$$

This is shown in Figure 7.40 where the implied pairs (b,d) and (c,e) are placed in the square for state pair (a,b). An *implied pair* of states indicates that equivalence is implied for the pair but has not yet been verified. During later steps in the procedure, implied pairs will be shown to be either equivalent or nonequivalent. Since this is a

Moore machine, the value of x_1 is not significant when comparing states for equivalence. Next, state pair (a,c) is considered. Output $z_1 = 0$ for states a and c; therefore,

$$a \equiv c \text{ if and only if } b \equiv f \text{ and } c \equiv g$$

This fact is indicated in the table for state pair (a,c). The process continues for each square in the implication table. When two states are not equivalent, an $\times$ is placed in the square that corresponds to the state pair under consideration, for example, (a,h), where the outputs differ. When two states are equivalent, this is indicated by placing the equivalence symbol $\equiv$ within the matrix square that corresponds to the state pair under consideration. For example, states h and m are equivalent, because output $z_1 = 1$ for both states and both states have state a as their next state. Therefore, the equivalence symbol $\equiv$ is placed in state pair (h,m).

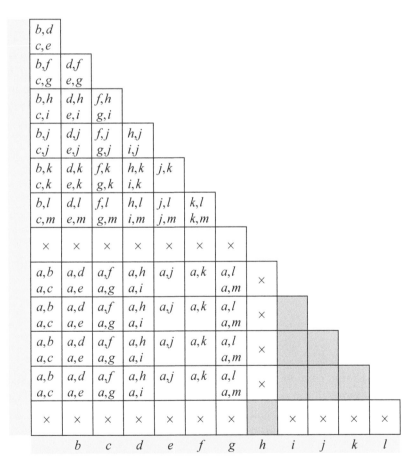

Figure 7.40   Implication table for the Moore machine of Table 7.16 after the first pass. The symbol $\times$ indicates nonequivalent states; the symbol $\equiv$ indicates equivalent states.

Self-implied pairs are redundant and need not be inserted. A self-implied pair is a state pair within a matrix square that is the same as the state pair under consideration. For example, in the entry shown below for state pair (s,x), the entry (s,x) in the matrix square can be eliminated, since (s,x) is a self-implied pair. Thus, $s \equiv x$ if and only if $s \equiv x$ and $r \equiv t$ can be reduced to $s \equiv x$ if and only if $r \equiv t$.

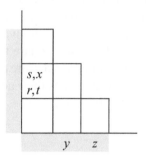

When all state pairs in the row-column coordinates have been compared, each square in the implication table contains implied pairs, the symbol $\times$ indicating that the state pair is not equivalent, or the symbol $\equiv$ indicating that the state pair is equivalent. The implied pairs in each square are now checked for equivalence. If one of the implied pairs in a square is not equivalent, then the state pair under consideration is not equivalent. For example, using state pair (s,x) shown below, if implied pair (h,m) is not equivalent, then $s \neg \equiv x$, where the symbol $\neg \equiv$ is read as "not equivalent."

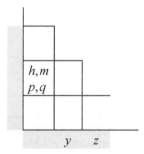

The second pass to find equivalent states is now performed. Beginning with state pair (a,b), continue down column a until each state pair (a,b) through (a,m) has been found either equivalent or not equivalent. It is not always possible, however, to immediately determine equivalency. For example,

$$a \equiv b \text{ if and only if } b \equiv d \text{ and } c \equiv e$$

At this point, the equivalency of state pairs (b,d) and (c,e) is not immediately evident; thus, their equivalency will be resolved at a later step.

Next, state pair (a,c) is checked for equivalence. Since

$$a \equiv c \text{ if and only if } b \equiv f \text{ and } c \equiv g$$

and since it is unknown at this time whether $b \equiv f$ or $c \equiv g$, the process proceeds to state pair (a,d), which specifies that

$$a \equiv d \text{ if and only if } b \equiv h \text{ and } c \equiv i$$

Since $b \neg \equiv h$ (the outputs are different), therefore, $a \neg \equiv d$ and an $\times$ is placed in the square for state pair (a,d). Continue in this manner for all columns in Figure 7.40. Notice that state pair (d,e) specifies that

$$d \equiv e \text{ if and only if } h \equiv j \text{ and } i \equiv j$$

This statement is true only if both state pairs (h,j) and (i,j) are equivalent. If either state pair is not equivalent, then $d \neg \equiv e$. Although state pair (i,j) is equivalent, state pair (h,j) is not equivalent. Therefore, $d \neg \equiv e$. The process of finding equivalent states continues until no more $\times$s can be inserted; that is, until no more nonequivalent states can be found. The result of the second pass is shown in Figure 7.41.

Now begin a third pass, using the implication table of Figure 7.41, to determine equivalency between the remaining state pairs. The results are presented in Figure 7.42 Finally, a fourth pass yields Figure 7.43, which illustrates the following equivalent states:

$$e \equiv f$$
$$h \equiv m$$
$$i \equiv j \equiv k \equiv l$$

The results obtained by the implication table are identical to those obtained by the row-matching procedure. Although the implication table method is tedious, it guarantees a reduced state diagram.

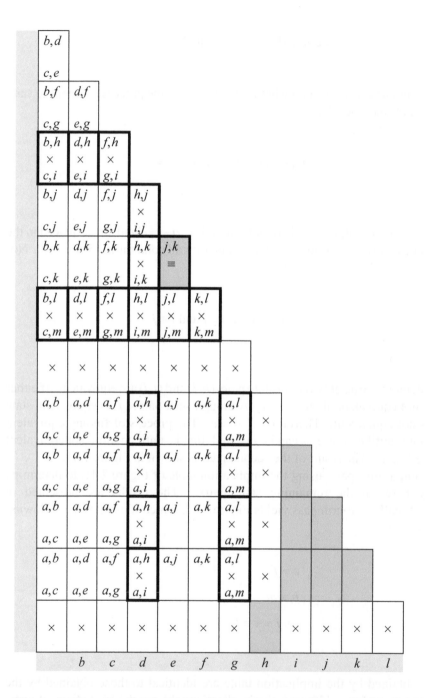

Figure 7.41   Implication table for the Moore machine of Table 7.16 after the second pass. The results of the second pass are shown in bold-lined squares. The symbol × indicates nonequivalent states; the symbol ≡ indicates equivalent states.

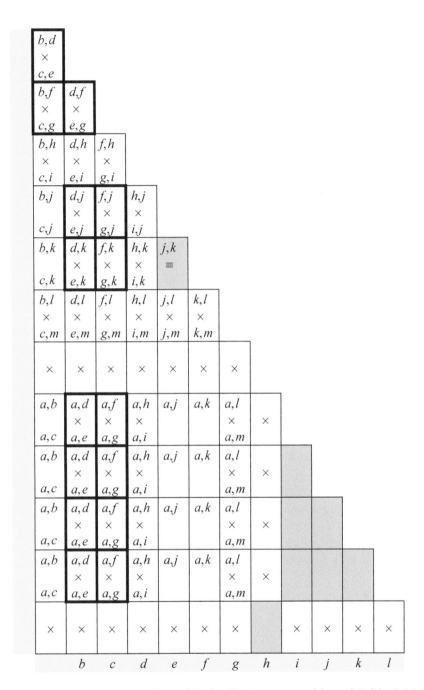

Figure 7.42   Implication table for the Moore machine of Table 7.16 after the third pass. The results of the third pass are shown in bold-lined squares. The symbol × indicates nonequivalent states; the symbol ≡ indicates equivalent states.

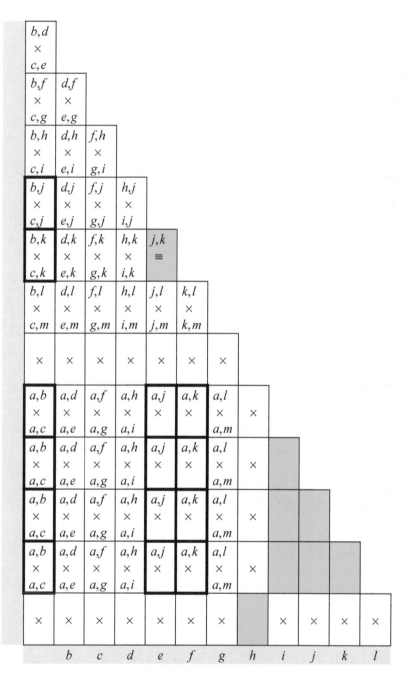

Figure 7.43 Implication table for the Moore machine of Table 7.16 after the fourth pass. The results of the fourth pass are shown in bold-lined squares. The symbol × indicates nonequivalent states; the symbol ≡ indicates equivalent states. Equivalent states are (e, f), (h, m), and (i, j, k, l).

Although the above two procedures were described for a Moore machine, the methods work equally well for a Mealy machine. When a row-column pair is compared for a Mealy machine, the outputs must be the same for each same input value.

There is one additional comment regarding row-column equivalent pairs; that is, interdependence. In the partial implication table shown below, $a \equiv i$. Also, $j \equiv b$ if $c \equiv k$.

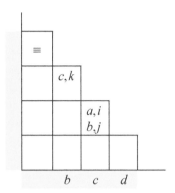

Examining state pair (c,k) for equivalence, we see that $c \equiv k$ if $a \equiv i$ and $b \equiv j$. Since it is already known that $a \equiv i$, the equivalence of c and k depends only on the equivalence of b and j. Conversely the equivalence of b and j depends only on the equivalence of c and k. It can be stated, therefore, that $c \equiv k$ if $b \equiv j$ if $c \equiv k$. This is an equivalence condition that is based upon a mutual dependence and is called *interdependence*.

Thus, there is an interdependence between state-pairs (b,j) and (c,k). Therefore, $b \equiv j$ and $c \equiv k$. This can be verified by examining the next-state table of the machine and observing that the output sequence beginning in state b will be the same as the output sequence beginning in state j for the same input sequence. The same rationale is true for states c and k.

7.2.2 Synchronous Registers

An ordered set of storage elements and associated combinational logic, where each cell performs an identical or similar operation, is called a *register*. Each cell of a register stores one bit of binary information. There are many different types of synchronous registers, including parallel-in, parallel-out; parallel-in, serial-out; serial-in, parallel-out; and serial-in, serial-out registers. In the synthesis of synchronous registers, it is not always necessary to use the formalized design procedure previously described. Usually, intuitive reasoning and experience are sufficient requisites for the synthesis of these elementary storage devices.

The next state of a register is usually a direct correspondence to the input vector, whose binary variables connect to the flip-flop data inputs, either directly or through δ next-state logic. Most registers are used primarily for temporary storage of binary data, either signed or unsigned, and do not modify the data internally; that is, the state of the register is unchanged until the next active clock transition. Other registers may modify the data in some elementary manner such as, shifting left or shifting right, where a left shift of one bit corresponds to a multiply-by-two operation and a right shift of one bit corresponds to a divide-by-two operation.

An n-bit register requires n storage elements, either SR latches, D flip-flops, or JK flip-flops. There are 2^n different states in an n-bit register, where each n-tuple corresponds to a unique state of the register.

Parallel-in, parallel-out registers   The simplest register, and the most prevalent, is the *parallel-in, parallel-out* (PIPO) register used for temporary storage of binary data. The synthesis procedure is not required for this type of register. Figure 7.44 illustrates a p-bit register containing only storage elements. There is a one-to-one correspondence between the input alphabet X, the state alphabet Y, and the output alphabet Z. The values of the present inputs $X_{i(t)}$ become the next state $Y_{k(t+1)}$ of the register at the next active clock transition.

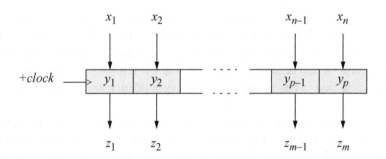

Figure 7.44   Block diagram for a parallel-in, parallel-out register.

A typical application for a PIPO register is for a general-purpose register (GPR) in an arithmetic and logic unit (ALU). General-purpose registers are used for temporary storage of data and for storing a base address or an index to be utilized in determining a memory location. They are also used as a memory address register (MAR) to address memory and as a memory data register (MDR) to contain a word of information that is sent to or received from memory.

An alternative approach to a PIPO is where the storage elements are JK flip-flops and the δ next-state logic not only provides the next state for the register by gating the input vector X_i, but also controls the time at which the register changes state. In this design, the register is clocked continuously by the system clock, which is a free-running astable multivibrator. The register is loaded, however, only when the *load* control signal is active. When the *load* input is inactive, the data inputs of each flip-flop are $JK = 00$, which causes no change to the state of the machine. Thus, the register remains in its present state until the *load* input changes to an active level. The new input vector X_i then replaces the previous state of the register.

Parallel-in, serial-out registers   A *parallel-in, serial-out* (PISO) register accepts binary input data in parallel and generates binary output data in serial form. The binary data can be shifted either left or right under control of a shift direction signal and a clock pulse, which is applied to all flip-flops simultaneously. The register shifts left or right 1 bit position at each active clock transition. Bits shifted out of one end of the register are lost unless the register is cyclic, in which case, the bits are shifted (or rotated) into the other end.

If the register is a PISO right-shift device, then two conditions determine the value of the bits shifted into the vacated positions on the left. If the binary data represents an unsigned number, then 0s are shifted into the vacated positions. If the binary data represents a signed number — with the high-order bit specified as the sign of the number, where a 0 bit represents a positive number and a 1 bit represents a negative number — then the sign bit extends right 1 bit position for each active clock transition.

The procedure for the synthesis of a PISO register is relatively straightforward and the formalized method can be circumvented. An example is a 4-bit register that receives a parallel input vector in the form of binary bits x_1, x_2, x_3, and x_4. This operand is stored in four flip-flops y_1, y_2, y_3, and y_4, as shown in Figure 7.45. Upon application of a clock signal, the operand shifts right 1 bit position. The serial output z_1 is generated from the output of flip-flop y_4. Zeroes fill the vacated positions on the left.

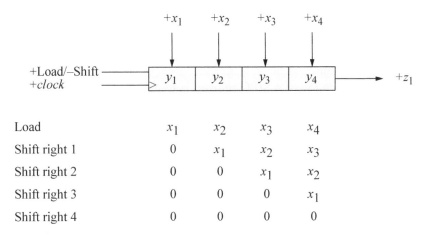

Load	x_1	x_2	x_3	x_4
Shift right 1	0	x_1	x_2	x_3
Shift right 2	0	0	x_1	x_2
Shift right 3	0	0	0	x_1
Shift right 4	0	0	0	0

Figure 7.45   Block diagram for a 4-bit parallel-in, serial-out shift register.

Upon completion of the load cycle, $y_i = x_i$. During the shift sequence, $y_i = y_{i-1}$ or 0, depending on the shift count. After four shift cycles, the state of the register is $y_1 y_2 y_3 y_4 = 0000$, and the process repeats with a new input vector X_i. One application of a PISO register is to convert data from a parallel bus into serial data for use by a single-track device, such as a disk drive. The serialization process occurs during a write operation.

Serial-in, parallel-out registers The *serial-in, parallel-out* (SIPO) register is another typical synchronous iterative network containing p identical cells. Data enters the register from the left and shifts serially to the right through all p stages, 1 bit position per clock pulse. After p shifts, the register is fully loaded and the bits are transferred in parallel to the destination.

An example of a 4-bit SIPO register is shown in Figure 7.46, in which four bits of serial data, x_1, x_2, x_3, and x_4 are shifted into the register from the left, where x_4 is the first bit entered. The initial state of the register is either unknown or reset to $y_1 y_2 y_3 y_4 = 0000$. During the shift sequence, $y_1 = x_i$ and $y_i = y_{i-1}$. After four shift cycles, the state of the register is $y_1 y_2 y_3 y_4 = x_1 x_2 x_3 x_4$ and the 4-bit word is transferred in parallel to a destination.

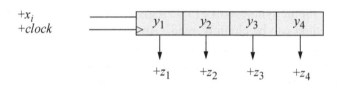

Figure 7.46   Block diagram for a 4-bit serial-in, parallel-out register.

A typical application of a serial-in, parallel-out register is to deserialize binary data from a single-track peripheral subsystem. The resulting word of parallel bits is placed on the system data bus.

A second useful application of a SIPO register is to generate a sequence of non-overlapping pulses for system timing. This provides a simple, yet effective state machine, where each pulse represents a different state. The machine is initially reset to $y_1 y_2 y_3 y_4 = 0000$. Whenever $y_1 y_2 y_3 = 000$, a 1 bit will be shifted into flip-flop y_1 at the next positive clock transition. If either y_1, y_2, or $y_3 = 1$, then a 0 bit will be shifted into flip-flop y_1, and $y_i = y_{i-1}$ at the next positive clock transition. Thus, the required four nonoverlapping pulses are generated.

Serial-in, serial-out registers The synthesis of a *serial-in, serial-out* (SISO) register is identical to that of a SIPO register, with the exception that only one output is required. The rightmost flip-flop provides the single output for the register as shown in Figure 7.47.

An important application of a SISO register is to deserialize data from a disk drive. A serial bit stream is read from a disk drive and converted into parallel bits by means of a SIPO register. When 8 bits have been shifted into the register, the bytes are shifted in parallel into a matrix of SISO registers, where each bit is shifted into a particular column. The SISO register, in this application, performs the function of a first-in, first-out (FIFO) queue and acts as a buffer between the disk drive and the system input/output (I/O) data bus.

Information is read from a disk sector into the FIFO and then transferred to a destination by means of the data bus. The destination may be a CPU register or a storage location if direct-memory access is implemented. The mode of transfer is bit parallel, byte serial and is either synchronous, where the transfer rate is determined by a system clock, or asynchronous, where the transfer rate is determined by the disk control unit.

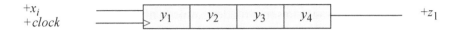

$+x_i$
$+clock$ $\quad\quad y_1 \quad y_2 \quad y_3 \quad y_4 \quad\quad\quad +z_1$

Figure 7.47   Block diagram for a 4-bit serial-in, serial-out register.

The mode of data transfer between a disk subsystem and a destination is usually in *burst mode*, in which the disk control unit remains logically connected to the system bus for the entire data transfer sequence; that is, until the complete sector has been transferred. Some systems, however, do not allow burst mode transfer, because this would prevent other peripheral devices from gaining control of the bus. This presents no problem when the disk control unit contains a FIFO. In this situation, data continues to be read from the disk and is transferred to the FIFO, where the bytes are retained until the disk control unit again gains control of the bus. The FIFO prevents data from being lost while the control unit is arbitrating for bus control.

The same implementation of a SISO register matrix can be used as an instruction queue in a CPU instruction pipeline. The CPU prefetches instructions from memory during unused memory cycles and stores the instructions in the FIFO queue. Thus, an instruction stream can be placed in the instruction queue to wait for decoding and execution by the processor. Instruction queueing provides an effective method to increase system throughput.

7.2.3 Synchronous Counters

Counters are essential devices used in the design of digital systems. Counters have a finite number of states and represent simple Moore machines in most cases. The λ output logic is usually a function of the present state only; that is, $\lambda(Y_{j(t)})$. The state of the counter is interpreted as an integer with respect to a modulus. A number A modulo n is defined as the remainder after dividing A by n. Some counters contain a set of binary input variables from which the counter achieves an initial state.

A clock input signal causes the counter flip-flops to react only at selected discrete intervals of time; in some cases, the clock input occurs randomly. Using the clock pulses to initiate state changes, the machine counts in either an ascending or descending sequence of states. Other counters have no inputs except a clock pulse and are usually reset to an initial state of $y_1 y_2 \ldots y_p = 00 \ldots 0$. In general, a p-stage counter counts modulo 2^p.

This section will discuss only synchronous counters; asynchronous counters are inherently slow, because of the ripple effect caused by the output of stage y_i functioning as the clock input for stage y_{i+1}. The maximum time for an asynchronous counter to change state occurs when all flip-flops are set and the count increments from $2^p - 1$ to zero.

The synchronous sequential machines in this section are associated with a set of transformations on a set of states and follow a prescribed sequence of states under control of a clock input signal. When the active clock transition occurs at the input, the state of the machine changes to some predetermined value as defined by the machine specifications. Counting sequentially is completely arbitrary, although the next state is usually an increment or decrement by one, or a state in which only one flip-flop changes state, as in a Gray code counter.

Modulo-10 counter   Modulo-10 counters are extensively used in digital computers when counting is required in radix 10. A modulo-10, or binary-coded decimal (BCD) decade counter, generates ten states in the following sequence: 0000, 0001, 0010, 0011, 0100, 0101, 0110, 0111, 1000, 1001, 0000, ... Thus, each decade requires four flip-flops.

The synthesis of a modulo-10 counter is relatively straightforward. The counter is initially reset to $y_1 y_2 y_3 y_4 = 0000$, then increments by one at each active clock transition until a state code of $y_1 y_2 y_3 y_4 = 1001$ is reached. At the next active clock transition, the counter sequences to state $y_1 y_2 y_3 y_4 = 0000$.

There are, however, six unused states, 1010 through 1111, that represent invalid numbers for BCD. These unused states can be regarded as "don't care" states for the purpose of minimizing the δ next-state logic, unless the counter is self-starting, in which case, all unused states contain entries which cause the counter to proceed to a predetermined state at the next active clock transition. The synthesis procedure begins below, using D flip-flops.

The state diagram for the modulo-10 counter is shown in Figure 7.48. The counter will not be self-starting; that is, all invalid BCD states will be considered as "don't care" or unused states. It is unlikely that the machine will enter an unused state; however, the possibility does exist. Digital systems enter unused states only under adverse environmental conditions such as, electrical noise, power supply voltage outside the specified operating range, or a hardware malfunction. If any of these situations occur, then the performance of the entire system is in jeopardy, not just the counter. The unused states correspond to minterms 10 through 15.

Every state in Figure 7.48 is unique. The outputs correspond to the state code of the individual states for this type of Moore machine. Therefore, no equivalent states exist. The state codes are assigned in sequence, $y_1 y_2 y_3 y_4 = 0000$ through 1001 for the valid digits. The invalid digits of $y_1 y_2 y_3 y_4 = 1010$ through 1111 constitute unused states.

The next-state table for this simple modulo-10 counter is not necessary — the state diagram is sufficient to obtain the input maps.

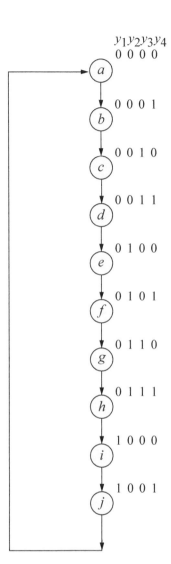

Figure 7.48   State diagram for a modulo-10 counter.

The input maps of Figure 7.49 can be derived either from the next-state table or from the state diagram. For example, using the state diagram, the entry for Dy_1 for location $y_1y_2y_3y_4 = 0111$ is obtained as follows: Flip-flop y_1 changes from 0 to 1 as the machine progresses from state h to state i. In the map for Dy_2 in minterm location $y_1y_2y_3y_4 = 0011$, a 1 is entered because flip-flop y_2 changes from 0 to 1 as the machine progresses from state d to state e. Four input maps are necessary, one for each D flip-flop. The input equations are presented in Equation 7.21.

Dy_1

y_1y_2 \ y_3y_4	00	01	11	10
00	0 (0)	0 (1)	0 (3)	0 (2)
01	0 (4)	0 (5)	1 (7)	0 (6)
11	– (12)	– (13)	– (15)	– (14)
10	1 (8)	0 (9)	– (11)	– (10)

Dy_2

y_1y_2 \ y_3y_4	00	01	11	10
00	0 (0)	0 (1)	1 (3)	0 (2)
01	1 (4)	1 (5)	0 (7)	1 (6)
11	– (12)	– (13)	– (15)	– (14)
10	0 (8)	0 (9)	– (11)	– (10)

Dy_3

y_1y_2 \ y_3y_4	00	01	11	10
00	0 (0)	1 (1)	0 (3)	1 (2)
01	0 (4)	1 (5)	0 (7)	1 (6)
11	– (12)	– (13)	– (15)	– (14)
10	0 (8)	0 (9)	– (11)	– (10)

Dy_4

y_1y_2 \ y_3y_4	00	01	11	10
00	1 (0)	0 (1)	0 (3)	1 (2)
01	1 (4)	0 (5)	0 (7)	1 (6)
11	– (12)	– (13)	– (15)	– (14)
10	1 (8)	0 (9)	– (11)	– (10)

Figure 7.49   Input maps for the modulo-10 counter of Figure 7.48.

$$Dy_1 = y_1y_4' + y_2y_3y_4$$

$$Dy_2 = y_2y_3' + y_2y_4' + y_2'y_3y_4$$

$$Dy_3 = y_1'y_3'y_4 + y_3y_4'$$

$$Dy_4 = y_4' \qquad\qquad (7.21)$$

No output maps are required for a single 4-bit modulo-10 counter. If, however, the counter is one decade of a multi-decade counter, then λ output logic is necessary to indicate when decade$_i$ has attained a count of $y_1 y_2 y_3 y_4 = 1001$. The next active clock transition will reset decade$_i$ and increment by one the next higher-order decade$_{i+1}$. The design of an n-digit BCD counter is constructed from n 4-bit modulo-10 counters.

The logic diagram of Figure 7.50 is derived from the D flip-flop input equations of Equation 7.21. The counter is reset initially to $y_1 y_2 y_3 y_4 = 0000$. The timing diagram is illustrated in Figure 7.51. The clock signal is supplied to all flip-flops simultaneously. State changes occur only on the positive clock transition. Using either Equation 7.21 or the logic diagram of Figure 7.50, the counter can be shown to increment through the modulo-10 counting sequence, then return to 0000 at the next positive clock transition.

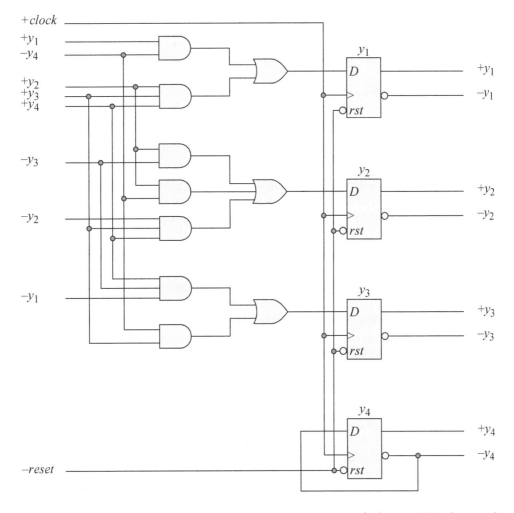

Figure 7.50   Logic diagram for the modulo-10 counter of Figure 7.48, where y_4 is the low-order flip-flop.

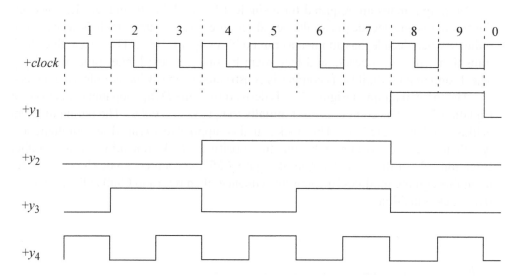

Figure 7.51   Timing diagram for the modulo-10 counter of Figure 7.50.

Johnson counter   The modulo-10 counter previously described had a counting sequence that increased in a binary manner from zero to nine; however, it used only ten combinations of 16 possible states. Other counters are frequently utilized in digital computers. These are down-counters that count in a descending sequence from some preset state to zero. The state diagram contains state codes that decrement by one for each succeeding state. The synthesis of these counters is identical to the method previously presented.

Still other counters can be designed for a unique application in which the counting sequence is neither entirely up nor entirely down. These have a nonsequential counting sequence that is prescribed by external requirements. Such a counter is shown in the state diagram of Figure 7.52, in which the counting sequence is $y_1y_2y_3 = 000, 100,$ 110, 111, 011, 001, 000, ... The counter is reset initially to $y_1y_2y_3 = 000$. For six of the eight possible states for three variables, the state transitions are completely defined. The remaining two states are unspecified and can be regarded as "don't care" states in order to minimize the δ next-state logic. This presents no problem under normal operating conditions where the environment is free from electrical interference.

The counter of Figure 7.52 represents a *Johnson counter* in which any two contiguous state codes (or code words) differ by only one variable. The Johnson counter is also referred to as a "Möbius counter," because the output of the last stage is inverted and fed back to the first stage. August F. Möbius was a German mathematician who discovered a one-sided surface that is constructed from a rectangle by holding one end fixed, rotating the opposite end through 180 degrees, and applying it to the first end.

It is similar, in this respect, to a Gray code counter. The Gray code concept is used in Karnaugh maps. Any physically adjacent minterm locations are also logically adjacent because they differ in only one variable and, therefore, can be combined into a term with fewer variables. Contiguous code words that are logically adjacent is an important consideration in eliminating output glitches and will be elaborated in detail in a later section.

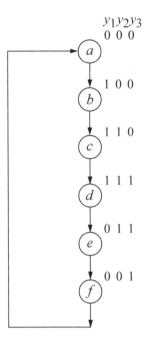

$y_1 y_2 y_3$
0 0 0

1 0 0

1 1 0

1 1 1

0 1 1

0 0 1

Figure 7.52   State diagram for a Johnson counter with a nonsequential counting sequence. There are two unused states: $y_1 y_2 y_3 = 010$ and 101.

Using the synthesis procedure previously described, the input maps are obtained using D flip-flops, as shown in Figure 7.53. The maps can be derived directly from the state diagram without the necessity of generating a next-state table. For example, from state b ($y_1 y_2 y_3 = 100$), the machine sequences to state c ($y_1 y_2 y_3 = 110$) where the next state for flip-flop y_1 is 1. Thus, a 1 is entered in minterm location $y_1 y_2 y_3 = 100$ for flip-flop y_1.

Likewise, from state c the machine proceeds to state d where the next state for y_1 is 1; therefore, a 1 is entered in minterm location $y_1 y_2 y_3 = 110$ for flip-flop y_1. In a similar manner, the remaining entries are obtained for the input map for y_1, as well as for the input maps for y_2 and y_3. The input equations are listed in Equation 7.22.

The logic diagram is shown in Figure 7.54. The counting sequence is easily verified by asserting the appropriate input logic levels to the flip-flop D inputs for each state of the counter and then applying the active clock transition. The timing diagram is shown in Figure 7.55.

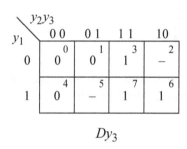

$$Dy_1$$

$$Dy_2$$

$$Dy_3$$

Figure 7.53   Input maps for the Johnson counter of Figure 7.52 using D flip-flops.
The unused states are $y_1y_2y_3 = 010$ and 101.

$$Dy_1 = y_3'$$

$$Dy_2 = y_1$$

$$Dy_3 = y_2 \tag{7.22}$$

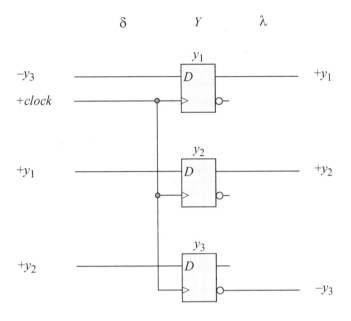

Figure 7.54   Logic diagram for the Johnson counter of Figure 7.52 using D flip-flops. The counting order is nonsequential. Flip-flop y_3 is the low-order stage.

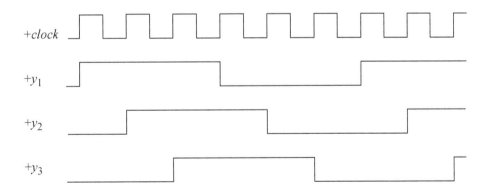

Figure 7.55   Timing diagram for the Johnson counter of Figure 7.54.

Pulse generator counter   Another interesting counter is a pulse generator counter in which the counter generates a series of nonoverlapping pulses as shown in the state diagram of Figure 7.56. These four disjoint pulses are a function of the counting sequence and can be used for system timing. The counter provides a simple, yet effective state machine, where each pulse represents a different state.

The pulses can be used in a digital system to define time slots in which different operations take place. For example, a peripheral control unit may have three different time slots: one to interface with the input/output channel of the computer; one dedicated to internal control unit operations; and one to interface with the peripheral device.

The D flip-flop input maps are shown in Figure 7.57 in which the unused states are treated as "don't care" states. The input equations are shown in Equation 7.23, which are used to design the logic diagram of Figure 7.58. The timing diagram is shown in Figure 7.59.

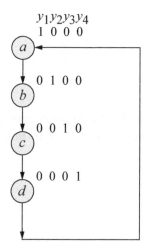

Figure 7.56   State diagram for a pulse generator counter.

y_1y_2 \ y_3y_4	0 0	0 1	1 1	1 0
0 0	– (0)	1 (1)	– (3)	0 (2)
0 1	0 (4)	– (5)	– (7)	– (6)
1 1	– (12)	– (13)	– (15)	– (14)
1 0	0 (8)	– (9)	– (11)	– (10)

Dy_1

y_1y_2 \ y_3y_4	0 0	0 1	1 1	1 0
0 0	– (0)	0 (1)	– (3)	0 (2)
0 1	0 (4)	– (5)	– (7)	– (6)
1 1	– (12)	– (13)	– (15)	– (14)
1 0	1 (8)	– (9)	– (11)	– (10)

Dy_2

Figure 7.57   Input maps for the pulse generator counter of Figure 7.56.

y_1y_2 \ y_3y_4	0 0	0 1	1 1	1 0
0 0	− (0)	0 (1)	− (3)	0 (2)
0 1	1 (4)	− (5)	− (7)	− (6)
1 1	− (12)	− (13)	− (15)	− (14)
1 0	0 (8)	− (9)	− (11)	− (10)

Dy_3

y_1y_2 \ y_3y_4	0 0	0 1	1 1	1 0
0 0	− (0)	0 (1)	− (3)	1 (2)
0 1	0 (4)	− (5)	− (7)	− (6)
1 1	− (12)	− (13)	− (15)	− (14)
1 0	0 (8)	− (9)	− (11)	− (10)

Dy_4

Figure 7.57   (Continued)

$$Dy_1 = y_4 \quad Dy_2 = y_1 \quad Dy_3 = y_2 \quad Dy_4 = y_3 \qquad (7.23)$$

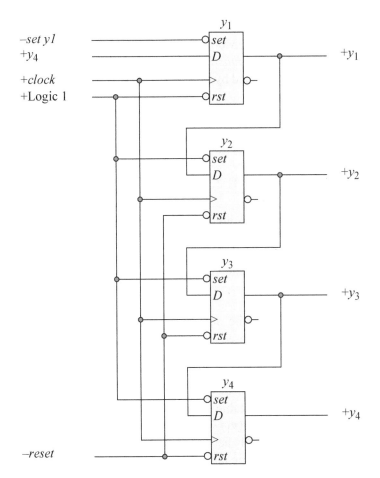

Figure 7.58   Logic diagram for the pulse generator counter of Figure 7.56.

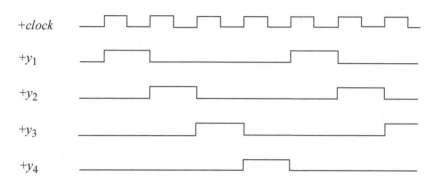

Figure 7.59   Timing diagram for the pulse generator counter of Figure 7.56.

7.2.4 Moore Machines

This section extends the concepts of Moore machines that were introduced previously and presents a procedure for synthesizing Moore machines. The primary focus of this section will be on the synthesis of *deterministic synchronous sequential machines*, in which the next state is uniquely determined by the present state $Y_{j(t)}$ and the present inputs $X_{i(t)}$.

The Moore model of sequential machines is the result of a paper by E.F. Moore in 1956. A Moore machine was formally defined in a previous section as a synchronous sequential machine characterized by the following 5-tuple:

$$M = (X,\ Y,\ Z,\ \delta,\ \lambda)$$

where X is the input alphabet, Y is the state alphabet, and Z is the output alphabet. The next-state function δ maps the Cartesian product of X and Y into Y, and thus, is determined by both the present inputs and the present state. The output function λ maps Y into Z such that the output vector is a function of the present state only and is independent of the external inputs.

Example 7.6   A Moore machine will be designed that operates according to the state diagram of Figure 7.60. There are three inputs x_1, x_2, and x_3; three *JK* state flip-flops y_1, y_2, and y_3; and two outputs z_1 that is asserted in state e ($y_1 y_2 y_3 = 111$) and z_2 that is asserted in state $f(y_1 y_2 y_3 = 110)$.

The *JK* flip-flop functional characteristics are shown in Table 7.17 and the excitation table in Table 7.18. The next-state table is shown in Table 7.19 and is generated directly from the state diagram. In state a ($y_1 y_2 y_3 = 000$), only x_1 is a contributing factor; in state b ($y_1 y_2 y_3 = 001$), x_2 is the only active input, and in state c ($y_1 y_2 y_3 = 010$), x_3 is the only active input. Therefore, some inputs are indicated as "don't care" inputs in the next-state table.

For example, consider the present state a ($y_1 y_2 y_3 = 000$). The next state for flip-flop y_1 is always $y_1 = 0$, regardless of the value of x_1. Therefore, the state transition sequence for y_1 is $0 \rightarrow 0$, yielding $Jy_1 = 0$ and $Ky_1 = -$ from Table 7.18.

In state a ($y_1 y_2 y_3 = 000$), if $x_1 = 1$, flip-flop y_2 has a next value of $y_2 = 0$ when the machine sequences to state b ($y_1 y_2 y_3 = 001$). Therefore, flip-flop y_2 sequences from $0 \rightarrow 0$. If $x_1 = 0$ in state a, flip-flop y_2 has a next value of $y_2 = 1$ when the machine sequences to state c ($y_1 y_2 y_3 = 010$); therefore, the state transition sequence is $0 \rightarrow 1$. From Table 7.18, the input values are $Jy_2 = x_1'$ and $Ky_2 = -$.

In state a ($y_1 y_2 y_3 = 000$), if $x_1 = 1$, flip-flop y_3 has a next value of 1 when the machine sequences to state b ($y_1 y_2 y_3 = 001$). Therefore, the state transition sequence for flip-flop y_3 is $0 \rightarrow 1$. If $x_1 = 0$ in state a ($y_1 y_2 y_3 = 000$), then flip-flop y_3 has a next value of 0 when the machine sequences to state c ($y_1 y_2 y_3 = 010$), providing a state transition sequence of $0 \rightarrow 0$. From Table 7.18, this yields input values $Jy_3 = x_1$ and $Ky_3 = -$. In a similar manner, the remaining entries are determined for the next-state table.

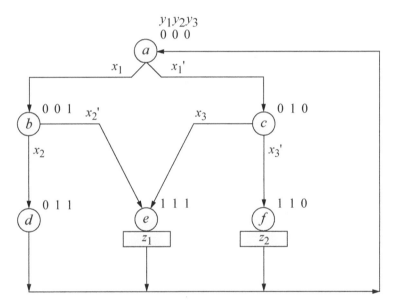

Figure 7.60   State diagram for the Moore machine of Example 7.6.

Table 7.17 *JK* Flip-Flop Functional Characteristics

J K	Function
0 0	No change
0 1	Reset
1 0	Set
1 1	Toggle

Table 7.18 Excitation Table for a *JK* Flip-Flop

Present State $Y_{j(t)}$	Next State $Y_{k(t+1)}$	Data Inputs J K
0	0	0 –
0	1	1 –
1	0	– 1
1	1	– 0

Table 7.19 Next-State Table for the Moore Machine of Example 7.6

y_1	y_2	y_3	x_1	x_2	x_3	y_1	y_2	y_3	Jy_1	Ky_1	Jy_2	Ky_2	Jy_3	Ky_3	z_1	z_2
\	Present State		\	Inputs		\	Next State		\	Flip-Flop Inputs					\	Outputs
0	0	0	0	–	–	0	1	0	0	–	1	–	0	–	0	0
0	0	0	1	–	–	0	0	1	0	–	0	–	1	–	0	0
0	0	1	–	0	–	1	1	1	1	–	1	–	–	0	0	0
0	0	1	–	1	–	0	1	1	0	–	1	–	–	0	0	0
0	1	0	–	–	0	1	1	0	1	–	–	0	0	–	0	0
0	1	0	–	–	1	1	1	1	1	–	–	0	1	–	0	0
0	1	1	–	–	–	0	0	0	0	–	–	1	–	1	0	0
1	0	0	–	–	–	–	–	–	–	–	–	–	–	–	–	–
1	0	1	–	–	–	–	–	–	–	–	–	–	–	–	–	–
1	1	0	–	–	–	0	0	0	–	1	–	1	0	–	0	1
1	1	1	–	–	–	0	0	0	–	1	–	1	–	1	1	0

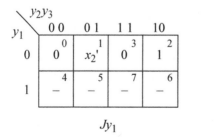

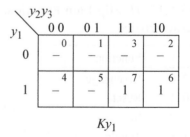

Figure 7.61   Input maps for the Moore machine of Example 7.6.

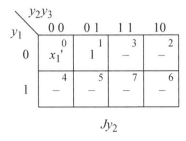

Jy2 map

y_1 \ y_2y_3	00	01	11	10
0	x_1' (0)	1 (1)	– (3)	– (2)
1	– (4)	– (5)	– (7)	– (6)

Jy_2

Ky2 map

y_1 \ y_2y_3	00	01	11	10
0	– (0)	– (1)	1 (3)	0 (2)
1	– (4)	– (5)	1 (7)	1 (6)

Ky_2

Jy3 map

y_1 \ y_2y_3	00	01	11	10
0	x_1 (0)	– (1)	– (3)	x_3 (2)
1	– (4)	– (5)	– (7)	0 (6)

Jy_3

Ky3 map

y_1 \ y_2y_3	00	01	11	10
0	– (0)	0 (1)	1 (3)	– (2)
1	– (4)	– (5)	1 (7)	– (6)

Ky_3

Figure 7.61   (Continued)

The input equations are listed in Equation 7.24 as obtained from the input maps. Consider the map for Jy_1. Minterm locations 1 and 5 combine to yield the term $y_2'y_3x_2'$; minterm locations 2 and 6 combine to yield y_2y_3'. The map for Ky_1 yields the term $Ky_1 = 1$, because every location contains either an entry of 1 or "don't care."

Consider the map for Jy_2. Every location contains either x_1' or a "don't care," because the entry of 1 in minterm location $1 = x_1 + x_1'$. Therefore, the first term for Jy_2 is x_1'. Now the entry of 1 in minterm location 1 combines with minterm locations 3, 5, and 7 to yield y_3. The equation for Ky_2 is $Ky_2 = y_3 + y_1$.

Consider the map for Jy_3. Minterm locations 0, 1, 4, and 5 combine to yield $y_2'x_1$. Minterms locations 2 and 3 combine to yield $y_1'y_2x_3$. The equation for Ky_3 is $Ky_3 = y_2$. The logic diagram is shown in Figure 7.62 and is designed using the equations of Equation 7.24.

$$Jy_1 = y_2'y_3x_2' + y_2y_3' \qquad Ky_1 = 1$$

$$Jy_2 = x_1' + y_3 \qquad Ky_2 = y_3 + y_1$$

$$Jy_3 = y_2'x_1 + y_1'y_2x_3 \qquad Ky_3 = y_2 \qquad (7.24)$$

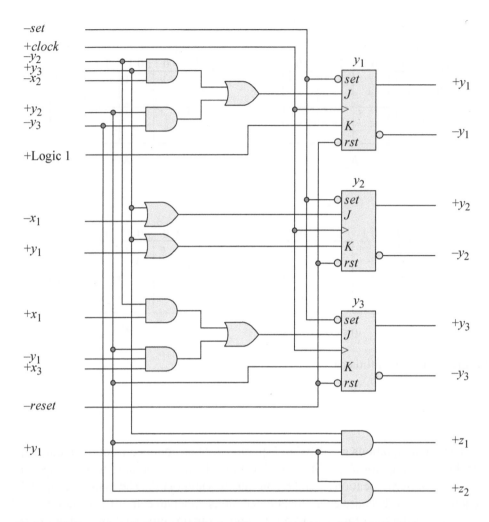

Figure 7.62   Logic diagram for the Moore machine of Example 7.6.

Example 7.7   A Moore machine will be designed that operates according to the state diagram of Figure 7.63. There are two inputs x_1, and x_2; three D state flip-flops y_1, y_2, and y_3; and three outputs z_1 that is asserted in state a ($y_1 y_2 y_3 = 000$), z_2 that is asserted in state c ($y_1 y_2 y_3 = 010$), and z_3 that is asserted in state d ($y_1 y_2 y_3 = 100$). There are three unused states: $y_1 y_2 y_3 = 001$, 101, and 110.

State codes were deliberately inserted that would cause glitches on the outputs; therefore, all inputs are asserted at time t_2 and deasserted at time t_3. This ensures that there will be no glitches on the outputs for any state transition sequence. For example, for a state transition from state b ($y_1 y_2 y_3 = 011$) to state d ($y_1 y_2 y_3 = 100$), all flip-flops change state. Therefore, all flip-flops could conceivably momentarily pass through state a ($y_1 y_2 y_3 = 000$) — depending on the propagation delays of the flip-flops — and assert output z_1.

It is possible that there could be three glitches on output z_2 for the following state transition sequences: state a ($y_1y_2y_3 = 000$) to state b ($y_1y_2y_3 = 011$) if y_2 sets before y_3 sets; from state b ($y_1y_2y_3 = 011$) to state d ($y_1y_2y_3 = 100$) if y_3 resets before y_1 and y_2 change state; and from state e ($y_1y_2y_3 = 111$) to state a ($y_1y_2y_3 = 000$) if y_1 and y_3 reset before y_2 resets. Output z_3 may produce a glitch when the machine sequences from state e ($y_1y_2y_3 = 111$) to state a ($y_1y_2y_3 = 000$) if y_2 and y_3 change state before y_1 changes state.

Asserting all outputs at time t_2 and deasserting all outputs at time t_3 negates the possibility of a glitch. Output glitches can occur at the active transition of the clock when the flip-flops are changing state. This is a very small interval of time and is referred to as the Δt time.

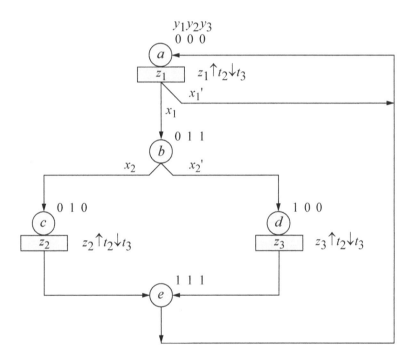

Figure 7.63   State diagram for the Moore machine of Example 7.7.

Since D flip-flops are used, there is no need for a next-state table — the input maps can be derived directly from the state diagram and are shown in Figure 7.64. For example, in state a ($y_1y_2y_3 = 000$), flip-flop y_1 always has a next value of 0 regardless of the path taken; therefore, a 0 is inserted in minterm location 0 of the map for Dy_1. In state b ($y_1y_2y_3 = 011$), flip-flop y_1 has a next value of 1 if input x_2 is deasserted; therefore, x_2' is inserted in minterm location 3 of the map for Dy_1. In a similar manner, the remaining entries in the map for Dy_1 and the maps for Dy_2 and Dy_3 are derived. The D input equations are shown in Equation 7.25 as derived from the input maps; the logic diagram is shown in Figure 7.65 and is designed from D the input equations.

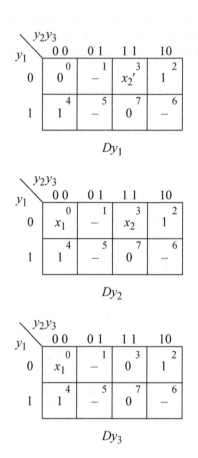

Figure 7.64   Input maps for the Moore machine of Example 7.7.

$$Dy_1 = y_1'y_3x_2' + y_2y_3' + y_1y_2'$$

$$Dy_2 = y_2'x_1 + y_1y_2' + y_1'y_3x_2 + y_2y_3'$$

$$Dy_3 = y_2'x_1 + y_1y_2' + y_2y_3' \tag{7.25}$$

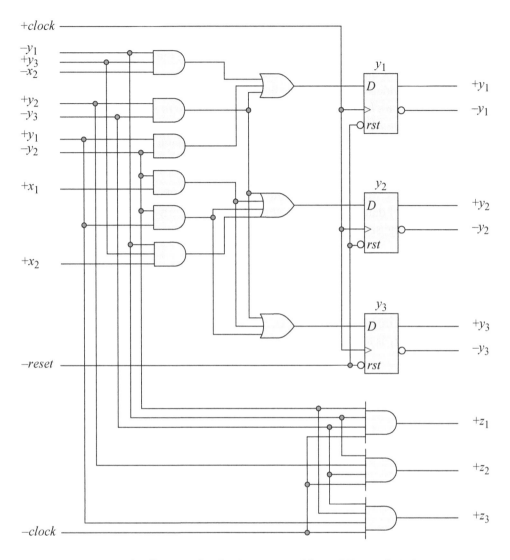

Figure 7.65   Logic diagram for the Moore machine of Example 7.7.

7.2.5 Mealy Machines

A Mealy machine is also an important class of finite-state machine and represents an alternative model that is more widely used than a Moore machine. The Mealy class of synchronous sequential machines is the result of a paper by G. H. Mealy in 1955 on the synthesis of sequential circuits. A Mealy machine is defined as a synchronous sequential machine characterized by the following 5-tuple:

$$M = (X, Y, Z, \delta, \lambda)$$

where X is the input alphabet, Y is the state alphabet, and Z is the output alphabet.

The next-state function δ maps the Cartesian product of X and Y into Y, and thus, is determined by both the present inputs and the present state. The output function λ maps the Cartesian product of X and Y into Z such that, the output vector is a function of both the present inputs and the present state.

This is the underlying difference between Moore and Mealy machines — *the outputs of a Moore machine are directly related to the present state only, whereas the outputs of a Mealy machine are a function of both the present state and the present inputs.* Examples will now be presented that illustrate the design procedure for Mealy synchronous sequential machines.

Example 7.8   A Mealy synchronous sequential machine generates an output z_1 whenever a serial data input line x_1 contains a 3-bit word with an odd number of 1s. The format for the serial data line is shown below.

$$x_1 = \left| \begin{array}{ccc} b_1 & b_2 & b_3 \end{array} \right| \begin{array}{ccc} b_1 & b_2 & b_3 \end{array} \left| \begin{array}{ccc} b_1 & b_2 & b_3 \end{array} \right| \cdots$$

where $b_i = 0$ or 1. There is no space between words. Output z_1 is asserted during the first half of the third bit time b_3; that is, $z_1 \uparrow t_1 \downarrow t_2$. The chart shown below lists the state codes in which z_1 is asserted for an odd number of 1s.

$x_1 =$	0	0	0		
	0	0	1	$\rightarrow$	z_1
	0	1	0	$\rightarrow$	z_1
	0	1	1		
	1	0	0	$\rightarrow$	z_1
	1	0	1		
	1	1	0		
	1	1	1	$\rightarrow$	z_1

The state diagram shown in Figure 7.66 in which the state codes have been selected so that there will be no glitches on output z_1 for any state transition sequence. There are three unused states: $y_1 y_2 y_3 = 101$, 110, and 111. If two or more flip-flops change value for a state transition sequence, then the machine may momentarily pass through either an unused state or a state in which there is no output — in both cases there will be no glitch on output z_1.

For example, for a state transition from state b ($y_1 y_2 y_3 = 001$) to state e ($y_1 y_2 y_3 = 100$), the machine may pass through state $y_1 y_2 y_3 = 101$ if flip-flop y_1 sets before flip-flop y_3 resets; however, state $y_1 y_2 y_3 = 101$ is an unused state. In a similar manner, for a transition from state c ($y_1 y_2 y_3 = 011$) to state a ($y_1 y_2 y_3 = 000$) with $x_1 = 0$, the machine may pass through state b ($y_1 y_2 y_3 = 001$) if y_2 resets before y_3 resets; however, state b has no output.

The D flip-flop input maps are shown in Figure 7.67 and the equations are shown in Equation 7.26. The map for output z_1 is shown in Figure 7.68 and the equation is shown in Equation 7.27. The logic diagram is shown in Figure 7.69.

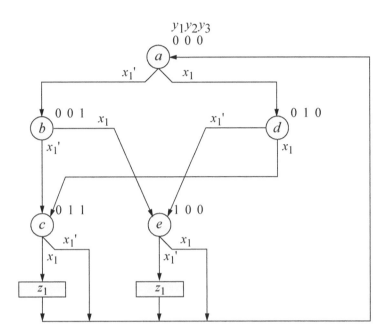

Figure 7.66   State diagram for the Mealy machine of Example 7.8.

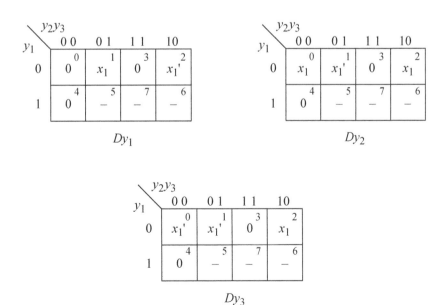

Figure 7.67   Input maps for the Mealy machine of Example 7.8.

$$Dy_1 = y_2'y_3x_1 + y_2y_3'x_1'$$

$$Dy_2 = y_1'y_3'x_1 + y_2'y_3x_1'$$

$$Dy_3 = y_1'y_2'x_1' + y_2y_3'x_1 \qquad\qquad (7.26)$$

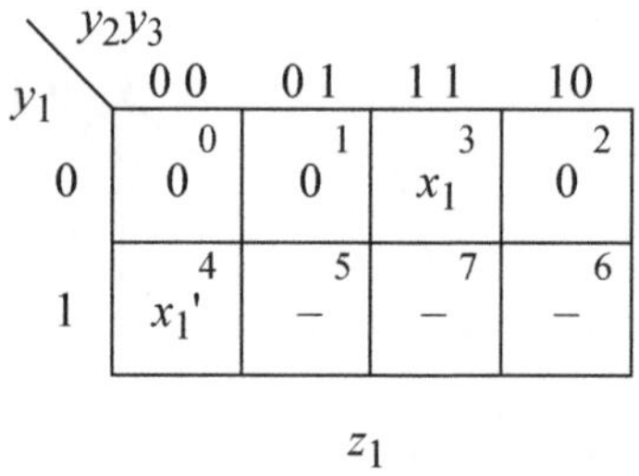

y_1 \ y_2y_3	0 0	0 1	1 1	1 0
0	0 (0)	0 (1)	x_1 (3)	0 (2)
1	x_1' (4)	– (5)	– (7)	– (6)

z_1

Figure 7.68   Output map for the Mealy machine of Example 7.8.

$$z_1 = y_1x_1' + y_2y_3x_1 \qquad\qquad (7.27)$$

Example 7.9   The state diagram for a Mealy synchronous sequential machine is shown in Figure 7.70 and will be implemented with *JK* flip-flops. There are three inputs x_1, x_2, and x_3 and one output z_1. There are two state flip-flops y_1 and y_2 that are reset to state a ($y_1y_2 = 11$) and one unused state $y_1y_2 = 00$.

The next-state table is shown in Table 7.20 and is obtained directly from the state diagram. For example, consider state a ($y_1y_2 = 11$). Input x_3 does not contribute to a state transition to state b ($y_1y_2 = 10$) or to state c ($y_1y_2 = 01$); therefore, input x_3 is entered as a "don't care" value in the next-state table. If $x_1 = 0$ in state a ($y_1y_2 = 11$), then the machine proceeds to state c ($y_1y_2 = 01$); therefore, the next-state table contains a next state of $y_1y_2 = 01$ whenever $x_1 = 0$.

If $x_1x_2 = 10$ in state a, then the machine remains in state a. If $x_1x_2 = 11$ in state a, then the machine proceeds to state b ($y_1y_2 = 10$), where output z_1 is asserted if $x_3 = 1$, then sequences to state a; otherwise, the machine proceeds to state a without asserting output z_1.

The input maps for the *JK* flip-flops are shown in Figure 7.71 and the equations are shown in Equation 7.28. Figure 7.72 contains the Karnaugh map for output z_1 with the equation in Equation 7.29. The input and output equations are used to design the logic diagram of Figure 7.73.

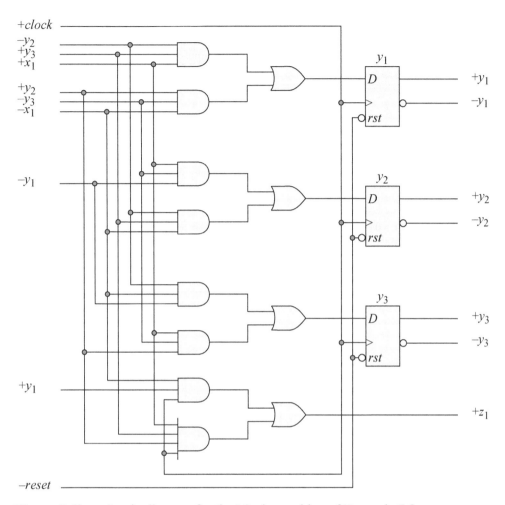

Figure 7.69   Logic diagram for the Mealy machine of Example 7.8.

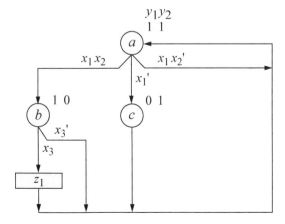

Figure 7.70   State diagram for the Mealy machine of Example 7.9.

Table 7.20 Next-State Table for the Mealy Machine of Example 7.9

| Present State | Inputs | Next State | Flip-Flop Inputs | | Output |
y_1 y_2	x_1 x_2 x_3	y_1 y_2	Jy_1 Ky_1	Jy_2 Ky_2	z_1
0 0	– – –	– –	– –	– –	–
0 1	– – –	1 1	1 –	– 0	0
1 0	– – 0	1 1	– 0	1 –	0
	– – 1	1 1	– 0	1 –	1
1 1	0 0 –	0 1	– 1	– 0	0
	0 1 –	0 1	– 1	– 0	0
	1 0 –	1 1	– 0	– 0	0
	1 1 –	1 0	– 0	– 1	0

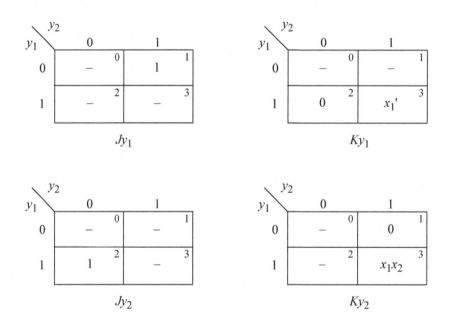

Figure 7.71   Input maps for the Mealy machine of Example 7.9.

$$Jy_1 = 1 \qquad\qquad Ky_1 = y_2 x_1'$$

$$Jy_2 = 1 \qquad\qquad Ky_2 = y_1 x_1 x_2 \qquad\qquad (7.28)$$

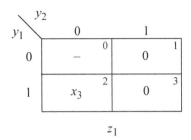

Figure 7.72   Output map for the Mealy machine of Example 7.9.

$$z_1 = y_2' x_3 \tag{7.29}$$

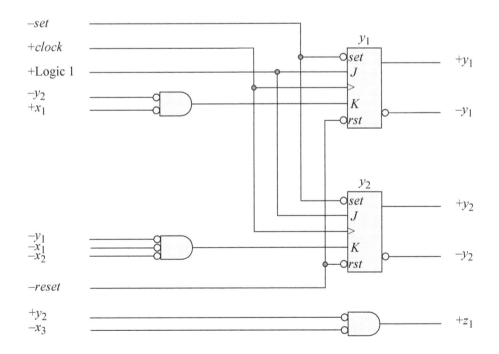

Figure 7.73   Logic diagram for the Mealy machine of Example 7.9.

7.2.6 Output Glitches

A *glitch* in synchronous sequential machines is any false or spurious electronic signal. These narrow, unwanted pulses wreak havoc in digital systems if the glitch occurs on an output signal. For example, the output signal of a logic circuit may connect to the clock input of a counter and produce erroneous results. Therefore, eliminating output glitches is extremely important, even at the expense of additional logic.

In synchronous systems, glitches can occur in the time period between the active clock transition and circuit stabilization. This is shown in Figure 7.74 and is indicated by the time duration Δt. It is during this time, when the machine is changing states, that the outputs are susceptible to glitches.

Due to varying propagation delays of the internal logic of the storage elements, the machine may enter an unstable, or transient, state. Although momentary in duration, this transient state can cause an output glitch in both Moore and Mealy machines if the output is decoded directly from the p-tuple state codes. If the outputs are enabled at time t_2, then glitches that occur during the period of instability are of no consequence — the machine has long since stabilized. Techniques will now be presented to eliminate output glitches.

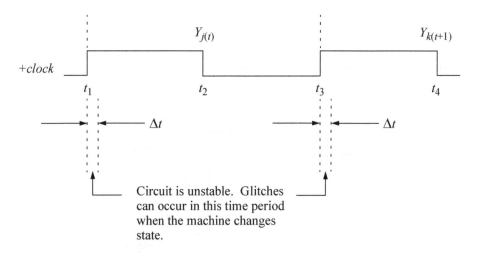

Figure 7.74   Illustration of the time intervals in which glitches are possible for synchronous sequential machines.

Glitch elimination using state code assignment   It is often possible to reassign state codes to avoid output glitches. This may result in additional δ next-state logic, however, if state code adjacency cannot be maintained. Whenever possible, state codes should be assigned such that there are a maximal number of adjacent 1s in the flip-flop input maps. This allows more minterm locations to be combined, resulting in minimized input equations in a sum-of-products form. State codes are adjacent if they differ in only one variable. For example, state codes $y_1y_2y_3 = 101$ and 100 are adjacent because only y_3 changes. Thus, minterm locations 101 and 100 can be combined into one term. However, state codes $y_1y_2y_3 = 101$ and 110 are not adjacent because two variables change: flip-flops y_2 and y_3.

However, if the primary concern is to eliminate output glitches, then additional δ next-state logic is inconsequential in order to produce a more reliable machine. Output glitches can occur when two or more flip-flops change state and only when the machine specifications require that the outputs be asserted at time t_1.

States which have the same output should have adjacent state code assignments; that is, if states Y_i and Y_j both have z_1 as an output, then Y_i and Y_j should be adjacent. This allows for a larger grouping of 1s in the output map. If there is a contention between minimizing the δ next-state logic or the λ output logic — where it is possible to make either two nonoutput states adjacent or two output states adjacent, but not both — then a greater savings in logic is usually realized by minimizing the δ next-state input logic rather than the λ output logic.

An example of a Moore machine in which output glitches may occur is shown in the state diagram of Figure 7.75(a). In analyzing the state diagram, three paths are found which may generate glitches on outputs z_1 or z_2, depending on the propagation delays of flip-flops y_1, y_2, and y_3. Figure 7.75(b) resolves the problem by reassigning state codes.

In Figure 7.75(a), the path from state b ($y_1y_2y_3 = 001$) to state e ($y_1y_2y_3 = 111$) may pass through state d ($y_1y_2y_3 = 101$) if flip-flop y_1 sets before flip-flop y_2 sets. This will assert output z_1 for a path that does not involve z_1.

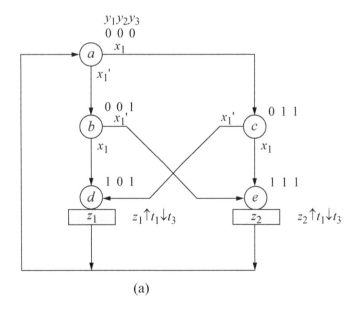

(a)

(Continued on next page)

Figure 7.75   State diagram for a Moore machine: (a) may produce glitches on output z_1 and (b) glitch-free outputs due to state code reassignment.

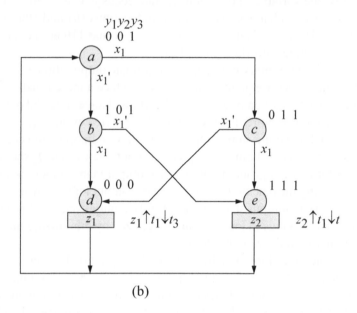

(b)

Figure 7.75   (Continued)

A path that may result in a glitch on z_2 is the path from state c $(y_1y_2y_3 = 011)$ to state d $(y_1y_2y_3 = 101)$, which may pass through state e $(y_1y_2y_3 = 111)$ if flip-flop y_1 sets before flip-flop y_2 resets. The path from state e $(y_1y_2y_3 = 111)$ to state a $(y_1y_2y_3 = 000)$ may pass through state d $(y_1y_2y_3 = 101)$ and result in a glitch on output z_1 if flip-flop y_2 resets before flip-flops y_1 and y_3 reset.

Glitch elimination using storage elements   It may not be possible to reassign state codes to eliminate output glitches. For example, Figure 7.76 presents a state diagram for a Moore machine in which any combination of state code assignments may still result in an output glitch. This can be verified by permuting the state codes so that all state transitions have been examined for all possible sets of codes.

In this machine, reassignment of state codes to eliminate output glitches is not possible, because all $2^p = 2^2 = 4$ state codes have been assigned and parallel paths exist for state transitions b, c, a and b, d, a. Thus, unused states cannot be utilized to avoid a state transition in which both flip-flops change state. A glitch may occur as the machine moves from state b $(y_1y_2 = 01)$ to state d $(y_1y_2 = 10)$ if flip-flop y_1 sets before flip-flop y_2 resets.

If the machine specifications require that $z_1\uparrow t_1\downarrow t_3$, then a D flip-flop in the λ output logic will satisfy this requirement, as shown in the logic diagram of Figure 7.77. A delayed clock signal is used to clock the D input of flip-flop z_1. The delay circuit delays the clock past the time (Δt) when a glitch could occur. Thus, although the output of the AND gate that decodes z_1 may glitch, the active clock transition for flip-flop z_1 will not occur until after the glitch has returned to a logic 0. The clock delay is a very small delay, thus the assertion/deassertion is still considered to be $\uparrow t_1 \downarrow t_3$. Delay circuits usually require a separate driver and receiver.

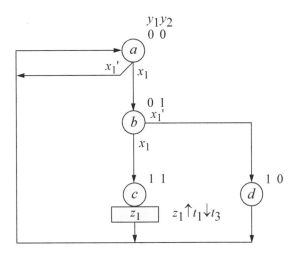

Figure 7.76   State diagram for a Moore machine with an inherent output glitch on z_1 for a state transition from state b to state d.

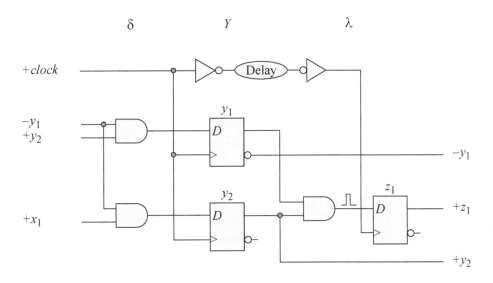

Figure 7.77   Logic diagram for the Moore machine of Figure 7.76 using a delayed clock to prevent a glitch on output z_1.

Glitch elimination using complemented clock   The simplest and most inexpensive method of eliminating output glitches is to include the complement of the machine clock in the implementation of the λ output logic. The logic that generates the output will consist of an AND gate which decodes the p-tuple state codes. One input of the AND gate is connected to the complement of the machine clock; that is, the negation of the clock signal which drives the state flip-flops. This will generate an output signal that is only one-half the duration of the clock cycle, but guarantees that the output is free from any erroneous assertions. The output is asserted at time t_2 and deasserted at time t_3.

Figure 7.78 shows a general block diagram for a Moore machine using the complement of the machine clock as an output gating function. A glitch that is caused by a state transition in which two or more flip-flops change state has no effect on the output — the glitch has returned to a logic 0 before the active level of the complemented clock occurs. Since the output assertion is for only one-half a clock period, storage elements are not required for the λ output logic. This method also applies to Mealy machines; the inputs, however, must remain active during the last half of the clock cycle.

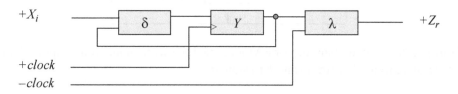

Figure 7.78   A generalized Moore machine which uses the complement of the machine clock as a gating function to eliminate output glitches.

Glitch elimination using delayed clock   One final technique is presented to circumvent the negative effects of glitches. If the machine specifications require that outputs be asserted at time t_1 and deasserted at t_2, then glitches are again possible, because output assertion occurs at the active clock transition. This technique applies to both Moore and Mealy machines.

Figure 7.79 shows a general block diagram for a Mealy machine which uses the active level of the delayed machine clock to enable the λ output logic. The state flip-flops are clocked by the $+clock$ signal, whereas the λ output logic is enabled by the $+clock$ delayed signal. The duration of the delay circuit must be equal to or greater than the time Δt — the time when glitches can occur. The machine has stabilized at the termination of the Δt period. The delay circuit can be either a delay element with a dedicated driver and receiver or simply an even number of inverters.

Since the outputs are asserted for one-half a clock period, storage elements are not required. Storage elements are necessary only when the outputs are active for a duration of one clock period, either $\uparrow t_1 \downarrow t_3$ or $\uparrow t_2 \downarrow t_4$. Since both clock and complemented clock are active for only one-half a clock cycle, they cannot be used to enable combinational λ output logic for one clock period.

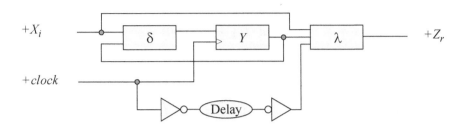

Figure 7.79   A general Mealy machine using a delayed clock as an output gating function, where $Z_r \uparrow t_1 \downarrow t_2$.

Glitches and output maps   Care must be taken when using "don't care" states in an output map in order to minimize the λ output logic. Any state transition that passes through an unused state — in which the beginning state and the destination state have the same output values — must have that same value placed in the unused state location. Since the unused state is a transient state, the output will glitch as the machine passes through that unused state en route to the destination state if a value is inserted that is different than the beginning and destination output value.

Consider the state diagram and output map for the Moore machine shown in Figure 7.80. The input variables have been omitted since they are not pertinent to the present discussion. When a reduced state diagram has been derived, the next step is to assign state codes. The state codes should be assigned so that no outputs will glitch as the machine progresses through the required state transition sequences. The state code assignment of Figure 7.80 results in an output that is free from transient assertions. This can be verified by checking all possible state transitions.

Referring to Figure 7.80(b), the equation for output z_1 can be minimized by combining the 1 in minterm 6 with one of the unused states in minterm 2, 4, or 7. Before using a "don't care" state, however, the state diagram must be checked to be certain that no state transition will occur that causes the machine to pass through the unused state en route to a destination state for a path that does not include z_1. Every path in the state diagram must be checked for this possible occurrence. If one path is found that may cause the machine to pass through an unused state that does not involve z_1, then a 0 must be inserted in the corresponding unused state.

If the "don't care" in minterm location 7 ($y_1 y_2 y_3 = 111$) combines with 1 entry in minterm location 6 ($y_1 y_2 y_3 = 110$) to yield $z_1 = y_1 y_2$, then the path from state c ($y_1 y_2 y_3 = 101$) to state d ($y_1 y_2 y_3 = 011$) may pass through the unused state in minterm location 7 and generate a glitch on output z_1 if flip-flop y_2 sets before flip-flop y_1 resets.

In a similar manner, combining minterm locations 2 and 6 to yield $z_1 = y_2 y_3'$ may produce a glitch on output z_1 for a transition from state d ($y_1 y_2 y_3 = 011$) to state a ($y_1 y_2 y_3 = 000$) if flip-flop y_3 resets before flip-flop y_2 resets. The problem of output glitches for this machine can be resolved by reassigning state codes as follows:

$$a: \ (y_1 y_2 y_3 = 000)$$
$$b: \ (y_1 y_2 y_3 = 001)$$
$$c: \ (y_1 y_2 y_3 = 101)$$
$$d: \ (y_1 y_2 y_3 = 100)$$
$$e: \ (y_1 y_2 y_3 = 010)$$

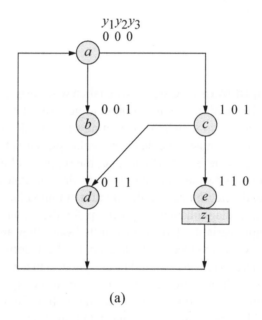

(a)

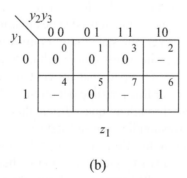

(b)

Figure 7.80   Moore machine: (a) state diagram and (b) output map.

7.3 Analysis of Asynchronous Sequential Machines

Another type of finite-state machine is an *asynchronous sequential machine*, where state changes occur on the application of input signals only — there is no machine clock. Like synchronous machines, the outputs of asynchronous machines are a function of either the present state only, or of the present state and the present inputs, corresponding to Moore and Mealy machines, respectively. The present state is directly related to the preceding sequence of inputs and states.

When an input variable changes in an asynchronous machine, the machine begins sequencing to a new state immediately. The input change may cause more than one storage element to alter its state, in which case the machine may sequence through various transient states before entering a stable state. This is caused by different propagation delays in the logic gates and storage elements.

The operational speed of a synchronous machine is regulated by, and is coincident with, a machine clock. Thus, the machine can change states only on the active transition of the clock signal. Because the speed of an asynchronous machine is not limited by a clock frequency, the operating characteristics are usually faster than those of a corresponding synchronous machine. The operational speed is limited only by the propagation delay of the longest path. The procedure for synthesizing a reliable asynchronous sequential machine is much more difficult and challenging than for a comparable synchronous sequential machine.

In the operation of synchronous sequential machines, the effect of transient output signals caused by varying circuit delays was negated by selecting an appropriate clock signal — either true or complemented — for the λ output logic. This is not possible, however, for asynchronous machines, because there is no system clock. The values placed in the output maps, therefore, must be carefully chosen.

Synchronous sequential machines occur more frequently in digital systems than their asynchronous counterparts, because of the ease with which the machines can be synthesized and the reliability imposed by a regularly occurring clock signal. In some cases, however, a machine clock may not be available, thus necessitating an asynchronous design. For extremely large synchronous machines using high-speed logic, the time required for the clock signal to propagate along the network of wires may be inordinately long. Thus, it is difficult to ensure the simultaneous arrival of the clock signal at all flip-flop clock inputs, which may hinder a reliable synchronous operation. In this situation also, an asynchronous design may be more appropriate.

Asynchronous machines are used extensively in the interface control logic of peripheral devices which attach to asynchronous interfaces between a channel (or I/O processor) and the device control unit. Since the interface contains no clock signal, the control unit utilizes randomly occurring interface control signals to transfer the data. During a write operation, the data bytes are transferred from the channel to the control unit interface logic which then synchronizes the data transfer rate to the speed of the peripheral device. During a read operation, the procedure is reversed.

Asynchronous sequential machines are implemented with set/reset (*SR*) latches as the storage elements. Thus, at least one *feedback* path is required in the synthesis of asynchronous machines. Figure 7.81(a) illustrates an asynchronous machine which contains one feedback path from output z_1 to the input logic. Asynchronous machines are implemented in a sum-of-products form. Figure 7.81(b) shows the same circuit re-drawn in the conventional latch configuration.

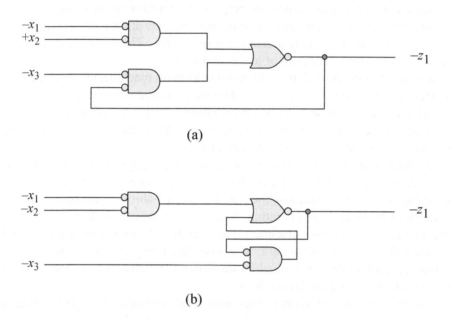

(a)

(b)

Figure 7.81   Asynchronous sequential machine with one feedback path: (a) sum-of-products and (b) conventional latch configuration.

7.3.1 Fundamental-Mode Model

Both synchronous and asynchronous sequential machines use feedback paths in their implementation. The feedback in synchronous machines combines with the input variables to form the δ next-state logic, which provides the next state for the machine at the next active clock transition. The feedback in asynchronous machines also combines with the δ next-state logic; however, this formation provides the necessary conditions to implement SR latches, and thus, generates a stable next state.

The primary approach to modeling and analyzing asynchronous sequential machines is to use a single delay element that represents the total delay of the entire machine and is placed in series with the feedback path. This simplifies the analysis procedure by specifying zero delay for all logic gates and interconnecting wires.

A general block diagram for an asynchronous sequential machine is shown in Figure 7.82. The input alphabet X consists of binary input variables $x_1, x_2, \cdots, x_n$ that can change value at any time and are represented as voltage levels rather than pulses. The state alphabet Y is characterized by p storage elements, where $Y_{1e}, Y_{2e}, \cdots, Y_{pe}$

are the *excitation* variables and $y_{1f}, y_{2f}, \cdots, y_{pf}$ are the *feedback* or *secondary* variables. The output alphabet Z is represented by $z_1, z_2, \cdots, z_m$.

Both the δ next-state logic and the λ output logic are composed of combinational logic circuits. The delay element in Figure 7.82 represents the total delay of the machine from the time an input changes until the machine has stabilized in the next state, and is represented as a time delay of Δt. The time correlation between the excitation variables Y_{ie} and the feedback variables y_{if} is specified by Equation 7.30.

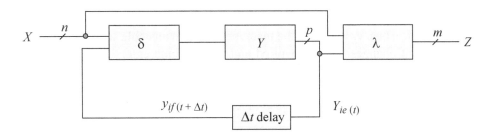

Figure 7.82   General block diagram of an asynchronous sequential machine.

$$y_{if(t + \Delta t)} = Y_{ie(t)} \tag{7.30}$$

For a given set of inputs, the machine will be in a stable state if and only if $y_{if} = Y_{ie}$, for $i = 1, 2, \ldots, p$; that is, after an appropriate delay, the feedback variables will be equal to the excitation variables, indicating that the machine has entered a *stable state*. When an input changes value, the δ next-state combinational logic produces a new set of values for the excitation variables $Y_{1e}, Y_{2e}, \cdots, Y_{pe}$. The machine then sequences through one or more unstable states. When the values of the feedback variables $y_{1f}, y_{2f}, \cdots, y_{pf}$ are equal to the values of the excitation variables $Y_{1e}, Y_{2e}, \cdots, Y_{pe}$, then the machine is stable in that state.

The transition from one stable state to another stable state is the result of a single change to an input variable. To ensure a *deterministic operation*, it is assumed that only one input changes at a time. If two inputs change simultaneously, then this will result in a *race* condition in which the machine may sequence to an incorrect next state. Race conditions are discussed in a later section.

Postulating that only one input changes value at a time should present no undo rigor in the synthesis procedure. It is always possible to affect a modification to the external circuitry so that only a single input changes value. This can be accomplished without altering the machine specifications. Also, when a binary input variable changes, the new value must be of sufficient duration to cause the machine to change state and proceed to a stable state. It is also assumed that no other input will change until the machine has entered a stable state. The previous two conditions specify a *fundamental-mode* operation. Thus, a fundamental-mode model has the following characteristics:

1.   Only one input changes at a time.

2.   No other input will change until the machine has sequenced to a stable state.

Whenever a single input changes, a new input vector is generated, which in turn produces a new set of excitation variables. At that instant in time, however, the feedback variables are not yet equal to the excitation variables because the Δt delay has not yet occurred. Thus, $y_{if} \neq Y_{ie}$ and the machine enters an unstable state. After a delay of Δt, all circuit changes have transpired and the machine enters a stable state where $y_{if} = Y_{ie}$. The machine remains in this stable state until another input change causes a new set of excitation variables to be generated. The machine will exit a stable state only when an input variable changes value.

7.3.2 Methods of Analysis

This section presents mechanisms for analyzing asynchronous sequential machines. These analysis techniques are similar in concept to those used for analyzing synchronous sequential machines. The behavior of an asynchronous machine can be completely specified by an excitation map, a next-state table, a state diagram, and a flow table. Also included in the analysis are the excitation and output equations.

In general, the excitation variables are a Boolean function of the inputs and the feedback variables. A Karnaugh map is a convenient method of representing the excitation variables. An excitation map is a Karnaugh map in which the columns are specified by the input variables and the rows by the feedback variables. The entries in the minterm locations represent the values of the excitation variables. The formats for representative excitation maps are shown in Figure 7.83 for one and two feedback variables.

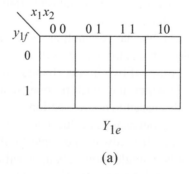

(a)

Figure 7.83   Formats for representative excitation maps: (a) two inputs and one excitation variable and (b) two inputs and two excitation variables.

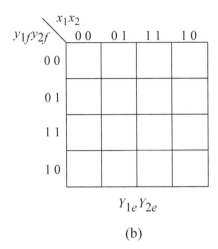

(b)

Figure 7.83   (Continued)

Example 7.10   Figure 7.84 illustrates a simple asynchronous sequential machine which will be used as an introductory example for analysis. The output of the circuit is the excitation variable Y_{1e}, which also corresponds to the machine output z_1. After a delay of Δt, the feedback variable y_{1f} becomes equal to the excitation variable Y_{1e} and the machine is stable.

The equation for Y_{1e} is obtained directly from the logic diagram as a function of the input variables x_1 and x_2 and of the feedback variable y_{1f}, as shown in Equation 7.31. The excitation variable is usually portrayed in the sum-of-products format. In this simple asynchronous machine, output z_1 has the same characteristics as the excitation variable Y_{1e}. Thus, the machine operates as a Moore model, because the output is a function of the state alphabet only and is independent of the input alphabet.

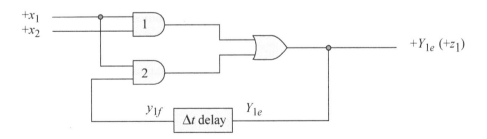

Figure 7.84   Asynchronous sequential machine with one feedback path.

The excitation map is obtained by plotting the equation for Y_{1e} in a Karnaugh map, as shown in Figure 7.85(a). The input variables x_1 and x_2 specify the columns of the map and are enumerated in the Gray code format in the same manner as for all Karnaugh maps of two or more variables. The feedback variable y_{1f} defines the values of the rows.

$$Y_{1e} = x_1 x_2 + x_1 y_{1f}$$

$$z_1 = Y_{1e} \tag{7.31}$$

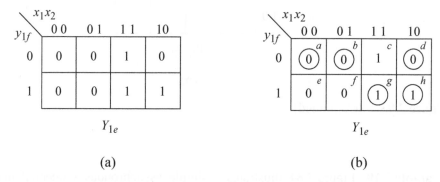

Figure 7.85   Karnaugh maps for the Moore machine of Figure 7.84: (a) excitation map and (b) excitation map showing stable states, which are indicated by circled entries.

In order to characterize the operation of the machine, the stable states must be identified. Recall that an asynchronous sequential machine is stable in a particular state only if the feedback variable is equal to the excitation variable; that is, $y_{1f} = Y_{1e}$. By convention, a *stable state* is indicated by circling the corresponding map entry, as shown in Figure 7.85(b). Since the values inserted in the minterm locations correspond to the values of Y_{1e}, any map entry that is equal to the row value of y_{1f} will indicate a stable state.

To facilitate the analysis of the machine, state names are inserted in the minterm locations together with the value of the excitation variable. When specifying a stable state, a circle is placed around the state variable or state name. Thus, in the row corresponding to $y_{1f} = 0$, states ⓐ, ⓑ, and ⓓ are stable, because in all three states $y_{1f} = Y_{1e} = 0$. Similarly, in the row corresponding to $y_{1f} = 1$, states ⓖ and ⓗ are stable, because in both states $y_{1f} = Y_{1e} = 1$.

The state transition sequences for the machine can now be determined. An asynchronous machine will remain in a stable state until an input changes. A change of a single input causes a horizontal movement in the excitation map. Assume that the machine is presently in stable state ⓐ ($y_{1f}x_1 x_2 = 000$) and that input x_2 changes from 0 to 1. The machine will proceed horizontally to state ⓑ ($y_{1f}x_1 x_2 = 001$) which is also stable, because $y_{1f} = Y_{1e} = 0$. The machine will remain in state ⓑ until an input again changes.

Similarly, if the machine is in state (a) and x_1 changes from 0 to 1, then the machine sequences horizontally in row $y_{1f} = 0$ and enters state (d). Since the excitation value does not change, the feedback value will not change. Thus, the entire operation takes place only in row $y_{1f} = 0$. Note that it is not possible to sequence from state (d) ($y_{1f}x_1x_2 = 010$) to state (h) ($y_{1f}x_1x_2 = 110$) directly, because inputs x_1 and x_2 remain unchanged. In order to leave a stable state, an input must change value. State (h) can be entered only from state (g). A state transition will occur from state (g) to state (h) if x_2 changes from 1 to 0. The machine will remain in state (h) until another change of input occurs, either to x_1 or x_2.

It is also not possible for the machine to sequence from state (b) to state (d) directly, because this transition requires a change of two input variables, a situation that is not allowed in asynchronous sequential machine operation. That is, $x_1x_2 = 01$ cannot change to $x_1x_2 = 10$ directly, or conversely. For the same reason, a simultaneous change from $x_1x_2 = 00$ to $x_1x_2 = 11$, or conversely, is not allowed.

When a single input changes, the machine can sequence to only those states that are contained in the column specified by the new input vector. Therefore, in state (b), when input x_1 changes from 0 to 1, the machine moves horizontally to unstable state c, then vertically in column $x_1x_2 = 11$, and settles in state (g). The state transition sequence for this input change is $(b) \rightarrow c \rightarrow (g)$. In unstable states, the inputs do not change — only the feedback variables change. To cause a feedback variable to change, the machine must pass through an unstable state.

If the next state is unstable, then the machine is in a transient state in the same feedback row and in the column specified by the new input values. There will be no vertical movement until the feedback variable begins to change in order to assume the value of the excitation variable. After a delay of Δt, the machine will move vertically in the same column and enter a row in which the value of the feedback variable is equal to the value of the previous excitation variable.

An alternative method of illustrating the flow of the machine is to tabulate the stable and unstable states as shown in the flow table of Figure 7.86. The flow table can be generated directly from the excitation map of Figure 7.85. The state names are chosen arbitrarily and only the stable states are assigned unique names. The flow table directly corresponds to the excitation map of Figure 7.85(b). The excitation variables, however, are replaced with state names, allowing for easier interpretation of machine operation.

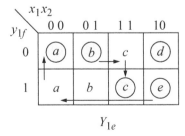

Figure 7.86   Flow table for the Moore machine of Figure 7.84.

The flow table provides a convenient notational technique for analyzing machine operation. The table specifies the state transition sequences resulting from an input change. Thus, the behavior of the machine is completely specified by the flow table. The flow table is an extremely useful instrument for analysis and will be used extensively in the synthesis of asynchronous sequential machines.

The circled states in Figure 7.86 represent stable states; the uncircled states represent unstable states and indicate the name of the destination stable state to which the machine will sequence after a delay of Δt. Thus, if the machine is presently in state $\textcircled{b}$ and input x_1 changes from 0 to 1, then the operation proceeds horizontally in row $y_{1f} = 0$ and the machine will sequence through transient state c and proceed to stable state $\textcircled{c}$ after a delay of Δt.

Similarly, the transition from state $\textcircled{e}$ to state $\textcircled{a}$ is the result of input x_1 changing from 1 to 0, which causes the operation to move horizontally in row $y_{1f} = 1$ to unstable state a. After a delay of Δt, the machine will pass through transient state a and enter state $\textcircled{a}$. All state transition sequences can be observed in the flow table, in which any transition is the result of a single input change.

Example 7.11 The logic diagram for a Moore asynchronous sequential machine is shown in Figure 7.87. The machine will be analyzed by obtaining the excitation map and the flow table. The excitation map is shown in Figure 7.88 and is constructed by applying all combinations of the inputs to the machine in conjunction with the feedback variable.

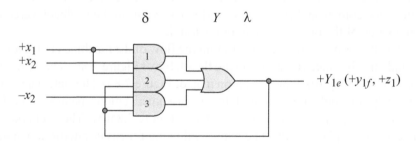

Figure 7.87   Logic diagram for the Moore machine of Example 7.11.

	$x_1 x_2$			
y_{1f}	0 0	0 1	1 1	1 0
0	$\textcircled{0}$	$\textcircled{0}$	1	$\textcircled{0}$
1	$\textcircled{1}$	0	$\textcircled{1}$	$\textcircled{1}$

Y_{1e}

Figure 7.88   Excitation map for the Moore machine of Example 7.11.

For example, if the feedback variable $y_{1f} = 0$, then the only condition that will cause the excitation variable Y_{1e} to be equal to 1 is when $x_1 x_2 = 11$; otherwise, $Y_{1e} = 0$. In a similar manner, if $y_{1f} = 1$, then the only condition that will cause the excitation variable Y_{1e} to be equal to 0 is when $x_1 x_2 = 01$; in all other cases, $Y_{1e} = 1$.

The flow table is shown in Figure 7.89 in which the state names for the stable states are arbitrarily chosen and correspond to the stable state locations in the excitation map. Assume that the machine is stable in state ⓑ where $y_{1f} = 0$ and input x_1 changes from 0 to 1. In Figure 7.87, gate 1 now has both inputs at a logic 1, which will cause Y_{1e} to be equal to a logic 1 indicated by unstable state e in Figure 7.89. After a delay of Δt, the feedback variable $y_{1f} = 1$ and the machine enters stable state ⓔ.

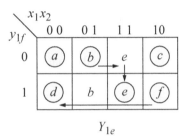

Figure 7.89   Flow table for the Moore machine of Example 7.11.

Consider the case where the machine is stable in state ⓒ with $x_1 x_2 = 10$. If x_2 changes from 0 to 1, then gate 1 now has both inputs at a logic 1, which will cause Y_{1e} to be equal to a logic 1. As before, the machine passes through unstable state e and enters stable state ⓔ.

Now assume that the machine is in stable state ⓕ where $x_1 x_2 = 10$ and $Y_{1e} = 1$. The feedback paths through gate 2 and gate 3 maintain $Y_{1e} = 1$. If input x_1 changes from 1 to 0, then the outputs of gate 1 and gate 2 become a logic 0; however, since $x_2 = 0$, the feedback path through gate 3 maintains the excitation variable $Y_{1e} = 1$ and the machine sequences to stable state ⓓ.

7.3.3 Hazards

The signal propagation delay must be considered in the analysis and synthesis of asynchronous sequential machines. When an input variable changes value, varying propagation delays caused by logic gates, wires, and different path lengths can produce erroneous transient signals on the outputs. These spurious signals are referred to as *hazards*. If the hazard occurs in the feedback path, then an incorrect state transition sequence may result.

Hazards are not usually a concern in synchronous sequential machines because the clock negates the effects of a hazard. A hazard may appear at the input of a flip-flop, but is deasserted before the machine changes state at the next active clock transition. In asynchronous sequential machines, however, where the next state is not synchronized with a clock signal, a hazard can cause an incorrect next state or an erroneous output, and must be eliminated.

These transitory signals generate a condition which is not specified in the expression for the machine, because Boolean algebra does not take into account the propagation delay of switching circuits. In this section, hazards will be examined and methods presented for detecting and correcting these transient phenomena so that correct operation of an asynchronous sequential machine can be assured.

Static hazards   Figure 7.90 illustrates a classic example of a combinational circuit with an inherent hazard. Although the network is combinational in structure, it may be an integral part of the δ next-state logic for an asynchronous sequential machine. The Karnaugh map which represents the circuit is shown in Figure 7.91 and the equation for output z_1 is shown in Equation 7.32. The map entries indicate that z_1 is asserted in minterm locations 1, 5, 6, and 7; thus, the function $z_1(x_1, x_2, x_3) = \Sigma_m(1, 5, 6, 7)$. In particular, z_1 is active if $x_1 x_2 x_3 = 111$ or 101; that is, output z_1 is set to a value of 1 regardless of the value of input x_2.

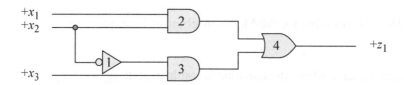

Figure 7.90   Combinational logic circuit that contains a potential static hazard.

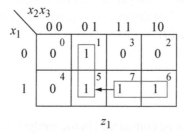

Figure 7.91   Karnaugh map for the logic diagram of Figure 7.90.

$$z_1 = x_1 x_2 + x_2' x_3 \qquad (7.32)$$

Thus, if x_2 changes from 0 to 1 or from 1 to 0, the state of z_1 should remain constant (or static) during the transition. Any deviation from this static condition is referred to as a "static hazard." If an input variable changes value causing an output to be deasserted momentarily when the output should remain asserted, then this is classified as a *static-1 hazard.* Conversely, when an input change causes an output to become asserted momentarily when it should remain deasserted, a *static-0 hazard* results. Figure 7.92 illustrates the two types of static hazards. In general, a static hazard is a singularity in which a single change to an input vector causes a momentary output transition to an incorrect state.

(a)   (b)

Figure 7.92   Static hazards: (a) static-1 hazard and (b) static-0 hazard.

If two adjacent minterm locations both contain a value of 1 and are covered by the same prime implicant, then a single change within that grouping of 1s will not produce a hazard. For example, in the Karnaugh map of Figure 7.91, changing the input vector from $x_1 x_2 x_3 = 111$ to 110 will not cause a static hazard because the corresponding 1s are covered by the term $x_1 x_2$. Thus, when x_3 changes from 1 to 0, the $x_1 x_2$ term maintains the output at a logic 1. Therefore, all groups of 1s in a Karnaugh map must be connected by redundant prime implicants (or consensus terms) to avoid possible hazards.

Referring to the logic diagram of Figure 7.90, when input x_2 changes from 1 to 0, the change is transmitted to the output along two paths: logic blocks 2 and 4; and logic blocks 1, 3, and 4. If the delay of block 2 is less than the combined delay of blocks 1 and 3, then all inputs to block 4 will be at a low level for a short duration. If the block 4 inputs remain at a low level for a sufficient duration, then output z_1 will generate a static-1 hazard.

Figure 7.93 illustrates an approach to analyzing the circuit of Figure 7.90 for hazards. For input x_2, the top row of +/– symbols refers to the logic levels produced as a result of input x_2 changing from a high to a low level. The bottom row of +/– symbols specify the logic levels which result when x_2 changes from a low to a high level.

Refer to the top row of +/– symbols for the discussion which follows. Assume that x_2 changes from a high to a low level. The first column in each of the seven sets of columns specifies a stable state, where $x_1 x_2 x_3 = 111$. The second column represents the state of the circuit after one gate delay, in which the inverter and gate 2 are in the same time relationship. The third column indicates the state of the circuit after two gate delays, in which gate 3 and gate 4 (from the top input) produce concurrent delays. At this time, output z_1 changes to a low voltage level. Finally, after three gate delays — the inverter, gate 3, and gate 4 (from the bottom input) — the circuit resumes a stable state, where z_1 returns to a high level.

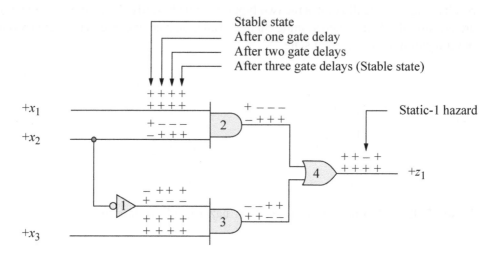

Figure 7.93   A method for analyzing a logic circuit for static hazards.

Now consider the case where input x_2 changes from a low to a high logic level. If the initial state of the circuit is $x_1 x_2 x_3 = 101$ and x_2 changes from a low to a high level, then no hazard will occur on output z_1, as indicated by the bottom row of +/− symbols. There will never be a time period when both inputs of the OR gate are at a low logic level. Thus, the circuit will not generate a static-1 hazard on z_1.

Because the map of Figure 7.91 specifies that output z_1 should remain at a constant high level when $x_1 x_3 = 11$, the effects of the hazard can be eliminated by adding a third term to the equation for z_1. The output can be made independent of the value of x_2 by including the redundant prime implicant $x_1 x_3$, which covers both the initial and terminal state of the transition. The redundant prime implicant will maintain the output at a constant high level during the transition.

The term $x_1 x_3$ is called a *hazard cover*, since it covers the detrimental effects of the hazard. The effects of a static hazard can be negated by combining adjacent groups of 1s in a Karnaugh map as shown in Figure 7.94 for the Karnaugh map of Figure 7.91. The new equation for z_1 is shown in Equation 7.33. The hazard will still occur at the OR gate inputs, but its effect will be negated due to the addition of gate 5, as shown in Figure 7.95. Although the circuit no longer contains a minimal number of gates, the operation is reliable.

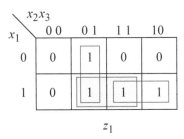

$$z_1$$

Figure 7.94   Negating the effects of a static hazard by combining adjacent groups of 1s with a redundant prime implicant term.

$$z_1 = x_2' x_3 + x_1 x_2 + x_1 x_3 \qquad\qquad (7.33)$$

Hazard cover ⟶

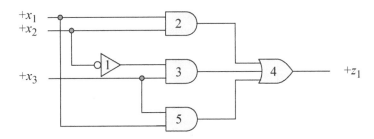

Figure 7.95   Eliminating the effects of a static-1 hazard for Figure 7.90 by adding the redundant prime implicant term $x_1 x_3$ to the equation for z_1 represented by gate 5.

Static-0 hazards can be analyzed and eliminated by using the same technique, except that adjacent groups of 0s are combined. Consider the Karnaugh map of Figure 7.96 in which 0s are grouped to form the product-of-sums expression shown in Equation 7.34. In minterm location $x_1 x_2 x_3 = 100$, if x_2 changes from 0 to 1, then the machine will generate a static-0 hazard as it sequences to minterm location $x_1 x_2 x_3 = 110$, as shown in Figure 7.97. However, the map indicates that z_1 should not change value during this transition; that is, z_1 should remain at a value of 0.

As in the sum-of-products implementation, the hazard can be eliminated by adding a redundant term. The additional sum term $(x_1' + x_3)$ is appended to the product-of-sums expression as shown in Equation 7.35. Applying the product terms $x_1'x_2'$ or x_2x_3 to Equation 7.35 will yield a value of 1 for z_1 as specified by the Karnaugh map of Figure 7.96 by combining groups of 1s.

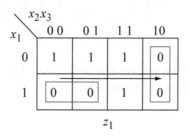

Figure 7.96   Karnaugh map for a product-of-sums with a potential static-0 hazard.

$$z_1 = (x_1' + x_2)(x_2' + x_3) \tag{7.34}$$

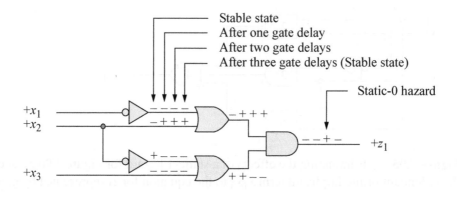

Figure 7.97   A static-0 hazard exhibited on output z_1 for the product-of-sums implementation of Figure 7.96.

$$z_1 = (x_1' + x_2)(x_2' + x_3)(x_1' + x_3) \tag{7.35}$$

Hazard cover

In order for there to be a potential static hazard, the groups of 1s or 0s must be adjacent. For example, the following equation will be used to generate a Karnaugh map which will be analyzed for possible static-1 and static-0 hazards:

$$z_1 = x_1' x_2 x_3 + x_1 x_3' + x_1' x_2' x_3$$

The Karnaugh map is shown in Figure 7.98, which clearly shows that the groups of 1s and 0s are not adjacent to other groups of 1s or 0s.

x_1 \ $x_2 x_3$	0 0	0 1	1 1	1 0
0	0 _(0)_	1 _(1)_	1 _(3)_	0 _(2)_
1	1 _(4)_	0 _(5)_	0 _(7)_	1 _(6)_

z_1

Figure 7.98   Karnaugh map which has no static-1 or static-0 hazards.

Dynamic hazards   Dynamic hazards, like static hazards, may also cause erroneous outputs in combinational circuits. If the combinational network is incorporated in the δ next-state logic of an asynchronous sequential machine, then an incorrect next state may result. The same rationale applies to dynamic hazards in the λ output combinational logic. A *dynamic hazard* is characterized by multiple output pulses resulting from a single change to the input vector.

When a single input variable changes value, an odd number of transitions may occur on the output signal, where the number of transitions is greater than one. Thus, the input change propagates toward the output signal along 3, 5, 7, ... paths. In order for a dynamic hazard to be realized, at least three different path lengths must be encountered. The first path will constitute a minimum propagation delay and cause the output to change value; the second path will cause an intermediate delay in which the output will return to its previous value; and the third path results in a maximum delay which causes the output to make a third transition. The last two transitions also generate a static hazard. Figure 7.99 illustrates two typical types of dynamic hazards.

Figure 7.99   Dynamic hazards with three transitions.

Dynamic hazards can be eliminated in a manner analogous to that used for static hazards; that is, by adding redundant terms to the output equation. It may also be possible to change the form of the equation to eliminate a dynamic hazard. Figure 7.100 illustrates a nonminimized combinational circuit with an inherent dynamic hazard. The equation for output z_1 is shown in Equation 7.36. Let x_1, x_3, and x_4 be active low inputs and x_2 be active high. With all inputs active at their respective levels, the circuit is stable and z_1 is at a high logic level.

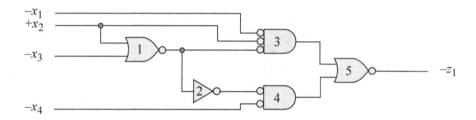

Figure 7.100   Combinational circuit with an inherent dynamic hazard.

$$z_1 = x_1 x_2'(x_2 + x_3') + x_2' x_3 x_4 \qquad (7.36)$$

The timing diagram of Figure 7.101 depicts the sequence of events that occur when input x_2 changes from a high to a low logic level. The waveforms labeled OR1, INV2, AND3, AND4, and OR5 refer to the outputs of their respective logic blocks. The deassertion of x_2 is immediate. The new value of x_2 propagates to the output terminal along three paths.

The first path involves two gate delays: gates 3 and 5. After one gate delay, the output of gate 3 changes from a low to a high level. This change is reflected on output z_1 after a second gate delay through gate 5, at which time z_1 changes from a high to a low level.

The second path involves three gate delays: gates 1, 3, and 5. After one gate delay, the output of gate 1 changes from a low to a high level. This change is reflected on the output of gate 3 after a second delay (through gate 3) which now changes from a high to a low level. Both inputs to gate 5 are now at a low logic level. Therefore, after a third delay (through gate 5), output z_1 changes from a low to a high level.

The third path entails four gate delays: gate 1, the inverter, gate 4, and gate 5. After one gate delay, the output of gate 1 changes from a low to a high level. This change is manifested on the output of the inverter, which changes from a high to a low level. Both inputs to gate 4 are now at a low logic level. Therefore, after a third delay (through gate 4), the output of gate 4 changes from a low to a high level. This change encounters the fourth delay for this path as the signal propagates through gate 5 causing output z_1 to change from a high to a low logic level. The output eventually changes state, as it should, but only after the occurrence of two extra transitions. The resulting effect on z_1 from this single input change is a dynamic hazard with a triple change of state.

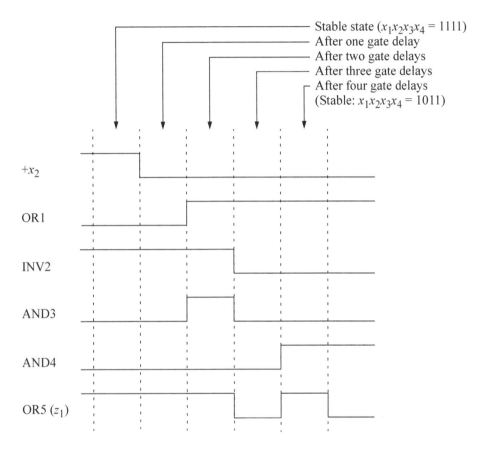

Figure 7.101 Timing diagram for the circuit of Figure 7.100 illustrating a dynamic hazard on output z_1.

Dynamic hazards, which are less common than static hazards, result from a single change to the input vector in which the change propagates to the output terminal along three or more different path lengths. Reconfiguring the logic equation, without changing the functionality of the circuit, will eliminate dynamic hazards. Dynamic hazards can also be removed as a direct result of eliminating static hazards, in which redundant terms are added to the output equation.

Essential hazards An asynchronous sequential machine may be free of static and dynamic hazards in the δ next-state logic and the λ output logic, but may still sequence to an incorrect next state due to inordinately long propagation delays in certain circuit elements. The propagation delay through a logic gate may increase due to temperature change, power supply output variation, or component aging.

An *essential hazard* is caused not by an incorrectly synthesized network, but by the operation of the circuit itself; that is, it is inherent in the machine's design. Thus, the essential hazard cannot be eliminated by adding redundant terms in the network equation or by changing the form of the equation. Since an essential hazard is the result of excessive propagation delay in certain network elements, the effect of the hazard can be nullified by introducing appropriate delays in other components, such that the cause of the hazard will be negated.

Essential hazards occur specifically in fundamental-mode machines and are characterized by two propagation paths: one path affecting storage element y_j, the other path affecting storage element y_k. Essential hazards can be detected by analyzing the excitation map or flow table of an asynchronous sequential machine.

The machine has a possible essential hazard if, beginning in a stable state, input x_i changes value and sequences the machine to a stable state that is different than the stable state reached after three successive changes to x_i. That is, an asynchronous sequential machine contains a possible essential hazard if a single change to an input variable x_i results in a transition from state S_j to state S_k, whereas three consecutive changes to x_i results in a state transition sequence which terminates in state S_l, where $S_k \neq S_l$.

Thus, when the machine is implemented, there may be a series of propagation delays that will cause the machine to sequence to an incorrect next stable state resulting from a single change to input x_i. This incorrect state change occurs because the change to x_i propagates to different circuit elements at different times. Consider the asynchronous sequential machine specified by the excitation maps and flow table of Figure 7.102 and Equation 7.37.

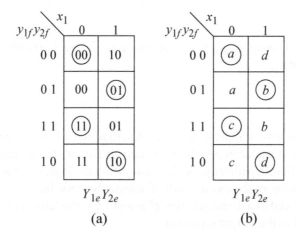

Figure 7.102   Karnaugh maps for an asynchronous sequential machine containing a possible essential hazard: (a) combined excitation map and (b) flow table.

$$Y_{1e} = y_{2f}'x_1 + y_{1f}x_1'$$

$$Y_{2e} = y_{1f}x_1' + y_{2f}x_1 \qquad\qquad (7.37)$$

The logic diagram is shown in Figure 7.103. If the machine is stable in state $\textcircled{a}$ ($y_{1f}y_{2f}x_1 = 000$) and input x_1 changes from a low to a high logic level, then an essential hazard is possible if the delay through the inverter is excessive; that is,

Inverter delay > delay of gate 1 + delay of gate 2 + delay of gate 4

The machine should sequence from state $\textcircled{a}$ ($y_{1f}y_{2f}x_1 = 000$) to state $\textcircled{d}$ ($y_{1f}y_{2f}x_1 = 101$) if the gate delays are within their respective specified range. However, the sequence described below is possible if the delay through the inverter greatly exceeds its maximum propagation delay.

Refer to the logic diagram of Figure 7.103, which is designed from the equations of Equation 7.37. If the machine is stable in state $\textcircled{a}$ ($y_{1f}y_{2f}x_1 = 000$) and input x_1 changes from a low to a high level, then the output of gate 1 changes to a low level after an appropriate delay. Latch y_1 will then set after a delay through gate 2. The circuit is now in a transient state of $y_{1f}y_{2f}x_1 = 101$.

The $+y_{1f}$ output of gate 2 connects to the input of gate 4. The inverter would normally have disabled gate 4 by this time and the machine would have stabilized in state $y_{1f}y_{2f}x_1 = 101$. The output of the inverter, however, is still at a high logic level due to excessive propagation delay. Therefore, the output of gate 4 changes to a low level and sets latch y_2 after a delay through gate 5. During this time period, the inverter propagates the change incurred by input x_1.

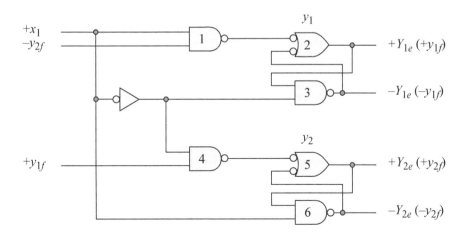

Figure 7.103   An asynchronous sequential machine with a possible essential hazard.

Even though the inverter has disabled the output of gate 4, latch y_2 remains set as a result of the feedback path $(+y_{2f})$ and the connection between the output of gate 6 and the input to gate 5. The inverter output also applies a low level to reset latch y_1. Since the output of gate 1 is disabled because of the $-y_{2f}$ signal from latch y_2, therefore latch y_1 is reset. After all propagation delays have elapsed, the machine is stable in state ⓑ $(y_{1f}y_{2f}x_1 = 011)$, which is an incorrect terminal state for the state transition sequence ⓐ → ⓓ.

The hazard thus realized is essential; that is, the design of the machine cannot be changed to eliminate the hazard and still conform to the machine specifications. The essential hazard occurs because the change to input x_1 is transmitted to different parts of the circuit at different times. The change of state for y_1 is received at gate 4 before the inverter has propagated the change from x_1. The inverter output would normally have applied a low level to the input of gate 4, effectively preventing latch y_2 from being set.

Essential hazards can be eliminated in an asynchronous sequential machine — without affecting the logical operation of the machine — by inserting appropriate delay circuits in strategic locations within the machine. For example, if sufficient delay was added to the output of latch y_1, then the change to input x_1 would propagate to all the appropriate gates before the change to y_1 was received at those gates. Thus, the essential hazard is eliminated and proper operation of the machine is assured.

7.3.4 Oscillations

An *oscillation* occurs in an asynchronous sequential machine when a single input change results in an input vector in which there is no stable state. Consider the excitation map of Figure 7.104 for Y_{1e}. There are two input variables x_1 and x_2 and one feedback variable y_{1f}. If the machine is in stable state ⓑ $(y_{1f}x_1x_2 = 001)$ where $y_{1f} = Y_{1e} = 0$ and x_1 changes from 0 to 1, then the machine sequences to transient state c $(y_{1f}x_1x_2 = 011)$. In state c, the excitation variable $Y_{1e} = 1$ and the feedback variable $y_{1f} = 0$.

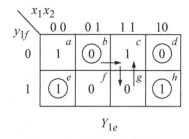

Y_{1e}

Figure 7.104   Excitation map for an asynchronous sequential machine containing an oscillation.

Thus, after a delay of Δt, the feedback variable becomes equal to the excitation variable and the machine proceeds to state g, where the feedback variable $y_{1f} = 1$. In state g, however, the excitation variable $Y_{1e} = 0$, designating state g as an unstable (or transient) state, because $y_{1f} \neq Y_{1e}$. After a further delay of Δt, the feedback variable becomes equal to the excitation variable and the machine sequences to state c, where the feedback variable $y_{1f} = 0$. Since the input vector $x_1x_2 = 11$ provides no stable state, the machine will oscillate between transient states c and g.

Another example depicting oscillations is shown in the excitation map for excitation variables Y_{1e} and Y_{2e} of Figure 7.105. All state transition sequences must be considered when analyzing the machine for oscillations. In this asynchronous sequential machine, there are multiple oscillations.

$y_{1f}y_{2f}$ \ x_1x_2	0 0	0 1	1 1	1 0
0 0	*a* 01	*b* 01	*c* 01	*d* 01
0 1	*e* (01)	*f* (01)	*g* 00	*h* 11
1 1	*i* 01	*j* (11)	*k* 01	*l* 10
1 0	*m* 11	*n* 11	*o* (10)	*p* (10)

$Y_{1e}Y_{2e}$

Figure 7.105 Excitation map containing multiple oscillations.

One oscillation occurs when starting in stable state (*f*) when x_1 changes from 0 to 1. The machine sequences to unstable state g, where $Y_{1e}Y_{2e} = 00$. After a delay of Δt, the feedback variables become equal to the excitation variables and the machine enters unstable state c, where $Y_{1e}Y_{2e} = 01$. After a further delay of Δt, the machine enters transient state g and the process repeats, causing the machine to oscillate between state g and state c. This oscillation is represented as (*f*) $\rightarrow g \leftrightarrow c$.

A second oscillation occurs when beginning in state (*j*) and x_1 changes from 0 to 1. The machine sequences to unstable state k, where $Y_{1e}Y_{2e} = 01$. After a delay of Δt, $y_{1f}y_{2f} = Y_{1e}Y_{2e} = 01$ and the machine proceeds to state g. The machine then oscillates between unstable states *g and c*. This oscillation is represented as (*j*) $\rightarrow k \rightarrow g \leftrightarrow c$.

In the synthesis of asynchronous sequential machines, the oscillation phenomenon should be avoided. The machine specifications can be modified slightly such that every input vector will provide at least one stable state. This modification should not drastically alter the general functional operation of the machine.

7.3.5 Races

In the analysis of synchronous sequential machines, if a change of state occurs between two states with nonadjacent state codes, then the machine may sequence through a transient state before entering the destination stable state. If the transient state contains a Moore-type output, then a transitory erroneous signal may be generated on the output. This glitch results from two or more variables changing state in a single state transition sequence in which the variables change values at different times.

A similar situation occurs in asynchronous sequential machines. If a single change to the input vector causes two or more excitation variables to change state, then multiple paths exist from the source stable state to the destination stable state. This is called a *race* condition.

There are two types of race conditions: noncritical and critical. A *noncritical race* is one that may cycle through one or more transient states before entering the correct destination stable state. A *critical race* is one that terminates in an incorrect destination stable state. Figure 7.106 shows a Karnaugh map for excitation variables Y_{1e} and Y_{2e} in which there is one noncritical race and one critical race.

$x_1 x_2$

$y_{1f} y_{2f}$	0 0	0 1	1 1	1 0
0 0	*a* 11	*b* (00)	*c* 11	*d* 11
0 1	*e* (01)	*f* 11	*g* 11	*h* 00
1 1	*i* 10	*j* (11)	*k* (11)	*l* 10
1 0	*m* (10)	*n* 00	*o* 11	*p* (10)

$Y_{1e} Y_{2e}$

Figure 7.106   Excitation map for $Y_{1e} Y_{2e}$ that has one noncritical race and one critical race.

The noncritical race is described first. Beginning in stable state ⓑ ($Y_{1e} Y_{2e} = 00$), if x_1 changes from 0 to 1, the machine proceeds to transient state c ($Y_{1e} Y_{2e} = 11$). Since both excitation variables change state, this constitutes a race condition. If Y_{1e} sets before Y_{2e} sets; that is, $Y_{1e} Y_{2e} = 10$, then after a delay of Δt, $y_{1f} y_{2f} = 10$ and the machine enters transient state o where $Y_{1e}Y_{2e} = 11$. After a further delay of Δt, $y_{1f}y_{2f} = 11$ and the machine enters and remains in stable state $\textcircled{k}$.

A second path for the race condition is when Y_{2e} sets before Y_{1e} sets; that is, $Y_{1e}Y_{2e} = 01$. After a delay of Δt, the feedback variables become equal to the excitation variables and the machine enters transient state g ($Y_{1e}Y_{2e} = 11$). After a further delay of Δt, $y_{1f}y_{2f} = 11$ and the machine enters and remains in stable state $\textcircled{k}$.

A third path for the race condition is when both excitation variables change simultaneously from $Y_{1e}Y_{2e} = 00$ to $Y_{1e}Y_{2e} = 11$. After a delay of Δt, the feedback variables become equal to the excitation variables and the machine proceeds directly to the destination stable state $\textcircled{k}$. The race conditions are summarized as follows and Figure 7.107 provides more detail for the intermediate steps:

$$\textcircled{b} \rightarrow c \rightarrow o \rightarrow \textcircled{k}$$
$$\textcircled{b} \rightarrow c \rightarrow g \rightarrow \textcircled{k}$$
$$\textcircled{b} \rightarrow c \rightarrow \textcircled{k}$$

$$\textcircled{b} \rightarrow c \rightarrow o \rightarrow \textcircled{k}$$

$$\textcircled{b}\ Y_{1e}Y_{2e} = 00 \rightarrow c\ (Y_{1e}Y_{2e} = 11)$$
$$\downarrow$$
$$Y_{1e}Y_{2e} = 10\ (Y_{1e}\text{sets before } Y_{2e} \text{ sets})$$
$$\downarrow \Delta t_1$$
$$y_{1f}y_{2f} = 10\ (o\colon Y_{1e}Y_{2e} = 11)$$
$$\downarrow \Delta t_2$$
$$y_{1f}y_{2f} = 11\ (\textcircled{k}\colon Y_{1e}Y_{2e} = 11)$$

$$\textcircled{b} \rightarrow c \rightarrow g \rightarrow \textcircled{k}$$

$$\textcircled{b}\ Y_{1e}Y_{2e} = 00 \rightarrow c\ (Y_{1e}Y_{2e} = 11)$$
$$\downarrow$$
$$Y_{1e}Y_{2e} = 01\ (Y_{2e} \text{ sets before } Y_{1e} \text{ sets})$$
$$\downarrow \Delta t_1$$
$$y_{1f}y_{2f} = 01\ (g\colon Y_{1e}Y_{2e} = 11)$$
$$\downarrow \Delta t_2$$
$$y_{1f}y_{2f} = 11\ (\textcircled{k}\colon Y_{1e}Y_{2e} = 11)$$

(Continued on next page)

Figure 7.107   Three possible paths for the state transition sequence $\textcircled{b} \rightarrow \textcircled{k}$.

$$\textcircled{b} \rightarrow c \rightarrow \textcircled{k}$$

$\textcircled{b}$ $Y_{1e}Y_{2e} = 00 \rightarrow c$ ($Y_{1e}Y_{2e} = 11$; Y_{1e} and Y_{2e} change state simultaneously)

$$\downarrow \Delta t$$
$$y_{1f}y_{2f} = 11 \ (\textcircled{k}: Y_{1e}Y_{2e} = 11)$$

Figure 7.107   (Continued)

Thus, three state transition sequences are possible, where each sequence results from a different propagation delay time through the excitation variable storage elements. Although a race condition exists, stable state $\textcircled{k}$ is the destination state for all three sequences. This is referred to as a *noncritical race* condition, since the destination stable state is the same regardless of the path taken.

There is also one critical race condition in Figure 7.106 when — beginning in stable state $\textcircled{b}$ — x_2 changes from 1 to 0. The machine then proceeds to transient state a ($Y_{1e}Y_{2e} = 11$). If Y_{1e} sets before Y_{2e} sets; that is, $Y_{1e}Y_{2e} = 10$, then after a delay of Δt, $y_{1f}y_{2f} = 10$ and the machine enters stable state $\textcircled{m}$ where $Y_{1e}Y_{2e} = 10$.

A second path for the race condition is when Y_{2e} sets before Y_{1e} sets; that is, $Y_{1e}Y_{2e} = 01$. After a delay of Δt, the feedback variables become equal to the excitation variables and the machine enters stable state $\textcircled{e}$ ($Y_{1e}Y_{2e} = 01$).

A third path for the race condition is when both excitation variables change simultaneously from $Y_{1e}Y_{2e} = 00$ to $Y_{1e}Y_{2e} = 11$. After a delay of Δt, the feedback variables become equal to the excitation variables and the machine proceeds to transient state i ($Y_{1e}Y_{2e} = 10$). After a further delay of Δt, the feedback variables become equal to the excitation variables and the machine enters stable state $\textcircled{m}$ ($Y_{1e}Y_{2e} = 10$). The race conditions are summarized as follows:

$$\textcircled{b} \rightarrow a \rightarrow \textcircled{m}$$
$$\textcircled{b} \rightarrow a \rightarrow \textcircled{e}$$
$$\textcircled{b} \rightarrow a \rightarrow i \rightarrow \textcircled{m}$$

The single change to the input vector results in a race condition because both excitation variables change value ($Y_{1e}Y_{2e} = 00$ to 11). However, since the destination stable state cannot be predicted, the machine will proceed to either stable state $\textcircled{m}$ or to stable state $\textcircled{e}$. This is termed a *critical race* condition and must be avoided.

Races can be avoided when it is possible to direct the machine through intermediate unstable states before reaching the destination stable state. This can be achieved by utilizing some of the unspecified entries in the excitation map. Also, it may be possible to add rows to the excitation map without increasing the number of excitation and feedback variables.

7.4 Synthesis of Asynchronous Sequential Machines

The synthesis (design) of asynchronous sequential machines is one of the most interesting and certainly the most challenging concepts of sequential machine design. In many situations, a synchronous clock is not available. The interface between an input/ output processor (IOP) — or channel — and an input/output (I/O) subsystem control unit is an example of an asynchronous condition. Many large computer channels communicate with an I/O subsystem by means of a signal interlocking protocol on the interface.

The control unit requests a word of data during a write operation by asserting an identifying signal called a "tag-in signal." The channel then places the word on the data bus and asserts an acknowledging tag-out signal. The device control unit accepts the data then deasserts the in-tag, allowing the channel to deassert the corresponding out-tag, completing the data transfer sequence for one word. An analogous situation occurs for a read operation in which the tag-in signal now indicates that a word is available on the data bus for the channel. The channel accepts the word and responds with the tag-out signal.

The data transfer sequence for the write and read operations was initiated, executed, and completed without utilizing a synchronizing clock signal. This technique permits not only a higher data transfer rate between the channel and an I/O device, but also allows the channel to communicate with I/O devices having a wide range of data transfer rates. The interface control logic in the device control unit is usually implemented as an asynchronous sequential machine. Even in large synchronous systems, it is often advantageous to allow certain subsystems to operate in an asynchronous manner, thereby increasing the overall speed of the system.

Transient signals are handled differently in the synthesis procedure for asynchronous machines. Techniques will be presented in this chapter to synthesize fundamental-mode asynchronous sequential machines irrespective of the varying delays of circuit components.

7.4.1 Synthesis Procedure

The synthesis procedure for asynchronous sequential machines is similar in many respects to that described for synchronous sequential machines. This section develops a systematic method for the synthesis of fundamental-mode asynchronous sequential machines using *SR* latches as the storage elements. The machine operation is specified as a timing diagram and/or verbal statements.

The synthesis procedure is summarized below. The key step in the synthesis of synchronous sequential machines is the derivation of the state diagram, whereas the key step in the synthesis of asynchronous sequential machines is the derivation of the primitive flow table. The synthesis procedure will result in a machine that operates according to the prescribed specifications. The solution, however, may not be unique.

1. **State diagram** The machine specifications are converted into a state diagram. A timing diagram and/or a verbal statement of the machine specifications is converted into a precise delineation which specifies the machine's operation for all applicable input sequences. This step is not a necessary requirement and is usually omitted; however, the state diagram characterizes the machine's operation in a graphical representation and adds completeness to the synthesis procedure.

2. **Primitive flow table** The machine specifications are converted to a state transition table called a "primitive flow table." This is the least methodical step in the synthesis procedure and the most important. The primitive flow table depicts the state transition sequences and output assertions for all valid input vectors. The flow table must correctly represent the machine's operation for all applicable input sequences, even those that are not initially apparent from the machine specifications.

3. **Equivalent states** The primitive flow table may have an inordinate number of rows. The number of rows can be reduced by finding equivalent states and then eliminating redundant states. If the machine's operation is indistinguishable whether commencing in state Y_i or state Y_j, then one of the states is redundant and can be eliminated. The flow table thus obtained, is a *reduced primitive flow table*.

4. **Merger diagram** The merger diagram graphically portrays the result of the merging process in which an attempt is made to combine two or more rows of the reduced primitive flow table into a single row. The result of the merging technique is analogous to that of finding equivalent states; that is, the merging process can also reduce the number of rows in the table and hence, reduce the number of feedback variables that are required. Fewer feedback variables will result in a machine with less logic and, therefore, less cost.

5. **Merged flow table** The merged flow table is constructed from the merger diagram. The table represents the culmination of the merging process in which two or more rows of a primitive flow table are replaced by a single equivalent row which contains one stable state for each merged row.

6. **Excitation maps and equations** An excitation map is generated for each excitation variable. Then the transient states are encoded, where applicable, to avoid critical race conditions. Appropriate assignment of the excitation variables for the transient states can minimize the δ next-state logic for the excitation variables. The operational speed of the machine can also be established at this step by reducing the number of transient states through which the machine must sequence during a cycle. Then the excitation equations are derived from the excitation maps. All static-1 and static-0 hazards are eliminated

from the network for a sum-of-products or product-of-sums implementation, respectively.

7. **Output maps and equations**   An output map is generated for each machine output. Output values are assigned for all nonstable states so that no transient signals will appear on the outputs. In this step, the speed of circuit operation can also be established. Then the output equations are derived from the output maps ensuring that all outputs will be free of momentary false outputs.

8. **Logic diagram**   The logic diagram is implemented from the excitation and output equations using an appropriate logic family.

7.4.2 Synthesis Examples

This section presents examples illustrating the synthesis of asynchronous sequential machines in their entirety. The machine specifications for these examples will consist of a verbal description only or a verbal description in conjunction with a representative timing diagram.

If the verbal description is the only means of conveying the operational characteristics of the machine, then the description must be comprehensive and precise. Sufficient detail must be provided in the description so that no additional information is required to elucidate the machine specifications. If the machine specifications are delineated in sufficient detail, then a timing diagram can be created as a further aid, if necessary.

When defining the operational characteristics by means of a timing diagram with two or more input variables and at least one output variable, a comprehensive representation of machine characteristics is usually prohibitive. This is due, in part, to space limitations for the diagram in order to show the arrangement and relationship of all combinations of the input and output binary variables.

In the following examples, the solution may not be unique. The synthesis specifications will stipulate whether the λ output logic is to be as fast as possible, as slow as possible, or in minimal form. In all examples, however, there must be no race conditions, no static-1 or static-0 hazards, and no momentary false outputs.

Example 7.12   This design has two inputs x_1 and x_2 and one output z_1, as shown in the timing diagram of Figure 7.108. Input x_1 acts as a gate for x_2; that is, x_2 will be gated to output z_1 only if x_1 precedes the assertion of x_2. If x_1 becomes deasserted while x_2 is asserted, then the full width of the x_2 pulse will appear on z_1 — the width of the x_2 pulse will not be decreased.

Primitive flow table   The primitive flow table transforms the machine specifications into a tabular representation which specifies the state transition sequences for all valid input combinations. The primitive flow table is characterized by having only one stable state in each row. This allows the output for that row to be uniquely specified as a function of a particular stable state. The stable states are indicated by circled entries, whereas the unstable, or transient states, are uncircled. The transient states specify the next stable state for a particular state transition sequence. The unspecified entries are indicated by a dash (–) and are the result of the fundamental-mode operation or an input vector that is invalid from a particular stable state.

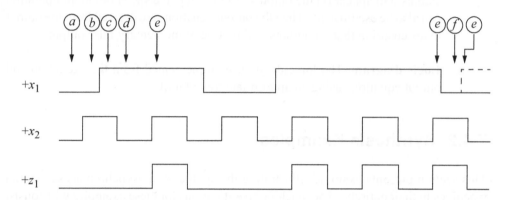

Figure 7.108   Timing diagram for the asynchronous sequential machine of Example 7.12.

Constructing the primitive flow table is a two-pass procedure. The first pass lists all stable states and their associated next states as obtained from the machine specifications such as, a timing diagram and/or a verbal description. These are the more obvious entries. Refer to the partial primitive flow table of Figure 7.109 for the discussion which follows.

Beginning at the leftmost section of the timing diagram, where the initial conditions are specified, proceed left to right assigning a stable state to each different combination of the input vector. The column headings represent the input vector. The table entries specify the stable states, transient states, invalid state transitions which are represented as dashes, and outputs.

The machine begins in stable state $@$ where $x_1 x_2 z_1 = 000$. Input x_2 is then asserted, which takes the machine through transient state b, terminating in stable state $\textcircled{b}$, as shown in the partial primitive flow table. The next change is the assertion of x_1. Since the assertion of x_1 did not precede the assertion of x_2, output z_1 remains deasserted and the machine enters transient state c and ends in stable state $\textcircled{c}$, where $x_1 x_2 z_1 = 110$.

The next change occurs when x_2 becomes deasserted, causing the machine to enter stable state $\textcircled{d}$ where output z_1 remains deasserted. The second x_2 pulse is then asserted, which sequences the machine to state $\textcircled{e}$. Because the assertion of x_1 preceded the assertion of x_2, output z_1 is asserted. If x_1 is still asserted when x_2 becomes deasserted, then this is equivalent to state $\textcircled{d}$, where $x_1 x_2 z_1 = 100$; therefore, the machine proceeds to state $\textcircled{d}$.

Consider stable state $\textcircled{e}$ again. If x_1 becomes deasserted while x_2 is asserted, then this represents a new state in which z_1 remains asserted due to the machine specifications. This new state is labeled stable state $\widehat{f}$. This completes the first pass through the timing diagram and is shown in the partial primitive flow table of Figure 7.109.

x_1x_2 0 0	0 1	1 1	1 0	z_1
$\widehat{a}$	b	–		0
	$\widehat{b}$	c	–	0
–		$\widehat{c}$	d	0
	–	e	$\widehat{d}$	0
–	f	$\widehat{e}$	d	1
	$\widehat{f}$		–	1

Figure 7.109 Partial primitive flow table for the asynchronous sequential machine of Example 7.12.

The second pass establishes the entries for any unspecified transient states and may necessitate creating additional rows. The second pass also establishes subsequences that are not expressly specified by the timing diagram. Careful consideration must be given to these subsequences to ensure that the machine operates according to the functional specifications. The complete primitive flow table is shown in Figure 7.110 and its construction is described in the following paragraphs.

x_1x_2 0 0	0 1	1 1	1 0	z_1
$\widehat{a}$	b	–	d	0
a	$\widehat{b}$	c	–	0
–	b	$\widehat{c}$	d	0
a	–	e	$\widehat{d}$	0
–	f	$\widehat{e}$	d	1
a	$\widehat{f}$	e	–	1

Figure 7.110 Complete primitive flow table for Example 7.12.

In stable state (a), if x_1 becomes asserted while x_2 remains deasserted, then this is equivalent to state (d), where z_1 is deasserted. Therefore, an entry of unstable state d is entered in column $x_1 x_2 = 10$. Also in state (a), a fundamental-mode model does not allow simultaneous changes to the inputs; that is, a change from $x_1 x_2 = 00$ to 11. Therefore, a "don't care" entry is inserted in column $x_1 x_2 = 11$.

In state (b) of the second row of the partial primitive flow table, if x_2 becomes deasserted, then this represents the same conditions as shown in state (a); therefore, an entry of a is inserted in that location. Due to a double change of inputs from $x_1 x_2 = 01$ to 10, a "don't care" entry is inserted in column $x_1 x_2 = 10$.

In state (c) of the third row of the Figure 7.109, if x_1 becomes deasserted, then this is equivalent to state (b), where $x_1 x_2 z_1 = 010$; therefore, an entry of b is inserted in column $x_1 x_2 = 01$. A "don't care" is inserted in column $x_1 x_2 = 00$ due to a simultaneous change of inputs.

In state (d) of the fourth row, if x_1 becomes deasserted, then this is equivalent to the conditions for stable state (a), which places an entry of a in column $x_1 x_2 = 00$ together with a "don't care" in column $x_1 x_2 = 01$. The fifth row containing stable state (e) has already been defined except for the "don't care" entry in column $x_1 x_2 = 00$.

Consider now stable state (f) in the sixth row in conjunction with the timing diagram. If x_2 becomes deasserted, then this is equivalent to state (a), where $x_1 x_2 z_1 = 000$. All situations must be considered when constructing a primitive flow table. Therefore, in state (f), it is possible for x_1 to become asserted while x_2 is still asserted. The machine then returns to state (e) and maintains output z_1 active.

Generation of the primitive flow table is the most critical step in the synthesis of asynchronous sequential machines. If the primitive flow table is correctly constructed, then the remaining steps in the synthesis procedure, if properly executed, will result in a machine that meets the performance criteria of the machine specifications. If, however, the primitive flow table does not delineate all transitions for every possible input sequence, then the remaining steps, although correct in themselves, will result in a machine that does not operate according to the machine specifications.

Equivalent states   Eliminating redundant states from a primitive flow table will generate a reduced primitive flow table which is equivalent to the original table. A primitive flow table — unless already containing a minimal number of rows — can be reduced to provide a table that contains fewer rows than the original table and yet completely characterizes the operation of the machine.

The definition of equivalence can be stated as shown below for asynchronous sequential machines. Two stable states Y_i and Y_j in a primitive flow table are defined to be equivalent if and only if all of the following rules apply:

1. The stable states have the same input vector; that is, the states are in the same column.

2. The outputs associated with Y_i and Y_j have the same value; that is, $Z_r(Y_i) = Z_r(Y_j)$.

3. The next states for Y_i and Y_j are the same or equivalent for every column in the two rows of the primitive flow table. That is, for each input combination in the rows of the two stable states, the following is observed:

   (a) Identical or equivalent state names, or
   (b) Two dashes.

In Figure 7.110, the only states that are potentially equivalent are:

   ⓑ and ⓕ

   ⓒ and ⓔ

Stable states ⓑ and ⓕ have different outputs, contradicting Rule 2 for equivalence. Therefore, states ⓑ and ⓕ are not equivalent. The same is true for states ⓒ and ⓔ; therefore, the primitive flow of Figure 7.110 is also the reduced primitive flow table. Each row of the primitive flow table corresponds to a different combination of feedback variables.

Merger diagram   Merging is a process of combining two or more rows of a reduced primitive flow table into a single row. The merging process reduces the number of rows in the flow table and, thus, may reduce the number of feedback variables. A reduction of feedback variables decreases the number of storage elements in the machine. The stable states in the rows that are merged are entered in the same location in the single merged row. When merging, the outputs associated with each row are disregarded. Thus, two rows of a reduced primitive flow table can merge, regardless of the output values of the rows under consideration.

Two rows can merge into a single row if the entries in the same column of each row satisfy the requirements of one of the following three merging rules:

1. Identical state entries, either stable or unstable
2. A state entry and a "don't care"
3. Two "don't care" entries

Merging rule 1 specifies that there must be no conflict in state name entries in the same column of the two rows under consideration. That is, two different states cannot both be active for the same input vector and the same combination of feedback variables. Three or more rows can merge into a single row if and only if all pairs of rows satisfy the conditions of the merging rules.

A stable state entry and an unstable state entry of the same name in the same column of two different rows are merged as the stable state entry, since the resultant state must be stable. Thus, in many cases, the merging process eliminates the transient unstable states. Two identical unstable states merge as an unstable state. Both stable and unstable states merge with a "don't care" entry as a stable and unstable state, respectively.

The merging process is facilitated by means of a merger diagram. The merger diagram depicts all rows of the reduced primitive flow table in a graphical representation. Each row of the table is portrayed as a vertex in the merger diagram. The rows in which merging is possible are connected by lines.

A set of rows in a reduced primitive flow table can merge into a single row if and only if the rows are strongly connected. That is, every row in the set must merge with all other rows in the set. For example, Figure 7.111 illustrates strongly connected sets of three, four, and five rows each. Each set in Figure 7.111 is a maximal compatible set, in which compatible pairs $\{@, ⓑ\}$, and $\{ⓑ, ©\}$ implies the compatibility of $\{@, ©\}$. Thus, the transitive property applies to maximal compatible sets. Adding another state to a maximal compatible set negates the transitive property on the new set, unless the set remains strongly connected.

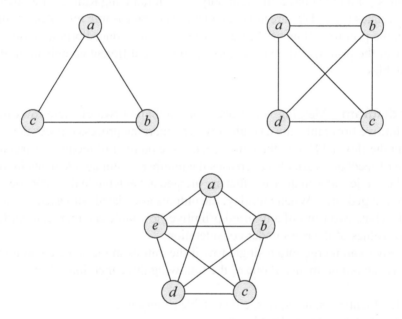

Figure 7.111   Strongly connected sets in which each row in the set can merge with all other rows in the set.

The objective of merging is to combine the maximal number of rows into a single merged row while maintaining the fewest number of merged rows. All strongly connected sets can be combined into single individual rows, one set per row. Figure 7.112 shows the merger diagram for the asynchronous sequential machine of Example 7.12. Using the rules for merging in Figure 7.110, it is seen that rows $@$ and $ⓑ$ can merge because there is no conflict in state names. Rows $@$ and $©$ can also merge as well as rows $\textcircled{a}$ and $\textcircled{d}$. Therefore, rows $\textcircled{a}$, $\textcircled{b}$, and $\textcircled{c}$ can merge into a single row. Row $\textcircled{a}$ cannot merge with any other row. In a similar manner, it is evident that rows $\textcircled{d}$, $\textcircled{e}$, and $\textcircled{f}$ can also merge into a single row.

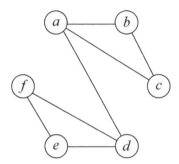

Figure 7.112   Merger diagram for the asynchronous sequential machine of Example 7.12.

Merged flow table   The next step in the synthesis procedure is the generation of a merged flow table. The merged flow table specifies the operational characteristics of the machine in a manner analogous to that of the primitive flow table and the reduced primitive flow table, but in a more compact form. Each row in a merged flow table represents a set of maximal compatible rows.

Most unspecified entries in the reduced primitive flow table are replaced with either a stable or an unstable state entry in the merged flow table. Since more than one stable state is usually present in each row of a merged flow table, many state transition sequences do not cause a change to the feedback variables. Thus, faster operational speed is realized.

The merged flow table is derived from the merger diagram in conjunction with the reduced primitive flow table. The partition of sets of maximal compatible rows obtained from the merger diagram dictates the minimal number of rows in the merged flow table.

To merge rows $\textcircled{a}$, $\textcircled{b}$, and $\textcircled{c}$ into a single row, simply transcribe each row, one row at a time, from the reduced primitive flow table to the merged flow table. For example, row $\textcircled{a}$ transfers as shown in Figure 7.113(a). Then transfer row $\textcircled{b}$ to the same row in the merged flow table, superimposing row $\textcircled{b}$ on row $\textcircled{a}$, as shown in Figure 7.113(b). Finally, transfer row $\textcircled{c}$ to the merged flow table, superimposing row $\textcircled{c}$ on previously transferred rows $\textcircled{a}$ and $\textcircled{b}$, as shown in Figure 7.113(c). Notice that no conflict in state names occurs during the merging of rows $\textcircled{a}$, $\textcircled{b}$ and $\textcircled{c}$ into a single merged row. This is a necessary requirement and exemplifies the rationale for merging. The merged flow table is shown in Figure 7.114.

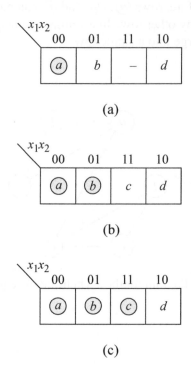

Figure 7.113   Top row of the merged flow table illustrating the transcribing of rows ⓐ, ⓑ, and ⓒ singly, from the reduced primitive flow table to the merged flow table: (a) row ⓐ transferred; (b) row ⓑ transferred and superimposed on row ⓐ; and (c) row ⓒ transferred and superimposed on rows ⓐ and ⓑ.

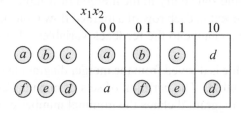

Figure 7.114   Merged flow table constructed from the merger diagram and the reduced primitive flow table.

Excitation map and equation   The merged flow table is the foundation from which the excitation maps are derived. The excitation map directly formulates the equations that are necessary to implement the logic for the excitation variables.

Each row of the merged flow table is assigned a unique combination of values for the feedback (or secondary) variables. The values of the feedback variables for each row then determine the values of the excitation variables for the stable state entries in the corresponding row of the excitation map.

Recall that a machine is stable in a particular state when the feedback variables are equal to the excitation variables. Thus, the values assigned to the excitation variables in each stable state of the excitation map are identical to those of the feedback variables for that row. The placement of each stable state of the excitation map represents a one-to-one mapping of the stable states in the corresponding locations of the merged flow table. The entries in the excitation map that correspond to unstable states in the merged flow table specify the next state to which the machine will sequence due to a change in the input vector. The excitation map for Y_{1e} is shown in Figure 7.115 and the equation for Y_{1e} is shown in Equation 7.38.

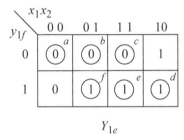

$$Y_{1e}$$

Figure 7.115 Excitation map for the asynchronous sequential machine of Example 7.12.

$$Y_{1e} = x_1 x_2' + x_2 y_{1f} + x_1 y_{1f} \qquad (7.38)$$

$$\text{Hazard cover} \quad \underline{\qquad}$$

Output map and equation The next step in the synthesis procedure is to assign values to the output variables. A Karnaugh map — referred to as an output map — facilitates this process. This step utilizes the results of two previous operations: the merged flow table and the reduced primitive flow table.

The merged flow table indicates the location of all stable states. Since the outputs are associated with stable states, the merged flow table indicates the location of the stable state output variables in their respective output maps. The merged flow table defines the format of the output map, which is constructed directly from the merged flow table and the reduced primitive flow table. The reduced primitive flow table specifies the output values of the corresponding stable states in the output map. The merged flow table, the excitation map, and the output map have the same number of inputs — and thus, the same number of columns — and the same number of feedback variables, necessitating the same number of rows. The values assigned to the input variables and the feedback variables are the same in both the excitation map and the output map.

The speed of circuit operation can be established during this phase. If different output values are associated with the initial and destination stable states for a state transition with only one intermediate state, then the intermediate unstable state can be assigned a value that is equal to either the initial or the destination stable state output value.

By assigning appropriate values to the unstable states in the output map, the outputs can be made to change value as soon as possible or as late as possible for a particular state transition. If the output value of the initial stable state is assigned to the intermediate state, then the change to the output is delayed until the machine enters the destination stable state. If, however, the output value of the destination stable state is assigned to the intermediate state, then the output value changes before the machine reaches the destination stable state.

The output map for z_1 is shown in Figure 7.116. Since stable state $\textcircled{c}$ sequences to stable state $\textcircled{d}$ and both states have an output value of $z_1 = 0$ — as shown in the primitive flow table — the intermediate state of $x_1x_2y_{1f} = 100$ must contain a value of 0; otherwise, there would be a glitch on output z_1. The same is true for the sequence $\textcircled{d}$ Æ $\textcircled{a}$, necessitating a 0 in location $x_1x_2y_{1f} = 001$. The equation for output z_1 is shown in Equation 7.39. The logic diagram is constructed from the excitation equation as a sum of products and the output equation, which is one of the product terms from the d next-state logic.

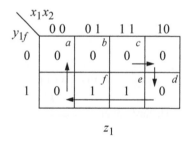

Figure 7.116   Output map for z_1 for the asynchronous sequential machine of Example 7.12.

$$z_1 = x_2 y_{1f} \tag{7.39}$$

Example 7.13   A timing diagram for an asynchronous sequential machine that has one input x_1 and two outputs z_1 and z_2 is shown in Figure 7.117. Output z_1 is asserted for the duration of every second x_1 pulse; output z_2 is asserted for the duration of every second z_1 pulse. The outputs are to respond as fast as possible to changes that occur on input x_1.

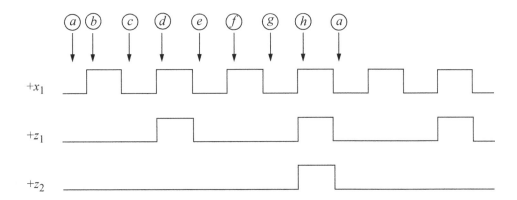

Figure 7.117   Timing diagram for the asynchronous sequential machine of Example 7.13.

The primitive flow table is shown in Figure 7.118. There are no equivalent states and no rows can merge. Therefore, the primitive flow table is also the reduced primitive flow table and the merged flow table. Since there are eight rows, there are three feedback variables y_{1f}, y_{2f}, and y_{3f}. The combined excitation map for excitation variables Y_{1e}, Y_{2e}, and Y_{3e} is shown in Figure 7.119.

x_1	0	1	z_1	z_2
ⓐ	b	0	0	
c	ⓑ	0	0	
ⓒ	d	0	0	
e	ⓓ	1	0	
ⓔ	f	0	0	
g	ⓕ	0	0	
ⓖ	h	0	0	
a	ⓗ	1	1	

Figure 7.118   Primitive flow table for Example 7.13.

$$x_1$$

$y_{1f}y_{2f}y_{3f}$	0	1
0 0 0	(000)ᵃ	001
0 0 1	011	(001)ᵇ
0 1 1	(011)ᶜ	010
0 1 0	110	(010)ᵈ
1 1 0	(110)ᵉ	111
1 1 1	101	(111)ᶠ
1 0 1	(101)ᵍ	100
1 0 0	000	(100)ʰ

$$Y_{1e}Y_{2e}Y_{3e}$$

Figure 7.119   Combined excitation map for the asynchronous sequential machine of Example 7.13.

The individual excitation maps are shown in Figure 7.120 and are derived directly from the combined excitation maps by copying the individual columns of the corresponding excitation variables to the appropriate map. The equations for the excitation variables are shown in Equation 7.40 and include redundant prime implicants to negate any potential hazards.

The output maps are shown in Figure 7.121 and are derived from the reduced primitive flow table. In state $\textcircled{c}$, output $z_1 = 0$; in state $\textcircled{d}$, output $z_1 = 1$. Therefore, as the machine sequences from state $\textcircled{c}$ to state $\textcircled{d}$ transient state d can have an output value for z_1 that is "don't care" — either 0 or 1. However, the machine specifications stipulate that the outputs must respond to input changes as quickly as possible. Therefore, an entry of 1 is inserted in location $y_{1f}y_{2f}y_{3f}x_1 = 0111$ in order to assert output z_1 as soon as possible. The same is true for z_1 in location $y_{1f}y_{2f}y_{3f}x_1 = 0100$ by inserting a value of $z_1 = 0$ and location $y_{1f}y_{2f}y_{3f}x_1 = 1000$ by inserting a 0.

The same rationale applies to output z_2 also in location $y_{1f}y_{2f}y_{3f}x_1 = 1011$ by inserting a value of 1 and in location $y_{1f}y_{2f}y_{3f}x_1 = 1000$ by inserting a value of 0. The logic diagram is shown in Figure 7.122 using AND gates and OR gates. Since all AND gates have inputs connected to the feedback variables, it is appropriate to connect an active-low reset signal to the AND gates; otherwise, the latches may assume a random state when power is turned on. A reset pulse guarantees that the machine will commence operation in state $\textcircled{a}$.

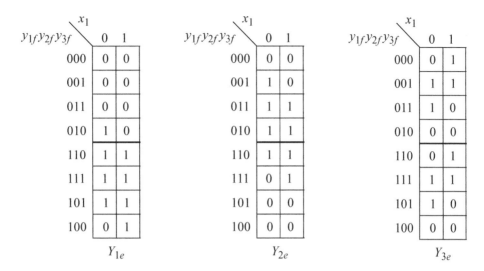

Figure 7.120   Individual excitation maps for the asynchronous sequential machine of Example 7.13.

$$Y_{1e} = y_{1f}x_1 + y_{1f}y_{3f} + y_{2f}y_{3f}'x_1' + y_{1f}y_{2f}$$

$$Y_{2e} = y_{2f}y_{3f}' + y_{2f}x_1 + y_{1f}'y_{3f}x_1' + y_{1f}'y_{2f}$$

$$Y_{3e} = y_{3f}x_1' + y_{1f}'y_{2f}'x_1 + y_{1f}y_{2f}x_1 + y_{1f}'y_{2f}'y_{3f} + y_{1f}y_{2f}y_{3f} \qquad (7.40)$$

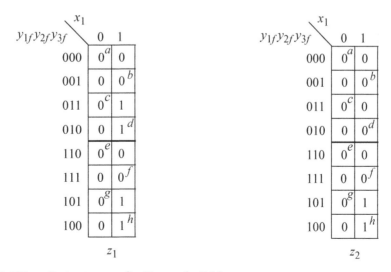

Figure 7.121   Output maps for Example 7.13.

$$z_1 = y_{1f}'y_{2f}x_1 + y_{1f}y_{2f}'x_1$$
$$z_2 = y_{1f}y_{2f}'x_1 \qquad (7.41)$$

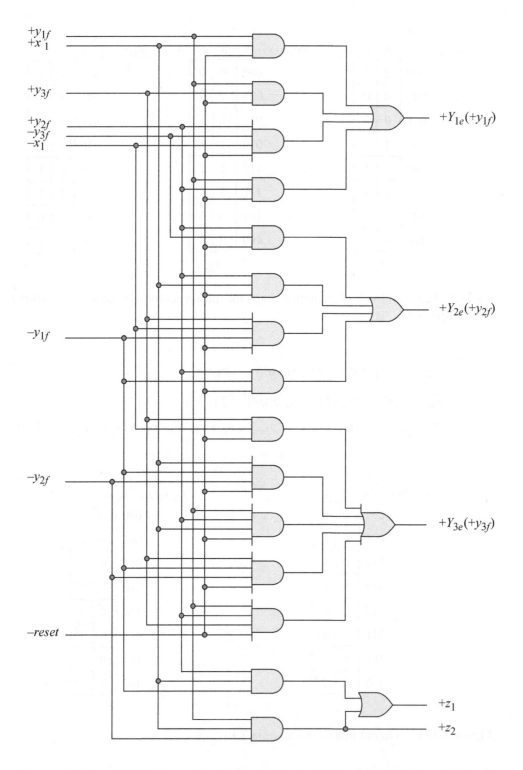

Figure 7.122   Logic diagram for the asynchronous sequential machine of Example 7.13.

Example 7.14 The timing diagram for an asynchronous sequential machine is shown in Figure 7.123 with one input x_1 and two outputs z_1 and z_2. The outputs will be implemented with the least amount of logic which may not necessarily be the fastest response time.

The primitive flow table is shown in Figure 7.124 in which there are no equivalent states, because each state has different outputs; therefore, this is also the reduced primitive flow table. The primitive flow table is generated directly from the timing diagram beginning in stable state $\textcircled{a}$, where $x_1 z_1 z_2 = 000$. A new stable state is inserted for each new combination of the variables $x_1 z_1 z_2$. The merger diagram is shown in Figure 7.125 in which no rows can merge because both columns for x_1 have at least one different state in each selected row. The combined excitation map — obtained from the primitive flow table — and the individual excitation maps are shown in Figure 7.126. The excitation equations are shown in Equation 7.42. The combined output map and the individual output maps are shown in Figure 7.127. The output equations are shown in Equation 7.43. The logic diagram is shown in Figure 7.128.

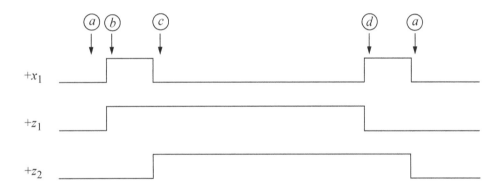

Figure 7.123 Timing diagram for the asynchronous sequential machine of Example 7.14.

x_1	0	1	z_1	z_2
	$\textcircled{a}$	b	0	0
	c	$\textcircled{b}$	1	0
	$\textcircled{c}$	d	1	1
	a	$\textcircled{d}$	0	1

Figure 7.124 Primitive flow table for Example 7.14.

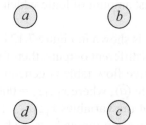

Figure 7.125   Merger diagram for Example 7.14.

$y_{1f}y_{2f}$ \ x_1	0	1
0 0	⓪⓪ ᵃ	01
0 1	11	⑩1 ᵇ
1 1	⑪1 ᶜ	10
1 0	00	⑩0 ᵈ

$Y_{1e} Y_{2e}$

$y_{1f}y_{2f}$ \ x_1	0	1
0 0	0	0
0 1	1	0
1 1	1	1
1 0	0	1

Y_{1e}

$y_{1f}y_{2f}$ \ x_1	0	1
0 0	0	1
0 1	1	1
1 1	1	0
1 0	0	0

Y_{2e}

Figure 7.126   Combined excitation map and individual excitation maps for Example 7.14.

$$Y_{1e} = x_1'y_{2f} + x_1y_{1f} + y_{1f}y_{2f}$$
$$Y_{2e} = x_1'y_{2f} + x_1y_{1f}' + y_{1f}'y_{2f} \tag{7.42}$$

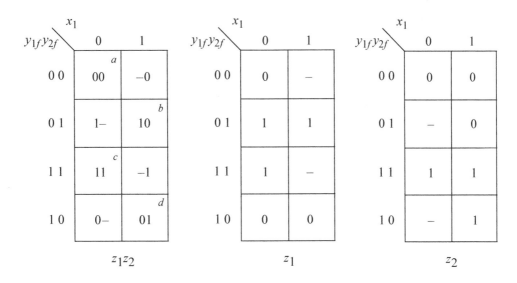

Figure 7.127   Combined output map and individual output maps for Example 7.14.

$$z_1 = y_{2f}$$
$$z_2 = y_{1f} \qquad\qquad (7.43)$$

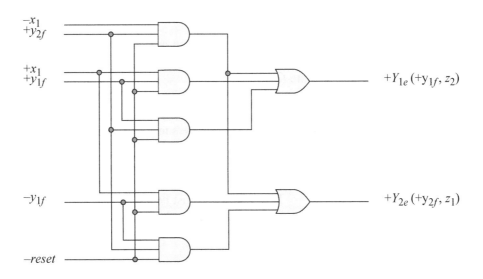

Figure 7.128   Logic diagram for Example 7.14.

Example 7.15   This example will concentrate on constructing a transition diagram. Methods will now be presented for assigning values to the feedback variables so that each state transition sequence will involve a change of only one excitation variable between logically contiguous rows in the cycle. Two changes of excitation variables will still be allowed, however, provided that the resulting race condition does not generate a critical race. Each transition in the merged flow table must be examined to ensure that the assigned feedback values differ by a change of only one variable between the row containing the beginning stable state and each successive pair of rows in the cycle, including the row containing the destination stable state. That is, the feedback variables must be assigned adjacent p-tuples for each successive row in the cycle.

Consider the merged flow table of Figure 7.129 consisting of three rows labeled 1, 2, and 3. Inspection of column $x_1 x_2 = 00$ indicates that the feedback values assigned to row 1 must be adjacent to those assigned to row 2, to provide a race-free cycle from state $\textcircled{b}$ or $\textcircled{g}$ through unstable state a to state $\textcircled{a}$. The same adjacency requirement is observed in column $x_1 x_2 = 11$.

$x_1 x_2$	00	01	11	10
1	$\textcircled{a}$	b	$\textcircled{e}$	c
2	a	$\textcircled{b}$	e	$\textcircled{g}$
3	$\textcircled{f}$	b	$\textcircled{d}$	$\textcircled{c}$

Figure 7.129   Merged flow table for an asynchronous sequential machine for Example 7.15.

Examination of column $x_1 x_2 = 10$ identifies an adjacency requirement between rows 1 and 3. This requirement accommodates a transition from either state $\textcircled{a}$ or $\textcircled{e}$ through unstable state c to state $\textcircled{c}$.

Similarly, rows 2 and 3 must have adjacent feedback values to realize a transition from state $\textcircled{d}$ or $\textcircled{f}$ through unstable state b to state $\textcircled{b}$. The observation of the above adjacency requirements for race-free operation are summarized as follows:

> Column $x_1 x_2 = 00$: Rows 1 and 2 must be adjacent.
> Column $x_1 x_2 = 01$: Rows 1 and 2 must be adjacent.
> Rows 2 and 3 must be adjacent.
> Column $x_1 x_2 = 11$: Rows 1 and 2 must be adjacent.
> Column $x_1 x_2 = 10$: Rows 1 and 3 must be adjacent.

In column $x_1 x_2 = 01$, only one stable state is specified; therefore, a critical race condition is impossible. A noncritical race, however, may occur from row 1 or 3 to row 2, depending on the assigned feedback values. Since noncritical races do not present a problem in the deterministic operation of an asynchronous sequential machine, the assignment of values for the feedback variables is not crucial.

The preceding requirements listed for each column specify the adjacencies that are needed to establish race-free operation for the indicated transitions. The same information is portrayed graphically in the *transition diagram* of Figure 7.130. For a merged flow table containing three rows, the rows are listed in a triangular arrangement. Each row is represented by a vertex. Each pair of rows, for which adjacency is required, is connected by a line. The connecting line indicates a requisite transition between the pair of rows.

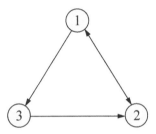

Figure 7.130   Transition diagram for the merged flow table of Figure 7.129.

There is no need to specify the direction of the lines, since adjacency is the only information that is relevant. However, directed lines in a transition diagram may be advantageous in visualizing the sequence of transitions.

The next step is to assign values to the vertices which will represent the values of the feedback variables. Each pair of logically adjacent rows must be assigned adjacent state codes. The transition diagram of Figure 7.130 illustrates that all three rows of the merged flow table must be adjacent. It is obviously not possible to assign 2-tuples to the three rows so that each row is adjacent to all other rows.

To illustrate this impracticability, observe the transition diagram of Figure 7.131. The codes assigned to the state variables in Figure 7.131(a) produce a noncritical race condition for a transition between rows 2 and 3. Figure 7.131(b) generates a critical race for a transition between rows 1 and 3. If the transition diagram contained four row vertices, then a 2-tuple Gray code assignment would realize a race-free operation.

The vertices of triangular or other polygons with an odd number of sides cannot be encoded with adjacent p-tuples for every row. The transition diagram must be altered so that triangular polygons do not appear. Modifying the transition diagram in this way may require more than a minimal number of state variables.

Since a merged flow table containing three rows requires two feedback variables, the addition of a fourth row will not increase the number of feedback variables. The number of rows in a table satisfies the expression of Equation 7.44, where r is the number of rows and p is the number of states, or feedback variables.

$$r = 2^p \tag{7.44}$$

Whether $r = 3$ or 4, the number of feedback variables remains the same. Therefore, a fourth row consisting of unspecified entries will be appended to the bottom row of the merged flow table. The additional row is not associated with any stable state. The unspecified entries will be used, where applicable, to establish intermediate unstable states which will direct the machine to the appropriate destination stable state. Two state variables are required to encode four rows.

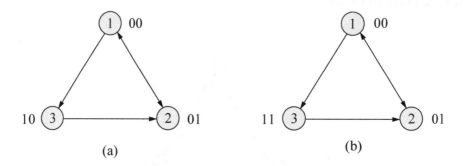

(a)   (b)

Figure 7.131 Transition diagrams illustrating state code assignments for a 3-row merged flow table containing noncritical and critical races: (a) a noncritical race between rows 2 and 3 and (b) a critical race between rows 1 and 3.

Figure 7.132 depicts the augmented merged flow table containing the original three rows plus a fourth row of unspecified entries. The transition diagram for the augmented flow table is shown in Figure 7.133. The transition from row 1 to row 3 is replaced by an equivalent sequence from row 1 to row 4 and then to row 3, as shown by the arrows in Figure 7.132 and Figure 7.133. This sequence represents a transition from stable state $\textcircled{a}$ or $\textcircled{e}$ through transient state c in column $x_1 x_2 = 10$, then to the unspecified entry in row 4, then to state $\textcircled{c}$ in row 3. All other transitions involve a change of only one feedback variable.

$x_1 x_2$	00	01	11	10
1	$\textcircled{a}$	b	$\textcircled{e}$	c
2	a	$\textcircled{b}$	e	$\textcircled{g}$
3	$\textcircled{f}$	b	$\textcircled{d}$	$\textcircled{c}$
4	—	—	—	—

Figure 7.132 Augmented merged flow table for Example 7.15.

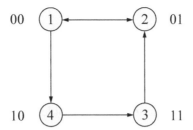

Figure 7.133 Transition diagram for the augmented merged flow table of Figure 7.132.

The codes next to each row vertex indicate the assigned values for the feedback variables. The choice of state codes is arbitrary in this context. Any assignment of sequential Gray code 2-tuples is a suitable choice. Each state name is now replaced by its corresponding assigned 2-tuple.

The excitation map for the augmented merged flow table is shown in Figure 7.134. This is a combined excitation map for excitation variables Y_{1e} and Y_{2e}. The feedback variables are $y_{1f}y_{2f}$. The stable states in each row are assigned excitation values that are equal to the feedback values of the corresponding row. The unstable states are assigned excitation values that direct the machine to the destination stable state.

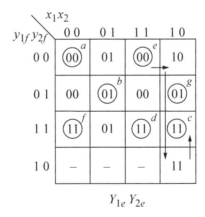

Figure 7.134 Combined excitation map for the augmented merged flow table of Figure 7.132.

For example, the transition from state $\textcircled{a}$ to state $\textcircled{b}$ is specified by $\textcircled{a} \rightarrow b \rightarrow \textcircled{b}$. Thus, $Y_{1e}Y_{2e} = \textcircled{00} \rightarrow 01 \rightarrow \textcircled{01}$. The entry $Y_{1e}Y_{2e} = 01$ for unstable state b in row $y_{1f}y_{2f} = 00$, column $x_1x_2 = 01$, specifies the excitation values which will become the values of the feedback variables after a delay of Δt, directing the machine to state $\textcircled{b}$ in row $y_{1f}y_{2f} = 01$, column $x_1x_2 = 01$.

Likewise, the transition from state $\textcircled{g}$ to state $\textcircled{e}$ requires excitation values of $Y_{1e}Y_{2e} = 00$ to be entered in unstable state e in row $y_{1f}y_{2f} = 01$, column $x_1x_2 = 11$. Thus, after a delay of Δt, the feedback variables become equal to the excitation variables and the machine enters state $\textcircled{e}$. The "don't care" entries in row 4 can be used as intermediate transient states to introduce cycles which direct the machine to the desired stable state. Code assignments must be avoided that would cause the machine to cycle continuously between unstable states.

The transition from state $\textcircled{e}$ to state $\textcircled{c}$ necessitates the values for $Y_{1e}Y_{2e}$ as shown in Figure 7.135. The sequence from state $\textcircled{e}$ to state $\textcircled{c}$ is graphically illustrated by the arrows in Figure 7.134. A similar path is realized for a transition from state $\textcircled{a}$ to state $\textcircled{c}$. Thus, the addition of the fourth row in the excitation map resolves the adjacency problem and thus, the possible critical race condition of the original merged flow table of Figure 7.129. Other arrangements are possible for the four rows of the augmented merged flow table.

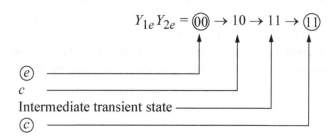

Figure 7.135   State transition sequence depicting the transition from stable state $\textcircled{e}$ to stable state $\textcircled{c}$.

The combined excitation map of Figure 7.134 is now separated into its constituent parts to obtain individual excitation maps for Y_{1e} and Y_{2e}, as shown in Figure 7.136. To obtain the excitation map for Y_{1e}, simply transfer the values for Y_{1e} from the minterm locations in the combined map to the same squares in the individual map. Repeat the process to obtain the excitation map for Y_{2e}.

The equations for the excitation variables are derived directly from the excitation maps. The equations can be specified in either a sum-of-products form or a product-of-sums form. Regardless of the form used, the Boolean equations must be free of static-1 and static-0 hazards. The sum-of-products notation is shown in Equation 7.45 for Y_{1e} and Y_{2e}; the product-of-sums form is shown in Equation 7.46 for each excitation variable. All equations are free of static hazards. In some instances, the equations may be manipulated — using the laws of Boolean algebra — to obtain a network with fewer logic gates.

$y_{1f}y_{2f}$ \ x_1x_2	0 0	0 1	1 1	1 0
0 0	0	0	0	1
0 1	0	0	0	0
1 1	1	0	1	1
1 0	–	–	–	1

Y_{1e}

$y_{1f}y_{2f}$ \ x_1x_2	0 0	0 1	1 1	1 0
0 0	0	1	0	0
0 1	0	1	0	1
1 1	1	1	1	1
1 0	–	–	–	1

Y_{2e}

Figure 7.136 Individual excitation maps obtained from Figure 7.134 for the asynchronous sequential machine of Example 7.15.

$$Y_{1e} = y_{1f}x_1 + y_{1f}x_2' + y_{2f}'x_1x_2'$$

$$Y_{2e} = y_{1f} + x_1'x_2 + y_{2f}x_1x_2' \qquad (7.45)$$

$$Y_{1e} = (x_1 + x_2')(y_{1f} + y_{2f}')(y_{1f} + x_1)(y_{1f} + x_2')$$

$$Y_{2e} = (y_{1f} + x_1 + x_2)(y_{1f} + x_1' + x_2')(y_{1f} + y_{2f} + x_1') \qquad (7.46)$$
$$(y_{1f} + y_{2f} + x_2)$$

Hazard cover

Example 7.16 A merged flow table is shown in Figure 7.137 in which multiple adjacencies are required. The state transitions can be determined with reference to individual rows in the merged flow table. This is accomplished by means of a transition diagram, which clearly identifies all race conditions. The four rows are listed as shown in the transition diagram of Figure 7.138.

Row 1 proceeds to row 2 by the following two transitions: $\textcircled{a} \to f \to \textcircled{f}$ and $\textcircled{b} \to c \to \textcircled{c}$. This is illustrated by the directed line connecting vertices 1 and 2 . A transition occurs from row 2 to rows 3 and 4 by state transitions $\textcircled{c} \to d \to \textcircled{d}$ and $\textcircled{f} \to g \to \textcircled{g}$, respectively, as indicated by the lines connecting row 2 to row 3 and row 4.

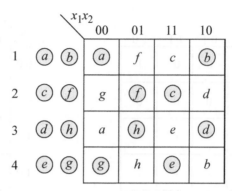

Figure 7.137   Merged flow table for Example 7.16.

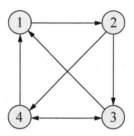

Figure 7.138   Transition diagram obtained from the merged flow table of Figure 7.137.

A transition is realized between rows 3 and 1 from state $\widehat{h}$ when the input vector changes from $x_1 x_2 = 01$ to 00 or from state $\widehat{d}$ when the input vector changes from $x_1 x_2 = 10$ to 00. Also, a transition can emanate from row 3 to row 4 from stable states $\widehat{h}$ or $\widehat{d}$ when the inputs change from $x_1 x_2 = 01$ to 11 or from $x_1 x_2 = 10$ to 11, respectively. Finally, a transition can occur from row 4 to row 3 by the sequence $\widehat{g} \rightarrow h \rightarrow \widehat{h}$ and from row 4 to row 1 by the sequence $\widehat{e} \rightarrow b \rightarrow \widehat{b}$.

Since it is impossible to assign two-tuple state variable codes so that all four rows are adjacent, it is necessary to introduce an additional state variable. Adding a state variable produces an augmented merged flow table containing eight rows as shown in Figure 7.139. The top four rows are unchanged from the original table. The bottom four rows contain unspecified entries in every column. The lower rows will be used to insert intermediate transient states, where applicable. An intermediate state will be inserted in the cycle whenever a state transition sequence causes two or more excitation variables to change state simultaneously. The values for the feedback variables will be obtained from a transition diagram.

x_1x_2

	00	01	11	10
1	ⓐ	f	c	ⓑ
2	g	Ⓕ	Ⓒ	d
3	a	Ⓗ	e	Ⓓ
4	Ⓖ	h	Ⓔ	b
5	–	–	–	–
6	–	–	–	–
7	–	–	–	–
8	–	–	–	–

Figure 7.139 Augmented merged flow table for the asynchronous sequential machine of Example 7.16.

The transition diagram for eight rows is shown in Figure 7.140. This is a cube in which the top surface is assigned the 3-tuple Gray code 000, 001, 011, 010 for row vertices 1, 2, 6, 4, respectively. The bottom surface is assigned Gray code values of 100, 101, 111, 110 for row vertices 3, 5, 8, 7, respectively.

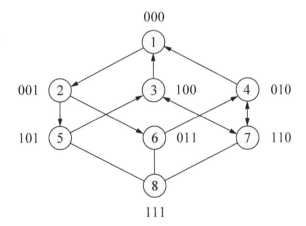

Figure 7.140 Transition diagram for the augmented merged flow table of Figure 7.139.

Any state transition that results in two or more state variables changing values simultaneously can be rerouted through unspecified entries in rows 5 through 8, such that each individual transition in the cycle takes place between adjacent state variable codes. For example, the transition $\textcircled{f} \rightarrow g \rightarrow \textcircled{g}$ from row 2 to row 4 can be directed first to row 6 and then to row 4. Each step of the cycle proceeds through contiguous vertices that involve a change of only one state variable. The transitions in the merged flow table of Figure 7.137 in which race conditions exist are shown redirected in the augmented merged flow table of Figure 7.139.

The transition from state $\textcircled{f}$ (row 2) to state $\textcircled{g}$ (row 4) passes through an intermediate state in row 6, as shown by the arrows. The transition from state $\textcircled{c}$ (row 2) to state $\textcircled{d}$ (row 3) causes two state variables to change (from 001 to 100), as shown in the transition diagram. To avoid this race condition, the transition from row 2 is directed first to row 5 and then to row 3. This modified state transition is

$$2\,(001) \rightarrow 5\,(101) \rightarrow 3\,(100); \qquad \textcircled{c} \rightarrow d \rightarrow \textcircled{d}$$

In a similar manner, the remaining transitions are modified in which race conditions are present. The arrows in Figure 7.139 illustrate the redirected transitions, in which each contiguous vertex (row) in the cycle involves a change of only one state variable. The modified row transitions are listed in Figure 7.141 together with their corresponding state transitions.

$$2\,(001) \rightarrow 6\,(011) \rightarrow 4\,(010); \qquad \textcircled{f} \rightarrow g \rightarrow \textcircled{g}$$
$$2\,(001) \rightarrow 5\,(101) \rightarrow 3\,(100); \qquad \textcircled{c} \rightarrow d \rightarrow \textcircled{d}$$
$$3\,(100) \rightarrow 7\,(110) \rightarrow 4\,(010); \qquad \textcircled{h} \rightarrow e \rightarrow \textcircled{e}$$
$$3\,(100) \rightarrow 7\,(110) \rightarrow 4\,(010); \qquad \textcircled{d} \rightarrow e \rightarrow \textcircled{e}$$
$$4\,(010) \rightarrow 7\,(110) \rightarrow 3\,(100); \qquad \textcircled{g} \rightarrow h \rightarrow \textcircled{h}$$
$$4\,(010) \rightarrow 7\,(110) \rightarrow 3\,(100); \qquad \textcircled{e} \rightarrow h \rightarrow \textcircled{h}$$

7.5   Analysis of Pulse-Mode Asynchronous Sequential Machines

Many situations are encountered in digital engineering where the input signals occur as pulses and in which there is no periodic clock signal to synchronize the operation of the sequential machine. Typical examples which use the principles of *pulse-mode* techniques are vending machines, demand-access road intersections, and automatic toll booths.

In the presentation of synchronous sequential machines, the data input signals were asserted as voltage levels. A periodic clock input was also required, such that state changes occurred on the active clock transition. In the synthesis of asynchronous sequential machines, the input variables were also considered as voltage levels; however, there was no machine clock to synchronize state changes. The operation of pulse-mode machines is similar, in some respects, to both synchronous and asynchronous sequential machines. State changes occur on the application of input pulses which trigger the storage elements, rather than on a clock signal. The input pulses, however, occur randomly in an asynchronous manner.

In pulse-mode sequential machines, each variable of the input alphabet X is active in the form of a pulse. The duration of the pulse is less than the propagation delay of the storage elements and associated logic gates. Thus, an input pulse will initiate a state change, but the completion of the change will not take place until after the corresponding input has been deasserted. Multiple inputs cannot be active simultaneously. There is no separate clock input in pulse-mode machines.

The storage elements in pulse-mode machines are usually level-sensitive rather than edge-triggered devices. Thus, *Set/Reset* (*SR*) latches using NAND or NOR logic are typically used in the implementation of pulse-mode machines. In order for the operation of the machine to be deterministic, some restrictions apply to the input pulses:

1. Input pulses must be of sufficient duration to trigger the storage elements.

2. The time duration of the pulses must be shorter than the minimal propagation delay through the combinational input logic and the storage elements, so that the pulses are deasserted before the storage elements can again change state.

3. The time duration between successive input pulses must be sufficient to allow the machine to stabilize before application of the next pulse.

4. Only one input pulse can be active at a time.

If the input pulse is of insufficient duration, then the storage elements may not be triggered and the machine will not sequence to the next state. If the pulse duration is too long, then the pulse will still be active when the machine changes from the present state $Y_{j(t)}$ to the next state $Y_{k(t+1)}$. The storage elements may then be triggered again and sequence the machine to an incorrect next state. If the time between consecutive pulses is too short, then the machine will be triggered while in an unstable condition, resulting in unpredictable behavior.

Since pulse inputs cannot occur simultaneously, a pulse-mode machine with n input signals can have only $n+1$ combinations of the input alphabet, instead of 2^n combinations as in synchronous sequential machines and asynchronous sequential machines that are not inherently characterized by pulse-mode operation. For example, for a pulse-mode machine with two inputs x_1 and x_2, three possible valid combinations can occur: $x_1 x_2 = 00$, 10, or 01. However, since no changes are initiated by an input vector of $x_1 x_2 = 00$, it is necessary to consider only the vectors $x_1 x_2 = 10$ and 01 when analyzing or designing a pulse-mode asynchronous sequential machine.

Similarly, for a machine with three inputs x_1, x_2, and x_3, only the following input vectors need be considered: $x_1 x_2 x_3 = 100$, 010, and 001. The absence of a clock signal implies that state transitions occur only when an input is asserted.

7.5.1 Analysis Procedure

Pulse-mode machines respond immediately to the assertion of an input signal without waiting for a clock signal. Analysis of pulse-mode sequential machines — as for synchronous sequential machines — consists of deriving a next-state table, input maps and equations, output maps and equations, and a state diagram for a given logic diagram. A timing diagram may also prove useful in analysis.

***SR* latches as storage elements**   The analysis proceeds in a manner analogous to that described for synchronous sequential machines. The predominant differences are the absence of a clock signal and the input restrictions mentioned previously. A Moore pulse-mode asynchronous sequential machine is shown in Figure 7.141, and will be analyzed with respect to a next-state table together with input and output maps.

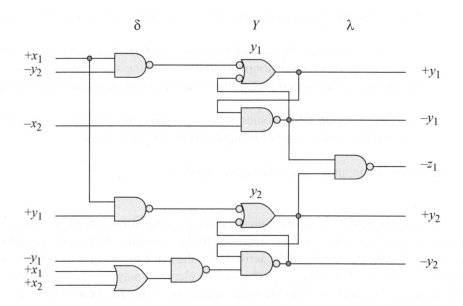

Figure 7.141   Pulse-mode Moore machine using latches for storage elements.

Assume that the machine is reset initially such that, $y_1 y_2 = 00$. Inputs x_1 and x_2 are assigned values of $x_1 x_2 = 10$ and 01 for each specified state. The equations to set and reset y_1 and y_2 are derived from the logic diagram and are shown in Equation 7.47.

If x_1 is pulsed when the machine is in an initial reset condition, then latch y_1 will set. Input pulse x_1 must be deasserted, however, before latch y_1 stabilizes in a set condition, otherwise the conditions to set latch y_2 would be established. If y_2 sets before x_1 is deasserted, then an erroneous next state will be generated because the input pulse must be deasserted before the state change is received at the input logic. Assuming that the input restrictions are met, latch y_2 will remain in a reset state. Thus, from an initial condition of $y_1 y_2 = 00$, a pulse on input x_1 causes the machine to proceed to state $y_1 y_2 = 10$, as shown in the first row of Table 7.21. Output z_1 remains inactive.

If input x_2 is pulsed in state $y_1 y_2 = 00$, then both latches will receive a reset pulse coincident with the assertion of x_2; the pulse will be active for the duration of x_2. Therefore, latches y_1 and y_2 will remain reset, as shown in the second row of Table 7.21. Output z_1 remains inactive.

$$Sy_1 = y_2' x_1$$

$$Ry_1 = x_2$$

$$Sy_2 = y_1 x_1$$

$$Ry_2 = y_1' (x_1 + x_2) \tag{7.47}$$

Table 7.21 Next-State Table for the Pulse-Mode Moore Machine of Figure 7.141

State Name	Present State $y_1\ y_2$	Inputs $x_1\ x_2$	Next State $y_1\ y_2$	Output z_1
a	0 0	1 0	1 0	0
	0 0	0 1	0 0	0
b	1 0	1 0	1 1	0
	1 0	0 1	0 0	0
c	1 1	1 0	1 1	0
	1 1	0 1	0 1	0
d	0 1	1 0	0 0	1
	0 1	0 1	0 0	1

Assume that the machine is now in state $y_1 y_2 = 10$ and that the input vectors $x_1 x_2 = 10$ and 01 are applied in sequence. Using either Equation 7.47 or the logic diagram, it is evident that a pulse on x_1 will cause y_1 to remain set and y_2 to be set. If x_2 is pulsed, then y_1 will be reset and y_2 will remain reset.

Let the present state be $y_1 y_2 = 11$ and the input vectors $x_1 x_2 = 10$ and 01 be applied in sequence. When x_1 is pulsed, no change occurs to latch y_1 because

$Sy_1 = y_2'x_1 = 01 = 0$. Thus, y_1 remains set. Latch y_2 will also remain set because $Sy_2 = y_1x_1 = 11 = 1$. When x_2 is asserted, y_1 will be reset and y_2 will remain unchanged at $y_2 = 1$ because $Ry_2 = y_1'(x_1 + x_2) = 0(0 + 1) = 0$.

Finally, when the machine is in state $y_1y_2 = 01$, output z_1 is asserted unconditionally, due to the Moore characteristics of the output. If input x_1 is pulsed, then no change occurs to latch y_1 because $Sy_1 = y_2'x_1 = 01 = 0$. There is also no set pulse to latch y_2 because $Sy_2 = y_1x_1 = 01 = 0$. However, a reset pulse is generated for y_2 when x_1 is asserted, since $Ry_2 = y_1'(x_1 + x_2) = 1(1 + 0) = 1$. A pulse on x_2 provides a reset pulse to y_1; thus, y_1 remains reset. Latch y_2 is also reset by an x_2 pulse because $Ry_2 = y_1'(x_1 + x_2) = 1(0 + 1) = 1$. Therefore, the assertion of x_1 or x_2 in state $y_1y_2 = 01$ returns the machine to the initial state of $y_1y_2 = 00$. Table 7.21 lists all possible states of two storage elements with the associated input vectors and the corresponding next states. Column z_1 lists the output values for the present state.

The input maps contain the same information as the next-state table, but in a different format. The maps for pulse-mode machines are slightly different than those for synchronous sequential machines because the input pulses are exclusive. Since the inputs cannot be asserted simultaneously, each latch requires n input maps, one map for each input $x_1, x_2, \cdots, x_n$. In this example, each latch requires two input maps, one each for x_1 and x_2, as shown in Figure 7.142. The maps for each latch are in the same row. Each column of maps corresponds to a separate input. The map entries are defined as follows:

> S indicates that the latch will be set.
> s indicates that the latch will remain set.
> R indicates that the latch will be reset.
> r indicates that the latch will remain reset.

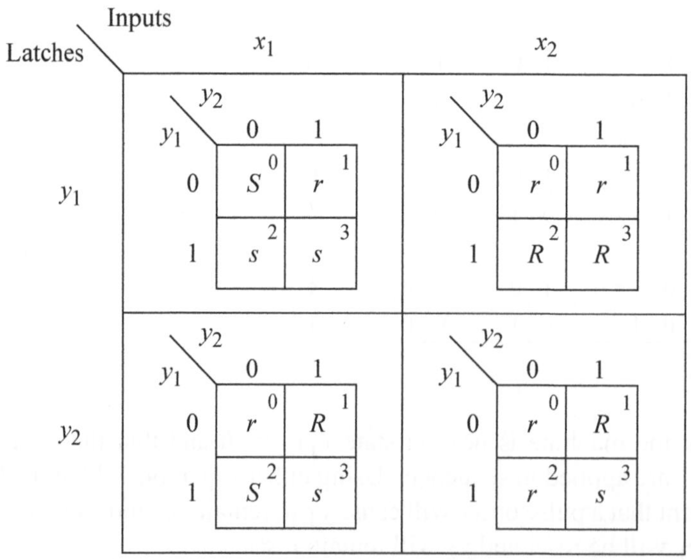

Figure 7.142   Input maps for the pulse-mode sequential machine of Figure 7.141.

In the presentation which follows, the set and reset equations of Equation 7.47 in conjunction with the next-state table of Table 7.21 will be utilized in deriving the entries for the input maps. The four maps will be constructed in parallel by considering each stable state in turn for each input vector.

Consider the input map in row y_1, column x_1 for an initial condition of $y_1 y_2 = 00$ when x_1 is pulsed. The next state for latch y_1 will be 1 for an input vector of $x_1 x_2 = 10$. That is, y_1 will change from $y_1 = 0$ to 1, which represents a set condition. Thus, the letter S is inserted in minterm location 0 of the map in row y_1, column x_1. The map in row y_2, column x_1 specifies the next states for latch y_2 when x_1 is pulsed. In state $y_1 y_2 = 00$, the set equation for latch y_2 is $Sy_2 = y_1 x_1 = 01 = 0$. Thus, there will be no change to the state of y_2, which remains reset, as indicated by the letter r.

Consider the map for y_1 in row y_1, column x_2 for an initial condition of $y_1 y_2 = 00$. The assertion of x_2 provides a reset pulse to latch y_1. Since y_1 remains in a reset state, the letter r is inserted in minterm location 0. Similarly, the letter r is entered in minterm location 0 of the map for y_2 in row y_2, column x_2, because $Ry_2 = y_1'(x_1 + x_2) = 1(0 + 1) = 1$.

Consider row y_1, column x_1, for a present state of $y_1 y_2 = 01$. When x_1 is asserted, there will be no change to latch y_1, because $Sy_1 = y_2' x_1 = 01 = 0$; that is, x_1 will not produce a set pulse to y_1. Thus, y_1 remains in a reset state, as indicated by the letter r. There is also no set pulse for y_2 when x_1 is asserted, because $Sy_2 = y_1 x_1 = 01 = 0$. Latch y_2, however, receives a reset pulse when x_1 is asserted because $Ry_2 = y_1'(x_1 + x_2) = 1(1 + 0) = 1$. Since y_2 changes from $y_2 = 1$ to 0; therefore, the letter R is entered in minterm location 1.

For a present state of $y_1 y_2 = 01$ in column x_2, the letters r and R are inserted in minterm location 1 for y_1 and y_2, respectively. Input x_2 provides a reset pulse to latch y_1, which remains reset. Input x_2 also provides a reset pulse for y_2 as specified by $Ry_2 = y_1'(x_1 + x_2) = 1(0 + 1) = 1$. Since the state of y_2 changes from 1 to 0, the letter R is entered in minterm location 1 of row y_2, column x_2.

In a similar manner, the remaining entries are derived. In column x_1, a pulse on input x_1 will provide either a set or reset pulse or leave the latches unchanged. In column x_2, a pulse on input x_2 will either reset the latches or leave them unchanged. Comparison of the entries in Table 7.21 with the entries in the input maps show a one-to-one correspondence for each state for identical input vectors.

When deriving the equations from the input maps, only the uppercase letters need be considered. The lowercase letters are treated as "don't care" entries and are used only if they aid in minimizing the equation. The equations derived from the input maps are identical to those shown in Equation 7.47. For example, consider the map in row y_1, column x_1. Minterms 0 and 2 can be combined, since both represent a set condition. Minterms 0 and 2 have common variables y_2' and x_1. Therefore, $Sy_1 = y_2' x_1$. The map in row y_1, column x_2, contains a reset entry in every location, indicating that x_2 will either reset latch y_1 or leave the latch in a reset state. Therefore, $Ry_1 = x_2$.

The map in row y_2, column x_1 contains the letters S and s in minterm locations 2 and 3, respectively, where s is considered a "don't care" entry. Since both squares possess the common variables y_1 and x_1, the set condition for latch y_2 is $Sy_2 = y_1 x_1$. The map also contains entries of r and R in minterm locations 0 and 1, respectively, where r is considered a "don't care" entry. The same values exist in the same locations

for the map in row y_2, column x_2. Therefore, the reset equation for latch y_2 is
$Ry_2 = y_1'x_1 + y_1'x_2 = y_1'(x_1 + x_2)$.

Sequential machines that operate in pulse mode will not have race conditions — either noncritical or critical — because only one input is active at a time and the machine is stable even in the absence of input pulses. Each product term in the set and reset equations contains at least one input variable. Also, no input variable appears in a complemented form because the complement of a variable corresponds to an inactive input signal.

SR latches with D flip-flops as storage elements　A pulse-mode machine is shown in the logic diagram of Figure 7.143. The machine has two inputs x_1 and x_2 and one output z_1. The storage elements consist of *SR* latches and *D* flip-flops. The output of each latch connects to the *D* input of the associated flip-flop forming a master-slave relationship. Since the *D* flip-flops are clocked on the trailing edge of the positive input pulses, state changes are not fed back to the δ next-state logic until the active input has been deasserted. Clocking the flip-flops on the negative edge of the positive input pulses delays the next state from affecting the input logic while an input pulse is still active. Thus, the machine operates in a deterministic manner.

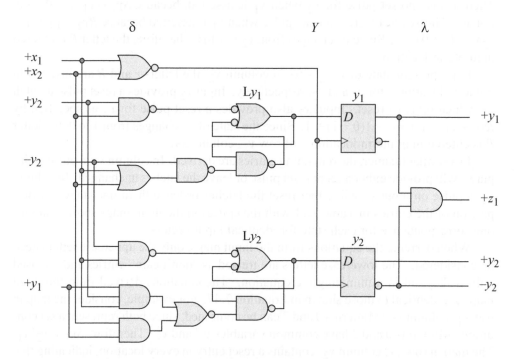

Figure 7.143　Logic diagram for a pulse-mode machine using latches and *D* flip-flops in a master-slave relationship.

The input equations are obtained directly from the logic diagram and are shown in Equation 7.48. Latch Ly_1 will be set if flip-flop y_2 is set and x_2 is pulsed. Latch Ly_1

will be reset if y_2 is reset and either x_1 or x_2 is pulsed. Thus, the set and reset conditions for latch Ly_1 are $SLy_1 = y_2x_2$ and $RLy_1 = y_2'(x_1 + x_2)$, respectively.

$$SLy_1 = y_2x_2$$

$$RLy_1 = y_2'x_1 + y_2'x_2 = y_2'(x_1 + x_2)$$

$$SLy_2 = y_2'x_1$$

$$RLy_2 = y_1y_2x_1 + y_1x_2 \tag{7.48}$$

Similarly, the set and reset conditions for latch Ly_2 are obtained. Latch Ly_2 will be set if flip-flop y_2 is reset and x_1 is pulsed. Latch Ly_2 will be reset if flip-flops y_1 and y_2 are both set and x_1 is pulsed, or if flip-flop y_1 is set and x_2 is pulsed. These conditions yield the set and reset equations $SLy_2 = y_2'x_1$ and $RLy_2 = y_1y_2x_1 + y_1x_2$, respectively.

The input equations for Dy_1 and Dy_2 are not required, since the state of each latch is transferred to its associated flip-flop on the negative transition of the active input variable.

Using the equations of Equation 7.48, the next-state table can now be constructed. If the machine is in the initial reset state of $y_1y_2 = 00$ and input x_1 is pulsed, then latch Ly_1 remains reset and latch Ly_2 is set, as shown in the first row of the next-state table of Table 7.22. If x_2 is pulsed in the initial reset condition, then both latches remain reset. Output z_1 remains inactive.

Table 7.22 Next-State Table for the Pulse-Mode Moore Machine of Figure 7.143

State Name	Present State $y_1\ y_2$	Inputs $x_1\ x_2$	Next State $y_1\ y_2$	Output z_1
a	0 0	1 0	0 1	0
	0 0	0 1	0 0	0
b	0 1	1 0	0 1	0
	0 1	0 1	1 1	0
c	1 1	1 0	1 0	0
	1 1	0 1	1 0	0
d	1 0	1 0	0 1	1
	1 0	0 1	0 0	1

In state $y_1 y_2 = 01$, if x_1 is asserted, then latches Ly_1 and Ly_2 will remain reset and set, respectively, and the machine will not change state. If x_2 is asserted, however, then latch Ly_1 will set and latch Ly_2 will remain set, because $SLy_1 = y_2 x_2 = 11 = 1$ and $Ry_2 = y_1 y_2 x_1 + y_1 x_2 = 010 + 01 = 0$, resulting in a transition from state $y_1 y_2 = 01$ to state $y_1 y_2 = 11$. Output z_1 remains inactive.

In state $y_1 y_2 = 11$, if x_1 is pulsed, then latch Ly_1 remains set while latch Ly_2 resets. The same set and reset conditions apply if x_2 is pulsed. Thus, a state transition will occur from $y_1 y_2 = 11$ to 10 if either x_1 or x_2 is pulsed.

Finally, in state $y_1 y_2 = 10$, output z_1 is asserted unconditionally, due to the Moore characteristics of the output variable. If x_1 is pulsed in state $y_1 y_2 = 10$, then latch Ly_1 is reset, whereas latch Ly_2 is set and the machine proceeds to state $y_1 y_2 = 01$. If x_2 is pulsed, then latch Ly_1 is reset and latch Ly_2 remains reset, causing a transition to the initial state of $y_1 y_2 = 00$.

The input maps can be derived directly from the next-state table or from the input equations. The input maps are shown in Figure 7.144, where the letters S and s indicate that the latch will be set or remain set, respectively, and the letters R and r indicate that the latch will be reset or remain reset, respectively.

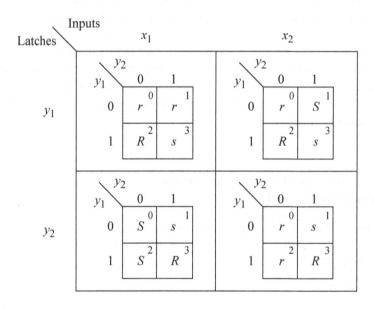

Figure 7.144   Input maps for the pulse-mode sequential machine of Figure 7.143.

The map entries correlate directly to the entries in the next-state table. For example, in state $y_1 y_2 = 10$, latch Ly_1 will be reset if x_1 is pulsed, as indicated by the letter R in minterm location 2 of the map in row y_1, column x_1. Also, in state $y_1 y_2 = 10$, latch Ly_2 will be set if x_1 is pulsed, as indicated by the letter S in minterm location 2 of the map in row y_2, column x_1.

Now consider the effect when input x_2 is asserted in state $y_1 y_2 = 10$. When x_2 is activated, a reset pulse is applied to latch Ly_1, as shown in the logic diagram and the next-state table. Since latch Ly_1 was set, the letter R is entered in minterm location 2 of the map for latch Ly_1 in row y_1, column x_2. The assertion of x_2 will also generate a reset pulse to latch Ly_2. However, since latch Ly_2 is already reset, an x_2 pulse causes latch Ly_2 to remain in a reset state, as specified by the letter r in minterm location 2 of the map for latch Ly_2 in row y_2, column x_2.

7.6   Synthesis of Pulse-Mode Asynchronous Sequential Machines

Due to the stringent requirements of input pulse characteristics, pulse-mode machines are less preferred than synchronous sequential machines, especially when the machine is implemented with subnanosecond logic. The synthesis concepts, however, are of sufficient importance to warrant a detailed presentation of the synthesis procedure.

Reliability of pulse-mode machines can be increased by inserting delay circuits of an appropriate duration in the output networks of the storage elements. The aggregate delay of the storage elements and the delay circuit must be of sufficient duration so that the input pulse will be deasserted before the storage element output signals arrive at the δ next-state logic.

7.6.1   Synthesis Procedure

Three techniques are commonly used to insert delays in the storage element outputs: An even number of inverters are connected in series with each latch output; a linear delay circuit is connected in series with each latch output; or an edge-triggered D flip-flop is connected in series with each latch output. The flip-flops are set to the state of the latches, but are triggered on the trailing edge of the input pulses. Thus, the flip-flop outputs — and therefore the state of the machine as represented by the SR latch outputs — are received at the δ next-state logic only when the active input pulse has been deasserted. The SR latches and the D flip-flops constitute a master-slave relationship and will be the primary means to implement pulse-mode asynchronous sequential machines in this section.

SR latches with D flip-flops as storage elements   The pulse width restrictions that are dominant in pulse-mode sequential machines can be eliminated by including D flip-flops in the feedback path from the SR latches to the δ next-state logic. Providing edge-triggered D flip-flops as a constituent part of the implementation negates the requirement of precisely-controlled input pulse durations. This is by far the most reliable means of synthesizing pulse-mode machines. The SR latches, in conjunction with the D flip-flops, form a master-slave configuration as shown in Figure 7.145.

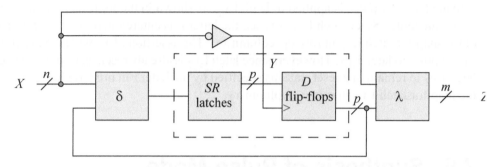

Figure 7.145   General block diagram of a pulse-mode asynchronous sequential machine using SR latches and D flip-flops in a master-slave configuration.

The output of each latch connects to the D input of its associated flip-flop which in turn connects to the δ next-state logic. The flip-flops are clocked on the complemented trailing edge of the active input variable. For example, if the input pulses are active high, then the D flip-flops are triggered on the inverted negative transition of the active input pulse; that is, the flip-flops are triggered on a positive transition, as shown in Figure 7.145. JK flip-flops may also be used as the slave storage elements. The output alphabet Z of pulse-mode machines can be generated as either levels for Moore machines or as pulses for Mealy machines.

Example 7.17   A Mealy pulse-mode asynchronous sequential machine will be synthesized which has two pulse input variables x_1 and x_2 and one output z_1. Output z_1 is asserted coincident with every second x_2 pulse, if and only if the pair of x_2 pulses is immediately preceded by an x_1 pulse. The machine will be implemented using NOR logic and inverters for the logic primitives. The storage elements will consist of SR latches and positive-edge-triggered D flip-flops in a master-slave configuration. A representative timing diagram is shown in Figure 7.146.

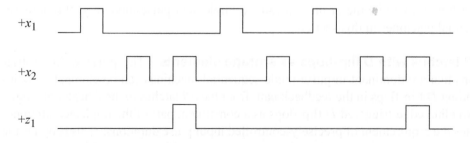

Figure 7.146   Representative timing diagram for the Mealy pulse-mode asynchronous sequential machine of Example 7.17.

The state diagram for the Mealy machine is shown in Figure 7.147; notice that only one input is asserted for any state transition. The input maps are shown in Figure 7.148. There is one unused state $y_1 y_2 = 11$. Beginning in state $\textcircled{a}$ ($y_1 y_2 = 00$) of the state diagram, if x_1 is asserted, then flip-flop y_1 remains reset; therefore, an entry of r is inserted in minterm location 0 of the input map for flip-flop y_1 in column x_1.

In state $\textcircled{c}$ ($y_1 y_2 = 10$), if x_1 is asserted, then flip-flop y_1 is reset; therefore, an entry of R is inserted in minterm location 2 of the map for flip-flop y_1 in column x_1. In state $\textcircled{b}$ ($y_1 y_2 = 01$), if x_2 is asserted, then flip-flop y_2 is reset; therefore, an entry of R is inserted in minterm location 1 of the map for flip-flop y_2 in column x_2. The remaining entries are inserted in a similar manner.

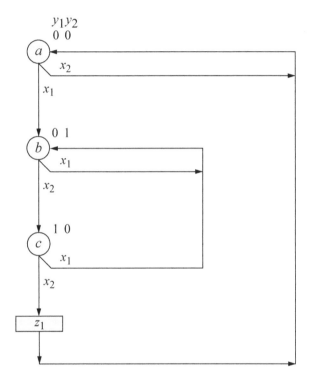

Figure 7.147   State diagram for the Mealy pulse-mode asynchronous sequential machine of Example 7.17.

The input equations are shown in Equation 7.49. The logic diagram is shown in Figure 7.149 using NOR logic for the logic primitives and the latches. The output from the latches connects directly to the D inputs of the clocked storage elements which are positive-edge-triggered D flip-flops. Output z_1 is asserted in state $\textcircled{c}$ if x_2 = 1. The output map for z_1 is shown in Figure 7.150 and the output equation is shown in Equation 7.50.

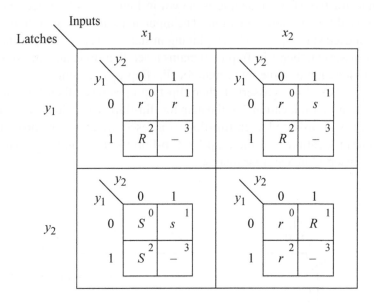

Figure 7.148   Input maps for the Mealy pulse-mode asynchronous sequential machine of Example 7.17.

$$SLy_1 = y_2x_2 \qquad RLy_1 = x_1 + y_1x_2 \qquad\qquad SLy_2 = x_1 \qquad RLy_2 = x_2 \qquad (7.49)$$

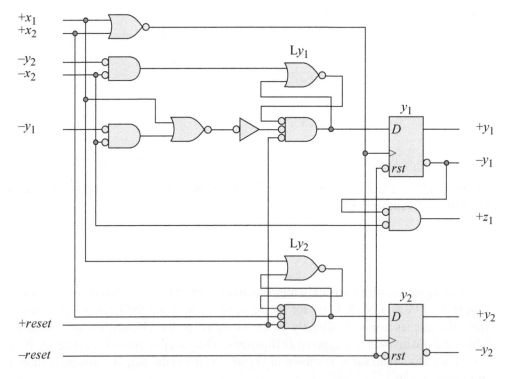

Figure 7.149   Logic diagram for the Mealy machine of Example 7.17.

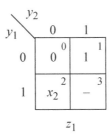

Figure 7.150   Output map for z_1 for the Mealy machine of Example 7.17.

$$z_1 = y_1 x_2 \tag{7.50}$$

Example 7.18   A Moore pulse-mode asynchronous sequential machine will be designed that has two inputs x_1 and x_2 and one output z_1. The negative transition of every second consecutive x_1 pulse will assert output z_1 as a level. The output will remain set for all following contiguous x_1 pulses. The output will be deasserted at the negative transition at the second of two consecutive x_2 pulses. The machine will be implemented with NAND logic for the δ next-state logic. The storage elements will be NAND SR latches and positive-edge-triggered D flip-flops in a master-slave configuration.

A representative timing diagram is shown in Figure 7.151 and the state diagram is shown in Figure 7.152 depicting all possible state transition sequences that conform to the functional specifications. The input maps are shown in Figure 7.153 using S (set), s (remains set), R (reset), and r (remains reset) entries. The corresponding set and reset equations for the SR latches are listed in Equation 7.51.

The output map for z_1 is shown in Figure 7.154 and the output equation appears in Equation 7.52. The logic diagram is shown in Figure 7.155 using only NAND logic — no inverters — for the δ next-state logic and SR latches together with positive-edge-triggered D flip-flops.

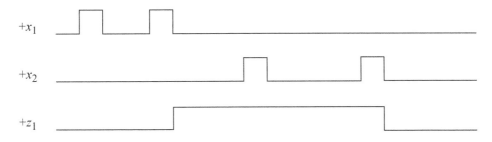

Figure 7.151   Representative timing diagram for the Moore pulse-mode asynchronous sequential machine of Example 7.18.

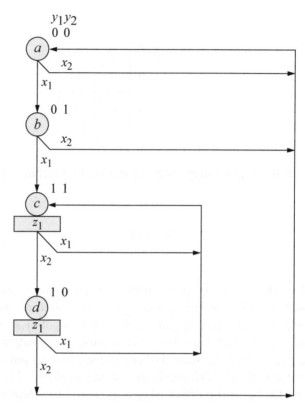

Figure 7.152   State diagram for the Moore pulse-mode asynchronous sequential machine of Example 7.18.

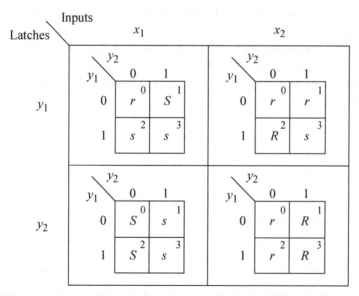

Figure 7.153   Input maps for the Moore pulse-mode asynchronous sequential machine of Example 7.18.

$$SLy_1 = y_2 x_1 \qquad RLy_1 = y_2' x_2$$

$$SLy_2 = x_1 \qquad RLy_2 = x_2 \qquad (7.51)$$

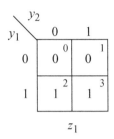

Figure 7.154   Output map for z_1 for the Moore pulse-mode asynchronous sequential machine of Example 7.18.

$$z_1 = y_1 \qquad (7.52)$$

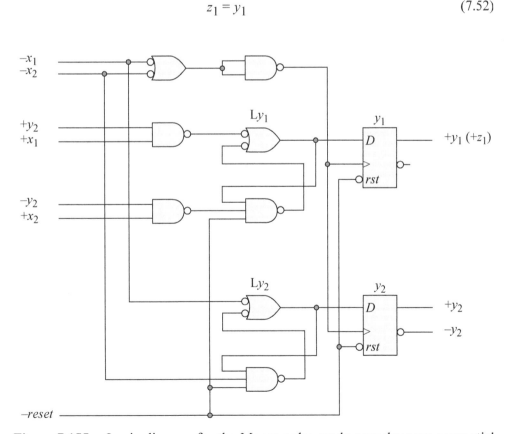

Figure 7.155   Logic diagram for the Moore pulse-mode asynchronous sequential machine of Example 7.18.

The logic diagram is implemented directly from Equation 7.51 and Equation 7.52. The state flip-flops y_1 and y_2 are clocked on the trailing transition of input x_1 or x_2 as shown in the timing diagram and the logic diagram. Because this is a Moore machine, the outputs are a function of the inputs only; therefore, when flip-flop y_1 is set, output z_1 is asserted, as shown in the state diagram.

The concept of pulse-mode asynchronous sequential machines is frequently encountered in a variety of common applications such as vending machines and traffic control for demand-access intersections. Because of the stringent timing requirements on input pulse duration associated with pulse-mode machines, only the *SR* latches with edge-triggered *D* flip-flops offer a high degree of reliability.

7.7 Problems

7.1   Obtain the state diagram for the Mealy synchronous sequential machine shown below. The machine is reset to $y_1 y_2 = 11$.

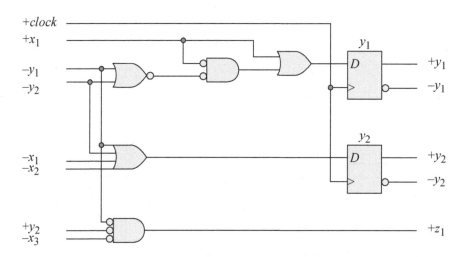

7.2   Obtain the state diagram for the Moore synchronous sequential machine shown below. The machine is reset to $y_1 y_2 = 00$.

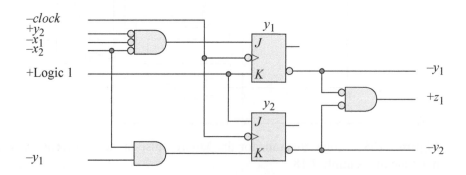

7.3 Obtain the state diagram for a Moore synchronous sequential machine that re-ceives 4-bit words over a serial data line x_1 then generates a fifth bit z_1, which maintains odd parity over all 5 bits. There is 1 bit space words.

7.4 The λ output logic for a synchronous sequential machine is shown below for three separate cases. Specify the assertion/deassertion statement for each of the three cases. The state flip-flops are clocked on the positive transition of the clock.

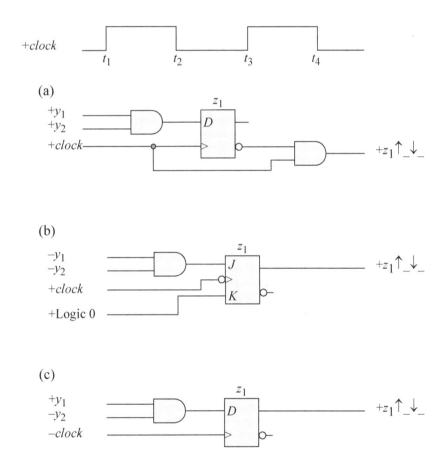

7.5 Select state codes for states b and e to eliminate all output glitches for the state diagram shown below. The Moore outputs z_1, z_2, and z_3 are asserted for the entire state time; that is, $\uparrow t_1 \downarrow t_3$.

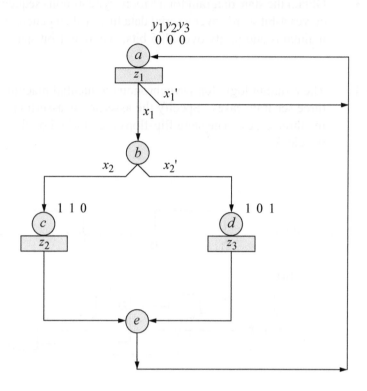

7.6   Find all equivalent states for the next-state table shown below.

Present State	Input x_1	Next State	Output z_1
a	0	b	0
	1	g	0
b	0	g	0
	1	a	1
c	0	h	0
	1	g	1
d	0	c	0
	1	a	1
e	0	h	0
	1	c	0

(Continued on next page)

Present State	Input x_1	Next State	Output z_1
f	0	c	0
	1	e	1
g	0	d	0
	1	g	1
h	0	c	0
	1	a	1

7.7   Determine the counting sequence for the counter shown below. The counter is reset initially to $y_1 y_2 = 00$, where y_2 is the low-order flip-flop.

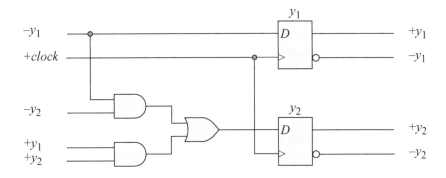

7.8   Determine the counting sequence for the counter shown below. The counter is reset initially to $y_1 y_2 = 00$, where y_2 is the low-order flip-flop.

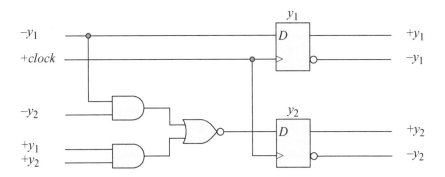

7.9   Determine the counting sequence for the counter shown below. The counter is reset initially to $y_1 y_2 y_3 = 000$, where y_3 is the low-order flip-flop.

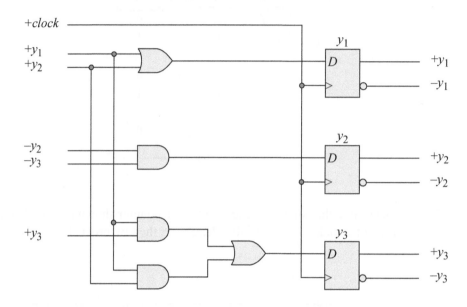

7.10   Design a synchronous modulo-11 counter with no self-starting state using *JK* flip-flops.

7.11   Design a counter using *D* flip-flops that counts in the following sequence:

000, 111, 001, 110, 010, 101, 011, 100, 000, . . .

7.12   Obtain the input equations for flip-flops y_1 and y_4 only, for a BCD counter that counts in the sequence shown below. The equations are to be in minimum form. Use *JK* flip-flops. There is no self-starting state.

y_1	y_2	y_3	y_4
0	0	0	0
0	0	0	1
0	0	1	0
0	0	1	1
0	1	0	0
0	1	0	1
0	1	1	0
0	1	1	1
1	0	0	0
1	0	0	1
0	0	0	0

7.13   Given the state diagram shown below for a Moore synchronous sequential machine with three inputs x_1, x_2, and x_3, derive the input maps for flip-flops y_1, y_2, and y_3. Then design the machine using linear-select multiplexers and logic primitives for the δ next-state logic. Use D flip-flops as the storage elements. Outputs z_1 and z_2 have the following assertion/deassertion times: $\uparrow t_2 \downarrow t_3$.

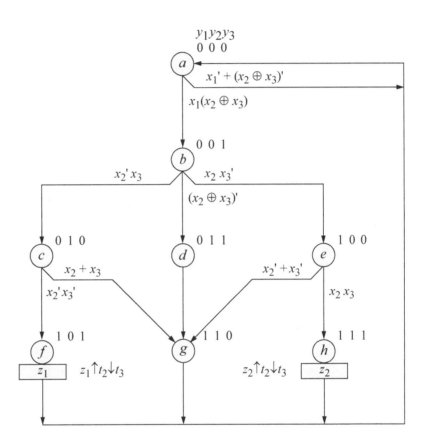

7.14   Given the state diagram shown below for a Moore synchronous sequential machine, implement the design using nonlinear-select multiplexers for the δ next-state logic together with logic primitives. Let flip-flops $y_1 y_3 = s_1 s_0$. Use positive-edge-triggered D flip-flops for the storage elements and a decoder for the λ output logic. Since the state codes may cause glitches on the outputs for some state transition sequences, z_1, z_2, and z_3 should be asserted at time t_2 and deasserted at time t_3.

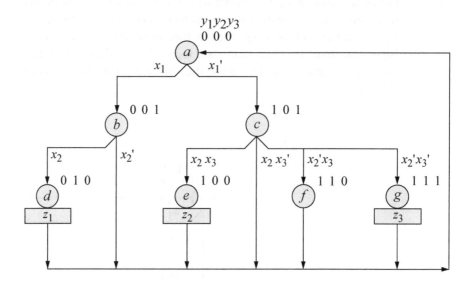

7.15   Design a Moore machine that has four parallel inputs x_1, x_2, x_3, and x_4 and two outputs z_1 and z_2. Output $z_1 \uparrow t_1 \downarrow t_3$; output $z_2 \uparrow t_2 \downarrow t_3$. The inputs constitute a 4-bit word. There is 1 bit space between words. Use positive-edge-triggered D flip-flops as the storage elements. The machine operates as follows:

(a) Output $z_1 = 1$ if the 4-bit word contains an odd number of 1s.
(b) Output $z_2 = 1$ if the 4-bit word contains an even number of 1s, but not zero 1s.

7.16   Given the Karnaugh shown below for function z_1, analyze all transitions resulting from a single change to the input vector and identify all static-1 and static-0 hazards. The inputs are active high. Obtain the sum-of-products equation containing the hazard cover and the product-of-sums equation containing the hazard cover.

x_1x_2 \ x_3x_4	0 0	0 1	1 1	1 0
0 0	1	1	0	0
0 1	1	1	0	1
1 1	0	0	0	1
1 0	0	0	0	0

z_1

7.17 Given the equation for z_1 shown below, identify all static-1 hazards and determine the hazard covers.

$$z_1 = x_1'x_3'x_5' + x_2x_3'x_4x_5' + x_2x_3x_4x_5 + x_1x_3'x_4x_5$$

7.18 Given the excitation map shown below for excitation variables Y_{1e} and Y_{2e}, obtain all possible state transition sequences that result in one or more oscillations. Indicate the starting stable state and the path that is taken when a single input changes value; for example, $\textcircled{q} \rightarrow r \leftrightarrow s$.

$Y_{1e} Y_{2e}$

7.19 Given the excitation map shown below for an asynchronous sequential machine, list all possible transitions (paths) that could occur for noncritical races, critical races, and oscillations. Indicate the starting stable state and the path that is taken when a single input changes.

$Y_{1e} Y_{2e}$

7.20   Analyze the Moore asynchronous sequential machine shown below by generating an excitation map and a flow table. Let the machine be initially reset to $y_{1f}y_{2f}x_1x_2 = 0000$. Determine the shortest input sequence which will assert output z_1.

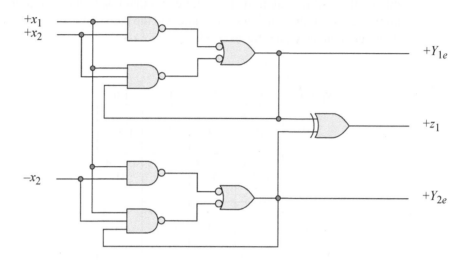

7.21   The equations for excitation variables Y_{1e}, Y_{2e}, and output z_1 are shown below for a Mealy asynchronous sequential machine.

$$Y_{1e} = x_1 x_2' + y_{1f} y_{2f}' x_2 + y_{2f} x_1' x_2$$
$$Y_{2e} = x_1' x_2 + y_{2f} x_2$$
$$z_1 = y_{1f} y_{2f} x_2 + y_{1f}' y_{2f}' x_2$$

(a) Obtain the excitation maps for the individual excitation variables.
(b) Obtain the combined excitation map and indicate the stable states.
(c) Obtain the output map for z_1.

7.22   Obtain the primitive flow table — not reduced — for an asynchronous sequential machine that has two inputs x_1 and x_2 and two outputs z_1 and z_2. If x_1 is asserted before x_2, then z_1 is asserted and remains active until z_2 is asserted. If x_2 is asserted before x_1, then z_2 is asserted and remains active until z_1 is asserted. Except for a reset condition, the outputs are mutually exclusive. The timing diagram shown below represents some typical input sequences.

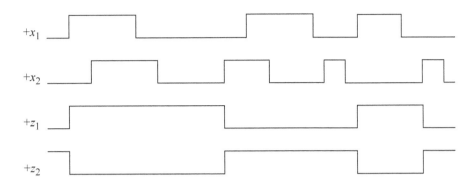

7.23   Obtain the primitive flow table for an asynchronous sequential machine that has two inputs x_1 and x_2 and one output z_1. Input x_1 is a periodic clock signal for a synchronous sequential machine under control of an asynchronous sequential machine. Input x_2 is used to control a single-step operation; that is, when x_2 is asserted, a single full width x_1 pulse — represented by output z_1 — is transferred to the synchronous sequential machine.

Input x_2 must be asserted before x_1 in order to generate a corresponding pulse on output z_1. If x_2 is deasserted before x_1 is deasserted, then output z_1 maintains its active state for the duration of x_1. A representative timing diagram is shown below.

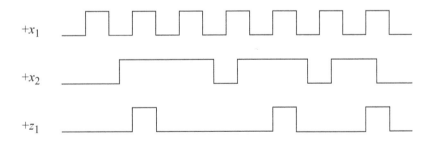

7.24   Identify equivalent states in the primitive flow table shown below. Then eliminate redundant states and obtain a reduced primitive flow table.

x_1x_2 00	01	11	10	z_1
(a)	b	–	c	0
a	(b)	d	–	1
f	–	e	(c)	1
–	h	(d)	g	0
–	h	(e)	g	0
(f)	b	–	c	0
a	–	d	(g)	0
f	(h)	e	–	0

7.25   Identify equivalent states in the primitive flow table shown below. Then eliminate redundant states and obtain a reduced primitive flow table.

x_1x_2 00	01	11	10	z_1	z_2
(a)	b	–	f	0	0
a	(b)	e	–	1	0
(c)	d	–	f	0	0
a	(d)	e	–	1	0
–	g	(e)	f	0	1
c	–	e	(f)	1	1
c	(g)	e	–	1	1

7.26   Derive the merger diagram from the reduced primitive flow table below.

x_1x_2	00	01	11	10	z_1	z_2
	ⓐ	b	—	d	0	0
	a	ⓑ	c	—	1	1
	—	j	ⓒ	d	1	0
	a	—	c	ⓓ	0	1
	ⓘ	j	—	d	0	1
	i	ⓙ	c	—	1	0

7.27   Given the merger diagram shown below, obtain a partition which contains sets of maximal compatible rows.

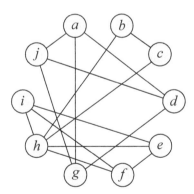

7.28   Derive the merger diagram from the reduced primitive flow table shown below.

x_1x_2 00	01	11	10	z_1	z_2
ⓐ	b	–	d	0	0
a	ⓑ	c	–	0	1
–	–	ⓒ	d	0	0
a	–	–	ⓓ	1	0
–	f	ⓔ	d	1	0
g	Ⓕ	c	–	1	0
ⓖ	h	–	d	0	1
g	ⓗ	e	–	1	1

7.29   Derive the transition diagram for the merged flow table shown below. Resolve all adjacency conflicts, then obtain the excitation maps and equations in a sum-of-products notation with no static-1 hazards.

x_1x_2 00	01	11	10
1	ⓐ	ⓑ	g	f
2	–	d	ⓒ	e
3	a	ⓓ	c	ⓔ
4	a	d	ⓖ	Ⓕ

7.30   Derive the transition diagram for the merged flow table shown below. Resolve all adjacency conflicts, then obtain the excitation maps and equations in a sum-of-products notation with no static-1 hazards.

x_1x_2	00	01	11	10
1	$\textcircled{a}$	$\textcircled{b}$	c	$\textcircled{g}$
2	–	e	$\textcircled{c}$	g
3	a	–	f	$\textcircled{d}$
4	a	$\textcircled{e}$	$\textcircled{f}$	d

7.31   Synthesize an asynchronous sequential machine which has two inputs x_1 and x_2 and one output z_1. Output z_1 is asserted for the duration of x_2 if and only if x_1 is already asserted. Assume that the initial state of the machine is $x_1x_2z_1 = 000$. A representative timing diagram is shown below. Obtain the excitation equations in both a sum-of-products and a product-of-sums form with no static-1 or static-0 hazards.

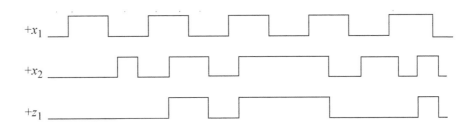

7.32   Synthesize an asynchronous sequential machine which has two inputs x_1 and x_2 and one output z_1. Output z_1 will be asserted coincident with the assertion of the first x_2 pulse and will remain active for the duration of the first x_2 pulse. The output will be asserted only if the assertion of x_1 precedes the assertion of x_2. Input x_1 will not become deasserted while x_2 is asserted. The λ output logic must have a minimal number of logic gates. A representative timing diagram is shown below.

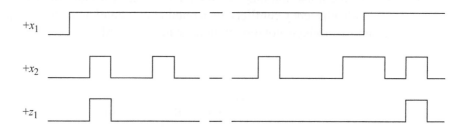

7.33   An asynchronous sequential machine has two inputs x_1 and x_2 and one output z_1. The machine operates according to the following specifications:

> If $x_1 x_2 = 00$, then the state of z_1 is unchanged.
> If $x_1 x_2 = 01$, then z_1 is deasserted.
> If $x_1 x_2 = 10$, then z_1 is asserted.
> If $x_1 x_2 = 11$, then z_1 changes state.

Derive the logic diagram using AND gates, OR gates, and inverters. The inputs are available in both high and low assertion. Assume that the initial conditions are $x_1 x_2 z_1 = 000$. The output must change as fast as possible. There must be no output glitches.

7.34   Analyze the Mealy pulse-mode sequential machine shown below. Obtain the next-state table, the input maps and equations, the output map and equation, and the state diagram.

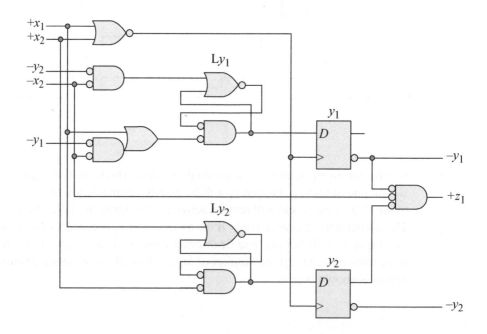

7.35   Analyze the Moore pulse-mode sequential machine shown below. Obtain the next-state table, the input maps and equations, the output map and equation, and the state diagram.

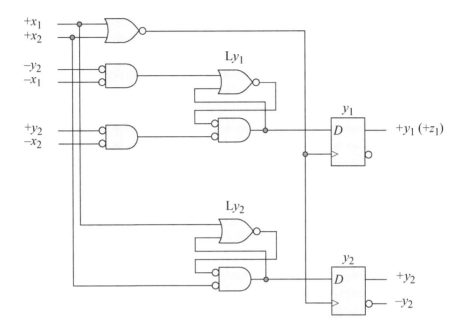

7.36   Synthesize a Moore pulse-mode sequential machine which has three inputs x_1, x_2, and x_3 and one output z_1. Output z_1 will be asserted coincident with the assertion of the x_3 pulse if and only if the x_3 pulse was preceded by an x_1 pulse followed by an x_2 pulse. That is, the input vector must be $x_1 x_2 x_3 = 100$, 000, 010, 000, 001 to assert z_1. Output z_1 will be deasserted at the next active x_1 pulse or x_2 pulse. Use NOR SR latches and positive-edge-triggered D flip-flops as the storage elements.

7.37   Given the state diagram shown below for a Moore pulse-mode asynchronous sequential machine, implement the machine using NAND gates for the SR latches and D flip-flops as the storage elements in a master-slave configuration. Use any type of gates for the logic primitives.

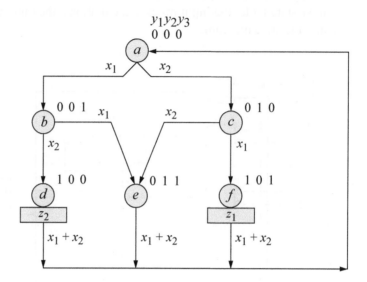

8

Sequential Logic Design Using Verilog HDL

This chapter includes numerous examples, including some of the design examples from Chapter 7 for a comparative study of the design methodologies used in Chapter 7 with the design methodologies used in the modeling constructs of Verilog HDL. Behavioral and structural modeling constructs will be used to design synchronous sequential machines, asynchronous sequential machines, and pulse-mode asynchronous sequential machines. Dataflow modeling will be used to design asynchronous sequential machines and pulse-mode asynchronous sequential machines in conjunction with structural modeling. Some of the examples will use all three modeling constructs.

Dataflow modeling uses the *continuous assignment* statement to design combinational logic without using gates and interconnecting nets. Continuous assignment statements provide a Boolean correspondence between the right-hand side expression and the left-hand side target. The continuous assignment statement uses the keyword **assign** and has the following syntax with optional drive strength and delay:

assign [drive_strength] [delay] left-hand side target = right-hand side expression

The continuous assignment statement assigns a value to a net (**wire**) that has been previously declared — it cannot be used to assign a value to a register. Therefore, the left-hand target must be a scalar or vector net or a concatenation of scalar and vector nets. The operands on the right-hand side can be registers, nets, or function calls. The registers and nets can be declared as either scalars or vectors.

Behavioral modeling describes the *behavior* of a digital system and is not concerned with the direct implementation of logic gates but more with the architecture of the system. This is an algorithmic approach to hardware implementation and represents a higher level of abstraction than other modeling constructs — the logic details and organization are left to the synthesis tool.

Verilog contains two structured procedure statements or behaviors: **initial** and **always**. A behavior may consist of a single statement or a block of statements delimited by the keywords **begin . . . end**. A module may contain multiple **initial** and **always** statements. These statements are the basic statements used in behavioral modeling and execute concurrently starting at time zero in which the order of execution is not important. All other behavioral statements are contained inside these structured procedure statements.

Structural modeling consists of instantiation of one or more of the following design objects:

- Built-in primitives
- User-defined primitives (UDPs)
- Design modules

Instantiation means to use one or more lower-level modules — including logic primitives — that are interconnected in the construction of a higher-level structural module. A module can be a logic gate, an adder, a multiplexer, a counter, or some other logical function. The objects that are instantiated are called *instances*. Structural modeling is described by the interconnection of these lower-level logic primitives or modules. The interconnections are made by wires that connect primitive terminals or module ports.

8.1 Synchronous Sequential Machines

Several examples of Moore and Mealy finite-state synchronous sequential machines will be designed using Verilog HDL in this section. The designs will utilize behavioral and structural modeling constructs and consist of sequence detectors and counters of various moduli.

Example 8.1   A Moore machine has one input x_1 and one output z_1. Output z_1 is asserted whenever the input contains a sequence $x_1 = 1101$; overlapping sequences are allowed. An example of correct sequences is shown below. The assertion/deassertion statement for z_1 is $\uparrow t_2 \downarrow t_3$.

$$x_1 = 0\ \underline{1\ 1\ 0\ 1}\ 0\ 1\ 0\ \underline{1\ 1\ 0\ 1}\ 1\ 0\ 1\ 0$$

Assert z_1

The design will use behavioral modeling with the **case** statement together with the **assign** statement for output z_1; therefore, a logic diagram is not necessary. The state diagram is shown in Figure 8.1 indicating the sequences to assert output z_1 and overlapping sequences. The module is shown in Figure 8.2 using behavioral modeling. The test bench is shown in Figure 8.3 and the outputs in Figure 8.4. The waveforms of Figure 8.5 show the input sequences to assert output z_1 and follow the state diagram exactly.

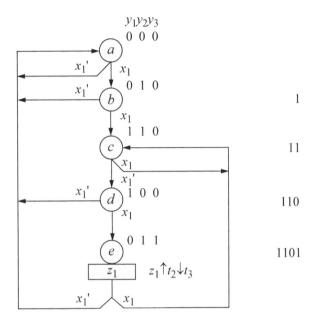

Figure 8.1   State diagram for the Moore machine of Example 8.1.

```
//behavioral moore synchronous sequential machine
module moore_ssm20 (rst_n, clk, x1, y, z1);

input rst_n, clk, x1;
output [1:3] y;
output z1;

reg [1:3] y, next_state;
wire z1;

//assign state codes
parameter    state_a = 3'b000,
             state_b = 3'b010,
             state_c = 3'b110,
             state_d = 3'b100,
             state_e = 3'b011;        //continued on next page
```

Figure 8.2   Behavioral module for the Moore machine of Example 8.1.

```verilog
always @ (posedge clk)         //set next state
begin
    if (~rst_n)
       y <= state_a;
    else
       y <= next_state;
end

assign   z1 = y[3] & ~clk;     //define output z1

always @ (y or x1)             //determine next state
begin
    case (y)
        state_a:
            if (x1==0)
               next_state = state_a;
            else
               next_state = state_b;

        state_b:
            if (x1==0)
               next_state = state_a;
            else
               next_state = state_c;

        state_c:
            if (x1==0)
               next_state = state_d;
            else
               next_state = state_c;

        state_d:
            if (x1==0)
               next_state = state_a;
            else
               next_state = state_e;

        state_e:
            if (x1==0)
               next_state = state_a;
            else
               next_state = state_c;

        default:next_state = state_a;
    endcase
end
endmodule
```

Figure 8.2   (Continued)

```
//test bench for moore synchronous sequential machine
module moore_ssm20_tb;

reg rst_n, clk, x1;
wire [1:3] y;
wire z1;

//display variables
initial
$monitor ("x1 = %b, state = %b, z1 = %b",
           x1, y, z1);

//define clock
initial
begin
   clk = 1'b0;
   forever
      #10clk = ~clk;
end

//define input sequence
initial
begin
   #0 rst_n = 1'b0;
      x1 = 1'b0;

   #5 rst_n = 1'b1;

   x1 = 1'b0;@ (posedge clk)  //remain in state_a (000)
   x1 = 1'b1;@ (posedge clk)  //go to state_b (010)
   x1 = 1'b0;@ (posedge clk)  //go to state_a (000)
   x1 = 1'b1;@ (posedge clk)  //go to state_b (010)
   x1 = 1'b1;@ (posedge clk)  //go to state_c (110)
   x1 = 1'b0;@ (posedge clk)  //go to state_d (100)
   x1 = 1'b1;@ (posedge clk)  //go to state_e (011); assert z1
   x1 = 1'b0;@ (posedge clk)  //go to state_a (000)
   x1 = 1'b1;@ (posedge clk)  //go to state_b (010)
   x1 = 1'b1;@ (posedge clk)  //go to state_c (110)
   x1 = 1'b0;@ (posedge clk)  //go to state_d (100)
   x1 = 1'b1;@ (posedge clk)  //go to state_e (011); assert z1
   x1 = 1'b1;@ (posedge clk)  //go to state_c (110); assert z1
   x1 = 1'b0;@ (posedge clk)  //go to state_d (100); assert z1
   x1 = 1'b1;@ (posedge clk)  //go to state_e (011); assert z1
   x1 = 1'b0;@ (posedge clk)  //go to state_a (000)

   #10    $stop;
end                          //continued on next page
```

Figure 8.3   Test bench for the Moore machine of Example 8.1.

```
//instantiate the module into the test bench
moore_ssm20 inst1 (
   .rst_n(rst_n),
   .clk(clk),
   .x1(x1),
   .y(y),
   .z1(z1)
   );

endmodule
```

Figure 8.3   (Continued)

```
x1 = 0, state = xxx, z1 = x    x1 = 1, state = 010, z1 = 0
x1 = 1, state = 000, z1 = 0    x1 = 0, state = 110, z1 = 0
x1 = 0, state = 010, z1 = 0    x1 = 1, state = 100, z1 = 0
x1 = 1, state = 000, z1 = 0    x1 = 1, state = 011, z1 = 0
x1 = 1, state = 010, z1 = 0    x1 = 1, state = 011, z1 = 1
x1 = 0, state = 110, z1 = 0    x1 = 0, state = 110, z1 = 0
x1 = 1, state = 100, z1 = 0    x1 = 1, state = 100, z1 = 0
x1 = 0, state = 011, z1 = 0    x1 = 0, state = 011, z1 = 0
x1 = 0, state = 011, z1 = 1    x1 = 0, state = 011, z1 = 1
x1 = 1, state = 000, z1 = 0    x1 = 0, state = 000, z1 = 0
```

Figure 8.4   Outputs for the Moore machine of Example 8.1.

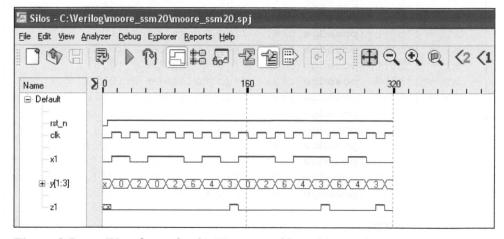

Figure 8.5   Waveforms for the Moore machine of Example 8.1.

Example 8.2   The same Moore synchronous sequential machine that was designed in Example 8.1 using behavioral modeling will now be designed using structural modeling.   This will provide a comparison between behavioral modeling and structural modeling for the same machine — the results should be identical.   Structural modeling usually requires algorithmic equations or a logic diagram as the basis for the module. The structural module will use built-in primitives and instantiated D flip-flops.   The test bench that was utilized in Example 8.1 will also be used in this example.

For convenience, the state diagram is reproduced in Figure 8.6.   The input maps are obtained directly from the state diagram and are shown in Figure 8.7; the D input equations are shown in Equation 8.1.   The logic diagram is shown in Figure 8.8 and depicts the instantiation names for the built-in primitives, the net names, and the D flip-flop instantiations.

The structural module, test bench, outputs, and waveforms are shown in Figure 8.9, Figure 8.10, Figure 8.11, and Figure 8.12, respectively.

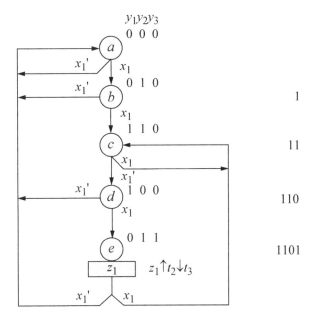

Figure 8.6   State diagram for the Moore machine of Example 8.2.

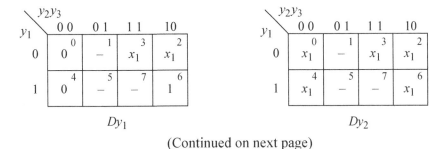

(Continued on next page)

Figure 8.7   Input maps for the D flip-flops of Example 8.2.

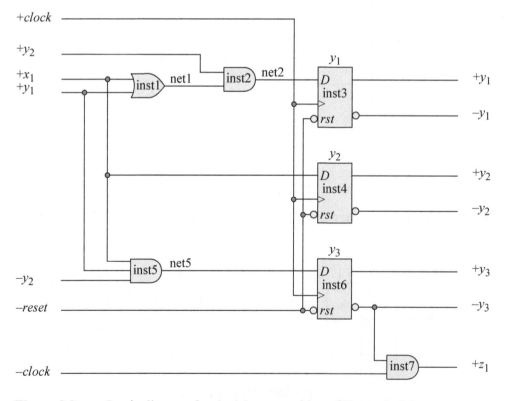

$y_2 y_3$			
y_1	0 0	0 1	1 1	1 0

Karnaugh map with cells: top row $y_1 = 0$: cell 0 = 0, cell 1 = –, cell 3 = 0, cell 2 = 0; bottom row $y_1 = 1$: cell 4 = x_1, cell 5 = –, cell 7 = –, cell 6 = 0.

Dy_3

Figure 8.7   (Continued)

$$Dy_1 = y_2 x_1 + y_1 y_2 = y_2 (x_1 + y_1)$$

$$Dy_2 = x_1$$

$$Dy_3 = y_1 y_2' x_1 \tag{8.1}$$

Figure 8.8   Logic diagram for the Moore machine of Example 8.2.

```verilog
//structural moore synchronous sequential machine
module moore_ssm20a (rst_n, clk, x1, y, z1);

input rst_n, clk, x1;
output [1:3] y;
output z1;

wire rst_n, clk, x1;
wire [1:3] y;
wire z1;

//define internal wires
wire net1, net2, net5;

//instantiate the logic for flip-flop y[1]
or     inst1 (net1, x1, y[1]);
and    inst2 (net2, y[2], net1);

//instantiate the D flip-flop for y[1]
d_ff_bh inst3 (
   .rst_n(rst_n),
   .clk(clk),
   .d(net2),
   .q(y[1])
   );

//instantiate the D flip-flop for y[2]
d_ff_bh inst4 (
   .rst_n(rst_n),
   .clk(clk),
   .d(x1),
   .q(y[2])
   );

//instantiate the logic for flip-flop y[3]
and    inst5 (net5, x1, y[1], ~y[2]);

//instantiate the D flip-flop for y[3]
d_ff_bh inst6 (
   .rst_n(rst_n),
   .clk(clk),
   .d(net5),
   .q(y[3])
   );

and    inst7 (z1, y[3], ~clk); //instantiate the logic for z1
endmodule
```

Figure 8.9   Structural module for the Moore machine of Example 8.2.

```verilog
//test bench for moore synchronous sequential machine
module moore_ssm20a_tb;

reg rst_n, clk, x1;
wire [1:3] y;
wire z1;

//display variables
initial
$monitor ("x1 = %b, state = %b, z1 = %b",
            x1, y, z1);

//define clock
initial
begin
   clk = 1'b0;
   forever
      #10clk = ~clk;
end

//define input sequence
initial
begin
   #0 rst_n = 1'b0;
      x1 = 1'b0;

   #5 rst_n = 1'b1;

   x1 = 1'b0;@ (posedge clk)   //remain in state_a (000)
   x1 = 1'b1;@ (posedge clk)   //go to state_b (010)
   x1 = 1'b0;@ (posedge clk)   //go to state_a (000)
   x1 = 1'b1;@ (posedge clk)   //go to state_b (010)
   x1 = 1'b1;@ (posedge clk)   //go to state_c (110)
   x1 = 1'b0;@ (posedge clk)   //go to state_d (100)
   x1 = 1'b1;@ (posedge clk)   //go to state_e (011); assert z1
   x1 = 1'b0;@ (posedge clk)   //go to state_a (000)
   x1 = 1'b1;@ (posedge clk)   //go to state_b (010)
   x1 = 1'b1;@ (posedge clk)   //go to state_c (110)
   x1 = 1'b0;@ (posedge clk)   //go to state_d (100)
   x1 = 1'b1;@ (posedge clk)   //go to state_e (011); assert z1
   x1 = 1'b1;@ (posedge clk)   //go to state_c (110); assert z1
   x1 = 1'b0;@ (posedge clk)   //go to state_d (100); assert z1
   x1 = 1'b1;@ (posedge clk)   //go to state_e (011); assert z1
   x1 = 1'b0;@ (posedge clk)   //go to state_a (000)

   #10    $stop;
end                             //continued on next page
```

Figure 8.10   Test bench for the Moore machine of Example 8.2.

```
//instantiate the module into the test bench
moore_ssm20a inst1 (
    .rst_n(rst_n),
    .clk(clk),
    .x1(x1),
    .y(y),
    .z1(z1)
    );

endmodule
```

Figure 8.10   (Continued)

```
x1 = 0, state = 000, z1 = 0      x1 = 1, state = 010, z1 = 0
x1 = 1, state = 000, z1 = 0      x1 = 0, state = 110, z1 = 0
x1 = 0, state = 010, z1 = 0      x1 = 1, state = 100, z1 = 0
x1 = 1, state = 000, z1 = 0      x1 = 1, state = 011, z1 = 0
x1 = 1, state = 010, z1 = 0      x1 = 1, state = 011, z1 = 1
x1 = 0, state = 110, z1 = 0      x1 = 0, state = 110, z1 = 0
x1 = 1, state = 100, z1 = 0      x1 = 1, state = 100, z1 = 0
x1 = 0, state = 011, z1 = 0      x1 = 0, state = 011, z1 = 0
x1 = 0, state = 011, z1 = 1      x1 = 0, state = 011, z1 = 1
x1 = 1, state = 000, z1 = 0      x1 = 0, state = 000, z1 = 0
```

Figure 8.11   Outputs for the Moore machine of Example 8.2.

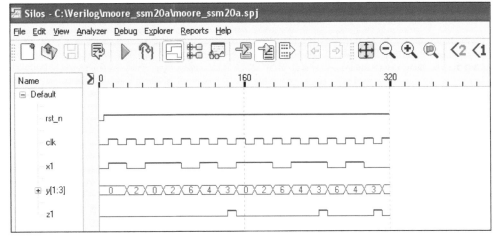

Figure 8.12   Waveforms for the Moore machine of Example 8.2.

Example 8.3   The state diagram for a Moore synchronous sequential machine is shown in Figure 8.13 which has two inputs x_1 and x_2 together with two outputs z_1 and z_2, both of which have the following assertion/deassertion: $\uparrow t_2 \downarrow t_3$. Behavioral modeling will be used in the implementation; therefore, the machine can be designed directly from the state diagram.

The behavioral module using the **case** statement is shown in Figure 8.14 — outputs z_1 and z_2 are defined using the continuous assignment statement of dataflow modeling. The test bench module is shown in Figure 8.15 using the **$random** system task to generate a random value for certain inputs when their value can be considered a "don't care" — either 0 or 1. The outputs and waveforms are shown in Figure 8.16 and Figure 8.17, respectively.

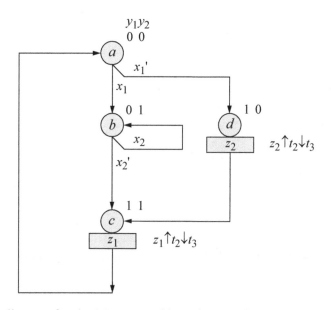

Figure 8.13   State diagram for the Moore machine of Example 8.3.

```
//behavioral moore synchronous sequential machine
module moore_ssm14 (clk, rst_n, x1, x2, y, z1, z2);

input clk, rst_n, x1, x2;
output [1:2] y;
output z1, z2;

reg [1:2] y, next_state;
wire z1, z2;

//continued on next page
```

Figure 8.14   Behavioral module for the Moore machine of Example 8.3.

```verilog
//assign state codes
parameter    state_a = 2'b00,
             state_b = 2'b01,
             state_c = 2'b11,
             state_d = 2'b10;

//set next state
always @ (posedge clk)
begin
   if (~rst_n)
      y <= state_a;
   else
      y <= next_state;
end

//define outputs
assign    z1 = (y[1] & y[2] & ~clk);
assign    z2 = (y[1] & ~y[2] & ~clk);

//determine next state
always @ (y or x1 or x2)
begin
   case (y)
      state_a:
         if (x1==0)
            next_state = state_d;
         else
            next_state = state_b;

      state_b:
         if (x2==0)
            next_state = state_c;
         else
            next_state = state_b;

      state_c: next_state = state_a;

      state_d: next_state = state_c;

      default: next_state = state_a;

   endcase
end

endmodule
```

Figure 8.14   (Continued)

```verilog
//test bench for moore synchronous sequential machine
module moore_ssm14_tb;

reg clk, rst_n, x1, x2;
wire [1:2] y;
wire z1, z2;

//display variables
initial
$monitor ("x1x2=%b, state=%b, z1z2=%b",
          {x1, x2}, y, {z1, z2});

//define clock
initial
begin
   clk = 1'b0;
   forever
      #10clk = ~clk;
end

//define input sequence
initial
begin
   #0 rst_n = 1'b0;      //reset to state_a
      x1 = 1'b0;
      x2 = 1'b0;
   #5 rst_n = 1'b1;

      x1 = 1'b1;   x2 = $random;
      @ (posedge clk)

      x1 = 1'b1;   x2 = $random;
      @ (posedge clk)   //go to state_b (01)

      x1 = $random;   x2 = 1'b1;
      @ (posedge clk)   //remain in state_b (01)

      x1 = $random;   x2 = 1'b0;
      @ (posedge clk)   //go to state_c (11)

      x1 = $random;   x2 = $random;
      @ (posedge clk)   //go to state_a (00)

      x1 = 1'b0;   x2 = $random;
      @ (posedge clk)   //go to state_d (10)

//continued on next page
```

Figure 8.15   Test bench for the Moore machine of Example 8.3.

```
      x1 = $random;   x2 = $random;
      @ (posedge clk)   //go to state_c (11)

      x1 = $random;   x2 = $random;
      @ (posedge clk)   //go to state_a (00)
   #10    $stop;
end

moore_ssm14 inst1 (   //instantiate the module
   .clk(clk),
   .rst_n(rst_n),
   .x1(x1),
   .x2(x2),
   .y(y),
   .z1(z1),
   .z2(z2)
   );
endmodule
```

Figure 8.15   (Continued)

```
x1x2=00, state=xx, z1z2=xx      x1x2=11, state=11, z1z2=10
x1x2=10, state=xx, z1z2=xx      x1x2=01, state=00, z1z2=00
x1x2=11, state=00, z1z2=00      x1x2=01, state=10, z1z2=00
x1x2=11, state=01, z1z2=00      x1x2=01, state=10, z1z2=01
x1x2=10, state=01, z1z2=00      x1x2=10, state=11, z1z2=00
x1x2=11, state=11, z1z2=00      x1x2=10, state=11, z1z2=10
                                x1x2=10, state=00, z1z2=00
```

Figure 8.16   Outputs for the Moore machine of Example 8.3.

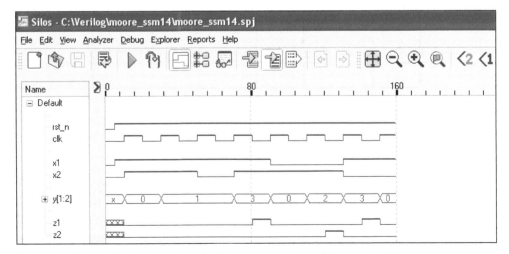

Figure 8.17   Waveforms for the Moore machine of Example 8.3.

Example 8.4   A Mealy synchronous sequential machine will be designed using behavioral modeling with the **case** statement and the **parameter** keyword. Output z_1 is asserted whenever a serial input data line contains a 3-bit word with an odd number of 1s. The format for the serial data line is shown below,

$$x_1 = \begin{vmatrix} b_1 & b_2 & b_3 \end{vmatrix} b_1 \quad b_2 \quad b_3 \begin{vmatrix} b_1 & b_2 & b_3 \end{vmatrix} \ldots$$

where $b_i = 0$ or 1. There is no space between words. Output z_1 is asserted during the first half of the third bit time b_3; that is, $z_1 \uparrow t_1 \downarrow t_2$.

The state diagram is shown in Figure 8.18 in which the state codes have been selected so that there will be no glitches on output z_1 for any state transition sequence. There are three unused states: $y_1 y_2 y_3 = 101, 110,$ and 111. If two or more flip-flops change value for a state transition sequence, then the machine may momentarily pass through either an unused state or a state in which there is no output — in both cases there will be no glitch on output z_1.

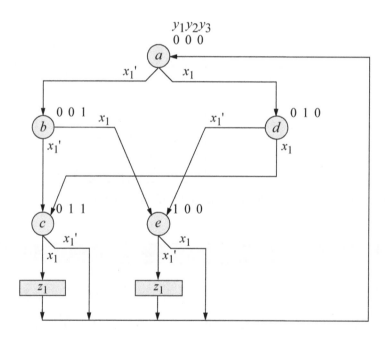

Figure 8.18   State diagram for the Mealy machine of Example 8.4.

The behavioral module is shown in Figure 8.19. The **default** keyword should be inserted at the end of a block containing a **case** statement. The test bench module is shown in Figure 8.20. Every state transition sequence requires a specific value for input x_1; therefore, the **$random** system task cannot be used. The outputs and waveforms are shown in Figure 8.21 and Figure 8.22, respectively.

```
//behavioral mealy synchronous sequential machine
module mealy_ssm7 (rst_n, clk, x1, y, z1);

input rst_n, clk, x1;
output [1:3] y;
output z1;

reg [1:3] y, next_state;
wire z1;

//assign state codes
parameter   state_a = 3'b000,
            state_b = 3'b001,
            state_c = 3'b011,
            state_d = 3'b010,
            state_e = 3'b100;

//set next state
always @ (posedge clk)
begin
   if (~rst_n)
      y <= state_a;
   else
      y <= next_state;
end

//define output
assign z1 = (y[1] & ~x1 & clk) | (y[2] & y[3] & x1 & clk);

//determine next state
always @ (y or x1)
begin
   case (y)
      state_a:
         if (x1==0)
            next_state = state_b;
         else
            next_state = state_d;

      state_b:
         if (x1==0)
            next_state = state_c;
         else
            next_state = state_e;

//continued on next page
```

Figure 8.19 Behavioral module for the Mealy machine of Example 8.4.

```
        state_c: next_state = state_a;

        state_d:
           if (x1==0)
              next_state = state_e;
           else
              next_state = state_c;

        state_e: next_state = state_a;

        default: next_state = state_a;
     endcase
  end
endmodule
```

Figure 8.19   (Continued)

```
//test bench for mealy synchronous sequential machine
module mealy_ssm7_tb;

reg rst_n, clk, x1;
wire [1:3] y;
wire z1;

//display variables
initial
$monitor ("x1=%b, state= %b, z1=%b",
           x1, y, z1);

//define clock
initial
begin
   clk = 1'b0;
   forever
      #10   clk = ~clk;
end

//define input sequence
initial
begin
   #0 rst_n = 1'b0;
      x1 = 1'b0;

   #5 rst_n = 1'b1;
//continued on next page
```

Figure 8.20   Test bench for the Mealy machine of Example 8.4.

```
        x1 = 1'b0; @ (posedge clk) //go to state_b (001)
        x1 = 1'b0; @ (posedge clk) //go to state_c (011)
        x1 = 1'b1; @ (posedge clk) //go to state_a (000)
        x1 = 1'b0; @ (posedge clk) //go to state_b (001)
        x1 = 1'b1; @ (posedge clk) //go to state_e (100); set z1
        x1 = 1'b0; @ (posedge clk) //go to state_a (000)
        x1 = 1'b1; @ (posedge clk) //go to state_d (010)
        x1 = 1'b0; @ (posedge clk) //go to state_e (100)
        x1 = 1'b1; @ (posedge clk) //go to state_a (000)
        x1 = 1'b1; @ (posedge clk) //go to state_d (010)
        x1 = 1'b1; @ (posedge clk) //go to state_c (011); set z1
        x1 = 1'b1; @ (posedge clk) //go to state_a (000)
        x1 = 1'b1; @ (posedge clk) //go to state_b (001)

    #10    $stop;
end

//instantiate the module into the test bench
mealy_ssm7 inst1 (
    .rst_n(rst_n),
    .clk(clk),
    .x1(x1),
    .y(y),
    .z1(z1)
    );
endmodule
```

Figure 8.20   (Continued)

```
x1=0, state= xxx, z1=0
x1=0, state= 000, z1=0
x1=1, state= 001, z1=0
x1=0, state= 100, z1=1
x1=0, state= 100, z1=0
x1=1, state= 000, z1=0
x1=0, state= 010, z1=0
x1=1, state= 100, z1=0
x1=0, state= 000, z1=0
x1=1, state= 001, z1=0
x1=1, state= 100, z1=0
x1=1, state= 000, z1=0
x1=1, state= 010, z1=0
x1=1, state= 011, z1=1
x1=1, state= 011, z1=0
x1=1, state= 000, z1=0
```

Figure 8.21   Outputs for the Mealy machine of Example 8.4.

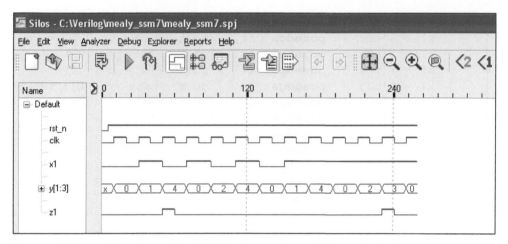

Figure 8.22   Waveforms for the Mealy machine of Example 8.4.

Example 8.5   A counter will be designed using structural modeling that counts in the sequence shown in the chart below. The storage elements will be positive-edge-triggered D flip-flops. The input maps are shown in Figure 8.23 and the input equations are shown in Equation 8.4. The logic diagram is shown in Figure 8.24.

y_1	y_2	y_3
0	0	0
1	1	1
0	0	1
1	1	0
0	1	0
1	0	1
0	1	1
1	0	0
0	0	0

The structural module is shown in Figure 8.25 using module instantiation of logic primitives that were previously designed and a positive-edge-triggered D flip-flop. The test bench is shown in Figure 8.26. The only inputs are the clock and reset variables. The duration of simulation is 160 time units, which is sufficient to show the complete counting sequence. The outputs are shown in Figure 8.27 which displays the counting sequence. The waveforms are shown in Figure 8.28.

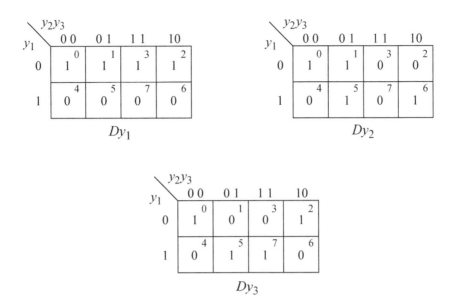

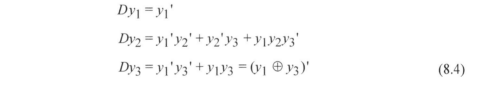

Figure 8.23 Input maps for the nonsequential counter of Example 8.5.

$$Dy_1 = y_1'$$

$$Dy_2 = y_1'y_2' + y_2'y_3 + y_1y_2y_3'$$

$$Dy_3 = y_1'y_3' + y_1y_3 = (y_1 \oplus y_3)' \tag{8.4}$$

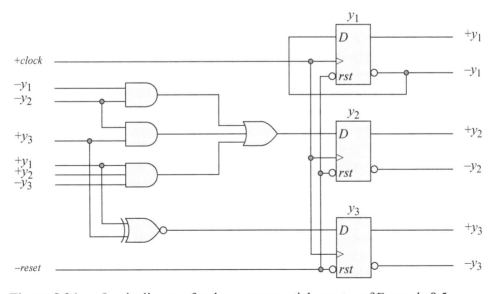

Figure 8.24 Logic diagram for the nonsequential counter of Example 8.5.

```verilog
//structural nonsequential counter
module ctr_non_seq4 (rst_n, clk, y);

input rst_n, clk;
output [1:3] y;

wire [1:3] y;

//define internal nets
wire net2, net3, net4, net5, net7;

//instantiate the flip-flop for y[1]
d_ff_bh inst1 (
   .rst_n(rst_n),
   .clk(clk),
   .d(~y[1]),
   .q(y[1])
   );

//instantiate the logic and D flip-flop for y[2]
and2_df inst2 (
   .x1(~y[1]),
   .x2(~y[2]),
   .z1(net2)
   );

and2_df inst3 (
   .x1(~y[2]),
   .x2(y[3]),
   .z1(net3)
   );

and3_df inst4 (
   .x1(y[1]),
   .x2(y[2]),
   .x3(~y[3]),
   .z1(net4)
   );

or3_df inst5 (
   .x1(net2),
   .x2(net3),
   .x3(net4),
   .z1(net5)
   );

//continued on next page
```

Figure 8.25   Structural module for the nonsequential counter of Example 8.5.

```
d_ff_bh inst6 (
   .rst_n(rst_n),
   .clk(clk),
   .d(net5),
   .q(y[2])
   );

//instantiate the logic and D flip-flop for y[3]
xnor2_df inst7 (
   .x1(y[1]),
   .x2(y[3]),
   .z1(net7)
   );

d_ff_bh inst8 (
   .rst_n(rst_n),
   .clk(clk),
   .d(net7),
   .q(y[3])
   );

endmodule
```

Figure 8.25   (Continued)

```
//test bench for nonsequential counter
module ctr_non_seq4_tb;

reg rst_n, clk;
wire [1:3] y;

//display count
initial
$monitor ("count = %b", y);

//define reset
initial
begin
   #0 rst_n = 1'b0;
   #5 rst_n = 1'b1;
end

//continued on next page
```

Figure 8.26   Test bench for the nonsequential counter of Example 8.5.

```
//define clock
initial
begin
   clk = 1'b0;
   forever
       #10clk = ~clk;
end

//define simulation time
initial
begin
   #160 $finish;
end

ctr_non_seq4 inst1 (          //instantiate the module
   .rst_n(rst_n),
   .clk(clk),
   .y(y)
   );
endmodule
```

Figure 8.26   (Continued)

```
count = 000                count = 101
count = 111                count = 011
count = 001                count = 100
count = 110                count = 000
count = 010                count = 111
```

Figure 8.27   Outputs for the nonsequential counter of Example 8.5.

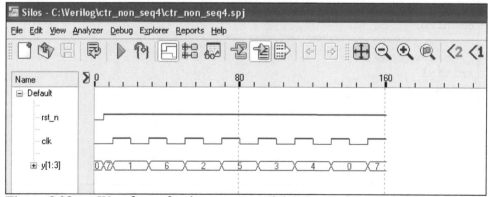

Figure 8.28   Waveforms for the nonsequential counter of Example 8.5.

Example 8.6   A Moore synchronous sequential machine will be designed using structural modeling that has four parallel inputs x_1, x_2, x_3, and x_4 and two outputs z_1 and z_2, as shown below.

$$
\begin{array}{cccc}
\left|\begin{array}{c} x_1 \\ x_2 \\ x_3 \\ x_4 \end{array}\right|
& \left|\begin{array}{c} x_1 \\ x_2 \\ x_3 \\ x_4 \end{array}\right|
& \left|\begin{array}{c} x_1 \cdots \rightarrow \\ x_2 \cdots \rightarrow \\ x_3 \cdots \rightarrow \\ x_4 \cdots \rightarrow \end{array}\right|
& \boxed{\begin{array}{c} \text{Moore} \\ \text{synchronous} \\ \text{sequential} \\ \text{machine} \end{array}}
\begin{array}{c} \rightarrow z_1 \\ \rightarrow z_2 \end{array}
\end{array}
$$

The inputs constitute a 4-bit word. There is one bit space between words. The machines operates as follows:

(a) Output $z_1 = 1$ if the 4-bit word contains an odd number of 1s.
(b) Output $z_2 = 1$ if the 4-bit word contains an even number of 1s, but not zero 1s.

The state diagram is shown in Figure 8.29 depicting one path for output z_1 and one path for output z_2. The assertion/deassertion statements for the outputs are: $z_1 \uparrow t_1 \downarrow t_3$ and $z_2 \uparrow t_2 \downarrow t_3$. The truth table for the machine is shown in Table 8.1, indicating the words that contain an even or odd number of 1s. The equation for an odd number of 1s — obtained from the truth table — is shown in Equation 8.5; the equation for an even number of 1s — but not zero 1s — is shown in Equation 8.6.

The logic diagram is shown in Figure 8.30 as obtained from Equation 8.5 and Equation 8.6, where flip-flop y_1 represents an even number of 1s and flip-flop y_2 represents an odd number of 1s. The output logic is obtained from the state diagram. The structural design module and the test bench module are shown in Figure 8.31 and Figure 8.32, respectively. The outputs and waveforms are shown in Figure 8.33 and Figure 8.34, respectively.

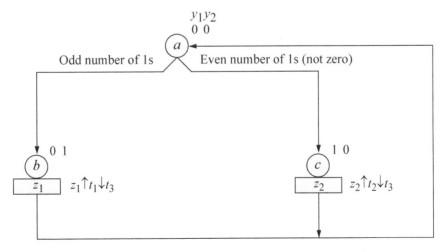

Figure 8.29   State diagram for the Moore machine of Example 8.6.

Table 8.1 Table Showing Words Containing an Odd or Even Number of 1s

x_1	x_2	x_3	x_4	
0	0	0	0	
	0	0	1	Odd number of 1s
0	0	1	0	
0	0	1	1	
	1	0	0	
0	1	0	1	
0	1	1	0	
	1	1	1	
	0	0	0	
1	0	0	1	
1	0	1	0	
	0	1	1	
1	1	0	0	
	1	0	1	
	1	1	0	
1	1	1	1	

$$
\begin{aligned}
\text{Odd} = {} & x_1'x_2'x_3'x_4 + x_1'x_2'x_3x_4' + x_1'x_2x_3'x_4' + x_1'x_2x_3x_4 + \\
& x_1x_2'x_3'x_4' + x_1x_2'x_3x_4 + x_1x_2x_3'x_4 + x_1x_2x_3x_4' \\
= {} & x_1'x_2'(x_3 \oplus x_4) + x_1'x_2(x_3 \oplus x_4)' + x_1x_2(x_3 \oplus x_4) + x_1x_2'(x_3 \oplus x_4)' \\
= {} & (x_3 \oplus x_4)(x_1 \oplus x_2)' + (x_3 \oplus x_4)'(x_1 \oplus x_2) \\
= {} & (x_1 \oplus x_2) \oplus (x_3 \oplus x_4)
\end{aligned}
\tag{8.5}
$$

$$
\begin{aligned}
\text{Even} = {} & x_1'x_2'x_3x_4 + x_1'x_2x_3'x_4 + x_1'x_2x_3x_4' + x_1x_2'x_3'x_4 + \\
& x_1x_2'x_3x_4' + x_1x_2x_3'x_4' + x_1x_2x_3x_4 \\
= {} & x_1'x_2(x_3 \oplus x_4) + x_1x_2(x_3 \oplus x_4)' + x_1x_2'(x_3 \oplus x_4) + x_1'x_2'x_3x_4 \\
= {} & (x_1 \oplus x_2)(x_3 \oplus x_4) + x_1x_2(x_3 \oplus x_4)' + x_1'x_2'x_3x_4
\end{aligned}
\tag{8.6}
$$

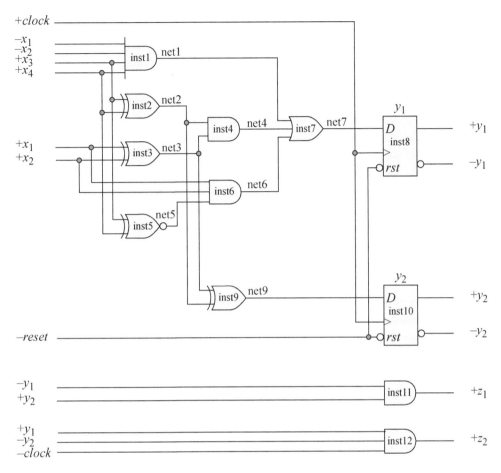

Figure 8.30 Logic diagram for the Moore machine of Example 8.6.

```
//structural moore synchronous sequential machine
module moore_ssm19a (rst_n, clk, x, y, z1, z2);

input rst_n, clk;
input [1:4] x;
output [1:2] y;
output z1, z2;

wire rst_n, clk;
wire [1:4] x;
wire [1:2] y;
wire z1, z2;

//continued on next page
```

Figure 8.31 Structural module for the Moore machine of Example 8.6.

```
//define internal wires
wire   net1, net2, net3, net4, net5,
       net6, net7, net9;

//instantiate the logic for flip-flop y[1]
and4_df inst1 (
   .x1(~x[1]),
   .x2(~x[2]),
   .x3(x[3]),
   .x4(x[4]),
   .z1(net1)
   );

xor2_df inst2 (
   .x1(x[3]),
   .x2(x[4]),
   .z1(net2)
   );

xor2_df inst3 (
   .x1(x[1]),
   .x2(x[2]),
   .z1(net3)
   );

and2_df inst4 (
   .x1(net2),
   .x2(net3),
   .z1(net4)
   );

xnor2_df inst5 (
   .x1(x[3]),
   .x2(x[4]),
   .z1(net5)
   );

and3_df inst6 (
   .x1(x[1]),
   .x2(x[2]),
   .x3(net5),
   .z1(net6)
   );

//continued on next page
```

Figure 8.31   (Continued)

```
or3_df inst7 (
   .x1(net1),
   .x2(net4),
   .x3(net6),
   .z1(net7)
   );

d_ff_bh inst8 (
   .rst_n(rst_n),
   .clk(clk),
   .d(net7),
   .q(y[1])
   );

//instantiate the logic for flip-flop y[2]
xor2_df inst9 (
   .x1(net3),
   .x2(net2),
   .z1(net9)
   );

d_ff_bh inst10 (
   .rst_n(rst_n),
   .clk(clk),
   .d(net9),
   .q(y[2])
   );

//instantiate the logic for outputs z1 and z2
and2_df inst11 (
   .x1(~y[1]),
   .x2(y[2]),
   .z1(z1)
   );

and3_df inst12 (
   .x1(y[1]),
   .x2(~y[2]),
   .x3(~clk),
   .z1(z2)
   );

endmodule
```

Figure 8.31   (Continued)

```
//test bench for moore synchronous sequential machine 19a
module moore_ssm19a_tb;

reg rst_n, clk;
reg [1:4] x;
wire [1:2] y;
wire z1, z2;

//display variables
initial
$monitor ("x = %b, state = %b, z1z2 = %b",
          x, y, {z1, z2});

//define clock
initial
begin
   clk = 1'b0;
   forever
      #10clk = ~clk;
end

//define input sequence
initial
begin
   #0 rst_n = 1'b0;
      x = 4'b0000;

   #5 rst_n = 1'b1;

      x = 4'b0001;
      @ (posedge clk)    //go to state_b (01); assert z1

      x = 4'b0000;
      @ (posedge clk)    //go to state_a (00)

      x = 4'b1010;
      @ (posedge clk)    //go to state_c (10); assert z2

      x = 4'b0000;
      @ (posedge clk)    //go to state_a (00)

      x = 4'b0111;
      @ (posedge clk)    //go to state_b (01); assert z1

      x = 4'b0000;
      @ (posedge clk)    //go to state_a (00)
//continued on next page
```

Figure 8.32   Test bench module for the Moore machine of Example 8.6.

```
      x = 4'b1111;
      @ (posedge clk)   //go to state_c (10); assert z2

      x = 4'b0000;
      @ (posedge clk)   //go to state_a (00)
   #10   $stop;
end

moore_ssm19a inst1 (    //instantiate the module
   .rst_n(rst_n),
   .clk(clk),
   .x(x),
   .y(y),
   .z1(z1),
   .z2(z2)
   );
endmodule
```

Figure 8.32 (Continued)

```
x = 0000, state = 00, z1z2 = 00 | x = 0111, state = 00, z1z2 = 00
x = 0001, state = 00, z1z2 = 00 | x = 0000, state = 01, z1z2 = 10
x = 0000, state = 01, z1z2 = 10 | x = 1111, state = 00, z1z2 = 00
x = 1010, state = 00, z1z2 = 00 | x = 0000, state = 10, z1z2 = 00
x = 0000, state = 10, z1z2 = 00 | x = 0000, state = 10, z1z2 = 01
x = 0000, state = 10, z1z2 = 01 | x = 0000, state = 00, z1z2 = 00
```

Figure 8.33 Outputs for the Moore machine of Example 8.6.

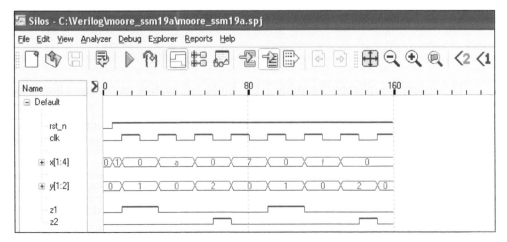

Figure 8.34 Waveforms for the Moore machine of Example 8.6.

8.2 Asynchronous Sequential Machines

Asynchronous sequential machines are another class of finite-state machines, where state changes occur on the application of input signals only — there is no machine clock. Like synchronous machines, the outputs of asynchronous machines are a function of either the present state only, or of the present state and the present inputs, corresponding to Moore and Mealy machines, respectively.

This section presents a variety of asynchronous sequential machines that are designed using Verilog HDL. These state machines were designed in Chapter 7 using traditional design methodologies and are presented in this section as a means to compare the two design techniques.

Example 8.7   An asynchronous sequential machine will be designed that has one input x_1 and two outputs z_1 and z_2. Output z_1 is asserted for the duration of every second x_1 pulse. Output z_2 is asserted for the duration of every second z_1 pulse. The outputs are to respond as fast as possible to changes in the input variable. A representative timing diagram is shown in Figure 8.35.

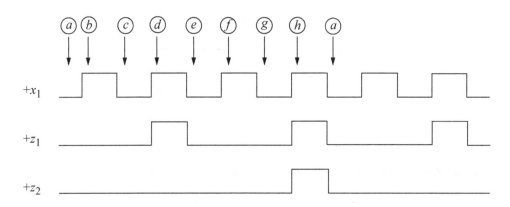

Figure 8.35   Representative timing diagram for the asynchronous sequential machine of Example 8.7.

This example replicates Example 7.13. Since this section emphasizes the design of asynchronous sequential machines using Verilog HDL; therefore, only the excitation equations, the output equations, and the logic diagram are presented. Refer to Example 7.13 for a detailed presentation of the steps required to design this asynchronous machine.

The excitation equations for Y_{1e}, Y_{2e}, and Y_{3e} are shown in Equation 8.7; the output equations for z_1 and z_2 are shown in Equation 8.8 and represent the fastest possible output changes. The logic diagram is shown in Figure 8.36, which depicts the logic primitives and associated net names. The machine will be designed using the continuous assignment statement **assign** for dataflow modeling.

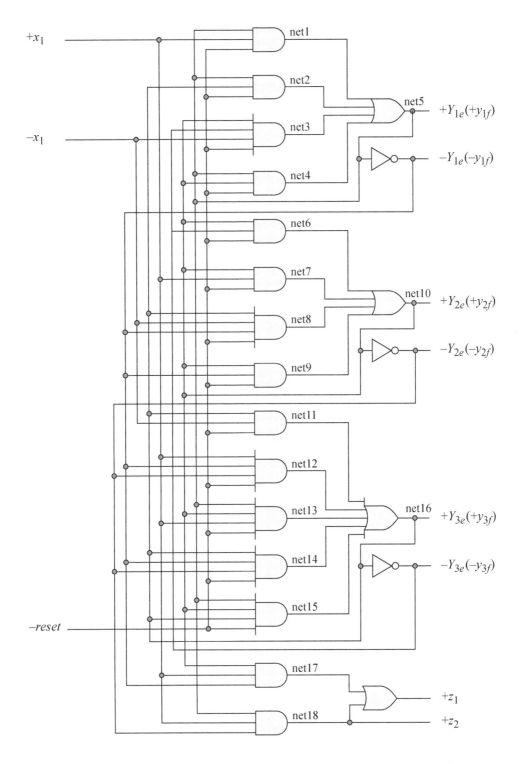

Figure 8.36 Logic diagram for the asynchronous sequential machine of Example 8.7.

The dataflow module and test bench are shown in Figure 8.37 and Figure 8.38, respectively. The test bench sequences the machine through the paths illustrated in the primitive flow table. The outputs and waveforms are shown in Figure 8.39 and Figure 8.40, respectively.

```verilog
//dataflow asynchronous sequential machine
module asm13 (rst_n, x1, z1, z2);

input rst_n, x1;
output z1, z2;

//define internal nets
wire   net1, net2, net3, net4, net5, net6,
       net7, net8, net9, net10, net11, net12,
       net13, net14, net15, net16, net17, net18;

//design the logic for excitation variable Y1e
assign   net1  = rst_n & net5 & x1,
         net2  = rst_n & net5 & net16,
         net3  = rst_n & net10 & ~net16 & ~x1,
         net4  = rst_n & net5 & net10,
         net5  = net1 | net2 | net3 | net4;

//design the logic for excitation variable Y2e
assign   net6  = rst_n & net10 & ~net16,
         net7  = rst_n & net10 & x1,
         net8  = rst_n & net16 & ~x1 & ~net5,
         net9  = rst_n & net10 & ~net5,
         net10 = net6 | net7 | net8 | net9;

//design the logic for excitation variable Y3e
assign   net11 = rst_n & net16 & ~x1,
         net12 = rst_n & x1 & ~net5 & ~net10,
         net13 = rst_n & net5 & net10 & x1,
         net14 = rst_n & net16 & ~net5 & ~net10,
         net15 = rst_n & net5 & net10 & net16,
         net16 = net11 | net12 | net13 | net14 | net15;

//design the logic for output z1
assign   net17 = net10 & x1 & ~net5,
         net18 = net5 & x1 & ~net10,
         z1    = net17 | net18;

//design the logic for output z2
assign   z2    = net18;

endmodule
```

Figure 8.37   Dataflow module for the asynchronous sequential machine of Example 8.7.

```
//test bench for asynchronous sequential machine
module asm13_tb;

reg rst_n, x1;
wire z1, z2;

//display variables
initial
$monitor ("x1 = %b, z1 = %b, z2 = %b", x1, z1, z2);

//apply input values
initial
begin
   #0    rst_n = 1'b0;
         x1 = 1'b0;   //reset to state_a
   #5    rst_n = 1'b1;

   #10   x1 = 1'b1;   //go to state_b
   #10   x1 = 1'b0;   //go to state_c
   #10   x1 = 1'b1;   //go to state_d; assert z1
   #10   x1 = 1'b0;   //go to state_e
   #10   x1 = 1'b1;   //go to state_f
   #10   x1 = 1'b0;   //go to state_g
   #10   x1 = 1'b1;   //go to state_h; assert z1 and z2
   #10   x1 = 1'b0;   //go to state_a
   #10   x1 = 1'b1;   //go to state_b
   #10   $stop;
end

//instantiate the module into the test bench
asm13 inst1 (
   .rst_n(rst_n),
   .x1(x1),
   .z1(z1),
   .z2(z2)
   );
endmodule
```

Figure 8.38   Test bench for the asynchronous sequential machine of Example 8.7.

```
x1 = 0, z1 = 0, z2 = 0        x1 = 1, z1 = 0, z2 = 0
x1 = 1, z1 = 0, z2 = 0        x1 = 0, z1 = 0, z2 = 0
x1 = 0, z1 = 0, z2 = 0        x1 = 1, z1 = 1, z2 = 1
x1 = 1, z1 = 1, z2 = 0        x1 = 0, z1 = 0, z2 = 0
x1 = 0, z1 = 0, z2 = 0        x1 = 1, z1 = 0, z2 = 0
```

Figure 8.39   Outputs for the asynchronous sequential machine of Example 8.7.

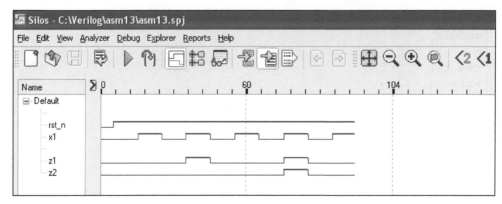

Figure 8.40   Waveforms for the asynchronous sequential machine of Example 8.7.

The next three examples will design the same asynchronous sequential machine using three different modeling techniques: dataflow modeling, behavioral modeling, and structural modeling. These three examples will offer a comparative study of three different modeling constructs for the same machine. The same test bench will be used for all three examples and will produce similar outputs and waveforms.

Example 8.8   An asynchronous sequential machine will be designed using dataflow modeling that has two inputs x_1 and x_2 and one output z_1. Output z_1 will be asserted coincident with the assertion of the first x_2 pulse and will remain active for the duration of the first x_2 pulse. The output will be asserted only if the assertion of x_1 precedes the assertion of x_2. Input x_1 will not become deasserted while x_2 is asserted. The λ output logic must have a minimal number of logic gates. A representative timing diagram is shown in Figure 8.41.

The primitive flow table is shown in Figure 8.42 in which there are no equivalent states. The merger diagram, the merged flow table, and the transition diagram are shown in Figure 8.43, Figure 8.44, and Figure 8.45, respectively.

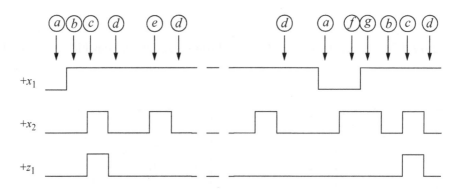

Figure 8.41   An asynchronous sequential machine for Example 8.8.

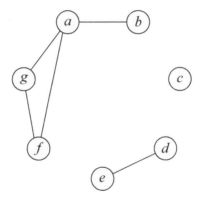

x_1x_2	00	01	11	10	z_1
	(a)	f	–	b	0
	a	–	c	(b)	0
	–	–	(c)	d	1
	a	–	e	(d)	0
	–	–	(e)	d	0
	a	(f)	g	–	0
	–	–	(g)	b	0

Figure 8.42 Primitive flow table for the asynchronous sequential machine of Example 8.8.

Figure 8.43 Merger diagram for the asynchronous sequential machine of Example 8.8.

				x_1x_2 00	01	11	10
1	(a)	(f)	(g)	(a)	(f)	(g)	b
2			(b)	a	–	c	(b)
3			(c)	–	–	(c)	d
4	(d)	(e)		a	–	(e)	(d)

Figure 8.44 Merged flow table for the asynchronous machine of Example 8.8.

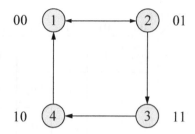

Figure 8.45   Transition diagram for the machine of Example 8.8.

The combined excitation map for Y_{1e} and Y_{2e} is shown in Figure 8.46 and the individual excitation maps are shown in Figure 8.47. The equations for Y_{1e} and Y_{2e} are shown in Equation 8.9. The output map for z_1 is shown in Figure 8.48 and the equation for z_1 is shown in Equation 8.10. The logic diagram — derived from the excitation and output equations — is shown in Figure 8.49.

$y_{1f} y_{2f}$ \ $x_1 x_2$	0 0	0 1	1 1	1 0
0 0	(00) ^{a}	(00) ^{f}	(00) ^{g}	01
0 1	00	–	11	(01) ^{b}
1 1	–	–	(11) ^{c}	10
1 0	00	–	(10) ^{e}	(10) ^{d}

$$Y_{1e}\ Y_{2e}$$

Figure 8.46   Combined excitation map for the asynchronous sequential machine of Example 8.8.

$y_{1f} y_{2f}$ \ $x_1 x_2$	0 0	0 1	1 1	1 0
0 0	0	0	0	0
0 1	0	–	1	0
1 1	–	–	1	1
1 0	0	–	1	1

$$Y_{1e}$$

$y_{1f} y_{2f}$ \ $x_1 x_2$	0 0	0 1	1 1	1 0
0 0	0	0	0	1
0 1	0	–	1	1
1 1	–	–	1	0
1 0	0	–	0	0

$$Y_{2e}$$

Figure 8.47   Individual excitation maps for the machine of Example 8.8.

$$Y_{1e} = y_{1f}x_1 + y_{2f}x_2$$

$$Y_{2e} = y_{2f}x_2 + y_{1f}'x_1x_2' + y_{1f}'y_{2f}x_1 \tag{8.9}$$

$y_{1f}y_{2f}$ x_1x_2	0 0	0 1	1 1	1 0
0 0	a 0	f 0	g 0	0
0 1	0	–	–	b 0
1 1	–	–	c 1	–
1 0	0	–	e 0	d 0

$$z_1$$

Figure 8.48   Output map for the asynchronous sequential machine of Example 8.8.

$$z_1 = y_{1f}y_{2f} \tag{8.10}$$

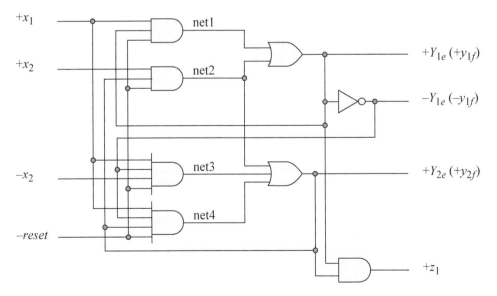

Figure 8.49   Logic diagram for the asynchronous sequential machine of Example 8.8.

The dataflow design module is shown in Figure 8.50 using the **assign** statement for all the logic primitive gates. The test bench — shown in Figure 8.51 — takes the machine through the state transition sequences shown in the primitive flow table and the timing diagram. The outputs are shown in Figure 8.52 and the waveforms in Figure 8.53.

```
//dataflow asynchronous sequential machine 16a
module asm16a (rst_n, x1, x2, y1e, y2e, z1);

input rst_n, x1, x2;
output y1e, y2e, z1;

//define internal nets
wire net1, net2, net3, net4;

//design the logic for excitation variable Y1e
assign    net1 = x1 & y1e & rst_n,
          net2 = x2 & y2e & rst_n,
          y1e = net1 | net2;

//design the logic for excitation variable Y2e
assign    net3 = x1 & ~x2 & ~y1e & rst_n,
          net4 = x1 & y2e & ~y1e & rst_n,
          y2e = net2 | net3 | net4;

//design the logic for output z1
assign    z1 = y1e & y2e;

endmodule
```

Figure 8.50   Dataflow module for the asynchronous sequential machine of Example 8.8.

```
//test bench for asynchronous sequential machine 16a
//use the primitive flow table or the timing diagram to follow
//the input sequence for the output and the waveforms
module asm16a_tb;

reg rst_n, x1, x2;
wire y1e, y2e, z1;

//display variables
initial
$monitor ("x1x2 = %b, state = %b, z1 = %b",  //next page
```

Figure 8.51   Test bench for the asynchronous sequential machine of Example 8.8.

```
//define input vectors
initial
begin
   #0    rst_n = 1'b0;
         x1 = 1'b0;
         x2 = 1'b0;

   #5    rst_n = 1'b1;

   #10   x1=1'b1;    x2=1'b0;    //go to state_b
   #10   x1=1'b1;    x2=1'b1;    //go to state_c; assert z1
   #10   x1=1'b1;    x2=1'b0;    //go to state_d
   #10   x1=1'b1;    x2=1'b1;    //go to state_e
   #10   x1=1'b1;    x2=1'b0;    //go to state_d
   #10   x1=1'b0;    x2=1'b0;    //go to state_a
   #10   x1=1'b0;    x2=1'b1;    //go to state_f
   #10   x1=1'b1;    x2=1'b1;    //go to state_g
   #10   x1=1'b1;    x2=1'b0;    //go to state_b
   #10   x1=1'b1;    x2=1'b1;    //go to state_c; assert z1
   #10   x1=1'b1;    x2=1'b0;    //go to state_d
   #10   x1=1'b0;    x2=1'b0;    //go to state_a

   #10   $stop;
end

//instantiate the module into the test bench
asm16a inst1 (
   .rst_n(rst_n),
   .x1(x1),
   .x2(x2),
   .y1e(y1e),
   .y2e(y2e),
   .z1(z1)
   );
endmodule
```

Figure 8.51   (Continued)

```
x1x2 = 00, state = 00, z1 = 0 | x1x2 = 01, state = 00, z1 = 0
x1x2 = 10, state = 01, z1 = 0 | x1x2 = 11, state = 00, z1 = 0
x1x2 = 11, state = 11, z1 = 1 | x1x2 = 10, state = 01, z1 = 0
x1x2 = 10, state = 10, z1 = 0 | x1x2 = 11, state = 11, z1 = 1
x1x2 = 11, state = 10, z1 = 0 | x1x2 = 10, state = 10, z1 = 0
x1x2 = 10, state = 10, z1 = 0 | x1x2 = 00, state = 00, z1 = 0
x1x2 = 00, state = 00, z1 = 0 |
```

Figure 8.52   Outputs for the asynchronous sequential machine of Example 8.8.

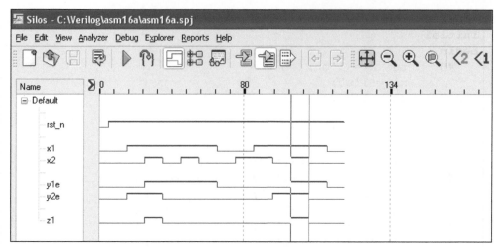

Figure 8.53   Waveforms for the asynchronous sequential machine of Example 8.8.

Example 8.9   This example repeats Example 8.8, but uses behavioral modeling to design the asynchronous sequential machine. The primitive flow table is duplicated in Figure 8.54 and shows the state code assignments that will be used in the behavioral module.

x_1x_2	00	01	11	10	z_1	
	(a)	f	$-$	b	0	(a) = 000
	a	$-$	c	(b)	0	(b) = 001
	$-$	$-$	(c)	d	1	(c) = 011
	a	$-$	e	(d)	0	(d) = 010
	$-$	$-$	(e)	d	0	(e) = 110
	a	(f)	g	$-$	0	(f) = 111
	$-$	$-$	(g)	b	0	(g) = 101

Figure 8.54   Primitive flow table for the asynchronous sequential machine of Example 8.9.

There is no need to obtain the merger diagram, the merged flow table, the transition diagram, the excitation maps, the output map, or the logic diagram. The behavior of the machine is specified and the logic design is accomplished by the synthesis tool.

The behavioral module is shown in Figure 8.55, in which state codes are assigned using the **parameter** keyword. The output is defined for each state and the **case** statement is used to determine the next state. The test bench is shown in Figure 8.56 — the primitive flow table can be used with the test bench to follow the machine through all of the state transition sequences.

The outputs are shown in Figure 8.57 and indicate the 3-bit state codes for the storage elements. The waveforms are shown in Figure 8.58 and display the same information as the waveforms of Figure 8.53, but in a slightly different form.

```verilog
//behavioral asynchronous sequential machine
module asm16 (rst_n, x1, x2, y, z1);

input rst_n, x1, x2;
output [1:3] y;
output z1;

wire rst_n, x1, x2;
reg [1:3] y, next_state;
reg z1;

//assign state codes
parameter    state_a = 3'b000,
             state_b = 3'b001,
             state_c = 3'b011,
             state_d = 3'b010,
             state_e = 3'b110,
             state_f = 3'b111,
             state_g = 3'b101;

//latch next state
always @ (x1 or x2 or rst_n)
begin
   if (~rst_n)
      y <= state_a;
   else
      y <= next_state;
end

//continued on next page
```

Figure 8.55   Behavioral module for the asynchronous sequential machine of Example 8.9.

```
//define output z1
always @ (x1 or x2 or y)
begin
   if (y == state_a)
      z1 = 1'b0;

   if (y == state_b)
      z1 = 1'b0;

   if (y == state_c)
      z1 = 1'b1;

   if (y == state_d)
      z1 = 1'b0;

   if (y == state_e)
      z1 = 1'b0;

   if (y == state_f)
      z1 = 1'b0;

   if (y == state_g)
      z1 = 1'b0;
end

//determine next state
always @ (x1 or x2)
begin
   case (y)
      state_a:    //== is logical equality; && is logical AND
         if ((x1==1'b1) && (x2==1'b0))
            next_state = state_b;
         else if ((x1==1'b0) && (x2==1'b1))
            next_state = state_f;
         else
            next_state = state_a;

      state_b:
         if ((x1==1'b0) && (x2==1'b0))
            next_state = state_a;
         else if ((x1==1'b1) && (x2==1'b1))
            next_state = state_c;
         else
            next_state = state_b;

//continued on next page
```

Figure 8.55   (Continued)

```
      state_c:
         if ((x1==1'b1) && (x2==1'b0))
            next_state = state_d;
         else
            next_state = state_c;


      state_d:
         if ((x1==1'b0) && (x2==1'b0))
            next_state = state_a;
         else if ((x1==1'b1) && (x2==1'b1))
            next_state = state_e;
         else
            next_state = state_d;


      state_e:
         if ((x1==1'b1) && (x2==1'b0))
            next_state = state_d;
         else
            next_state = state_e;


      state_f:
         if ((x1==1'b0) && (x2==1'b0))
            next_state = state_a;
         else if ((x1==1'b1) && (x2==1'b1))
            next_state = state_g;
         else
            next_state = state_f;


      state_g:
         if ((x1==1'b1) && (x2==1'b0))
            next_state = state_b;
         else
            next_state = state_g;

      default: next_state = state_a;

   endcase
end

endmodule
```

Figure 8.55   (Continued)

```
//test bench for asynchronous sequential machine
module asm16_tb;

reg rst_n, x1, x2;
wire [1:3] y;
wire z1;

//display variables
initial
$monitor ("x1x2 = %b, y = %b, z1 = %b",
          {x1, x2}, y, z1);

//define input vectors
initial
begin
   #0    rst_n = 1'b0;
         x1 = 1'b0;
         x2 = 1'b0;

   #5    rst_n = 1'b1;

   #10   x1=1'b1;    x2=1'b0;    //go to state_b
   #10   x1=1'b1;    x2=1'b1;    //go to state_c; assert z1
   #10   x1=1'b1;    x2=1'b0;    //go to state_d
   #10   x1=1'b1;    x2=1'b1;    //go to state_e
   #10   x1=1'b1;    x2=1'b0;    //go to state_d
   #10   x1=1'b0;    x2=1'b0;    //go to state_a
   #10   x1=1'b0;    x2=1'b1;    //go to state_f
   #10   x1=1'b1;    x2=1'b1;    //go to state_g
   #10   x1=1'b1;    x2=1'b0;    //go to state_b
   #10   x1=1'b1;    x2=1'b1;    //go to state_c; assert z1
   #10   x1=1'b1;    x2=1'b0;    //go to state_d
   #10   x1=1'b0;    x2=1'b0;    //go to state_a

   #10   $stop;
end

//instantiate the module into the test bench
asm16 inst1 (
   .rst_n(rst_n),
   .x1(x1),
   .x2(x2),
   .y(y),
   .z1(z1)
   );

endmodule
```

Figure 8.56   Test bench for the asynchronous sequential machine of Example 8.9.

```
x1x2 = 00, y = 000, z1 = 0
x1x2 = 10, y = 001, z1 = 0
x1x2 = 11, y = 011, z1 = 1
x1x2 = 10, y = 010, z1 = 0
x1x2 = 11, y = 110, z1 = 0
x1x2 = 10, y = 010, z1 = 0
x1x2 = 00, y = 000, z1 = 0
x1x2 = 01, y = 111, z1 = 0
x1x2 = 11, y = 101, z1 = 0
x1x2 = 10, y = 001, z1 = 0
x1x2 = 11, y = 011, z1 = 1
x1x2 = 10, y = 010, z1 = 0
x1x2 = 00, y = 000, z1 = 0
```

Figure 8.57   Outputs for the asynchronous sequential machine of Example 8.9.

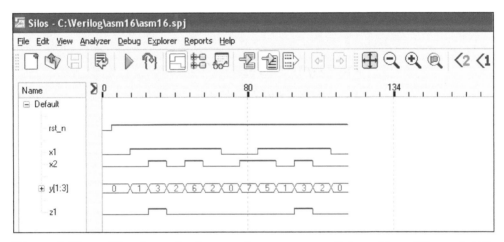

Figure 8.58   Waveforms for the asynchronous sequential machine of Example 8.9.

Example 8.10   This example repeats the asynchronous sequential machine of Example 8.8, but uses structural modeling in the implementation. The primitive flow table is reproduced in Figure 8.59 so that the operation of the machine can be easily followed as it proceeds through the state transition sequences. The various design steps will not be reproduced — they can be examined by referring to Example 8.8.

The excitation equations for Y_{1e} and Y_{2e} and the output equation for z_1 are replicated in Equation 8.9 and Equation 8.10, respectively. The logic diagram is shown in Figure 8.60 and displays the instantiation names and the net names which will be used in the structural module.

x_1x_2	00	01	11	10	z_1
	ⓐ	f	–	b	0
	a	–	c	ⓑ	0
	–	–	ⓒ	d	1
	a	–	e	ⓓ	0
	–	–	ⓔ	d	0
	a	ⓕ	g	–	0
	–	–	ⓖ	b	0

Figure 8.59   Primitive flow table for the asynchronous sequential machine of Example 8.10.

$$Y_{1e} = y_{1f}x_1 + y_{2f}x_2$$

$$Y_{2e} = y_{2f}x_2 + y_{1f}'x_1x_2' + y_{1f}'y_{2f}x_1 \qquad (8.11)$$

$$z_1 = y_{1f}y_{2f} \qquad (8.12)$$

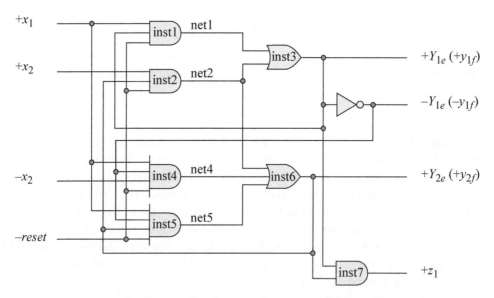

Figure 8.60   Logic diagram for the asynchronous machine of Example 8.10.

The structural module is shown in Figure 8.61 using dataflow logic primitives that were previously designed. The instantiation names and net names correlate to the instantiation names and net names indicated in the logic diagram. The test bench module is shown in Figure 8.62. The input sequence takes the machine through the state transitions shown in the primitive flow table. The outputs and waveforms are shown in Figure 8.63 and Figure 8.64, respectively.

```verilog
//structural asynchronous sequential machine 16b
module asm16b (rst_n, x1, x2, y1e, y2e, z1);

input rst_n, x1, x2;
output y1e, y2e;
output z1;

//define internal nets
wire net1, net2, net4, net5;

//instantiate the logic for latch y1e
and3_df inst1 (
   .x1(y1e),
   .x2(x1),
   .x3(rst_n),
   .z1(net1)
   );

and3_df inst2 (
   .x1(y2e),
   .x2(x2),
   .x3(rst_n),
   .z1(net2)
   );

or2_df inst3 (
   .x1(net1),
   .x2(net2),
   .z1(y1e)
   );

//instantiate the logic for latch y2e
and4_df inst4 (
   .x1(~y1e),
   .x2(x1),
   .x3(~x2),
   .x4(rst_n),
   .z1(net4)
   );                  //continued on next page
```

Figure 8.61   Structural module for the asynchronous machine of Example 8.10.

```
and4_df inst5 (
   .x1(~y1e),
   .x2(x1),
   .x3(y2e),
   .x4(rst_n),
   .z1(net5)
   );

or3_df inst6 (
   .x1(net2),
   .x2(net4),
   .x3(net5),
   .z1(y2e)
   );

//instantiate the logic for output z1
and2_df inst7 (
   .x1(y1e),
   .x2(y2e),
   .z1(z1)
   );

endmodule
```

Figure 8.61   (Continued)

```
//test bench for asynchronous sequential machine 16b
//use the primitive flow table to follow the input sequence
module asm16b_tb;

reg rst_n, x1, x2;
wire y1e, y2e;
wire z1;

initial      //display variables
$monitor ("x1x2 = %b, state = %b, z1 = %b",
          {x1, x2}, {y1e, y2e}, z1);

initial      //define input vectors
begin
   #0    rst_n = 1'b0;
         x1 = 1'b0;
         x2 = 1'b0;
   #5    rst_n = 1'b1;
//continued on next page
```

Figure 8.62   Test bench for the asynchronous machine of Example 8.10.

```
   #10    x1=1'b1;    x2=1'b0;    //go to state_b
   #10    x1=1'b1;    x2=1'b1;    //go to state_c; assert z1
   #10    x1=1'b1;    x2=1'b0;    //go to state_d
   #10    x1=1'b1;    x2=1'b1;    //go to state_e
   #10    x1=1'b1;    x2=1'b0;    //go to state_d
   #10    x1=1'b0;    x2=1'b0;    //go to state_a
   #10    x1=1'b0;    x2=1'b1;    //go to state_f
   #10    x1=1'b1;    x2=1'b1;    //go to state_g
   #10    x1=1'b1;    x2=1'b0;    //go to state_b
   #10    x1=1'b1;    x2=1'b1;    //go to state_c; assert z1
   #10    x1=1'b1;    x2=1'b0;    //go to state_d
   #10    x1=1'b0;    x2=1'b0;    //go to state_a

   #10    $stop;
end

//instantiate the module into the test bench
asm16b inst1 (
   .rst_n(rst_n),
   .x1(x1),
   .x2(x2),
   .y1e(y1e),
   .y2e(y2e),
   .z1(z1)
   );

endmodule
```

Figure 8.62   (Continued)

```
x1x2 = 00, state = 00, z1 = 0
x1x2 = 10, state = 01, z1 = 0
x1x2 = 11, state = 11, z1 = 1
x1x2 = 10, state = 10, z1 = 0
x1x2 = 11, state = 10, z1 = 0
x1x2 = 10, state = 10, z1 = 0
x1x2 = 00, state = 00, z1 = 0
x1x2 = 01, state = 00, z1 = 0
x1x2 = 11, state = 00, z1 = 0
x1x2 = 10, state = 01, z1 = 0
x1x2 = 11, state = 11, z1 = 1
x1x2 = 10, state = 10, z1 = 0
x1x2 = 00, state = 00, z1 = 0
```

Figure 8.63   Outputs for the asynchronous machine of Example 8.10.

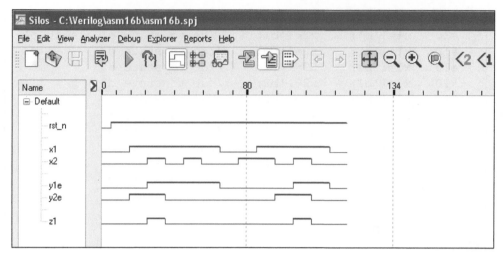

Figure 8.64   Waveforms for the asynchronous machine of Example 8.10.

8.3   Pulse-Mode Asynchronous Sequential Machines

Before beginning the design examples of pulse-mode asynchronous sequential machines, a brief review will be presented. As mentioned previously, each variable of the input alphabet is in the form of a pulse. The duration of each pulse is less than the propagation delay of the storage elements and the associated logic gates. Thus, an input pulse will initiate a state change, but the completion of the change will not take place until after the corresponding input has been deasserted. Multiple inputs cannot be active simultaneously. There is no separate clock input in pulse-mode machines.

Unlike a system clock, which has a specified frequency, the input pulses can occur randomly and more than one input pulse can generate an output. If the input pulse is of insufficient duration, then the storage elements may not be triggered and the machine will not sequence to the next state. If the pulse duration is too long, then the pulse will still be active when the machine changes from the present state $Y_{j(t)}$ to the next state $Y_{k(t+1)}$. The storage elements may then be triggered again and sequence the machine to an incorrect next state. If the time between consecutive pulses is too short, then the machine will be triggered while in an unstable condition, resulting in unpredictable behavior.

The pulse width restrictions that are dominant in pulse-mode sequential machines can be eliminated by including D flip-flops in the feedback path from the SR latches to the δ next-state logic. Providing edge-triggered D flip-flops as a constituent part of the implementation negates the requirement of precisely controlled input pulse durations. This is by far the most reliable means of synthesizing pulse-mode machines. The SR latches — in conjunction with the D flip-flops — form a master-slave configuration.

Example 8.11 A Mealy pulse-mode sequential machine will be designed which has two inputs x_1 and x_2 and one output z_1. Output z_1 is asserted coincident with every second x_2 pulse, if and only if the pair of x_2 pulses is immediately preceded by an x_1 pulse.

A combination of dataflow modeling — using the continuous assignment statement — and structural modeling will be used in the implementation. The storage elements will consist of SR latches and D flip-flops in a master-slave configuration. The design will be implemented with NOR logic for the SR latches and the logic primitives. A representative timing diagram is shown in Figure 8.65 and the state diagram is shown in Figure 8.66.

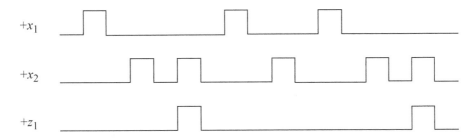

Figure 8.65 Representative timing diagram for the Mealy pulse-mode asynchronous sequential machine of Example 8.11.

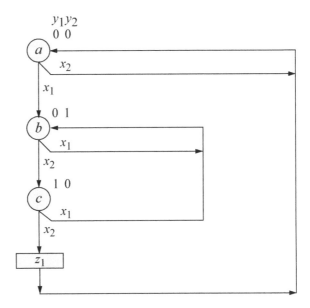

Figure 8.66 State diagram for the Mealy pulse-mode asynchronous sequential machine of Example 8.11.

The input maps and input equations are shown in Figure 8.67 and Equation 8.13, respectively. The output map and output equation are shown in Figure 8.68 and Equation 8.14, respectively. The logic diagram is shown in Figure 8.69, which instantiates positive-edge-triggered D flip-flops that were previously designed using behavioral modeling.

The mixed-design module — using dataflow and structural modeling is shown in Figure 8.70. The test bench is shown in Figure 8.71, which sequences the machine through all the state transitions shown in the state diagram. The outputs and waveforms are shown in Figure 8.72 and Figure 8.73, respectively.

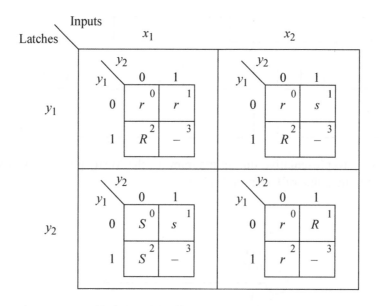

Figure 8.67   Input maps for the Mealy pulse-mode asynchronous sequential machine of Example 8.11.

$$SLy_1 = y_2 x_2$$

$$RLy_1 = x_1 + y_1 x_2$$

$$SLy_2 = x_1$$

$$RLy_2 = x_2 \qquad\qquad (8.13)$$

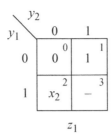

Figure 8.68 Output map for the Mealy pulse-mode asynchronous sequential machine of Example 8.11.

$$z_1 = y_1 x_2 \qquad (8.14)$$

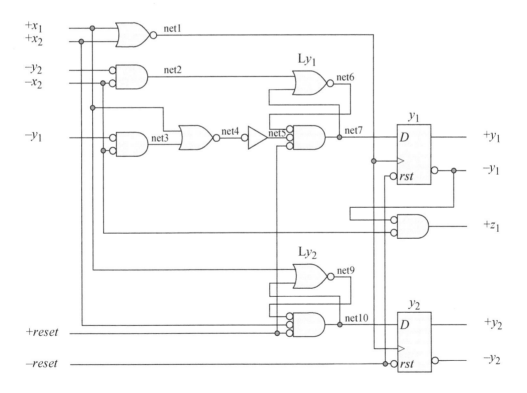

Figure 8.69 Logic diagram for the Mealy pulse-mode asynchronous sequential machine of Example 8.11.

```
//dataflow/structural mealy pulse-mode
//asynchronous sequential machine
module pm_asm7 (rst_n, rst, x1, x2, y1, y2, z1);

input rst_n, rst, x1, x2;
output y1, y2, z1;

//define internal nets
wire   net1, net2, net3, net4, net5,
       net6, net7, net9, net10;

//design the D flip-flop clock
assign net1 = ~(x1 | x2);

//design for latch Ly1
assign     net2 = ~(~y2 | ~x2),
           net3 = ~(~y1 | ~x2),
           net4 = ~(x1 | net3),
           net5 = ~net4,
           net6 = ~(net2 | net7),
           net7 = ~(net6 | net5 | rst);

//instantiate the D flip-flop for y1
d_ff_bh inst1 (
   .rst_n(rst_n),
   .clk(net1),
   .d(net7),
   .q(y1)
   );

//design for latch Ly2
assign     net9 = ~(x1 | net10),
           net10 = ~(net9 | x2 | rst);

//instantiate the D flip-flop for y2
d_ff_bh inst2 (
   .rst_n(rst_n),
   .clk(net1),
   .d(net10),
   .q(y2)
   );

//design the logic for output z1
assign     z1 = ~(~y1 | ~x2);

endmodule
```

Figure 8.70   Mixed-design module for the Mealy pulse-mode asynchronous sequential machine of Example 8.11.

```
//test bench for mealy pulse-mode asynchronous machine
module pm_asm7_tb;

reg rst_n, rst, x1, x2;
wire y1, y2, z1;

initial                        //display inputs and output
$monitor ("x1x2 = %b, state = %b, z1 = %b",
          {x1, x2}, {y1, y2}, z1);

initial                        //define input sequence
begin
   #0    rst_n = 1'b0;        //reset to state_a (00)
         rst = 1'b1;
         x1 = 1'b0;   x2 = 1'b0;
   #5    rst_n = 1'b1;
         rst = 1'b0;

   #10   x1 = 1'b0;   x2 = 1'b1;
   #10   x1 = 1'b0;   x2 = 1'b0;   //remain in state_a (00)

   #10   x1 = 1'b1;   x2 = 1'b0;
   #10   x1 = 1'b0;   x2 = 1'b0;   //go to state_b (01)

   #10   x1 = 1'b1;   x2 = 1'b0;
   #10   x1 = 1'b0;   x2 = 1'b0;   //remain in state_b (01)

   #10   x1 = 1'b0;   x2 = 1'b1;
   #10   x1 = 1'b0;   x2 = 1'b0;   //go to state_c (10)

   #10   x1 = 1'b0;   x2 = 1'b1;
   #10   x1 = 1'b0;   x2 = 1'b0;   //go to state_a (00); set z1

   #10   x1 = 1'b1;   x2 = 1'b0;
   #10   x1 = 1'b0;   x2 = 1'b0;   //go to state_b (01)

   #10   x1 = 1'b0;   x2 = 1'b1;
   #10   x1 = 1'b0;   x2 = 1'b0;   //go to state_c (10)

   #10   x1 = 1'b0;   x2 = 1'b1;
   #10   x1 = 1'b0;   x2 = 1'b0;   //go to state_a (00); set z1

   #10   x1 = 1'b0;   x2 = 1'b1;
   #10   x1 = 1'b0;   x2 = 1'b0;   //remain in state_a (00)
   #10   $stop;
end                            //continued on next page
```

Figure 8.71   Test bench for the Mealy pulse-mode asynchronous sequential machine of Example 8.11.

```
//instantiate the module into the test bench
pm_asm7 inst1 (
    .rst_n(rst_n),
    .rst(rst),
    .x1(x1),
    .x2(x2),
    .y1(y1),
    .y2(y2),
    .z1(z1)
    );
endmodule
```

Figure 8.71   (Continued)

```
x1x2 = 00, state = 00, z1 = 0   x1x2 = 00, state = 00, z1 = 0
x1x2 = 01, state = 00, z1 = 0   x1x2 = 10, state = 00, z1 = 0
x1x2 = 00, state = 00, z1 = 0   x1x2 = 00, state = 01, z1 = 0
x1x2 = 10, state = 00, z1 = 0   x1x2 = 01, state = 01, z1 = 0
x1x2 = 00, state = 01, z1 = 0   x1x2 = 00, state = 10, z1 = 0
x1x2 = 10, state = 01, z1 = 0   x1x2 = 01, state = 10, z1 = 1
x1x2 = 00, state = 01, z1 = 0   x1x2 = 00, state = 00, z1 = 0
x1x2 = 01, state = 01, z1 = 0   x1x2 = 01, state = 00, z1 = 0
x1x2 = 00, state = 10, z1 = 0   x1x2 = 00, state = 00, z1 = 0
x1x2 = 01, state = 10, z1 = 1
```

Figure 8.72   Outputs for the Mealy pulse-mode asynchronous sequential machine of Example 8.11.

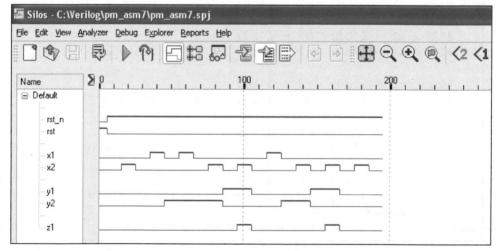

Figure 8.73   Waveforms for the Mealy pulse-mode asynchronous sequential machine of Example 8.11.

Example 8.12 A Moore pulse-mode asynchronous sequential machine will be designed which has three inputs x_1, x_2, and x_3 and one output z_1. Output z_1 will be asserted coincident with the assertion of the x_3 pulse if and only if the x_3 pulse was preceded by an x_1 pulse followed by an x_2 pulse. That is, the input vector must be $x_1 x_2 x_3 = 100, 000, 010, 000, 001$ to assert z_1. Output z_1 will be deasserted at the next active x_1 pulse or x_2 pulse. A representative timing diagram is shown in Figure 8.74.

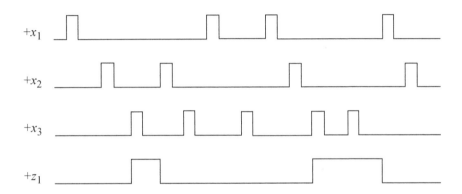

Figure 8.74 Timing diagram for the Moore pulse-mode asynchronous sequential machine of Example 8.12.

The machine will be implemented with NOR *SR* latches and positive-edge-triggered *D* flip-flops as the storage elements in a master-slave configuration. The design module will consist of dataflow modeling and structural modeling. The dataflow construct uses the continuous assignment statement **assign** for the combinational logic primitives, which consist of AND gates, OR gates, and NOR gates. The structural modeling method will instantiate positive-edge-triggered *D* flip-flops that were previously designed using behavioral modeling.

The state diagram is shown in Figure 8.75. The machine remains in state *a* until the first x_1 pulse occurs, which is the correct input to begin a sequence to assert output z_1. As mentioned previously, the state diagram does not contain any complemented variables because a pulse-mode machine with three inputs can have only

$$n + 1 = 3 + 1 = 4$$

combinations of the input variables, including $x_1 x_2 x_3 = 000$, instead of the 2^n combinations as in synchronous sequential machines.

The input maps are shown in Figure 8.76, where *S* indicates that the latch will be set; *s* indicates that the latch will remain set; *R* indicates that the latch will be reset; and *r* indicates that the latch will remain reset. The input equations are shown in Equation 8.13.

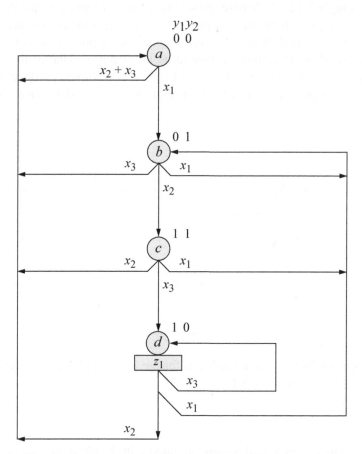

Figure 8.75   State diagram for the Moore pulse-mode machine of Example 8.12.

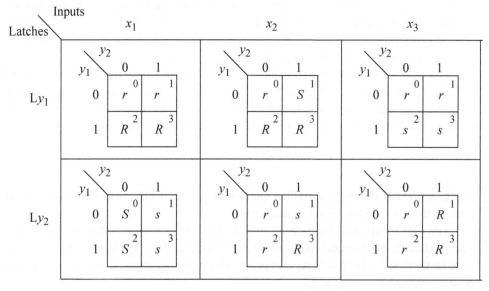

Figure 8.76   Input maps for the Moore pulse-mode machine of Example 8.12.

$$SLy_1 = y_1'y_2x_2 \qquad\qquad RLy_1 = x_1 + y_1x_2$$

$$SLy_2 = x_1 \qquad\qquad RLy_2 = y_1x_2 + x_3 \qquad\qquad (8.15)$$

The output map and equation are shown in Figure 8.77 and Equation 8.16, respectively. The logic diagram is shown in Figure 8.78.

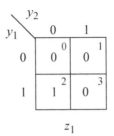

Figure 8.77   Output map for the Moore pulse-mode machine of Example 8.12.

$$z_1 = y_1y_2' \qquad\qquad (8.16)$$

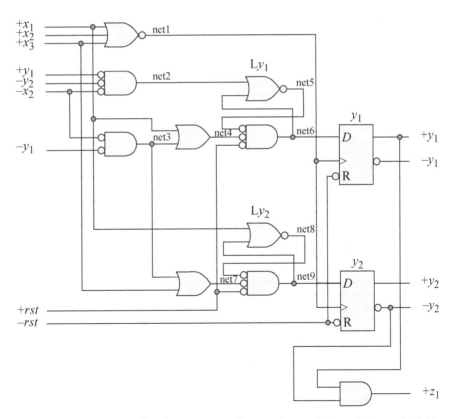

Figure 8.78   Logic diagram for the Moore pulse-mode machine of Example 8.12.

The mixed-design module is shown in Figure 8.79, where the continuous assignment statement is used to design the primitive logic gates by assigning the logic function to the net names. The test bench is shown in Figure 8.80, which sequences the machine through the different paths in the state diagram. The outputs and waveforms are shown in Figure 8.81 and Figure 8.82, respectively.

```
//dataflow/structural moore pulse-mode
//asynchronous sequential machine
module pm_asm11 (rst_n, rst, x1, x2, x3, y, z1);

input rst_n, rst, x1, x2, x3;
output [1:2] y;
output z1;

//define internal nets
wire   net1, net2, net3, net4, net5, net6,
       net7, net8, net9;

//design for the D flip-flops clock
assign net1 = ~(x1 | x2 | x3);

//design for latch Ly1
assign   net2 = ~(y[1] | ~y[2] | ~x2),
         net3 = ~(~y[1] | ~x2),
         net4 = (net3 | x1),
         net5 = ~(net2 | net6),
         net6 = ~(net5 | net4 | rst);

//instantiate the D flip-flop for y[1]
d_ff_bh inst1 (
   .rst_n(rst_n),
   .clk(net1),
   .d(net6),
   .q(y[1])
   );

//design for latch for Ly2
assign   net7 = (net3 | x3),
         net8 = ~(x1 | net9),
         net9 = ~(net8 | net7 | rst);

//continued on next page
```

Figure 8.79   Mixed-design module for the Moore pulse-mode machine of Example 8.12.

```
//instantiate the D flip-flop for y[2]
d_ff_bh inst2 (
   .rst_n(rst_n),
   .clk(net1),
   .d(net9),
   .q(y[2])
   );

//design the logic for output z1
assign   z1 = y[1] & ~y[2];

endmodule
```

Figure 8.79   (Continued)

```
//test bench for moore pulse-mode
//asynchronous sequential machine
module pm_asm11_tb;

reg rst_n, rst, x1, x2, x3;
wire [1:2] y;
wire z1;

//display inputs and output
initial
$monitor ("x1x2x3 = %b, state = %b, z1 = %b",
          {x1, x2, x3}, y, z1);

//define input sequence
initial
begin
   #0    rst_n = 1'b0;      //reset to state_a (00)
         rst = 1'b1;
         x1 = 1'b0;
         x2 = 1'b0;
         x3 = 1'b0;

   #5    rst_n = 1'b1;
         rst = 1'b0;

   #10   x1 = 1'b0;   x2 = 1'b1;   x3 = 1'b0;
   #10   x1 = 1'b0;   x2 = 1'b0;   x3 = 1'b0; //to state_a

//continued on next page
```

Figure 8.80   Test bench for the Moore pulse-mode machine of Example 8.12.

```
      #10    x1 = 1'b0;   x2 = 1'b0;   x3 = 1'b1;
      #10    x1 = 1'b0;   x2 = 1'b0;   x3 = 1'b0; //to state_a

      #10    x1 = 1'b1;   x2 = 1'b0;   x3 = 1'b0;
      #10    x1 = 1'b0;   x2 = 1'b0;   x3 = 1'b0; //to state_b

      #10    x1 = 1'b1;   x2 = 1'b0;   x3 = 1'b0;
      #10    x1 = 1'b0;   x2 = 1'b0;   x3 = 1'b0; //to state_b

      #10    x1 = 1'b0;   x2 = 1'b0;   x3 = 1'b1;
      #10    x1 = 1'b0;   x2 = 1'b0;   x3 = 1'b0; //to state_a

      #10    x1 = 1'b1;   x2 = 1'b0;   x3 = 1'b0;
      #10    x1 = 1'b0;   x2 = 1'b0;   x3 = 1'b0; //to state_b

      #10    x1 = 1'b0;   x2 = 1'b1;   x3 = 1'b0;
      #10    x1 = 1'b0;   x2 = 1'b0;   x3 = 1'b0; //to state_c

      #10    x1 = 1'b1;   x2 = 1'b0;   x3 = 1'b0;
      #10    x1 = 1'b0;   x2 = 1'b0;   x3 = 1'b0; //to state_b

      #10    x1 = 1'b0;   x2 = 1'b1;   x3 = 1'b0;
      #10    x1 = 1'b0;   x2 = 1'b0;   x3 = 1'b0; //to state_c

      #10    x1 = 1'b0;   x2 = 1'b1;   x3 = 1'b0;
      #10    x1 = 1'b0;   x2 = 1'b0;   x3 = 1'b0; //to state_a

      #10    x1 = 1'b1;   x2 = 1'b0;   x3 = 1'b0;
      #10    x1 = 1'b0;   x2 = 1'b0;   x3 = 1'b0; //to state_b

      #10    x1 = 1'b0;   x2 = 1'b1;   x3 = 1'b0;
      #10    x1 = 1'b0;   x2 = 1'b0;   x3 = 1'b0; //to state_c

      #10    x1 = 1'b0;   x2 = 1'b0;   x3 = 1'b1;
      #10    x1 = 1'b0;   x2 = 1'b0;   x3 = 1'b0; //to state_d); z1

      #10    x1 = 1'b1;   x2 = 1'b0;   x3 = 1'b0;
      #10    x1 = 1'b0;   x2 = 1'b0;   x3 = 1'b0; //to state_b

      #10    x1 = 1'b0;   x2 = 1'b1;   x3 = 1'b0;
      #10    x1 = 1'b0;   x2 = 1'b0;   x3 = 1'b0; //to state_c

      #10    x1 = 1'b0;   x2 = 1'b0;   x3 = 1'b1;
      #10    x1 = 1'b0;   x2 = 1'b0;   x3 = 1'b0; //to state_d; z1

//continued on next page
```

Figure 8.80   (Continued)

```
    #10     x1 = 1'b0;   x2 = 1'b0;   x3 = 1'b1;
    #10     x1 = 1'b0;   x2 = 1'b0;   x3 = 1'b0;   //to state_d; z1

    #10     x1 = 1'b0;   x2 = 1'b0;   x3 = 1'b1;
    #10     x1 = 1'b0;   x2 = 1'b0;   x3 = 1'b0;   //to state_d; z1

    #10     x1 = 1'b0;   x2 = 1'b1;   x3 = 1'b0;
    #10     x1 = 1'b0;   x2 = 1'b0;   x3 = 1'b0;   //to state_a
    #20     $stop;
end

//instantiate the module into the test bench
pm_asm11 inst1 (
    .rst_n(rst_n),
    .rst(rst),
    .x1(x1),
    .x2(x2),
    .x3(x3),
    .y(y),
    .z1(z1)
    );
endmodule
```

Figure 8.80   (Continued)

```
x1x2x3=000, state=00, z1=0      x1x2x3=000, state=00, z1=0
x1x2x3=010, state=00, z1=0      x1x2x3=100, state=00, z1=0
x1x2x3=000, state=00, z1=0      x1x2x3=000, state=01, z1=0
x1x2x3=001, state=00, z1=0      x1x2x3=010, state=01, z1=0
x1x2x3=000, state=00, z1=0      x1x2x3=000, state=11, z1=0
x1x2x3=100, state=00, z1=0      x1x2x3=001, state=11, z1=0
x1x2x3=000, state=01, z1=0      x1x2x3=000, state=10, z1=1
x1x2x3=100, state=01, z1=0      x1x2x3=100, state=10, z1=1
x1x2x3=000, state=01, z1=0      x1x2x3=000, state=01, z1 0
x1x2x3=001, state=01, z1=0      x1x2x3=010, state=01, z1=0
x1x2x3=000, state=00, z1=0      x1x2x3=000, state=11, z1=0
x1x2x3=100, state=00, z1=0      x1x2x3=001, state=11, z1=0
x1x2x3=000, state=01, z1=0      x1x2x3=000, state=10, z1=1
x1x2x3=010, state=01, z1=0      x1x2x3=001, state=10, z1=1
x1x2x3=000, state=11, z1=0      x1x2x3=000, state=10, z1=1
x1x2x3=100, state=11, z1=0      x1x2x3=001, state=10, z1=1
x1x2x3=000, state=01, z1=0      x1x2x3=000, state=10, z1=1
x1x2x3=010, state=01, z1=0      x1x2x3=010, state=10, z1=1
x1x2x3=000, state=11, z1=0      x1x2x3=000, state=00, z1=0
x1x2x3=010, state=11, z1=0
```

Figure 8.81   Outputs for the Moore pulse-mode machine of Example 8.12.

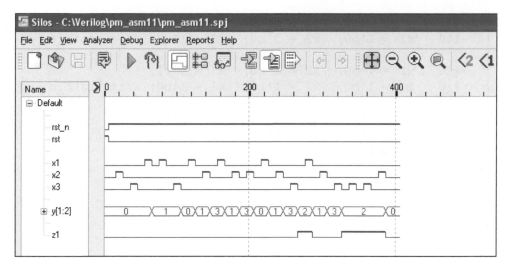

Figure 8.82   Waveforms for the Moore pulse-mode machine of Example 8.12.

Example 8.13   This example repeats Example 8.12, but uses only structural modeling and instantiates previously designed dataflow modules for the logic primitives and latches plus a behavioral module *D* flip-flop for the positive-edge-triggered *D* flip-flops.

The timing diagram, state diagram, input maps and equations, and the output map and equation are the same as those shown in Example 8.12 and will not be replicated. The logic diagram is reproduced in Figure 8.83 and shows the instantiation names and the net names that are used in the structural design module of Figure 8.84.

The test bench is identical except for the module name and the instantiated name of the design module. The outputs are also identical; therefore, the test bench and outputs will not be shown. The waveforms are the same as those of Example 8.12 and are shown in Figure 8.85 for comparison.

This section has presented an important class of asynchronous sequential machines in which pulses trigger state changes. The *SR* latch and *D* flip-flop in a master-slave relationship is by far the most reliable method to implement pulse-mode asynchronous sequential machines.

Other methods may also be used such as *SR* latches as storage elements without *D* flip-flops. As stated previously, the critical factor in the synthesis of pulse-mode sequential machines is controlling the pulse duration of the input signals. If the pulse duration is less than the minimal width to trigger a latch or greater than the maximal width that ensures only one state change, then the machine will not function reliably in a deterministic manner.

Another method to design pulse-mode machines is to use level-sensitive *T* flip-flops as storage elements that are designed with *SR* latches. A *T* flip-flop has one input *T* and two complementary outputs. If the flip-flop is reset, then an active pulse on the *T* input will toggle the flip-flop to the set state; if the flip-flop is set, then a pulse on the *T* input will toggle the flip-flop to the reset state.

Another flip-flop that is occasionally used in pulse-mode machines is the *SR-T* flip-flop. The *SR-T* flip-flop possesses the combined operational characteristics of both the *SR* latch and the *T* flip-flop. The *SR-T* flip-flop contains three inputs, set (S), reset (R), and toggle (T). If the flip-flop is reset, then a pulse on either the S input, the T input, or both will set the flip-flop. A pulse on the R input will have no effect. If the flip-flop is set, then a pulse on either the R input, the T input, or both will reset the flip-flop. A pulse on the S input will cause no change.

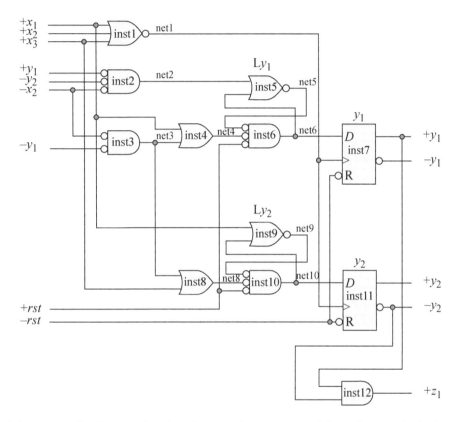

Figure 8.83   Logic diagram for the Moore pulse-mode machine of Example 8.13.

```
//structural moore pulse-mode
//asynchronous sequential machine
module pm_asm9a (rst_n, rst, x1, x2, x3, y, z1);

input rst_n, rst, x1, x2, x3;
output [1:2] y;
output z1;
//continued on next page
```

Figure 8.84   Structural module for the Moore pulse-mode machine of Example 8.13.

```verilog
//define internal nets
wire   net1, net2, net3, net4, net5, net6,
       net8, net9, net10;

//design for the D flip-flops clock
nor3_df inst1 (
   .x1(x1),
   .x2(x2),
   .x3(x3),
   .z1(net1)
   );

//instantiate the logic for latch Ly1
nor3_df inst2 (
   .x1(y[1]),
   .x2(~y[2]),
   .x3(~x2),
   .z1(net2)
   );

nor2_df inst3 (
   .x1(~y[1]),
   .x2(~x2),
   .z1(net3)
   );

or2_df inst4 (
   .x1(x1),
   .x2(net3),
   .z1(net4)
   );

nor2_df inst5 (
   .x1(net2),
   .x2(net6),
   .z1(net5)
   );

nor3_df inst6 (
   .x1(net5),
   .x2(net4),
   .x3(rst),
   .z1(net6)
   );

//continued on next page
```

Figure 8.84   (Continued)

```
//instantiate the D flip-flop for y[1]
d_ff_bh inst7 (
    .rst_n(rst_n),
    .clk(net1),
    .d(net6),
    .q(y[1])
    );

//instantiate the logic for latch Ly2
or2_df inst8 (
    .x1(x3),
    .x2(net3),
    .z1(net8)
    );

nor2_df inst9 (
    .x1(x1),
    .x2(net10),
    .z1(net9)
    );

nor3_df inst10 (
    .x1(net9),
    .x2(net8),
    .x3(rst),
    .z1(net10)
    );

//instantiate the D flip-flop for y[2]
d_ff_bh inst11 (
    .rst_n(rst_n),
    .clk(net1),
    .d(net10),
    .q(y[2])
    );

//instantiate the logic for output z1
and2_df inst12 (
    .x1(y[1]),
    .x2(~y[2]),
    .z1(z1)
    );

endmodule
```

Figure 8.84   (Continued)

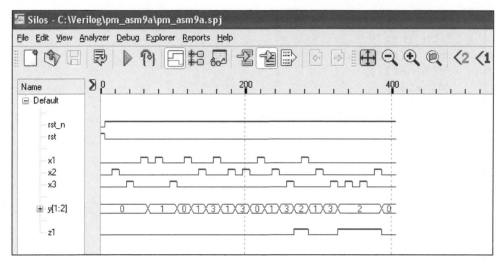

Figure 8.85   Waveforms for the Moore pulse-mode machine of Example 8.13.

8.4 Problems

8.1   Given the state diagram shown below for a Moore synchronous sequential machine, obtain the input maps for negative-edge-triggered JK flip-flops and the logic diagram. Then implement the machine using structural modeling. Obtain the test bench, the outputs, and the waveforms.

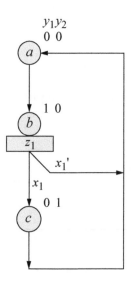

8.2   Use structural modeling to design the Mealy synchronous sequential machine that is represented by the state diagram shown below. Use a positive-edge-triggered *JK* flip-flop and additional AND gates and OR gates. Obtain the structural module, the test bench module, the outputs, and the waveforms.

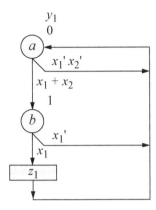

8.3   A state diagram is shown below for a synchronous sequential machine with both Moore and Mealy outputs. Use behavioral modeling with the **parameter** keyword and the **case** statement to design the machine. The outputs are implemented with the continuous assignment construct.

Obtain the behavioral module using negative-edge-triggered flip-flops, the test bench module, the outputs, and the waveforms.

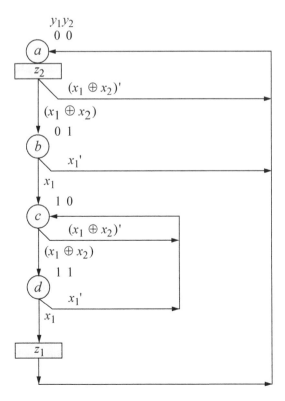

8.4   Design a modulo-10 counter with a self-starting state of $y_1 y_2 y_3 y_4 = 0000$; that is, any count that is not part of the modulo-10 counting sequence has a next state of $y_1 y_2 y_3 y_4 = 0000$. The counting sequence is as follows: $y_1 y_2 y_3 y_4 = 0000, 0001, 0010, 0011, 0100, 0101, 0110, 0111, 1000, 1001, 0000$. Obtain the input maps, the input equations, and the logic diagram.

Obtain the structural module using dataflow logic primitives of AND gates and OR gates plus positive-edge-triggered D flip-flops as the storage elements that are defined using behavioral modeling. Obtain the test bench module, the outputs, and waveforms.

8.5   Design a Moore synchronous sequential machine that generates the following sequence: $y_1 y_2 y_3 y_4 = 1000, 0100, 0010, 0001, 1000, \ldots$. The unused states can be treated as "don't care" states. Obtain the input maps, the input equations, and the logic diagram. Obtain the structural module using positive-edge-triggered D flip-flops with active-low set inputs and active-low reset inputs. Obtain the test bench module, the outputs, and the waveforms.

8.6   Given the state diagram shown below for a synchronous sequential machine, obtain the next-state table, the input maps for JK flip-flops, the input equations, and the logic diagram using AND gates and OR gates and positive-edge-triggered JK flip-flops.

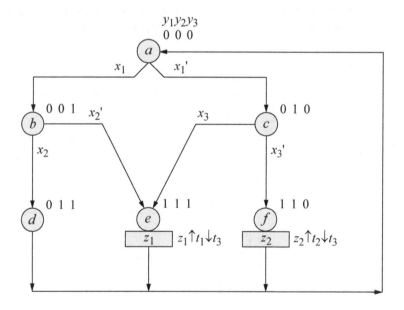

Since there may be a glitch on output z_2 for a transition from state c ($y_1 y_2 y_3 = 010$) to state e ($y_1 y_2 y_3 = 111$), the assertion/deassertion is $z_2 \uparrow t_2 \downarrow t_3$. Obtain the structural design module by instantiating dataflow modules for the logic primitives and a behavioral module for the positive-edge-triggered JK flip-flops. Obtain the test bench module, the outputs, and the waveforms.

8.7 Use behavioral modeling to design the Moore synchronous sequential machine shown in the state diagram below. All outputs are asserted at time t_2 and deasserted at time t_3. Obtain the test bench, the outputs, and the waveforms.

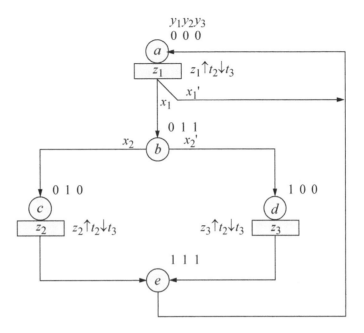

8.8 Given the state diagram shown below for a Mealy synchronous sequential machine, obtain next-state table, the input maps for *JK* flip-flops, the input equations, the output map, the output equation, and the logic diagram using NOR logic for the logic gates and positive-edge-triggered *JK* flip-flops for the storage elements.

Then design the machine using structural modeling and instantiate logic primitives that were designed using dataflow modeling. The *JK* flip-flops are designed using behavioral modeling. Obtain the test bench, the outputs, and the waveforms.

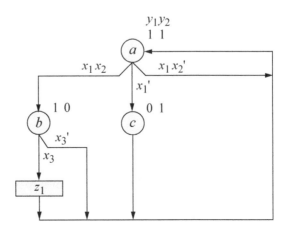

8.9   Determine the counting sequence for the counter shown below by implementing the machine in a structural module. The counter is reset initially to y_1y_2 =
00, where y_2 is the low-order flip-flop. Obtain the test bench, outputs, and
waveforms.

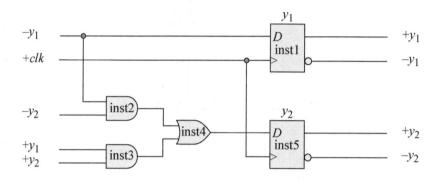

8.10   Determine the counting sequence for the counter shown below. The counter
is reset initially to $y_1y_2y_3$ = 000, where y_3 is the low-order flip-flop. Use
structural modeling in the implementation. Obtain the test bench, the outputs,
and the waveforms.

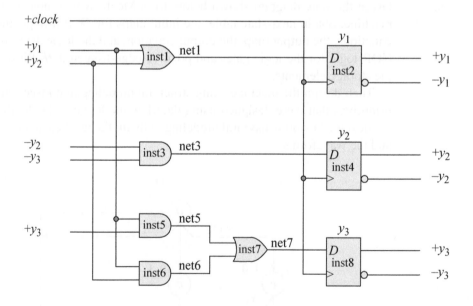

8.11   Given the state diagram shown below for a Moore synchronous sequential
machine with three inputs x_1, x_2, and x_3, derive the input maps for flip-flops
y_1, y_2, and y_3. Obtain the logic diagram using linear-select multiplexers,

logic primitives, and positive-edge-triggered D flip-flops as the storage elements.

The multiplexers are designed using behavioral modeling; the logic primitives are designed using dataflow modeling. The D flip-flops are designed using behavioral modeling. Outputs z_1 and z_2 have the following assertion/deassertion times: $\uparrow t_2 \downarrow t_3$. Obtain the structural module, the test bench module, the outputs, and the waveforms.

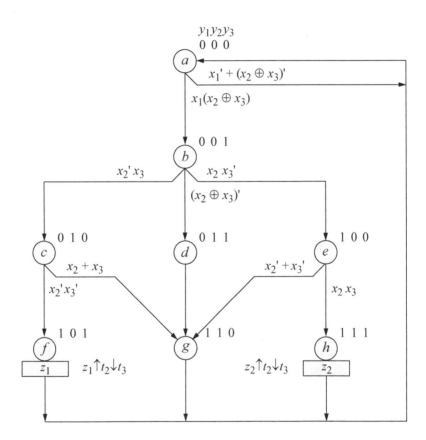

8.12   Given the state diagram shown below for a Moore synchronous sequential machine, implement the design using nonlinear-select multiplexers for the δ next-state logic together with logic primitives. Let flip-flops $y_1 y_3 = s_1 s_0$. Use positive-edge-triggered D flip-flops for the storage elements and a decoder for the λ output logic. Since the state codes may cause glitches on the outputs for some state transition sequences, z_1, z_2, and z_3 should be asserted at time t_2 and deasserted at time t_3. Obtain the structural module, the test bench module, the outputs, and the waveforms. Design the 4:1 multiplexer using built-in primitives; design the D flip-flops using behavioral modeling; design the decoder using dataflow modeling.

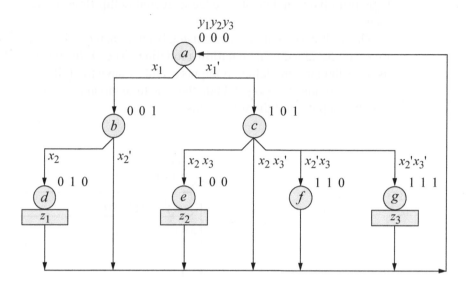

8.13   Design a Moore machine that has four parallel inputs x_1, x_2, x_3, and x_4 and
two outputs z_1 and z_2. Output $z_1\uparrow t_1 \downarrow t_3$; output $z_2 \uparrow t_2 \downarrow t_3$. The inputs constitute a 4-bit word $x[1:4]$. There is one bit space between words. The ma-
chine operates as follows:

   (a) Output $z_1 = 1$ if the 4-bit word contains an odd number of 1s.
   (b) Output $z_2 = 1$ if the 4-bit word contains an even number of 1s,
       including zero 1s.

Use behavioral modeling to implement the machine, then obtain the test
bench, the outputs, and the waveforms.

8.14   Design a Mealy asynchronous sequential machine using structural modeling
that generates an output z_1 whenever a serial data line x_1 contains a sequence
of three consecutive 1s. Overlapping sequences are allowed. Obtain the state
diagram, the input maps, and the logic diagram using AND gates and OR
gates for the logic primitives and positive-edge-triggered D flip-flops as the
storage elements. Obtain the structural module, the test bench module showing overlapping sequences, the outputs, and the waveforms.

8.15   This design replicates the asynchronous sequential machine of Example 7.12
and is to be designed using dataflow modeling. The machine has two inputs
x_1 and x_2 and one output z_1, as shown in the timing diagram below. Input x_1
acts as a gate for x_2; that is, x_2 will be gated to output z_1 only if x_1 precedes the
assertion of x_2. If x_1 becomes deasserted while x_2 is asserted, then the full
width of the x_2 pulse will appear on z_1 — the width of the x_2 pulse will not be
decreased.

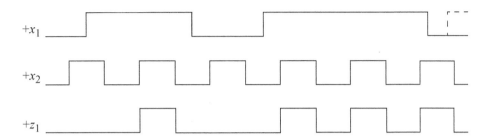

Refer to Example 7.12 for the design procedure for this problem. The resulting logic diagram is shown below with one excitation variable Y_{1e}. Obtain the design module using dataflow modeling, the test bench module, the outputs, and the waveforms.

8.16 Refer to Example 7.14 for the design procedure for this problem. The timing diagram for the asynchronous sequential machine is shown below with one input x_1 and two outputs z_1 and z_2. The outputs will be implemented with the least amount of logic which may not necessarily be the fastest response time.

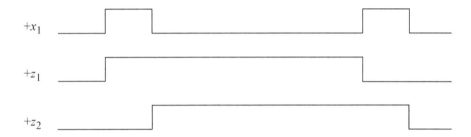

The primitive flow table and the logic diagram are shown below. The logic diagram is to be implemented using structural modeling. The logic primitives are to be designed using dataflow modeling. Obtain the structural design module, the test bench module, the outputs, and the waveforms.

x_1	0	1	z_1	z_2
	ⓐ	b	0	0
	c	ⓑ	1	0
	ⓒ	d	1	1
	a	ⓓ	0	1

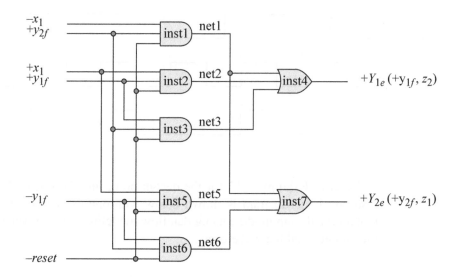

8.17   Design an asynchronous sequential machine which has two inputs x_1 and x_2 and one output z_1. Output z_1 is asserted for the duration of x_2 if and only if x_1 is already asserted. Assume that the initial state of the machine is $x_1 x_2 z_1$ = 000. A representative timing diagram is shown below. Design the machine using dataflow modeling, and obtain the test bench, the outputs, and the waveforms.

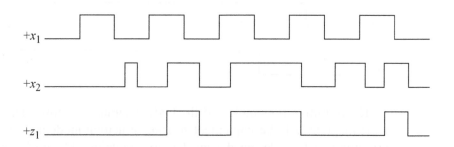

8.18   An asynchronous sequential machine has two inputs x_1 and x_2 and one output z_1. The machine operates according to the following specifications:

If $x_1 x_2$ = 00, then the state of z_1 is unchanged.
If $x_1 x_2$ = 01, then z_1 is deasserted.
If $x_1 x_2$ = 10, then z_1 is asserted.
If $x_1 x_2$ = 11, then z_1 changes state.

Design the machine using dataflow modeling. The inputs are available in both high and low assertion. Assume that the initial conditions are $x_1 x_2 z_1$ = 000. The output must change as fast as possible. There must be no output glitches. A representative timing diagram is shown below. Obtain the dataflow design module, the test bench module, the outputs, and the waveforms.

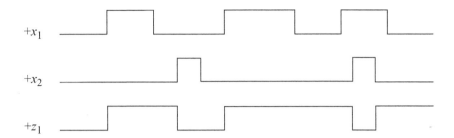

8.19   Given the state diagram shown below, design a Moore pulse-mode asynchronous sequential machine using *SR* latches and *D* flip-flops in a master-slave configuration. Obtain the input maps, the input equations, and the logic diagram. Use structural modeling for the design module by instantiating dataflow modules for the logic primitives and a behavioral module for the *D* flip-flops. Obtain the test bench, the outputs, and the waveforms.

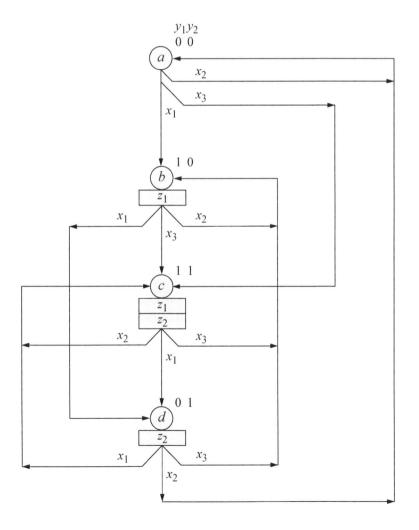

8.20   Design a Mealy machine that has three pulse input variables x_1, x_2, and x_3 and one output z_1 that is asserted coincident with x_3 whenever the sequence $x_1 x_2 x_3 = 100$, 010, 001 occurs. The storage elements will consist of SR latches and positive-edge-triggered D flip-flops in a master-slave configuration.

A representative timing diagram displaying valid input sequences and corresponding outputs is shown below. State code assignment is arbitrary for the state diagram, since input pulses trigger all state transitions and the machine does not begin to sequence to the next state until the input pulse, which initiated the transition, has been deasserted. Thus, output z_1 will not glitch. Obtain the input maps, the input equations, and the logic diagram. Then obtain the structural module using dataflow modeling for the logic primitives. Use NOR logic that is designed using dataflow modeling for the SR latches. Use behavioral modeling for the D flip-flop. Obtain the test bench, the outputs, and the waveforms.

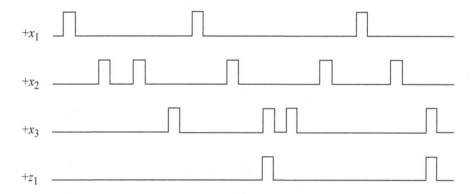

8.21   Design a Moore pulse-mode asynchronous sequential machine that has two inputs x_1 and x_2 and one output z_1. The negative transition of every second consecutive x_1 pulse will assert output z_1 as a level. The output will remain set for all following contiguous x_1 pulses. The output will be deasserted at the negative transition of the second of two consecutive x_2 pulses. The machine will be implemented with NAND logic for the δ next-state logic. The storage elements will be NAND SR latches and positive-edge-triggered D flip-flops in a master-slave configuration.

A representative timing diagram is shown. Generate a state diagram that depicts all possible state transition sequences that conform to the functional specifications. Obtain the input maps, the input equations, and the logic diagram.

Then design the structural module using dataflow modeling for the logic primitives and behavioral modeling for the positive-edge-triggered D flip-flops. Use dataflow modeling for the implementation of output z_1. Obtain the test bench, the outputs, and the waveforms.

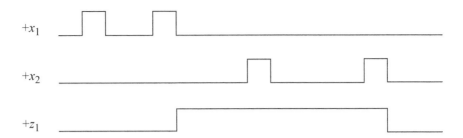

8.22 Given the state diagram shown below for a Moore pulse-mode asynchronous sequential machine, implement the machine using NAND gates for the *SR* latches and *D* flip-flops as the storage elements in a master-slave configuration. Use any type of gates for the logic primitives.

Derive the input maps and input equations, then obtain the logic diagram. Design the machine using dataflow and structural modeling. Use the continuous assignment construct for the logic primitives and latches; instantiate positive-edge-triggered *D* flip-flops that were designed using behavioral modeling.

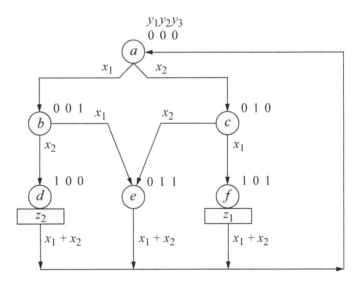

9

Programmable Logic Devices

There are three main types of programmable logic devices (PLDs): *programmable read-only memories* (PROMs), *programmable array logic* (PAL) devices, and *programmable logic array* (PLA) devices.

A related device is a *field-programmable gate array* (FPGA) which implements complex logic functions by means of a configuration program stored in an on-chip memory. The program establishes the interconnection of internal logic blocks and I/O blocks, thus allowing the same FPGA to be used in many different applications.

Another related device is the *application-specific integrated circuit* (ASIC), which is used to design a specific product or application. ASICs can be customized for a particular use such as a cell phone or other hand-held device, rather than for general-purpose use.

Current ASICs contain processors and memory and are designed using a hardware description language (HDL). These HDL software systems simplify the task of logic design using PLDs and also perform logic minimization and test vector generation for system simulation. The ultimate goal is to specify the input/output characteristics of a machine in a high-level language. The hardware-software system will then synthesize the machine to yield a minimized, functionally tested unit.

Programmable logic devices can be used in the synthesis (design) of both combinational and sequential logic networks. PLDs are prefabricated integrated circuits (ICs) in which fused and hard-wired interconnections are used and implement 2-level switching functions by means of an AND array and an OR array.

The basic organization of a PLD consists of an AND array driving an OR array as shown in Figure 9.1. There is a set of inputs X_i containing n input signals and a set of outputs Z_i containing m output signals. The amount of programming capability depends upon the type of PLD that is used. For example, a PROM contains a fixed AND array and a programmable OR array; a PAL contains a programmable AND array and a fixed OR array; a PLA contains both a programmable AND array and a programmable OR array. Both PAL and PLA architectures have versions which contain storage elements in conjunction with combinational logic.

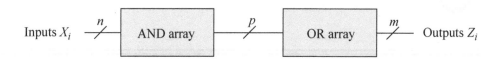

Figure 9.1   Basic organization of a programmable logic device.

The following sections illustrate the use of PLDs in the synthesis of combinational logic and synchronous sequential machines. The PLDs that will be presented are PROMs, PAL devices, PLAs, and FPGAs.

9.1 Programmable Read-Only Memory

A PROM is a storage device in which the information is permanently stored; that is, the data remains valid even after power is turned off. PROMs are used for application programs, tables, code conversion, control store for microprogram sequencers, and other functions in which the stored data is not changed. The organization of a PROM is essentially the same as that for other PLDs: an input vector (an address) connects to an AND array which in turn connects to an OR array which generates the output vector (or word) for the PROM.

The concept of read-only memories (ROMs) for sequential machine synthesis and processor control is quite common. ROMs are also used extensively in developing new microprogram-controlled systems. Various versions of programmable ROM (PROM) devices are available which significantly reduce the turnaround time required for modifications to existing machines.

Erasable PROMs (EPROMs) can be reprogrammed at the user's location, thereby enabling the designer to modify the stored program on site. There are two main types of EPROMs, distinguished by the method used for erasure: the electrically erasable PROM (EEPROM) in which the program is erased upon application of specific electrical signals and the ultraviolet erasable PROM (UVEPROM) in which the program is erased when exposed to ultraviolet radiation.

In general, a PROM contains n inputs and m outputs. Because the inputs function as an address, there are 2^n unique addresses to select one of 2^n words. The AND array decodes the address to select a specific word in memory. Thus, the interconnections in the AND array are fixed by the manufacturer and cannot be programmed by the user. The OR array, however, is programmable. The interconnections in the OR array are programmed by the user using special internal circuitry and a programming device to indicate the bit configuration of each word in memory. Each interconnection functions as a fuse; thus, the fuse can be left intact (indicating a logic 1) or opened (indicating a logic 0).

Figure 9.2 illustrates the internal organization of a PROM with two address inputs x_1 and x_2 and four outputs z_1, z_2, z_3, and z_4. Inputs x_1 and x_2 select one of four words using the AND array decoder: word 0, 1, 2, or 3 that corresponds to $x_1 x_2 = 00, 01, 10$, or 11, respectively. Thus, the AND array cannot be programmed, as indicated by the "hardwired" connection symbol " ● ." The OR array, however, is programmable. The symbol "×" indicates an intact fuse at the intersection of the AND gate product term and the OR gate input and provides a logic 1 to the specified OR gate input. The absence of an × indicates an open fuse, which provides a logic 0 to the OR gate input.

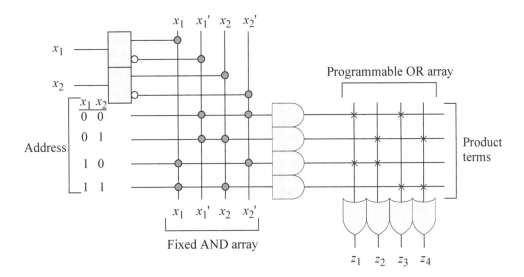

Figure 9.2   PROM organization with two address inputs x_1 and x_2 and four outputs z_1, z_2, z_3, and z_4.

9.1.1 Combinational Logic

All unused inputs in the AND array correspond to an open or *floating* input. Thus, all unused AND gate inputs must generate a logic 1 so that the output of the AND gate will be a function of the "hard-wired" connections only. The PROM illustrated in Figure 9.2 contains four words with four bits per word as shown in Table 9.1. When the

PROM address is $x_1 x_2 = 10$, the output word is $z_1 z_2 z_3 z_4 = 1100$, which may be a data constant depending on the application of the PROM at that address.

Table 9.1 PROM Program

Address $x_1 x_2$	Outputs $z_1 z_2 z_3 z_4$
0 0	1 0 1 0
0 1	0 1 0 1
1 0	1 1 0 0
1 1	0 0 1 1

Although a PROM is primarily used to store application programs, code conversions, tables, and data constants, it can also be used to implement logic functions. For example, output z_1 of Figure 9.2 is asserted when $z_1 = x_1' x_2' + x_1 x_2'$. The number of outputs can be increased without increasing the number of inputs. The two input variables in Figure 9.2 are decoded into four lines by means of four AND gates that represent the AND array. By convention, only one line is drawn as the input to each AND gate. This single line, however, represents four lines, one line each for x_1, x_1', x_2, and x_2'. The input buffer drivers generate the true and complement of each input variable. Each AND gate output represents one of the four minterms of two variables. The four outputs of the AND array decoder are connected through fuses to each OR gate. A PROM generates outputs that are in a sum-of-minterms form.

Example 9.1   A PROM will be used to implement combinational logic in a sum-of-minterms form. Consider the truth table of Table 9.2. There are two inputs that select one of four words, where each word consists of three bits z_1, z_2, and z_3. Each output can be represented as a sum-of-minterms. Although the functions can be more easily implemented with discrete gates, this example serves to illustrate the technique for PROM programming to implement combinational logic functions.

Table 9.2 PROM Programming for Example 9.1

Address $x_1 x_2$	Outputs $z_1 z_2 z_3$
0 0	1 1 0
0 1	0 1 1
1 0	1 0 1
1 1	0 0 0

Figure 9.3 shows the PROM organization and programming for Example 9.1. The fuses are shown either intact or open to correspond to the appropriate entries in the truth table, where a 1 and 0 specify an intact or open fuse, respectively. For example, z_1 is connected by intact fuses to AND gate outputs corresponding to $x_1'x_2' + x_1x_2'$. Any unused OR gate input must not contribute to the generated function; therefore, all open-fused OR gate inputs must provide a logic 0 to the gate. This example demonstrates the general procedure for implementing combinational logic in a PROM. The number of inputs and outputs determines the size of the PROM. The truth table is then generated, which specifies the programming requirements of the device. No minimization is necessary, since the size of the PROM is determined by the number of inputs and outputs.

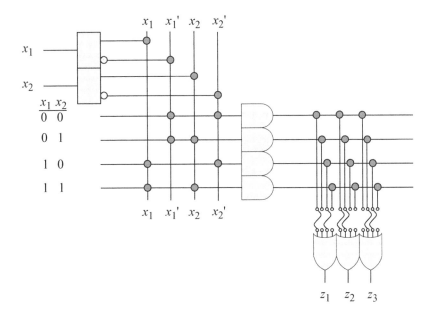

Figure 9.3   PROM organization and programming to implement the functions shown in Equation 9.1.

$$z_1 = x_1'x_2' + x_1x_2'$$

$$z_2 = x_1'x_2' + x_1'x_2$$

$$z_3 = x_1'x_2 + x_1x_2' \tag{9.1}$$

Example 9.2   The truth table to implement a sum-of-minterms combinational circuit is shown in Table 9.3. The equations that represent the sum of minterms are shown in Equation 9.2. The PROM organization and programming are shown in Figure 9.4.

The 4-bit output word can be easily expanded to a larger word by simply concatenating additional OR gates in the OR array. The size (or capacity) of a PROM is characterized by the expression $2^n \times m$, where n represents the number of inputs and m specifies the number of outputs. Therefore, n inputs requires 2^n AND gates and m outputs requires m OR gates.

Table 9.3 Truth Table for the PROM of Example 9.2

Address Inputs $x_1\ x_2$	Outputs $z_1\ z_2\ z_3\ z_4$
0 0	1 0 1 0
0 1	0 1 0 1
1 0	1 1 0 0
1 1	0 0 1 1

$$z_1\,(x_1, x_2) = \Sigma\,(0, 2)$$

$$z_2\,(x_1, x_2) = \Sigma\,(1, 2)$$

$$z_3\,(x_1, x_2) = \Sigma\,(0, 3)$$

$$z_4\,(x_1, x_2) = \Sigma\,(1, 3) \tag{9.2}$$

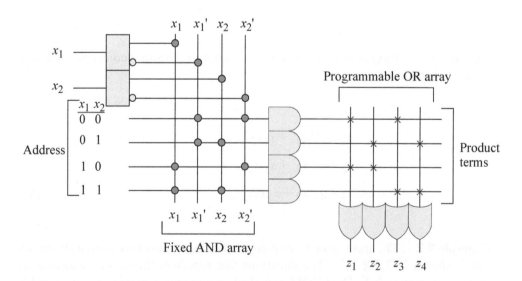

Figure 9.4   PROM organization and programming for Example 9.2.

9.1.2 Sequential Logic

PROMs provide a simple, yet elegant method of implementing synchronous sequential machines. A general block diagram of a PROM-controlled Moore sequential machine is shown in Figure 9.5. A Mealy machine can be portrayed by inserting λ output logic as a function of the input alphabet X. The PROM address inputs are comprised of the machine inputs $x_1, x_2, \dots, x_n$ concatenated with the outputs of the state flip-flops $y_1, y_2, \dots, y_p$. The outputs of the PROM connect to an $(m + p)$-stage register, which is clocked on the positive transition of the machine clock. The outputs, therefore, are asserted at time t_1 and deasserted at t_3. Other versions of output assertion and deassertion are possible by employing the complement of the machine clock or by including λ output logic which is enabled by a particular clock phase; for example, $z_i \uparrow t_2 \downarrow t_3$.

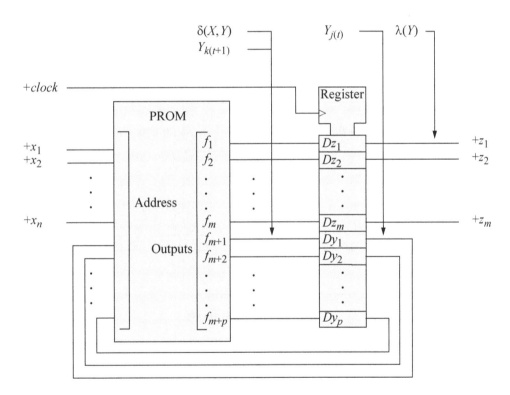

Figure 9.5   General block diagram of a $2^n \times (m + p)$ PROM-controlled Moore synchronous sequential machine.

The next state $Y_{k(t+1)}$ is generated by a unique word in the PROM as a function of the present inputs $X_{i(t)}$ and the present state $Y_{j(t)}$. The word that is addressed by $X_{i(t)} \cdot Y_{j(t)}$ appears at the outputs of the OR array as $f_1, f_2, \dots, f_m, f_{m+1}, \dots, f_{m+p}$. These outputs connect to the D inputs of the register such that f_i connects to Dz_i and f_{m+i} connects to Dy_i. The register is loaded from the PROM on the positive clock transition.

The present state is fed back to the PROM address inputs and, after an appropriate propagation delay through the PROM arrays, the next state and next outputs are available at the D inputs of the register. At the next positive clock transition, the next state is loaded into the register and becomes the present state.

Example 9.3   Given the state diagram of Figure 9.6, a Moore machine will be synthesized using a PROM for the δ next-state logic and a register for the state flip-flops and outputs, plus any additional logic that is necessary for the λ output logic. The machine has two serial inputs x_1 and x_2 and three outputs z_1, z_2, and z_3, which are asserted and deasserted as shown.

When the sequence $x_1 x_2 = 00$ occurs, z_1 is asserted; when the sequence $x_1 x_2 = 11$ occurs, z_2 is asserted. Output z_3 is asserted during the fourth clock period, regardless of the path taken, indicating the end of any sequence. Figure 9.7 depicts valid sequences to assert z_1 and z_2 and a sequence in which only z_3 is asserted.

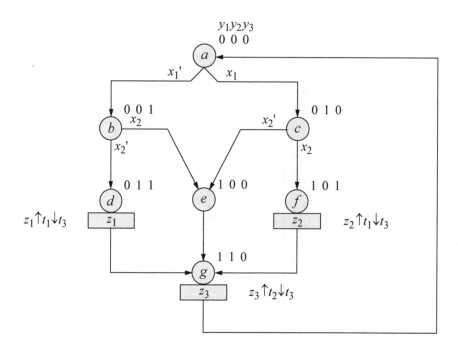

Figure 9.6   State diagram for a Moore synchronous sequential machine to detect a sequence of $x_1 x_2 = 00$ or 11 to assert output z_1 or z_2, respectively, and then to assert output z_3. There is one unused state: $y_1 y_2 y_3 = 111$.

The necessary conditions for two states to be equivalent are summarized as follows:

  1. The two states under consideration must have the same outputs.
  2. Both states must have the same or equivalent next state.

Both conditions must be true for equivalence to exist. The only potential equivalence is between states b and c. However, both states have different next states, depending upon the value of input x_2. Therefore, the state diagram contains no equivalent states.

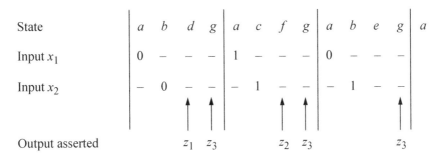

State	a	b	d	g	a	c	f	g	a	b	e	g	a
Input x_1	0	–	–	–	1	–	–	–	0	–	–	–	–
Input x_2	–	0	–	–	–	1	–	–	–	1	–	–	–
Output asserted			z_1	z_3			z_2	z_3				z_3	

Figure 9.7   Illustration of valid sequences to assert outputs z_1 and z_2 and a sequence where only z_3 is asserted.

The assignment of state codes is arbitrary in this example, since there is no need to minimize the δ next-state logic using state code adjacency — the input logic is implemented by the PROM program. Also, the λ output logic does not require specific state codes for minimization, since the outputs are a function of the PROM program, and are obtained directly from the register.

Table 9.4 specifies the PROM program for the Moore machine of Figure 9.6. The table represents the next-state table of the synthesis procedure. The table also substitutes as the input maps and output maps, respectively. Since there are two inputs and three storage elements, the PROM address consists of five input lines: x_1, x_2, y_1, y_2, and y_3.

Six outputs are required: three for the next state y_1, y_2, and y_3 and three for the present outputs z_1, z_2 and z_3. The size of the PROM, therefore, is $2^n \times m = 2^5 \times 6$; that is, there are 32 words with 6 bits per word. Only the first 28 words, however, need to be programmed, since $y_1 y_2 y_3 = 111$ is an unused state, providing four unused PROM words: $y_1 y_2 y_3 x_1 x_2 = 11100$ through 11111.

Output glitches are not possible in this type of PROM implementation, because the PROM outputs are stable at the D inputs of the register long before the positive clock transition occurs. Thus, the output flip-flops will never enter a metastable condition.

The PROM is programmed directly from the state diagram. For example, refer to Table 9.4 in conjunction with the state diagram. In state a ($y_1 y_2 y_3 = 000$), the next state is b ($y_1 y_2 y_3 = 001$) if $x_1 = 0$, and c ($y_1 y_2 y_3 = 010$) if $x_1 = 1$. State a provides no outputs. In state c ($y_1 y_2 y_3 = 010$), the next state is e ($y_1 y_2 y_3 = 100$) if $x_2 = 0$, and f ($y_1 y_2 y_3 = 101$) if $x_2 = 1$. State c does not generate an output. The next state for state d ($y_1 y_2 y_3 = 011$) is state g ($y_1 y_2 y_3 = 110$), regardless of the values of x_1 and x_2. In state d, however, output z_1 is asserted while z_2 and z_3 are both inactive.

Table 9.4 PROM Program for the Moore Machine of Figure 9.6

State Name	PROM Address Present State $y_1\ y_2\ y_3$	Present Inputs $x_1\ x_2$	PROM Outputs Next State $y_1\ y_2\ y_3$	Present Outputs $z_1\ z_2\ z_3$
a	0 0 0	0 0	0 0 1	0 0 0
	0 0 0	0 1	0 0 1	0 0 0
	0 0 0	1 0	0 1 0	0 0 0
	0 0 0	1 1	0 1 0	0 0 0
b	0 0 1	0 0	0 1 1	0 0 0
	0 0 1	0 1	1 0 0	0 0 0
	0 0 1	1 0	0 1 1	0 0 0
	0 0 1	1 1	1 0 0	0 0 0
c	0 1 0	0 0	1 0 0	0 0 0
	0 1 0	0 1	1 0 1	0 0 0
	0 1 0	1 0	1 0 0	0 0 0
	0 1 0	1 1	1 0 1	0 0 0
d	0 1 1	0 0	1 1 0	1 0 0
	0 1 1	0 1	1 1 0	1 0 0
	0 1 1	1 0	1 1 0	1 0 0
	0 1 1	1 1	1 1 0	1 0 0
e	1 0 0	0 0	1 1 0	0 0 0
	1 0 0	0 1	1 1 0	0 0 0
	1 0 0	1 0	1 1 0	0 0 0
	1 0 0	1 1	1 1 0	0 0 0
f	1 0 1	0 0	1 1 0	0 1 0
	1 0 1	0 1	1 1 0	0 1 0
	1 0 1	1 0	1 1 0	0 1 0
	1 0 1	1 1	1 1 0	0 1 0
g	1 1 0	0 0	0 0 0	0 0 1
	1 1 0	0 1	0 0 0	0 0 1
	1 1 0	1 0	0 0 0	0 0 1
	1 1 0	1 1	0 0 0	0 0 1

A standard 32×8 PROM is sufficient for this application. There will be four un-used words and two unused PROM outputs. When the PROM program has been established, the device is then programmed by means of a PROM programmer — a unit that addresses each fuse location in turn and either opens the fuse or leaves it intact, depending upon the PROM output values specified in Table 9.4.

The logic diagram is shown in Figure 9.8. Two registers are depicted to illustrate the separate functions of the state register and the output register, although only a single 6-bit register is sufficient. The state flip-flop outputs are fed back to the PROM inputs to combine with the machine inputs to form the address function.

Outputs z_1 and z_2 are produced directly from the output register, providing the requisite assertion and deassertion of t_1 and t_3, respectively, whereas z_3 is generated by an AND gate using the $-clock$ signal as an enabling factor to assert z_3 at time t_2 and deassert z_3 at t_3. The timing diagram for a representative sequence to assert z_2 and z_3 is shown in Figure 9.9.

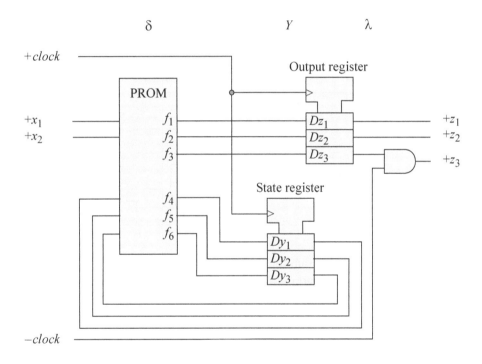

Figure 9.8   Logic diagram for the Moore machine of Figure 9.6 using a PROM for the δ next-state logic which is programmed according to the contents of Table 9.4.

9.2 Programmable Array Logic

A PAL device confirms to the general structure of a PLD. The number of AND gates and OR gates is variable, depending on the part number of the commercially available PAL. Some PALs contain ten dedicated inputs and eight outputs, some of which have dual input/output (I/O) functions. Each output section consists of eight programmable AND gates connected to a fixed OR gate. There are eight output sections for a total of 64 AND gates. In many cases, the outputs are also fed back through separate buffers (drivers) to the programmable AND array.

The basic organization of a PAL device is shown in Figure 9.10. The AND array is programmable and the OR array is fixed. The requirement that a PROM must have 2^n AND gates in the AND array for n inputs is not a restriction for a PAL — the number of AND gates in the AND array is not a function of the number of inputs.

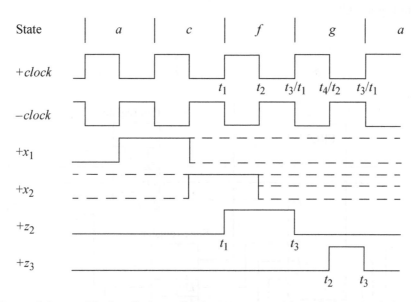

Figure 9.9   Timing diagram for the Moore machine of Figure 9.6 showing a valid sequence of $x_1 x_2 = 11$ to assert output z_2. Output z_3 is asserted in state g.

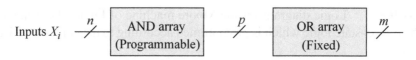

Figure 9.10   Block diagram for the basic organization of a PAL device.

By convention, the logic symbols in the AND array use a single horizontal line connected to the gate input, where the single line represents all of the inputs to the gate. The intersecting vertical lines in the AND array represent the device inputs, both true and complemented.

9.2.1 Combinational Logic

When designing with a PAL, the Boolean expressions should be minimized, if necessary, so that the number of product terms in each expression does not exceed the number of AND gates in each AND-OR structure. If the number of product terms is too large, then the function can still be realized by utilizing two AND-OR structures. Feedback paths can also be utilized from the outputs to drivers which connect to the AND array.

Example 9.4 The following Boolean expressions will be implemented using a PAL device:

$$z_1(x_1, x_2, x_3) = \Sigma_m(1, 2, 6)$$

$$z_2(x_1, x_2, x_3) = \Sigma_m(0, 1, 5, 6, 7)$$

$$z_3(x_1, x_2, x_3) = \Sigma_m(1, 2, 4, 6, 7)$$

Using Boolean algebra or Karnaugh maps, the above functions convert to the following sum-of-products forms:

$$z_1 = x_1'x_2'x_3 + x_2x_3'$$

$$z_2 = x_1'x_2' + x_1x_2 + x_2'x_3$$

$$z_3 = x_1'x_2'x_3 + x_2x_3' + x_1x_2 + x_1x_3'$$

$$= z_1 + x_1x_2 + x_1x_3'$$

Figure 9.11 illustrates a PAL device consisting of three inputs and three outputs that implements the three Boolean functions of Example 9.4. Each input is connected to a buffer-driver which generates both true and complemented outputs of the corresponding input. The device consists of nine AND gates forming the programmable AND array and three OR gates which form the fixed OR array. Each AND gate has eight fused programmable inputs as shown by the eight vertical lines intersecting each horizontal line. The horizontal line is called the *product line* and symbolizes the multiple-input configuration of the AND gate. The output of each AND gate is the corresponding *product term*.

Function z_1 contains only two terms; thus, the bottom AND gate is not required. To ensure that the output of the lower AND gate does not contribute to output z_1, the fuses for inputs x_1 and x_1' are left intact. Thus, the AND gate receives both the true and complemented values of input x_1, so that the output of the gate is $x_1x_1' = 0$.

Function z_2 is programmed in a similar manner. In this case, all three AND gates are utilized. Function z_3 requires four product terms; however, the first two terms are the same as the two terms that represent z_1. Therefore, using output z_1, function z_3 can be reduced to three terms, as shown in Figure 9.11.

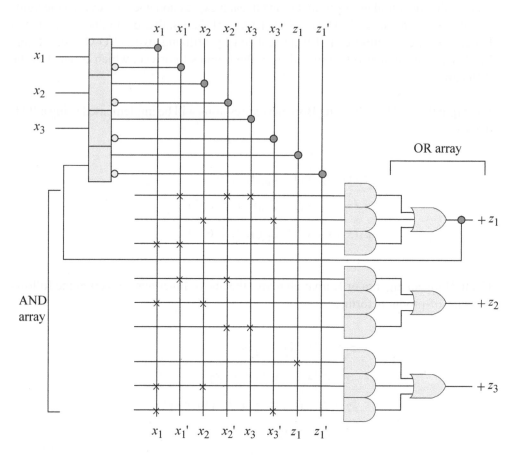

Figure 9.11   A PAL device using three inputs and three outputs to implement the Boolean expressions of Example 9.4.

9.2.2 Sequential Logic

The AND array allows product terms to be programmed which then connect to a predefined OR array. The restriction of the prewired OR array is compensated by the wide variety of available PAL configurations. Specifying the configuration of the OR array, therefore, is simply a matter of device selection.

Each AND gate in a PAL has $2n$ inputs, where n is the number of device inputs. The symbol "×" indicates an intact fuse, which connects a unique variable — either true or complemented — to one of the six AND gate inputs. The absence of an × indicates an open fuse, which supplies a logic 1 to the AND gate. Thus, the product terms consist only of the input variables specified by an ×.

Some PALs contain pins that provide both input and output functions. The PAL shown in Figure 9.12 contains a 3-state output driver that is enabled by a product term. When the product term is active, the sum-of-products function is transferred to output z_1. This function is also fed back to the input logic, providing a necessary requirement for synthesizing SR latches. When the 3-state driver is disabled, the output network of the driver operates as an input source. This bidirectional capability of the $-z_1(+x_4)$ signal is useful for shifting left or right when the PAL is used as the δ next-state logic for one stage of a shift register.

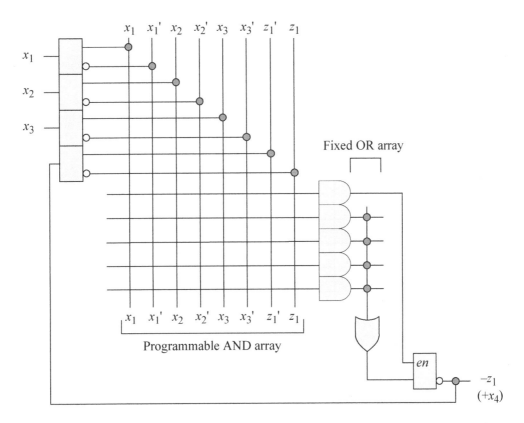

Figure 9.12 PAL circuit in which the 3-state output driver is enabled by a product term in the AND array. The output of the 3-state driver is bidirectional.

More complex PALs contain not only the basic AND-OR array organization, but also additional circuitry for feedback signals and output registers. One stage of a typical PAL of this type is shown in Figure 9.13. The active-high output of the D flip-flop for y_1 is connected to a 3-state driver (buffer). Output z_1 is transmitted to external logic by applying a high voltage level to the $+enable$ $(+en)$ signal of the PAL input buffer. The active-low output of the storage element is fed back to the δ input logic.

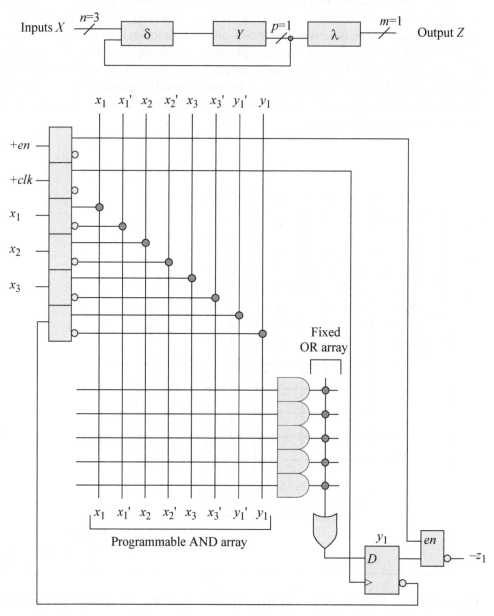

Figure 9.13   Typical PAL stage showing the AND-OR arrays with output flip-flop and feedback signal.

It is interesting to note that the basic organization of the PAL in Figure 9.13 conforms to the general structure of a synchronous sequential machine. The input drivers, together with the AND-OR arrays, constitute the δ next-state logic; flip-flop y_1 is the storage element; and the 3-state driver represents the λ output logic.

The input equations obtained from input maps are in a sum-of-products form, which is the requisite format for programming the AND array. Thus, PALs and other PLDs are ideally suited for synthesizing synchronous sequential machines. To illustrate the synthesis procedure for machines that are implemented with PAL devices, two examples will be presented. The first is a machine depicting a Gray code counter; the second is a sequence detector with Mealy outputs.

Example 9.5   A 3-bit Gray code counter will be synthesized that counts in the following sequence: 000, 001, 011, 010, 110, 111, 101, 100, 000. A single PAL device will be used for both the δ next-state logic and the storage elements, which consist of positive-edge-triggered D flip-flops.

The state diagram is shown in Figure 9.14 and the input maps in Figure 9.15. The input equations for Dy_1, Dy_2, and Dy_3 are listed in Equation 9.3 in a sum-of-products form. If the state flip-flops do not have a reset input, then each term of Dy_1, Dy_2, and Dy_3 must contain a third variable — a *reset* input x_1 — as shown in Equation 9.4. Thus, the counter operates only when the reset input is inactive.

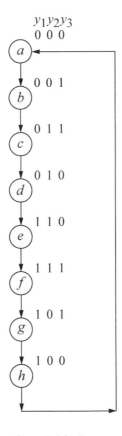

Figure 9.14   State diagram for a 3-bit Gray code counter.

y_1 \ y_2y_3	00	01	11	10
0	0	0	0	1
1	0	1	1	1

Dy_1

y_1 \ y_2y_3	00	01	11	10
0	0	1	1	1
1	0	0	0	1

Dy_2

y_1 \ y_2y_3	00	01	11	10
0	1	1	0	0
1	0	0	1	1

Dy_3

Figure 9.15   Input maps for the Gray code counter of Figure 9.14.

$$Dy_1 = y_1y_3 + y_2y_3'$$
$$Dy_2 = y_1'y_3 + y_2y_3'$$
$$Dy_3 = y_1'y_2' + y_1y_2$$

(9.3)

$$Dy_1 = y_1y_3x_1' + y_2y_3'x_1'$$
$$Dy_2 = y_1'y_3x_1' + y_2y_3'x_1'$$
$$Dy_3 = y_1'y_2'x_1' + y_1y_2x_1'$$

(9.4)

The logic diagram using PAL technology is shown in Figure 9.16 and is programmed directly from the sum-of-products input equations of Equation 9.3. The only inputs to the counter are the +*clock* signal and the +*enable* signal. All remaining AND array inputs are feedback signals from the state flip-flops through drivers with inverting and noninverting outputs.

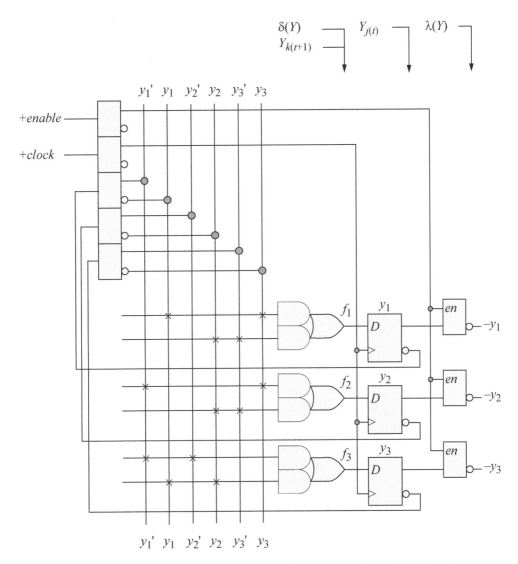

Figure 9.16 Logic diagram for the Gray code counter using a PAL device.

The complemented output of flip-flop y_i is fed back to an input driver. Thus, when flip-flop y_i is set (=1), the complemented output of the driver generates a positive voltage level which is available to all AND gate inputs. If the input equation for y_i contains a flip-flop variable in its true form, then the complemented output of the driver is programmed to connect to the appropriate positive-input AND gate in the AND array.

The δ next-state logic consists of the feedback drivers, the programmable AND array, and the fixed OR array.

Outputs f_1, f_2, and f_3 of the OR gates connect to Dy_1, Dy_2, and Dy_3, respectively. The present state $Y_{j(t)}$ is fed back through drivers to the AND array and, after an appropriate propagation delay, appears at the flip-flop D inputs as the next state $Y_{k(t+1)}$. When the +*enable* signal is active, the state of the counter is available to external devices as active-low voltage levels.

Example 9.6   PAL devices can be used to implement any synchronous sequential machine, including a sequence detector. A Mealy machine will be synthesized which checks for the sequence 01111110 (7E hexadecimal) on a serial input line x_1. This bit configuration is used in the High-Level Data Link Control (HDLC) protocol as a flag character to detect the beginning and the end of a frame. The format for an HDLC frame is shown in Figure 9.17.

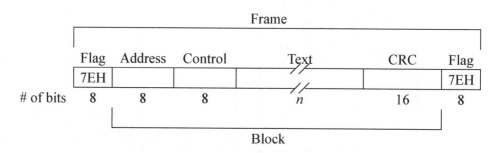

Figure 9.17   Format for the HDLC protocol.

The *address* field identifies the primary or secondary destination for the frame transmission. The *control* field contains commands, responses and sequence numbers used to control transmission and defines the functions of the frame such as last frame begin transferred. The *text* field contains the data that is being transmitted to the receiver. The *cyclic redundancy check* (CRC) character is used to detect and correct any errors that occurred during transmission. The HDLC method of transmitting data is *self-clocking*, which is explained in detail in Chapter 11.

In order to avoid ambiguity between a flag character and the bit sequence of 01111110 in the text field, a 0 bit is inserted in the data (text) field after every occurrence of five consecutive 1s as the frame is being transmitted. The receiver then deletes the 0 bit that occurs after five consecutive 1s. This is shown in Figure 9.18, where a flag of 7E hexadecimal is part of the block.

The state diagram for this Mealy machine is graphically depicted in Figure 9.19. The machine remains in state *a* until the start of transmission is indicated by a high-to-low transition on input x_1. The bit sequence is then received beginning with bit 0, one bit per clock period. When the first 1 bit has been detected in state *b*, any subsequent 0 bit that occurs before six consecutive 1s returns the machine to state *b* to begin checking for a new valid sequence. Similarly, seven consecutive 1s returns the machine to state a to begin checking for a new valid sequence. Only when $x_1 = 01111110$ is output z_1 asserted. Changes to x_1 occur at time t_2.

Data to be sent:	0	1	1	1	1	1	1	0		
Data that is sent:	0	1	1	1	1	1	**0**	1	0	

Data to be sent:	0	1	1	1	1	1	0	
Data that is sent:	0	1	1	1	1	1	**0**	0

Figure 9.18   HDLC protocol format to avoid ambiguity between a flag character and a text sequence of 01111110.

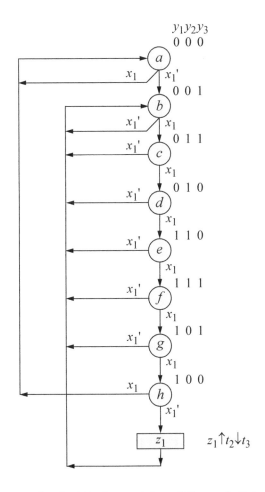

Figure 9.19   State diagram for the Mealy machine of Example 9.6.

The input maps, obtained from the state diagram, are shown in Figure 9.20. In state a ($y_1 y_2 y_3 = 000$), the next states for flip-flops y_1 and y_2 are 0 and 0, respectively, regardless of the value of x_1. The next state for y_3, however, is determined by the value of x_1; if $x_1 = 0$, then the next state for y_3 is 1, otherwise the next state is 0. In states d, e, f, and g, flip-flop y_1 has a next state of 1 only if $x_1 = 1$. Therefore, x_1 is entered in the map as a map-entered variable. Likewise, in states b, c, d, and e, flip-flop y_2 has a next value of 1 only if $x_1 = 1$.

In the input map for flip-flop y_3, the next state is never an unconditional 0, irrespective of the path taken. The next state will be either an unconditional 1 or a value dependent upon x_1; that is, if $x_1 = 0$, then $y_3 = 1$. Since a logic $1 = x_1 + x_1'$; therefore, every minterm location in the map can be given a value of x_1'. This accounts for the x_1' term in the equation for Dy_3 of Equation 9.5. The x_1 term of the logic 1 expression must now be taken into account. This is very easily accomplished by reverting to the minterm value of 1, which generates the two remaining terms for Dy_3.

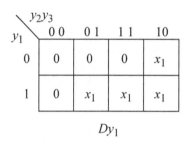

Dy_1

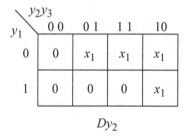

Dy_2

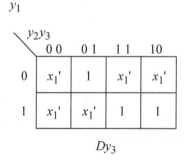

Dy_3

Figure 9.20   Input maps for the Mealy machine of Figure 9.19.

The input equations are listed in Equation 9.5 in a sum-of-products form for implementation using a PAL device. The term $y_2 y_3' x_1$ in the equations for Dy_1 and Dy_2 cannot be shared in order to minimize logic as is possible when using discrete logic gates. There is no internal connection from the product term to the AND array.

$$Dy_1 = y_1 y_3 x_1 + y_2 y_3' x_1$$

$$Dy_2 = y_1' y_3 x_1 + y_2 y_3' x_1$$

$$Dy_3 = x_1' + y_1' y_2' y_3 + y_1 y_2 \qquad (9.5)$$

The output map is shown in Figure 9.21, as specified by the state diagram. The equation for output z_1 is shown in Equation 9.6. The PAL programming and logic diagram are illustrated in Figure 9.22. The PAL is programmed directly from the input and output equations of Equation 9.5 and Equation 9.6, respectively.

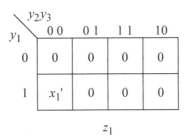

$$z_1$$

Figure 9.21 Output map for the Mealy machine of Figure 9.19.

$$z_1 = y_1 y_2' y_3' x_1' \qquad (9.6)$$

There is no need to enable output z_1 with the $-clock$ signal to conform to the assertion/deassertion statement of $z_1 \uparrow t_2 \downarrow t_3$. The output assertion and deassertion times are determined inherently by the changes to the input variable x_1, which occur at time t_2. For example, in state h, if x_1 remains asserted at time t_2, then the machine proceeds to state a. If x_1 changes from a value of 1 to a value of 0 in state h, the change occurs at t_2, in which case, output z_1 will be asserted at t_2 and deasserted at t_3 when the machine leaves state h. The machine then proceeds to state b. A representative timing diagram for a valid input sequence of $x_1 = 01111110$ is shown in Figure 9.23.

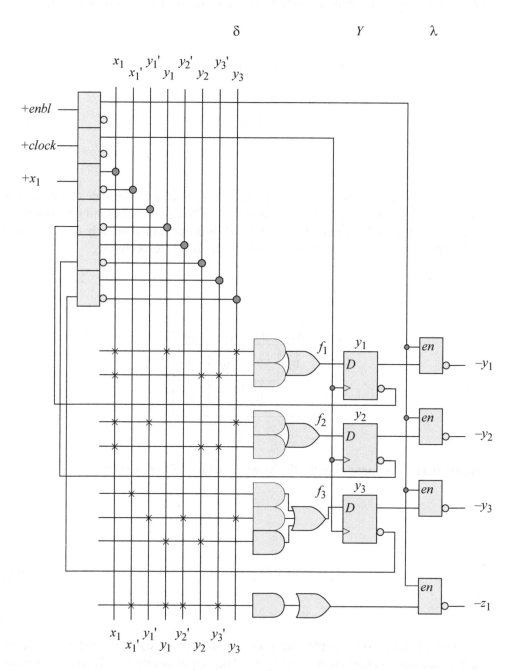

Figure 9.22   Logic diagram for the Mealy machine of Figure 9.19 using a PAL device.

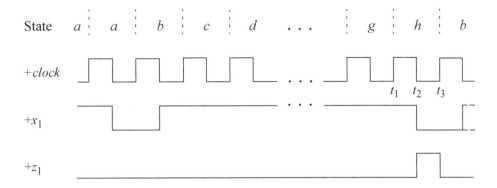

Figure 9.23   Timing diagram for the Mealy machine of Figure 9.19.

9.3 Programmable Logic Arrays

The basic organization of a PLA is shown in Figure 9.24. Both the AND array and the OR array are programmable. Since both arrays are programmable, the PLA has more programming capability and thus, more flexibility than the PROM or PAL.

The output function in a PLA is limited only by the number of AND gates in the AND array, since all AND gates can be programmed to connect to all OR gates. The output function in a PAL, however, is restricted not only by the number of AND gates in the AND array, but also by the fixed connections from the AND array outputs to the OR array.

9.3.1 Combinational Logic

The PLA in Figure 9.24 is programmed to generate the following Boolean functions for z_1, z_2, z_3, and z_4:

$$z_1 = x_1 x_2' + x_1' x_2$$

$$z_2 = x_1 x_3 + x_1' x_3'$$

$$z_3 = x_1 x_2' + x_1 x_3 + x_1' x_2' x_3'$$

$$z_4 = \text{Logic 1}$$

Output z_4 is a logic 1, because the inputs to the AND gate whose output connects to the OR gate have all fuses open. Thus, the inputs to the AND gate become high voltage levels due to internal circuitry, and generate a high level on the output.

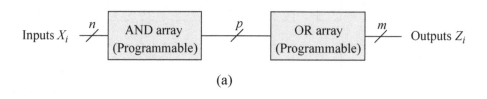

(a)

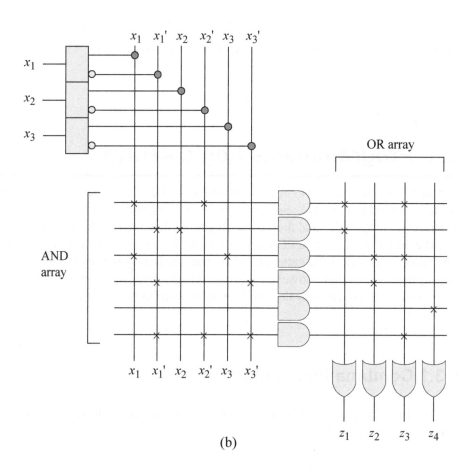

(b)

Figure 9.24   Basic organization of a PLA; (a) block diagram; and (b) implementation using three inputs and four outputs.

As can be seen from the above examples, the output function may not use all of the input variables x_i. Combinational logic networks can be characterized in terms of fundamental logic operations such as, AND, OR, and NOT. This also includes the functionally complete set of logic gates, NAND and NOR. The output symbols z_i may be asserted either high or low. Combinational logic is used extensively to represent the next-state function and the output function for both synchronous and asynchronous sequential machines.

9.3.2 Sequential Logic

Programmable logic arrays offer a high degree of flexibility, because both the AND array and the OR array are programmable. A PLA has n input variables, $x_1, x_2, ..., x_n$ and m output functions, $z_1, z_2, ..., z_m$. The number of AND gates — which determines the number of product terms — is variable, depending upon the PLA version.

Like the previous PLDs, each input x_i connects to a buffer which increases the driving capability of the input and also provides both true and complemented values of the input signal. The OR array permits each OR gate to access any product term. Thus, the programmable OR array allows all OR gates to access the same product terms simultaneously. Each output z_i is generated from a sum-of-product expression which is a function of the input variables.

PLAs are characterized by the number of input variables, the number of product terms, and the number of output functions. All Boolean expressions can be decomposed into a sum-of-products representation. For example, the exclusive-OR function $x_1 \oplus x_2$ equates to $x_1 x_2' + x_1' x_2$. The multiple-input, multiple-output logic circuit shown in Figure 9.25(a) is illustrated by an equivalent programmable representation in Figure 9.25(b) using a PLA device to implement the Boolean equations. The symbol $\times$ indicates an intact fuse; the absence of an $\times$ indicates an open fuse, where the unconnected input assumes a logic 1 voltage level for an AND gate input and a logic 0 voltage level for an OR gate input.

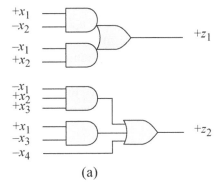

(a)

Figure 9.25   Implementation of a multiple-input, multiple-output 2-level logic circuit: (a) conventional sum-of-products representation and (b) equivalent circuit using a PLA device.

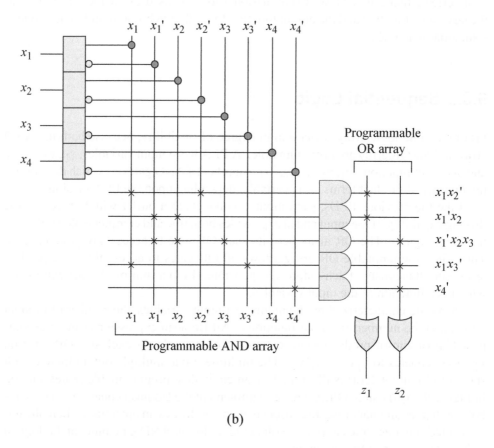

(b)

Figure 9.25   (Continued)

Multiple levels of logic can be reduced to a 2-level sum-of-products form. Although this implementation may require more logic gates, the high integration of PLAs easily absorbs the additional logic. The sum-of-products form permits the propagation delay through the PLA to be reduced considerably, resulting in a higher-speed function. Programmable devices usually reduce the amount of board space required, as well as the number of interconnections and the power supply size, thereby reducing the cost of the machine.

Example 9.7   The synthesis of synchronous sequential machines can be realized using PLAs in a manner analogous to that used for PROMs and PALs. A judicious choice of state codes guarantees that there will be no output glitches for any state transition sequence. The rules shown below are useful in assigning state codes such that there will be a maximal number of 1s in adjacent squares of the input maps, thus minimizing the δ next-state logic. It should be noted, however, that these rules do not guarantee a minimum solution with the fewest number of terms and the fewest number of variables per term. There may be several sets of state code assignments that meet the adjacency requirements, but not all will result in a minimally reduced set of input equations.

1. When a state has two possible next states, then the two next states should be adjacent; that is, if an input causes a state transition from state Y_i to either Y_j or Y_k, then Y_j and Y_k should be assigned adjacent state codes.

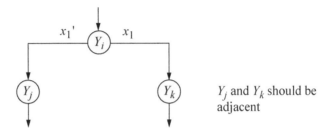

Y_j and Y_k should be adjacent

2. When two states have the same next state, the two states should be adjacent; that is, if Y_i and Y_j both have Y_k as a next state, then Y_i and Y_j should be assigned adjacent state codes.

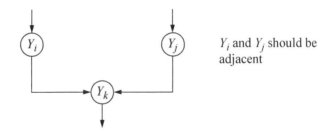

Y_i and Y_j should be adjacent

3. A third rule is useful in minimizing the λ output logic. States which have the same output should have adjacent state code assignments; that is, if states Y_i and Y_j both have z_1 as an output, then Y_i and Y_j should be adjacent. This allows for a larger grouping of 1s in the output map.

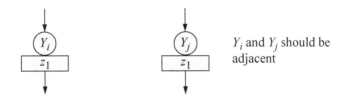

Y_i and Y_j should be adjacent

Figure 9.26 illustrates a state diagram for a Moore machine with one input x_1 and two outputs z_1 and z_2. The machine will be implemented using a PLA and three positive-edge-triggered D flip-flops. Because the outputs are asserted at time t_1 and deasserted at t_3, the λ output logic simply decodes states c ($y_1 y_2 y_3 = 111$) and f ($y_1 y_2 y_3 = 100$), asserting outputs z_1 and z_2, respectively.

Using the rules for state code adjacency stated above, states a and f should be adjacent, because the next state for both is state a (Rule 2). Also, states c and d should be adjacent, because they are both possible next states for state b (Rule 1) and have the same next state, e (Rule 2). The state assignment shown in Figure 9.26 precludes the possibility of glitches on outputs z_1 and z_2. This can be verified by checking all paths to determine if any state transition produces a transient state that is identical to the state codes for states c or f. This condition can occur only if two or more flip-flops change state for a particular state transition.

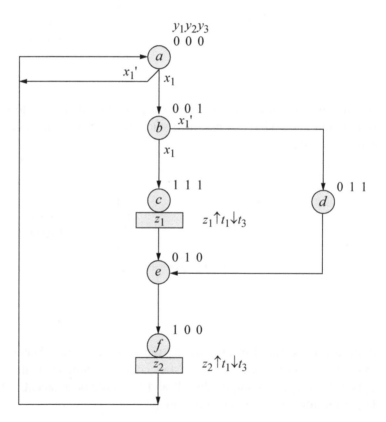

Figure 9.26   State diagram for the Moore synchronous sequential machine of Example 9.7. Unused states are: $y_1 y_2 y_3 = 101$ and 110.

The path from state a to state b produces a change of state for only one flip-flop (y_3). Similarly, the path from state b to state d produces only one change — y_2 changes from 0 to 1. Although the transition from state b to state c produces two changes, flip-flop y_3 remains set — y_3 must be reset for the machine to enter state f and assert output z_2. The path from state d to state e results in only one change of flip-flop variable (y_3). Both y_1 and y_3 change state when the machine proceeds from state c to state e; however, y_2 remains set, thus negating a glitch on z_2 in state f. The path from state e to state f produces a change to both y_1 and y_2, but flip-flop y_3 remains reset — y_3 must be set for the machine to enter state c and assert output z_1. Finally, the path from state f to state a occurs when only y_1 changes state. Thus, no state transition will produce a glitch on either output z_1 or z_2.

The input maps are shown in Figure 9.27 as obtained from the state diagram using input x_1 as a map-entered variable. The Karnaugh maps yield minimum sum-of-products expressions, as shown in Equation 9.7.

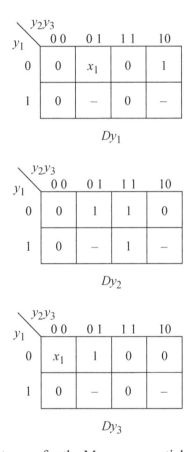

Figure 9.27   Input maps for the Moore sequential machine of Figure 9.26.

$$Dy_1 = y_2'y_3x_1 + y_2y_3'$$

$$Dy_2 = y_3$$

$$Dy_3 = y_1'y_2'x_1 + y_2'y_3 \tag{9.7}$$

The output maps are shown in Figure 9.28. The output equations can be minimized if the minterms for z_1 and z_2 are combined with unused minterm $y_1 y_2 y_3 = 101$ or 110. If these unused states are to be used for minimization, however, they must not function as transient states for any state sequence that does not include the corresponding output.

The only transition that may pass through unused state $y_1 y_2 y_3 = 101$ is the path from state b to state c. This presents no hazard for output z_1, however, because this sequence includes z_1. Output z_1 may be asserted slightly early, but no glitch will be generated. Therefore, a 1 can be inserted in $y_1 y_2 y_3 = 101$ in order to minimize the equation for z_1.

Since z_1 is already minimized, it is immaterial whether $y_1 y_2 y_3 = 110$ could be used to minimize z_1. This unused state could not be used in any case, however, because the path from state e to state f may pass through transient state $y_1 y_2 y_3 = 110$, and this sequence does not involve z_1.

Now consider the unused states to minimize z_2. The path from state b to state c does not include output z_2 in either the initial state or the destination state; therefore, a 0 must be inserted in state $y_1 y_2 y_3 = 101$ in the output map for z_2. Similarly, the path from state c to state e may pass through unused state $y_1 y_2 y_3 = 110$; therefore, a 0 must be inserted in minterm location $y_1 y_2 y_3 = 110$. The equations for output z_1 and z_2 are shown in Equation 9.8.

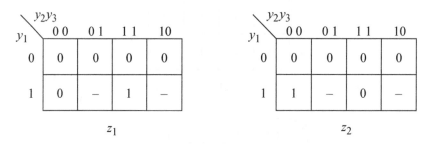

Figure 9.28   Output maps for the Moore sequential machine of Figure 9.26.

$$z_1 = y_1 y_3$$

$$z_2 = y_1 y_2' y_3' \tag{9.8}$$

The logic is shown in Figure 9.29 using a PLA with positive-edge-triggered D flip-flops. To obtain the logic function for Dy_1, Dy_2, and Dy_3, the AND array is programmed according to the product terms of Equation 9.7 and the OR array is programmed to obtain the appropriate sum-of-products for the respective Dy_i input. In the same manner, the AND and OR arrays are programmed to generate outputs z_1 and z_2 according to Equation 9.8.

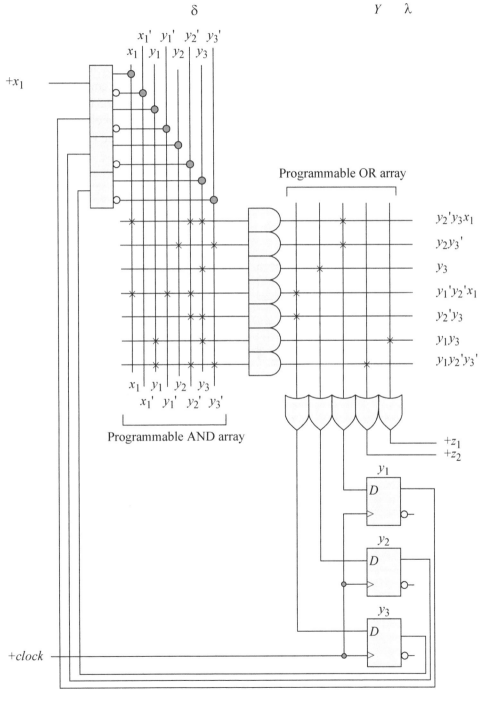

Figure 9.29   Implementation of the Moore machine of Figure 9.26 using a PLA device.

9.4 Field-Programmable Gate Arrays

Several versions of field-programmable gate arrays (FPGAs) are available, depending on the manufacturer. This section describes the organization and operation of a generalized FPGA and concludes with a synthesis example for a Moore machine.

FPGAs are high-density gate arrays that can be configured by the user for specific applications. The organization of a typical FPGA, as shown in Figure 9.30, consists of a matrix of identical logic block elements encompassed by a perimeter of input/output (I/O) blocks. Each logic block element contains a programmable combinational logic function, storage elements, and output logic, as shown in Figure 9.31. The storage elements feed back to the combinational logic. Thus, the three major elements that comprise a synchronous sequential machine are available in this type of FPGA: δ next-state logic, storage elements, and λ output logic.

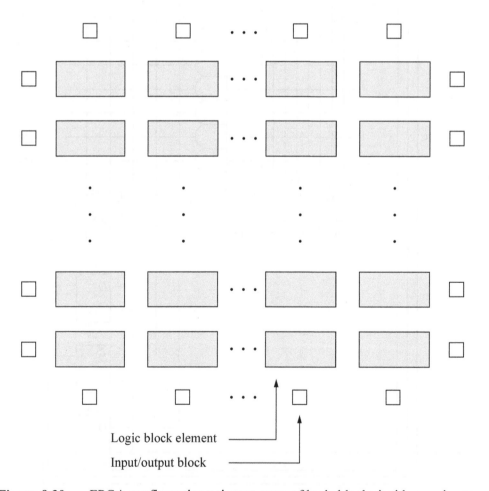

Logic block element ————
Input/output block ————

Figure 9.30   FPGA configuration using an array of logic blocks inside a perimeter of input/output blocks.

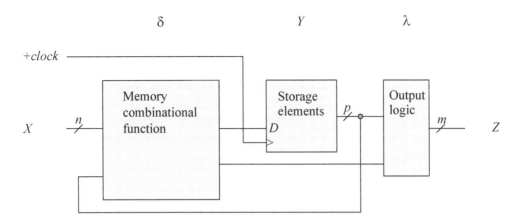

Figure 9.31   Typical logic block element of a programmable configurable FPGA.

The combinational function utilizes a memory table-lookup technique which can be programmed to implement any Boolean function of the input variables. The output of the combinational function connects to the storage elements. The output signals can be programmed to be a function of either the storage elements only or the storage elements and the present inputs, providing characteristics that are consistent with both Moore and Mealy machines.

The fused PLDs discussed in previous sections provide a fixed organization, whereas an FPGA provides an uncommitted organization that is configured by a resident program. The configuration program is loaded into a resident memory on the FPGA during the power-up sequence. The program conditions the combinational function in each array element to provide the required input logic for the storage elements and the output logic for the I/O blocks. The program also establishes the I/O blocks as input, output, or bidirectional circuits. The storage elements can be programmed to operate as SR latches or edge-triggered D flip-flops. The I/O blocks provide an interface between the internal logic blocks and the external pins of the array.

Interconnection between logic blocks and between logic blocks and I/O blocks is accomplished by vertical and horizontal signal paths between the rows and columns of the logic blocks. Programmable switching matrices are provided at the intersection of the vertical and horizontal interconnect lines to provide a network for interblock communication.

This type of FPGA relies heavily on the software configuration program and, therefore, provides a general organization that can be adapted to meet the individual requirements of the user. The configuration program in the array's resident memory completely defines the operation of each logic block element.

Assume that the memory combinational function of a logic block has five inputs, providing 5-variable minterms for any combinational function. The memory consists of a 32-by-1 array of storage cells. Thus, the five input variables form an address that accesses a unique 1-bit word in memory to produce a single output f_1. All address minterms can be programmed to assert f_1. A second output f_2 is available when the memory is partitioned into two segments; that is, two outputs are generated from two separate 4-variable functions. In this case, the addressing restriction requires that one input address variable x_i be common to both functions.

Figure 9.32 shows the two configurations of function generation. In Figure 9.32(a) all five inputs form an address minterm to generate output $f_1(x_1, x_2, x_3, x_4, x_5)$. Figure 9.32(b) illustrates a 2-function implementation. If input x_1 is the common input variable, then output f_1 is a function of x_1 and any 3-variable combination of x_2, x_3, x_4, and x_5. Similarly, f_2 is a function of x_1 and any different 3-variable combination of x_2, x_3, x_4, and x_5.

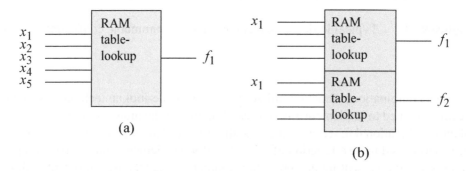

Figure 9.32   RAM table-lookup function generator: (a) one 5-variable function f_1 and (b) two 4-variable functions f_1 and f_2.

Example 9.8   Figure 9.33 illustrates a state diagram for a Moore machine containing two inputs x_1 and x_2, and two outputs z_1 and z_2. The machine will have a self-starting state of $y_1y_2y_3 = 000$. Thus, if either of the unused states ($y_1y_2y_3 = 100$ or 111) is entered due to noise or machine malfunction, then the outputs are made inactive and the machine returns to state a. Three flip-flops are required for this 6-state synchronous sequential machine.

Recall that outputs may produce erroneous values when the machine changes state at the active clock transition. Since both outputs are asserted at time t_1 and deasserted at t_3, state codes must be assigned such that no state transition will cause a glitch on z_1 or z_2. It is unlikely that a glitch will filter through the memory address decoder to the output; however, the exercise is worthwhile.

The state codes shown in Figure 9.33 satisfy this requirement. The absence of glitches can be verified by checking all ten paths in Figure 9.33 for multiple changes of flip-flop states and observing whether state $d\,(y_1y_2y_3 = 101)$ or state $f\,(y_1y_2y_3 = 110)$ is entered as a transient state.

The input maps, obtained from the state diagram, are shown in Figure 9.34. Refer to the state diagram in conjunction with the input map for flip-flop y_1. In state b, the machine proceeds to state d only if $x_1x_2 = 11$. Thus, x_1x_2 is entered in minterm location $y_1y_2y_3 = 001$ in the map for flip-flop y_1. The same is true for the transition from state c to state f, necessitating an entry of x_1x_2 in minterm location $y_1y_2y_3 = 010$. With the exception of the unused states, all other map entries are assigned a value of 0, since the next state for y_1 in states a, d, e, and f is 0.

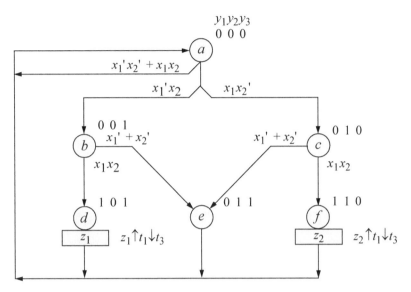

Figure 9.33   State diagram for the Moore machine of Example 9.8.

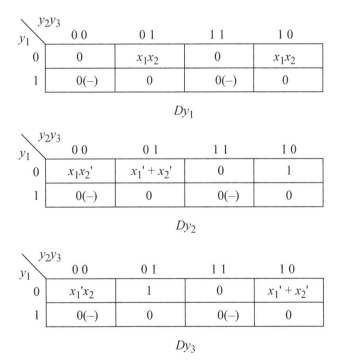

Figure 9.34   Input maps for the Moore machine of Figure 9.33.

In constructing the input map for flip-flop y_2, three paths are possible from state a; however, only one path — from state a to state c — results in a next state of 1 for flip-flop y_2. This transition occurs if $x_1 x_2 = 10$. Thus, the expression $x_1 x_2'$ is entered in minterm location $y_1 y_2 y_3 = 000$. The path from state b to state e results in a next state of 1 for y_2 if either $x_1 = 0$ or $x_2 = 0$. Thus, the expression $x_1' + x_2'$ is entered in minterm location 001 in the input map for y_2. In state c, the next state for y_2 is 1, regardless of the path taken. In the remaining three states, d, e, and f, the machine proceeds to state a, where $y_2 = 0$. In the same manner, the input map for flip-flop y_3 is constructed.

The equations for Dy_1, Dy_2, and Dy_3 are shown in Equation 9.9. The unused states for y_1, y_2, and y_3 are set to 0, because minimization of the input equations is not necessary. The equations do not have to be minimized for this type of memory-configurable FPGA, since a minimal set of terms would not reduce the size of the memory unit.

$$Dy_1 = y_1' y_2' y_3 x_1 x_2 + y_1' y_2 y_3' x_1 x_2$$

$$Dy_2 = y_1' y_2' y_3' x_1 x_2' + y_1' y_2' y_3 x_1' + y_1' y_2' y_3 x_2' + y_1' y_2 y_3'$$

$$Dy_3 = y_1' y_2' y_3' x_1' x_2 + y_1' y_2' y_3 + y_1' y_2 y_3' x_1' + y_1' y_2 y_3' x_2' \qquad (9.9)$$

The output maps for z_1 and z_2 are shown in Figure 9.35. Since the Moore outputs are a function of the present state only, there can be no more than three variables in the output equations. Thus, a single logic block element can accommodate the two outputs. The combinational memory unit is configured to generate functions f_1 and f_2, which correspond to z_1 and z_2, respectively. The output equations do not require minimization, since no reduction in internal logic is possible. The output equations are listed in Equation 9.10.

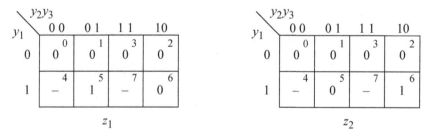

Figure 9.35   Output maps for the Moore machine of Figure 9.33.

$$z_1 = y_1 y_2' y_3$$

$$z_2 = y_1 y_2 y_3' \qquad (9.10)$$

This type of FPGA operates in a manner analogous to that of a PROM-controlled synchronous sequential machine. That is, a resident program completely determines the machine's behavior as a function of the present state and the present inputs. The memory table-lookup program for this Moore machine is shown in Table 9.5, which is a tabular representation of the three input maps and the two output maps.

Table 9.5   Memory Table-Lookup Program for the Moore Machine of Figure 9.33

Memory Inputs $y_1 y_2 y_3 x_1 x_2$	Memory Outputs $f_1(Dy_1)$	$f_1(Dy_2)$	$f_1(Dy_3)$	$f_1(z_1)$	$f_2(z_2)$
0 0 0 0 0	0	0	0	0	0
0 0 0 0 1	0	0	1	0	0
0 0 0 1 0	0	1	0	0	0
0 0 0 1 1	0	0	0	0	0
0 0 1 0 0	0	1	1	0	0
0 0 1 0 1	0	1	1	0	0
0 0 1 1 0	0	1	1	0	0
0 0 1 1 1	1	0	1	0	0
0 1 0 0 0	0	1	1	0	0
0 1 0 0 1	0	1	1	0	0
0 1 0 1 0	0	1	1	0	0
0 1 0 1 1	1	1	0	0	0
0 1 1 0 0	0	0	0	0	0
0 1 1 0 1	0	0	0	0	0
0 1 1 1 0	0	0	0	0	0
0 1 1 1 1	0	0	0	0	0
1 0 0 0 0	0	0	0	0	0
1 0 0 0 1	0	0	0	0	0
1 0 0 1 0	0	0	0	0	0
1 0 0 1 1	0	0	0	0	0
1 0 1 0 0	0	0	0	1	0
1 0 1 0 1	0	0	0	1	0
1 0 1 1 0	0	0	0	1	0
1 0 1 1 1	0	0	0	1	0
1 1 0 0 0	0	0	0	0	1
1 1 0 0 1	0	0	0	0	1
1 1 0 1 0	0	0	0	0	1
1 1 0 1 1	0	0	0	0	1
1 1 1 0 0	0	0	0	0	0
1 1 1 0 1	0	0	0	0	0
1 1 1 1 0	0	0	0	0	0
1 1 1 1 1	0	0	0	0	0

Four logic block elements are required, one each for Dy_1, Dy_2, Dy_3, and z_1z_2. Each logic block contains one memory unit configured as 32 one-bit words. The table is established from the input and output equations of Equation 9.9 and Equation 9.10, respectively. For example, the equation for Dy_1 contains two 5-variable terms, $y_1'y_2'y_3x_1x_2$ and $y_1'y_2y_3'x_1x_2$. Thus, these two minterms address the memory unit associated with y_1 and each generates a logic 1 for Dy_1 when the appropriate inputs are active, as is evident in rows 8 and 12 of Table 9.5.

The equation for Dy_2 contains one 5-variable term, two 4-variable terms, and one 3-variable term, each of which generates a logic 1. The first term generates a logic 1 when $y_1y_2y_3x_1x_2 = 00010$. The second term produces two logic 1 entries, one each for minterms $y_1'y_2'y_3x_1'x_2$ and $y_1'y_2'y_3x_1'x_2'$. Although the variable x_2 is eliminated from both minterms by the distributive law, the product term must still generate a logic 1 for the two address minterms.

The same rationale is true for the third term of Dy_2. The fourth term, $y_1'y_2y_3'$, generates four logic 1 entries that correspond to the four minterms, $y_1'y_2y_3'x_1'x_2'$, $y_1'y_2y_3'x_1'x_2$, $y_1'y_2y_3'x_1x_2'$, and $y_1'y_2y_3'x_1x_2$. Some of the product terms for Dy_2 share common minterms, resulting in eight logic 1 entries for Dy_2. Eight logic 1 entries are also required for the sum-of-product expression for Dy_3. Outputs z_1 and z_2 are both 3-variable functions. This results in a logic 1 entry for the four minterm combinations where $y_1y_2y_3 = 101$ for z_1 and where $y_1y_2y_3 = 110$ for z_2. During the power-up sequence, the configuration program is loaded into the memory unit of each logic block element according to the format of Table 9.5.

The logic diagram is illustrated in Figure 9.36. Each state flip-flop is connected to the f_1 output of its corresponding memory element, which characterizes the operation of the flip-flop. Outputs z_1 and z_2 are asserted from the combinational logic functions generated by the configuration program in two segments of a single logic block element.

Refer to the state diagram of Figure 9.33, the table-lookup program of Table 9.5, and the logic diagram of Figure 9.36 for the discussion of machine operation which follows. The machine is reset to state a, then the values of inputs x_1 and x_2 are examined. If x_1 and x_2 are equal in value, either 00 or 11, then the machine remains in state a. This is represented by the expression $(x_1 \oplus x_2)' = x_1'x_2' + x_1x_2$. However, when the expression $x_1 \oplus x_2$ is true, the machine proceeds to state b if $x_1x_2 = 01$ or to state c if $x_1x_2 = 10$. If $y_1y_2y_3x_1x_2 = 00010$ in state a, then this minterm addresses a unique 1-bit word in the memory table-lookup unit for flip-flop y_2. The $f_1(Dy_2)$ output assumes a logic 1 value. At the next positive clock transition, flip-flop y_2 is set to 1 and the state code is $y_1y_2y_3 = 010$. In a similar manner, the machine progresses through the remaining states.

In state d, each flip-flop memory unit is addressed by $y_1y_2y_3 = 101$ concatenated with the present input values. One of four memory words is selected, depending on the values of x_1 and x_2. Each 1-bit word, however, provides a logic 0 to Dy_1, Dy_2, and Dy_3 such that, the next state is a ($y_1y_2y_3 = 000$). Also, in state d, the combinational memory unit for output z_1 is programmed to generate a logic 1 for $f_1(z_1)$, asserting z_1 at time t_1. Output z_1 is deasserted when the machine changes to state a at t_3. Similarly, output z_2 is asserted at time t_1 in state f and deasserted when the machine sequences to state a.

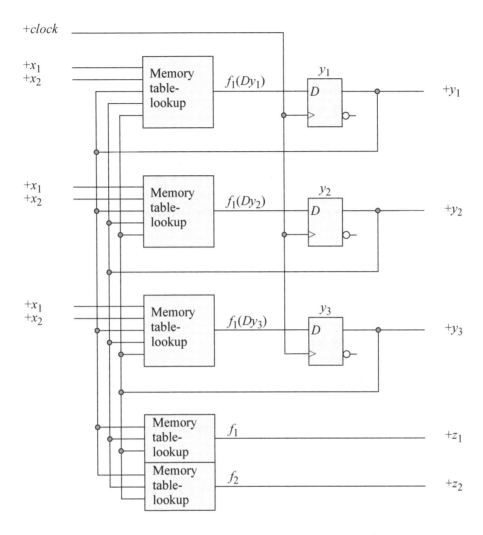

Figure 9.36   Logic diagram for the Moore machine of Figure 9.33 utilizing four logic block elements in a FPGA.

9.5 Problems

9.1   For a PROM, indicate which of the following statements are true:

(a) The AND array is fixed.
(b) The OR array is programmable.
(c) The OR array is fixed.
(d) The AND array is programmable.
(e) Both the AND array and the OR array are fixed.
(f) Both the AND array and the OR array are programmable.
(g) The AND array is fixed; the OR array is programmable.
(h) The AND array is programmable; the OR array is fixed.

9.2   For a PAL, indicate which of the following statements are true:

(a) The AND array is fixed.
(b) The OR array is programmable.
(c) The OR array is fixed.
(d) The AND array is programmable.
(e) Both the AND array and the OR array are fixed.
(f) Both the AND array and the OR array are programmable.
(g) The AND array is fixed; the OR array is programmable.
(h) The AND array is programmable; the OR array is fixed.

9.3   For a PLA, indicate which of the following statements are true:

(a) The AND array is fixed.
(b) The OR array is programmable.
(c) The OR array is fixed.
(d) The AND array is programmable.
(e) Both the AND array and the OR array are fixed.
(f) Both the AND array and the OR array are programmable.
(g) The AND array is fixed; the OR array is programmable.
(h) The AND array is programmable; the OR array is fixed.

9.4   Given the state diagram shown below for a Moore synchronous sequential machine, implement the machine using a PROM for the δ next-state logic and positive-edge-triggered D flip-flops for the storage elements.

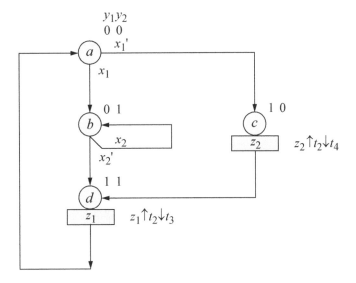

9.5   Given the state diagram shown below for a Moore machine, implement the machine using a PROM for the δ next-state logic and a parallel-in, parallel-out register for the storage elements. The machine has a self-starting state of $y_1y_2y_3 = 111$, which is also a terminal state.

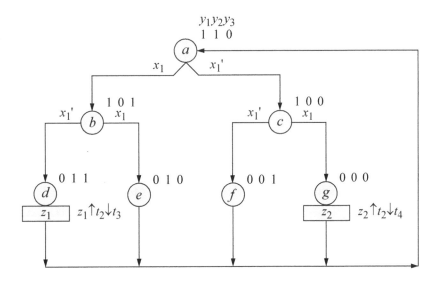

9.6   Given the state diagram shown below with both Moore- and Mealy-type outputs, implement the machine using a PROM for the δ next-state logic and a parallel-in, parallel-out register for the storage elements. Outputs z_1 and z_2 asserted at time t_2 and deasserted at t_3 in their respective states, depending on the value of x_1. Output z_3 is asserted at time t_2 and deasserted at t_3. The machine has a self-starting state of $y_1y_2y_3 = 000$.

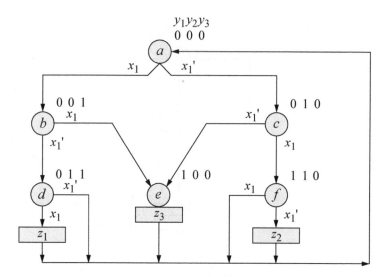

9.7   Synthesize a Moore sequential machine that receives a 4-bit word on a parallel input bus. The input bus is designated $x_1 x_2 x_3 x_4$ as shown below. The parallel bus specifies two 2-bit operands, where $x_1 x_2$ is the multiplicand and $x_3 x_4$ is the multiplier. Inputs x_2 and x_4 are the low-order bits of the multiplicand and multiplier, respectively.

There are four outputs $z_1 z_2 z_3 z_4$ which indicate the product when the multiplicand $x_1 x_2$ is multiplied by the multiplier $x_3 x_4$. Output z_4 is the low-order product bit. All outputs are asserted at time t_2 and deasserted at t_3. Design the Moore machine using a PROM for the δ next-state logic and parallel-in, parallel-out registers for the state flip-flops and the λ output logic.

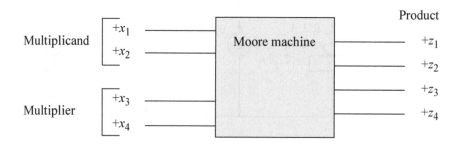

9.8   Design a 3-bit Johnson counter using a PAL device for the δ next-state logic, the D flip-flop storage elements, and the λ output logic. The counter counts in the following sequence: 000, 100, 110, 111, 011, 001, 000.

9.9 Design a 4-bit Moore synchronous sequential machine that operates according to the following sequence: $y_1y_2y_3y_4 = 0000, 1000, 1100, 1110, 1111, 0000$. Output z_1 is asserted unconditionally in state $y_1y_2y_3y_4 = 1111$. Use a PAL device for the δ next-state logic, the positive-edge-triggered D flip-flops, and for the λ output logic. Use clock delayed as a clock input for the D flip-flop that generates output z_1.

9.10 Design a Moore machine whose operation is defined by the state diagram shown below. Use a PAL device for the δ next-state logic, the D flip-flop storage elements, and the λ output logic. Use x_1, x_2, and x_3 as map-entered variables.

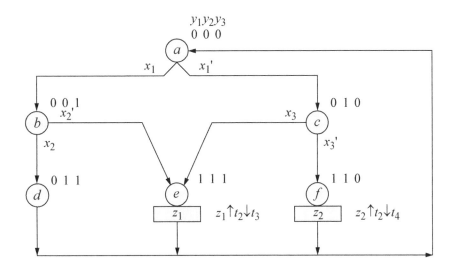

9.11 This problem repeats Problem 9.7, but uses a PAL device instead of a PROM. Synthesize a Moore sequential machine that receives a 4-bit word on a parallel input bus. The input bus is designated $x_1x_2x_3x_4$. The parallel bus specifies two 2-bit operands, where x_1x_2 is the multiplicand and x_3x_4 is the multiplier. Inputs x_2 and x_4 are the low-order bits of the multiplicand and multiplier, respectively.

There are four outputs $z_1z_2z_3z_4$ which indicate the product when the multiplicand x_1x_2 is multiplied by the multiplier x_3x_4. Output z_4 is the low-order product bit. All outputs are asserted at time t_2 and deasserted at t_3. Design the Moore machine using a PAL device for the δ next-state, the state D flip-flops, and the λ output logic. A state diagram is shown below.

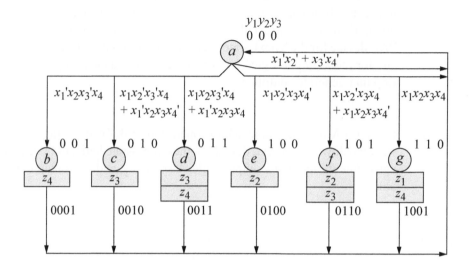

9.12   Implement the state diagram shown below for a Moore machine using a PLA device for the δ next-state logic, the state flip-flops, and the λ output logic. Use x_1 and x_2 as map-entered variables. The outputs are asserted at time t_1 and deasserted at t_3. Before implementing the design, be certain that no state transition sequence passes through states a, c, or d for a path that does not include the corresponding output.

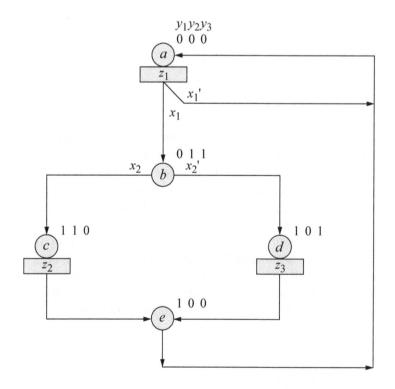

9.13 Design a synchronous counter that counts according to the specifications shown below. Input x_1 is a mode control line such that,

If $x_1 = 0$, then count up
If $x_1 = 1$, then count down

The counter has an initial state of $y_1 y_2 y_3 = 000$. Use a PLA device for the δ next-state logic, the state flip-flops, and the λ output logic. Use input x_1 as a map-entered variable. The counting sequence is shown below.

$y_1\ y_2\ y_3$	x_1
0 0 0	0
0 0 1	0
0 1 0	0
0 1 1	0
1 0 0	0
1 0 1	0
1 1 0	0
1 1 1	0
0 0 0	0
. . .	

$y_1\ y_2\ y_3$	x_1
1 1 1	1
1 1 0	1
1 0 1	1
1 0 0	1
0 1 1	1
0 1 0	1
0 0 1	1
0 0 0	1
1 1 1	1
. . .	

9.14 Implement the Boolean functions shown below by programming a PLA. Do not minimize the functions before programming the PLA. Use the functions as shown in a sum-of-minterms form.

$$z_1 (x_1, x_2, x_3, x_4) = \Sigma_m (4, 5, 10, 11, 12)$$

$$z_2 (x_1, x_2, x_3, x_4) = \Sigma_m (0, 1, 3, 4, 11)$$

9.15 The fuse connections for a PLA are shown below. Generate a truth table for a PROM that would implement the same logic function for output z_1 as the PLA.

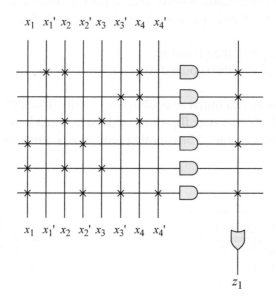

10

Digital and Analog Conversion

Many systems handle both linear (analog) and digital data. Digital-to-analog conversion and analog-to-digital conversion are required in many digital systems when the digital system must be interfaced to the analog system. A digital system has only two distinct voltage levels, corresponding to binary values of logic 0 and logic 1. In contrast, an analog system has voltages that vary continuously from a minimum voltage to a maximum voltage. For example, a 4-bit digital device has a range from 0000 to 1111, providing 16 discrete values. The analog interpretation of this range represents an infinite set of values that range from 0000 (0 volts) to 1111 (+15 volts).

10.1 Operational Amplifier

Before proceeding to the topic of converting between digital and analog systems, a brief introduction to operational amplifiers will be presented. The *operational amplifier* (*op amp*) is an integral device that is used in converting from digital to analog and from analog to digital. The op amp is a linear amplifier that has two inputs and one output, as shown in Figure 10.1. A *linear* circuit contains only linear elements and independent sources. This section on the operational amplifier will concentrate on its functional operation and not on the detailed design of the device.

The op amp is a high gain amplifier that functions as a voltage-controlled voltage source. It has a high input impedance and a low output impedance. The two inputs are labeled V_1 (inverting input indicated by a –) and V_2 (noninverting input indicated by a +); the output is labeled V_{out}. There are also power supply inputs of + and – voltages, that are usually not shown.

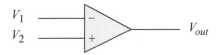

Figure 10.1   Block diagram symbol for an operational amplifier.

In *single-ended* operation, only one signal is amplified — one of the inputs is set to zero. If $V_2 = 0$, then the output V_{out} is out of phase with the input signal V_1. This is called the "inverting mode" and V_1 is the inverting input. When $V_1 = 0$, the output signal V_{out} is in phase with V_2. This is called the "noninverting mode" and V_2 is the noninverting input. The output V_{out} is given by Equation 10.1,

$$V_{out} = A_v (V_2 - V_1) \tag{10.1}$$

where A_v is the voltage gain of the op amp
V_1 is the inverting input
V_2 is the noninverting input

Feedback amplifier   The output V_{out} is proportional to the negative of V_1; thus, the signal that is applied to the V_1 input is shifted in phase by 180° — it is inverted. The gain of op amps of the same type can vary widely. Therefore, the amplifying circuit — which is part of the op amp — should be designed so that the overall gain is independent of the gain of the op amp. This a form of *feedback amplifier* that incorporates negative feedback and is shown in Figure 10.2, where R_F is the feedback resistor.

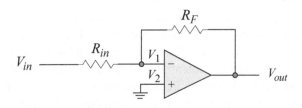

Figure 10.2   Feedback inverting amplifier.

The voltage gain of the overall circuit is defined as shown in Equation 10.2 and the output voltage is shown in Equation 10.3. The output voltage is opposite in polarity to that of the input voltage.

$$A_v = V_{out} / V_{in} = (R_F / R_{in}) \tag{10.2}$$

$$V_{out} = - (R_F / R_{in})\, V_{in} \tag{10.3}$$

Analog summing circuit   More inputs can be added to the feedback amplifier, resulting in a *summing amplifier*. A summing circuit algebraically adds several input signals; thus, the output of the summing amplifier is proportional to the algebraic sum of the inputs. Each input generates a constituent part of the output that is independent of the other inputs. A summing amplifier is shown in Figure 10.3 in which the op amp is configured as an inverting-mode circuit and represents a voltage-controlled voltage source.

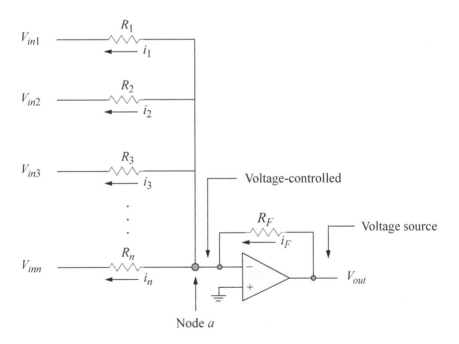

Figure 10.3   An inverting summing amplifier.

By applying Kirchhoff's current law on *nodal analysis*, the equation for V_{out} can be obtained. *Kirchhoff's current law* states that the algebraic sum of the currents entering a node is zero; therefore, the sum of the currents entering node a in Figure 10.3 is equal to the sum of the currents leaving node a. The steps to derive the equation for V_{out} are shown below; the equation for V_{out} is shown in Equation 10.4.

$$i_F = i_1 + i_2 + i_3 + \ldots + i_n$$

and since $i_1 = V_{in1} / R_1$, therefore,

$$-(V_{out} / R_F) = (V_{in1} / R_1) + (V_{in2} / R_2) + (V_{in3} / R_3) + \ldots + (V_{inn} / R_n)$$

$$V_{out} = -R_F [(V_{in1} / R_1) + (V_{in2} / R_2) + (V_{in3} / R_3) + \ldots + (V_{inn} / R_n)] \qquad (10.4)$$

10.2 Digital-to-Analog Conversion

A summing amplifier with a binary-weighted resistor ladder network is used for digital-to-analog conversion. Discrete bits of information in the form of logic 1s and 0s represent the input signals in the form of digital data. This is converted to an analog signal whose magnitude is related to some physical quantity such as voltage. Some applications include: a driver for plotters, an interface for numerical-controlled machine; and a digital-to-audio interface.

Resolution   The resolution is an indication of the number of possible analog output levels expressed as the number of input bits that the converter can accept and is usually specified as a percentage. For example, a 4-bit digital-to-analog converter has 2^4 levels and a resolution of one part in $2^4 - 1$; that is, one part in 15. Expressed as a percentage, the resolution is $(1/15)(100) = 6.67\%$.

For an n-bit converter, the number of discrete steps is $2^n - 1$, where n is the number of bits. An 8-bit converter has $2^8 = 256$ possible output levels, including zero, as shown in Figure 10.4, where the resolution is one part in $2^8 - 1$. Resolution can also be expressed as the number of bits that are converted.

Accuracy   Accuracy is a measure of the deviation of the analog output level from its expected level for any input combination. It is expressed as a percentage of the full-scale output voltage.

Settling time   Settling time is the total time measured from a digital input change to the time that the analog output reaches its new value within $\pm 1/2$ of the low-order bit.

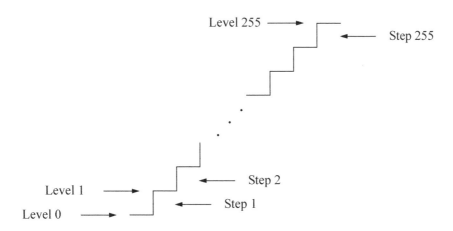

Figure 10.4 Output levels for an 8-bit digital-to-analog converter.

10.2.1 Binary-Weighted Resistor Network Digital-to-Analog Converter

This section will design a 4-bit digital-to-analog converter using a binary-weighted resistor network, which is simply a summing amplifier with weighted inputs. The values of the input resistors are chosen to be inversely proportional to the values of the binary weights; that is, the lowest-valued resistor R is equivalent to the highest binary weight of 2^3, and the highest-valued resistor $8R$ is equivalent to the lowest binary weight of 2^0. The high-order binary input b_3 connects to the smallest-valued resistor R. The circuit converts a 4-bit binary value

$$b_3\, b_2\, b_1\, b_0$$

which connect to resistors R, $2R$, $4R$, and $8R$, respectively. The 4-bit binary inputs will generate an equivalent analog voltage with 16 discrete voltage levels.

The binary inputs connect to four single-pole, double-throw switches; the single-pole of each switch connects to one of the binary-weighted resistors. The switches are analog field-effect transistor (FET) switches that act as either a short circuit or an open circuit for the binary data.

Each input resistor has either no current (connected to ground) or a current dependent on the input voltage level and the value of the input resistor. The amount of current in each resistor depends on the value of the input resistor. A connection to the voltage indicates a logic 1 level; a connection to ground indicates a logic 0 level. All of the input currents are added algebraically and go through the feedback resistor R_F. The binary-weighted resistor network digital-to-analog converter is shown in Figure 10.5.

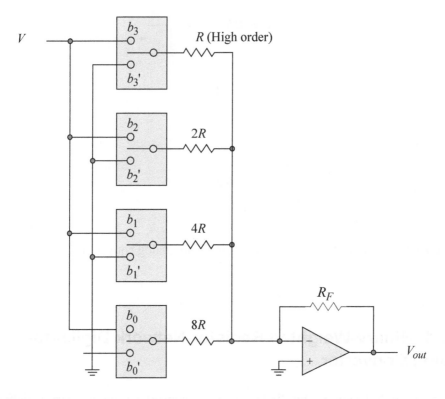

Figure 10.5   Binary-weighted resistor network for a digital-to-analog converter.

Using Equation 10.4, the voltage for V_{out} can be obtained as shown in Equation 10.5, where the (V / iR) terms represent currents.

$$V_{out} = -R_F \left[(V / R) + (V / 2R) + (V / 4R) + (V / 8R)\right] \tag{10.5}$$

However, the smallest resistor R corresponds to b_3; $2R$ corresponds to b_2; $4R$ corresponds to b_1; and $8R$ corresponds to b_0. Therefore, V_{out} can be represented by Equation 10.6, where the $b_i = 1$ or 0. Note that the term $(b_2 / 2)$ is one-half the current of b_3. Rearranging Equation 10.6 yields Equation 10.7.

$$V_{out} = -\left[(VR_F / R)\left[b_3 + (b_2 / 2) + (b_1 / 4) + (b_0 / 8)\right]\right] \tag{10.6}$$

$$V_{out} = -(VR_F / 8R)(8b_3 + 4b_2 + 2b_1 + b_0) \tag{10.7}$$

By adjusting ($VR_F / 8R$), the level of the output voltage V_{out} can be adjusted. Since the b_is are either 1s or 0s, the output voltage V_{out} will be proportional to the binary number $b_3 b_2 b_1 b_0$; that is, there will be 16 unique values. The binary-weighted resistors produce a binary-weighted set of voltages.

If $b_3 b_2 b_1 b_0 = 0000$, then from Equation 10.7,

$$V_{out} = - (VR_F / 8R) (0 + 0 + 0 + 0)$$
$$= 0$$

If $b_3 b_2 b_1 b_0 = 0001$, then from Equation 10.7,

$$V_{out} = - (VR_F / 8R) (0 + 0 + 0 + 1)$$
$$= - (VR_F / 8R) 1$$

If $b_3 b_2 b_1 b_0 = 1000$, then from Equation 10.7, V_{out} is 8 times more negative than $b_3 b_2 b_1 b_0 = 0001$, as shown below.

$$V_{out} = - (VR_F / 8R) [8(1) + 0 + 0 + 0]$$
$$= - (VR_F / 8R) 8$$

If $b_3 b_2 b_1 b_0 = 1100$, then from Equation 10.7, V_{out} is 12 times more negative than $b_3 b_2 b_1 b_0 = 0001$, as shown below.

$$V_{out} = - (VR_F / 8R) [8(1) + 4(1) + 0 + 0]$$
$$= - (VR_F / 8R) 12$$

Finally, if $b_3 b_2 b_1 b_0 = 1111$, then from Equation 10.7, V_{out} is 15 times more negative than $b_3 b_2 b_1 b_0 = 0001$, as shown below.

$$V_{out} = - (VR_F / 8R) [8(1) + 4(1) + 2(1) + 1(1)]$$
$$= - (VR_F / 8R) 15$$

Examples that show the relationship between the binary input data and the analog output voltage are shown in Figure 10.6, Figure 10.7, Figure 10.8, Figure 10.9, and Figure 10.10. All voltages that are shown in the figures in this chapter are the actual voltages as measured with a digital voltmeter.

Figure 10.6 shows a 4-bit binary input of $b_3 b_2 b_1 b_0 = 0101$ (5_{10}), yielding an output voltage of -6.250 volts, which is the actual voltage as measured with a voltmeter. Figure 10.7 shows a 4-bit binary input of $b_3 b_2 b_1 b_0 = 1010$ (10_{10}), yielding an output voltage of -12.500 volts, which is twice the voltage of $b_3 b_2 b_1 b_0 = 0101$.

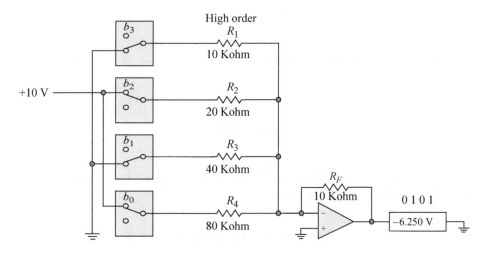

Figure 10.6　　Digital-to-analog converter with a binary input of $b_3 b_2 b_1 b_0 = 0101$.

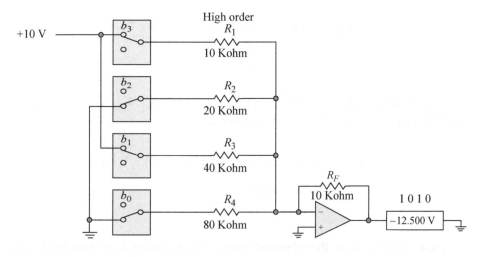

Figure 10.7　　Digital-to-analog converter with a binary input of $b_3 b_2 b_1 b_0 = 1010$.

Figure 10.8 shows a 4-bit binary input of $b_3 b_2 b_1 b_0 = 0001$ (1_{10}), yielding an output voltage of -1.250 volts. Figure 10.8 shows a 4-bit binary input of $b_3 b_2 b_1 b_0 = 1100$ (12_{10}), yielding an output voltage of -15.000 volts, which is 12 times the voltage of $b_3 b_2 b_1 b_0 = 0001$.

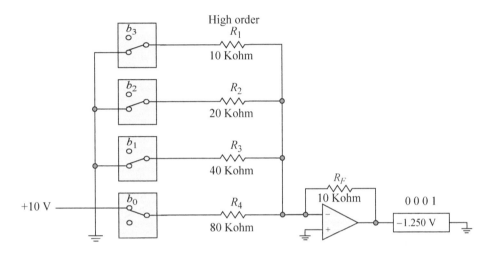

Figure 10.8   Digital-to-analog converter with a binary input of $b_3 b_2 b_1 b_0 = 0001$.

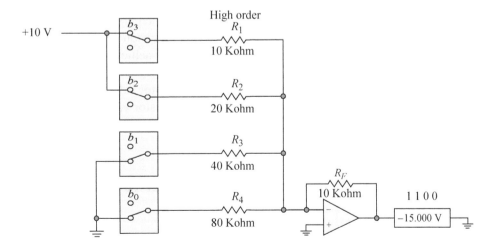

Figure 10.9   Digital-to-analog converter with a binary input of $b_3 b_2 b_1 b_0 = 1100$.

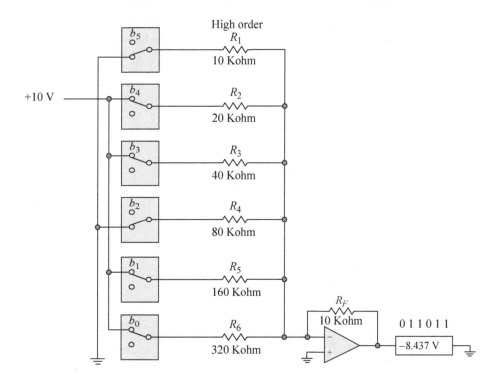

Figure 10.10   Digital-to-analog converter with a binary input of $b_3\ b_2\ b_1\ b_0 =$ 011011.

There is a problem with the binary-weighted resistor network in that the precision of the digital-to-analog converter is limited by the precision of the resistors. If n binary-weighted resistors are used, then it is difficult to achieve tight tolerances for large-valued resistors if $n \geq 5$.

This problem can be resolved by using an $R - 2R$ resistor ladder network. Since all of the resistors have a value of either R or $2R$, large resistor values are not required. Closer tolerance is also possible.

10.2.2 $R - 2R$ Resistor Network Digital-to-Analog Converter

A digital-to-analog converter that is more reasonable for converting digital numbers with a large number of bits uses a ladder network in which there are only two values of resistors: R and $2R$. This solves the problem associated with binary-weighted resistors because there are only two values of resistors regardless of the number of binary inputs. The derivation of the $R - 2R$ resistor ladder network is described below.

The equivalent resistance for the two $2R$ resistors in parallel shown in Figure 10.11 is

$$[(2R)(2R)] / (2R + 2R) = 4R^2 / 4R = R$$

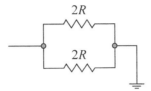

Figure 10.11   Parallel resistors with an equivalent resistance of R.

Now the lower $2R$ resistor is replaced with the equivalent network shown in Figure 10.12. The equivalent resistance of the network is still R.

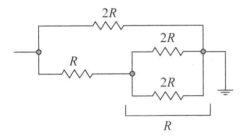

Figure 10.12   Next step in deriving the $R - 2R$ ladder network.

Now the network is redrawn as a ladder network as shown in Figure 10.13, where the equivalent resistance of the entire network is still R.

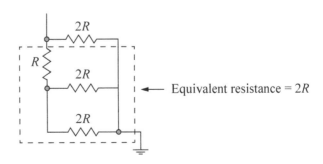

Figure 10.13   Next step in deriving the $R - 2R$ ladder network.

Now the lower $2R$ resistor is replaced with the equivalent network shown in Figure 10.14. The equivalent resistance of the network is still R. This process can be continued to any size ladder. A 4-bit $R-2R$ ladder digital-to-analog converter is shown in Figure 10.15.

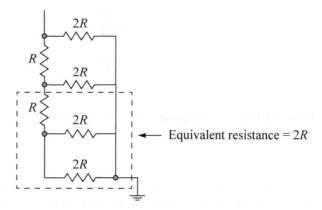

Figure 10.14   Next step in deriving the $R-2R$ ladder network.

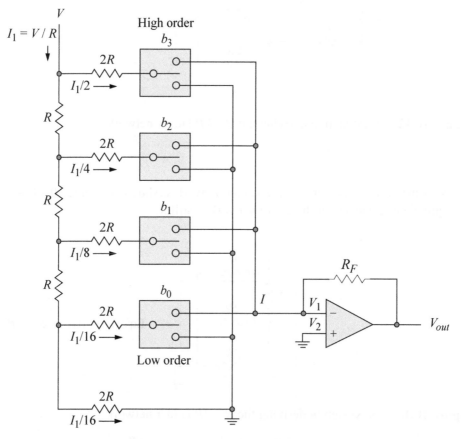

Figure 10.15   A 4-bit $R-2R$ digital-to-analog converter.

When the binary input $b_i = 1$, the switch is in the "up" position; when the binary input $b_i = 0$, the switch is in the "down" position. When a switch is in the "up" position, then its current contributes to the total current I, otherwise it does not. The total current is specified in Equation 10.8, where R is the equivalent resistance of the ladder network and V is the input voltage.

$$I = [(V - V_1) / R] [1/2(b_3) + 1/4(b_2) + 1/8(b_1) + 1/16(b_0)] \qquad (10.8)$$

However, I is also equal to $I = (V_1 - V_{out}) / R_F$. Therefore, Equation 10.9 states

$$(V_1 - V_{out}) / R_F = [(V - V_1) / R] [1/2(b_3) + 1/4(b_2) + 1/8(b_1) + 1/16(b_0)] \qquad (10.9)$$

Assuming that the gain A_v of the op amp is very large (50,000 or more) so that V_{out} $\gg V_1$ yields Equation 10.10 from which the final equation for V_{out} is shown in Equation 10.11.

$$- V_{out} / R_F = (V / R) [1/2(b_3) + 1/4(b_2) + 1/8(b_1) + 1/16(b_0)] \qquad (10.10)$$

$$V_{out} = - [(V R_F) / (16R)] (8b_3 + 4b_2 + 2b_1 + b_0) \qquad (10.11)$$

Thus, the desired digital-to-analog conversion has been achieved using only two values of resistors (R and $2R$) in the ladder network. The binary-weighted currents produce a binary-weighted set of voltages on V_{out}.

Examples of $R - 2R$ resistor ladder networks for digital-to-analog conversion are shown in Figure 10.16 and Figure 10.17 for 4-bit converters. The binary input data for Figure 10.16 is $b_3 b_2 b_1 b_0 = 0101$ (5_{10}), which equates to a voltage of -2.844 volts. The binary input data for Figure 10.17 is $b_3 b_2 b_1 b_0 = 1010$ (10_{10}), which equates to a voltage of -5.687 volts, twice the voltage of Figure 10.16. The values can be verified by substituting the variables from Figure 10.16 into Equation 10.11, as shown below.

$$V_{out} = - [(V R_F) / (16R)] (8b_3 + 4b_2 + 2b_1 + b_0)$$

$$= - [(10)(9,100) / (16)(10,000)] [8(0) + 4(1) + 2(0) + 1(1)]$$

$$= - 0.56875 (4 + 1)$$

$$= - 2.84375 \text{ volts}$$

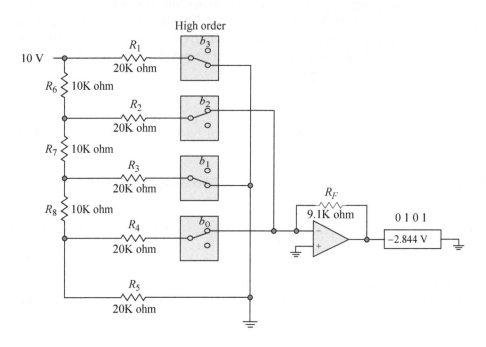

Figure 10.16   An $R-2R$ resistor ladder network for a digital-to-analog converter for a binary input of $b_3 b_2 b_1 b_0 = 0101$.

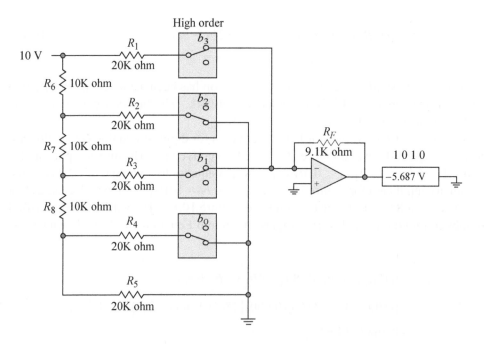

Figure 10.17   An $R-2R$ resistor ladder network for a digital-to-analog converter for a binary input of $b_3 b_2 b_1 b_0 = 1010$.

10.3 Analog-to-Digital Conversion

The function of an analog-to-digital (A/D) converter is to measure the amplitude of an analog input voltage and produce a binary output that represents the value of the analog input voltage. The binary output is proportional to the analog input. Three types of analog-to-digital converters will be presented in this section: counter A/D converter; successive approximation A/D converter; and a simultaneous (flash) A/D converter. All three types use comparators in the conversion process.

10.3.1 Comparators

A comparator is an operational amplifier that compares the amplitude of one voltage with another voltage; that is, a comparator determines if an input signal level is greater than or less than a reference voltage level or whether the input signal is outside the range of two voltage levels. It provides a logic output that indicates the amplitude relationship between two analog signals. A comparator is basically an operational amplifier that operates in *open-loop* mode; that is, there is no feedback from the output to either of the two inputs.

The symbol for a comparator is shown in Figure 10.18. The symbol for a comparator is the same as that for an operational amplifier; thus, the symbol is labelled *comp* (comparator) to avoid confusion. The input voltage V_{in} is compared with the reference voltage V_{ref}. The input voltage V_{in} can also be connected to *comp+* and V_{ref} can also be connected to *comp−*. The input currents of a comparator are zero.

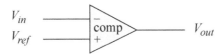

Figure 10.18   An operational amplifier used as a comparator.

Using the configuration shown in Figure 10.18,

$$\text{If } V_{in} < V_{ref}, \text{ then } V_{out} = \text{logic 1}$$
$$\text{If } V_{in} > V_{ref}, \text{ then } V_{out} = \text{logic 0}$$

This can be restated as follows: the output will be at a logic 1 voltage level whenever the difference voltage between the noninverting input and the inverting input is positive; that is

$$\text{If } (V_+ - V_-) > 0, \text{ then } V_{out} = \text{logic 1}$$

$$\text{If } (V_+ - V_-) < 0, \text{ then } V_{out} = \text{logic 0}$$

The input/output characteristics are shown in Figure 10.19. The output of the comparator changes state when $V_+ = V_-$. A single comparator can be considered to be a 1-bit analog-to-digital converter.

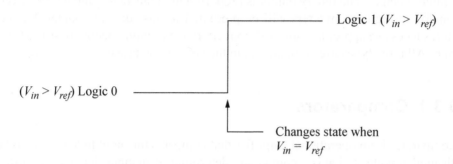

Figure 10.19   Input/output characteristics of a comparator.

A comparator can be connected to the input voltage and the reference voltage two ways, as shown in Figure 10.20. In Figure 10.20(a), the following statements are valid:

$$(V_{in} < V_{ref}) = \text{logic } 1; (V_{in} > V_{ref}) = \text{logic } 0$$

$$(V_+ - V_-) > 0 = \text{logic } 1; (V_+ - V_-) < 0 = \text{logic } 0$$

In Figure 10.20(b), the following statements are valid:

$$(V_{in} > V_{ref}) = \text{logic } 1; (V_{in} < V_{ref}) = \text{logic } 0$$

$$(V_+ - V_-) > 0 = \text{logic } 1; (V_+ - V_-) < 0 = \text{logic } 0$$

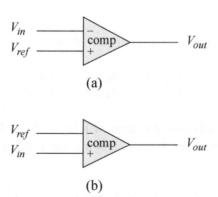

Figure 10.20   Two ways to connect a comparator.

10.3.2 Counter Analog-to-Digital Converter

This approach uses a count-up binary counter whose output is connected to a digital-to-analog (D/A) converter and a register. The analog output of the D/A converter is used as a reference voltage which connects to the inverting input of a comparator. An input voltage connects to the noninverting input of the comparator and is compared to the reference voltage that is generated by the D/A converter. The input voltage V_{in} is sampled and held so that V_{in} remains constant during the conversion process.

A 4-bit counter A/D converter is shown in Figure 10.21. The counter is initially reset; therefore, $V_{in} > V_{ref}$ (logic 1). The logic 1 output of the comparator is ANDed with the clock pulses and the counter continues to count. When the output of the comparator changes from a logic 1 ($V_{in} > V_{ref}$) to a logic 0 ($V_{in} \leq V_{ref}$), the clock for the counter is disabled and the binary count is loaded into the register. The output of the D/A converter is proportional to the count.

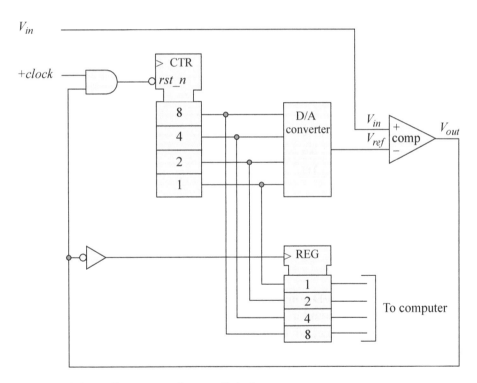

Figure 10.21   Counter analog-to-digital converter.

An example demonstrating the operation of the counter A/D converter is shown in Table 10.1 in which the analog input voltage V_{in} corresponds to binary 1100. The counter increments from 0000 to 1011, during which time $V_{in} > V_{ref}$, maintaining V_{out} at a logic 1 value. When the count equals 1100, $V_{in} \leq V_{ref}$ and V_{out} changes from a logic 1 to a logic 0. This disables the clock to the counter and loads the binary value that corresponds to V_{in} into the register. This binary count is the digital representation of the analog input V_{in}. The number of cycles may be 2^n, where n is the number of bits in the counter. The counter is then reset for the next operation and the process repeats.

**Table 10.1 Example Illustrating the
Operation of the Counter A/D Converter**

Binary Counter 8 4 2 1	$V_{in} > V_{ref}$	$V_{in} \leq V_{ref}$	V_{out}
0 0 0 0	Yes	No	1
0 0 0 1	Yes	No	1
0 0 1 0	Yes	No	1
0 0 1 1	Yes	No	1
0 1 0 0	Yes	No	1
0 1 0 1	Yes	No	1
0 1 1 0	Yes	No	1
0 1 1 1	Yes	No	1
1 0 0 0	Yes	No	1
1 0 0 1	Yes	No	1
1 0 1 0	Yes	No	1
1 0 1 1	Yes	No	1
1 1 0 0	**No**	**Yes**	**0**

10.3.3 Successive Approximation Analog-to-Digital Converter

Successive approximation A/D conversion is a popular method of conversion and is relatively fast compared to counter A/D conversion and also uses a D/A converter in its implementation. The successive approximation method has a fixed conversion time that is proportional to the number of bits in the binary data. The *flash* converter — which is presented in Section 10.3.4 — is the fastest A/D converter.

Successive approximation is similar to the binary search technique used in programming. For example, assume that a number between 0 and 15 is to be found by asking the least number of questions that can be answered by "yes" or "no," where a "yes" equals 1 and a "no" equals 0. The technique is shown below for the number 10. Thus, it is possible to find one number out of 16 by asking four questions (binary search). Similarly, to find one number between 0 and 255 requires eight questions. The block diagram for a successive approximation analog-to-digital converter is shown in Figure 10.22 and is controlled by a state machine.

Is the number	≥ 8	≥ 12	≥ 10	= 11
	Yes	No	Yes	No
	1	0	1	0

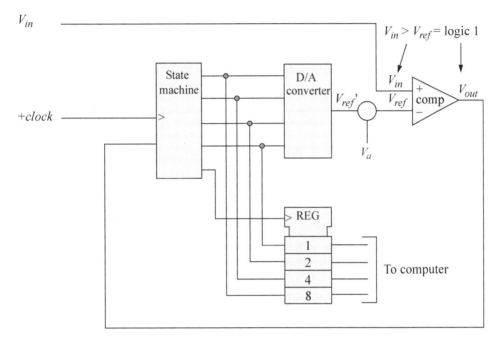

Figure 10.22   Successive approximation analog-to-digital converter.

The output of the D/A converter V_{ref}' is reduced by V_a by an amount that is 1/2 of the voltage change in the output of the D/A converter due to a change in the low-order bit of the binary data, as shown in Figure 10.23. Thus, V_a is a bias voltage.

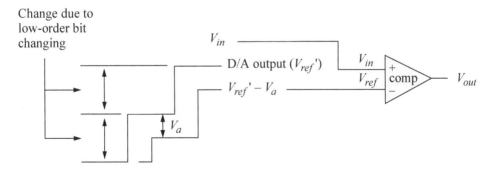

Figure 10.23   Illustration of V_{ref}' minus the bias voltage V_a to produce V_{ref}.

The reason for using the bias voltage is so that there will be only two states:

$$\text{One state for } V_{in} > V_{ref}$$
$$\text{One state for } V_{in} < V_{ref}$$

Three states are not necessary; that is, there should be no state for $V_{in} = V_{ref}$. The steps shown below will go through a sequence of approximations to obtain a digital representation of V_{in}. The input voltage V_{in} is obtained from a sample-and-hold circuit so that V_{in} is constant during the A/D conversion. In this description, four bits will represent the voltage. The value of the bits is unknown; therefore, they will be represented by x_i. The four bits are as follows, where b_0 is the low-order bit:

	8	4	2	1
Bit	b_3	b_2	b_1	b_0
	x_3	x_2	x_1	x_0

Step 1   Set the high-order bit $x_3 = 1$ and the remaining bits $x_2 x_1 x_0 = 000$; that is try 1000. This number (1000) is sent to the D/A converter. The output of the D/A converter V_{ref}' is modified by V_a to produce V_{ref} and is compared to V_{in}. If $V_{in} < V_{ref}$, then 1000 is too large. Thus, $(V_+ - V_-) < 0$ and $V_{out} = $ logic 0 making the high-order bit $x_3 = 0$. If $V_{in} > V_{ref}$, then 1000 is too small. Thus, $(V_+ - V_-) > 0$ and $V_{out} = $ logic 1 making $x_3 = 1$. Therefore, the bit under test is determined by the following:

$$\text{If } V_{out} = 0, \text{ then } b_i = 0$$
$$\text{If } V_{out} = 1, \text{ then } b_i = 1$$

Step 2   The value of the high-order bit b_3 is now known; therefore, the four bits are $b_3 x_2 x_1 x_0$. Now set $x_2 = 1$ and $x_1 x_0 = 00$; that is, try $b_3 100$ — the value of b_3 was determined by step 1. The number $b_3 100$ is sent to the D/A converter whose output is modified by V_a and compared to the input voltage V_{in}. If $V_{in} < V_{ref}$, then $b_3 100$ is too large. Thus, $(V_+ - V_-) < 0$ and $V_{out} = $ logic 0 making bit $x_2 = 0$. If $V_{in} > V_{ref}$, then $b_3 100$ is too small. Thus, $(V_+ - V_-) > 0$ and $V_{out} = $ logic 1 making $x_2 = 1$.

The process repeats until the value of all bits is determined. The state machine then generates a pulse to the clock input of the register and the binary (digital) data is loaded into the register. As V_{ref} approaches V_{in} from above or below, then when V_{ref} crosses V_{in}, V_{out} changes state.

The successive approximation method determines each bit of the binary data sequentially, beginning with the high-order bit. The hardware utilized in this technique is similar to that used in the counter A/D approach, except that a state machine controls the operation. Binary data of n bits requires n clock cycles. A state diagram for the operation of the state machine is shown in Figure 10.24. An example of successive approximation A/D conversion is shown in Figure 10.25 for an input voltage of $V_{in} = 1010$.

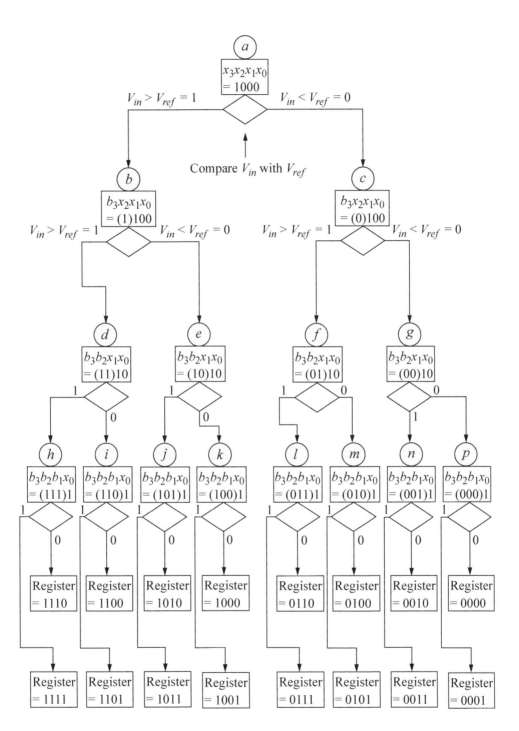

Figure 10.24   State diagram illustrating the sequence of operations for the successive approximation analog-to-digital converter.

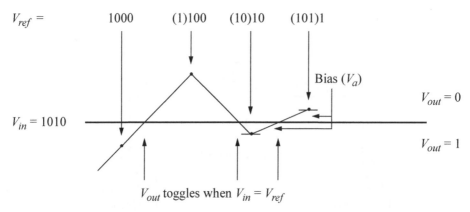

$V_{ref} =$ 1000   (1)100   (10)10   (101)1

Bias (V_a)

$V_{out} = 0$

$V_{in} = 1010$

$V_{out} = 1$

V_{out} toggles when $V_{in} = V_{ref}$

Figure 10.25   Example of successive approximation A/D conversion for $V_{in} = 1010$.

10.3.4 Simultaneous Analog-to Digital Converter

A simultaneous (or flash) A/D converter is one of the fastest converters, since the conversion process takes place by comparators operating in parallel. It does, however, require a large amount of hardware because many comparators must be utilized. The analog input voltage V_{in} is compared to a reference voltage V_{ref}. When $V_{in} > V_{ref}$ for a particular comparator, the output voltage V_{out} for that comparator is a logic 1.

The output of a simultaneous converter is usually connected to the inputs of a priority encoder or a ROM that generates an n-bit binary code. The reference voltage for each comparator is established by a resistor voltage divider as shown in Figure 10.26.

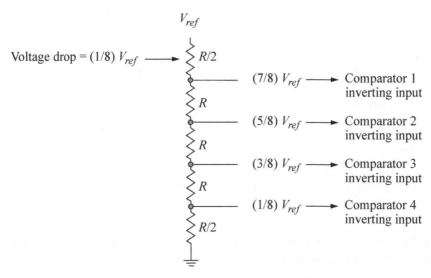

V_{ref}

Voltage drop = $(1/8) V_{ref}$ $\longrightarrow$ $R/2$

$(7/8) V_{ref} \longrightarrow$ Comparator 1 inverting input

R

$(5/8) V_{ref} \longrightarrow$ Comparator 2 inverting input

R

$(3/8) V_{ref} \longrightarrow$ Comparator 3 inverting input

R

$(1/8) V_{ref} \longrightarrow$ Comparator 4 inverting input

$R/2$

Figure 10.26   Voltage divider for a 4-comparator flash A/D converter.

The reference voltage (7/8) V_{ref}, which is connected to the inverting input of comparator 1 of Figure 10.26, is the result of a voltage drop of (1/8) V_{ref} across the top resistor $R/2$ and is derived as follows:

$$\frac{R/2}{R/2 + R + R + R + R/2} = \frac{R/2}{4R/1} = R/2 \times 1/4R = 1/8$$

That is, 1/8 of the voltage V_{ref} is dropped across the top resistor $R/2$, which leaves (7/8) V_{ref} to be applied to the inverting input of comparator 1. In a similar manner, (5/8) V_{ref} is obtained by

$$\frac{R/2 + R}{R/2 + R + R + R + R/2} = \frac{3R/2}{4R/1} = 3R/2 \times 1/4R = 3/8$$

That is, 3/8 of the voltage V_{ref} is dropped across the top resistors $R/2$ and R, which leaves (5/8) V_{ref} to be applied to the inverting input of comparator 2. In a similar manner, the remaining fractions of V_{ref} are obtained for their respective comparators. A 4-comparator simultaneous A/D converter is shown in Figure 10.27.

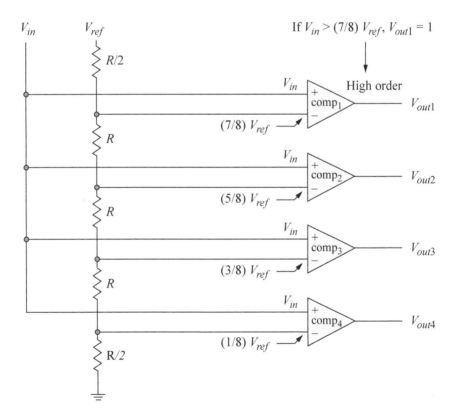

Figure 10.27   Simultaneous A/D converter.

The resistor network forms a voltage divider that establishes the reference voltage to each comparator: $(7/8)V_{ref}$, $(5/8)V_{ref}$, $(3/8)V_{ref}$, $(1/8)V_{ref}$. The following statements hold true:

$$\text{If } V_{in} > V_{ref}, \text{ then } V_{out} = 1$$
$$\text{If } V_{in} \leq V_{ref}, \text{ then } V_{out} = 0$$

For example, if $V_{in} > (5/8)V_{ref}$, then $V_{out2} = 1$. If $V_{out1} V_{out2} V_{out3} V_{out4} = 0011$, then $(3/8)V_{ref} < V_{in} \leq (5/8)V_{ref}$ and the resolution is $(1/4)V_{ref}$. Also, if the comparator outputs are $V_{out1} V_{out2} V_{out3} V_{out4} = 0001$, then $(1/8)V_{ref} < V_{in} \leq (3/8)V_{ref}$ and the resolution is $(1/4)V_{ref}$.

With four comparators, the resolution is $(1/4)V_{ref}$; with eight comparators, the resolution is $(1/8)V_{ref}$. When A/D comparators are designed, there is a *resolution* error because a finite number of binary bits is used to represent an infinite set of input voltage levels. The error that is produced when attempting to represent an infinite set by a finite set is called a *quantization* error.

The 4-bit simultaneous A/D comparator can produce five possible output levels (states). If binary numbers are used to represent the five possible levels, then Table 10.2 is obtained. The comparator outputs are used as inputs to a priority encoder or ROM, which then produce the desired binary numbers.

The simultaneous A/D converter output occurs after 1 unit of time; the successive approximation A/D converter output occurs after n units of time; the counter A/D converter output occurs after 2^n units of time.

Table 10.2 Comparator Outputs for a Simultaneous A/D Converter

	Comparator Outputs				Binary Outputs of ROM or Priority Encoder
	V_{out1}	V_{out2}	V_{out3}	V_{out4}	
$V_{in} \leq (1/8)V_{ref} =$	0	0	0	0	0 0 0
$(1/8)V_{ref} < V_{in} \leq (3/8)V_{ref} =$	0	0	0	1	0 0 1
$(3/8)V_{ref} < V_{in} \leq (5/8)V_{ref} =$	0	0	1	1	0 1 0
$(5/8)V_{ref} < V_{in} \leq (7/8)V_{ref} =$	0	1	1	1	0 1 1
$(7/8)V_{ref} < V_{in} =$	1	1	1	1	1 0 0

Examples of simultaneous analog-to-digital converters are shown in Figure 10.28 and Figure 10.29. Figure 10.28 is designed using four comparators, providing digital outputs of $comp_1 comp_2 comp_3 comp_4 = 0011$, in which the voltage levels correspond to transistor-transistor logic (TTL) levels. Figure 10.29 uses five comparators to generate digital data of $comp_1 comp_2 comp_3 comp_4 comp_5 = 00111$. Figure 10.29 also shows the reference voltage V_{ref} at each comparator inverting input. If $V_{in} > V_{ref}$, then $V_{out} =$ a logic 1; if $V_{in} \leq V_{ref}$, then $V_{out} =$ a logic 0. Note that when the first logic 1 appears in the output, all following outputs are a logic 1.

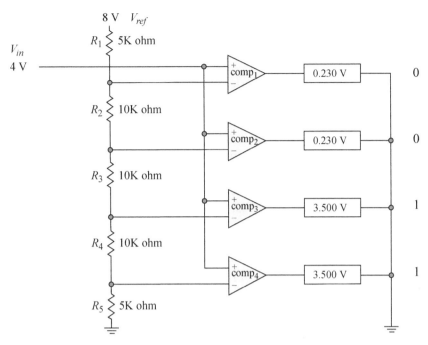

Figure 10.28   A 4-comparator simultaneous A/D converter.

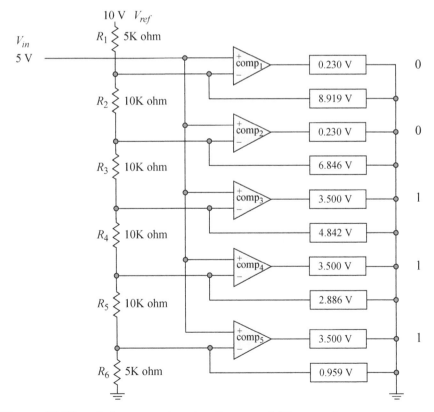

Figure 10.29   A 5-comparator simultaneous A/D converter.

The outputs of Figure 10.28 can be easily verified by calculating the reference voltage at the inverting input of each comparator, then comparing that voltage with the input voltage.

$$\text{If } V_{in} > V_{ref}, \text{ then } V_{out} = 1$$
$$\text{If } V_{in} \leq V_{ref}, \text{ then } V_{out} = 0$$

The input voltage $V_{in} = 4$ volts; the reference voltage $V_{ref} = 8$ volts. Using the calculations for Figure 10.26 in conjunction with Figure 10.27, the reference voltage at the inverting inputs of each comparator is as follows:

$$comp_1 = 7 \text{ volts; therefore } V_{out} = 0$$
$$comp_2 = 5 \text{ volts; therefore } V_{out} = 0$$
$$comp_3 = 3 \text{ volts; therefore } V_{out} = 1$$
$$comp_4 = 1 \text{ volts; therefore } V_{out} = 1$$

which agrees with the voltage outputs of Figure 10.28, when interpreted as TTL voltage levels, where a low voltage level is typically $+0.2$ volts and a high voltage level is typically $+3.4$ volts.

10.4 Problems

10.1   Determine the resolution of a 4-bit digital-to-analog (D/A) converter in terms of percentage.

10.2   Each increment in the binary count of a 5-bit D/A converter will increase the output voltage by what fraction of the maximum output voltage?

10.3   Determine the resolution of a 5-bit D/A converter in terms of percentage.

10.4   Using the operational amplifier shown below,

    (a)   Determine V_{out} for the following conditions:
        $A_v = 0.533$
        $V_{in} = 3.75$ volts

    (b)   Determine V_{out} for the following conditions:
        $V_{in} = 1.5$ volts
        $R_{in} = 5 \text{ K}\Omega$
        $R_F = 10 \text{ K}\Omega$

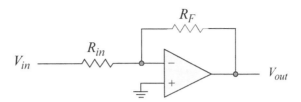

10.5   The input voltage to an operational amplifier is 10 millivolts generating an output of 2 volts. Determine the closed-loop voltage gain.

10.6   Given a closed-loop voltage gain of 330 for an inverting operational amplifier, determine the value of the feedback resistor if R_{in} is 1 KΩ.

10.7   A D/A converter uses a binary-weighted, four-resistor network with a smallest resistor value of 25 KΩ, a feedback resistor R_F of 10 KΩ, and a voltage of +5 volts. The operational amplifier provides a very high impedance load to the resistor network, and the inverting input is a "virtual" ground, so that the output is proportional to the current through the feedback resistor R_F — the sum of the input currents. Almost all of the current is through R_F and into the low impedance output of the operational amplifier. The inverting input is approximately 0 volts.

The inverting input of the operational amplifier is at 0 volts (virtual ground). Assuming that a binary 1 corresponds to +5 volts, then the current through any of the input resistors is 5 volts divided by the resistance value.

Almost none of this current goes into the inverting input, because of its extremely high impedance. Therefore, all of the current goes through the feedback resistor. Since one end of R_F is at 0 volts, the voltage drop across R_F equals the output voltage

(a)   Determine the output voltage V_{out} when each input is connected in turn (one at a time) to +5 volts.

(b)   Determine V_{out} for an input binary number of $b_3\ b_2\ b_1\ b_0 = 1011$.

10.8   A D/A converter used a binary-weighted resistor network of 150 KΩ , 75 KΩ, 37.5 KΩ, and 18.75 KΩ (high order). The feedback resistor $R_F = 20$ KΩ. The input voltage $V_{in} = +3$ volts. Determine the voltage increment in V_{out} that is caused by a change in the low-order bit of the binary input.

10.9   Given the initial equation shown below for a binary-weighted D/A converter, determine the output of the D/A converter shown on the following page at time t_4 when the waveforms on the following page are applied to the inputs, where a high level represents +5 volts and a low level represents 0 volts:

$$V_{out} = -R_f(V/R_1 + V/2R_1 + V/4R_1 + V/8R_1 + \ldots)$$

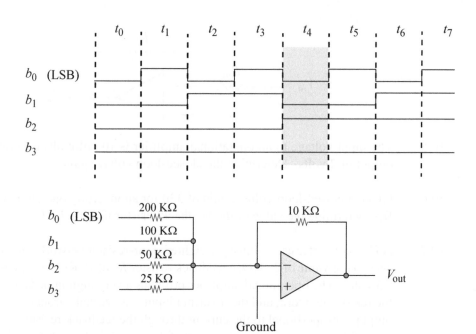

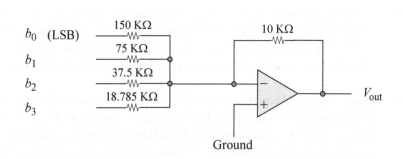

10.10   Determine the output voltage V_{out} of a D/A converter which uses a binary-weighted resistor network, where

$V = +5$ volts
$R_F = 10 \text{ K}\Omega$
$R_1 = 10 \text{ K}\Omega$ (low-value resistor)

and the binary input data $= 110110$ (low-order)

10.11   A binary-weighted resistor network D/A converter is shown below. Four input switches b_0 through b_3 connect to 0 volts or $+3$ volts. With the switches set as follows, determine the answers for parts (a) through (c):

$b_0 = 0$ volts, $b_1 = 0$ volts $b_2 = +3$ volts $b_3 = +3$ volts

(a) R_{in} (b) A_v (c) V_{out}

10.12   The binary-weighted resistor network of a D/A converter consists of the resistors shown below. Calculate R_{in} for the ladder network.

$$R_1 = 150 \text{ K}\Omega$$
$$R_2 = 75 \text{ K}\Omega$$
$$R_3 = 37.5 \text{ K}\Omega$$
$$R_4 = 18.75 \text{ K}\Omega$$

10.13   A binary-weighted resistor network D/A converter is shown below. Four input switches b_0 through b_3 connect to 0 volts or +3 volts. With the switches set to $b_3 b_2 b_1 b_0 = 0101$, determine the value of R_{in}, A_v, and V_{out}.

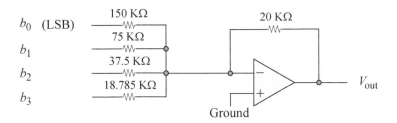

10.14   A binary-weighted resistor D/A converter has $R = 18.75$ KΩ, $R_F = 30$ KΩ, and $V = +5$ volts. Obtain V_{out} when the digital inputs are $b_4 b_3 b_2 b_1 b_0 = 11001$ (low order), where a 1 indicates a closed switch connecting to +5 volts.

10.15   Determine the output voltage V_{out} of a D/A converter which uses an $R-2R$ resistor network, where the supply voltage $V = +5$ volts, $R_F = 5$ KΩ, and $R = 5$ KΩ. The binary data input is 110110 (low order).

10.16   A binary-coded decimal (BCD) $R-2R$ D/A converter has eight inputs that represent a voltage whose maximum value is 99 volts. The inputs are labeled as follows:

$$b_{30}\, b_{20}\, b_{10}\, b_{00} \qquad b_3\, b_2\, b_1\, b_0$$
$$.$$

where b_{30} and b_0 are the low-order bit positions of the tens and units digits, respectively. Write the equation for V_{out}.

10.17   Determine the output voltage V_{out} of a D/A converter that uses an $R-2R$ resistor network, where $V = +12$ volts, $R_F = 10$ KΩ, and $R = 5$ KΩ. The binary input is 100110 (low order).

10.18   A counter A/D converter is shown below. The analog input voltage V_{in} has a range of 0 to +3 volts. The output of the D/A converter has a range of 0 to +3 volts. The counter is initially reset to 0000. With the analog input voltage V_{in} equal to +1.95 volts, determine the following:

(a) The binary output.
(b) The number of clock pulses required to obtain the binary output.

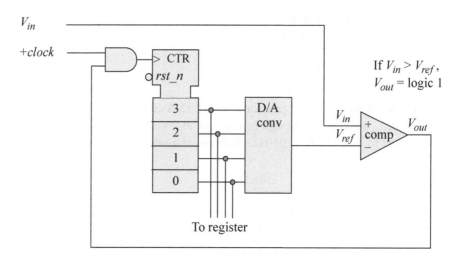

10.19   Draw the state diagram for the state machine in a successive approximation analog-to-digital converter, where the analog input corresponds to the binary number 10110. Draw only the path that corresponds to the binary number 10110.

10.20   Design a simultaneous/flash comparator using five comparators that will generate a digital output of 00001. Let $V_{in} = 2$ volts and $V_{ref} = 10$ volts.

10.21   Show all possible output binary words of a simultaneous/flash comparator analog-to-digital converter that uses eight comparators *comp1* through *comp8*, where *comp8* is the low-order bit.

10.22   Design a simultaneous/flash comparator using five comparators that will generate a digital output of 00111. Let $V_{in} = 5$ volts and $V_{ref} = 8$ volts.

11

Magnetic Recording Fundamentals

This chapter covers the reading and writing of magnetic recording media as applied to magnetic tapes, disks, and other magnetic systems. In order to achieve high density — and thus high capacity — a binary bit must be recorded in the smallest possible space on the magnetic medium and as close as possible to adjacent bits. There are two main types of magnetic recording media: flexible media such as tapes and hard media such as disks and drums.

Figure 11.1 depicts a write head that records binary data on a magnetic recording medium. The *flying height* of the head h is the distance between the write head and the recording medium. The fringe field is maximum at $x = 0$ and diminishes as x increases or decreases. A write current i_{write} passes through the write coil to write the binary bits. The current flows in one of two directions, depending on whether a 1 bit or a 0 bit is being written. The narrower the air gap and the smaller the distance h between the write head and the recording medium, the smaller the bit, and thus the higher the density. Erasing the magnetic medium is accomplished by applying a strong high frequency over the magnetic surface.

Over the past several years, numerous different types of codes have been used to store binary data. The code that was chosen for an application depended on fundamental requirements such as density and the complexity of the read circuitry. The following sections present various coding techniques that are used in writing binary data on a magnetic surface.

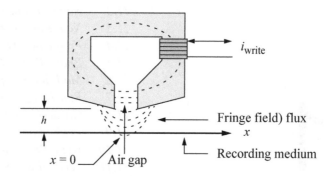

Figure 11.1   Magnetic write head.

11.1 Return to Zero

The *return to zero* (RZ) method is currently rarely used due to its inherent low density capability. This is a result of having two transitions per bit cell and the return to zero magnetization (demagnetized state) during the second half of the bit cell. These demagnetized regions require considerable space and reduce bit density.

A major consideration when choosing a particular coding technique is the ability for the code to be self-clocking. The term *self-clocking* is used to indicate a code in which the readback data can be used to generate clock pulses to sample the data; that is, the clock pulses are synchronized with the data. If the selected code is not self-clocking, then an external clock must be utilized. The problem with using an external clock is that the clock may become unsynchronized with the data due to varying media speeds and drift in the clocking circuits. Thus, it is desirable to have a code that provides a means to synchronize — or trigger — the strobe for the read signal. The RZ code is self-clocking.

For the RZ code, two transitions are available for every stored bit. When reading a 1 bit, a positive transition is followed by a negative transition. When reading a 0 bit, a negative transition is followed by a positive transition, as shown in Figure 11.2 for an arbitrary sequence of 001101. This bit sequence was chosen because it represents all combinations of bit-to-bit sequences: $0 \rightarrow 0$, $0 \rightarrow 1$, $1 \rightarrow 1$, and $1 \rightarrow 0$.

During readback, a positive transition generates a positive read voltage pulse and a negative transition generates a negative read voltage pulse. The clock pulses are obtained by means of a peak detector, whose output generates a narrow signal that is coincident with the read voltage pulses and samples the corresponding read voltage pulse. Only the pulses at the beginning of a bit cell are sampled. That is why it is imperative that the clock pulses are synchronized to the speed of the input/output device. If a positive read voltage pulse is sampled, then the stored bit is a 1; if a negative read voltage pulse is sampled, then the stored bit is a 0. The read voltage pulses in the middle of the bit cells are ignored.

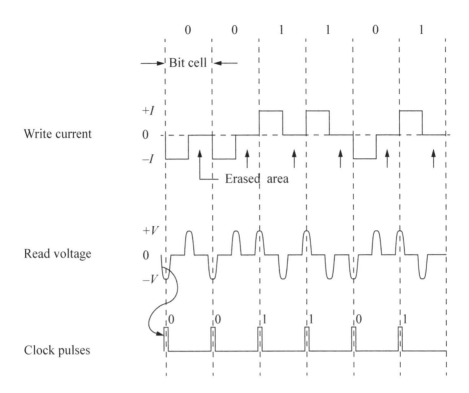

Figure 11.2   Waveforms for the RZ code for the bit pattern 001101.

11.2 Nonreturn to Zero

The density of the RZ code can be increased by eliminating the demagnetized area between bit cells so that the transitions do not return to zero magnetization, thus generating a *nonreturn to zero* (NRZ) code. There is a change in magnetization only when the bit pattern changes from 0 to 1 or from 1 to 0. When the bit pattern changes from 0 to 1, a positive pulse is generated; when the bit pattern changes from 1 to 0, a negative pulse is generated.

There is a major problem with the NRZ code in that a signal is generated in the read head only when a transition occurs between 0 and 1 or between 1 and 0. Therefore, the stored data cannot be used to generate clock pulses; thus, the NRZ code is not self-clocking.

This is evident when long strings of 0s or 1s are encountered in the stored data. An external clock must be used that is resynchronized at transitions. A clock pulse and a negative read pulse equates to a 0 bit; a clock pulse and a positive read pulse equates to a 1 bit; a clock pulse and no read pulse means that the bit is the same as indicated by the immediately previous pulse.

There is a second major problem in which an erroneous pulse that is caused by noise or a read error will cause all following bits to be in error until the next read pulse occurs. If a string of 0s is injected with a spurious pulse, then the following 0s in the string will be interpreted as 1s. A third problem occurs because a medium dropout — an absence of recording material — cannot be distinguished from a stored bit with no transition. An example of the NRZ code is shown in Figure 11.3 using the bit pattern 001101, in which the leading 0 is preceded by a 1 bit in order to generate a transition.

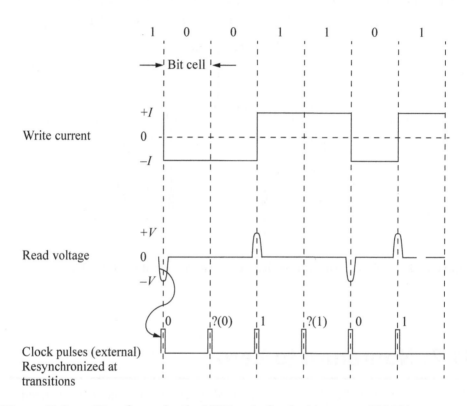

Figure 11.3   Waveforms for the NRZ code for the bit pattern 001101.

11.3 Nonreturn to Zero Inverted

The lack of self-clocking for the NRZ code can be partially resolved with a slight modification to the NRZ code, where a transition occurs for each stored 1 bit and no transition occurs for each stored 0 bit. The resulting waveform is comparable to the inverted waveform of the NRZ code, and is therefore referred to as the *nonreturn to zero inverted* (NRZI) code.

For strings of 0s, a clock cannot be generated from the stored data; therefore, the NRZI code is not self-clocking, necessitating the need for an external clock. Positive and negative read voltage pulses occur for contiguous 1s in the stored data, giving an unambiguous distinction between 1s and 0s in the stored data. The waveforms for the NRZI code are shown in Figure 11.4 using the binary data 001101 with an implied 0 to the right of the rightmost 1 bit, thus generating no transition for that bit cell.

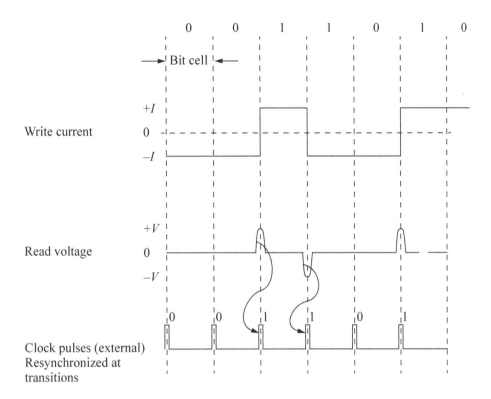

Figure 11.4   Waveforms for the NRZI code for the bit pattern 001101.

Neither NRZ nor NRZI codes are self-clocking for single-track systems such as disk drives. However, NRZI provides a pulse for each 1 bit; thus, multi-track systems such as tape drives which use odd parity across a character ensures at least one read pulse for each character. Skew can be a limiting factor for multi-track systems which use parity for clocking. Tape systems generally use phase encoding which provides self-clocking within each bit cell or group-coded recording which is also self-clocking. Both of those techniques are covered in later sections.

11.4 Frequency Modulation

The lack of self-clocking for NRZ and NRZI codes can be resolved by inserting a clock transition at the beginning of each bit cell. This can be a positive or negative transition and represents a built-in clock pulse. This method of recording binary data on a recording surface is called *frequency modulation* (FM). Each bit cell must now be capable of two transitions; thus, this recording technique is also referred to as *double frequency* (DF). This is not, however, double density — that method is covered in a later section.

A 1 bit is recorded by a transition — either positive of negative — in the middle of the bit cell; a 0 bit contains no transition in the middle of the bit cell. The clock pulses are generated from the read voltage pulses as in previous methods. The clock pulses then generate a strobe pulse in the middle of the corresponding bit cell to sample the state of the read pulse in the middle of the bit cell. If the strobe pulse samples a read voltage pulse — either positive or negative — then the bit cell contains a 1 bit; the absence of a read voltage pulse indicates a 0 bit. An example of FM encoding is shown in Figure 11.5 using the bit pattern 001101.

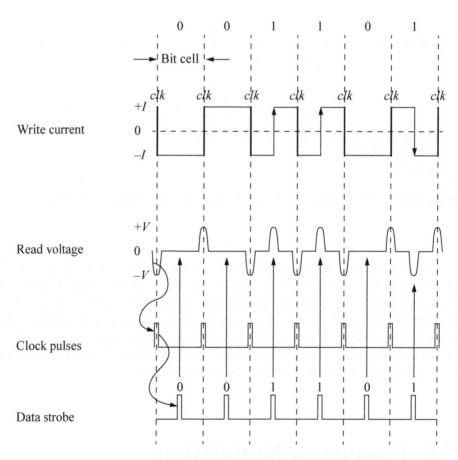

Figure 11.5   Waveforms for the FM code for the bit pattern 001101.

11.5 Phase Encoding

In the *phase encoding* (PE) method, a transition is generated at the beginning of every bit cell to indicate the state of the bit in that cell: a positive transition indicates a 1 bit; a negative transition indicates a 0 bit. This provides a way to obtain self-clocking. Phase encoding is also referred to as *phase modulation*.

Phase encoding increases the frequency of transitions within a bit cell for contiguous strings of 1s or 0s, because each bit cell must contain two transitions. The read voltage pulses that are produced at the beginning of each bit cell are used to generate clock pulses that strobe the read voltage pulses to determine the value of the bit. The frequency of PE is similar to FM. Figure 11.6 shows an example of PE using the bit pattern 001101.

Notice that the write current waveform for the rightmost bit stops halfway through the bit cell. This is because the next bit is unknown. If it is a 1 bit, then a negative transition must take place so that a positive transition can occur at the beginning of the next bit cell to indicate a 1 bit. If it is a 0 bit, then there is no transition in the middle so that a negative transition can occur at the beginning of the next bit cell to indicate a 0 bit. Pulses that occur in the middle of a bit cell are ignored.

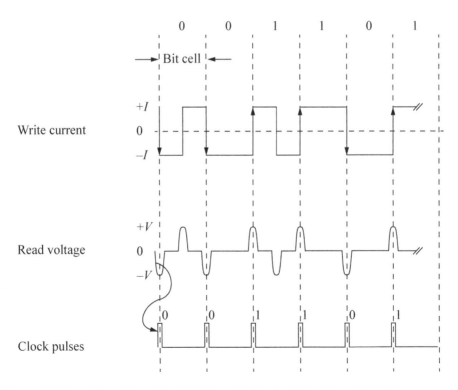

Figure 11.6   Waveforms for the PE code for the bit pattern 001101.

11.6 Modified Frequency Modulation

It is advantageous to have as few transitions as possible yet still maintain adequate self-clocking. This can be accomplished by using the frequency modulation (FM) method and providing clock transitions only where they are needed; that is, the FM method is modified to obtain the *modified frequency modulation* (MFM) code.

In the FM code, a clock transition is inserted at the beginning of every bit cell. Every 1 bit has a transition in the middle of the bit cell, whereas a 0 bit has no transitions in the middle of the bit cell. Therefore, the clock transitions can be eliminated for 1 bits, but are necessary for contiguous 0 bits.

At most there will be a 2-bit-cell spacing in a bit pattern of 101 in which there are no transitions; that is, a transition in the middle of the first 1-bit cell, no transition for the 0-bit cell, and a transition in the middle of the second 1-bit cell. The clock circuitry will not drift significantly during 2 bit cells in which there are no transitions; therefore, self-clocking is achieved.

Figure 11.7 shows the derivation of the MFM code from the FM code. Assume that a 0 bit precedes the leftmost 0 bit. Since the frequency of transitions for the MFM code is one-half the frequency of transitions for the FM code, the MFM code can be compressed to one-half the distance and still maintain the same frequency as the FM code — this is referred to as *double density*. Figure 11.8 shows an additional example of the MFM code.

11.7 Run-Length Limited

The *run-length limited* (RLL) code limits the number of contiguous 0s in a sequence and extends the concept presented in the section on the MFM code in which certain clock transitions (pulses) were removed. The MFM code was obtained from the FM code by removing certain clock transitions and retaining the remaining clock transitions.

For the RLL code, all clock transitions are removed except those between contiguous 0s. Let the removed clock transitions be equal to a 0 bit and the retained clock transitions be equal to a 1 bit. Then the data bits are interleaved with the clock bits to obtain a bit configuration that represents the stored data in which there is no distinction between data and clock transitions.

The stored data is then written on the recording medium using the NRZI encoding technique. This guarantees a minimum of one 0 between two successive 1s and a maximum of three 0s between two successive 1s. The minimum number ensures a minimum transition spacing to control pulse crowding. The maximum number maintains clock synchronization for self-clocking — the clock circuitry will not drift significantly in 4 bit cells.

The transformation is controlled by an algorithm during writing and an inverse algorithm during reading. In order to meet the constraints of the RLL code , m bits of data are mapped into n bits of code by the algorithm, where $n > m$. Figure 11.9 illustrates the concept of the RLL code using the bit configuration 001101.

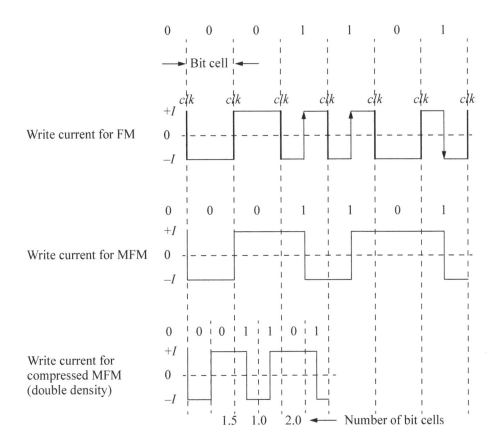

Figure 11.7   Waveforms for the MFM code for the bit pattern 001101.

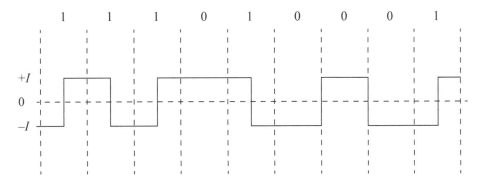

Figure 11.8   Another example of MFM encoding for the bit configuration 111010001.

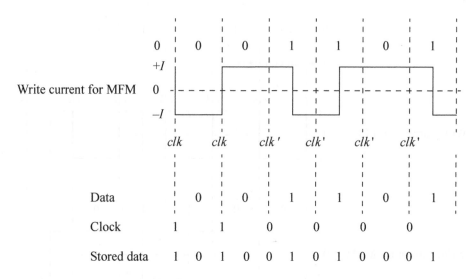

Figure 11.9   Illustration of the RLL code for the bit configuration 001101.

11.8 Group-Coded Recording

The *group-coded recording* (GCR) encoding method writes the data using the NRZI format, but allows no more than two contiguous 0s in the sequence. This guarantees that at least one transition will occur in every 3 bit cells. Thus, self-clocking can be realized. Each byte is partitioned into 4 bits. Each 4-bit group is translated into 5 bits such that there will be no more than two contiguous 0s in the sequence, either within a 5-bit group or between contiguous 5-bit groups. During a read operation, the 5-bit code is translated in reverse into 4 bits.

Out of $2^5 = 32$ combinations of 5 bits, only seventeen 5-bit codes are usable, as shown in Table 11.1. Of these, any code word with three of more 0s, or two 0s on either end of the 5-bit character, as shown below, cannot be used.

10001

11100

00111

The 5-bit code word 11100 cannot be used because it may be left-adjacent to a code word of 0xxxx, thus allowing three or more 0s to be adjacent. Likewise, the 5-bit code 00111 cannot be used because it may be right-adjacent to a code word of xxxx0, thus allowing three or more 0s to be adjacent. Therefore, the following concatenated 5-bit code words are invalid because the code word 11100 is invalid:

11100 11010

Table 11.2 lists the 5-bit codes that are translated from the 4-bit data words.

Table 11.1 Valid and Invalid 5-Bit Codes for GCR

5-Bit GCR Code		5-Bit GCR Code	
0 0 0 0 0		1 0 0 0 0	
0 0 0 0 1		1 0 0 0 1	
0 0 0 1 0		1 0 0 1 0	Valid
0 0 0 1 1		1 0 0 1 1	Valid
0 0 1 0 0		1 0 1 0 0	
0 0 1 0 1		1 0 1 0 1	Valid
0 0 1 1 0		1 0 1 1 0	Valid
0 0 1 1 1		1 0 1 1 1	Valid
0 1 0 0 0		1 1 0 0 0	
0 1 0 0 1	Valid	1 1 0 0 1	Valid
0 1 0 1 0	Valid	1 1 0 1 0	Valid
0 1 0 1 1	Valid	1 1 0 1 1	Valid
0 1 1 0 0		1 1 1 0 0	
0 1 1 0 1	Valid	1 1 1 0 1	Valid
0 1 1 1 0	Valid	1 1 1 1 0	Valid
0 1 1 1 1	Valid	1 1 1 1 1	Valid

Table 11.2 Translation Table for GCR

4-Bit Data	5-Bit GCR Code
0 0 0 0	1 1 0 0 1
0 0 0 1	1 1 0 1 1
0 0 1 0	1 0 0 1 0
0 0 1 1	1 0 0 1 1
0 1 0 0	1 1 1 0 1
0 1 0 1	1 0 1 0 1
0 1 1 0	1 0 1 1 0
0 1 1 1	1 0 1 1 1
1 0 0 0	1 1 0 1 0
1 0 0 1	0 1 0 0 1
1 0 1 0	0 1 0 1 0
1 0 1 1	0 1 0 1 1
1 1 0 0	1 1 1 1 0
1 1 0 1	0 1 1 0 1
1 1 1 0	0 1 1 1 0
1 1 1 1	0 1 1 1 1

Examples of GCR translation is shown below.

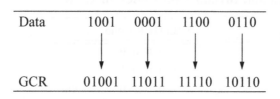

Data	1001	0001	1100	0110
GCR	01001	11011	11110	10110

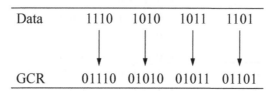

Data	1110	1010	1011	1101
GCR	01110	01010	01011	01101

The GCR format for 9-track tape drives is shown below. Each track of the tape is partitioned into 4-bit groups, then translated into 5-bit groups of the GCR code. Every 7 bytes of data on a track are followed by an error checking and correction (ECC) byte.

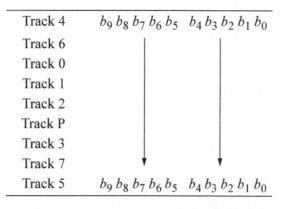

The 5-bits of GCR data are sent to the tape drive and are encoded using the NRZI format. The parity track P and the other most frequently used tracks are placed in the center region of the tape, away from the edges. This ensures that the parity and information bits will not be lost if the edges become damaged.

11.9 Peak Shift

Peak shift is caused when two or more transitions occur sequentially; for example, 0110 and 01110. This bit crowding causes the induced voltage peaks to shift away from each other and can occur in both writing and reading. In the NRZI format, a 1 bit is represented by a transition, either positive or negative, as shown in Figure 11.10. The distance between the pulses is indicated by the letter D. For low density, D is large and bit crowding is not a factor. For high density, the distance between successive 1s should be as small as possible. As D becomes smaller, peak shifting can occur, as shown in Figure 11.11.

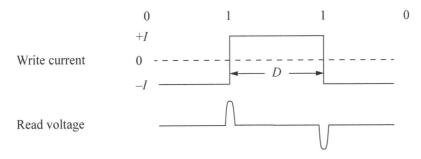

Figure 11.10   NRZI encoding for the bit sequence 0110.

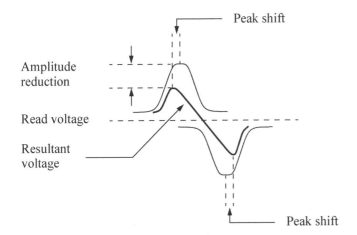

Figure 11.11   Example illustrating peak shift caused by bit crowding.

11.10 Write Precompensation

For high density, as the distance D is decreased, the two pulses in Figure 11.11 begin to overlap and subtract from each other. This overlap and subtraction generates a reduction in amplitude and a shift in position (time) where the peaks occur; that is, peak shift. Peak shift can be reduced during writing by a process called *write precompensation* in which the bits are shifted in the opposite direction to the expected shift. For example, the bit pattern 0110 would cause the third bit from the left to be detected later than its normal position. The peripheral control unit would cause the bit to be written earlier so that the bit would be closer to its normal position when read. Conversely, a bit that would appear early is written later.

An example showing peak shift for the bit pattern 01111110 is demonstrated in Figure 11.12 using the NRZI encoding method. The figure depicts the ideal read voltage waveforms and the actual read voltage waveforms that result from peak shift. In this case, peak shift occurs mainly at the extremities of the bit pattern. Since symmetry occurs in the middle of the bit pattern, there is no peak shift. Peak shift is caused by a characteristic of magnetic media to expand areas of high density into areas of low density. Figure 11.13 shows a similar example using MFM, but uses write precompensation to nullify the effects of peak shift.

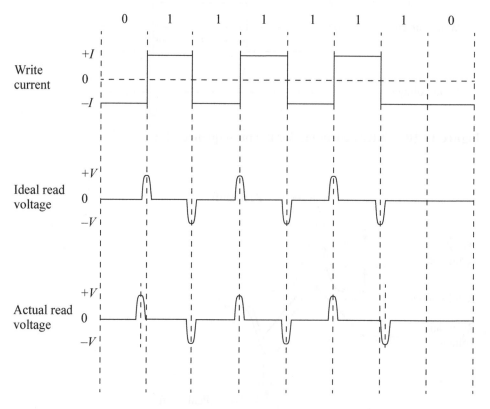

Figure 11.12   An illustration showing the effects of peak shift for the bit pattern 01111110.

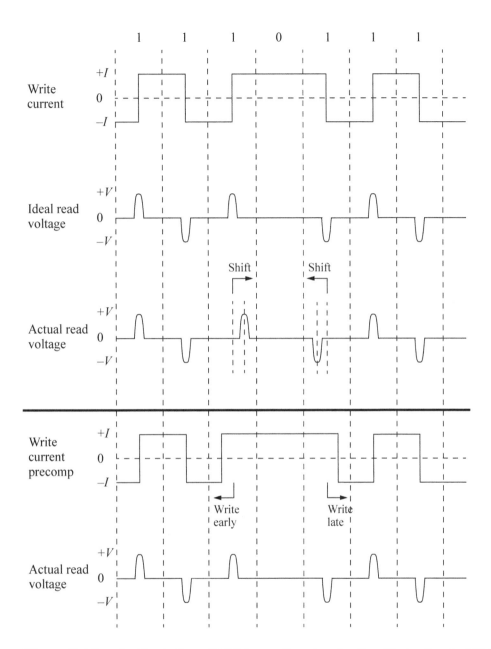

Figure 11.13   An illustration of MFM encoding showing the effects of peak shift for the bit pattern 1110111 and the use of write precompensation to nullify the effects of peak shift.

The amount of bit shift depends on the density of the preceding and following bits. The write precompensation technique requires a lookahead register of 4 to 8 bits in order to determine which bits must be written early or late.

Unwanted signals from adjacent tracks or from previous data that was not completely erased is referred to as *crosstalk*. The amount of *flux changes per inch* (FCI) is the number of transitions per inch that a particular encoding technique requires. For example, the maximum FCI for the NRZI encoding method is equal to the bits per inch (BPI). The maximum FCI for the PE and FM encoding methods is 2 BPI, as shown in Figure 11.14.

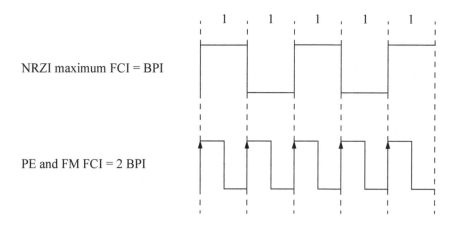

Figure 11.14   Maximum number of flux changes per inch for NRZI, PE, and FM.

11.11 Vertical Recording

Horizontal (longitudinal) recording writes magnetic bits horizontally on the recording medium by aligning the magnetic domains of the north and south poles of the bits longitudinally; that is, the magnetic orientation of the data bits is aligned horizontally parallel to the recording surface. In order to increase bit density, *vertical* (perpendicular or monopole) recording minimizes the bit area by writing bits vertically on the recording medium by aligning the north and south poles of the bits perpendicular to the recording surface; that is, the magnetic orientation of the data bits is aligned vertically.

In vertical recording, the head and recording surface are part of a single magnetic circuit in which the gap is formed by the *flying height* of the head, which is the distance between the read/write head and the recording medium. The flying height of most current disk read/write heads is approximately 0.5 microinches.

Figure 11.15 shows a basic disk read/write head used in vertical recording. The recording surface is a high coercivity layer with a high resistance to flux; the polarization is vertical. The high permeability layer provides a return path for the magnetic flux. The area labelled A in Figure 11.15 is the return pole and is approximately 1,000 times the area of the head probe to provide highly dispersed flux. The air gap is formed by the flying height of the head.

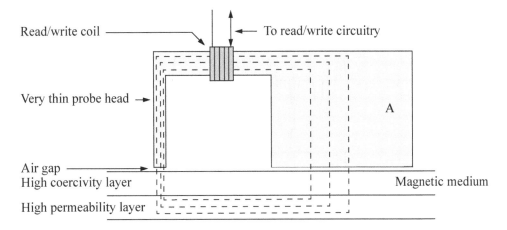

Read/write coil ⎯⎯⎯⎯⎯ ⎯ To read/write circuitry

Very thin probe head →

A

Air gap

High coercivity layer

Magnetic medium

High permeability layer

Figure 11.15 Read/write head for a disk drive.

The track edges in vertical recording are less noisy and better defined due to the vertical pole head configuration. Sharp track edges allow higher track density and a smaller bit aspect ratio. Vertical recording has increased the areal density in current disk drives to approximately 350 gigabits per square inch and is expected to reach 500 gigabits per square inch.

11.12 Problems

11.1   Indicate the data bit sequence for the read voltage waveforms shown below.

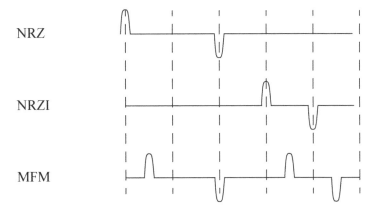

NRZ

NRZI

MFM

11.2   Give the data bit sequence for each of the following read voltage waveforms:

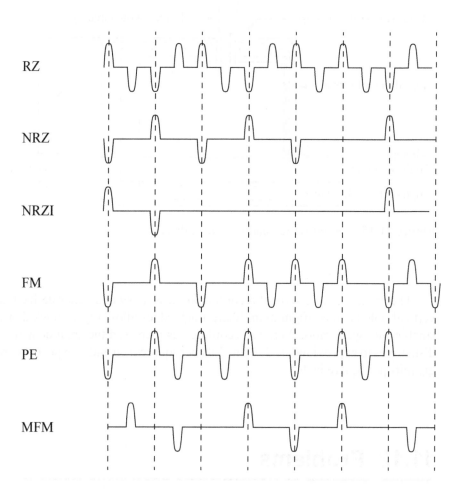

11.3   Given the bit pattern 011110, obtain the write current waveform using MFM encoding and show the actual read voltage waveform without write precompensation. Start the write current waveform at $+I$.

11.4   Give the data bit sequence for each of the following read voltage waveforms:

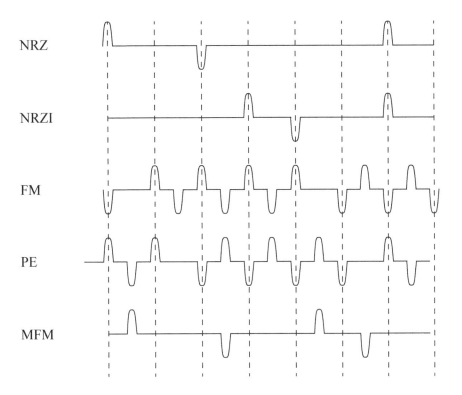

11.5   Show the write current waveforms for the bit sequence 1110010 for each of the following encoding methods: RZ, NRZ, NRZI, FM, PE, and MFM. Begin the NRZI, FM, and MFM waveforms at a +*I* current level.

11.6   Show the write current waveform for the bit sequence 0001101 for RZ, NRZ, NRZI, FM, PE, and MFM. Begin the NRZI, FM, and MFM waveforms at a –*I* current level.

11.7   Given the bit pattern 11100101, convert the byte of data to the GCR format, then draw the write current waveform for the GCR code using the NRZI encoding technique. Begin the waveform with a +*I* level.

11.8   Given the following byte of data, show the stored data for the RLL code using the MFM encoding method: 1000 1101.

11.9   Each of the data shown below is obtained from two 4-bit segments using the GCR translation table. Indicate which data is valid or invalid for GCR encoding.

10011	00110	Valid ___	Invalid ___
00111	11010	Valid ___	Invalid ___
10100	11001	Valid ___	Invalid ___
01010	10001	Valid ___	Invalid ___
10110	00110	Valid ___	Invalid ___

11.10   Give the data bit sequence for the following read voltage waveform:

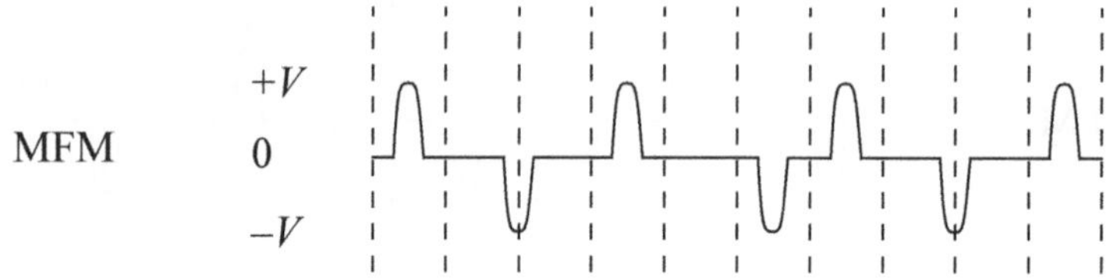

11.11   Translate the following GCR-encoded bit patterns into the corresponding data bytes:

GCR Encoded		Data Byte
01110	01001	
11001	01011	
01101	10110	

11.12   The NRZI format can be made self-clocking for single-track systems by using _____ or _____ .

11.13   Unwanted signals from adjacent tracks or from previous data that was not completely erased are called _____ .

11.14   Indicate which of the 5-bit groups shown below could be used as the valid stored code for GCR.

	Can be Used	Cannot be Used
00111		
01000		
10001		
11100		
10010		
01010		
11010		
10110		
01100		
00110		

11.15   Each of the data shown below represents 1 byte, which was obtained by translating two 4-bit segments into two 5-bit segments. Indicate which data is valid or invalid for GCR encoding.

10011	00110	Valid _____	Invalid _____
00111	11010	Valid _____	Invalid _____
10100	11001	Valid _____	Invalid _____
01010	10001	Valid _____	Invalid _____
10110	00110	Valid _____	Invalid _____
01100	11001	Valid _____	Invalid _____
01001	01101	Valid _____	Invalid _____
10010	01001	Valid _____	Invalid _____
01101	00110	Valid _____	Invalid _____
11100	11111	Valid _____	Invalid _____

11.16   Determine the maximum number of flux changes per inch (FCI) for records written at 6,250 bits per inch for the following encoding methods:

RZ, NRZ, NRZI, FM, PE, MFM

11.17   Determine which of the following encoding methods are self-clocking for single-track systems:

RZ, NRZ, NRZI, DF, FM, PE, PM, MFM

12

Additional Topics in Digital Design

This chapter presents additional topics in digital design that do not fit into a common category. Boolean functions can be minimized and represented in a sum-of-products form or in a product-of-sums form. Functional decomposition of Boolean functions is a process in which a large function is decomposed into smaller functions that produce the same results. Examples of functional decomposition are presented and are designed using Verilog HDL. Iterative networks are useful in many applications such as adders, comparators, array multipliers, and single-bit detection circuits; the theory and design of such circuits are discussed and then designed using Verilog HDL.

The Hamming code error detection and correction method was discussed in an earlier chapter and is presented in this chapter in more detail and designed using Verilog HDL. The cyclic redundancy check (CRC) code for error detection and correction is also covered; these unique devices are implemented with linear feedback shift registers.

Residue checking and parity prediction are presented which are useful techniques in detecting errors in adders. Residue checking involves the parity of the augend, the addend, and the sum. Parity prediction compares the parity of the augend, the addend, and the carries — the predicted parity — with the actual parity of the sum. A method to determine the condition codes of sum < 0, sum $= 0$, and sum > 0 before the actual sum is obtained is also introduced.

A 4-function arithmetic and logic unit (ALU) is designed using structural modeling and an 8-function ALU is designed using behavioral modeling. A final topic describing how memories are designed using Verilog HDL is also included.

921

12.1 Functional Decomposition

Functional decomposition has emerged as an important technique in automatic logic design such as field-programmable gate arrays (FPGAs). A function may be composed of several simpler functions; thus, large and complex functions can be decomposed into a system of smaller functions. These smaller functions — or subfunctions — are more easily manipulated and provide a solution to the overall design.

One approach of functional decomposition uses multiplexers. Multiplexer implementation for a logic function will now be reviewed prior to their use in functional decomposition. Equation 12.1 illustrates an expression for a function $z_1(x_1, x_2, x_3)$, where x_3 is the low-order bit, and is represented by the Karnaugh map of Figure 12.1. The Karnaugh map is already organized for a 4:1 multiplexer of the type shown in Figure 12.2 and will be used to implement the function z_1, where the select inputs $s_1 s_0 = x_2 x_3$. The general equation for a 4:1 multiplexer is shown in Equation 12.2.

$$z_1(x_1, x_2, x_3) = x_1 x_2' + x_2 x_3' + x_1' x_2 \tag{12.1}$$

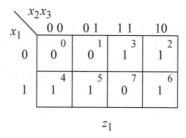

Figure 12.1   Karnaugh map for the function z_1 shown in Equation 12.1.

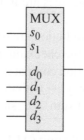

Figure 12.2   A 4:1 multiplexer to implement the function for z_1 shown in Equation 12.1.

$$z_1 = s_1's_0'd_0 + s_1's_0d_1 + s_1s_0'd_2 + s_1s_0d_3 \qquad (12.2)$$

Using the Karnaugh map of Figure 12.1, the equation for z_1 is shown in Equation 12.3 and the multiplexer implementation is shown in Figure 12.3, with the select inputs defined as follows: $s_1s_0 = x_2x_3$.

$$z_1 = x_2'x_3'(x_1) + x_2'x_3(x_1) + x_2x_3'(1) + x_2x_3(x_1') \qquad (12.3)$$

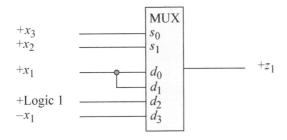

Figure 12.3   Multiplexer implementation for the function z_1 of Equation 12.3.

Example 12.1   Equation 12.4 shows an expression for a function $z_1(x_1, x_2, x_3, x_4, x_5)$, which will be implemented in Verilog HDL using dataflow modeling. The AND gates and OR gates are designed with the continuous assignment statement **assign**. The interconnecting wires are assigned net names as follows: $net1 = x_4'x_5'$ through $net7 = x_1x_2'x_3x_4x_5$. Two more examples will then be presented for the same function using structural modeling: one using a 4:1 multiplexer and one using functional decomposition and multiplexers. In all three implementations, the results for the function z_1 will be identical.

$$z_1 = x_4'x_5' + x_1'x_2'x_4x_5 + x_1'x_3x_4x_5 + x_2x_3'x_5'$$
$$+ x_1x_2x_5' + x_1x_2'x_3'x_5' + x_1x_2'x_3x_4x_5 \qquad (12.4)$$

The dataflow module is shown in Figure 12.4 and the test bench module is shown in Figure 12.5. The outputs are shown in Figure 12.6 which lists all 32 combinations of the five variables x_1, x_2, x_3, x_4, and x_5; the waveforms are shown in Figure 12.7.

```
//dataflow functional decomposition
module func_decomp1 (x1, x2, x3, x4, x5, z1);

input x1, x2, x3, x4, x5;   //define inputs and output
output z1;

wire x1, x2, x3, x4, x5;    //define inputs and output as wire
wire z1;

//define internal nets
wire net1, net2, net3, net4, net5, net6, net7;

//define logic using continuous assignment
assign    net1 = ~x4 & ~x5,
          net2 = ~x1 & ~x2 & x4 & x5,
          net3 = ~x1 & x3 & x4 & x5,
          net4 = x2 & ~x3 & ~x5,
          net5 = x1 & x2 & ~x5,
          net6 = x1 & ~x2 & ~x3 & ~x5,
          net7 = x1 & ~x2 & x3 & x4 & x5;
assign    z1 = net1 | net2 | net3 | net4 | net5 | net6 | net7;
endmodule
```

Figure 12.4   Dataflow module for the function $z_1(x_1, x_2, x_3, x_4, x_5)$.

```
//test bench for functional decomposition
module func_decomp1_tb;

reg x1, x2, x3, x4, x5;
wire z1;

initial      //apply input vectors and display variables
begin: apply_stimulus
   reg [5:0] invect;
   for (invect = 0; invect < 32; invect = invect + 1)
      begin
         {x1, x2, x3, x4, x5} = invect [5:0];
         #10 $display ("x1 x2 x3 x4 x5 = %b z1 = %b",
                       {x1, x2, x3, x4, x5}, z1);
      end
end          //continued on next page
```

Figure 12.5   Test bench for the dataflow module of Figure 12.4.

```
//instantiate the module into the test bench
func_decomp1 inst1 (
   .x1(x1),
   .x2(x2),
   .x3(x3),
   .x4(x4),
   .x5(x5),
   .z1(z1)
   );
endmodule
```

Figure 12.5   (Continued)

```
x1 x2 x3 x4 x5 = 00000 z1 = 1 | x1 x2 x3 x4 x5 = 10000 z1 = 1
x1 x2 x3 x4 x5 = 00001 z1 = 0 | x1 x2 x3 x4 x5 = 10001 z1 = 0
x1 x2 x3 x4 x5 = 00010 z1 = 0 | x1 x2 x3 x4 x5 = 10010 z1 = 1
x1 x2 x3 x4 x5 = 00011 z1 = 1 | x1 x2 x3 x4 x5 = 10011 z1 = 0
x1 x2 x3 x4 x5 = 00100 z1 = 1 | x1 x2 x3 x4 x5 = 10100 z1 = 1
x1 x2 x3 x4 x5 = 00101 z1 = 0 | x1 x2 x3 x4 x5 = 10101 z1 = 0
x1 x2 x3 x4 x5 = 00110 z1 = 0 | x1 x2 x3 x4 x5 = 10110 z1 = 0
x1 x2 x3 x4 x5 = 00111 z1 = 1 | x1 x2 x3 x4 x5 = 10111 z1 = 1
x1 x2 x3 x4 x5 = 01000 z1 = 1 | x1 x2 x3 x4 x5 = 11000 z1 = 1
x1 x2 x3 x4 x5 = 01001 z1 = 0 | x1 x2 x3 x4 x5 = 11001 z1 = 0
x1 x2 x3 x4 x5 = 01010 z1 = 1 | x1 x2 x3 x4 x5 = 11010 z1 = 1
x1 x2 x3 x4 x5 = 01011 z1 = 0 | x1 x2 x3 x4 x5 = 11011 z1 = 0
x1 x2 x3 x4 x5 = 01100 z1 = 1 | x1 x2 x3 x4 x5 = 11100 z1 = 1
x1 x2 x3 x4 x5 = 01101 z1 = 0 | x1 x2 x3 x4 x5 = 11101 z1 = 0
x1 x2 x3 x4 x5 = 01110 z1 = 0 | x1 x2 x3 x4 x5 = 11110 z1 = 1
x1 x2 x3 x4 x5 = 01111 z1 = 1 | x1 x2 x3 x4 x5 = 11111 z1 = 0
```

Figure 12.6   Outputs for the function $z_1(x_1, x_2, x_3, x_4, x_5)$.

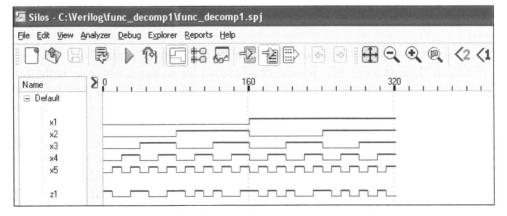

Figure 12.7   Waveforms for the function $z_1(x_1, x_2, x_3, x_4, x_5)$.

Example 12.2 The same function, $z_1(x_1, x_2, x_3, x_4, x_5)$, of Example 12.1 will now be implemented using a 4:1 multiplexer. Equation 12.4 is plotted on a Karnaugh map as shown in Figure 12.8 using x_5 as a map-entered variable, in which the select inputs for the multiplexer are $s_1 s_0 = x_3 x_4$.

x_1x_2 \ x_3x_4	0 0	0 1	1 1	1 0
0 0	x_5'	x_5	x_5	x_5'
0 1	x_5'	x_5'	x_5	x_5'
1 1	x_5'	x_5'	x_5'	x_5'
1 0	x_5'	x_5'	x_5	x_5'

z_1

Figure 12.8 Karnaugh map for Equation 12.4.

From the Karnaugh map of Figure 12.8, the equation for function z_1 is obtained as shown in Equation 12.5 with reference to the multiplexer select inputs. The logic diagram is shown in Figure 12.9 using a 4:1 multiplexer and logic primitive gates.

$$\begin{aligned} z_1 &= x_3'x_4'(x_5') + x_3'x_4[(x_1'x_2'x_5) + (x_2x_5') + (x_1x_5')] \\ &+ x_3x_4'(x_5') + x_3x_4[(x_1'x_5) + (x_2'x_5) + (x_1x_2x_5')] \end{aligned} \tag{12.5}$$

The dataflow module for the 4:1 multiplexer is shown in Figure 12.10. The select inputs are defined as *[1:0]s*, which specifies two inputs $s_1 s_0$, where s_0 is the low-order select input; therefore, $s_1 = x_3$ and $s_0 = x_4$. The multiplexer data inputs are defined as *[3:0]d*, which species four inputs d_3, d_2, d_1, d_0, where d_0 is the low-order data input. When specifying the select and data inputs in the structural module, the variables must be in the same order as indicated in the multiplexer dataflow module.

The structural module for the logic diagram is shown in Figure 12.11 in which there is a direct correspondence between the instantiation names and the net names in the logic diagram with those in the structural design module. The AND gates, the OR gates, and the multiplexer are designed using dataflow modeling and then instantiated into the structural module. The test bench module is shown in Figure 12.12. The outputs and waveforms are shown in Figure 12.13 and Figure 12.14, respectively.

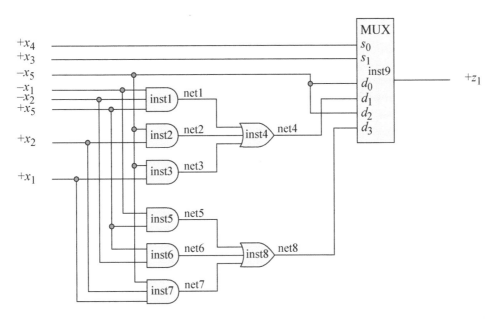

Figure 12.9 Logic diagram for the function $z_1(x_1, x_2, x_3, x_4, x_5)$ of Example 12.2.

```
//dataflow 4:1 mux
module mux4a_df (s, d, z1);

input [1:0] s;
input [3:0] d;
output z1;

wire [1:0] s;
wire [3:0] d;
wire z1;

assign    z1 =  (~s[1] & ~s[0] & d[0]) |
                (~s[1] &  s[0] & d[1]) |
                ( s[1] & ~s[0] & d[2]) |
                ( s[1] &  s[0] & d[3]);

endmodule
```

Figure 12.10 Dataflow module for a 4:1 multiplexer.

```
//structural functional decomposition
module func_decomp2 (x1, x2, x3, x4, x5, z1);


//define inputs and output
input x1, x2, x3, x4, x5;
output z1;


//define internal nets
wire net1, net2, net3, net4, net5, net6, net7, net8;


//instantiate the logic gates
and3_df inst1 (
    .x1(~x1),
    .x2(~x2),
    .x3(x5),
    .z1(net1)
    );

and2_df inst2 (
    .x1(~x5),
    .x2(x2),
    .z1(net2)
    );

and2_df inst3 (
    .x1(~x5),
    .x2(x1),
    .z1(net3)
    );

or3_df inst4 (
    .x1(net1),
    .x2(net2),
    .x3(net3),
    .z1(net4)
    );

and2_df inst5 (
    .x1(~x1),
    .x2(x5),
    .z1(net5)
    );

//continued on next page
```

Figure 12.11   Structural module for Figure 12.9.

```
and2_df inst6 (
   .x1(x5),
   .x2(~x2),
   .z1(net6)
   );

and3_df inst7 (
   .x1(~x5),
   .x2(x2),
   .x3(x1),
   .z1(net7)
   );

or3_df inst8 (
   .x1(net5),
   .x2(net6),
   .x3(net7),
   .z1(net8)
   );

//instantiate the 4:1 multiplexer
mux4a_df inst9 (
   .s({x3, x4}),
   .d({net8, ~x5, net4, ~x5}),
   .z1(z1)
   );
endmodule
```

Figure 12.11   (Continued)

```
//test bench for functional decomposition
module func_decomp2_tb;

reg x1, x2, x3, x4, x5;
wire z1;

initial     //apply input vectors and display variables
begin: apply_stimulus
   reg [5:0] invect;
   for (invect = 0; invect < 32; invect = invect + 1)
      begin
         {x1, x2, x3, x4, x5} = invect [5:0];
         #10 $display ("x1 x2 x3 x4 x5 = %b z1 = %b",
                       {x1, x2, x3, x4, x5}, z1);
      end
end          //continued on next page
```

Figure 12.12   Test bench for the structural module of Figure 12.11.

```
func_decomp2 inst1 (      //instantiate the module
   .x1(x1),
   .x2(x2),
   .x3(x3),
   .x4(x4),
   .x5(x5),
   .z1(z1)
   );
endmodule
```

Figure 12.12   (Continued)

```
x1 x2 x3 x4 x5 = 00000 z1 = 1 | x1 x2 x3 x4 x5 = 10000 z1 = 1
x1 x2 x3 x4 x5 = 00001 z1 = 0 | x1 x2 x3 x4 x5 = 10001 z1 = 0
x1 x2 x3 x4 x5 = 00010 z1 = 0 | x1 x2 x3 x4 x5 = 10010 z1 = 1
x1 x2 x3 x4 x5 = 00011 z1 = 1 | x1 x2 x3 x4 x5 = 10011 z1 = 0
x1 x2 x3 x4 x5 = 00100 z1 = 1 | x1 x2 x3 x4 x5 = 10100 z1 = 1
x1 x2 x3 x4 x5 = 00101 z1 = 0 | x1 x2 x3 x4 x5 = 10101 z1 = 0
x1 x2 x3 x4 x5 = 00110 z1 = 0 | x1 x2 x3 x4 x5 = 10110 z1 = 0
x1 x2 x3 x4 x5 = 00111 z1 = 1 | x1 x2 x3 x4 x5 = 10111 z1 = 1
x1 x2 x3 x4 x5 = 01000 z1 = 1 | x1 x2 x3 x4 x5 = 11000 z1 = 1
x1 x2 x3 x4 x5 = 01001 z1 = 0 | x1 x2 x3 x4 x5 = 11001 z1 = 0
x1 x2 x3 x4 x5 = 01010 z1 = 1 | x1 x2 x3 x4 x5 = 11010 z1 = 1
x1 x2 x3 x4 x5 = 01011 z1 = 0 | x1 x2 x3 x4 x5 = 11011 z1 = 0
x1 x2 x3 x4 x5 = 01100 z1 = 1 | x1 x2 x3 x4 x5 = 11100 z1 = 1
x1 x2 x3 x4 x5 = 01101 z1 = 0 | x1 x2 x3 x4 x5 = 11101 z1 = 0
x1 x2 x3 x4 x5 = 01110 z1 = 0 | x1 x2 x3 x4 x5 = 11110 z1 = 1
x1 x2 x3 x4 x5 = 01111 z1 = 1 | x1 x2 x3 x4 x5 = 11111 z1 = 0
```

Figure 12.13   Outputs for the structural module of Figure 12.11.

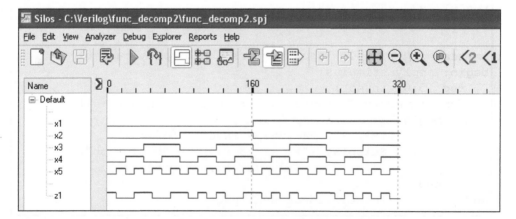

Figure 12.14   Waveforms for the structural module of Figure 12.11.

Example 12.3 This example presents an alternative method for designing the function that was given in the two previous examples by utilizing the technique of *functional decomposition*. The function will be decomposed into two smaller functions and then implemented with two multiplexers connected in series. The equation for function z_1 is reproduced in Equation 12.6.

$$z_1 = x_4' x_5' + x_1' x_2' x_4 x_5 + x_1' x_3 x_4 x_5 + x_2 x_3' x_5'$$
$$+ x_1 x_2 x_5' + x_1 x_2' x_3' x_5' + x_1 x_2' x_3 x_4 x_5 \qquad (12.6)$$

The function for z_1 is then decomposed into two functions as shown in Equation 12.7, which consists of an independent function $f(x_1, x_2, x_3)$ to be implemented by the first multiplexer, and a simple disjoint decomposition of z_1 to be implemented by the second multiplexer.

$$z_1(x_1, x_2, x_3, x_4, x_5) = z_1[f(x_1, x_2, x_3), x_4, x_5] \qquad (12.7)$$

Independent function ———
Simple disjoint decomposition ———

Equation 12.6 is then plotted on a 5-variable Karnaugh map, as shown in Figure 12.15. Each of the rows is a function of x_1, x_2, x_3, which have a value of 0, or 1, or a function $f(x_1, x_2, x_3)$, or a function $f'(x_1, x_2, x_3)$.

$x_4 x_5$ \ $x_1 x_2 x_3$	000	001	011	010	110	111	101	100	
0 0	1	1	1	1	1	1	1	1	1
0 1	0	0	0	0	0	0	0	0	0
1 1	1	1	1	0	0	0	1	0	$f(x_1, x_2, x_3)$
1 0	0	0	0	1	1	1	0	1	$f'(x_1, x_2, x_3)$

z_1

Figure 12.15 Karnaugh map for the function z_1.

From row $x_4 x_5 = 11$ of Figure 12.15, the expression for the independent function $f(x_1, x_2, x_3)$ can be derived by plotting the function on a Karnaugh map, as shown in Figure 12.16. This yields Equation 12.8. The 4:1 multiplexer for the independent function f can now be obtained by assigning the select inputs from the Karnaugh map of Figure 12.16 as follows: $s_1 s_0 = x_2 x_3$, as shown in Figure 12.17. The 4:1 multiplexer for the function z_1 can be obtained in a similar manner by assigning the select inputs from the Karnaugh map of Figure 12.15 as follows: $s_1 s_0 = x_4 x_5$ as shown in Figure 12.17.

Figure 12.16   Karnaugh map for the independent function $f(x_1, x_2, x_3)$.

$$f(x_1, x_2, x_3) = x_1' x_2' + x_1' x_3 + x_2' x_3 \qquad (12.8)$$

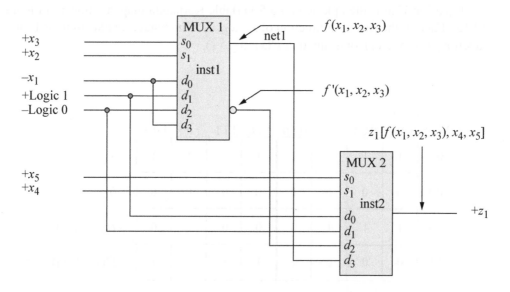

Figure 12.17   Logic diagram for the function $z_1(x_1, x_2, x_3, x_4, x_5)$ using functional decomposition.

The output of multiplexer 2 is $z_1[f(x_1, x_2, x_3), x_4, x_5]$, which is identical to the equation shown in Equation 12.4 and is derived as follows from Figure 12.17, resulting in Equation 12.9:

$$z_1(x_1, x_2, x_3, x_4, x_5) = z_1[f(x_1, x_2, x_3), x_4, x_5]$$

$$= x_4'x_5'(1) + x_4'x_5(0) + x_4x_5'[(x_1'x_2') + (x_1'x_3) + (x_2'x_3)]' +$$

$$x_4x_5[(x_1'x_2') + (x_1'x_3) + (x_2'x_3)]$$

$$= x_4'x_5' + x_4x_5'[(x_1 + x_2)(x_1 + x_3')(x_2 + x_3')] +$$

$$x_1'x_2'x_4x_5 + x_1'x_3x_4x_5 + x_2'x_3x_4x_5$$

$$= x_4'x_5' + x_4x_5'[(x_1 + x_1x_3' + x_1x_2 + x_2x_3')(x_2 + x_3')] +$$

$$x_1'x_2x_4x_5 + x_1'x_3x_4x_5 + x_2'x_3x_4x_5$$

$$= x_4'x_5' + x_1x_2x_4x_5' + x_1x_2x_3'x_4x_5' + x_1x_2x_4x_5' +$$

$$x_2x_3'x_4x_5' + x_1x_3'x_4x_5' + x_1x_2x_3'x_4x_5' + x_2x_3'x_4x_5' +$$

$$x_1'x_2'x_4x_5 + x_1'x_3x_4x_5 + x_2'x_3x_4x_5$$

$$= x_4'x_5' + x_1'x_2'x_4x_5 + x_1'x_3x_4x_5 + x_2x_3'x_5' + x_1x_2x_5' +$$

$$x_1x_2'x_3'x_5' + x_1x_2'x_3x_4x_5 \qquad (12.9)$$

The structural module for the functional decomposition method is shown in Figure 12.18 using only two 4:1 multiplexers that were designed using dataflow modeling. The test bench is shown in Figure 12.19. The outputs and waveforms are shown in Figure 12.20 and Figure 12.21, respectively, and are identical to those obtained in Example 12.1 and Example 12.2.

In Example 12.3, it is seen that functional decomposition has achieved a significant reduction in logic. Example 12.1 required eight logic gates; Example 12.2 required eight logic gates plus a 4:1 multiplexer; Example 12.3 required two 4:1 multiplexers.

```
//structural functional decomposition
module func_decomp3 (x1, x2, x3, x4, x5, z1);

input x1, x2, x3, x4, x5;   //define inputs and output
output z1;

//continued on next page
```

Figure 12.18 Structural module for the functional decomposition design of Figure 12.17.

```
//define internal wire
wire net1;

//instantiate the 4:1 multiplexer
//for the independent function f
mux4a_df inst1 (
   .s({x2, x3}),
   .d({~x1, 1'b0, 1'b1, ~x1}),
   .z1(net1)
   );

//instantiate the 4:1 multiplexer for the function z1
mux4a_df inst2 (
   .s({x4, x5}),
   .d({net1, ~net1, 1'b0, 1'b1}),
   .z1(z1)
   );

endmodule
```

Figure 12.18   (Continued)

```
//test bench for functional decomposition
module func_decomp3_tb;

reg x1, x2, x3, x4, x5;
wire z1;

//apply input vectors and display variables
initial
begin: apply_stimulus
   reg [5:0] invect;
   for (invect = 0; invect < 32; invect = invect + 1)
      begin
         {x1, x2, x3, x4, x5} = invect [5:0];
         #10 $display ("x1 x2 x3 x4 x5 = %b z1 = %b",
                        {x1, x2, x3, x4, x5}, z1);
      end
end

//continued on next page
```

Figure 12.19   Test bench for the functional decomposition module of Figure 12.18.

```
func_decomp3 inst1 (      //instantiate the module
   .x1(x1),
   .x2(x2),
   .x3(x3),
   .x4(x4),
   .x5(x5),
   .z1(z1)
   );
endmodule
```

Figure 12.19   (Continued)

```
x1 x2 x3 x4 x5 = 00000 z1 = 1 │ x1 x2 x3 x4 x5 = 10000 z1 = 1
x1 x2 x3 x4 x5 = 00001 z1 = 0 │ x1 x2 x3 x4 x5 = 10001 z1 = 0
x1 x2 x3 x4 x5 = 00010 z1 = 0 │ x1 x2 x3 x4 x5 = 10010 z1 = 1
x1 x2 x3 x4 x5 = 00011 z1 = 1 │ x1 x2 x3 x4 x5 = 10011 z1 = 0
x1 x2 x3 x4 x5 = 00100 z1 = 1 │ x1 x2 x3 x4 x5 = 10100 z1 = 1
x1 x2 x3 x4 x5 = 00101 z1 = 0 │ x1 x2 x3 x4 x5 = 10101 z1 = 0
x1 x2 x3 x4 x5 = 00110 z1 = 0 │ x1 x2 x3 x4 x5 = 10110 z1 = 0
x1 x2 x3 x4 x5 = 00111 z1 = 1 │ x1 x2 x3 x4 x5 = 10111 z1 = 1
x1 x2 x3 x4 x5 = 01000 z1 = 1 │ x1 x2 x3 x4 x5 = 11000 z1 = 1
x1 x2 x3 x4 x5 = 01001 z1 = 0 │ x1 x2 x3 x4 x5 = 11001 z1 = 0
x1 x2 x3 x4 x5 = 01010 z1 = 1 │ x1 x2 x3 x4 x5 = 11010 z1 = 1
x1 x2 x3 x4 x5 = 01011 z1 = 0 │ x1 x2 x3 x4 x5 = 11011 z1 = 0
x1 x2 x3 x4 x5 = 01100 z1 = 1 │ x1 x2 x3 x4 x5 = 11100 z1 = 1
x1 x2 x3 x4 x5 = 01101 z1 = 0 │ x1 x2 x3 x4 x5 = 11101 z1 = 0
x1 x2 x3 x4 x5 = 01110 z1 = 0 │ x1 x2 x3 x4 x5 = 11110 z1 = 1
x1 x2 x3 x4 x5 = 01111 z1 = 1 │ x1 x2 x3 x4 x5 = 11111 z1 = 0
```

Figure 12.20   Outputs for the functional decomposition module of Figure 12.18.

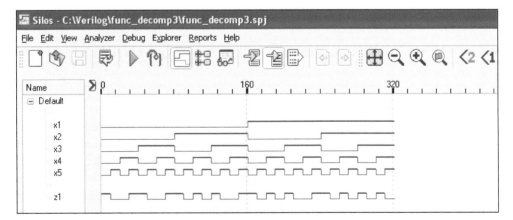

Figure 12.21   Waveforms for the functional decomposition module of Figure 12.18.

Determining whether a function has a simple disjoint decomposition is a relatively simple process. Using the original Karnaugh map that represents the function, if the rows or columns can be identified with the values 0, 1, a function $f(x_1, x_2, x_3, \ldots, x_n)$, or a function $f'(x_1, x_2, x_3, \ldots, x_n)$, then f is the independent function and functional decomposition is possible. If not, then the Karnaugh map is permuted and the process is repeated. Equation 12.10 will be plotted on the Karnaugh map shown in Figure 12.22 to assess whether functional decomposition is possible. The equation for function z_1, obtained from the Karnaugh map, is shown in Equation 12.11.

$$z_1(x_1, x_2, x_3, x_4) = \Sigma_m (0, 4, 9, 10, 13, 14) \tag{12.10}$$

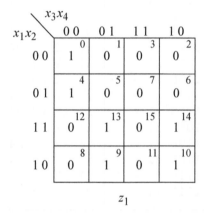

Figure 12.22   Karnaugh map that is generated from Equation 12.10.

$$z_1(x_1, x_2, x_3, x_4) = x_1'x_3'x_4' + x_1x_3'x_4 + x_1x_3x_4' \tag{12.11}$$

It is evident from examining the Karnaugh map that functional decomposition is not possible with this version of the map. Therefore, the Karnaugh map will be permuted to determine if functional decomposition is feasible using a different version of the map. The permuted version is shown in Figure 12.23. Rows $x_1x_3 = 00$ and 11 can be used for the independent function $f(x_2, x_4) = x_4'$; row $x_1x_3 = 10$ can be used for $f'(x_2, x_4) = x_4$; and row $x_1x_3 = 01$ can be used to represent a logic 0.

Using the select inputs of a 4:1 multiplexer defined as follows: $s_1s_0 = x_1x_3$, the equation for z_1 can be derived as shown in Equation 12.12. This design could use two multiplexers as before; however, since variable x_2 is not a constituent part of the equation for z_1, only one 4:1 multiplexer is necessary, as shown in Figure 12.24.

x_1x_3 \ x_2x_4	0 0	0 1	1 1	1 0	
0 0	1 (0)	0 (1)	0 (5)	1 (4)	$f(x_2, x_4) = x_4'$
0 1	0 (2)	0 (3)	0 (7)	0 (6)	0
1 1	1 (10)	0 (11)	0 (15)	1 (14)	$f(x_2, x_4) = x_4'$
1 0	0 (8)	1 (9)	1 (13)	0 (12)	$f'(x_2, x_4) = x_4$

z_1

Figure 12.23   The permuted Karnaugh map of Figure 12.22.

$$z_1(x_1, x_2, x_3, x_4) = z_1[f(x_3, x_4), x_1, x_2]$$
$$= x_1'x_3' f(x_2, x_4) + x_1'x_3(0) + x_1x_3'f'(x_2, x_4) +$$
$$x_1x_3 f(x_2, x_4)$$
$$= x_1'x_3'(x_4') + x_1'x_3(0) + x_1x_3'(x_4) + x_1x_3(x_4')$$
$$= x_1'x_3'x_4' + x_1x_3'x_4 + x_1x_3x_4' \qquad (12.12)$$

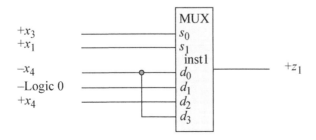

Figure 12.24   Logic diagram for Equation 12.11 using functional decomposition.

Two Verilog HDL modules will now be designed: a dataflow module for Equation 12.11 and a structural module for Figure 12.24. The outputs and waveforms should be the same for both implementations. The dataflow module is shown in Figure 12.25 and the test bench module is shown in Figure 12.26. The outputs and waveforms are shown in Figure 12.27 and Figure 12.28, respectively.

The structural module is shown in Figure 12.29 using a single 4:1 multiplexer and the test bench module is shown in Figure 12.30. The outputs and waveforms are shown in Figure 12.31 and Figure 12.32, respectively. The outputs and waveforms for both design methods are identical. As can be seen from this short presentation, functional decomposition may significantly decrease the amount of logic required to implement a logic function.

```
//dataflow functional decomposition
module func_decomp4 (x1, x2, x3, x4, z1);

//define inputs and output
input x1, x2, x3, x4;
output z1;

//define inputs and output as wire
wire x1, x2, x3, x4;
wire z1;

//define internal nets
wire net1, net2, net3;

//define logic using continuous assignment
assign    net1 = ~x1 & ~x3 & ~x4,
          net2 = x1 & ~x3 & x4,
          net3 = x1 & x3 & ~x4;

//define logic for output z1
assign    z1 = net1 | net2 | net3;

endmodule
```

Figure 12.25   Dataflow module for Equation 12.11.

```
//test bench for functional decomposition
module func_decomp4_tb;

reg x1, x2, x3, x4;
wire z1;
//continued on next  page
```

Figure 12.26   Test bench for Figure 12.25.

```
initial      //apply input vectors and display variables
begin: apply_stimulus
   reg [4:0] invect;
   for (invect = 0; invect < 16; invect = invect + 1)
      begin
         {x1, x2, x3, x4} = invect [4:0];
         #10 $display ("x1 x2 x3 x4 = %b, z1 = %b",
                  {x1, x2, x3, x4}, z1);
      end
end

func_decomp4 inst1 (    //instantiate the module
   .x1(x1),
   .x2(x2),
   .x3(x3),
   .x4(x4),
   .z1(z1)
   );
endmodule
```

Figure 12.26   (Continued)

```
x1 x2 x3 x4 = 0000, z1 = 1    x1 x2 x3 x4 = 1000, z1 = 0
x1 x2 x3 x4 = 0001, z1 = 0    x1 x2 x3 x4 = 1001, z1 = 1
x1 x2 x3 x4 = 0010, z1 = 0    x1 x2 x3 x4 = 1010, z1 = 1
x1 x2 x3 x4 = 0011, z1 = 0    x1 x2 x3 x4 = 1011, z1 = 0
x1 x2 x3 x4 = 0100, z1 = 1    x1 x2 x3 x4 = 1100, z1 = 0
x1 x2 x3 x4 = 0101, z1 = 0    x1 x2 x3 x4 = 1101, z1 = 1
x1 x2 x3 x4 = 0110, z1 = 0    x1 x2 x3 x4 = 1110, z1 = 1
x1 x2 x3 x4 = 0111, z1 = 0    x1 x2 x3 x4 = 1111, z1 = 0
```

Figure 12.27   Outputs for the dataflow module of Figure 12.25.

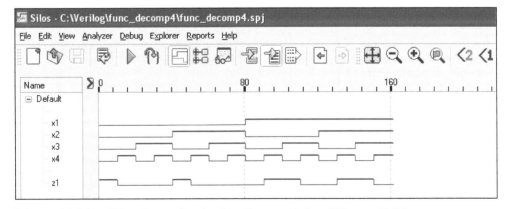

Figure 12.28   Waveforms for the dataflow module of Figure 12.25.

```
//structural functional decomposition
module func_decomp5 (x1, x2, x3, x4, z1);

//define inputs and output
input x1, x2, x3, x4;
output z1;

//instantiate the 4:1 multiplexer
mux4a_df inst1 (
   .s({x1, x3}),
   .d({~x4, x4, 1'b0, ~x4}),
   .z1(z1)
   );
endmodule
```

Figure 12.29   Structural module for the functional decomposition design of Equation 12.12.

```
//test bench for functional decomposition
module func_decomp5_tb;

reg x1, x2, x3, x4;
wire z1;

//apply input vectors and display variables
initial
begin: apply_stimulus
   reg [4:0] invect;
   for (invect = 0; invect < 16; invect = invect + 1)
      begin
         {x1, x2, x3, x4} = invect [4:0];
         #10 $display ("x1 x2 x3 x4 = %b, z1 = %b",
                       {x1, x2, x3, x4}, z1);
      end
end

//instantiate the module into the test bench
func_decomp5 inst1 (
   .x1(x1),
   .x2(x2),
   .x3(x3),
   .x4(x4),
   .z1(z1)
   );
endmodule
```

Figure 12.30   Test bench for the structural module of Figure 12.29.

```
x1 x2 x3 x4 = 0000, z1 = 1     x1 x2 x3 x4 = 1000, z1 = 0
x1 x2 x3 x4 = 0001, z1 = 0     x1 x2 x3 x4 = 1001, z1 = 1
x1 x2 x3 x4 = 0010, z1 = 0     x1 x2 x3 x4 = 1010, z1 = 1
x1 x2 x3 x4 = 0011, z1 = 0     x1 x2 x3 x4 = 1011, z1 = 0
x1 x2 x3 x4 = 0100, z1 = 1     x1 x2 x3 x4 = 1100, z1 = 0
x1 x2 x3 x4 = 0101, z1 = 0     x1 x2 x3 x4 = 1101, z1 = 1
x1 x2 x3 x4 = 0110, z1 = 0     x1 x2 x3 x4 = 1110, z1 = 1
x1 x2 x3 x4 = 0111, z1 = 0     x1 x2 x3 x4 = 1111, z1 = 0
```

Figure 12.31   Outputs for the structural module of Figure 12.29.

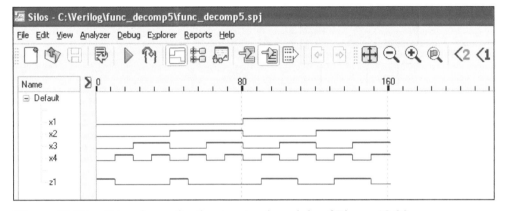

Figure 12.32   Waveforms for the structural module of Figure 12.29.

12.2 Iterative Networks

An *iterative machine* is an organization of identical cells (or elements) which are interconnected in an ordered manner. An iterative machine (or network) may consist of combinational logic arranged in a linear array in which signals between cells propagate in one direction only. A parity checker and comparator are examples of combinational iterative networks. Or, the iterative network may consist of sequential cells, such as found in shift registers and simple binary counters.

A general block diagram of a combinational iterative network is shown in Figure 12.33. There are k cells labeled $cell_1$, $cell_2$, ... , $cell_i$, ... , $cell_k$ and n external input variables for each cell, designated $x_{11}, x_{12}, ... , x_{1n}$ for $cell_1$, $x_{i1}, x_{i2}, ... , x_{in}$ for $cell_i$ and $x_{k1}, x_{k2}, ... , x_{kn}$ for $cell_k$. There are p internal cell inputs that are generated as outputs from the previous cell. For a typical $cell_i$, these internal interconnections are labeled $y_{(i-1)1}, y_{(i-1)2}, ... , y_{(i-1)p}$. The information received from the previous cell can be considered as intercell carry signals. The network also contains m output signals, designated $z_{11}, z_{12}, \ldots, z_{1m}$ for cell$_1$, $z_{i1}, z_{i2}, \ldots, z_{im}$ for cell$_i$, and $z_{k1}, z_{k2}, \ldots, z_{km}$ for cell$_k$. The network outputs can be generated from each cell or from only the rightmost cell$_k$. The number of cells in a combinational iterative network is equal to the number of input sets that are applied to the network, where a typical set of inputs is $x_{i1}, x_{i2}, x_{i3},$ $\ldots, x_{in}$.

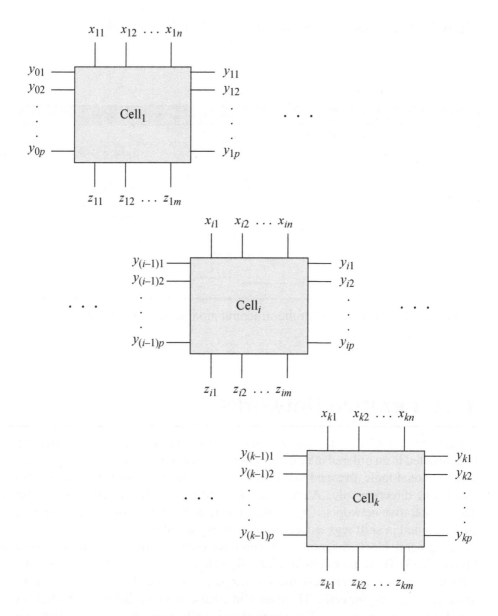

Figure 12.33   General block diagram of an iterative network.

An iterative network is a logical structure composed of identical cells; thus, it is a cascade of identical combinational or sequential circuits (cells) in which the first or last cells may be different than the other cells in the network. Since an iterative network consists of identical cells, it is only necessary to design a typical cell, and then to replicate that cell for the entire network.

Example 12.4   This example demonstrates a method to design a single-bit detection circuit. In this example, a typical cell will be designed, then instantiated four times into a higher-level module to detect a single bit in a 4-bit input vector $x[1:4]$. Figure 12.34 shows the block diagram of a typical cell and Figure 12.35 shows the internal logic of the cell, which will be instantiated four times into the higher-level circuit of Figure 12.36.

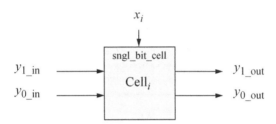

Figure 12.34   Typical cell for a single-bit detection circuit that will be instantiated four times into a higher-level structural module.

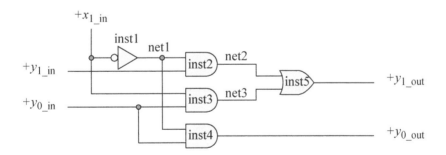

Figure 12.35   Internal logic for a typical cell for the single-bit detection circuit of Example 12.4.

The module for the typical cell is shown in Figure 12.37 using built-in logic primitives. The structural module for the single-bit detection circuit and the test bench module are shown in Figure 12.38 and Figure 12.39, respectively. The outputs and waveforms are shown in Figure 12.40 and Figure 12.41, respectively.

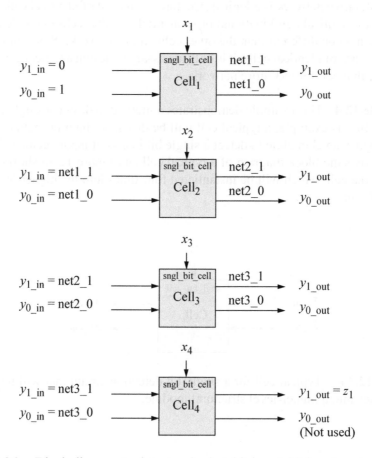

Figure 12.36   Block diagram to detect a single 1 bit in a 4-bit input vector $x[1:4]$.

In Figure 12.36, the input and output lines are defined as follows:

- y_{1_in} is an active-high input line indicating that a single 1 bit was detected up to that cell.

- y_{0_in} is an active-high input line indicating that no 1 bits were detected up to that cell.

- y_{1_out} is an active-high output line indicating that a single 1 bit was detected up to and including that cell.

- y_{0_out} is an active-high output line indicating that no 1 bits were detected up to and including that cell.

```
//typical cell for single-bit detection
module sngl_bit_cell (x1_in, y1_in, y0_in, y1_out, y0_out);

input x1_in, y1_in, y0_in;
output y1_out, y0_out;

not  inst1 (net1, x1_in);
and  inst2 (net2, net1, y1_in);
and  inst3 (net3, x1_in, y0_in);
and  inst4 (y0_out, net1, y0_in);
or   inst5 (y1_out, net2, net3);

endmodule
```

Figure 12.37   Typical cell that is instantiated four times into a structural module to detect a single 1 bit in an input vector *x[1:4]*.

```
//structural single-bit detection module
module sngl_bit_detect2 (x1, x2, x3, x4, z1);

input x1, x2, x3, x4;
output z1;

//instantiate the single-bit cell modules
//cell 1
sngl_bit_cell  inst1(
   .x1_in(x1),
   .y1_in(1'b0),
   .y0_in(1'b1),
   .y1_out(net1_1),
   .y0_out(net1_0)
   );

//cell 2
sngl_bit_cell  inst2(
   .x1_in(x2),
   .y1_in(net1_1),
   .y0_in(net1_0),
   .y1_out(net2_1),
   .y0_out(net2_0)
   );

//continued on next page
```

Figure 12.38   Structural module to detect a single 1 bit in a 4-bit input vector *x[1:4]* in which the typical cell of Figure 12.37 is instantiated four times.

```
//cell 3
sngl_bit_cell  inst3(
   .x1_in(x3),
   .y1_in(net2_1),
   .y0_in(net2_0),
   .y1_out(net3_1),
   .y0_out(net3_0)
   );

//cell 4
sngl_bit_cell  inst4(
   .x1_in(x4),
   .y1_in(net3_1),
   .y0_in(net3_0),
   .y1_out(z1)
   );
endmodule
```

Figure 12.38   (Continued)

```
//test bench for the single-bit detection module
module sngl_bit_detect2_tb;

reg x1, x2, x3, x4;
wire z1;

initial      //apply input vectors
begin: apply_stimulus
   reg [4:0] invect;
   for (invect=0; invect<16; invect=invect+1)
      begin
         {x1, x2, x3, x4} = invect [4:0];
         #10 $display ("x1x2x3x4 = %b, z1 = %b",
                       {x1, x2, x3, x4}, z1);
      end
end
//instantiate the module into the test bench
sngl_bit_detect2 inst1 (
   .x1(x1),
   .x2(x2),
   .x3(x3),
   .x4(x4),
   .z1(z1)
   );
endmodule
```

Figure 12.39   Test bench for the single-bit detection module of Figure 12.38.

```
x1x2x3x4 = 0000,  z1 = 0        x1x2x3x4 = 1000,  z1 = 1
x1x2x3x4 = 0001,  z1 = 1        x1x2x3x4 = 1001,  z1 = 0
x1x2x3x4 = 0010,  z1 = 1        x1x2x3x4 = 1010,  z1 = 0
x1x2x3x4 = 0011,  z1 = 0        x1x2x3x4 = 1011,  z1 = 0
x1x2x3x4 = 0100,  z1 = 1        x1x2x3x4 = 1100,  z1 = 0
x1x2x3x4 = 0101,  z1 = 0        x1x2x3x4 = 1101,  z1 = 0
x1x2x3x4 = 0110,  z1 = 0        x1x2x3x4 = 1110,  z1 = 0
x1x2x3x4 = 0111,  z1 = 0        x1x2x3x4 = 1111,  z1 = 0
```

Figure 12.40   Outputs for the single-bit detection module of Figure 12.38.

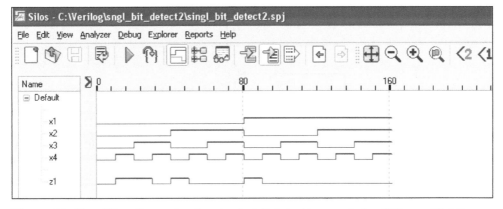

Figure 12.41   Waveforms for the single-bit detection module of Figure 12.38.

Example 12.5   The combinational iterative network of Example 12.4 will now be synthesized as a functionally equivalent sequential machine. The input vector $x_1, x_2,$ $\ldots, x_i, \ldots, x_k$ is applied to the machine serially such that one bit x_i is processed for each machine clock cycle. The usual design procedure for synchronous sequential machines will be followed. The state diagram is illustrated in Figure 12.42.

The machine is reset to state a ($y_1y_2 = 00$) and the input variable x_i is evaluated. The machine remains in state a as long as each bit $x_i = 0$. If $x_i = 1$, then the machine proceeds to state b ($y_1y_2 = 01$) where output z_1 is asserted. The assertion of z_1 in state b indicates that a single 1 bit has been detected thus far. The state of z_1 is valid only after k clock cycles when the k inputs have been sequenced into the machine and evaluated.

In state b, the next 1 bit sequences the machine to state c ($y_1y_2 = 11$) where z_1 is deasserted. State c is a terminal state in which the machine remains regardless of the subsequent bit pattern. Since the transition from state c ($y_1y_2 = 11$) to state a ($y_1y_2 = 00$) cannot occur, the machine will never generate state $y_1y_2 = 01$ as a transient state, which would produce a spurious output on z_1. Therefore, the λ output logic for z_1 is simply an AND gate to decode $y_1 y_2 = 01$, negating the necessity of a complemented clock input to eliminate a glitch. Output z_1 is asserted at time t_1 in state b and remains asserted for all subsequent 0 bits.

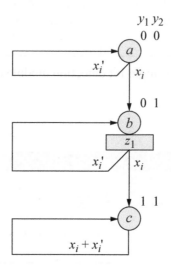

Figure 12.42   State diagram for a synchronous sequential machine to detect a single 1 bit in a serial bit stream.

The input maps of Figure 12.43 are derived from the state diagram as in previous examples. In state a ($y_1 y_2 = 00$), flip-flop y_1 has a next value of 0 regardless of the path taken. Flip-flop y_2 has a next value of 1 only if $x_i = 1$; therefore, input variable x_i is inserted in minterm location 0. Likewise, in state b ($y_1 y_2 = 01$), y_1 has a next value of 1 only if $x_i = 1$. Flip-flop y_2, however, has a next value of 1 regardless of the value of x_i. Therefore, $x_i + x_i' = 1$ is placed in minterm location 1 of the input map for y_2. In state c ($y_1 y_2 = 11$), both y_1 and y_2 have a next value of 1 for $x_i = 0$ or 1.

Consider the input map for flip-flop y_1. Input x_i is common to minterm locations 1 and 3, both of which are in column $y_2 = 1$, yielding the term $y_2 x_i$. The expression $x_i + x_i' = 1$ in minterm 3 combines with the unused state to yield the term y_1. Regarding the input map for flip-flop y_2, every square contains the term x_i. Therefore, x_i is a component of the equation for y_2. Since the expression $x_i + x_i' = 1$ is common to minterm locations 1 and 3, the term y_2 is also a constituent part of the equation for y_2. The input equations for flip-flops y_1 and y_2 are listed in Equation 12.13.

Output z_1 is asserted unconditionally in state b ($y_1 y_2 = 01$). Thus, this Moore-type output is a function of y_1 and y_2 only, and is independent of the value of x_i. The equation for z_1 is specified in Equation 12.13. The logic diagram is shown in Figure 12.44 displaying the instantiation names and net names and will be implemented using the structural module shown in Figure 12.45. The test bench module is shown in Figure 12.46. The outputs and waveforms are shown in Figure 12.47 and Figure 12.48, respectively.

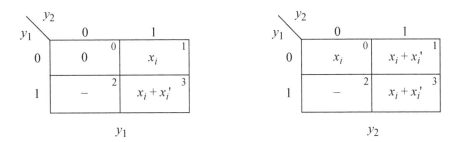

Figure 12.43 Input maps for the Moore machine of Figure 12.42.

$$y_1 = y_2 x_i + y_1$$
$$y_2 = x_i + y_2$$
$$z_1 = y_1' y_2 \qquad\qquad (12.13)$$

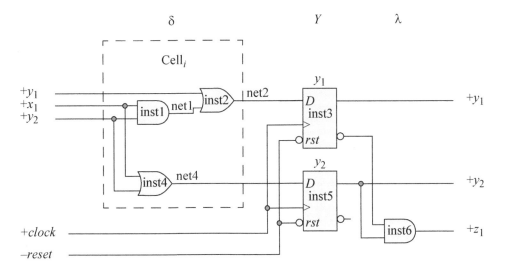

Figure 12.44 Logic diagram for a synchronous sequential machine to detect a single 1 bit in a serial bit stream.

```
//structural moore single-bit detector
module moore_ssm_sngl_bit (rst_n, clk, x1, y1, y2, z1);

//define inputs and output
input rst_n, clk, x1;
output y1, y2;
output z1;

//define internal nets
wire net1, net2, net4;

//instantiate the logic for flip-flop y1
and2_df inst1 (
    .x1(x1),
    .x2(y2),
    .z1(net1)
    );

or2_df inst2 (
    .x1(y1),
    .x2(net1),
    .z1(net2)
    );

//instantiate the D flip-flop for y1
d_ff_bh inst3 (
    .rst_n(rst_n),
    .clk(clk),
    .d(net2),
    .q(y1)
    );

//instantiate the logic for flip-flop y2
or2_df inst4 (
    .x1(x1),
    .x2(y2),
    .z1(net4)
    );

//instantiate the D flip-flop for y2
d_ff_bh inst5 (
    .rst_n(rst_n),
    .clk(clk),
    .d(net4),
    .q(y2)
    );
//continued on next page
```

Figure 12.45   Structural module for the single-bit detector of Example 12.5.

```
//instantiate the logic for output z1
and2_df inst6 (
   .x1(~y1),
   .x2(y2),
   .z1(z1)
   );

endmodule
```

Figure 12.45 (Continued)

```
//test bench for moore single-bit detector
module moore_ssm_sngl_bit_tb;

reg rst_n, clk, x1;
wire y1, y2, z1;

//display inputs and output
initial
$monitor ("x1 = %b, state = %b, z1 = %b",
           x1, {y1, y2}, z1);

//define clock
initial
begin
   clk = 1'b0;
   forever
      #10 clk = ~clk;
end

//define input vectors
initial
begin
   #0    rst_n = 1'b0;      //reset to state_a (00)
         x1 = 1'b0;
   #5    rst_n = 1'b1;

   @ (posedge clk)
      //if x1 = 1'b0 in state_a, go to state_a (00)

   x1 = 1'b1;      @ (posedge clk)
      //if x1 = 1'b1 in state_a, go to state_b (01)

//continued on next page
```

Figure 12.46 Test bench for the structural module of Figure 12.45.

```
   x1 = 1'b0;      @ (posedge clk)
      //if x1 = 1'b0 in state_b, go to state_b (01)

   x1 = 1'b0;      @ (posedge clk)
      //if x1 = 1'b0 in state_b, go to state_b (01)
//----------------------------------------------------
   #20    rst_n = 1'b0;      //reset to state_a (00)
   #25    rst_n = 1'b1;

   x1 = 1'b0;      @ (posedge clk)
      //if x1 = 1'b0 in state_a, go to state_a (00)

   x1 = 1'b1;      @ (posedge clk)
      //if x1 = 1'b1 in state_a, go to state_b (01)

   x1 = 1'b0;      @ (posedge clk)
      //if x1 = 1'b0 in state_b, go to state_b (01)

   x1 = 1'b0;      @ (posedge clk)
      //if x1 = 1'b0 in state_b, go to state_b (01)

   x1 = 1'b1;      @ (posedge clk)
      //if x1 = 1'b1 in state_b, go to state_c (11)

   x1 = 1'b0;      @ (posedge clk)
      //if x1 = 1'b0 in state_c, go to state_c (11)

   x1 = 1'b1;      @ (posedge clk)
      //if x1 = 1'b1 in state_c, go to state_c (11)

   x1 = 1'b1;      @ (posedge clk)
      //if x1 = 1'b1 in state_c, go to state_c (11)

   #10    $stop;
end

//instantiate the module into the test bench
moore_ssm_sngl_bit inst1 (
   .rst_n(rst_n),
   .clk(clk),
   .x1(x1),
   .y1(y1),
   .y2(y2),
   .z1(z1)
   );

endmodule
```

Figure 12.46   (Continued)

```
x1 = 0, state = 00, z1 = 0
x1 = 1, state = 00, z1 = 0
x1 = 0, state = 01, z1 = 1
x1 = 0, state = 00, z1 = 0
x1 = 1, state = 00, z1 = 0
x1 = 0, state = 01, z1 = 1
x1 = 1, state = 01, z1 = 1
x1 = 0, state = 11, z1 = 0
x1 = 1, state = 11, z1 = 0
```

Figure 12.47   Outputs for the structural module of Figure 12.45.

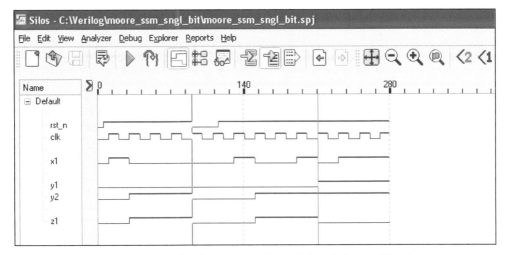

Figure 12.48   Waveforms for the structural module of Figure 12.45.

12.3 Hamming Code

This section extends the concept of Hamming code presented in Chapter 1 by providing more detailed information and presents the theory, logic design, and Verilog design of a Hamming code circuit. A *code word* contains n bits consisting of m message bits plus k parity check bits as shown in Figure 12.49. The m bits represent the information or message part of the code word; the k bits are used for detecting and correcting errors, where $k = n - m$. The *Hamming distance* of two code words X and Y is the number of bits in which the two words differ in their corresponding columns.

For example, the Hamming distance is three for code words X and Y as shown Figure 12.50. Since the minimum distance is three, single error detection and correction is possible.

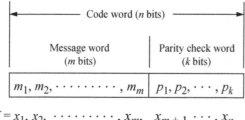

Code word $X = x_1, x_2, \cdots\cdots\cdots, x_m, \; x_{m+1}, \cdots, x_n$

Figure 12.49   Code word of n bits containing m message bits and k parity check bits.

$$
\begin{array}{ccccccccc}
X = & 1 & 0 & 1 & 1 & 1 & 1 & 0 & 1 \\
Y = & 1 & 0 & 1 & 0 & 1 & 1 & 1 & 0
\end{array}
$$

Figure 12.50   Two code words to illustrate a Hamming distance of three.

Let $X = x_1, x_2, \cdots x_n$ and $Y = y_1, y_2, \cdots y_n$ be the ordered n-tuples that represent two code words, where $x_i, y_i \in \{0, 1\}$ for all i. The Hamming distance is the number of x_i and y_i that are different and can be defined mathematically as

$$
H_d\,(x_i, y_i) = \sum_{i=1}^{n}\left(x_i \oplus y_i\right)
$$

The Hamming distance is characterized by the following mathematical properties:

1. $H_d\,(X, Y) = H_d\,(Y, X)$
2. $H_d\,(X, Y) = 0$ if and only if $x_i = y_i$ for $i = 1, 2, \cdots, n$
3. $H_d\,(X, Y) > 0$ if at least one $x_i \neq y_i$ for $i = 1, 2, \cdots, n$
4. $H_d\,(X, Y) + H_d\,(Y, Z) \geq H_d\,(X, Z)$

Since there can be an error in *any* bit position, including the parity check bits, there must be a sufficient number of k parity check bits to identify any of the $m + k$ bit positions. The parity check bits are normally embedded in the code word and are positioned in columns with column numbers that are a power of two, as shown below for a code word containing four message bits (m_3, m_5, m_6, m_7) and three parity bits (p_1, p_2, p_4).

Column number	1	2	3	4	5	6	7
Code word =	p_1	p_2	m_3	p_4	m_5	m_6	m_7

Each parity bit maintains odd parity over a unique group of bits as shown below for a code word of four message bits.

$E_1 =$	p_1	m_3	m_5	m_7
$E_2 =$	p_2	m_3	m_6	m_7
$E_4 =$	p_4	m_5	m_6	m_7

The placement of the parity bits in certain columns is not arbitrary. Each of the variables in group E_1 contain a 1 in column 1 (2^0) of the binary representation of the column number as shown below.

	8	4	2	1
Group E_1	2^3	2^2	2^1	2^0
p_1	0	0	0	1
m_3	0	0	1	1
m_5	0	1	0	1
m_7	0	1	1	1
$\cdots$				

Since p_1 has only a single 1 in the binary representation of column 1, p_1 can therefore be used as a parity check bit for a message bit in *any* column in which the binary representation of the column number has a 1 in column 1 (2^0). Thus, group E_1 can be expanded to include other message bits, as shown below.

$$p_1, m_3, m_5, m_7, m_9, m_{11}, m_{13}, m_{15}, m_{17}, \cdots$$

Each of the variables in group E_2 contain a 1 in column 2 (2^1) of the binary representation of the column number as shown below.

	8	4	2	1
Group E_2	2^3	2^2	2^1	2^0
p_2	0	0	1	0
m_3	0	0	1	1
m_6	0	1	1	0
m_7	0	1	1	1
. . .				

Since p_2 has only a single 1 in the binary representation of column 2, p_2 can therefore be used as a parity check bit for a message bit in *any* column in which the binary representation of the column number has a 1 in column 2 (2^1). Thus, group E_2 can be expanded to include other message bits, as shown below.

$$p_2, m_3, m_6, m_7, m_{10}, m_{11}, m_{14}, m_{15}, m_{18}, \cdots$$

Each of the variables in group E_4 contain a 1 in column 4 (2^2) of the binary representation of the column number as shown below.

	8	4	2	1
Group E_4	2^3	2^2	2^1	2^0
p_4	0	1	0	0
m_5	0	1	0	1
m_6	0	1	1	0
m_7	0	1	1	1
. . .				

Since p_4 has only a single 1 in the binary representation of column 4, p_4 can therefore be used as a parity check bit for a message bit in *any* column in which the binary representation of the column number has a 1 in column 4 (2^2). Thus, group E_4 can be expanded to include other message bits, as shown below.

$$p_4, m_5, m_6, m_7, m_{12}, m_{13}, m_{14}, m_{15}, m_{20}, \cdots$$

The format for embedding parity bits in the code word can be extended easily to any size message. For example, the code word for an 8-bit message is encoded as follows:

$$p_1, p_2, m_3, p_4, m_5, m_6, m_7, p_8, m_9, m_{10}, m_{11}, m_{12}$$

where $m_3, m_5, m_6, m_7, m_9, m_{10}, m_{11}, m_{12}$ are the message bits and p_1, p_2, p_4, p_8 are the parity check bits for groups E_1, E_2, E_4, E_8, respectively, as shown below.

Group E_1 =	p_1	m_3	m_5	m_7	m_9	m_{11}
Group E_2 =	p_2	m_3	m_6	m_7	m_{10}	m_{11}
Group E_4 =	p_4	m_5	m_6	m_7	m_{12}	
Group E_8 =	p_8	m_9	m_{10}	m_{11}	m_{12}	

For messages, the bit with the highest numbered subscript is the low-order bit. Thus, the low-order message bit is m_{12} for a byte of data that is encoded using the Hamming code. A 32-bit message requires six parity check bits:

$$p_1, p_2, p_4, p_8, p_{16}, p_{32}$$

There is only one parity bit in each group. The parity bits are independent and no parity bit checks any other parity bit. Consider the following code word for a 4-bit message:

$$p_1, p_2, m_3, p_4, m_5, m_6, m_7$$

The parity bits are generated so that there are an odd number of 1s in the following groups:

E_1 =	p_1	m_3	m_5	m_7
E_2 =	p_2	m_3	m_6	m_7
E_4 =	p_4	m_5	m_6	m_7

For example, the parity bits are generated as follows:

$$p_1 = (m_3 \oplus m_5 \oplus m_7)'$$
$$p_2 = (m_3 \oplus m_6 \oplus m_7)'$$
$$p_4 = (m_5 \oplus m_6 \oplus m_7)'$$

Example 12.6 A 4-bit message (0110) will be encoded using the Hamming code then transmitted. The message, transmitted code word, and received code word are shown below.

	p_1	p_2	m_3	p_4	m_5	m_6	m_7
Message to be sent			0		1	1	0
Code word sent	0	0	0	1	1	1	0
Code word received	0	0	0	1	0	1	0

From the received code word, it is seen that bit 5 is in error. When the code word is received, the parity of each group is checked using the bits assigned to that group, as shown below.

Group $E_1 =$	$p_1\, m_3\, m_5\, m_7 =$	$0\ 0\ 0\ 0 =$	Error $=\ 1$
Group $E_2 =$	$p_2\, m_3\, m_6\, m_7 =$	$0\ 0\ 1\ 0 =$	No error $=\ 0$
Group $E_4 =$	$p_4\, m_5\, m_6\, m_7 =$	$1\ 0\ 1\ 0 =$	Error $=\ 1$

A parity error is assigned a value of 1; no parity error is assigned a value of 0. The groups are then listed according to their binary weight. The resulting binary number is called the *syndrome word* and indicates the bit in error; in this case, bit 5. The bit in error is then complemented to yield a correct message of 0110.

	2^2	2^1	2^0
Groups	E_4	E_2	E_1
Syndrome word	1	0	1

Since there are three groups in this example, there are eight combinations. Each combination indicates the column of the bit in error, including the parity bits. A syndrome word of $E_4 E_2 E_1 = 000$ indicates no error in the received code word.

Double error detection and single error correction can be achieved by adding a parity bit for the entire code word. The format is shown below, where p_{cw} is the parity bit for the code word.

$$\text{Code word} = p_1\, p_2\, m_3\, p_4\, m_5\, m_6\, m_7\, p_{cw}$$

Several examples using the code word parity bit are shown in Table 12.1. The examples include both single error correction and double error detection. In the final example, the syndrome word is 000, indicating no error. However, the code word parity is incorrect; therefore, there must be no single error and also no double error — the error occurred in the code word parity bit. The logic diagram is shown in Figure 12.51.

Table 12.1 Examples of Single Error Correction and Double Error Detection

	Code Word Format							Syndrome	Code Word	Single	Double
	p_1	p_2	m_3	p_4	m_5	m_6	m_7	p_{cw}	$E_4E_2E_1$	Parity	Error	Error
Sent	0	0	1	1	0	0	0	1				
Received	0	0	0	1	0	0	0	1	0 1 1	Bad	Yes	No
Sent	0	0	0	1	1	1	0	0				
Received	0	1	0	1	1	1	0	0	0 1 0	Bad	Yes	No
Sent	1	0	0	1	1	0	1	1				
Received	1	1	0	1	0	0	1	1	1 1 1	Good	No	Yes
Sent	0	0	1	1	0	0	0	1				
Received	0	0	1	1	0	1	1	1	0 1 1	Good	No	Yes
Sent	0	0	1	1	0	0	0	1				
Received	0	0	1	1	0	0	0	0	0 0 0	Bad	No	No

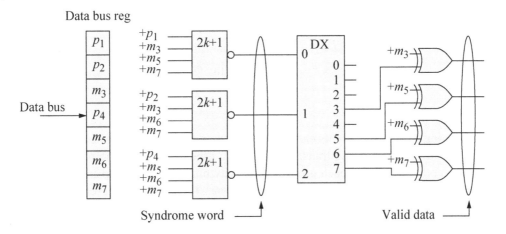

Figure 12.51 Hamming code error detection and correction.

The bit in error can be complemented (inverted) by using an exclusive-OR circuit. Any bit that is exclusive-ORed with a logic 1 is inverted as shown below. The inputs of the exclusive-OR circuits connect to the output of the decoder — which indicates the bit in error — and the error bit is obtained from the data register. If only the correct message is required, then the parity bits do not have to be corrected.

$$
\begin{aligned}
0 \oplus 0 &= 0 \\
0 \oplus 1 &= 1 \\
1 \oplus 0 &= 1 \\
1 \oplus 1 &= 0
\end{aligned}
$$

Example 12.7 This example encodes an 8-bit message in Hamming code for transmission. At the receiver, the message is checked for errors and any single-bit error is corrected. The code word, including the eight message bits and the four parity bits, is as follows:

Code word = p_1 p_2 m_3 p_4 m_5 m_6 m_7 p_8 m_9 m_{10} m_{11} m_{12}

The code word is partitioned into the following four groups that will be used to generate the syndrome word: E_1, E_2, E_4, and E_8. The four groups and their respective code word bits are shown below. The parity of each group is odd.

Group E_1 = p_1 m_3 m_5 m_7 m_9 m_{11}
Group E_2 = p_2 m_3 m_6 m_7 m_{10} m_{11}
Group E_4 = p_4 m_5 m_6 m_7 m_{12}
Group E_8 = p_8 m_9 m_{10} m_{11} m_{12}

The logic diagram for this example is shown in Figure 12.52. The m_i bits are the message bits that are being transmitted and are used to generate the parity bits p_i to maintain odd parity over each group. The parity bits for each group are generated in the test bench by a task labeled *task pbit_generate*. The mr_i bits represent the received message bits that may contain errors. Errors are injected into the received message by a task in the test bench labeled *task error_inject*. The *e1_err*, *e2_err*, *e4_err*, and *e8_err* outputs of the exclusive-NOR functions represent the syndrome word which connects to the inputs of the 4:16 decoder.

The decoder outputs specify the message bit of the received code word in error. The active decoder output is then exclusive-ORed with the corresponding received message bit to correct the bit in error. The valid message is represented by the mv_i bits.

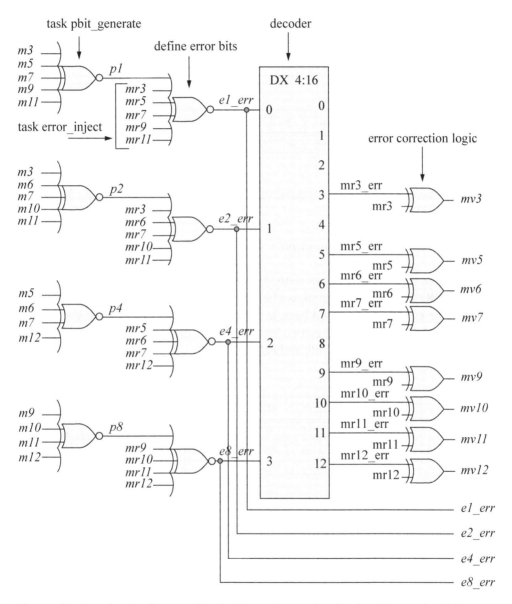

Figure 12.52 Logic diagram for the Hamming code circuit of Example 12.7.

The dataflow module, test bench module, and outputs are shown in Figure 12.53, Figure 12.54, and Figure 12.55, respectively. In the test bench, parity bits $p1$, $p2$, $p4$, and $p8$ are generated for each corresponding group by the task *task pbit_generate*. Single-bit errors are inserted in the received mr_i bits in the test bench by the task *task error_inject*. The outputs of Figure 12.55 indicate the message that was sent, the received message containing the error, the column number in which the error occurred, and the corrected message containing valid data. Since parity bits are not corrected, the byte of data specifies the message bits only. Recall that the column numbers for the message bits are m_3, m_5, m_6, m_7, m_9, m_{10}, m_{11}, and m_{12}. Thus, the high-order message bit (m_3) is the leftmost bit listed and the low-order message bit (m_{12}) is the rightmost bit listed.

Consider the case where an error is inserted into message bit *m3*. The code word, shown in the test bench, is formatted as follows:

Bit position =	11	10	9	8	7	6	5	4	3	2	1	0
Data =	*m3*	*m5*	*m6*	*m7*	*m9*	*m10*	*m11*	*m12*	*p1*	*p2*	*p4*	*p8*

The statement *error_inject (11)* passes the constant 11 to the task as the *bit_number*. Then the statement

$$bit\_position = 1'b1 << bit\_number$$

shifts a 1 bit eleven bit positions to the left to location *m3*. The message bit *m3* is exclusive-ORed with the 1 bit as shown below, thereby inverting *m3*. The message containing the error is then passed back to the task invocation as the received message, which is then corrected by the error correction logic. In a similar manner, errors are injected into message bits *m7*, *m9*, and *m12*.

Data =	*m3*	*m5*	*m6*	*m7*	*m9*	*m10*	*m11*	*m12*	*p1*	*p2*	*p4*	*p8*
XOR =	1	0	0	0	0	0	0	0	0	0	0	0
Received message =	*m3'*	*m5*	*m6*	*m7*	*m9*	*m10*	*m11*	*m12*	*p1*	*p2*	*p4*	*p8*

```
//dataflow module for Hamming code to encode an 8-bit message
module hamming_code (m3, m5, m6, m7, m9, m10, m11, m12,
                     p1, p2, p4,p8,
                     mv3, mv5, mv6, mv7, mv9, mv10, mv11, mv12,
                     e1_err, e2_err, e4_err, e8_err);

input m3, m5, m6, m7, m9, m10, m11, m12;
input p1, p2, p4, p8;
//continued on next page
```

Figure 12.53   Dataflow module to illustrate Hamming code error detection and correction.

```
output mv3, mv5, mv6, mv7, mv9, mv10, mv11, mv12;
output e1_err, e2_err, e4_err, e8_err;

wire mr3_err, mr5_err, mr6_err, mr7_err;
wire mr9_err, mr10_err, mr11_err, mr12_err;

//define the error bits
assign    e1_err = ~(p1 ^ m3 ^ m5 ^ m7 ^ m9 ^ m11),
          e2_err = ~(p2 ^ m3 ^ m6 ^ m7 ^ m10 ^ m11),
          e4_err = ~(p4 ^ m5 ^ m6 ^ m7 ^ m12),
          e8_err = ~(p8 ^ m9 ^ m10 ^ m11 ^ m12);

//design the decoder
assign    mr3_err= (~e8_err) & (~e4_err) & (e2_err)  & (e1_err),
       mr5_err= (~e8_err) & (e4_err)  & (~e2_err) & (e1_err),
       mr6_err= (~e8_err) & (e4_err)  & (e2_err)  & (~e1_err),
       mr7_err= (~e8_err) & (e4_err)  & (e2_err)  & (e1_err),
       mr9_err= (e8_err)  & (~e4_err) & (~e2_err) & (e1_err),
       mr10_err = (e8_err)  & (~e4_err) & (e2_err)  & (~e1_err),
       mr11_err = (e8_err)  & (~e4_err) & (e2_err)  & (e1_err),
       mr12_err = (e8_err)  & (e4_err)  & (~e2_err) & (~e1_err);

//design the correction logic
assign    mv3 = (mr3_err)  ^ (m3),
          mv5 = (mr5_err)  ^ (m5),
          mv6 = (mr6_err)  ^ (m6),
          mv7 = (mr7_err)  ^ (m7),
          mv9 = (mr9_err)  ^ (m9),
          mv10= (mr10_err) ^ (m10),
          mv11= (mr11_err) ^ (m11),
          mv12= (mr12_err) ^ (m12);
endmodule
```

Figure 12.53   (Continued)

```
//test bench for the Hamming code module
module hamming_code_tb;

reg m3, m5, m6, m7, m9, m10, m11, m12;
reg ms3, ms5, ms6, ms7, ms9, ms10, ms11, ms12;
reg mr3, mr5, mr6, mr7, mr9, mr10, mr11, mr12;
wire mv3, mv5, mv6, mv7, mv9, mv10, mv11, mv12;
wire e1_err, e2_err, e4_err, e8_err;

//continued on next page
```

Figure 12.54   Test bench for the Hamming code dataflow module of Figure 12.53.

```
reg p1, p2, p4, p8;


initial
$display ("bit_order = m3, m5, m6, m7, m9, m10, m11, m12");


initial
$monitor ("sent=%b, rcvd=%b, error=%b, valid=%b",
          {ms3, ms5, ms6, ms7, ms9, ms10, ms11, ms12},
          {mr3, mr5, mr6, mr7, mr9, mr10, mr11, mr12},
          {e8_err, e4_err, e2_err, e1_err},
          {mv3, mv5, mv6, mv7, mv9, mv10, mv11, mv12});


initial
begin
//-------------------------------------------------------------
   #0    {ms3,ms5,ms6,ms7,ms9,ms10,ms11,ms12}=8'b1010_1010;
         {m3,m5,m6,m7,m9,m10,m11,m12} =
         {ms3,ms5,ms6,ms7,ms9,ms10,ms11,ms12};
   pbit_generate(m3,m5,m6,m7,m9,m10,m11,m12);   //invoke task

         //no error injected
         {mr3, mr5, mr6, mr7, mr9, mr10, mr11, mr12} =
         {m3, m5, m6, m7, m9, m10, m11, m12};
//-------------------------------------------------------------
   #10   {ms3,ms5,ms6,ms7,ms9,ms10,ms11,ms12}=8'b1010_1010;
         {m3,m5,m6,m7,m9,m10,m11,m12} =
         {ms3,ms5,ms6,ms7,ms9,ms10,ms11,ms12};
   pbit_generate(m3,m5,m6,m7,m9,m10,m11,m12);   //invoke task

         //inject error into m3
   error_inject(11);                             //invoke task
         {mr3,mr5,mr6,mr7,mr9,mr10,mr11,mr12} =
         {m3,m5,m6,m7,m9,m10,m11,m12};
//-------------------------------------------------------------
   #10   {ms3,ms5,ms6,ms7,ms9,ms10,ms11,ms12}=8'b0101_0101;
         {m3,m5,m6,m7,m9,m10,m11,m12} =
         {ms3,ms5,ms6,ms7,ms9,ms10,ms11,ms12};
   pbit_generate(m3,m5,m6,m7,m9,m10,m11,m12);   //invoke task

         //inject error into m7
   error_inject(8);                              //invoke task
         {mr3,mr5,mr6,mr7,mr9,mr10,mr11,mr12} =
         {m3,m5,m6,m7,m9,m10,m11,m12};
//-------------------------------------------------------------

//continued on next page
```

Figure 12.54   (Continued)

```
//---------------------------------------------------------
   #10    {ms3,ms5,ms6,ms7,ms9,ms10,ms11,ms12}=8'b1111_0000;
          {m3,m5,m6,m7,m9,m10,m11,m12} =
          {ms3,ms5,ms6,ms7,ms9,ms10,ms11,ms12};
   pbit_generate(m3,m5,m6,m7,m9,m10,m11,m12);    //invoke task

          //inject error into m9
   error_inject(7);                               //invoke task
          {mr3,mr5,mr6,mr7,mr9,mr10,mr11,mr12} =
          {m3,m5,m6,m7,m9,m10,m11,m12};
//---------------------------------------------------------
   #10    {ms3,ms5,ms6,ms7,ms9,ms10,ms11,ms12}=8'b0110_1101;
          {m3,m5,m6,m7,m9,m10,m11,m12} =
          {ms3,ms5,ms6,ms7,ms9,ms10,ms11,ms12};
   pbit_generate(m3,m5,m6,m7,m9,m10,m11,m12);    //invoke task

          //inject error into m12
   error_inject(4);                               //invoke task
          {mr3,mr5,mr6,mr7,mr9,mr10,mr11,mr12} =
          {m3,m5,m6,m7,m9,m10,m11,m12};
//---------------------------------------------------------
   #10    $stop;
end

task pbit_generate;
input m3, m5, m6, m7, m9, m10, m11, m12;
begin
   p1 = ~(m3 ^ m5 ^ m7 ^ m9 ^ m11);
   p2 = ~(m3 ^ m6 ^ m7 ^ m10 ^ m11);
   p4 = ~(m5 ^ m6 ^ m7 ^ m12);
   p8 = ~(m9 ^ m10 ^ m11 ^ m12);
end
endtask

task error_inject;
   input [3:0] bit_number;
   reg [11:0] bit_position;
   reg [11:0] data;
   begin
      bit_position = 1'b1 << bit_number;

      data = {m3,m5,m6,m7,m9,m10,m11,m12,p1,p2,p4,p8};
      {m3,m5,m6,m7,m9,m10,m11,m12,p1,p2,p4,p8} =
         data ^ bit_position;
end
endtask
//continued on next page
```

Figure 12.54   (Continued)

```
//instantiate the module into the test bench
hamming_code inst1 (
    .m3(m3),
    .m5(m5),
    .m6(m6),
    .m7(m7),
    .m9(m9),
    .m10(m10),
    .m11(m11),
    .m12(m12),
    .p1(p1),
    .p2(p2),
    .p4(p4),
    .p8(p8),


    .mv3(mv3),
    .mv5(mv5),
    .mv6(mv6),
    .mv7(mv7),
    .mv9(mv9),
    .mv10(mv10),
    .mv11(mv11),
    .mv12(mv12),


    .e1_err(e1_err),
    .e2_err(e2_err),
    .e4_err(e4_err),
    .e8_err(e8_err)
    );

endmodule
```

Figure 12.54   (Continued)

```
bit_order = m3, m5, m6, m7, m9, m10, m11, m12

sent=10101010, rcvd=10101010, error=0000, valid=10101010
sent=10101010, rcvd=00101010, error=0011, valid=10101010
sent=10101010, rcvd=10111010, error=0111, valid=10101010
sent=11110000, rcvd=11111000, error=1001, valid=11110000
sent=01101101, rcvd=01101100, error=1100, valid=01101101
```

Figure 12.55   Outputs for the Hamming code test bench of Figure 12.54.

12.4 Cyclic Redundancy Check Code

The intent of this section is to provide an overview of some fundamental attributes of linear feedback shift registers and to illustrate their use as a pragmatic solution to error detection and correction. These devices are especially useful for large strings of serial binary data, such as found on single-track storage devices. When data bytes are stored on a disk sector, they are usually followed by a *cyclic redundancy check* (CRC) character. The CRC character is formed by passing the bits serially through a CRC generator, then appending the CRC character to the end of the bit stream, as shown in Figure 12.56. The binary data can be represented by a *binary polynomial P(X)*, where the data bits are the coefficients of the polynomial. For example, the data byte 1001 1011 can be written as a seventh degree polynomial $P(X)$ as shown in Equation 12.14.

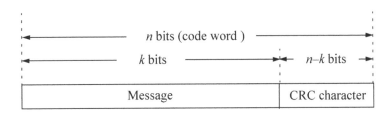

Figure 12.56   Format for an *n*-bit code word.

$$P(X) = 1x^7 + 0x^6 + 0x^5 + 1x^4 + 1x^3 + 0x^2 + 1x^1 + 1x^0 \qquad (12.14)$$

If the polynomial $P(X)$ is divided by a second polynomial, called a *generator polynomial G(X)*, then the result is a *quotient polynomial Q(X)* plus a *remainder polynomial R(X)*; that is,

$$\frac{P(X)}{G(X)} = Q(X) + R(X)$$

The CRC character that is added to the end of the binary bit stream is the remainder $R(X)$. The CRC generator performs a continuous division of the bit stream polynomial $P(X)$ by the generator polynomial $G(X)$. The degree (highest exponent value) of $G(X)$ is

$$0 < G(X) < P(X)$$

Also, $G(X)$ has a nonzero coefficient for the X^0 term.

Division of the data polynomial by the generator polynomial is performed by modulo-2 division. Since modulo-2 is a linear operation, the register used to generate and check CRC characters is called a *linear feedback shift register*. Consider the previous data stream of

$$
\begin{array}{llllllllll}
\text{Bit position} & 7 & 6 & 5 & 4 & 3 & 2 & 1 & 0 & \\
\text{Data} & 1 & 0 & 0 & 1 & 1 & 0 & 1 & 1 & (\text{LSB})
\end{array}
$$

which yields a data (or message) polynomial of $P(X) = 1x^7 + 0x^6 + 0x^5 + 1x^4 + 1x^3 + 0x^2 + 1x^1 + 1x^0$, or simply $x^7 + x^4 + x^3 + x + 1$ (LSB).

Assume a generator polynomial of $x^4 + x + 1$. The CRC check character is formed by dividing the shifted message polynomial by the generator polynomial. The code word that is transmitted to the destination during a write operation consists of n bits: a message segment of k bits, and a CRC check bit segment of $n - k$ bits, which is the remainder polynomial.

When the bit stream is received at the destination during a read operation, both the data bits and the CRC bits are shifted through an identical CRC generator/checker circuit. Because the received bit stream now includes the remainder polynomial, the division operation — produced by shifting — results in a remainder of zero if no errors occurred during transmission or reception; that is, $R(X) = 0$.

To encode a message polynomial $P(X)$, multiply $P(X)$ by $x(n - k)$, because the transmitted code word will have 12 bits: eight message bits and four check bits ($n - k = 12 - 8 = 4$). Thus, the exponent of each message bit will be increased by x^4. The code word is then divided by the generator polynomial $G(X)$ to yield a quotient $Q(X)$ and a remainder $R(X)$, which is appended to the end of the message bits. This encoding process is expressed in Equation 12.15.

$$
\frac{x^{n-k}P(X)}{G(X)} = Q(X) + \frac{R(X)}{G(X)} \tag{12.15}
$$

To elucidate the encoding method, assume a message of $P(X) = 1001\ 1011 = x^7 + x^4 + x^3 + x + 1$ and a generator polynomial of $G(X) = x^4 + x + 1$. The transmitted message is multiplied by x^4 — the high-order variable of the generator polynomial — so that the code word will contain n bits $[k + (n - k)]$. This corresponds to a left shift of $n - k$ bits. This new message of n bits is then divided by the generator polynomial to produce an n-bit code word containing a k-bit message and its corresponding $n - k$ check bits, which is the CRC character. The encoding technique is shown in Figure 12.57(a).

Encode the message

Message polynomial $P(X) = 1001\ 1011 = x^7 + x^4 + x^3 + x + 1$

Generator polynomial $G(X) = x^4 + x + 1$

Multiply the message polynomial by x^4

The shifted message polynomial $P'(X)$
$$= x^4(x^7 + x^4 + x^3 + x + 1)$$
$$= x^{11} + x^8 + x^7 + x^5 + x^4$$

Divide $P'(X)$ by $G(X)$.

$$
\begin{array}{r}
x^7 \quad + x \quad + 1 \\
x^4 + x + 1 \,\overline{\big)\, x^{11} \quad + x^8 \quad + x^7 \quad + x^5 \quad + x^4} \\
\underline{x^{11} \quad + x^8 \quad + x^7} \\
x^5 \quad + x^4 \\
\underline{x^5 \quad + x^2 \quad + x} \\
x^4 \quad + x^2 \quad + x \\
\underline{x^4 \qquad\quad + x \quad + 1} \\
+ x^2 \qquad\quad + 1
\end{array}
$$

Therefore, the transmitted code word $= x^{11} + x^8 + x^7 + x^5 + x^4 + x^2 + 1$

CRC check bits ⟶

(a)

Check the received code word

Received code word $= x^{11} + x^8 + x^7 + x^5 + x^4 + x^2 + 1$

Divide the received code word by the generator polynomial $G(X)$

$$
\begin{array}{r}
x^7 \quad + x \quad + 1 \\
x^4 + x + 1 \,\overline{\big)\, x^{11} \quad + x^8 \quad + x^7 \quad + x^5 \quad + x^4 \quad + x^2 \quad + 1} \\
\underline{x^{11} \quad + x^8 \quad + x^7} \\
+ x^5 \quad + x^4 \quad + x^2 \quad + 1 \\
\underline{+ x^5 \qquad\qquad + x^2 \quad + 1} \\
+ x^4 \quad + x \quad + 1 \\
\underline{+ x^4 \quad + x \quad + 1} \\
0
\end{array}
$$

(b)

Figure 12.57 Encoding the message polynomial $P(X) = x^7 + x^4 + x^3 + x + 1$ for transmission using the generator polynomial $G(X) = x^4 + x + 1$: (a) divide $P'(X)$ by $G(X)$ to yield $R(X)$ and (b) checking the received polynomial $C(X)$ for errors.

The remainder polynomial $R(X) = x^2 + 1$ is appended to the end of the message and $P'(X) + R(X)$ is transmitted. Thus, the transmitted code word $C(X)$ is

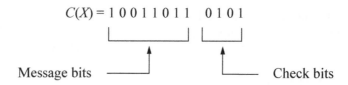

$$C(X) = 1\ 0\ 0\ 1\ 1\ 0\ 1\ 1 \quad 0\ 1\ 0\ 1$$

Message bits ——⌐   ⌐—— Check bits

When the code word is read back from the destination, the received polynomial $C(X)$ — containing the appended remainder — is divided by the generator polynomial $G(X)$. If no error has occurred in transmission, then the remainder $R(X)$ will be zero. This is illustrated in Figure 12.57(b).

The hardware for encoding a message polynomial is shown in Figure 12.58, and consists of a linear feedback shift register for use with the generator polynomial $G(X)$. Since modulo-2 addition is performed, the shift register contains two exclusive-OR gates. The message bits enter the register from the left beginning with the high-order bit. The gating logic shown at the output of flip-flop y_4 circulates the shifted polynomial through the feedback paths until the code word polynomial is generated, then allows the CRC character to be transmitted.

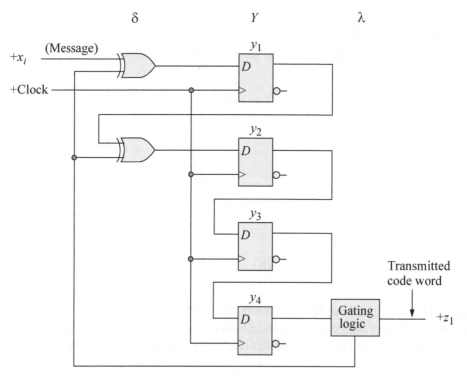

Figure 12.58   Linear feedback shift register for generator polynomial $x^4 + x + 1$.

Table 12.2 tabulates the sequence of shifts which will generate the remainder $R(X)$. The shift register is reset to zero initially. Recall that the message polynomial was shifted left $n - k = 12 - 8 = 4$ bit positions to allow four bits for the CRC character. This is represented by shift counts 9 through 12. After 12 shifts, the remainder polynomial $R(X) = x^2 + 1 = 0101$.

Table 12.2 Shift Sequence to Generate the Remainder Polynomial $R(X)$ for Message 1001 1011

Shift	Message $P'(X)$	y_1	y_2	y_3	y_4	(High Order)
		0	0	0	0	
1	1	1	0	0	0	
2	0	0	1	0	0	
3	0	0	0	1	0	
4	1	1	0	0	1	
5	1	0	0	0	0	
6	0	0	0	0	0	
7	1	1	0	0	0	
8	1	1	1	0	0	
9	0	0	1	1	0	
10	0	0	0	1	1	
11	0	1	1	0	1	
12	0	1	0	1	0	$= R(X) = x^2 + 1 = 0101$

When the code word is read back from the destination, it is processed through the same shift register. Because the remainder is also shifted through the register, the result will be zero if no error has occurred during transmission. The sequence of shifts is tabulated in Table 12.3 and indicates that the received code word — and thus the message — was free from errors.

Table 12.3 Tabulation of Shift Sequences for the Received Word Containing No Error

Shift	Received Code Word	y_1	y_2	y_3	y_4	(High Order)
		0	0	0	0	
1	1	1	0	0	0	
2	0	0	1	0	0	
3	0	0	0	1	0	
4	1	1	0	0	1	
5	1	0	0	0	0	
(Continued on next page)						

Table 12.3 Tabulation of Shift Sequences for the Received Word Containing No Error

Shift	Received Code Word	y_1	y_2	y_3	y_4	(High Order)
6	0	0	0	0	0	
7	1	1	0	0	0	
8	1	1	1	0	0	
9	0	0	1	1	0	
10	1	1	0	1	1	
11	0	1	0	0	1	
12	1	0	0	0	0	= No error

Assume that an error occurred in bit position 7 of the received code word, as shown below.

Bit position =	11	10	9	8	7	6	5	4		3	2	1	0
Received code word =	1	0	0	1	**0**	0	1	1		0	1	0	1
				Message						Check bits			

Therefore, The received code word polynomial is $x^{11} + x^8 + x^5 + x^4 + x^2 + 1$. Table 12.4 represents a shift sequence in which the received code word contained an error in bit position 7. The remainder is nonzero as indicated in shift count 12 and is expressed by the error polynomial $E(X)$. Using $E(X)$, a table lookup procedure specifies the bit in error. To correct the error in this example, the x^7 term is added to the received code word using modulo-2 addition, resulting in the original code word containing the message and check bits.

The primitive generator polynomial $x^4 + x + 1$ is only one of many primitive polynomials used for error detection and correction. A popular generator polynomial for disk drives and binary synchronous communication is $x^{16} + x^{15} + x^2 + 1$.

Table 12.4 Received Code Word Containing an Error

Shift	Received Code Word	y_1	y_2	y_3	y_4	(High Order)
		0	0	0	0	
1	1	1	0	0	0	
2	0	0	1	0	0	
3	0	0	0	1	0	
4	1	1	0	0	1	
5	0	1	0	0	0	
6	0	0	1	0	0	
7	1	1	0	1	0	

(Continued on next page)

Table 12.4 Received Code Word Containing an Error

Shift	Received Code Word	y_1	y_2	y_3	y_4	(High Order)
8	1	1	1	0	1	
9	0	1	0	1	0	
10	1	1	1	0	1	
11	0	1	0	1	0	
12	1	1	1	0	1	$= R(X) = E(X)x^7$

12.5 Residue Checking

This section presents a method of detecting errors in addition operations. When an integer A is divided by an integer m (modulus) a unique remainder R (residue) is generated, where $R < m$. If two integers A and B are divided by the same modulus m and produce the same remainder R, then A and B are *congruent*, as shown below.

$$A \equiv B \quad \text{mod-}m$$

Congruence is indicated either by the symbol $\equiv$ or by the symbol $\Rightarrow$. The residue R of a number A modulo-m is specified by Equation 12.16. The integers A and B are of the form shown in Equation 12.17, where a_i and b_i are the digit values and r is the radix.

$$R(A) \equiv A \bmod\text{-}m \tag{12.16}$$

$$A = a_{n-1}\, r^{n-1}\, a_{n-2}\, r^{n-2} \ldots a_1 r^1\, a_0 r^0$$

$$B = b_{n-1}\, r^{n-1}\, b_{n-2}\, r^{n-2} \ldots b_1 r^1\, b_0 r^0 \tag{12.17}$$

The congruence relationship can be obtained by dividing the integer A by an integer m (modulus) to produce Equation 12.18, where $0 \leq R < m$.

$$A = (quotient \times m) + R \tag{12.18}$$

Numbers that are congruent to the same modulus can be added and the result is a valid congruence, as shown in Equation 12.19.

$$R(A) \equiv A \text{ mod-}m$$

$$R(B) \equiv B \text{ mod-}m$$

$$R(A) + R(B) \equiv (A \text{ mod-}m) + (B \text{ mod-}m)$$

$$\equiv (A + B) \text{ mod-}m$$

$$\equiv R(A + B) \qquad (12.19)$$

The operations of subtraction and multiplication yield similar results; that is, the difference of the residues of two numbers is congruent to the residue of their difference and the product of the residues of two numbers is congruent to the residue of their products, as shown in Equation 12.20. A similar procedure is not available for division.

$$R(A) - R(B) \equiv R(A - B)$$

$$R(A) \times R(B) \equiv R(A \times B) \qquad (12.20)$$

For the radix 2 number system, modulo-3 residue can detect single-bit errors using only two check bits. The modulo-3 results are shown in Table 12.5 for binary data 0000 through 1111.

Table 12.5 Modulo 3 Residue

Binary Data				Modulo-3 Residue	
8	4	2	1	2	1
0	0	0	0	0	0
0	0	0	1	0	1
0	0	1	0	1	0
0	0	1	1	0	0
0	1	0	0	0	1
0	1	0	1	1	0
0	1	1	0	0	0
0	1	1	1	0	1
1	0	0	0	1	0
1	0	0	1	0	0
1	0	1	0	0	1
1	0	1	1	1	0
1	1	0	0	0	0
1	1	0	1	0	1
1	1	1	0	1	0
1	1	1	1	0	0

Examples will now be presented to illustrate error detection in addition using Equation 12.19. The *carry-in* and *carry-out* will also be considered.

Example 12.8 Let the augend A and the addend B be decimal values of 42 and 87, respectively, to yield a sum of 129. The *carry-in* = 0; the *carry-out* = 0.

$$
\begin{array}{lllll}
A = & 0010 \;\; 1010 & \rightarrow & \text{mod-3} = 0 \\
B = & 0101 \;\; 0111 & \rightarrow & \text{mod-3} = 0
\end{array} \quad \rightarrow \text{mod-3} = \mathbf{0}
$$

$$
\begin{array}{llll}
& & 0 \leftarrow & & \text{No error} \\
\hline
Sum = & 0 \leftarrow 1000 \;\; 0001 & \rightarrow & \rightarrow & \rightarrow \text{mod-3} = \mathbf{0}
\end{array}
$$

Example 12.9 Let the augend A and the addend B be decimal values of 42 and 87, respectively, with a *carry-in* = 1 and a *carry-out* = 0 to yield a sum of 130.

$$
\begin{array}{lllll}
A = & 0010 \;\; 1010 & \rightarrow & \text{mod-3} = 0 \\
B = & 0101 \;\; 0111 & \rightarrow & \text{mod-3} = 0
\end{array} \quad \rightarrow \text{mod-3} = 0 + cin = \mathbf{1}
$$

$$
\begin{array}{llll}
& & 1 & & \text{No error} \\
\hline
Sum = & 1000 \;\; 0010 & \rightarrow & \rightarrow & \rightarrow \text{mod-3} = \mathbf{1}
\end{array}
$$

Example 12.10 Let the augend A and the addend B be decimal values of 122 and 149, respectively, with a *carry-in* = 0 and a *carry-out* = 1 to yield a sum of 271.

$$
\begin{array}{lllll}
A = & 0111 \;\; 1010 & \rightarrow & \text{mod-3} = 2 \\
B = & 1001 \;\; 0101 & \rightarrow & \text{mod-3} = 2
\end{array} \quad \rightarrow \text{mod-3} = 1 + cin = \mathbf{1}
$$

$$
\begin{array}{llll}
& & 0 & & \text{No error} \\
\hline
Sum = & 1 \leftarrow 0000 \;\; 1111 & \rightarrow & \rightarrow & \rightarrow \text{mod-3} = \mathbf{1}
\end{array}
$$

Example 12.11   Let the augend A and the addend B be decimal values of 122 and 148, respectively, with a *carry-in* = 1 and a *carry-out* = 1 to yield a sum of 271.

$$
\begin{array}{lllll}
A = & 0111 \;\; 1010 & \rightarrow & \text{mod-3} = 2 \\
B = & 1001 \;\; 0100 & \rightarrow & \text{mod-3} = 1
\end{array} \quad \rightarrow \text{mod-3} = 0 + cin = \mathbf{1}
$$

$$
\begin{array}{llll}
& & 1 & & \text{No error} \\
\hline
Sum = & 1 \leftarrow 0000 \;\; 1111 & \rightarrow & \rightarrow & \rightarrow \text{mod-3} = \mathbf{1}
\end{array}
$$

Example 12.12 Let the augend A and the addend B be decimal values of 122 and 148, respectively, with a *carry-in* = 1 and a *carry-out* = 1 to yield a sum of 271. Assume that an error occurred in adding the low-order four bits, as shown below. The modulo-3 sum of the residues of the augend and addend is 1; however, the modulo-3 residue of the sum is 0, indicating an error.

$$
\begin{array}{llll}
A = & 0111 \quad 1010 & \rightarrow \quad \text{mod-3} = 2 \\
B = & 1001 \quad 0100 & \rightarrow \quad \text{mod-3} = 1
\end{array}
\quad \rightarrow \text{mod-3} = 0 + cin = \mathbf{1}
$$

$$
\begin{array}{llllll}
& \underline{\qquad\qquad\qquad 1 \qquad\qquad} & & & & \text{Error} \\
Sum = & 1 \leftarrow \ 0000 \quad 1110 \ \rightarrow & & \rightarrow & \rightarrow \text{mod-3} = \mathbf{0}
\end{array}
$$

The same technique can be used for detecting errors in binary-coded decimal (BCD) addition. However, each digit is treated separately for each operand. Then the modulo-3 sum of the residues of each digit determines the residue of the operand. Examples are shown below.

Example 12.13 The following two decimal numbers will be added using BCD addition: 452 and 120 to yield a sum of 572.

$$
\begin{array}{llllllll}
& & & & \multicolumn{3}{c}{\text{Residue of}} \\
& & & & \multicolumn{3}{c}{\text{segments}} \\
A = & 0100 & 0101 & 0010 & 1 & 2 & 2 & \rightarrow 2 \\
B = & \underline{0001} & \underline{0010} & \underline{0000} & 1 & 2 & 0 & \rightarrow 0 \\
Sum = & 0101 & 0111 & 0010 & 2 & 1 & 2 & \rightarrow
\end{array}
\qquad
\begin{array}{l}
\rightarrow \text{mod-3} = \mathbf{2} \\
\\
\rightarrow \text{mod-3} = \mathbf{2}
\end{array}
\qquad \text{No error}
$$

Example 12.14 The following two decimal numbers will be added using BCD addition: 676 and 738 to yield a sum of 1414.

$$
\begin{array}{llllllll}
& & & & \multicolumn{3}{c}{\text{Residue of}} \\
& & & & \multicolumn{3}{c}{\text{segments}} \\
A = & 0110 & 0111 & 0110 & 0 & 1 & 0 & \rightarrow 1 \\
B = & 0111 & 0011 & 1000 & 1 & 0 & 2 & \rightarrow 0 \\
& & & \underline{\ \ 1 \leftarrow \ 1110} \\
& & \underline{\ \ 1 \leftarrow \ 1011} & 0110 \\
& 1 \leftarrow \ 1110 & 0110 & 0100 \\
& \underline{0110} & \underline{0001} \\
& 0100
\end{array}
$$

$$
\rightarrow \text{mod-3} = \mathbf{1}
$$

No error

$$
\begin{array}{lllllllll}
& \downarrow & \downarrow & \downarrow & \downarrow \\
Sum = & 0001 \quad 0100 & 0001 & 0100 & 1 & 1 & 1 & 1 & \rightarrow \quad \rightarrow \text{mod-3} = \mathbf{1}
\end{array}
$$

Example 12.15   The following two decimal numbers will be added using BCD addition: 769 and 476 to yield a sum of 1245.

				Residue of segments					
$A =$	0111	0110	1001	1	0	0	$\rightarrow 1$	$\rightarrow$ mod-3 = **0**	
$B =$	0100	0111	0110	1	1	0	$\rightarrow 2$		

$$\begin{array}{cccc}
 & & 1 & \leftarrow \quad 1111 \\
 & 1 \leftarrow & 1110 & \quad 0110 \\
1 \leftarrow & 1100 & 0110 & \quad 0101 \\
 & 0110 & 0100 & \\
 & 0010 & & \quad \text{No error}
\end{array}$$

	$\downarrow$	$\downarrow$	$\downarrow$	$\downarrow$					
$Sum =$	0001 0010	0100	0101	1	2	1	2 $\rightarrow$	$\rightarrow$ mod-3 = 0	

Example 12.16   The following two decimal numbers will be added using BCD addition: 732 and 125 to yield a sum of 857. If no error occurred in the addition operation, then the sum of the residues of the augend and the addend is 2 and the residue of the sum is also 2. Assume that an error occurred in adding the middle four bits, as shown below. The modulo-3 sum of the residues of the augend and addend is 2; however, the modulo-3 residue of the sum is 1, indicating an error.

				Residue of segments					
$A =$	0111	0011	0010	1	0	2	$\rightarrow 0$	$\rightarrow$ mod-3 = **2**	
$B =$	0001	0010	0101	1	2	2	$\rightarrow 2$		Error
$Sum =$	1000	0100	0111	2	1	1	$\rightarrow$	$\rightarrow$ mod-3 = **1**	

The Verilog HDL dataflow module for modulo-3 residue checking is shown in Figure 12.59 using the concept of Equation 12.19, where the arithmetic operator % indicates the modulus (remainder/residue) operation. The test bench is shown in Figure 12.60 and the outputs are shown in Figure 12.61. The outputs show the residue of augend a (*residue_a*) and the addend b (*residue_b*), the residue of the residues of a and b (*residue_a_b*), and the residue of the sum (*residue_sum*).

```
//dataflow modulo-3 residue checking
module residue_chkg (a, b, cin, sum, cout,
                     residue_a, residue_b, residue_a_b,
                     residue_sum);
input [7:0] a, b;
input cin;
output [7:0] sum;
output cout;
output [1:0] residue_a, residue_b, residue_a_b, residue_sum;

//obtain sum
assign {cout, sum} = a + b + cin;

//obtain residues for a and b
assign   residue_a = a % 3,
         residue_b = b % 3;

//obtain the residue of the residues
assign   residue_a_b = (residue_a + residue_b + cin) % 3;

//obtain the residue of the sum
assign   residue_sum = {cout, sum} % 3;
endmodule
```

Figure 12.59   Dataflow module for modulo-3 residue checking.

```
//test bench for modulo-3 residue checking
module residue_chkg_tb;

reg [7:0] a, b;
reg cin;
wire [7:0] sum;
wire cout;

wire [1:0] residue_a, residue_b, residue_a_b, residue_sum;

//display variables
initial
$monitor ("a=%b, b=%b, cin=%b, sum=%b, cout=%b, residue_a=%b,
          residue_b=%b, residue_a_b=%b, residue_sum=%b",
          a, b, cin, sum, cout, residue_a, residue_b,
          residue_a_b, residue_sum);

//continued on next page
```

Figure 12.60   Test bench for the modulo-3 residue checking module.

```
//apply input vectors
initial
begin
   #0     a = 8'b0111_1010;
          b = 8'b0011_0101;
          cin = 1'b0;

   #10    a = 8'b0111_1010;
          b = 8'b1001_0100;
          cin = 1'b1;

   #10    a = 8'b0010_1010;
          b = 8'b0101_0111;
          cin = 1'b1;

   #10    $stop;
end

residue_chkg inst1 (     //instantiate the module
   .a(a),
   .b(b),
   .cin(cin),
   .sum(sum),
   .cout(cout),
   .residue_a(residue_a),
   .residue_b(residue_b),
   .residue_a_b(residue_a_b),
   .residue_sum(residue_sum)
   );
endmodule
```

Figure 12.59   (Continued)

```
a=01111010, b=00110101, cin=0,
sum=10101111, cout=0, residue_a=10, residue_b=10,
residue_a_b=01, residue_sum=01

a=01111010, b=10010100, cin=1,
sum=00001111, cout=1, residue_a=10, residue_b=01,
residue_a_b=01, residue_sum=01

a=00101010, b=01010111, cin=1,
sum=10000010, cout=0, residue_a=00, residue_b=00,
residue_a_b=01, residue_sum=01
```

Figure 12.61   Outputs for the modulo-3 residue checking module.

12.6 Parity Prediction

Parity prediction is an alternative way to check for errors in addition. It usually requires less hardware than the modulo-3 residue checking method. Parity prediction compares the predicted parity of the sum with the actual parity of the sum, as shown in Equation 12.21. The actual parity is denoted by P_{sum}. The predicted parity is specified by $P_a \oplus P_b \oplus P_{cy}$, where P_a is the parity of operand a (augend), P_b is the parity of operand b (addend), and P_{cy} is the parity of the carries. The block diagram for a parity-checked adder using parity prediction is shown in Figure 12.62.

$$P_{sum} = P_a \oplus P_b \oplus P_{cy} \qquad (12.21)$$

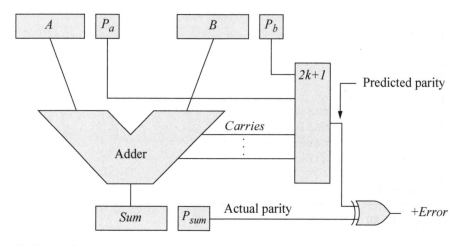

Figure 12.62   Block diagram for a parity-checked adder.

The block labeled $2k+1$ is an odd parity element, whose output is a logic 1 if the number of logic 1s on the inputs is odd. The output of the $2k+1$ predicted parity circuit is compared to the actual parity of the sum. An error will be undetected if there is an even number of errors in the operands. The carries are obtained from a carry-look-ahead technique using the *generate* and *propagate* functions. The truth table for stage $_i$ of a full adder is shown in Table 12.6, where the carry-in c_{i-1} to stage $_i$ is the carry-out from the previous lower-order stage $_{i-1}$. The Karnaugh map for the carry-out from stage $_i$ (c_i) is shown in Figure 12.63 and the equation for c_i is shown in Equation 12.22.

Table 12.6 Truth Table for a Full Adder

a_i	b_i	c_{i-1}	sum_i	c_i
0	0	0	0	0
0	0	1	1	0
0	1	0	1	0
0	1	1	0	1
1	0	0	1	0
1	0	1	0	1
1	1	0	0	1
1	1	1	1	1

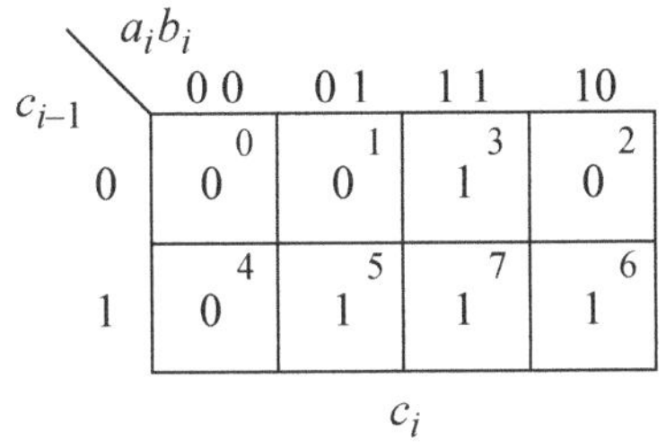

Figure 12.63   Karnaugh map for the carry-out of stage$_i$ of a full adder.

$$c_i = a_i b_i + a_i c_{i-1} + b_i c_{i-1}$$
$$= a_i b_i + (a_i + b_i) c_{i-1} \tag{12.22}$$

Two auxiliary functions can now be defined: *generate* and *propagate*, as shown below.

$$\text{Generate: } G_i = a_i b_i$$

$$\text{Propagate: } P_i = a_i + b_i$$

Equation 12.22 can now be rewritten in terms of the generate and propagate functions, as shown in Equation 12.23 for adder stage$_i$. Equation 12.23 can be used recursively to obtain the carries for parity prediction, as shown in Equation 12.24, where c_{-1} is the carry-in to the low-order stage of the adder.

$$c_i = G_i + P_i c_{i-1} \tag{12.23}$$

$$c_0 = G_0 + P_0 c_{-1}$$

$$c_1 = G_1 + P_1 c_0$$

$$= G_1 + P_1(G_0 + P_0 c_{-1})$$

$$= G_1 + P_1 G_0 + P_1 P_0 c_{-1}$$

$$c_2 = G_2 + P_2 G_1 + P_2 P_1 G_0 + P_2 P_1 P_0 c_{-1} \qquad (12.24)$$

. . .

Examples for parity prediction are shown below.

$a =$	0	1	1	1	$P_a =$	0	
+) $b =$	0	0	1	1	$P_b =$	1	$\oplus = 1$
Carries =	1	1	1	0 ← (c_{in})	$P_{cy} =$	0	
Sum =	1	0	1	0	$P_{sum} =$	1	

$a =$	0	1	0	1	$P_a =$	1	
+) $b =$	0	0	1	0	$P_b =$	0	$\oplus = 0$
Carries =	0	0	0	0 ← (c_{in})	$P_{cy} =$	1	
Sum =	0	1	1	1	$P_{sum} =$	0	

$a =$	0	1	1	0	$P_a =$	1	
+) $b =$	0	1	1	0	$P_b =$	1	$\oplus = 1$
Carries =	1	1	0	0 ← (c_{in})	$P_{cy} =$	1	
Sum =	1	1	0	0	$P_{sum} =$	1	

$$
\begin{array}{lcccccl}
a = & 0 & 1 & 1 & 0 & & P_a = & 1 \\
+)\quad b = & 0 & 1 & 1 & 0 & & P_b = & 1 \quad \oplus = \mathbf{0} \\
\text{Carries} = & 1 & 1 & 0 & 1 & \leftarrow (c_{in}) & P_{cy} = & 0 \\
\text{Sum} = & 1 & 1 & 0 & 1 & & P_{sum} = & \mathbf{0}
\end{array}
$$

$$
\begin{array}{lcccccl}
a = & 1 & 1 & 1 & 1 & & P_a = & 1 \\
+)\quad b = & 1 & 1 & 1 & 1 & & P_b = & 1 \quad \oplus = \mathbf{1} \\
\text{Carries} = & 1 & 1 & 1 & 1 & \leftarrow (c_{in}) & P_{cy} = & 1 \\
\text{Sum} = & 1 & 1 & 1 & 1 & & P_{sum} = & \mathbf{1}
\end{array}
$$

$$
\begin{array}{lcccccl}
a = & 1 & 1 & 1 & 1 & & P_a = & 1 \\
+)\quad b = & 1 & 1 & 1 & 1 & & P_b = & 1 \quad \oplus = \mathbf{0} \\
\text{Carries} = & 1 & 1 & 1 & 0 & \leftarrow (c_{in}) & P_{cy} = & 0 \\
\text{Sum} = & 1 & 1 & 1 & 0 & & P_{sum} = & \mathbf{0}
\end{array}
$$

As a final example, a byte adder will be examined together with the carries and the carry-in to the adder for the example shown below. Then the logic diagram will be designed to illustrate the operation of the parity prediction hardware. The logic diagram is shown in Figure 12.64. The logic blocks labeled $2k+1$ are the general symbol to generate appropriate outputs if there are an odd number of logic 1s on the input.

The inputs $c_{in}, c_0, c_1, c_2, c_3, c_4, c_5, c_6$ represent the carries. The outputs of the logic block labeled 1 represents the parity of the carries P_{cy}, which, in conjunction with the parity of operand $a(P_a)$ and the parity of operand $b(P_b)$, denote the predicted parity. The predicted parity is then compared with the actual parity of the sum to indicate whether an error occurred. Assume that an error occurred during addition and bits 3 2 1 0 = 0011, making the parity of the sum a logic 1. The inputs to the exclusive-OR circuit would then be 0 and 1, generating a logic 1 output indicating an error.

$$
\begin{array}{lccccccccl}
\text{Bit \#} & 7 & 6 & 5 & 4 & 3 & 2 & 1 & 0 & \\
a = & 0 & 1 & 1 & 1 & 0 & 0 & 0 & 0 & & P_a = & 0 \\
+)\quad b = & 0 & 0 & 1 & 0 & 0 & 0 & 1 & 0 & & P_b = & 1 \quad \oplus = \mathbf{0} \\
\text{Carries} = & 1 & 1 & 0 & 0 & 0 & 0 & 0 & 0 & \leftarrow (c_{in}) & P_{cy} = & 1 \\
\text{Sum} = & 1 & 0 & 0 & 1 & 0 & 0 & 1 & 0 & & P_{sum} = & \mathbf{0}
\end{array}
$$

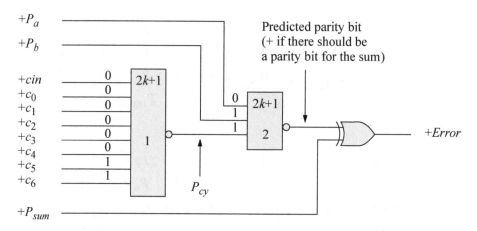

Figure 12.64   Logic diagram for parity prediction.

12.7 Condition Codes for Addition

For high-speed addition, it is desirable to have the following condition codes generated in parallel with the add operation: sum = 0; sum < 0; sum > 0; and overflow. This negates the necessity of waiting to obtain the condition codes until after the sum is available. The symbol for a full adder for stage i is shown in Figure 12.65 and the truth table for the sum s_i is shown in Table 12.7, indicating that s_i has a value of 0 in four rows. The Karnaugh map for the sum s_i is shown in Figure 12.66 and the equation for s_i is shown in Equation 12.25.

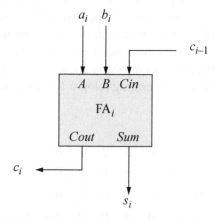

Figure 12.65   Symbol for a full adder.

Table 12.7 Truth Table for the Sum of a Full Adder

a_i	b_i	c_{i-1}	s_i
0	0	0	
0	1	0	1
1	0	0	1
1	1	0	
0	0	1	1
0	1	1	
1	0	1	
1	1	1	1

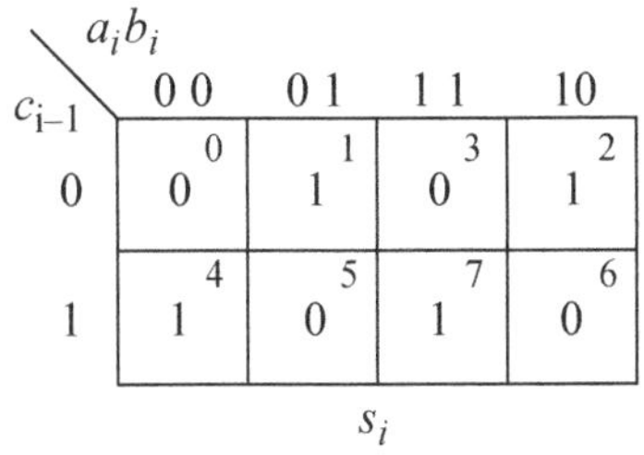

Figure 12.66   Karnaugh map for the sum of a full adder.

$$s_i = a_i'b_ic_{i-1}' + a_ib_i'c_{i-1}' + a_i'b_i'c_{i-1} + a_ib_ic_{i-1}$$
$$= c_{i-1}'(a_i \oplus b_i) + c_{i-1}(a_i \oplus b_i)'$$
$$= (a_i \oplus b_i) \oplus c_{i-1} \qquad (+ \text{ if } s_i \text{ is } 1; - \text{ if } s_i \text{ is } 0) \qquad (12.25)$$

Assume that the operands consist of 4-bytes (bits 0 through 31), using the format shown in Figure 12.67 and that the addition operation is an algebraic; that is, the operands are signed numbers in 2s complement representation. The operands consist of two parts: the numeric part (bits 0 through 30) and the sign part (bit 31). The logic diagram for the sum s_i together with the logic to indicate that a byte contains all zeroes is shown in Figure 12.68. Two exclusive-OR circuits are required for each bit and 16 exclusive-OR circuits for each byte.

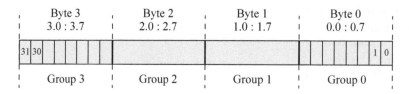

Figure 12.67   Format for the augend and addend.

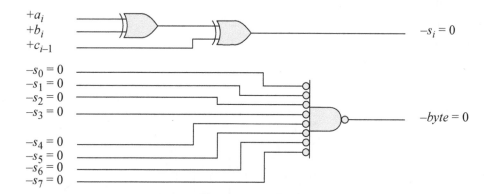

Figure 12.68   Logic diagram to indicate that the sum s_i is zero and that a byte is zero.

The logic to determine if the numeric part is zero or nonzero, the sum is zero or nonzero, and the sign is positive or negative is shown in Figure 12.69. The logic to determine the four condition codes is shown in Figure 12.70. Using this approach, the condition codes for the sum can be obtained independently and in parallel with the addition operation.

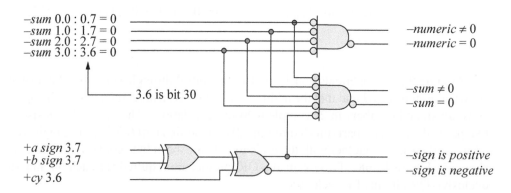

Figure 12.69   Logic diagram to determine whether the numeric part is zero or nonzero, the sum is zero or nonzero, and the sign is positive or negative.

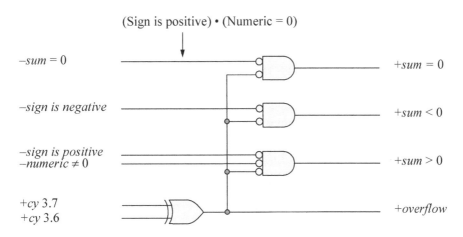

Figure 12.70 Logic diagram to generate the four condition codes: sum < 0; sum = 0; sum > 0; and overflow.

12.8 Arithmetic and Logic Unit

Arithmetic and logic units (ALUs) perform the arithmetic operations of addition, subtraction, multiplication, and division in fixed-point, decimal, and floating-point number representations. They also perform various logical operations such as AND, OR, exclusive-OR, exclusive-NOR, and complementation. An ALU is one of five main units of a computer: input, output, memory (storage), arithmetic and logic unit, and control (or sequencer), which is usually microprogrammed. The ALU is the core of the computer and, together with the control unit, form the processor.

There are two types of information in a computer: instructions and data. The instructions perform arithmetic and logic operations, transfer data, and execute programs from memory. Data are numbers or coded characters. There are four general categories of storage hierarchy in a computer: internal registers, cache memory, main memory, and peripheral memory. Internal registers are very high speed, but have very small capacity; cache memory is high speed — but slower than registers — and has a larger capacity; main memory is lower speed, but has a larger capacity; peripheral memory is the lowest speed, but has the largest capacity.

Most computer operations are executed by the ALU. For example, operands located in memory can be transferred to the ALU, where an operation is performed on the operands, then the result is stored in memory. The ALU performs calculations in different number representations; performs logical operations; performs comparisons, and performs shift operations such as shift right algebraic, shift right logical, shift left algebraic, and shift left logical. The algebraic shift operations refer to signed operands in 2s complement representation; the logical shift operations refer to unsigned operands. ALUs also perform operations on packed and unpacked decimal operands. This section will present two examples of designing arithmetic and logic units using Verilog HDL.

Example 12.17 A 4-function arithmetic and logic unit (ALU) will be designed in this example using the **case** construct. There are two 4-bit inputs: operands $A[3:0]$ and $B[3:0]$ and one 2-bit input $opcode[1:0]$. There is one 8-bit output $z[7:0]$ which contains the result of the operations and is declared to be eight bits to accommodate the $2n$-bit product. The **parameter** keyword will declare and assign values to the operation codes.

A block diagram is shown in Figure 12.71. The behavioral module is shown in Figure 12.72. The test bench is shown in Figure 12.73 in which input vectors are applied for each operation code. Notice that the results are correctly represented in 2s complement representation; thus, $4 - 8 = -4$ (11111100) and $6 - 15 = -9$ (11110111). The outputs and waveforms are shown in Figure 12.74 and Figure 12.75, respectively. The remainder of divide operations is truncated.

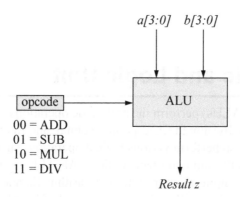

Figure 12.71   Block diagram for the 4-function ALU of Example 12.17.

```
//behavioral 4-bit ALU
module alu4 (a, b, opcode, z);

input [3:0] a, b;
input [1:0] opcode;
output [7:0] z;

wire [3:0] a, b;        //inputs are wire
wire [1:0] opcode;
reg [7:0] z;            //outputs are reg

//continued on next page
```

Figure 12.72   Behavioral module for the 4-function ALU of Example 12.17.

```
//define operation codes
parameter    addop = 2'b00,
             subop = 2'b01,
             mulop = 2'b10,
             divop = 2'b11;

//perform operations
always @(a or b or opcode)
begin
case (opcode)
      addop: z = a + b;
      subop: z = a - b;
      mulop: z = a * b;
      divop: z = a / b;
endcase
end

endmodule
```

Figure 12.72   (Continued)

```
//test bench for 4-bit ALU
module alu4_tb;

reg [3:0] a, b;
reg [1:0] opcode;
wire [7:0] z;

//display variables
initial
$monitor ("a=%b, b=%b, opcode=%b, result=%b",
            a, b, opcode, z);

//apply input vectors
initial
begin
//add operation
  #0   a=4'b0001; b=4'b0001; opcode=2'b00;   //sum = 2

  #10  a=4'b0010; b=4'b1101; opcode=2'b00;   //sum = 15(f)

  #10  a=4'b1111; b=4'b1111; opcode=2'b00;   //sum = 30(1e)

//continued on next page
```

Figure 12.73   Test bench for the 4-function ALU of Figure 12.72.

```
//subtract operation
  #10  a=4'b1000; b=4'b0100; opcode=2'b01;  //diff = 4

  #10  a=4'b1111; b=4'b0101; opcode=2'b01;  //diff = 10(a)

  #10  a=4'b1110; b=4'b0011; opcode=2'b01;  //diff = 11(b)

  #10  a=4'b0100; b=4'b1000; opcode=2'b01;  //diff = -4(fc)

  #10  a=4'b0110; b=4'b1111; opcode=2'b01;  //diff = -9(f7)


//multiply operation
  #10  a=4'b0100; b=4'b0111; opcode=2'b10;  //product = 28(1c)

  #10  a=4'b0101; b=4'b0011; opcode=2'b10;  //product = 15(f)

  #10  a=4'b1111; b=4'b1111; opcode= 'b10;  //product = 225(e1)


//divide operation
  #10  a=4'b1111; b=4'b0101; opcode=2'b11;  //quotient = 3

  #10  a=4'b1100; b=4'b0011; opcode=2'b11;  //quotient = 4

  #10  a=4'b1110; b=4'b0010; opcode=2'b11;  //quotient = 7

  #10  a=4'b0011; b=4'b1100; opcode=2'b11;  //quotient = 0

  #10  $stop;

end


//instantiate the module into the test bench
alu4 inst1 (
   .a(a),
   .b(b),
   .opcode(opcode),
   .z(z)
   );

endmodule
```

Figure 12.73   (Continued)

```
Add
a=0001, b=0001, opcode=00, result=00000010
a=0010, b=1101, opcode=00, result=00001111
a=1111, b=1111, opcode=00, result=00011110

Subtract
a=1000, b=0100, opcode=01, result=00000100
a=1111, b=0101, opcode=01, result=00001010
a=1110, b=0011, opcode=01, result=00001011
a=0100, b=1000, opcode=01, result=11111100
a=0110, b=1111, opcode=01, result=11110111

Multiply
a=0100, b=0111, opcode=10, result=00011100
a=0101, b=0011, opcode=10, result=00001111
a=1111, b=1111, opcode=10, result=11100001

Divide
a=1111, b=0101, opcode=11, result=00000011
a=1100, b=0011, opcode=11, result=00000100
a=1110, b=0010, opcode=11, result=00000111
a=0011, b=1100, opcode=11, result=00000000
```

Figure 12.74   Outputs for the 4-function ALU of Figure 12.72.

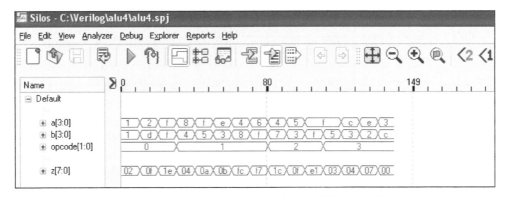

Figure 12.75   Waveforms for the 4-function ALU of Figure 12.72.

Example 12.18   This example presents the design of a 4-function ALU for two 4-bit operands $A[3:0]$ and $B[3:0]$. The operation codes are shown in Table 12.8, where c_1 and c_0 are two control inputs with c_0 being the low-order bit. The design will be a structural design instantiating the following modules: a 4-bit ripple adder, a 4:1 multiplexer, a 2-input AND gate, a 2-input OR gate, and an exclusive-OR circuit.

**Table 12.8 Operation Codes
for the 4-Function ALU**

c[1]	c[0]	Operation
0	0	Add
0	1	Subtract
1	0	And
1	1	Or

The logic diagram is shown in Figure 12.76 and represents only one of several methods of designing the ALU. This method illustrates the instantiation of several different modules into a structural top-level module. Figure 12.76 also shows the instantiation names and net names.

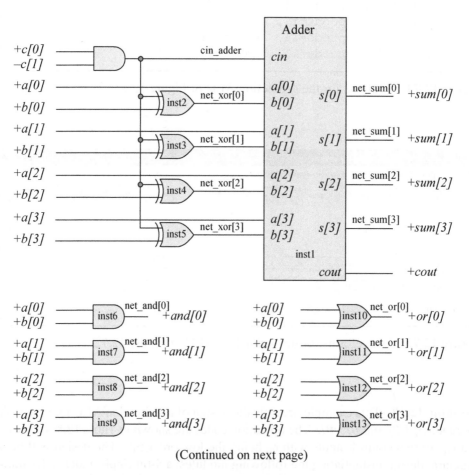

(Continued on next page)

Figure 12.76   Logic diagram for the 4-function ALU.

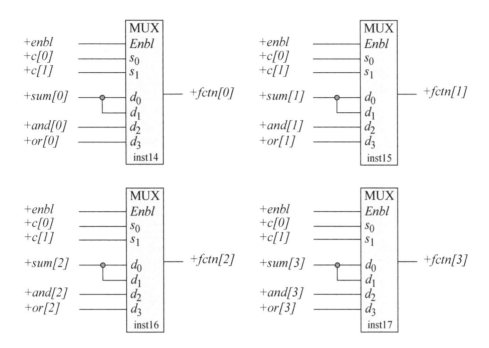

Figure 12.76   (Continued)

The dataflow modules for the 4-bit adder and the 4:1 multiplexer are shown Figure 12.77 and Figure 12.78, respectively. The structural module that instantiates the modules for the adder, exclusive-OR circuits, AND gates, OR gates, and multiplexers is shown in Figure 12.79. The test bench, outputs, and waveforms are shown in Figure 12.80, Figure 12.81, and Figure 12.82, respectively.

```verilog
//dataflow for a 4-bit adder
module adder4_df (a, b, cin, sum, cout);

//list inputs and outputs
input [3:0] a, b;
input cin;
output [3:0] sum;
output cout;

wire [3:0] a, b;   //define signals as wire for dataflow
wire cin, cout;    //(or default to wire)
wire [3:0] sum;

assign {cout, sum} = a + b + cin;
endmodule
```

Figure 12.77   Dataflow module for a 4-bit adder.

```
//dataflow 4:1 mux
module mux4_df (s, d, enbl, z1);


input [1:0] s;
input [3:0] d;
input enbl;
output z1;


wire [1:0] s;
wire [3:0] d;
wire enbl;
wire z1;


assign   z1 =   (~s[1] & ~s[0] & d[0] & enbl) |
                (~s[1] &  s[0] & d[1] & enbl) |
                ( s[1] & ~s[0] & d[2] & enbl) |
                ( s[1] &  s[0] & d[3] & enbl);


endmodule
```

Figure 12.78   Dataflow module for a 4:1 multiplexer.

```
//structural 4-function alu
module alu_4function (a, b, ctrl, enbl, cout, fctn);


input [3:0] a, b;
input [1:0] ctrl;
input enbl;
output [3:0] fctn;
output cout;


wire [3:0] a, b;
wire [1:0] ctrl;
wire enbl;
wire [3:0] fctn;
wire cout;


//continued on next page
```

Figure 12.79   Structural module for a 4-function ALU.

```verilog
//define internal nets
wire cin_adder;
wire [3:0] net_xor, net_sum, net_and, net_or;

assign cin_adder = (~ctrl [1] & ctrl [0]);

//instantiate the adder
adder4_df inst1 (
   .a(a),
   .b(net_xor),
   .cin(cin_adder),
   .sum(net_sum),
   .cout(cout)
   );

//instantiate the xor circuits
xor2_df inst2 (
   .x1(cin_adder),
   .x2(b[0]),
   .z1(net_xor[0])
   );

xor2_df inst3 (
   .x1(cin_adder),
   .x2(b[1]),
   .z1(net_xor[1])
   );

xor2_df inst4 (
   .x1(cin_adder),
   .x2(b[2]),
   .z1(net_xor[2])
   );

xor2_df inst5 (
   .x1(cin_adder),
   .x2(b[3]),
   .z1(net_xor[3])
   );

//instantiate the and gates
and2_df inst6 (
   .x1(a[0]),
   .x2(b[0]),
   .z1(net_and[0])
   );
//continued on next page
```

Figure 12.79   (Continued)

```
and2_df inst7 (
   .x1(a[1]),
   .x2(b[1]),
   .z1(net_and[1])
   );

and2_df inst8 (
   .x1(a[2]),
   .x2(b[2]),
   .z1(net_and[2])
   );

and2_df inst9 (
   .x1(a[3]),
   .x2(b[3]),
   .z1(net_and[3])
   );

//instantiate the or gates
or2_df inst10 (
   .x1(a[0]),
   .x2(b[0]),
   .z1(net_or[0])
   );

or2_df inst11 (
   .x1(a[1]),
   .x2(b[1]),
   .z1(net_or[1])
   );

or2_df inst12 (
   .x1(a[2]),
   .x2(b[2]),
   .z1(net_or[2])
   );

or2_df inst13 (
   .x1(a[3]),
   .x2(b[3]),
   .z1(net_or[3])
   );

//continued on next page
```

Figure 12.79   (Continued)

```
mux4_df inst14 (                //instantiate the multiplexers
   .s(ctrl),
   .d({net_or[0], net_and[0], net_sum[0], net_sum[0]}),
   .enbl(enbl),
   .z1(fctn[0])
   );

mux4_df inst15 (
   .s(ctrl),
   .d({net_or[1], net_and[1], net_sum[1], net_sum[1]}),
   .enbl(enbl),
   .z1(fctn[1])
   );

mux4_df inst16 (
   .s(ctrl),
   .d({net_or[2], net_and[2], net_sum[2], net_sum[2]}),
   .enbl(enbl),
   .z1(fctn[2])
   );

mux4_df inst17 (
   .s(ctrl),
   .d({net_or[3], net_and[3], net_sum[3], net_sum[3]}),
   .enbl(enbl),
   .z1(fctn[3])
   );
endmodule
```

Figure 12.79   (Continued)

```
//test bench for the 4-function alu
module alu_4function_tb;

reg [3:0] a, b;
reg [1:0] ctrl;
reg enbl;
wire [3:0] fctn;
wire cout;

initial            //display variables
$monitor ("ctrl = %b, a = %b, b = %b, cout=%b, fctn = %b",
        ctrl, a, b, cout, fctn);
//continued on next page
```

Figure 12.80   Test bench for the structural module for a 4-function ALU.

```verilog
//apply input vectors
initial
begin
//add operation
  #0  ctrl=2'b00;  a=4'b0000;  b=4'b0000;  enbl=1'b1;//sum=0000
  #10 ctrl=2'b00;  a=4'b0001;  b=4'b0011;  enbl=1'b1;//sum=0100
  #10 ctrl=2'b00;  a=4'b0111;  b=4'b0011;  enbl=1'b1;//sum=1010
  #10 ctrl=2'b00;  a=4'b1101;  b=4'b0110;  enbl=1'b1;//sum=10011
  #10 ctrl=2'b00;  a=4'b0011;  b=4'b1111;  enbl=1'b1;//sum=10010


//subtract operation
  #10 ctrl=2'b01;  a=4'b0111;  b=4'b0011;  enbl=1'b1;//diff=0100
  #10 ctrl=2'b01;  a=4'b1101;  b=4'b0011;  enbl=1'b1;//diff=1010
  #10 ctrl=2'b01;  a=4'b1111;  b=4'b0011;  enbl=1'b1;//diff=1100
  #10 ctrl=2'b01;  a=4'b1111;  b=4'b0001;  enbl=1'b1;//diff=1110
  #10 ctrl=2'b01;  a=4'b1100;  b=4'b0111;  enbl=1'b1;//diff=0101


//and operation
  #10 ctrl=2'b10;  a=4'b1100;  b=4'b0111;  enbl=1'b1;//and=0100
  #10 ctrl=2'b10;  a=4'b0101;  b=4'b1010;  enbl=1'b1;//and=0000
  #10 ctrl=2'b10;  a=4'b1110;  b=4'b0111;  enbl=1'b1;//and=0110
  #10 ctrl=2'b10;  a=4'b1110;  b=4'b1111;  enbl=1'b1;//and=1110
  #10 ctrl=2'b10;  a=4'b1111;  b=4'b0111;  enbl=1'b1;//and=0111


//or operation
  #10 ctrl=2'b11;  a=4'b1100;  b=4'b0111;  enbl=1'b1;//or=1111
  #10 ctrl=2'b11;  a=4'b1100;  b=4'b0100;  enbl=1'b1;//or=1100
  #10 ctrl=2'b11;  a=4'b1000;  b=4'b0001;  enbl=1'b1;//or=1001

  #10 $stop;
end


//instantiate the module into the test bench
alu_4function inst1 (
  .a(a),
  .b(b),
  .ctrl(ctrl),
  .enbl(enbl),
  .fctn(fctn),
  .cout(cout)
  );

endmodule
```

Figure 12.80   (Continued)

```
Add
ctrl = 00, a = 0000, b = 0000, cout=0, fctn = 0000
ctrl = 00, a = 0001, b = 0011, cout=0, fctn = 0100
ctrl = 00, a = 0111, b = 0011, cout=0, fctn = 1010
ctrl = 00, a = 1101, b = 0110, cout=1, fctn = 0011
ctrl = 00, a = 0011, b = 1111, cout=1, fctn = 0010

Subtract
ctrl = 01, a = 0111, b = 0011, cout=1, fctn = 0100
ctrl = 01, a = 1101, b = 0011, cout=1, fctn = 1010
ctrl = 01, a = 1111, b = 0011, cout=1, fctn = 1100
ctrl = 01, a = 1111, b = 0001, cout=1, fctn = 1110
ctrl = 01, a = 1100, b = 0111, cout=1, fctn = 0101

AND
ctrl = 10, a = 1100, b = 0111, cout=1, fctn = 0100
ctrl = 10, a = 0101, b = 1010, cout=0, fctn = 0000
ctrl = 10, a = 1110, b = 0111, cout=1, fctn = 0110
ctrl = 10, a = 1110, b = 1111, cout=1, fctn = 1110
ctrl = 10, a = 1111, b = 0111, cout=1, fctn = 0111

OR
ctrl = 11, a = 1100, b = 0111, cout=1, fctn = 1111
ctrl = 11, a = 1100, b = 0100, cout=1, fctn = 1100
ctrl = 11, a = 1000, b = 0001, cout=0, fctn = 1001
```

Figure 12.81   Outputs for the structural module for a 4-function ALU.

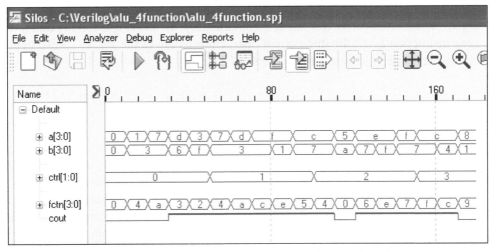

Figure 12.82   Waveforms for the structural module for a 4-function ALU.

12.9 Memory

Memory (or storage) is one of the main units of a computer from which instructions and data can be loaded directly into processor registers for subsequent execution or processing. There are two types of storage arrays that are internal in a computer: cache memory and main memory. As mentioned in a previous section, cache is smaller than main memory, but much faster; main memory is larger than cache, but much slower. The speed with which memory can be accessed is a major contributing factor in the execution speed of programs. It is for this reason that cache memories are a predominant unit in a memory system. Cache memory is an intermediate device that is logically situated between the processor and main memory, as shown in Figure 12.83.

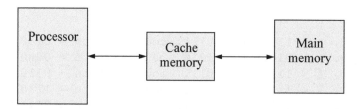

Figure 12.83   Cache memory and main memory relative to the processor.

Figure 12.84 lists the internal processor registers that are used to interface with memory. The memory address register (MAR) contains the memory address from where data is to be retrieved during a read operation or where data is to be stored during a write operation. The memory data register (MDR) contains the data for the read or write operation. The instruction register (IR) contains the current instruction, including the operation code, source address, and destination address.

The program counter (PC) contains the address of the next instruction to be executed. When an instruction is fetched from memory, the PC is updated to point to the next instruction by adding the current instruction length to the PC, as shown below, where the brackets indicate "the contents of."

$$PC \longleftarrow [PC] + \text{instruction length}$$

The general-purpose registers (GPRs) have multiple uses. They are used for temporary storage of data and for working storage. Another use is as an index register in which a constant is stored to generate the effective address of an operand. A GPR can also be used as a base register that is added to the index register to provide additional flexibility in generating an effective address — the contents of both registers can be changed by programming.

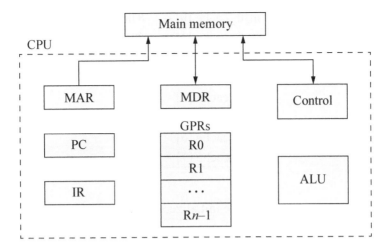

Figure 12.84   Internal registers for a central processing unit.

Memories can be represented in Verilog by an array of registers and are declared using a **reg** data type as follows:

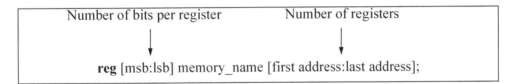

A 32-word register with one byte per word would be declared as follows:

reg [7:0] memory_name [0:31];

An array can have only two dimensions. Memories must be declared as **reg** data types, not as **wire** data types. A register can be assigned a value using one statement, as shown below.

reg [15:0] buff_reg;
buff_reg = 16'h7ab5;

Register *buff_reg* is assigned the value $0111\ 1010\ 1011\ 0101_2$. Values can be stored in memories by assigning a value to each word individually, as shown below for an instruction cache of eight registers with 8 bits per register.

reg [7:0] instr_cache [0:7];

instr_cache [0] =	8'h08;
instr_cache [1] =	8'h09;
instr_cache [2] =	8'h0a;
instr_cache [3] =	8'h0b;
instr_cache [4] =	8'h0c;
instr_cache [5] =	8'h0d;
instr_cache [6] =	8'h0e;
instr_cache [7] =	8'h0f;

Alternatively, memories can be initialized by means of one of the following system tasks:

$readmemb   for binary data
$readmemh   for hexadecimal data

A text file is prepared for the specified memory in either binary or hexadecimal format. The system task reads the file and loads the contents into memory. An example of loading the *instr_cache* memory described above with binary data is shown in Example 12.19.

Example 12.19 A block diagram of the instruction cache is shown in Figure 12.85 with a 3-bit program counter *pc [2:0]* as an input vector to address the instruction cache and an 8-bit *ic_data_out [7:0]* bus as an output vector. Each address in the cache is considered to be a cache line containing a block of data, in this case 1 byte. The contents of the cache represent the instructions to execute a computer program. The instructions are decoded to generate the operation code, source operand address, and destination operand address.

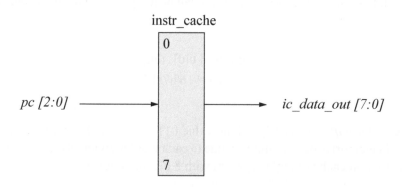

Figure 12.85 Block diagram of the instruction cache for Example 12.19.

The file shown below is created and saved as *icache.instr* (the addresses are shown for reference). The file is saved as a separate file in the project folder without the **.v** extension.

```
Address 0: 0000_1000
Address 1: 0000_1001
Address 2: 0000_1010
Address 3: 0000_1011
Address 4: 0000_1100
Address 5: 0000_1101
Address 6: 0000_1110
Address 7: 0000_1111
```

The contents of *icache.instr* are loaded into the memory *instr_cache* beginning at location 0, by the following two statements:

reg [7:0] instr_cache [0:7];

$readmemb ("icache.instr", instr_cache);

The Verilog code for the above procedure is shown in Figure 12.86 using binary data for the file called *icache.instr*. The test bench, the outputs, and the waveforms are shown in Figure 12.87, Figure 12.88, and Figure 12.89, respectively.

There are two ports in Figure 12.86: input port *pc* and output port *ic_data_out*, which is declared as **reg** because it operates as a storage element. The instruction cache is defined as an array of eight 8-bit registers by the following statement:

reg [7:0] instr_cache [0:7];

An **initial** procedural construct is used to load data from the *icache.instr* file into the instruction cache *instr_cache* by means of the system task **$readmemb**. The **initial** statement executes only once to initialize the instruction cache. An **always** procedural construct is then used to read the contents of the instruction cache based on the value of the program counter; that is, *ic_data_out* receives the contents of the instruction cache at the address specified by the program counter. The program counter is an event control used in the **always** statement — when the program counter changes, the statement in the **begin** . . . **end** block is executed.

```
//behavioral to load memory with
//binary data from file icache.instr
module mem_load (pc, ic_data_out);

//continued on next page
```

Figure 12.86   Module using **$readmemb** to load an instruction cache.

```
//list inputs and outputs
input pc;
output ic_data_out;

//list wire and reg
wire [2:0] pc;              //a program counter to address 8 words
reg [7:0] ic_data_out;

//define memory size
//instr_cache is an array of eight 8-bit regs
reg [7:0] instr_cache [0:7];

//define memory contents
//load instr_cache from file icache.instr
initial
begin
   $readmemb ("icache.instr", instr_cache);
end

//use a program counter to access the instr_cache
always @ (pc)
begin
   ic_data_out = instr_cache [pc];
end

endmodule
```

Figure 12.86   (Continued)

The test bench of Figure 12.87 has a 3-bit input vector *pc* declared as **reg** because it retains its value and an 8-bit output vector *ic_data_out* declared as **wire**. All eight combinations of the program counter are listed in 10-time-unit increments. Then the contents of the instruction cache are displayed by means of a **for** loop using **integer** *i* as a control variable.

```
//mem_load test bench
module mem_load_tb;

integer i;      //used to display contents

reg [2:0] pc;
wire [7:0] ic_data_out;
//continued on next page
```

Figure 12.87   Test bench for the *mem_load* behavioral module of Figure 12.86.

```
//assign values to the program counter
initial
begin
   #0    pc = 3'b000;
   #10   pc = 3'b001;
   #10   pc = 3'b010;
   #10   pc = 3'b011;
   #10   pc = 3'b100;
   #10   pc = 3'b101;
   #10   pc = 3'b110;
   #10   pc = 3'b111;

   #15   $stop;
end


//display the contents of the instruction cache
initial
begin
   for (i=0; i<8; i=i+1)
   begin
      #10 $display ("address %h = %b", i, ic_data_out);
   end
   #150 $stop;
end


//instantiate the module into the test bench
mem_load inst1 (
   .pc(pc),
   .ic_data_out(ic_data_out)
   );

endmodule
```

Figure 12.87 (Continued)

```
address 00000000 = 00001000
address 00000001 = 00001001
address 00000002 = 00001010
address 00000003 = 00001011
address 00000004 = 00001100
address 00000005 = 00001101
address 00000006 = 00001110
address 00000007 = 00001111
```

Figure 12.88 Outputs for the *mem_load* behavioral module of Figure 12.86.

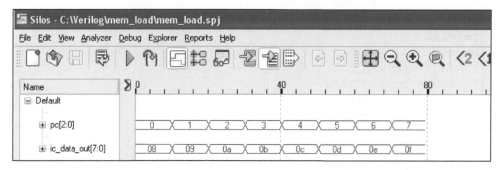

Figure 12.89   Waveforms for the *mem_load* behavioral module of Figure 12.86.

12.10 Problems

12.1   Given the Karnaugh map shown below, implement the logic using a 4:1 multiplexer with the select inputs assigned as follows: $s_1 s_0 = x_1 x_4$.

$x_3 x_4$

$x_1 x_2$	0 0	0 1	1 1	1 0
0 0	1 [0]	0 [1]	0 [3]	0 [2]
0 1	1 [4]	0 [5]	0 [7]	0 [6]
1 1	0 [12]	1 [13]	0 [15]	1 [14]
1 0	0 [8]	1 [9]	0 [11]	1 [10]

z_1

12.2   Implement the function shown below using one 4:1 multiplexer.

$$z_1(x_1, x_2, x_3) = x_1 x_2' x_3' + x_1' x_2 x_3' + x_1' x_2' x_3 + x_1 x_2 x_3$$

12.3   Implement the function shown below using a 4:1 multiplexer and additional logic gates, if necessary.

$$z_1(x_1, x_2, x_3, x_4) = \Sigma_m(4, 5, 7, 9, 13, 15)$$

12.4   Given the following function, use functional decomposition to implement the function using a 4:1 multiplexer:

$$z_1(x_1, x_2, x_3, x_4) = x_1' x_4' + x_1 x_3$$

12.5   Given the following function, use functional decomposition to implement the function using a 4:1 multiplexer:

$$z_1(x_1, x_2, x_3, x_4) = \Sigma_m(0, 3, 4, 5, 9, 10, 12, 13)$$

12.6   Given the following function, use functional decomposition to implement the function using 4:1 multiplexers, then implement the design using dataflow modeling for the sum-of-products expression and structural modeling for the multiplexer design:

$$z_1(x_1, x_2, x_3, x_4, x_5) = x_1' x_3' x_4' + x_1' x_2 + x_1' x_4' x_5' + x_1' x_3 x_4 x_5 +$$
$$x_1 x_2' x_3' x_4 + x_1 x_2' x_4 x_5' + x_1 x_2' x_3 x_4' x_5$$

12.7   Design a parity checker using an iterative network to determine the parity of a 4-bit word x_1, x_2, x_3, x_4. If the parity of the word is odd, then the output z_1 will be asserted high; otherwise, $z_1 = 0$. Design the network using dataflow modeling. Obtain the design module, the test bench module, the outputs, and the waveforms.

12.8   Design a 4-bit comparator as an iterative network using Verilog HDL. The inputs are $A[1:4]$ and $B[1:4]$. There are three scalar outputs that define the relationship between the two vectors: $z_1 = (A < B)$, $z_2 = (A = B)$, and $z_3 = (A > B)$. Design a typical cell using dataflow modeling, then instantiate the typical cell four times into a structural module for the 4-bit comparator. For the typical cell, obtain the dataflow module, the test bench module for several input vectors, and the outputs. For the comparator design, obtain the structural module, the test bench module for several input vectors, and the outputs.

12.9   Design a 3-bit Gray code counter as an iterative network. A gated D flip-flop will first be designed as a typical cell that has four data inputs x_1, x_2, x_3, and x_4. These inputs will be used to generate a sum-of-products expression that connects to the D input of the flip-flop. The typical cell will then be instantiated three times into a structural module to implement the Gray code counter. The sum-of-products expressions that will be used for the typical cells are obtained from the Karnaugh maps that represent the counting sequence of the Gray code.

    The typical cell is to be implemented with a D flip-flop that was designed using behavioral modeling; the logic primitive gates for the sum-of-products expression are to be designed using dataflow modeling. Obtain the structural module for the typical cell by instantiating the behavioral D flip-flop and the logic primitives. Obtain the test bench module, the outputs, and the waveforms for the typical cell. Then obtain the structural module for the Gray code counter by instantiating the typical cell three times, and obtain the test bench module, the outputs, and the waveforms.

12.10   Design a sequence detector that is implemented as a combinational iterative network that will generate an output z_1 if the pattern 0110 is detected anywhere in an 8-bit vector $x[1:8]$. Overlapping sequences are valid.

Obtain the next-state table, the Karnaugh maps, and the equations for a typical cell. Then design the typical cell using built-in primitives, obtain the test bench for several combinations of the inputs, and the outputs.

Then design the 8-bit sequence detector using structural modeling by instantiating the typical cell eight times into the structural module. Obtain the test bench for the sequence detector for several different input vectors, including overlapping sequences. Obtain the outputs and the waveforms.

12.11   Three code words, each containing a message of four bits which are encoded using the Hamming code, are received as shown below. Determine the correct 4-bit messages that were transmitted using odd parity.

	Received Code Words						
Bit Position	1	2	3	4	5	6	7
(a)	0	1	0	1	0	1	0
(b)	1	1	0	0	1	1	0
(c)	0	0	1	0	1	1	1

12.12   An 11-bit message is to be encoded using the Hamming code with odd parity. Write the equations for all of the groups that are required for the encoding process.

12.13   The 7-bit code words shown below are received using Hamming code with odd parity. Determine the syndrome word for each received code word.

(a)

Bit Position =	1	2	3	4	5	6	7
Received Code Word =	1	1	1	1	1	1	1

(b)

Bit Position =	1	2	3	4	5	6	7
Received Code Word =	0	0	0	0	0	0	0

12.14 A code word containing one 8-bit message, which is encoded using the Hamming code with odd parity, is received as shown below. Determine the 8-bit message that was transmitted.

Bit Position	1	2	3	4	5	6	7	8	9	10	11	12
Received Code Word	1	1	0	0	1	0	1	1	0	0	0	1

12.15 A code word containing one 8-bit message, which is encoded using the Hamming code with odd parity, is received as shown below. Determine the 8-bit message that was transmitted.

Bit Position	1	2	3	4	5	6	7	8	9	10	11	12
Received Code Word	0	1	0	1	1	0	0	1	0	0	1	1

12.16 A code word containing one 12-bit message, which is encoded using the Hamming code with odd parity, is received as shown below. Determine the syndrome word.

1	2	3	4	5	6	7	8	9	10	11	12	13	14	15	16	17
1	0	1	1	1	1	1	0	0	0	1	0	1	1	1	0	1

12.17 A code word containing one 12-bit message, which is encoded using the Hamming code with odd parity, is received as shown below. Determine the 12-bit message that was transmitted.

1	2	3	4	5	6	7	8	9	10	11	12	13	14	15	16	17
1	0	1	1	1	1	0	0	0	0	0	0	1	1	1	0	1

12.18 Given the 8-bit message 0 1 0 1 1 0 1 1, construct a code word to be transmitted that corrects single errors and detects double errors.

12.19 Obtain the code word using the Hamming code with odd parity for the following message word: 1 1 0 1 0 1 0 1 1 1 1.

12.20 Determine the number of parity bits that are required to encode messages of 16 and 48 bits using the Hamming code with odd parity.

12.21 Perform residue checking on the following BCD numbers: 1000 0111 0101 and 0110 0100 0010.

12.22   Perform residue checking on the following BCD numbers: 1001 1001 1001 and 1001 1001 1001.

12.23   Perform parity prediction on the following operands with a carry-in = 1: 0111 and 0111.

12.24   Perform parity prediction on the following operands with a carry-in = 0: 1010 and 1010.

12.25   Using mixed-design modeling, design a 4-function arithmetic and logic unit to execute the following four operations: add, multiply, AND, and OR. The operands are 8 bits defined as $A[7:0]$ and $B[7:0]$. The result of all operations will be 16 bits $rslt[15:0]$. The state of the following flags are to be determined:

The parity flag $pf = 1$ if the result has an even number of 1s; otherwise, $pf = 0$.

The sign flag sf represents the state of the leftmost result bit.

The zero flag $zf = 1$ if the result of an operation is zero; otherwise, $zf = 0$.

Obtain the design module, the test bench module using several values for each of the four operations, the outputs, and the waveforms.

12.26   Design an 8-function ALU for two 4-bit operands $A[3:0]$ and $B[3:0]$. The operation codes are shown below. The design will be a behavioral design using the **parameter** keyword and the **case** statement. Obtain the behavioral module, the test bench module for several different operands, the outputs, and the waveforms.

Operation	Operation Code
Add	000
Subtract	001
Multiply	010
Divide	011
AND	100
OR	101
Exclusive-OR	110
Exclusive-NOR	111

12.27   Design a 6-function ALU using behavioral modeling with the **parameter** keyword and the **case** statement to execute the 8-bit shift and rotate operations shown below. Obtain the design module, the test bench module for several different variables, the outputs, and the waveforms.

SRA	Shift right arithmetic (For signed operands)	The sign bit [7] is propagated right 1 bit position. All other bits shift right one bit position.
SRL	Shift right logical (For unsigned operands)	All bits shift right 1 bit position. A zero is shifted into the vacated bit [7] position.
SLA	Shift left arithmetic (For signed operands)	The sign bit [7] does not shift. One bit is shifted out of bit position [6]. All other bits shift left 1 bit position. A zero is shifted into bit [0].
SLL	Shift left logical (For unsigned operands)	All bits shift left 1 bit position. A zero is shifted into the vacated bit [0] position.
ROR	Rotate right	Bit [0] rotates to the bit [7] position. All other bits shift right 1 bit position.
ROL	Rotate left	Bit [7] rotates to the bit [0] position. All other bits shift left 1 bit position.

12.28   Using behavioral modeling, design a memory unit that consists of sixteen 8-bit words. Store the bytes in memory by assigning a value to each address location individually. Beginning with address 0, store the following hexadecimal bytes: 00H, 01H, 02H, 03H, 14H, 15H, 16H, 17H, 28H, 29H, 2AH, 2BH, 3CH, 3DH, 3EH, 3FH. Obtain the design module, the test bench module, the outputs, and the waveforms.

12.29   Using behavioral modeling, design a memory unit that consists of sixteen 8-bit words. Store the bytes in memory by creating a file that defines the data and save the file as a separate file in the project folder without the **.v** extension. Use the system task **$readmemb** to load the data into memory. Beginning with address 0, store the following hexadecimal bytes: 00H, 01H, 02H, 03H, 14H, 15H, 16H, 17H, 28H, 29H, 2AH, 2BH, 3CH, 3DH, 3EH, 3FH. Obtain the design module and the test bench module. Show the file to be loaded by the system task **$readmemb**. Obtain the outputs and the waveforms.

Appendix A

Event Queue

Event management in Verilog hardware description language (HDL) is controlled by an event queue. Verilog modules generate events in the test bench, which provide stimulus to the module under test. These events can then produce new events by the modules under test. Since the Verilog HDL Language Reference Manual (LRM) does not specify a method of handling events, the simulator must provide a way to arrange and schedule these events in order to accurately model delays and obtain the correct order of execution. The manner of implementing the event queue is vendor-dependent.

Time in the event queue advances when every event that is scheduled in that time step is executed. Simulation is finished when all event queues are empty. An event at time t may schedule another event at time t or at time $t + n$.

A.1 Event Handling for Dataflow Assignments

Dataflow constructs consist of continuous assignments using the **assign** statement. The assignment occurs whenever simulation causes a change to the right-hand side expression. Unlike procedural assignments, continuous assignments are order independent — they can be placed anywhere in the module.

Consider the logic diagram shown in Figure A.1 which is represented by the two dataflow modules of Figure A.2 and Figure A.3. The test bench for both modules is shown in Figure A.4. The only difference between the two dataflow modules is the reversal of the two **assign** statements. The order in which the two statements execute is not defined by the Verilog HDL LRM; therefore, the order of execution is indeterminate.

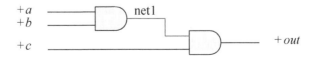

Figure A.1   Logic diagram to demonstrate event handling.

```
module dataflow (a, b, c, out);          module dataflow (a, b, c, out)

input a, b, c;                           input a, b, c;
output out;                              output out;

wire a, b, c;                            wire a, b, c;
wire out;                                wire out;

//define internal net                    //define internal net
wire net1;                               wire net1;

assign net1 = a & b;                     assign out = net1 & c;
assign out = net1 & c;                   assign net1 = a & b;

endmodule                                endmodule
```

Figure A.2   Dataflow module 1.   **Figure A.3**   Dataflow module 2.

```
module dataflow_tb;              end
                                //instantiate the module
reg test_a, test_b, test_c;     dataflow inst1
wire test_out;                      .a(test_a),
                                    .b(test_b),
initial                             .c(test_c),
begin                               .out(test_out)
   test_a = 1'b1;                   );
   test_b = 1'b0;
   test_c = 1'b0;              endmodule

   #10    test_b = 1'b1;
          test_c = 1'b1;
   #10    $stop;
```

Figure A.4   Test bench for Figure A.2 and Figure A.3.

Assume that the simulator executes the assignment order shown in Figure A.2 first. When the simulator reaches time unit #10 in the test bench, it will evaluate the right-hand side of *test_b = 1'b1;* and place its value in the event queue for an immediate scheduled assignment. Since this is a blocking statement, the next statement will not execute until the assignment has been made. Figure A.5 represents the event queue after the evaluation. The input signal *b* will assume the value of *test_b* through instantiation.

Event queue					
Scheduled event 5	Scheduled event 4	Scheduled event 3	Scheduled event 2	Scheduled event 1	Time unit
				test_b ← 1'b1 b ← 1'b1	t = #10
◄──────────────────────────────────				Order of execution	

Figure A.5   Event queue after execution of *test_b = 1'b1;*.

After the assignment has been made, the simulator will execute the *test_c = 1'b1;* statement by evaluating the right-hand side, and then placing its value in the event queue for immediate assignment. The new event queue is shown in Figure A.6. The entry that is not shaded represents an executed assignment.

Event queue					
Scheduled event 5	Scheduled event 4	Scheduled event 3	Scheduled event 2	Scheduled event 1	Time unit
			test_c ← 1'b1 c ← 1'b1	test_b ← 1'b1 b ← 1'b1	t = #10
◄──────────────────────────────────				Order of execution	

Figure A.6   Event queue after execution of *test_c = 1'b1;*.

When the two assignments have been made, time unit #10 will have ended in the test bench, which is the top-level module in the hierarchy. The simulator will then enter the instantiated dataflow module during this same time unit and determine that events have occurred on input signals *b* and *c* and execute the two continuous assignments. At this point, inputs *a*, *b*, and *c* will be at a logic 1 level. However, *net1* will still contain a logic 0 level as a result of the first three assignments that executed at time #0 in the test bench. Thus, the statement *assign out = net1 & c;* will evaluate to a logic 0, which will be placed in the event queue and immediately assigned to *out*, as shown in Figure A.7.

Event queue					
Scheduled event 5	Scheduled event 4	Scheduled event 3	Scheduled event 2	Scheduled event 1	Time unit
		out ← 1'b0 test_out ← 1'b0	test_c ← 1'b1 c ← 1'b1	test_b ← 1'b1 b ← 1'b1	$t = \#10$
◄───────────────────────────────────────				Order of execution	

Figure A.7   Event queue after execution of *assign out = net1 & c;*.

The simulator will then execute the *assign net1 = a & b;* statement in which the right-hand side evaluates to a logic 1 level. This will be placed on the queue and immediately assigned to *net1* as shown in Figure A.8.

Event queue					
Scheduled event 5	Scheduled event 4	Scheduled event 3	Scheduled event 2	Scheduled event 1	Time unit
	net1 ← 1'b1	out ← 1'b0 test_out ← 1'b0	test_c ← 1'b1 c ← 1'b1	test_b ← 1'b1 b ← 1'b1	$t = \#10$
◄───────────────────────────────────────				Order of execution	

Figure A.8   Event queue after execution of *assign net1 = a & b;*.

When the assignment has been made to *net1*, the simulator will recognize this as an event on *net1*, which will cause all statements that use *net1* to be reevaluated. The only statement to be reevaluated is *assign out = net1 & c;*. Since both *net1* and *c* equal a logic 1 level, the right-hand side will evaluate to a logic 1, resulting in the event queue shown in Figure A.9.

Event queue					
Scheduled event 5	Scheduled event 4	Scheduled event 3	Scheduled event 2	Scheduled event 1	Time unit
out ← 1'b1	net1 ← 1'b1	out ← 1'b0	test_c ← 1'b1	test_b ← 1'b1	$t = \#10$
test_out ← 1'b1		test_out ← 1'b0	c ← 1'b1	b ← 1'b1	
←				Order of execution	

Figure A.9   Event queue after execution of *assign out = net1 & c;*.

The test bench signal *test_out* must now be updated because it is connected to *out* through instantiation. Because the signal *out* is not associated with any other statements within the module, the output from the module will now reflect the correct output. Since all statements within the dataflow module have been processed, the simulator will exit the module and return to the test bench. All events have now been processed; therefore, time unit #10 is complete and the simulator will advance the simulation time.

Since the order of executing the **assign** statements is irrelevant, processing of the dataflow events will now begin with the *assign net1 = a & b;* statement to show that the result is the same. The event queue is shown in Figure A.10.

Event queue					
Scheduled event 5	Scheduled event 4	Scheduled event 3	Scheduled event 2	Scheduled event 1	Time unit
		net1 ← 1'b1	test_c ← 1'b1	test_b ← 1'b1	$t = \#10$
			c ← 1'b1	b ← 1'b1	
←				Order of execution	

Figure A.10   Event queue beginning with the statement *assign net1 = a & b;*.

Once the assignment to *net1* has been made, the simulator recognizes this as a new event on *net1*. The existing event on input *c* requires the evaluation of statement *assign out = net1 & c;*. The right-hand side of the statement will evaluate to a logic 1, and will be placed on the event queue for immediate assignment, as shown in Figure A.11.

Event queue					
Sched- uled event 5	Scheduled event 4	Scheduled event 3	Scheduled event 2	Scheduled event 1	Time unit
	out ← 1'b1 test_out ← 1'b1	net1 ← 1'b1	test_c ← 1'b1 c ← 1'b1	test_b ← 1'b1 b ← 1'b1	t = #10
				Order of execution	

Figure A.11   Event queue after execution of *assign out = net1 & c;*.

A.2 Event Handling for Blocking Assignments

The blocking assignment operator is the equal (=) symbol. A blocking assignment evaluates the right-hand side arguments and completes the assignment to the left-hand side before executing the next statement; that is, the assignment *blocks* other assignments until the current assignment has been executed.

Example A.1   Consider the code segment shown in Figure A.12 using blocking assignments in conjunction with the event queue of Figure A.13. There are no interstatement delays and no intrastatement delays associated with this code segment. In the first blocking assignment, the right-hand side is evaluated and the assignment is scheduled in the event queue. Program flow is blocked until the assignment is executed. This is true for all blocking assignment statements in this code segment. The assignments all occur in the same simulation time step *t*.

```
always @ (x2 or x3 or x5 or x7)
begin
   x1 = x2 | x3;
   x4 = x5;
   x6 = x7;
end
```

Figure A.12   Code segment with blocking assignments.

Event queue					
Scheduled event 5	Scheduled event 4	Scheduled event 3	Scheduled event 2	Scheduled event 1	Time unit
		$x6 \leftarrow x7\ (t)$	$x4 \leftarrow x5\ (t)$	$x1 \leftarrow x2 \mid x3\ (t)$	t
←				Order of execution	

Figure A.13   Event queue for Figure A.12.

Example A.2   The code segment shown in Figure A.14 contains an interstatement delay. Both the evaluation and the assignment are delayed by two time units. When the delay has taken place, the right-hand side is evaluated and the assignment is scheduled in the event queue as shown in Figure A.15. The program flow is blocked until the assignment is executed.

```
always @ (x2)
begin
   #2 x1 = x2;
end
```

Figure A.14   Blocking statement with interstatement delay.

Event queue					
Scheduled event 5	Scheduled event 4	Scheduled event 3	Scheduled event 2	Scheduled event 1	Time unit
					t
				$x1 \leftarrow x2\ (t+2)$	$t+2$
←				Order of execution	

Figure A.15   Event queue for Figure A.14.

Example A.3   The code segment of Figure A.16 shows three statements with interstatement delays of $t+2$ time units. The first statement does not execute until simulation time $t+2$ as shown in Figure A.17. The right-hand side $(x_2 \mid x_3)$ is evaluated at the current simulation time which is $t+2$ time units, and then assigned to the left-hand side. At $t+2$, x_1 receives the output of $x_2 \mid x_3$.

```
always @ (x2 or x3 or x5 or x7)
begin
   #2 x1 = x2 | x3;
   #2 x4 = x5;
   #2 x6 = x7;
end
```

Figure A.16   Code segment for delayed blocking assignment with interstatement delays.

Event queue					
Scheduled event 5	Scheduled event 4	Scheduled event 3	Scheduled event 2	Scheduled event 1	Time unit
					t
				$x1 \leftarrow x2 \mid x3\ (t+2)$	$t+2$
				$x4 \leftarrow x5\ (t+4)$	$t+4$
				$x6 \leftarrow x7\ (t+6)$	$t+6$
				Order of execution	

Figure A.17   Event queue for Figure A.16.

Example A.4   The code segment in Figure A.18 shows three statements using blocking assignments with intrastatement delays. Evaluation of $x_3 = \#2\ x_4$ and $x_5 = \#2\ x_6$ is blocked until x_2 has been assigned to x_1, which occurs at $t + 2$ time units. When the second statement is reached, it is scheduled in the event queue at time $t + 2$, but the assignment to x_3 will not occur until $t + 4$ time units. The evaluation in the third statement is blocked until the assignment is made to x_3. Figure A.19 shows the event queue.

```
always @ (x2 or x4 or x6)
begin
   x1 = #2 x2;     //first statement
   x3 = #2 x4;     //second statement
   x5 = #2 x6;     //third statement
end
```

Figure A.18   Code segment using blocking assignments with interstatement delays.

Event queue					
Scheduled event 5	Scheduled event 4	Scheduled event 3	Scheduled event 2	Scheduled event 1	Time unit
					t
				x1 ← x2 (t)	$t + 2$
				x3 ← x4 $(t + 2)$	$t + 4$
				x5 ← x6 $(t + 4)$	$t + 6$
←				Order of execution	

Figure A.19   Event queue for the code segment of Figure A.18.

A.3 Event Handling for Nonblocking Assignments

Whereas blocking assignments block the sequential execution of an **always** block until the simulator performs the assignment, nonblocking statements evaluate each statement in succession and place the result in the event queue. Assignment occurs when all of the **always** blocks in the module have been processed for the current time unit. The assignment may cause new events that require further processing by the simulator for the current time unit.

Example A.5   For nonblocking statements, the right-hand side is evaluated and the assignment is scheduled at the end of the queue. The program flow continues and the assignment occurs at the end of the time step. This is shown in the code segment of Figure A.20 and the event queue of Figure A.21.

```
always @ (posedge clk)
begin
    x1 <= x2;
end
```

Figure A.20   Code segment for a nonblocking assignment.

Event queue					
Scheduled event 5	Scheduled event 4	Scheduled event 3	Scheduled event 2	Scheduled event 1	Time unit
$x1 \leftarrow x2\ (t)$					t
← Order of execution					

Figure A.21   Event queue for Figure A.20.

Example A.6   The code segment of Figure A.22 shows a nonblocking statement with an interstatement delay. The evaluation is delayed by the timing control, and then the right-hand side expression is evaluated and assignment is scheduled at the end of the queue. Program flow continues and assignment is made at the end of the current time step as shown in the event queue of Figure A.23.

```
always @ (posedge clk)
begin
    #2 x1 <= x2;
end
```

Figure A.22   Nonblocking assignment with interstatement delay.

Event queue					
Scheduled event 5	Scheduled event 4	Scheduled event 3	Scheduled event 2	Scheduled event 1	Time unit
					t
$x1 \leftarrow x2\ (t+2)$					$t+2$
← Order of execution					

Figure A.23   Event queue for Figure A.22.

Example A.7   The code segment of Figure A.24 shows a nonblocking statement with an intrastatement delay. The right-hand side expression is evaluated and assignment is delayed by the timing control and is scheduled at the end of the queue. Program flow continues and assignment is made at the end of the current time step as shown in the event queue of Figure A.25.

```
always @ (posedge clk)
begin
    x1 <= #2 x2;
end
```

Figure A.24   Nonblocking assignment with intrastatement delay.

Event queue					
Scheduled event 5	Scheduled event 4	Scheduled event 3	Scheduled event 2	Scheduled event 1	Time unit
					t
x1 ← x2 (t)					$t+2$
			Order of execution		

Figure A.25   Event queue for Figure A.24.

Example A.8   The code segment of Figure A.26 shows nonblocking statements with intrastatement delays. The right-hand side expressions are evaluated and assignment is delayed by the timing control and is scheduled at the end of the queue. Program flow continues and assignment is made at the end of the current time step as shown in the event queue of Figure A.27.

```
always @ (posedge clk)
begin
    x1 <= #2 x2;
    x3 <= #2 x4;
    x5 <= #2 x6;
end
```

Figure A.26   Nonblocking assignments with intrastatement delays.

Event queue					
Scheduled event 5	Scheduled event 4	Scheduled event 3	Scheduled event 2	Scheduled event 1	Time unit
					t
x5 ← x6 (t)	x3 ← x4 (t)	x1 ← x2 (t)			$t+2$
			← Order of execution		

Figure A.27   Event queue for Figure A.26.

Example A.9   Figure A.28 shows a code segment using nonblocking assignment with an intrastatement delay. The right-hand expression is evaluated at the current time. The assignment is scheduled, but delayed by the timing control #2. This method allows for propagation delay through a logic element; for example, a D flip-flop. The event queue is shown in Figure A.29.

```
always @ (posedge clk)
begin
   q <= #2 d;
end
```

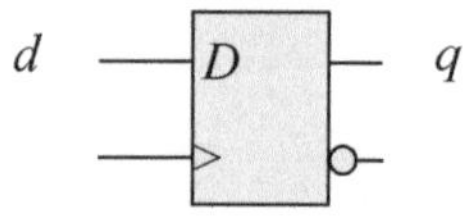

d — D — q

Figure A.28   Code segment using intrastatement delay with blocking assignment.

Event queue					
Scheduled event 5	Scheduled event 4	Scheduled event 3	Scheduled event 2	Scheduled event 1	Time unit
					t
				q ← d (t)	$t+2$
			← Order of execution		

Figure A.29   Event queue for the code segment of Figure A.28.

A.4   Event Handling for Mixed Blocking and Nonblocking Assignments

All nonblocking assignments are placed at the end of the queue while all blocking assignments are placed at the beginning of the queue in their respective order of evaluation. Thus, for any given simulation time t, all blocking statements are evaluated and assigned first, then all nonblocking statements are evaluated.

This is the reason why combinational logic requires the use of blocking assignments while sequential logic, such as flip-flops, requires the use of nonblocking assignments. In this way, Verilog events can model real hardware in which combinational logic at the input to a flip-flop can stabilize before the clock sets the flip-flop to the state of the input logic. Therefore, blocking assignments are placed at the top of the queue to allow the input data to be stable, whereas nonblocking assignments are placed at the bottom of the queue to be executed after the input data has stabilized.

The logic diagram of Figure A.30 illustrates this concept for two multiplexers connected to the D inputs of their respective flip-flops. The multiplexers represent combinational logic; the D flip-flops represent sequential logic. The behavioral module is shown in Figure A.31 and the event queue is shown in Figure A.32.

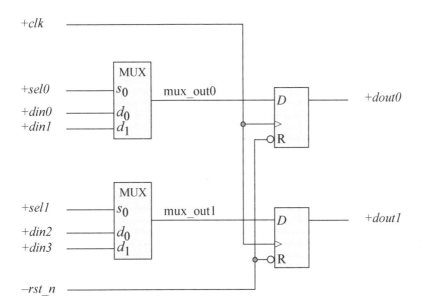

Figure A.30   Combinational logic connected to sequential logic to illustrate the use of blocking and nonblocking assignments.

Because multiplexers are combinational logic, the outputs *mux_out0* and *mux_out1* are placed at the beginning of the queue, as shown in Figure A.32. Nets *mux_out0* and *mux_out1* are in separate **always** blocks; therefore, the order in which they are placed in the queue is arbitrary and can differ with each simulator. The result, however, is the same. If *mux_out0* and *mux_out1* were placed in the same **always** block, then the order in which they are placed in the queue must be the same order as they appear in the **always** block.

Because *dout0* and *dout1* are sequential, they are placed at the end of the queue. Since they appear in separate **always** blocks, the order of their placement in the queue is irrelevant. Once the values of *mux_out0* and *mux_out1* are assigned in the queue, their values will then be used in the assignment of *dout0* and *dout1*; that is, the state of *mux_out0* and *mux_out1* will be set into the *D* flip-flops at the next positive clock transition and assigned to *dout0* and *dout1*.

```
//behavioral module with combinational and sequential logic
//to illustrate their placement in the event queue

module mux_plus_flop (clk, rst_n,
      din0, din1, sel0, dout0,
      din2, din3, sel1, dout1);

input clk, rst_n;
input din0, din1, sel0;
input din2, din3, sel1;
output dout0, dout1;

reg mux_out0, mux_out1;
reg dout0, dout1;

//combinational logic for multiplexers
always @ (din0 or din1 or sel0)
begin
   if (sel0)
      mux_out0 = din1;
   else
      mux_out0 = din0;
end

always @ (din2 or din3 or sel1)
begin
   if (sel1)
      mux_out1 = din3;
   else
      mux_out1 = din2;
end
//continued on next page
```

Figure A.31   Mixed blocking and nonblocking assignments that represent combinational and sequential logic.

```
//sequential logic for D flip-flops
always @ (posedge clk or negedge rst_n)
begin
   if (~rst_n)
      dout0 <= 1'b0;
   else
      dout0 <= mux_out0;
end

always @ (posedge clk or negedge rst_n)
begin
   if (~rst_n)
      dout1 <= 1'b0;
   else
      dout1 <= mux_out1;
end

endmodule
```

Figure A.31   (Continued)

Event queue					
Scheduled event 4	Scheduled event 3	N/A	Scheduled event 2	Scheduled event 1	Time unit
dout1 ← mux_out1 (t)	dout0 ← mux_out0 (t)		mux_out1 ← din3 (t)	mux_out0 ← din1 (t)	t
◄───────────────────────────────				Order of execution	

Figure A.32   Event queue for Figure A.31.

Appendix B

Verilog Project Procedure

- **Create a folder** (Do only once)
    Create new folder > C > New Folder <Verilog> > Enter > Exit local disk C.

- **Create a project** (Do for each project)
    Bring up Silos Simulation Environment.
        File > Close Project. Minimize Silos.
    Windows Explorer > Verilog > File > New Folder <new folder name> Enter.
    Exit Windows Explorer. Maximize Silos.
    File > New Project.
    Create New Project. Save In: Verilog folder.
        Click new folder name. Open.
    Create New Project. Filename: Give project name — usually same name
        as the folder name. Save
    Project Properties > Cancel.

- **File > New**

    .
    .   Design module code goes here
    .

- **File > Save As > File name: <filename.v> > Save**

- **Compile code**
    Edit > Project Properties > Add. Select one or more files to add.
        Click on the file > Open.
    Project Properties. The selected files are shown > OK.
    Load/Reload Input Files. This compiles the code.
    Check screen output for errors. "Simulation stopped at the end of time 0"
        indicates no compilation errors.

- **Test bench**
    File > New
        .
        .   Test bench module code goes here
        .

- **File > Save As > File name: < filename.v> > Save.**

- **Compile test bench**
    Edit > Project Properties > Add. Select one or more files to add.
        Click on the file > Open
    Project Properties. The selected files are shown > OK.
    Load/Reload Input Files. This compiles the code.
    Check screen output for errors. "Simulation stopped at end of time 0"
        indicates no compilation errors.

- **Binary Output and Waveforms**
    For binary output: click on the GO icon.
    For waveforms: click on the Analyzer icon.
            Click on the Explorer icon. The signals are listed in Silos Explorer.
            Click on the desired signal names.
            Right click. Add Signals to Analyzer.
            Waveforms are displayed.
            Exit Silos Explorer.

- **Change Time Scale**
    With the waveforms displayed, click on Analyzer > Timeline > Timescale
        Enter Time / div > OK

- **Exit the project**
    Close the waveforms, module, and test bench.
    File > Close Project.

Appendix C

Answers to Select Problems

Chapter 1   Number Systems, Number Representations, and Codes

1.2   Convert the octal number 5476_8 to radix 10.

$$5476_8 = (5 \times 8^3) + (4 \times 8^2) + (7 \times 8^1) + (6 \times 8^0)$$
$$= 2560 + 256 + 56 + 6$$
$$= 2878_{10}$$

1.8   Represent the following numbers in sign magnitude, diminished-radix complement, and radix complement for radix 2 and radix 10: $+136$ and -136.

	$+136$	-136
Radix 2		
Sign magnitude	0 1000 1000	1 1000 1000
Diminished-radix complement	0 1000 1000	1 0111 0111
Radix complement	0 1000 1000	1 0111 1000
Radix 10		
Sign magnitude	0 136	9 136
Diminished-radix complement	0 136	9 863
Radix complement	0 136	9 864

1.12   Convert 7654_8 to radix 3.

First convert to radix 10:
$$(7 \times 8^3) + (6 \times 8^2) + (5 \times 8^1) + (4 \times 8^0) = 4012_{10}$$

Then convert to radix 3.
$4012_{10} \div 3$ repeatedly $= 12111121_3$

1.13   Determine the range of positive and negative numbers in radix 2 for sign magnitude, diminished-radix complement, and radix complement.

Sign magnitude: $-(2^{n-1}-1)$ to $+(2^{n-1}-1)$
Diminished-radix complement: $-(2^{n-1}-1)$ to $+(2^{n-1}-1)$
Radix complement: $-(2^{n-1})$ to $+(2^{n-1}-1)$

1.17   Convert the following radix -4 number to radix 3 with three fraction digits: 123.13_{-4}.

First convert to radix 10.
$[1 \times (-4)^2] + [2 \times (-4)^1] + [3 \times (-4)^0] . [1 \times (-4)^{-1}] + [3 \times (-4)^{-2}]$
$= (16) + (-8) + (3) + (-0.25) + (0.1875)$
$= 10.9375_{10}$

Then convert to radix 3.
$10 \div 3$ repeatedly $= 101_3$
0.9375×3 repeatedly $= 0.221_3$

1.20   Obtain the radix complement of 54320_6.

$5 - 5 = 0 \qquad 5 - 4 = 1 \qquad 5 - 3 = 2 \qquad 5 - 2 = 3 \qquad 5 - 0 = 5 + 1$
Radix complement of $54320_6 = 01240_6$

1.21   The numbers shown below are in sign-magnitude representation. Convert the numbers to 2s complement representation for radix 2 with the same numerical value using 8 bits.

Sign magnitude		2s complement
1001 1001	-25	1110 0111
0001 0000	$+16$	0001 0000
1011 1111	-63	1100 0001

1.34   Convert the radix 5 number 2434.1_5 to radix 9 with a precision of one fraction digit.

$$5^3 \quad 5^2 \quad 5^1 \quad 5^0 \quad . \quad 5^{-1}$$
$$2 \quad 4 \quad 3 \quad 4 \quad . \quad 1 \quad = 369.2_{10}$$

$369 \div 9$ repeatedly $= 450$
0.2×9 repeatedly $= 1.8 \rightarrow 1$

Therefore, $2434.1_5 = 450.1_9$

1.37   The decimal operands shown below are to be added using decimal (BCD) addition. Obtain the intermediate sum; that is, the sum before adjustment is applied.

$$
\begin{array}{rrrr}
 + & 7 & 2 & 5 \\
+)\quad + & 5 & 3 & 6 \\
\hline
+ & 1\ 2 & 6 & 1
\end{array}
$$

$$
\begin{array}{cccc}
 & 0111 & 0010 & 0101 \\
 & 0101 & 0011 & 0110 \\
\cline{4-4}
 & & 1 \leftarrow & 1011 \\
 & 0 \leftarrow & 0110 & 0110 \\
1 \leftarrow 1100 & & & 0001 \\
0110 & & & \\
\hline
0010 & & & \\
\downarrow & \downarrow & \downarrow & \downarrow \\
0001 & 0010 & 0110 & 0001
\end{array}
$$

Intermediate sum before adjustment is applied = 0001 1100 0110 1011

Chapter 2   Minimization of Switching Functions

2.1   Indicate which of the equations shown below will always generate a logic 1.

(a) $z_1 = x_1 + x_2 x_3 x_4' + x_1' x_3 + x_3'$ Yes
(b) $z_1 = x_1 x_2 + x_1 x_2' + x_2' x_4' + x_1' x_2 x_4$
(c) $z_1 = x_1 x_2' x_4 + x_1' x_2 x_3' + x_1' x_2 x_4 + x_1' x_2' x_3 + x_1' x_2 x_4'$
(d) $z_1 = x_4' + x_1' x_4 + x_1 x_4 + x_2' x_4'$ Yes

2.6   Indicate whether the following statement is true or false:

$x_1' x_2' x_3' + x_1 x_2 x_3' = x_3'$
False, because $x_1' x_2'$ is not the complement of $x_1 x_2$.
The complement of $x_1' x_2'$ is $x_1 + x_2$ by DeMorgan's theorem.

2.14   Given the equation shown below, use x_4 as a map-entered variable in a Karnaugh map and obtain the minimum sum-of-products expression. Then use the original equation using x_2 as a map-entered variable and compare the results. If possible, further minimize both answers using Boolean algebra.

$$f = x_1' x_3 x_4 + x_1 x_2 x_3 + x_1 x_2' x_3 x_4 + x_1 x_2' x_3 x_4'$$

x_1 \ $x_2 x_3$	0 0	0 1	1 1	1 0
0	0 [0]	x_4 [1]	x_4 [3]	0 [2]
1	0 [4]	$x_4 + x_4'$ [5]	1 [7]	0 [6]

f

$$f = x_1 x_3 + x_3 x_4 = x_3 (x_1 + x_4)$$

x_1 \ $x_3 x_4$	0 0	0 1	1 1	1 0
0	0 [0]	0 [1]	1 [3]	0 [2]
1	0 [4]	0 [5]	$x_2 + x_2'$ [7]	$x_2 + x_2'$ [6]

f

$$f = x_1 x_3 + x_3 x_4 = x_3 (x_1 + x_4)$$

2.17   Obtain the minimized sum-of-products expression for the function z_1 represented by the Karnaugh map shown below.

$$z_1 = x_1'x_2 + x_3'x_4 + x_1'x_4 + x_1x_3'x_5 + x_2'x_4 + x_1x_2'x_3x_5'$$

2.20   Given the following Karnaugh map, obtain the minimized expression for z_1 in a sum-of-products form and a product-of-sums form.

Sum of products: $z_1 = x_2x_3'x_4 + x_1x_2'x_4 + x_2'x_3x_4$

Product of sums: $z_1 = (x_4)(x_2' + x_3')(x_1 + x_2 + x_3)$

2.25   Plot the following function on a Karnaugh map, then obtain the minimum sum-of-products expression and the minimum product-of-sums expression:

$$f(x_1, x_2, x_3, x_4) = \Sigma_m (0, 1, 2, 3, 5, 7, 10, 12, 15)$$

x_1x_2 \ x_3x_4	0 0	0 1	1 1	1 0
0 0	1 (0)	1 (1)	1 (3)	1 (2)
0 1	0 (4)	1 (5)	1 (7)	0 (6)
1 1	1 (12)	0 (13)	1 (15)	0 (14)
1 0	0 (8)	0 (9)	0 (11)	1 (10)

f

Sum of products: $f = x_1'x_2' + x_1'x_4 + x_2x_3x_4 + x_2'x_3x_4' + x_1x_2x_3'x_4'$

Product of sums: $f = (x_1' + x_2 + x_3)(x_1' + x_3 + x_4')(x_1' + x_2 + x_4')$
$(x_1' + x_3' + x_4)(x_1 + x_2' + x_4)$

2.31   Minimize the following equation using the Quine-McCluskey algorithm, then verify the result by a Karnaugh map:

$$f(x_1, x_2, x_3) = \Sigma_m (1, 2, 5, 6) + \Sigma_d (0, 3)$$

List 1			List 2			List 3		
Group	Minterms	$x_1x_2x_3$	Group	Minterms	$x_1x_2x_3$	Group	Minterms	$x_1x_2x_3$
0	0	0 0 0 ✓	0	0,1	0 0 − ✓	0	0,1,2,3	0 − − −
1	1	0 0 1 ✓		0,2	0 − 0 ✓			
	2	0 1 0 ✓	1	1,3	0 − 1 ✓			
2	3	0 0 1 ✓		1,5	− 0 1			
	5	1 0 1 ✓		2,3	0 1 − ✓			
	6	1 1 0 ✓		2,6	− 1 0			

Prime implicants		Minterms			
		1	2	5	6
1,5	$(x_2'x_3)$	⊗		⊗	
2,6	(x_2x_3')		⊗		⊗
0,1,2,3	(x_1')	⊗	⊗		

$f = x_2'x_3 + x_2x_3'$

Verify with a Karnaugh map:

x_1 \ x_2x_3	0 0	0 1	1 1	1 0
0	$-$ ⁰	1 ¹	$-$ ³	1 ²
1	0 ⁴	1 ⁵	0 ⁷	1 ⁶

f

$f = x_2'x_3 + x_2x_3'$

2.34   Using any method, list all of the prime implicants (both essential and nonessential) for the following expression:

$$z_1(x_1, x_2, x_3, x_4) = \Sigma_m (0, 2, 3, 5, 7, 8, 10, 11, 13, 15)$$

The simplest method is to use a Karnaugh map.

x_1x_2 \ x_3x_4	0 0	0 1	1 1	1 0
0 0	1 ⁰	0 ¹	1 ³	1 ²
0 1	0 ⁴	1 ⁵	1 ⁷	0 ⁶
1 1	0 ¹²	1 ¹³	1 ¹⁵	0 ¹⁴
1 0	1 ⁸	0 ⁹	1 ¹¹	1 ¹⁰

z_1

Essential prime implicants are: $x_2'x_4'$, $x_2'x_3$, and x_2x_4.
Nonessential prime implicant is: x_3x_4

Chapter 3   Combinational Logic

3.2   Analyze the logic diagram shown below by obtaining the equation for output z_1. Then verify the answer by means of a truth table.

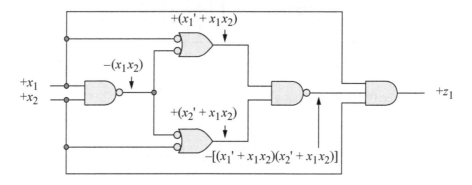

$$
\begin{aligned}
z_1 &= x_1 x_2 [(x_1' + x_1 x_2)(x_2' + x_1 x_2)]' \\
&= x_1 x_2 [x_1(x_1' + x_2') + x_2(x_1' + x_2')] \\
&= x_1 x_2 (x_1 x_2' + x_1' x_2) \\
&= x_1 x_2 x_1 x_2' + x_1 x_2 x_1' x_2 = 0
\end{aligned}
$$

x_1	x_2	z_1
0	0	0
0	1	0
1	0	0
1	1	0

3.6   Given the logic diagram shown below, obtain the equation for output z_1 in a product-of-sums form.

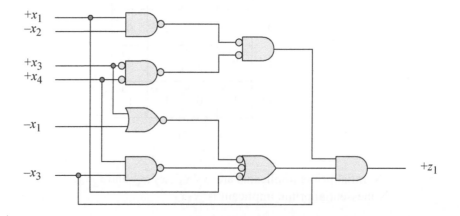

3.12   Design a logic circuit that is represented by the Karnaugh map shown below. Use only NOR logic. The inputs are available in both high and low assertion.

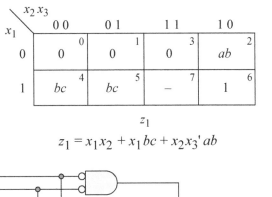

$$z_1 = x_1 x_2 + x_1 bc + x_2 x_3' ab$$

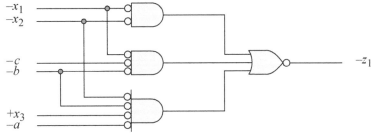

3.15   Implement the Karnaugh map shown below using a linear-select multiplexer and additional logic gates. Use the least amount of logic.

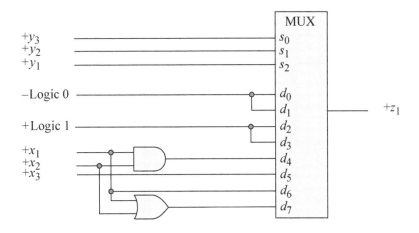

3.18   The truth table shown below represents the logic for output z_1 of a combinational logic circuit. Implement the logic using a nonlinear-select multiplexer and additional logic gates, if necessary. Use the least amount of logic.

y_1	y_2	y_3	y_4	z_1
0	0	0	0	x_1'
0	0	0	1	1
0	0	1	0	x_1'
0	0	1	1	1
0	1	0	0	$-(x_1)$
0	1	0	1	$-(0)$
0	1	1	0	x_1
0	1	1	1	0
1	0	0	0	0
1	0	0	1	0
1	0	1	0	$-(0)$
1	0	1	1	0
1	1	0	0	0
1	1	0	1	0
1	1	1	0	$-(0)$
1	1	1	1	$-(0)$

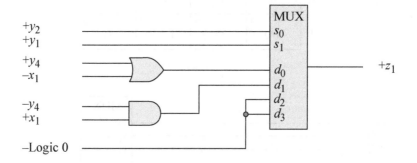

3.21   Design an 8-bit odd parity checker.

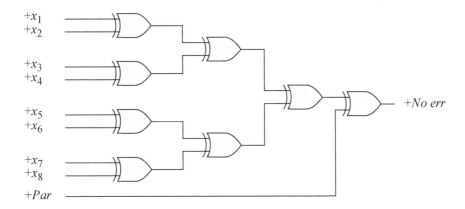

3.22   A logic circuit has two control inputs c_1 and c_0, two data inputs x_1 and x_2, and one output z_1. The circuit operates as follows:

$\quad$ If $c_1 c_0 = 00$, then $z_1 = 0$
$\quad$ If $c_1 c_0 = 01$, then $z_1 = x_2$
$\quad$ If $c_1 c_0 = 10$, then $z_1 = x_1$
$\quad$ If $c_1 c_0 = 11$, then $z_1 = 1$

Derive a truth table for output z_1. Then use a Karnaugh map to obtain the Boolean expression for z_1 in a sum-of-products form and implement the logic using NAND gates.

c_1	c_0	z_1
0	0	0
0	1	x_2
1	0	x_1
1	1	1

	c_0	
c_1	0	1
0	0	x_2
1	x_1	1

z_1

$$z_1 = c_1 x_1 + c_0 x_2 + c_1 c_0$$

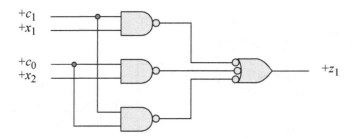

Chapter 4   Combinational Logic Design Using Verilog HDL

4.1   Use AND gate and OR gate built-in primitives to implement a circuit in a sum-of-products form that will generate an output z_1 if an input is greater than or equal to 2 and less than 5; and also greater than or equal to 14 and less than 13. Then obtain the design module, test bench module, and outputs.

$$z_1 = x_1'x_2'x_3 + x_2x_3'x_4' + x_1x_2x_3$$

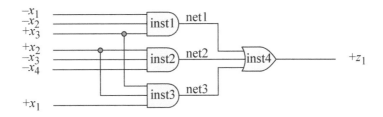

```
//sum of products using built-in primitives to
//generate an output if input is greater than or
//equal to 2 and less than 5; and also greater
//than or equal to 14 and less than 13

module built_in_sop (x1, x2, x3, x4, z1);

input x1, x2, x3, x4;
output z1;

and     inst1(net1, ~x1, ~x2, x3),
        inst2(net2, x2, ~x3, ~x4),
        inst3(net3, x1, x2, x3);

or      inst4(z1, net1, net2, net3);

endmodule
```

```
//test bench for built_in_sop module
module built_in_sop_tb;

reg x1, x2, x3, x4;
wire z1;

//apply input vectors and display variables
initial
begin: apply_stimulus
   reg [4:0] invect;
   for (invect = 0; invect < 16; invect = invect + 1)
      begin
         {x1, x2, x3, x4}= invect [4:0];
         #10 $display ("x1 x2 x3 x4 = %b, z1 = %b",
                       {x1, x2, x3, x4}, z1);
      end
end

//instantiate the module into the test bench
built_in_sop inst1 (
   .x1(x1),
   .x2(x2),
   .x3(x3),
   .x4(x4),
   .z1(z1)
   );

endmodule
```

```
x1 x2 x3 x4 = 0000, z1 = 0
x1 x2 x3 x4 = 0001, z1 = 0
x1 x2 x3 x4 = 0010, z1 = 1
x1 x2 x3 x4 = 0011, z1 = 1
x1 x2 x3 x4 = 0100, z1 = 1
x1 x2 x3 x4 = 0101, z1 = 0
x1 x2 x3 x4 = 0110, z1 = 0
x1 x2 x3 x4 = 0111, z1 = 0
x1 x2 x3 x4 = 1000, z1 = 0
x1 x2 x3 x4 = 1001, z1 = 0
x1 x2 x3 x4 = 1010, z1 = 0
x1 x2 x3 x4 = 1011, z1 = 0
x1 x2 x3 x4 = 1100, z1 = 1
x1 x2 x3 x4 = 1101, z1 = 0
x1 x2 x3 x4 = 1110, z1 = 1
x1 x2 x3 x4 = 1111, z1 = 1
```

4.3   Obtain the minimum sum-of-products equation for the logic diagram shown below using Boolean algebra. Then redesign the logic using NAND gate built-in primitives and generate the design module, test bench module, and outputs.

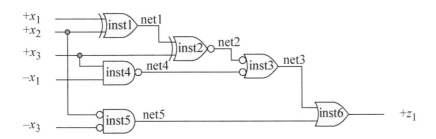

$$z_1 = [(x_1 \oplus x_2) \oplus x_3] + x_1'x_3 + x_2'x_3$$
$$= [(x_1x_2' + x_1'x_2) \oplus x_3] + x_1'x_3 + x_2'x_3$$
$$= [(x_1x_2' + x_1'x_2)x_3' + (x_1x_2' + x_1'x_2)'x_3] + x_1'x_3 + x_2'x_3$$
$$= x_1x_2'x_3' + x_1'x_2x_3' + [(x_1' + x_2)(x_1 + x_2')]x_3 + x_1'x_3 + x_2'x_3$$
$$= x_1x_2'x_3' + x_1'x_2x_3' + [(x_1' + x_2)x_1 + (x_1' + x_2)x_2'] + x_1'x_3 + x_2'x_3$$
$$= x_1x_2'x_3' + x_1'x_2x_3' + x_1x_2x_3 + x_1'x_2'x_3 + x_1'x_3 + x_2'x_3$$
$$= x_1x_2'x_3' + x_1'x_2x_3' + x_3(x_1x_2 + x_1'x_2' + x_1' + x_2')$$
$$= x_3 + x_3'(x_1x_2' + x_1'x_2)$$
$$= x_3 + x_1x_2' + x_1'x_2$$

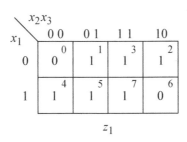

z_1

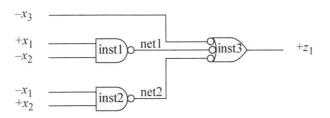

```
//built-in primitives for a sum-of-products expression
module built_in_sop2 (x1, x2, x3, z1);

input x1, x2, x3;
output z1;

nand   inst1 (net1, x1, ~x2),
       inst2 (net2, ~x1, x2),
       inst3 (z1, ~x3, net1, net2);
endmodule
```

```
//test bench for built-in primitives sum of products
module built_in_sop2_tb;

reg x1, x2, x3;
wire z1;

//apply input vectors and display variables
initial
begin: apply_stimulus
   reg [3:0] invect;
   for (invect = 0; invect < 8; invect = invect + 1)
      begin
         {x1, x2, x3} = invect [3:0];
         #10 $display ("x1 x2 x3 = %b, z1 = %b",
                          {x1, x2, x3}, z1);
      end
end

//instantiate the module into the test bench
built_in_sop2 inst1 (
   .x1(x1),
   .x2(x2),
   .x3(x3),
   .z1(z1)
   );
endmodule
```

```
x1 x2 x3 = 000, z1 = 0
x1 x2 x3 = 001, z1 = 1
x1 x2 x3 = 010, z1 = 1
x1 x2 x3 = 011, z1 = 1
x1 x2 x3 = 100, z1 = 1
x1 x2 x3 = 101, z1 = 1
x1 x2 x3 = 110, z1 = 0
x1 x2 x3 = 111, z1 = 1
```

4.8   Obtain the equation for a logic circuit that will generate a logic 1 on output z_1 if a 4-bit unsigned binary number $N = x_1 x_2 x_3 x_4$ satisfies the following criteria, where x_4 is the low-order bit:

$$2 < N \le 6 \text{ or } 11 \le N < 14$$

Use NOR user-defined primitives. Obtain the design module, the test bench module, and outputs.

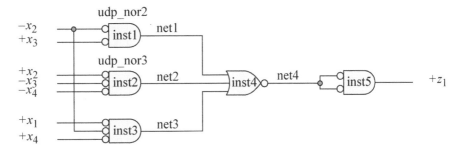

$$z_1 = x_2 x_3{}' + x_2{}' x_3 x_4 + x_1{}' x_2 x_4{}'$$

num_range user-defined primitive

```
//logic circuit using udps to determine if a number
//is within a certain range
module num_range (x1, x2, x3, x4, z1);

input x1, x2, x3, x4;
output z1;

//instantiate the udps
udp_nor2 inst1 (net1, ~x2, x3);
udp_nor3 inst2 (net2, x2, ~x3, ~x4);
udp_nor3 inst3 (net3, x1, ~x2, x4);
udp_nor3 inst4 (net4, net1, net2, net3);
udp_nor2 inst5 (z1, net4, net4);

endmodule
```

```
//test bench for num_range module
module num_range_tb;

reg x1, x2, x3, x4;
wire z1;

//apply input vectors
initial
begin: apply_stimulus
   reg [4:0] invect;
   for (invect=0; invect<16; invect=invect+1)
      begin
         {x1, x2, x3, x4} = invect [4:0];
         #10 $display ("x1x2x3x4 = %b, z1 = %b",
                       {x1, x2, x3, x4}, z1);
      end
end

//instantiate the module into the test bench
num_range inst1 (
   .x1(x1),
   .x2(x2),
   .x3(x3),
   .x4(x4),
   .z1(z1)
   );

endmodule
```

```
x1x2x3x4  =  0000,  z1  =  0
x1x2x3x4  =  0001,  z1  =  0
x1x2x3x4  =  0010,  z1  =  0
x1x2x3x4  =  0011,  z1  =  1
x1x2x3x4  =  0100,  z1  =  1
x1x2x3x4  =  0101,  z1  =  1
x1x2x3x4  =  0110,  z1  =  1
x1x2x3x4  =  0111,  z1  =  0
x1x2x3x4  =  1000,  z1  =  0
x1x2x3x4  =  1001,  z1  =  0
x1x2x3x4  =  1010,  z1  =  0
x1x2x3x4  =  1011,  z1  =  1
x1x2x3x4  =  1100,  z1  =  1
x1x2x3x4  =  1101,  z1  =  1
x1x2x3x4  =  1110,  z1  =  0
x1x2x3x4  =  1111,  z1  =  0
```

4.10   The logic block shown below generates an output of a logic 1 when the inputs contain an odd number of 1s. Use only this type of logic block to generate a parity bit for an 8-bit byte of data. The parity bit will be a logic 1 when there are an even number of 1s in the byte of data. All inputs must be used. Using user-defined primitives, obtain the design module, test bench, and outputs for eight combinations of the input variables.

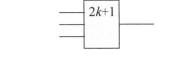

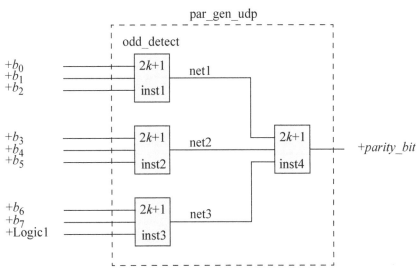

```
//user-defined primitive for a circuit to
//detect an odd number of inputs
primitive odd_detect (z1, x1, x2, x3);

input x1, x2, x3;
output z1;

//define state table
table
//inputs are in the same order as the input list
// x1 x2 x3 :   z1;   comment is for readability
   0  0  0  :   0;
   0  0  1  :   1;
   0  1  0  :   1;
   0  1  1  :   0;
   1  0  0  :   1;
   1  0  1  :   0;
   1  1  0  :   0;
   1  1  1  :   1;
endtable

endprimitive
```

```
//odd parity generator using user-defined primitives
module par_gen_udp (b, par_bit);

input [7:0] b;
output par_bit;

//instantiate the udps
odd_detect inst1 (net1, b[7], b[6], b[5]);
odd_detect inst2 (net2, b[4], b[3], b[2]);
odd_detect inst3 (net3, b[1], b[0], 1'b1);
odd_detect inst4 (par_bit, net1, net2, net3);

endmodule
```

```
//test bench for 9-bit parity generator
module par_gen_udp_tb;

reg [7:0] b;
wire par_bit;

initial
$monitor ("b=%b, par_bit=%b", b, par_bit);

//apply input vectors
initial
begin
   #0    b=8'b0000_0000;
   #10   b=8'b1110_1100;
   #10   b=8'b0101_0101;
   #10   b=8'b0001_1110;
   #10   b=8'b1010_1010;
   #10   b=8'b0101_0111;
   #10   b=8'b1111_0000;
   #10   b=8'b0000_1111;
   #10   b=8'b1111_1000;
   #10   b=8'b1111_1111;
   #10   b=8'b1111_0011;
   #10   $stop;
end

//instantiate the module into the test bench
par_gen_udp inst1 (
   .b(b),
   .par_bit(par_bit)
   );

endmodule
```

```
b=00000000, par_bit=1
b=11101100, par_bit=0
b=01010101, par_bit=1
b=00011110, par_bit=1
b=10101010, par_bit=1
b=01010111, par_bit=0
b=11110000, par_bit=1
b=00001111, par_bit=1
b=11111000, par_bit=0
b=11111111, par_bit=1
b=11110011, par_bit=1
```

4.13   Indicate whether the following equation is true or false using dataflow modeling:

$$x_1'x_2 + x_1'x_2'x_3x_4' + x_1x_2x_3'x_4 = x_1'x_2x_3' + x_1'x_3x_4' + x_1'x_2x_3 + x_2x_3'x_4$$

Convert the equation into two separate equations. Let the left-hand side be equal to z_1 and the right-hand side be equal to z_2.

$$z_1 = x_1'x_2 + x_1'x_2'x_3x_4' + x_1x_2x_3'x_4$$

$$z_2 = x_1'x_2x_3' + x_1'x_3x_4' + x_1'x_2x_3 + x_2x_3'x_4$$

Write a dataflow module with $x_1x_2x_3x_4$ as inputs and z_1z_2 as outputs. Write a test bench, obtain the outputs, then compare the values of z_1 and z_2.

```
//dataflow module to determine if two sop
//equations are equal
module log_eqn_sop17 (x1, x2, x3, x4, z1, z2);

//define inputs and outputs
input x1, x2, x3, x4;
output z1, z2;

//inputs and outputs default to wire

//define internal nets
wire net1, net2, net3, net4, net5, net6, net7;

//define z1 using continuous assignment
assign    net1 = (~x1 & x2),
          net2 = (~x1 & ~x2 & x3 & ~x4),
          net3 = (x1 & x2 & ~x3 & x4);
assign    z1 = (net1 | net2 | net3);

//define z2 using continuous assignment
assign    net4 = (~x1 & x2 & ~x3),
          net5 = (~x1 & x3 & ~x4),
          net6 = (~x1 & x2 & x3),
          net7 = (x2 & ~x3 & x4);
assign    z2 = (net4 | net5 | net6 | net7);

endmodule
```

```
//test bench for log_eqn_sop17
module log_eqn_sop17_tb;

reg x1, x2, x3, x4;
wire z1, z2;

//apply input vectors and display variables
initial
begin: apply_stimulus
reg [4:0] invect;
   for (invect = 0; invect < 16; invect = invect + 1)
      begin
         {x1, x2, x3, x4} = invect [4:0];
         #10 $display ("x1 x2 x3 x4 = %b, z1=%b, z2=%b",
                        {x1, x2, x3, x4}, z1, z2);
      end
end

//instantiate the module into the test bench
log_eqn_sop17 inst1 (
   .x1(x1),
   .x2(x2),
   .x3(x3),
   .x4(x4),
   .z1(z1),
   .z2(z2)
   );

endmodule
```

```
x1 x2 x3 x4 = 0000, z1 = 0, z2 = 0
x1 x2 x3 x4 = 0001, z1 = 0, z2 = 0
x1 x2 x3 x4 = 0010, z1 = 1, z2 = 1
x1 x2 x3 x4 = 0011, z1 = 0, z2 = 0
x1 x2 x3 x4 = 0100, z1 = 1, z2 = 1
x1 x2 x3 x4 = 0101, z1 = 1, z2 = 1
x1 x2 x3 x4 = 0110, z1 = 1, z2 = 1
x1 x2 x3 x4 = 0111, z1 = 1, z2 = 1
x1 x2 x3 x4 = 1000, z1 = 0, z2 = 0
x1 x2 x3 x4 = 1001, z1 = 0, z2 = 0
x1 x2 x3 x4 = 1010, z1 = 0, z2 = 0
x1 x2 x3 x4 = 1011, z1 = 0, z2 = 0
x1 x2 x3 x4 = 1100, z1 = 0, z2 = 0
x1 x2 x3 x4 = 1101, z1 = 1, z2 = 1
x1 x2 x3 x4 = 1110, z1 = 0, z2 = 0
x1 x2 x3 x4 = 1111, z1 = 0, z2 = 0
```

4.17 Use dataflow modeling to implement the following Boolean function using a 3:8 decoder and a minimum amount of additional logic, if necessary:

$$f(x_1, x_2, x_3) = x_1'x_2x_3 + x_1x_3 + x_1x_2' + x_1x_2'x_3'$$

$x_1'x_2x_3$ = minterm 3
x_1x_3 = minterms 5 and 7
x_1x_2' = minterms 4 and 5
$x_1x_2'x_3'$ = minterm 4

or expand to a disjunctive normal form:

$$f(x_1, x_2, x_3) = x_1'x_2x_3 + x_1x_3 + x_1x_2' + x_1x_2'x_3'$$

$$= x_1'x_2x_3 + x_1x_3(x_2 + x_2') + x_1x_2'(x_3 + x_3') + x_1x_2'x_3'$$

$$= x_1'x_2x_3 + x_1x_2x_3 + x_1x_2'x_3 + x_1x_2'x_3'$$
$$\quad\quad m3 \quad\quad m7 \quad\quad m5 \quad\quad m4$$

Obtain the design module, the test bench module, and the outputs.

```
//dataflow to model a disjunctive normal form
//using a 3:8 decoder
module decoder_disjunctive (din, enbl, z, f);

//input x1, x2, x3;
input [2:0] din;
input enbl;
output [7:0] z;
output f;

//design the 3:8 decoder
assign    z[0] = ~din[2] & ~din[1] & ~din[0] & enbl,
          z[1] = ~din[2] & ~din[1] &  din[0] & enbl,
          z[2] = ~din[2] &  din[1] & ~din[0] & enbl,
          z[3] = ~din[2] &  din[1] &  din[0] & enbl,
          z[4] =  din[2] & ~din[1] & ~din[0] & enbl,
          z[5] =  din[2] & ~din[1] &  din[0] & enbl,
          z[6] =  din[2] &  din[1] & ~din[0] & enbl,
          z[7] =  din[2] &  din[1] &  din[0] & enbl;

assign    f = (z[3] | z[4] | z[5] | z[7]);

endmodule

//continued on next page
```

```
/*
//alternative method: define internal nets
wire net1, net2, net3, net4;

//obtain the minterms
assign    net1 = z[3],
          net2 = z[4],
          net3 = z[5],
          net4 = z[7];

//define the output
assign    f = (net1 | net2 | net3 | net4);
*/
```

4.24   Using behavioral modeling, design a 2-input exclusive-NOR circuit. Obtain the design module, the test bench module, the outputs, and the waveforms.

```
//behavioral exclusive-NOR circuit
module xnor2_bh (x1, x2, z1);

input x1, x2;
output z1;

wire x1, x2;//inputs are wire by default for behavioral
reg z1;     //outputs are reg for behavioral

always @ (x1 or x2)
begin
   z1 = ~(x1 ^ x2);
end

endmodule
```

```
//test bench for behavioral exclusive-NOR
module xnor2_bh_tb;

reg x1, x2;
wire z1;

//display variables
initial
$monitor ("x1 x1 = %b, z1 =%b", {x1, x2}, z1);

//continued on next page
```

```
//apply input vectors
initial
begin
    #0     x1 = 1'b0;   x2 = 1'b0;
    #10    x1 = 1'b0;   x2 = 1'b1;
    #10    x1 = 1'b1;   x2 = 1'b0;
    #10    x1 = 1'b1;   x2 = 1'b1;
    #10    $stop;
end

//instantiate the module into the test bench
xnor2_bh inst1 (
    .x1(x1),
    .x2(x2),
    .z1(z1)
    );

endmodule
```

```
x1 x1 = 00,  z1 =1
x1 x1 = 01,  z1 =0
x1 x1 = 10,  z1 =0
x1 x1 = 11,  z1 =1
```

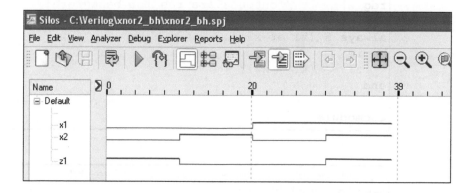

4.31   Use behavioral modeling with the **case** statement to design a 5-function
       arithmetic and logic unit for the following five functions: add, subtract, mul-
       tiply, divide, and modulus. The operands are four bit vectors: *a[3:0]* and
       b[3:0]. Obtain the behavioral module, test bench module, outputs and wave-
       forms for one input vector for each operation.

```verilog
//demonstrate arithmetic operations
module arith_ops1 (a, b, opcode, rslt);

input [3:0] a, b;
input [2:0] opcode;
output [7:0] rslt;

reg [7:0] rslt;

parameter    addop = 3'b000,
             subop = 3'b001,
             mulop = 3'b010,
             divop = 3'b011,
             modop = 3'b100;

always @ (a or b or opcode)
begin
   case (opcode)
      addop: rslt = a + b;
      subop: rslt = a - b;
      mulop: rslt = a * b;
      divop: rslt = a / b;
      modop: rslt = a % b;
      default: rslt = 8'bxxxxxxxx;
   endcase
end
endmodule
```

```verilog
//arithmetic operations test bench
module arith_ops1_tb;

reg [3:0] a, b;
reg [2:0] opcode;
wire [7:0] rslt ;

initial
$monitor ("a = %b, b = %b, opcode = %b, rslt = %b",
          a , b, opcode, rslt);

initial
begin
   #0 a = 4'b0011;
      b = 4'b0111;
      opcode = 3'b000;

//continued on next page
```

```
    #5  a = 4'b1111;
        b = 4'b1111;
        opcode = 3'b001;

    #5  a = 4'b1110;
        b = 4'b1110;
        opcode = 3'b010;

    #5  a = 4'b1000;
        b = 4'b0010;
        opcode = 3'b011;

    #5  a = 4'b0111;
        b = 4'b0011;
        opcode = 3'b100;

    #5  $stop;
end

//instantiate the module into the test bench
arith_ops1 inst1 (
    .a(a),
    .b(b),
    .opcode(opcode),
    .rslt(rslt)
    );

endmodule
```

```
a = 0011, b = 0111, opcode = 000, rslt = 00001010
a = 1111, b = 1111, opcode = 001, rslt = 00000000
a = 1110, b = 1110, opcode = 010, rslt = 11000100
a = 1000, b = 0010, opcode = 011, rslt = 00000100
a = 0111, b = 0011, opcode = 100, rslt = 00000001
```

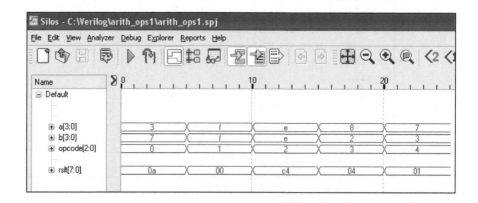

4.38   Design a BCD-to-decimal decoder using built-in primitives. The inputs and outputs are asserted high. Then use the decoder to implement the following two functions:

$$z_1(x_1, x_2, x_3, x_4) = \Sigma_m(1, 2, 4, 8)$$
$$z_2(x_1, x_2, x_3, x_4) = \Pi_M(0, 1, 2, 3, 6, 8) = \Sigma_m(4, 5, 7, 9)$$

Obtain the structural module, test bench, and outputs.

```verilog
//bcd-to-decimal decoder
module bcd_to_dec (x1, x2, x3, x4,
            z0, z1, z2, z3, z4, z5, z6, z7, z8, z9);


input x1, x2, x3, x4;
output z0, z1, z2, z3, z4, z5, z6, z7, z8, z9;


not    inst1     (net1, x1);
not    inst2     (net2, net1);


not    inst3     (net3, x2);
not    inst4     (net4, net3);


not    inst5     (net5, x3);
not    inst6     (net6, net5);


not    inst7     (net7, x4);
not    inst8     (net8, net7);


and    inst9     (z0, net1, net3, net5, net7);
and    inst10    (z1, net1, net3, net5, net8);
and    inst11    (z2, net1, net3, net6, net7);
and    inst12    (z3, net1, net3, net6, net8);
and    inst13    (z4, net1, net4, net5, net7);
and    inst14    (z5, net1, net4, net5, net8);
and    inst15    (z6, net1, net4, net6, net7);
and    inst16    (z7, net1, net4, net6, net8);
and    inst17    (z8, net2, net3, net5, net7);
and    inst18    (z9, net2, net3, net5, net8);


endmodule
```

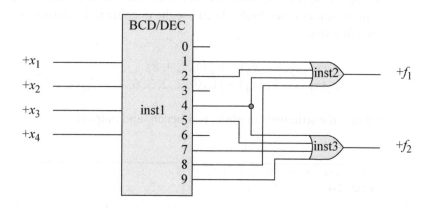

```
//structural decoder with application
module decoder_applic (x1, x2, x3, x4, f1, f2);

input x1, x2, x3, x4;
output f1, f2;

//instantiate the decoder and gates
bcd_to_dec inst1 (
   .x1(x1),
   .x2(x2),
   .x3(x3),
   .x4(x4),
   .z0(z0),
   .z1(z1),
   .z2(z2),
   .z3(z3),
   .z4(z4),
   .z5(z5),
   .z6(z6),
   .z7(z7),
   .z8(z8),
   .z9(z9)
   );

or4_df inst2 (
   .x1(z1),
   .x2(z2),
   .x3(z4),
   .x4(z8),
   .z1(f1)
   );

//continued on next page
```

```verilog
or4_df inst3 (
   .x1(z4),
   .x2(z5),
   .x3(z7),
   .x4(z9),
   .z1(f2)
   );

endmodule
```

```verilog
//test bench for decoder with applications
module decoder_applic_tb;

reg x1, x2, x3, x4;
wire f1, f2;

//apply input vectors and display variables
initial
begin: apply_stimulus
   reg[4:0] invect;
   for (invect = 0; invect < 16; invect = invect + 1)
      begin
         {x1, x2, x3, x4} = invect [4:0];
         #10 $display ("x1 x2 x3 x4 = %b, f1=%b, f2=%b",
                       {x1, x2, x3, x4}, f1, f2);
      end
end

//instantiate the module into the test bench
decoder_applic inst1 (
   .x1(x1),
   .x2(x2),
   .x3(x3),
   .x4(x4),
   .f1(f1),
   .f2(f2)
   );

endmodule
```

```
x1 x2 x3 x4 = 0000, f1 = 0, f2 = 0
x1 x2 x3 x4 = 0001, f1 = 1, f2 = 0
x1 x2 x3 x4 = 0010, f1 = 1, f2 = 0
x1 x2 x3 x4 = 0011, f1 = 0, f2 = 0
x1 x2 x3 x4 = 0100, f1 = 1, f2 = 1
x1 x2 x3 x4 = 0101, f1 = 0, f2 = 1
x1 x2 x3 x4 = 0110, f1 = 0, f2 = 0
x1 x2 x3 x4 = 0111, f1 = 0, f2 = 1
x1 x2 x3 x4 = 1000, f1 = 1, f2 = 0
x1 x2 x3 x4 = 1001, f1 = 0, f2 = 1
x1 x2 x3 x4 = 1010, f1 = 0, f2 = 0
x1 x2 x3 x4 = 1011, f1 = 0, f2 = 0
x1 x2 x3 x4 = 1100, f1 = 0, f2 = 0
x1 x2 x3 x4 = 1101, f1 = 0, f2 = 0
x1 x2 x3 x4 = 1110, f1 = 0, f2 = 0
x1 x2 x3 x4 = 1111, f1 = 0, f2 = 0
```

4.41   Design a structural module that will generate a high output z_1 if a 4-bit binary input $x[3:0]$ has a value less than or equal to five or greater than nine. Generate a Karnaugh map and obtain the equation for z_1 in a sum-of-products form. Instantiate dataflow modules into the structural module. Obtain the design module, the test bench module for all combinations of the inputs, and the outputs.

$$z_1 = x_3'x_2' + x_2x_1' + x_3x_1$$

```verilog
//structural module for 5 >= N > 9
module log_eqn_sop12 (x, z1);

input [3:0] x;
output z1;

wire net1, net2, net3;   //define internal nets

and2_df inst1 (    //instantiate the logic gates for z1
   .x1(~x[3]),
   .x2(~x[2]),
   .z1(net1)
   );

and2_df inst2 (
   .x1(x[2]),
   .x2(~x[1]),
   .z1(net2)
   );

and2_df inst3 (
   .x1(x[3]),
   .x2(x[1]),
   .z1(net3)
   );

or3_df inst4 (
   .x1(net1),
   .x2(net2),
   .x3(net3),
   .z1(z1)
   );

endmodule
```

```
//test bench for 5 >= N > 9
module log_eqn_sop12_tb;
reg [3:0] x;
wire z1;

//apply input vectors and display variables
initial
begin: apply_stimulus
   reg [4:0] invect;
   for (invect=0; invect<16; invect=invect+1)
      begin
         x = invect [4:0];
         #10 $display ("x=%b, z1=%b", x, z1);
      end
end

//instantiate the module into the test bench
log_eqn_sop12 inst1 (
   .x(x),
   .z1(z1)
   );

endmodule
```

```
x=0000, z1=1
x=0001, z1=1
x=0010, z1=1
x=0011, z1=1
x=0100, z1=1
x=0101, z1=1
x=0110, z1=0
x=0111, z1=0
x=1000, z1=0
x=1001, z1=0
x=1010, z1=1
x=1011, z1=1
x=1100, z1=1
x=1101, z1=1
x=1110, z1=1
x=1111, z1=1
```

Chapter 5   Computer Arithmetic

5.1   For $n = 8$, let A and B be two fixed-point binary numbers in 2s complement representation, where $A = 01101101$ and $B = 11001111$. Determine $A + B' + 1 + B$, where B' is the 1s complement.

$B' + 1 = $ 2s complement of $B = -B$. Therefore, $A + B' + 1 + B = A$

5.7   Obtain the sum of the following radix 5 numbers:

$$
\begin{array}{r}
1\ 2\ 3\ 4 \\
4\ 3\ 2\ 1 \\
3\ 3\ 3\ 3 \\
\hline
1 \leftarrow 4\ 4\ 4\ 3
\end{array}
$$

5.11   Add the following numbers, which are shown in radix complementation. Obtain the sum in radix 4: $0201_{10} + 321_4$.

$0201_{10} = 03021_4$ using the divide-by-4-repeatedly method.

$$
\begin{array}{r}
3\ \ 0\ \ 2\ \ 1_4 \\
+)\ \ 3\ \ 3\ \ 3\ \ 2\ \ 1_4 \\
\hline
0\ \ 3\ \ 0\ \ 0\ \ 2_4
\end{array}
$$

5.14   Use the paper-and-pencil method to multiply the following numbers, which are in 2s complement representation:

111111 (−1)
001011 (+11)

$$
\begin{array}{r}
1\ 1\ 1\ 1\ 1\ 1 \quad (-1) \\
\times)\ \ 0\ 0\ 1\ 0\ 1\ 1 \quad (+11) \\
\hline
1\ 1\ 1\ 1\ 1\ 1\ 1\ 1\ 1\ 1\ 1\ 1 \\
1\ 1\ 1\ 1\ 1\ 1\ 1\ 1\ 1\ 1\ 1 \\
0\ 0\ 0\ 0\ 0\ 0\ 1\ 1\ 1\ 0 \\
1\ 1\ 1\ 1\ 1\ 1\ 1\ 1\ 1 \\
0\ 0\ 0\ 0\ 0\ 0\ 0\ 0 \\
0\ 0\ 0\ 0\ 0\ 0\ 0 \\
\hline
1\ 1\ 1\ 1\ 1\ 1\ 1\ 1\ 0\ 1\ 0\ 1 \quad (-11)
\end{array}
$$

5.15   Use the sequential add-shift method to multiply the following operands, which are in 2s complement representation:

$$\text{Multiplicand } A = \quad 1111$$
$$\text{Multiplier } B = \quad 0111$$

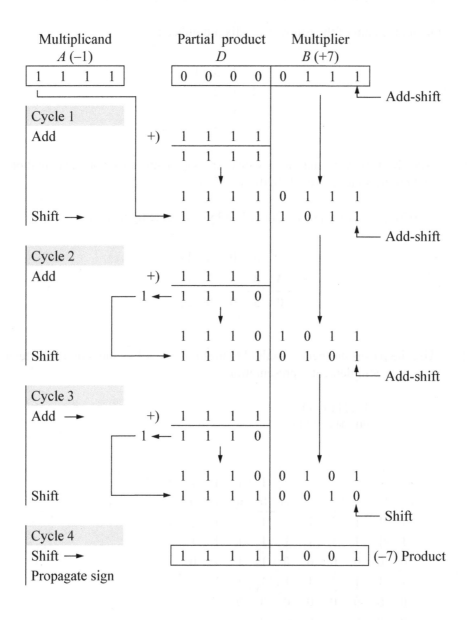

5.19   Use the Booth algorithm to multiply the following operands, which are in 2s complement representation:

Multiplicand $A =$ 110001
Multiplier $B =$ 100110

							Multiplicand		1	1	0	0	0	1		−15
Multiplier						×)	1	0	0	1	1	0	0	−26
															+390

Booth algorithm															
Multiplicand							1	1	0	0	0	1		−15	
Recoded multiplier					×)	−1	0	+1	0	−1	0			
0	0	0	0	0	0	0	0	0	0	0	0			
0	0	0	0	0	0	0	1	1	1	1				
1	1	1	1	1	0	0	0	1						
0	0	0	1	1	1	1								
0	0	1	1	0	0	0	0	1	1	0		+390		

5.22   Use bit-pair recoding to multiply the following operands, which are in 2s complement representation:

Multiplicand $A =$ 0100110
Multiplier $B =$ 1001101

				0	1 0	0 1	1 0	+38
		×)		1	0 0	1 1	0 1	−51
					↓			−1938
			0	1 0	0 1	1 0	
		1	0 0	1 1	0 1 0	
		−1	+1	−1	+1		

0 0	0 0	0 0	0 0	1 0	0 1	1 0
1 1	1 1	1 1	0 1	1 0	1 0	
0 0	0 0	1 0	0 1	1 0		
1 1	0 1	1 0	1 0			
1 1	0	0	0 1	1 0	1	0	−1938

5.28   Perform the indicated decimal operation on the operands shown below.

(+20) + (−32)

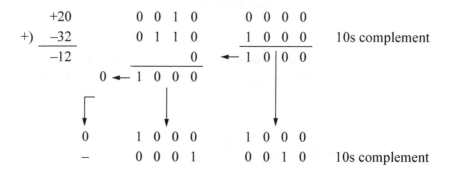

	+20	0 0 1 0	0 0 0 0	
+)	−32	0 1 1 0	1 0 0 0	10s complement
	−12	0	← 1 0│0 0	
		0 ← 1 0 0 0		

| | 0 | 1 0 0 0 | 1 0 0 0 | |
| | − | 0 0 0 1 | 0 0 1 0 | 10s complement |

5.32   Perform the operation listed below for normalized floating-point numbers using 8-bit fractions.

$$
\begin{array}{rr}
+ & 3\ 3 \\
+) \ - & 1\ 0 \\
\hline
+ & 2\ 3
\end{array}
$$

	0•1 0 0 0 0 1 0 0	× 2⁶	
−)	1•1 0 1 0 0 0 0 0	× 2⁴	
	0•1 0 0 0 0 1 0 0	× 2⁶	
−)	1•0 0 1 0 1 0 0 0	× 2⁶	Align fractions
	0•1 0 0 0 0 1 0 0	× 2⁶	
+)	1•1 1 0 1 1 0 0 0	× 2⁶	2s complement
1 ←	•0 1 0 1 1 1 0 0	× 2⁶	
	0•0 1 0 1 1 1 0 0	× 2⁶	
	0•1 0 1 1 1 0 0 0	× 2⁵	Normalize (SL1)
			Sum = +23

Chapter 6   Computer Arithmetic Design Using Verilog HDL

6.5   Use behavioral modeling to design a 5-function ALU for the following five functions: add, subtract, multiply, divide, and modulus. The operands are 4-bit vectors *A[3:0]* and *B[3:0]*. Obtain the behavioral module, the test bench module which applies one input vector for each operation, the outputs, and the waveforms.

```verilog
//behavioral arithmetic operations
module arith_ops1 (a, b, opcode, rslt);

input [3:0] a, b;
input [2:0] opcode;
output [7:0] rslt;

reg [7:0] rslt;

parameter    addop = 3'b000,
             subop = 3'b001,
             mulop = 3'b010,
             divop = 3'b011,
             modop = 3'b100;

always @ (a or b or opcode)
begin
   case (opcode)
      addop: rslt = a + b;
      subop: rslt = a - b;
      mulop: rslt = a * b;
      divop: rslt = a / b;
      modop: rslt = a % b;
      default: rslt = 8'bxxxxxxxx;
   endcase
end

endmodule
```

```verilog
//arithmetic operations test bench
module arith_ops1_tb;

reg [3:0] a, b;
reg [2:0] opcode;
wire [7:0] rslt ;

initial
$monitor ("a = %b, b = %b, opcode = %b, rslt = %b",
            a , b, opcode, rslt);

initial
begin
   #0 a = 4'b0011;
      b = 4'b0111;
      opcode = 3'b000;

   #5 a = 4'b1111;
      b = 4'b1111;
      opcode = 3'b001;

   #5 a = 4'b1110;
      b = 4'b1110;
      opcode = 3'b010;

   #5 a = 4'b1000;
      b = 4'b0010;
      opcode = 3'b011;

   #5 a = 4'b0111;
      b = 4'b0011;
      opcode = 3'b100;

   #5 $stop;
end

//instantiate the module into the test bench
arith_ops1 inst1 (
   .a(a),
   .b(b),
   .opcode(opcode),
   .rslt(rslt)
   );

endmodule
```

a = 0011,	b = 0111,	opcode = 000,	rslt = 00001010
a = 1111,	b = 1111,	opcode = 001,	rslt = 00000000
a = 1110,	b = 1110,	opcode = 010,	rslt = 11000100
a = 1000,	b = 0010,	opcode = 011,	rslt = 00000100
a = 0111,	b = 0011,	opcode = 100,	rslt = 00000001

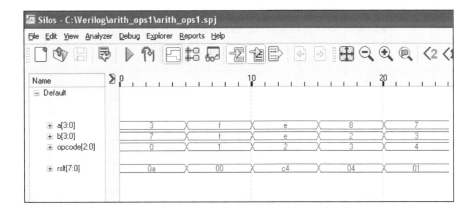

6.10   Shift registers are used in various arithmetic applications; for example, the sequential add-shift multiply algorithm and the shift-subtract division algorithm. Design a 4-bit shift register that operates as a parallel-in, parallel-out register and shifts the contents right 1 bit position. There is a scalar input called *fctn* that allows the register to load data (*fctn* = 1) and shift data (*fctn* = 0). Use the **parameter** and **case** keywords to define the load and shift right operations.

The concept of *clock* will be introduced in this problem and will be covered in more detail in the chapter on sequential logic. Clock pulses are synchronization signals provided by a clock, usually an astable multivibrator, which generates a periodic series of pulses. A clock signal can be generated by the following code segment in which *clk* changes state every 10 time units:

```
//define clock
    initial
    begin
        clk = 1'b0;
        forever
            #10  clk = ~clk;
    end
```

The keywords **posedge** and **negedge** refer to the rising and falling edge of the clock, respectively. Obtain the design module, the test bench module for four different operands, the outputs, and the waveforms.

```verilog
//mixed-design shift register
module shift_reg2 (clk, rst_n, data_in, fctn, q);

input clk, rst_n;
input [3:0] data_in;
input fctn;
output [3:0] q;

reg [3:0] q;
wire [3:0] shift_r_data;

//define internal register
reg [3:0] d;

parameter    ld  = 1'b1;
parameter    shr = 1'b0;

assign shift_r_data = q >> 1;

always @ (shift_r_data or fctn or data_in or q)
begin
   case (fctn)
      ld:  d = data_in;
      shr: d = shift_r_data;
      default: d = 4'b0000;
   endcase
end

always @ (posedge clk or negedge rst_n)
begin
   if (~rst_n)
      q <= 4'b0000;
   else
      q <= d;
end

endmodule
```

```verilog
//test bench for shift register
module shift_reg2_tb;

reg clk, rst_n;
reg [3:0] data_in;
reg fctn;
wire [3:0] q;

//continued on next page
```

```
//display variables
initial
$monitor ("data_in=%b, fctn=%b, q=%b",
            data_in, fctn, q);

//define clock
initial
begin
   clk =1'b0;
   forever
      #10 clk = ~clk;
end

//define operand and fctn
initial
begin
   #0    rst_n = 1'b0;
         data_in = 4'b0000;
         fctn = 1'b1;          //load

   #10    rst_n = 1'b1;
//-----------------------------------------------
//load data_in and shift right one bit position
   data_in = 4'b1111;
   fctn = 1'b1;
   @ (posedge clk)           //load q

   fctn = 1'b0;              //shift right
   @ (posedge clk)
//-----------------------------------------------
   data_in = 4'b1010;
   fctn = 1'b1;
   @ (posedge clk)           //load q

   fctn = 1'b0;              //shift right
   @ (posedge clk)
//-----------------------------------------------
   data_in = 4'b1100;
   fctn = 1'b1;
   @ (posedge clk)           //load q

   fctn = 1'b0;
   @ (posedge clk)           //shift right
//-----------------------------------------------

//continued on next page
```

```
//------------------------------------------------
   data_in = 4'b0110;
   fctn = 1'b1;
   @ (posedge clk)            //load q

   fctn = 1'b0;
   @ (posedge clk)            //shift right
//------------------------------------------------

   #20     $stop;
end

//instantiate the module into the test bench
shift_reg2 inst1 (
   .clk(clk),
   .rst_n(rst_n),
   .data_in(data_in),
   .fctn(fctn),
   .q(q)
   );
endmodule
```

```
data_in=0000, fctn=1, q=0000
data_in=1111, fctn=0, q=1111
data_in=1010, fctn=1, q=0111
data_in=1010, fctn=0, q=1010
data_in=1100, fctn=1, q=0101
data_in=1100, fctn=0, q=1100
data_in=0110, fctn=1, q=0110
data_in=0110, fctn=0, q=0110
data_in=0110, fctn=0, q=0011
```

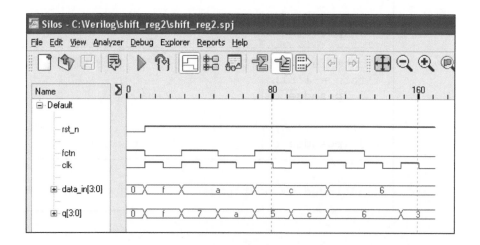

6.13   Use carry-save adders to add 7 bits from the same column. Check the design
for correct operation.

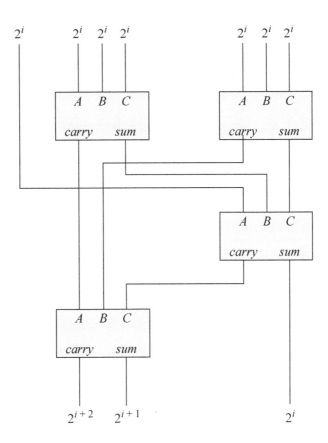

Chapter 7   Sequential Logic

7.2   Obtain the state diagram for the Moore synchronous sequential machine shown below. The machine is reset to $y_1y_2 = 00$.

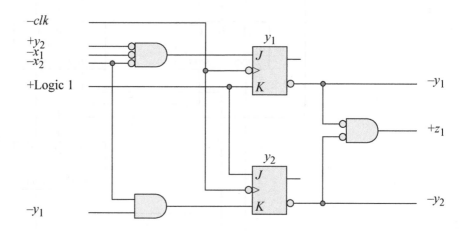

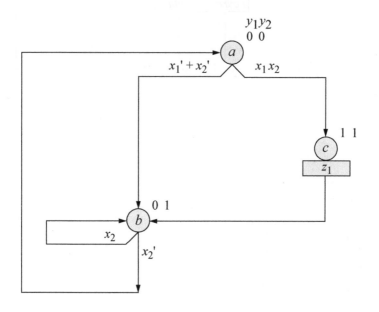

7.6   Find all equivalent states for the next-state table shown below.

Present State	Input x_1	Next State	Output z_1
a	0	b	0
	1	g	0
b	0	g	0
	1	a	1
c	0	h	0
	1	g	1
d	0	c	0
	1	a	1
e	0	h	0
	1	c	0
f	0	c	0
	1	e	1
g	0	d	0
	1	g	1
h	0	c	0
	1	a	1

$a \equiv e$ if $(b \equiv h) \cdot (g \equiv c)$
   $b \equiv h$ is true
   $g \equiv c$ is true

$f \equiv d$ if $a \equiv e$ (True)

Therefore, $a \equiv e$
       $b \equiv h \equiv d \equiv f$
       $c \equiv g$

7.12   Obtain the input equations for flip-flops y_1 and y_4 only, for a BCD counter that counts in the sequence shown below. The equations are to be in minimum form. Use *JK* flip-flops. There is no self-starting state.

y_1	y_2	y_3	y_4
0	0	0	0
0	0	0	1
0	0	1	0
0	0	1	1
0	1	0	0
0	1	0	1
0	1	1	0
0	1	1	1
1	0	0	0
1	0	0	1
0	0	0	0

Transition	J	K
$0 \rightarrow 0$	0	–
$0 \rightarrow 1$	1	–
$1 \rightarrow 0$	–	1
$1 \rightarrow 1$	–	0

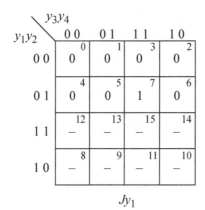

$$Jy_1 = y_2 y_3 y_4 \qquad Ky_1 = y_4$$
$$Jy_1 = 1 \qquad\qquad Ky_1 = 1$$

7.17   Given the equation for z_1 shown below, identify all static-1 hazards and determine the hazard covers.

$$z_1 = x_1'x_3'x_5' + x_2x_3'x_4x_5' + x_2x_3x_4x_5 + x_1x_3'x_4x_5$$

$x_5 = 0$

x_1x_2 \ x_3x_4	0 0	0 1	1 1	1 0
0 0	1 [0]	1 [2]	0 [6]	0 [4]
0 1	1 [8]	1 [10]	0 [14]	0 [12]
1 1	0 [24]	1 [26]	0 [30]	0 [28]
1 0	0 [16]	0 [18]	0 [22]	0 [20]

$x_5 = 1$

x_1x_2 \ x_3x_4	0 0	0 1	1 1	1 0
0 0	0 [1]	0 [3]	0 [7]	0 [5]
0 1	0 [9]	0 [11]	1 [15]	0 [13]
1 1	0 [25]	1 [27]	1 [31]	0 [29]
1 0	0 [17]	1 [19]	0 [23]	0 [21]

z_1

Sum-of-products equation with possible hazards:
$$z_1 = x_1'x_3'x_5' + x_2x_3'x_4x_5' + x_2x_3x_4x_5 + x_1x_3'x_4x_5$$

Sum-of-products equation with hazard covers:
$$z_1 = x_1'x_3'x_5' + x_2x_3'x_4x_5' + x_2x_3x_4x_5 + x_1x_3'x_4x_5$$
$$+ x_1x_2x_4x_5 + x_1x_2x_3'x_4$$

7.25   Identify equivalent states in the primitive flow table shown below. Then eliminate redundant states and obtain a reduced primitive flow table.

x_1x_2	00	01	11	10	z_1	z_2
	Ⓐ	b	–	f	0	0
	a	Ⓑ	e	–	1	0
	Ⓒ	d	–	f	0	0
	a	Ⓓ	e	–	1	0
	–	g	Ⓔ	f	0	1
	c	–	e	Ⓕ	1	1
	c	Ⓖ	e	–	1	1

Equivalent states are: $\textcircled{b} \equiv \textcircled{d}$

$\textcircled{a} \equiv \textcircled{c}$ if $\textcircled{b} \equiv \textcircled{d}$ (true), therefore,

$\textcircled{a} \equiv \textcircled{c}$

x_1x_2	00	01	11	10	z_1	z_2
$\textcircled{a}$	b	$-$	f	0	0
a	$\textcircled{b}$	e	$-$	1	0
$-$	g	$\textcircled{e}$	f	0	1
a	$-$	e	$\textcircled{f}$	1	1
a	$\textcircled{g}$	e	$-$	1	1

7.31   Synthesize an asynchronous sequential machine which has two inputs x_1 and x_2 and one output z_1. Output z_1 is asserted for the duration of x_2 if and only if x_1 is already asserted. Assume that the initial state of the machine is $x_1x_2z_1 = 000$. A representative timing diagram is shown below. Obtain the excitation equations in both a sum-of-products and a product-of-sums form with no static-1 or static-0 hazards.

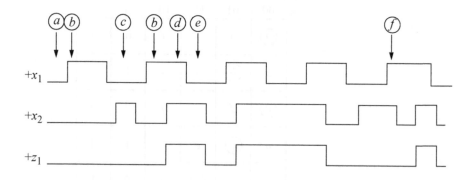

Primitive flow table

x_1x_2	00	01	11	10	z_1
	ⓐ	c	–	b	0
	a	–	d	ⓑ	0
	a	ⓒ	f	–	0
	–	e	ⓓ	b	1
	a	ⓔ	d	–	1
	–	c	ⓕ	b	0

Merger diagram

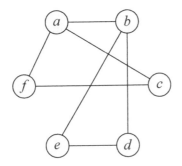

Merged flow table

			x_1x_2	00	01	11	10
ⓐ	ⓒ	ⓕ		ⓐ	ⓒ	ⓕ	b
ⓑ	ⓓ	ⓔ		a	ⓔ	ⓓ	ⓑ

Excitation map

y_{1f} \ x_1x_2	00	01	11	10
0	⓪ ^{a}	⓪ ^{c}	⓪ ^{f}	1
1	0	① ^{e}	① ^{d}	① ^{b}

$$Y_{1e}$$

$$Y_{1e} = y_{1f}x_2 + x_1x_2' + y_{1f}x_1$$

$$Y_{1e} = (y_{1f} + x_2')(x_1 + x_2)(y_{1f} + x_1)$$

Output map

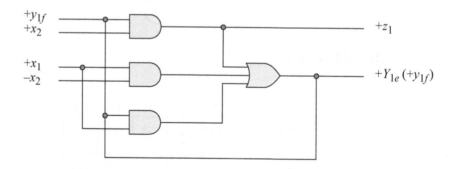

y_{1f} \ x_1x_2	00	01	11	10
0	a 0	c 0	f 0	0
1	0	e 1	d 1	b 0

Y_{1e}

The transition $\textcircled{a} \rightarrow \textcircled{b}$ must maintain output z_1 at a 0 level.

The transition $\textcircled{b} \rightarrow \textcircled{a}$ must maintain output z_1 at a 0 level.

Logic diagram

$+y_{1f}$
$+x_2$

$+x_1$
$-x_2$

$+z_1$

$+Y_{1e}\,(+y_{1f})$

7.36   Synthesize a Moore pulse-mode sequential machine which has three inputs $x_1, x_2,$ and x_3 and one output z_1. Output z_1 will be asserted coincident with the assertion of the x_3 pulse if and only if the x_3 pulse was preceded by an x_1 pulse followed by an x_2 pulse. That is, the input vector must be $x_1x_2x_3 = 100, 000, 010, 000, 001$ to assert z_1. Output z_1 will be deasserted at the next active x_1 pulse or x_2 pulse. Use NOR SR latches and positive-edge-triggered D flip-flops as the storage elements.

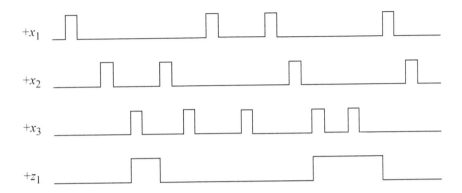

State diagram

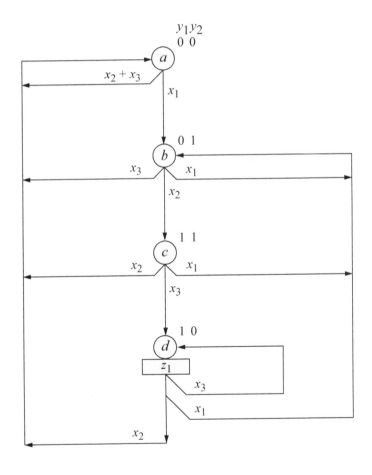

Input maps

Inputs / Latches	x_1	x_2	x_3
Ly_1	$y_1\backslash y_2$: $0,0\!=r$; $0,1\!=r$; $1,0\!=R$; $1,1\!=R$	$y_1\backslash y_2$: $0,0\!=r$; $0,1\!=S$; $1,0\!=R$; $1,1\!=R$	$y_1\backslash y_2$: $0,0\!=r$; $0,1\!=r$; $1,0\!=s$; $1,1\!=s$
Ly_2	$y_1\backslash y_2$: $0,0\!=S$; $0,1\!=s$; $1,0\!=S$; $1,1\!=s$	$y_1\backslash y_2$: $0,0\!=r$; $0,1\!=s$; $1,0\!=r$; $1,1\!=R$	$y_1\backslash y_2$: $0,0\!=r$; $0,1\!=R$; $1,0\!=r$; $1,1\!=R$

$$SLy_1 = y_1'y_2x_2 \qquad RLy_1 = x_1 + y_1x_2$$

$$SLy_2 = x_1 \qquad RLy_2 = y_1x_2 + x_3$$

Output map

$y_1\backslash y_2$	0	1
0	0	0
1	1	0

z_1

$$z_1 = y_1y_2'$$

Logic diagram

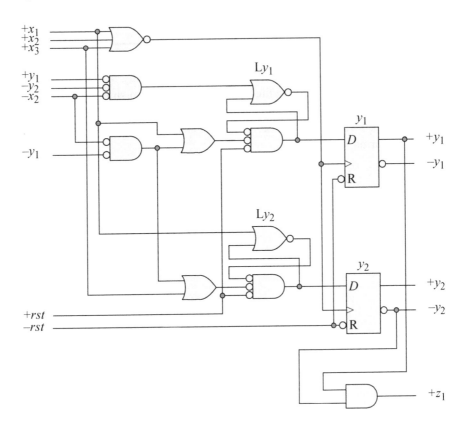

Chapter 8   Sequential Logic Design Using Verilog HDL

8.2   Use structural modeling to design the Mealy synchronous sequential machine that is represented by the state diagram shown below. Use a positive-edge-triggered JK flip-flop and additional AND gates and OR gates. Obtain the structural module, the test bench module, the outputs, and the waveforms.

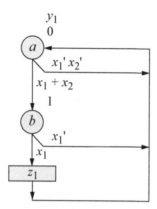

Table 8.1 Next-State Table for the Mealy Machine of Problem 8.2

State Name	Present State y_1	Inputs $x_1 x_2$	Flip-Flop Inputs $Jy_1 Ky_1$	Next State y_1	Output z_1
a	0	0 0	0 1	0	0
	0	0 1	1 1	1	0
	0	1 0	1 1	1	0
	0	1 1	1 1	1	0
b	1	0 0	0 1	0	0
	1	0 1	0 1	0	0
	1	1 0	0 1	0	1
	1	1 1	0 1	0	1

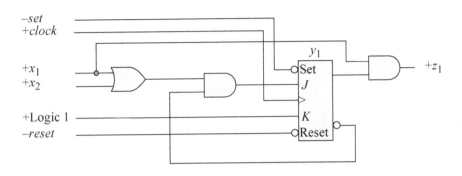

```
//structural mealy synchronous sequential machine
module mealy_ssm6 (clk, set_n, rst_n, x1, x2, y1, z1);

input set_n, rst_n, clk, x1, x2;
output y1, z1;

wire set_n, rst_n, clk, x1, x2;
wire y1, z1;

//define internal wires
wire net1, net2;

//instantiate the logic for flip-flop y1
or2_df inst1 (
    .x1(x1),
    .x2(x2),
    .z1(net1)
    );

and2_df inst2 (
    .x1(net1),
    .x2(~y1),
    .z1(net2)
    );

jkff inst3 (
    .clk(clk),
    .j(net2),
    .k(1'b1),
    .set_n(set_n),
    .rst_n(rst_n),
    .q(y1)
    );

//continued on next page
```

```
//instantiate the logic for output z1
and2_df inst4 (
   .x1(x1),
   .x2(y1),
   .z1(z1)
   );

endmodule
```

```
//test bench for mealy synchronous sequential machine
module mealy_ssm6_tb;

reg set_n, rst_n, clk, x1, x2;
wire y1, z1;

//display inputs and output
initial
$monitor ("rst=%b, clk=%b, state=%b, x1x2=%b z1=%b",
          rst_n, clk, y1, {x1, x2}, z1);

//define clock
initial
begin
   clk = 1'b0;
   forever
      #10 clk = ~clk;
end

//define input sequence
initial
begin
   #0 set_n = 1'b1;
      rst_n = 1'b0;
      x1 = 1'b0;
   #5 rst_n = 1'b1;

   x1 = 1'b1;   x2 = $random;
      @ (posedge clk)              //go to state_b (1)

   x1 = 1'b1;   x2 = $random;   //assert z1
      @ (posedge clk)              //go to state_a (0)

   x1 =1'b1;    x2 = $random;
      @ (posedge clk)              //go to state_b (1)

//continued on next page
```

```
   x1 = 1'b0;   x2 = $random;
      @ (posedge clk)            //go to state_a (0)


   x1 = 1'b0;   x2 = 1'b1;
      @ (posedge clk)            //go to state_b (1)


   x1 = 1'b0;   x2 = $random;
      @ (posedge clk)            //go to state_a (0)


   x1 = 1'b0;   x2 = 1'b0;
      @ (posedge clk)            //remain in state_a
   #10    $stop;
end

//instantiate the module into the test bench
mealy_ssm6 inst1 (
   .set_n(set_n),
   .rst_n(rst_n),
   .clk(clk),
   .x1(x1),
   .x2(x2),
   .y1(y1),
   .z1(z1)
   );

endmodule
```

```
rst=0, clk=0, state=0, x1x2=0x z1=0
rst=1, clk=0, state=0, x1x2=10 z1=0
rst=1, clk=1, state=1, x1x2=11 z1=1
rst=1, clk=0, state=1, x1x2=11 z1=1
rst=1, clk=1, state=0, x1x2=11 z1=0
rst=1, clk=0, state=0, x1x2=11 z1=0
rst=1, clk=1, state=1, x1x2=01 z1=0
rst=1, clk=0, state=1, x1x2=01 z1=0
rst=1, clk=1, state=0, x1x2=01 z1=0
rst=1, clk=0, state=0, x1x2=01 z1=0
rst=1, clk=1, state=1, x1x2=01 z1=0
rst=1, clk=0, state=1, x1x2=01 z1=0
rst=1, clk=1, state=0, x1x2=00 z1=0
rst=1, clk=0, state=0, x1x2=00 z1=0
rst=1, clk=1, state=0, x1x2=00 z1=0
```

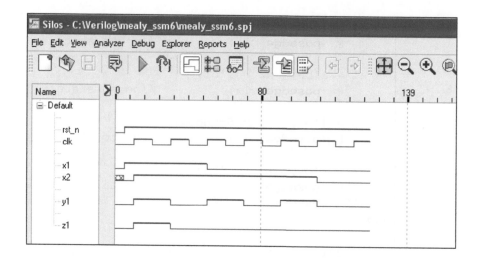

8.6   Given the state diagram shown below for a synchronous sequential machine, obtain the next-state table, the input maps for *JK* flip-flops, the input equations, and the logic diagram using AND gates and OR gates and positive-edge-triggered *JK* flip-flops.

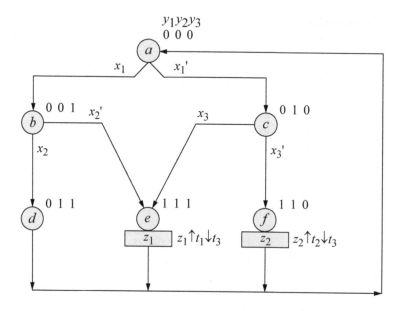

Present State	Inputs	Next State	Flip-Flop Inputs						Outputs
$y_1 y_2 y_3$	$x_1 x_2 x_3$	$y_1 y_2 y_3$	Jy_1	Ky_1	Jy_2	Ky_2	Jy_3	Ky_3	$z_1 z_2$
0 0 0	0 – –	0 1 0	0	–	1	–	0	–	0 0
0 0 0	1 – –	0 0 1	0	–	0	–	1	–	0 0
0 0 1	– 0 –	1 1 1	1	–	1	–	–	0	0 0
0 0 1	– 1 –	0 1 1	0	–	1	–	–	0	0 0
0 1 0	– – 0	1 1 0	1	–	–	0	0	–	0 0
0 1 0	– – 1	1 1 1	1	–	–	0	1	–	0 0
0 1 1	– – –	0 0 0	0	–	–	1	–	1	0 0
1 0 0	– – –	– – –	–	–	–	–	–	–	– –
1 0 1	– – –	– – –	–	–	–	–	–	–	– –
1 1 0	– – –	0 0 0	–	1	–	1	0	–	0 1
1 1 1	– – –	0 0 0	–	1	–	1	–	1	1 0

y_1 \ $y_2 y_3$	0 0	0 1	1 1	10
0	[0] 0	[1] x_2'	[3] 0	[2] 1
1	[4] –	[5] –	[7] –	[6] –

Jy_1

y_1 \ $y_2 y_3$	0 0	0 1	1 1	10
0	[0] –	[1] –	[3] –	[2] –
1	[4] –	[5] –	[7] 1	[6] 1

Ky_1

y_1 \ $y_2 y_3$	0 0	0 1	1 1	10
0	[0] x_1'	[1] 1	[3] –	[2] –
1	[4] –	[5] –	[7] –	[6] –

Jy_2

y_1 \ $y_2 y_3$	0 0	0 1	1 1	10
0	[0] –	[1] –	[3] 1	[2] 0
1	[4] –	[5] –	[7] 1	[6] 1

Ky_2

y_2y_3

y_1	0 0	0 1	1 1	1 0
0	x_1 ⁰	– ¹	– ³	x_3 ²
1	– ⁴	– ⁵	– ⁷	0 ⁶

Jy_3

y_2y_3

y_1	0 0	0 1	1 1	1 0
0	– ⁰	0 ¹	1 ³	– ²
1	– ⁴	– ⁵	1 ⁷	– ⁶

Ky_3

$$Jy_1 = y_2'y_3x_2' + y_2y_3' \qquad Ky_1 = 1$$
$$Jy_2 = x_1' + y_3 \qquad Ky_2 = y_3 + y_1$$
$$Jy_3 = y_2'x_1 + y_1'y_2x_3 \qquad Ky_3 = y_2$$

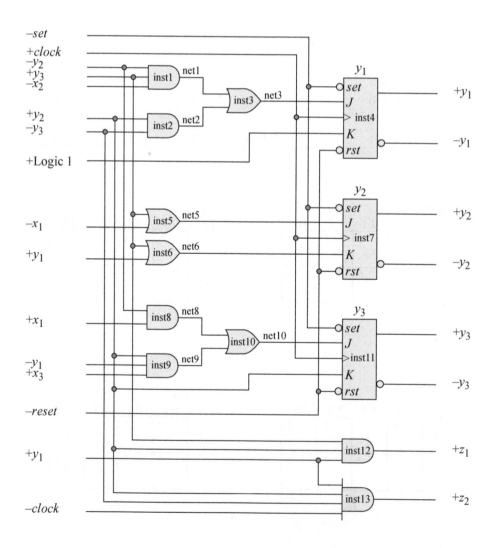

Since there may be a glitch on output z_2 for a transition from state c ($y_1 y_2 y_3$ = 010) to state e ($y_1 y_2 y_3$ = 111), the assertion/deassertion is $z_2 \uparrow t_2 \downarrow t_3$. Obtain the structural design module by instantiating dataflow modules for the logic primitives and a behavioral module for the positive-edge-triggered JK flip-flops. Obtain the test bench module, the outputs, and the waveforms.

```
//structural moore synchronous sequential machine
module moore_ssm15 (set_n, rst_n, clk, x1, x2, x3,
                    y, z1, z2);

input set_n, rst_n, clk, x1, x2, x3;
output [1:3] y;
output z1, z2;

//define internal wires
wire net1, net2, net3, net5, net7, net8, net9, net10;

//instantiate the logic for flip-flop y[1]
and3_df inst1 (
   .x1(~y[2]),
   .x2(y[3]),
   .x3(~x2),
   .z1(net1)
   );

and2_df inst2 (
   .x1(y[2]),
   .x2(~y[3]),
   .z1(net2)
   );

or2_df inst3 (
   .x1(net1),
   .x2(net2),
   .z1(net3)
   );

jkff inst4 (
   .set_n(1'b1),
   .rst_n(rst_n),
   .clk(clk),
   .j(net3),
   .k(1'b1),
   .q(y[1])
   );

//continued on next page
```

```
//instantiate the logic for flip-flop y[2]
or2_df inst5 (
   .x1(y[3]),
   .x2(~x1),
   .z1(net5)
   );

or2_df inst6 (
   .x1(y[3]),
   .x2(y[1]),
   .z1(net6)
   );

jkff inst7 (
   .set_n(1'b1),
   .rst_n(rst_n),
   .clk(clk),
   .j(net5),
   .k(net6),
   .q(y[2])
   );

//instantiate the logic for flip-flop y[3]
and2_df inst8 (
   .x1(~y[2]),
   .x2(x1),
   .z1(net8)
   );

and3_df inst9 (
   .x1(y[2]),
   .x2(~y[1]),
   .x3(x3),
   .z1(net9)
   );

or2_df inst10 (
   .x1(net8),
   .x2(net9),
   .z1(net10)
   );

jkff inst11 (
   .set_n(1'b1),
   .rst_n(rst_n),
   .clk(clk),
   .j(net10),
   .k(y[2]),
   .q(y[3])
   );
//continued on next page
```

```verilog
//instantiate the logic for outputs z1 and z2
and3_df inst12 (
   .x1(y[1]),
   .x2(y[2]),
   .x3(y[3]),
   .z1(z1)
   );

and4_df inst13 (
   .x1(y[1]),
   .x2(y[2]),
   .x3(~y[3]),
   .x4(~clk),
   .z1(z2)
   );

endmodule
```

```verilog
//test bench for the moore synchronous machine
module moore_ssm15_tb;

reg set_n, rst_n, clk, x1, x2, x3;
wire [1:3] y;
wire z1, z2;

//display inputs and outputs
initial
$monitor ("clk=%b, x1x2x3=%b, y=%b, z1z2=%b",
          clk, {x1, x2, x3}, y, {z1, z2});

//define clock
initial
begin
   clk = 1'b0;
   forever
      #10   clk = ~clk;
end

//define input sequence
initial
begin
   #0    rst_n = 1'b0;  //reset to state_a (000)
         x1 = 1'b0;
         x2 = 1'b0;
         x3 = 1'b0;

   #5    rst_n = 1'b1;
//continued on next page
```

```
               x1 = 1'b0;   x2 = $random;   x3 = $random;
               @ (posedge clk)   //go to state_c (010)

               x1 = $random;   x2 = $random; x3 = 1'b1;
               @ (posedge clk)   //go to state_e (111); set z1

               x1 = $random;   x2 = $random;   x3 = $random;
               @ (posedge clk)   //go to state_a (000)

               x1 = 1'b1;   x2 = $random;   x3 = $random;
               @ (posedge clk)   //go to state_b (001)

               x1 = $random;   x2 = 1'b1;   x3 = $random;
               @ (posedge clk)   //go to state_d (011)

               x1 = $random;   x2 = $random;   x3 = $random;
               @ (posedge clk)   //go to state_a (000)

               x1 = 1'b0;   x2 = $random;   x3 = $random;
               @ (posedge clk)   //go to state_c (010)

               x1 = $random;   x2 = $random;   x3 = 1'b0;
               @ (posedge clk)   //go to state_f (110); set z2

               x1 = $random;   x2 = $random;   x3 = $random;
               @ (posedge clk)   //go to state_a (000)

               x1 = 1'b1;   x2 = $random;   x3 = $random;
               @ (posedge clk)   //go to state_b (001)

               x1 = $random;   x2 = 1'b0;   x3 = $random;
               @ (posedge clk)   //go to state_e (111); set z1

               x1 = $random;   x2 = $random;   x3 = $random;
               @ (posedge clk)   //go to state_a (000)

               x1 = 1'b1;   x2 = $random;   x3 = $random;
               @ (posedge clk)   //go to state_b (001)

        #10    $stop;

end

//continued on next page
```

```
//instantiate the module into the test bench
moore_ssm15 inst1 (
    .set_n(set_n),
    .rst_n(rst_n),
    .clk(clk),
    .x1(x1),
    .x2(x2),
    .x3(x3),
    .y(y),
    .z1(z1),
    .z2(z2)
    );

endmodule
```

```
clk=0,  x1x2x3=000,  y=000,  z1z2=00
clk=0,  x1x2x3=001,  y=000,  z1z2=00
clk=1,  x1x2x3=111,  y=010,  z1z2=00
clk=0,  x1x2x3=111,  y=010,  z1z2=00
clk=1,  x1x2x3=111,  y=111,  z1z2=10
clk=0,  x1x2x3=111,  y=111,  z1z2=10
clk=1,  x1x2x3=101,  y=000,  z1z2=00
clk=0,  x1x2x3=101,  y=000,  z1z2=00
clk=1,  x1x2x3=110,  y=001,  z1z2=00
clk=0,  x1x2x3=110,  y=001,  z1z2=00
clk=1,  x1x2x3=110,  y=011,  z1z2=00
clk=0,  x1x2x3=110,  y=011,  z1z2=00
clk=1,  x1x2x3=010,  y=000,  z1z2=00
clk=0,  x1x2x3=010,  y=000,  z1z2=00
clk=1,  x1x2x3=100,  y=010,  z1z2=00
clk=0,  x1x2x3=100,  y=010,  z1z2=00
clk=1,  x1x2x3=110,  y=110,  z1z2=00
clk=0,  x1x2x3=110,  y=110,  z1z2=01
clk=1,  x1x2x3=110,  y=000,  z1z2=00
clk=0,  x1x2x3=110,  y=000,  z1z2=00
clk=1,  x1x2x3=000,  y=001,  z1z2=00
clk=0,  x1x2x3=000,  y=001,  z1z2=00
clk=1,  x1x2x3=101,  y=111,  z1z2=10
clk=0,  x1x2x3=101,  y=111,  z1z2=10
clk=1,  x1x2x3=111,  y=000,  z1z2=00
clk=0,  x1x2x3=111,  y=000,  z1z2=00
clk=1,  x1x2x3=111,  y=001,  z1z2=00
```

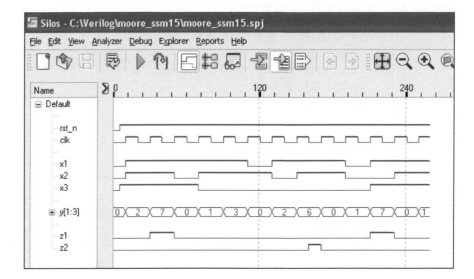

8.9   Determine the counting sequence of the counter shown below by implement-
        ing the machine in a structural module. The counter is reset initially to $y_1 y_2$
        = 00, where y_2 is the low-order flip-flop. Obtain the test bench, outputs, and
        waveforms.

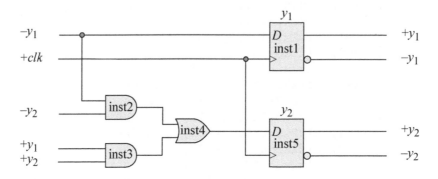

```
//structural nonsequential counter
module ctr_non_seq1 (rst_n, clk, y);

input rst_n, clk;
output [1:2] y;

wire [1:2] y;

//instantiate the D flip-flop for y[1]
d_ff_bh inst1 (
    .rst_n(rst_n),
    .clk(clk),
    .d(~y[1]),
    .q(y[1])
    );                //continued on next page
```

```
//instantiate the logic for flip-flop y[2]
and2_df inst2 (
   .x1(~y[1]),
   .x2(~y[2]),
   .z1(net2)
   );

and2_df inst3 (
   .x1(y[1]),
   .x2(y[2]),
   .z1(net3)
   );

or2_df inst4 (
   .x1(net2),
   .x2(net3),
   .z1(net4)
   );

//instantiate the D flip-flop for y[2]
d_ff_bh inst5 (
   .rst_n(rst_n),
   .clk(clk),
   .d(net4),
   .q(y[2])
   );

endmodule
```

```
//test bench for nonsequential counter
module ctr_non_seq1_tb;

reg rst_n, clk;
wire [1:2] y;

//display count
initial
$monitor ("count = %b", y);

//define reset
initial
begin
   #0 rst_n = 1'b0;
   #5 rst_n = 1'b1;
end

//continued on next page
```

```
//define clock
initial
begin
   clk = 1'b0;
   forever
      #10clk = ~clk;
end

//define simulation time
initial
begin
   #100 $finish;
end

//instantiate the module into the test bench
ctr_non_seq1 inst1 (
   .rst_n(rst_n),
   .clk(clk),
   .y(y)
   );

endmodule
```

```
count = 00
count = 11
count = 01
count = 10
count = 00
count = 11
count = 01
```

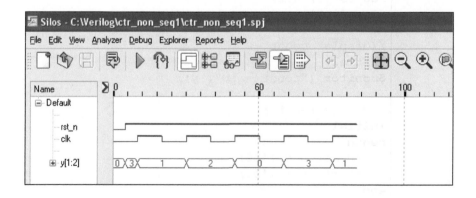

8.13   Design a Moore machine that has four parallel inputs x_1, x_2, x_3, and x_4 and two outputs z_1 and z_2. Output $z_1 \uparrow t_1 \downarrow t_3$; output $z_2 \uparrow t_2 \downarrow t_3$. The inputs constitute a 4-bit word $x[1:4]$. There is 1 bit space between words. The machine operates as follows:

    (a) Output $z_1 = 1$ if the 4-bit word contains an odd number of 1s.
    (b) Output $z_2 = 1$ if the 4-bit word contains an even number of 1s, including zero 1s.

Use behavioral modeling to implement the machine, then obtain the test bench, the outputs, and the waveforms.

    A random number is generated for $x[1:4]$ when the machine sequences from state b ($y_1 y_2 = 01$) or from state c ($y_1 y_2 = 10$) to state a ($y_1 y_2 = 00$).

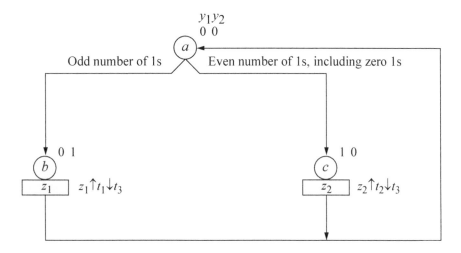

```
//behavioral moore synchronous sequential machine
module moore_ssm19 (rst_n, clk, x, y, z1, z2);

input rst_n, clk;
input [1:4] x;
output [1:2] y;
output z1, z2;

reg [1:2] y, next_state;
wire z1, z2;

//assign state codes
parameter    state_a = 2'b00,
             state_b = 2'b01,
             state_c = 2'b10;

//continued on next page
```

```verilog
//set next state
always @ (posedge clk)
begin
   if (~rst_n)
      y = state_a;
   else
      y = next_state;
end

//define outputs
assign z1 = (~y[1] & y[2]);
assign z2 = (y[1] & ~y[2] & ~clk);

//determine next state
always @ (y or x)
begin
   case (y)
         state_a:
            if (^x)                   //odd # of 1s
               next_state = state_b;//01; assert z1

            else                      //even # of 1s
               next_state = state_c;//10; assert z2

         state_b:
               next_state = state_a;//000

         state_c:
               next_state = state_a;//000

      default:
            next_state = state_a;
   endcase
end
endmodule
```

```verilog
//test bench for moore synchronous sequential machine
module moore_ssm19_tb;

reg rst_n, clk;
reg [1:4] x;
wire [1:2] y;
wire z1, z2;

//display variables
initial
$monitor ("x = %b, state = %b, z1z2 = %b",
            x, y, {z1, z2});  //continued on next page
```

```
//define clock
initial
begin
   clk = 1'b0;
   forever
       #10clk = ~clk;
end

//define input sequence
initial
begin
   #0 rst_n = 1'b0;
       x = 4'b0000;

   #5 rst_n = 1'b1;

       x = 4'b0001;
       @ (posedge clk)  //go to state_b (01); assert z1

       x = $random;
       @ (posedge clk)  //go to state_a (00)

       x = 4'b1010;
       @ (posedge clk)  //go to state_c (10); assert z2

       x = $random;
       @ (posedge clk)  //go to state_a (00)

       x = 4'b0111;
       @ (posedge clk)  //go to state_b (01); assert z1

       x = $random;
       @ (posedge clk)  //go to state_a (00)

       x = 4'b1111;
       @ (posedge clk)  //go to state_c (10); assert z2

       x = $random;
       @ (posedge clk)  //go to state_a (00)

   #40    $stop;
end

//continued on next page
```

```
//instantiate the module into the test bench
moore_ssm19 inst1 (
    .rst_n(rst_n),
    .clk(clk),
    .x(x),
    .y(y),
    .z1(z1),
    .z2(z2)
    );

endmodule
```

```
x = 0000, state = xx, z1z2 = xx
x = 0001, state = xx, z1z2 = xx
x = 0100, state = 00, z1z2 = 00
x = 1010, state = 01, z1z2 = 10
x = 0001, state = 00, z1z2 = 00
x = 0111, state = 01, z1z2 = 10
x = 1001, state = 00, z1z2 = 00
x = 1111, state = 10, z1z2 = 00
x = 1111, state = 10, z1z2 = 01
x = 0011, state = 00, z1z2 = 00
x = 0011, state = 10, z1z2 = 00
x = 0011, state = 10, z1z2 = 01
x = 0011, state = 00, z1z2 = 00
```

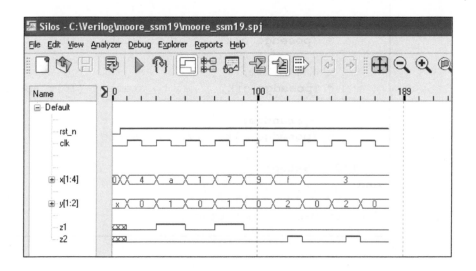

8.18   An asynchronous sequential machine has two inputs x_1 and x_2 and one output z_1. The machine operates according to the following specifications:

If $x_1 x_2 = 00$, then the state of z_1 is unchanged.
If $x_1 x_2 = 01$, then z_1 is deasserted.
If $x_1 x_2 = 10$, then z_1 is asserted.
If $x_1 x_2 = 11$, then z_1 changes state.

Design the machine using dataflow modeling. The inputs are available in both high and low assertion. Assume that the initial conditions are $x_1 x_2 z_1 =$ 000. The output must change as fast as possible. There must be no output glitches. A representative timing diagram is shown below. Obtain the dataflow design module, the test bench module, the outputs, and the waveforms.

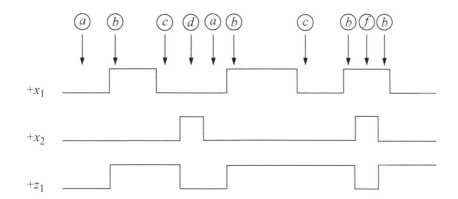

Primitive flow table

$x_1 x_2$ 00	01	11	10	z_1
ⓐ	d	–	b	0
c	–	f	ⓑ	1
ⓒ	d	–	b	1
a	ⓓ	e	–	0
–	d	ⓔ	b	1
–	d	ⓕ	b	0

There are no equivalent states.

Merger diagram

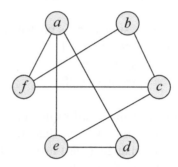

Merged flow table

			x_1x_2			
			00	01	11	10
ⓐ	ⓓ	ⓔ	ⓐ	ⓓ	ⓔ	b
ⓑ	ⓒ	ⓕ	ⓒ	d	ⓕ	ⓑ

Excitation map

y_{1f}	x_1x_2 00	01	11	10
0	0 ^{a}	0 ^{d}	0 ^{e}	1
1	1 ^{c}	0	1 ^{f}	1 ^{b}

Y_{1e}

$$Y_{1e} = x_1x_2' + y_{1f}x_1 + y_{1f}x_2'$$

Output map

y_{1f}	x_1x_2 00	01	11	10
0	0 ^{a}	0 ^{d}	1 ^{e}	1
1	1 ^{c}	0	0 ^{f}	1 ^{b}

z_1

$$z_1 = y_{1f}'x_1 + y_{1f}x_2' + x_1x_2'$$

Logic diagram

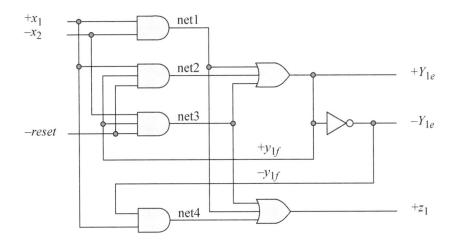

```verilog
//dataflow asynchronous sequential machine 17
module asm17 (rst_n, x1, x2, y1e, z1);


input rst_n, x1, x2;
output y1e, z1;


//define internal nets
wire net1, net2, net3, net4;


//design the logic for excitation variable Y1e
assign    net1 = x1 & ~x2,
          net2 = x1 & y1e & rst_n,
          net3 = y1e & ~x2 & rst_n,
          y1e = net1 | net2 | net3;


//design the logic for output z1
assign    net4 = ~y1e & x1,
          z1 = net1 | net3 | net4;

endmodule
```

```verilog
//test bench for asynchronous sequential machine
//use the primitive flow table to follow the input
//sequence for the output and waveforms
module asm17_tb;

reg rst_n, x1, x2;
wire y1e, z1;

initial            //display variables
$monitor ("x1x2 = %b, state = %b, z1 =%b",
           {x1, x2}, y1e, z1);

initial               //define input vectors
begin
   #0    rst_n = 1'b0;
         x1 = 1'b0;
         x2 = 1'b0;

   #5    rst_n = 1'b1;

   #10   x1 = 1'b1;   x2 = 1'b0;   //to state_b; set z1
   #10   x1 = 1'b0;   x2 = 1'b0;   //to state_c; set z1
   #10   x1 = 1'b0;   x2 = 1'b1;   //to state_d
   #10   x1 = 1'b0;   x2 = 1'b0;   //to state_a
   #10   x1 = 1'b1;   x2 = 1'b0;   //to state_b; set z1
   #20   x1 = 1'b0;   x2 = 1'b0;   //to state_c; set z1
   #10   x1 = 1'b1;   x2 = 1'b0;   //to state_b; set z1
   #10   x1 = 1'b1;   x2 = 1'b1;   //to state_f
   #10   x1 = 1'b1;   x2 = 1'b0;   //to state_b; set z1
   #10   x1 = 1'b1;   x2 = 1'b1;   //to state_f
   #10   x1 = 1'b1;   x2 = 1'b0;   //to state_b; set z1
   #10   x1 = 1'b0;   x2 = 1'b0;   //to state_c; set z1
   #10   x1 = 1'b0;   x2 = 1'b1;   //to state_d
   #10   x1 = 1'b1;   x2 = 1'b1;   //to state_e; set z1
   #10   x1 = 1'b1;   x2 = 1'b0;   //to state_b; set z1
   #10   x1 = 1'b0;   x2 = 1'b0;   //to state_c; set z1
   #10   x1 = 1'b0;   x2 = 1'b1;   //to state_d
   #10   x1 = 1'b0;   x2 = 1'b0;   //to state_a
   #10   $stop;
end

asm17 inst1 (      //instantiate the module
   .rst_n(rst_n),
   .x1(x1),
   .x2(x2),
   .y1e(y1e),
   .z1(z1)
   );
endmodule
```

```
x1x2 = 00, state = 0, z1 =0
x1x2 = 10, state = 1, z1 =1
x1x2 = 00, state = 1, z1 =1
x1x2 = 01, state = 0, z1 =0
x1x2 = 00, state = 0, z1 =0
x1x2 = 10, state = 1, z1 =1
x1x2 = 00, state = 1, z1 =1
x1x2 = 10, state = 1, z1 =1
x1x2 = 11, state = 1, z1 =0
x1x2 = 10, state = 1, z1 =1
x1x2 = 11, state = 1, z1 =0
x1x2 = 10, state = 1, z1 =1
x1x2 = 00, state = 1, z1 =1
x1x2 = 01, state = 0, z1 =0
x1x2 = 11, state = 0, z1 =1
x1x2 = 10, state = 1, z1 =1
x1x2 = 00, state = 1, z1 =1
x1x2 = 01, state = 0, z1 =0
x1x2 = 00, state = 0, z1 =0
```

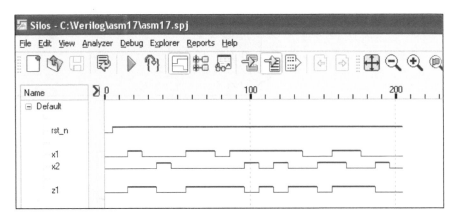

8.21   Design a Moore pulse-mode asynchronous sequential machine that has two inputs x_1 and x_2 and one output z_1. The negative transition of every second consecutive x_1 pulse will assert output z_1 as a level. The output will remain set for all following contiguous x_1 pulses. The output will be deasserted at the negative transition of the second of two consecutive x_2 pulses. The machine will be implemented with NAND logic for the δ next-state logic. The storage elements will be NAND SR latches and positive-edge-triggered D flip-flops in a master-slave configuration.

A representative timing diagram is shown. Generate a state diagram that depicts all possible state transition sequences that conform to the functional specifications. Obtain the input maps, the input equations, and the logic diagram.

Then design the structural module using dataflow modeling for the logic primitives and behavioral modeling for the positive-edge-triggered D flip-flops. Use dataflow modeling for the implementation of output z_1. Obtain the test bench, the outputs, and the waveforms.

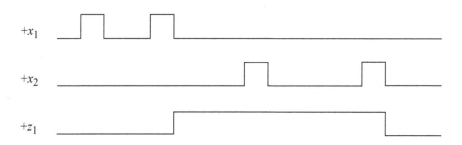

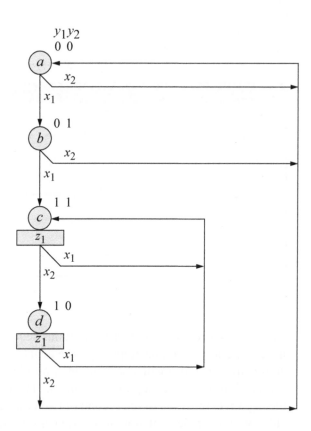

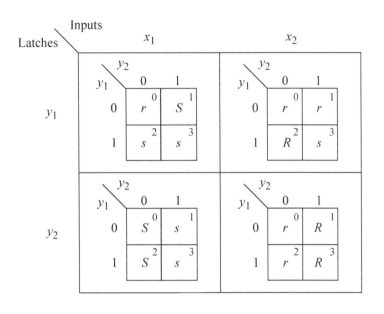

$$SLy_1 = y_2x_1 \qquad RLy_1 = y_2'x_2$$

$$SLy_2 = x_1 \qquad RLy_2 = x_2$$

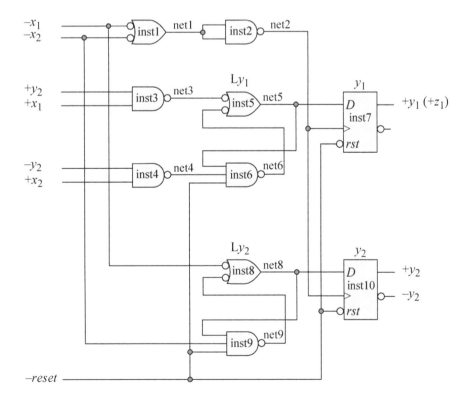

```verilog
//structural/dataflow moore pulse-mode asm
module pm_asm8 (rst_n, x1, x2, y1, y2, z1);

input rst_n, x1, x2;
output y1, y2, z1;

//define internal nets
wire   net1, net2, net3, net4,
       net5, net6, net8, net9;

//instantiate logic for the clock for the D flip-flops
nand2_df inst1 (
   .x1(~x1),
   .x2(~x2),
   .z1(net1)
   );

nand2_df inst2 (
   .x1(net1),
   .x2(net1),
   .z1(net2)
   );

//instantiate the logic for latch Ly1
nand2_df inst3 (
.x1(y2),
.x2(x1),
.z1(net3)
);

nand2_df inst4 (
.x1(~y2),
.x2(x2),
.z1(net4)
);

nand2_df inst5 (
   .x1(net3),
   .x2(net6),
   .z1(net5)
   );

nand3_df inst6 (
   .x1(net5),
   .x2(net4),
   .x3(rst_n),
   .z1(net6)
   );                 //continued on next page
```

```
//instantiate the D flip-flop for y1
d_ff_bh inst7 (
   .rst_n(rst_n),
   .clk(net2),
   .d(net5),
   .q(y1)
   );

//instantiate the logic for latch Ly2
nand2_df inst8 (
   .x1(~x1),
   .x2(net9),
   .z1(net8)
   );

nand3_df inst9 (
   .x1(net8),
   .x2(~x2),
   .x3(rst_n),
   .z1(net9)
   );

//instantiate the D flip-flop for y2
d_ff_bh inst10 (
   .rst_n(rst_n),
   .clk(net2),
   .d(net8),
   .q(y2)
   );

//design the logic for output z1
assign z1 = y1;
endmodule
```

```
//test bench for the pulse-mode asm
module pm_asm8_tb;

reg rst_n, x1, x2;
wire y1, y2, z1;

//display inputs and output
initial
$monitor ("x1x2 = %b, state = %b, z1 = %b",
            {x1, x2}, {y1, y2}, z1);

//continued on next page
```

```
//define input sequence
initial
begin
   #0     rst_n = 1'b0;              //reset to state_a
          x1 = 1'b0;
          x2 = 1'b0;

   #5     rst_n = 1'b1;

   #10    x1 = 1'b0;  x2 = 1'b1;
   #10    x1 = 1'b0;  x2 = 1'b0;  //to state_a

   #10    x1 = 1'b1;  x2 = 1'b0;
   #10    x1 = 1'b0;  x2 = 1'b0;  //to state_b

   #10    x1 = 1'b0;  x2 = 1'b1;
   #10    x1 = 1'b0;  x2 = 1'b0;  //to state_a

   #10    x1 = 1'b1;  x2 = 1'b0;
   #10    x1 = 1'b0;  x2 = 1'b0;  //to state_b

   #10    x1 = 1'b1;  x2 = 1'b0;
   #10    x1 = 1'b0;  x2 = 1'b0;  //to state_c; set z1

   #10    x1 = 1'b0;  x2 = 1'b1;
   #10    x1 = 1'b0;  x2 = 1'b0;  //to state_d; set z1

   #10    x1 = 1'b0;  x2 = 1'b1;
   #10    x1 = 1'b0;  x2 = 1'b0;  //to state_a; rst z1

   #10    x1 = 1'b1;  x2 = 1'b0;
   #10    x1 = 1'b0;  x2 = 1'b0;  //to state_b

   #10    x1 = 1'b1;  x2 = 1'b0;
   #10    x1 = 1'b0;  x2 = 1'b0;  //to state_c; set z1

   #10    x1 = 1'b1;  x2 = 1'b0;
   #10    x1 = 1'b0;  x2 = 1'b0;  //to state_c; set z1

   #10    x1 = 1'b0;  x2 = 1'b1;
   #10    x1 = 1'b0;  x2 = 1'b0;  //to state_d; set z1

   #10    x1 = 1'b1;  x2 = 1'b0;
   #10    x1 = 1'b0;  x2 = 1'b0;  //to state_c; set z1

   #10    x1 = 1'b0;  x2 = 1'b1;
   #10    x1 = 1'b0;  x2 = 1'b0;  //to state_d; set z1
//continued on next page
```

```
    #10    x1 = 1'b0;   x2 = 1'b1;
    #10    x1 = 1'b0;   x2 = 1'b0;   //to state_a; rst z1
    #10    $stop;
end

//instantiate the module into the test bench
pm_asm8 inst1 (
    .rst_n(rst_n),
    .x1(x1),
    .x2(x2),
    .y1(y1),
    .y2(y2),
    .z1(z1)
    );

endmodule
```

```
x1x2 = 00,  state = 00,  z1 = 0
x1x2 = 01,  state = 00,  z1 = 0
x1x2 = 00,  state = 00,  z1 = 0
x1x2 = 10,  state = 00,  z1 = 0
x1x2 = 00,  state = 01,  z1 = 0
x1x2 = 01,  state = 01,  z1 = 0
x1x2 = 00,  state = 00,  z1 = 0
x1x2 = 10,  state = 00,  z1 = 0
x1x2 = 00,  state = 01,  z1 = 0
x1x2 = 10,  state = 01,  z1 = 0
x1x2 = 00,  state = 11,  z1 = 1
x1x2 = 01,  state = 11,  z1 = 1
x1x2 = 00,  state = 10,  z1 = 1
x1x2 = 01,  state = 10,  z1 = 1
x1x2 = 00,  state = 00,  z1 = 0
x1x2 = 10,  state = 00,  z1 = 0
x1x2 = 00,  state = 01,  z1 = 0
x1x2 = 10,  state = 01,  z1 = 0
x1x2 = 00,  state = 11,  z1 = 1
x1x2 = 10,  state = 11,  z1 = 1
x1x2 = 00,  state = 11,  z1 = 1
x1x2 = 01,  state = 11,  z1 = 1
x1x2 = 00,  state = 10,  z1 = 1
x1x2 = 10,  state = 10,  z1 = 1
x1x2 = 00,  state = 11,  z1 = 1
x1x2 = 01,  state = 11,  z1 = 1
x1x2 = 00,  state = 10,  z1 = 1
x1x2 = 01,  state = 10,  z1 = 1
x1x2 = 00,  state = 00,  z1 = 0
```

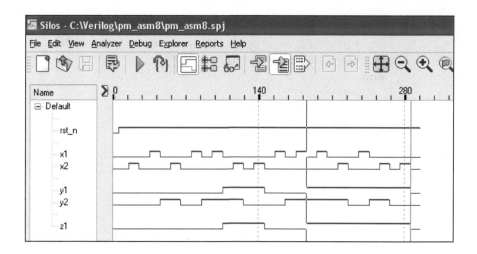

Chapter 9   Programmable Logic Devices

9.4   Given the state diagram shown below for a Moore synchronous sequential machine, implement the machine using a PROM for the δ next-state logic and positive-edge-triggered D flip-flops for the storage elements.

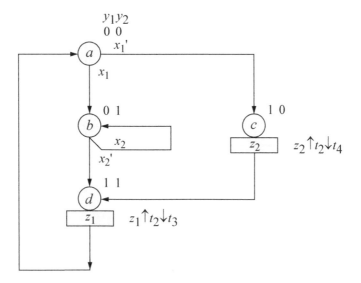

PROM program

PROM Address		PROM Outputs	
Present State $y_1\,y_2$	Inputs $x_1\,x_2$	Next State $y_1\,y_2$	Outputs $z_1\,z_2$
0 0	0 0	1 0	0 0
0 0	0 1	1 0	0 0
0 0	1 0	0 1	0 0
0 0	1 1	0 1	0 0
0 1	0 0	1 1	0 0
0 1	0 1	0 1	0 0
0 1	1 0	1 1	0 0
0 1	1 1	0 1	0 0
1 0	0 0	1 1	0 1
1 0	0 1	1 1	0 1
1 0	1 0	1 1	0 1
1 0	1 1	1 1	0 1
1 1	0 0	0 0	1 0
1 1	0 1	0 0	1 0
1 1	1 0	0 0	1 0
1 1	1 1	0 0	1 0

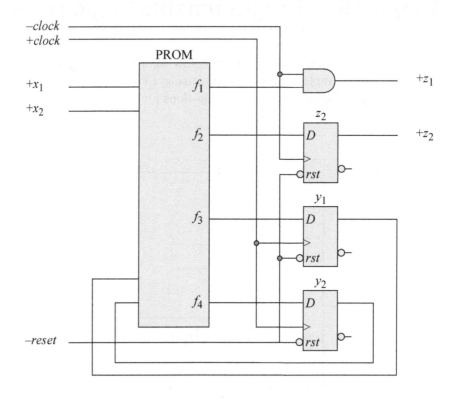

9.8   Design a 3-bit Johnson counter using a PAL device for the δ next-state logic, the D flip-flop storage elements, and the λ output logic. The counter counts in the following sequence: 000, 100, 110, 111, 011, 001, 000.

$y_2 y_3$

y_1	0 0	0 1	1 1	1 0
0	1 ⁰	0 ¹	0 ³	– ²
1	1 ⁴	– ⁵	0 ⁷	1 ⁶

Dy_1

$y_2 y_3$

y_1	0 0	0 1	1 1	1 0
0	0 ⁰	0 ¹	0 ³	– ²
1	1 ⁴	– ⁵	1 ⁷	1 ⁶

Dy_2

$y_2 y_3$

y_1	0 0	0 1	1 1	1 0
0	0 ⁰	0 ¹	1 ³	– ²
1	0 ⁴	– ⁵	1 ⁷	1 ⁶

Dy_3

$$Dy_1 = y_3'$$

$$Dy_2 = y_1$$

$$Dy_3 = y_2$$

Logic diagram

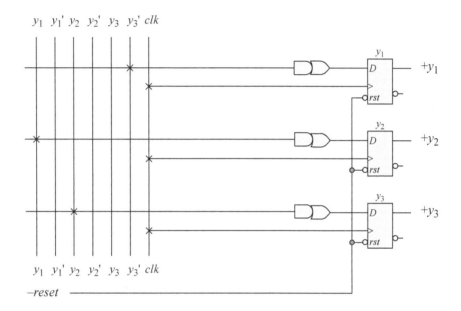

9.12   Implement the state diagram shown below for a Moore machine using a PLA device for the δ next-state logic, the state flip-flops, and the λ output logic. Use x_1 and x_2 as map-entered variables. The outputs are asserted at time t_1 and deasserted at t_3. Before implementing the design, be certain that no state transition sequence passes through states a, c, or d for a path that does not include the corresponding output.

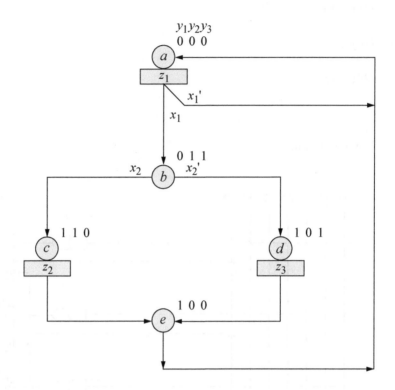

Input maps

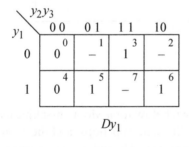

Dy_1 (left map):

y_1 \ y_2y_3	0 0	0 1	1 1	1 0
0	0	–	1	–
1	0	1	–	1

Dy_1

Dy_2 (right map):

y_1 \ y_2y_3	0 0	0 1	1 1	1 0
0	x_1	–	x_2	–
1	0	0	–	0

Dy_2

Dy_3:

y_1 \ y_2y_3	0 0	0 1	1 1	1 0
0	x_1	–	x_2'	–
1	0	0	–	0

Dy_3

Input equations

$$Dy_1 = y_2 + y_3$$

$$Dy_2 = y_1'y_2'x_1 + y_1'y_2x_2$$

$$Dy_3 = y_1'y_2'x_1 + y_1'y_2x_2'$$

Output maps

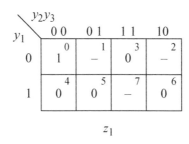

z_1

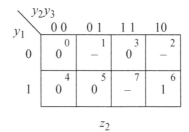

z_2

z_3

Output equations

$$z_1 = y_1'y_2'$$

$$z_2 = y_2y_3'$$

$$z_3 = y_2'y_3$$

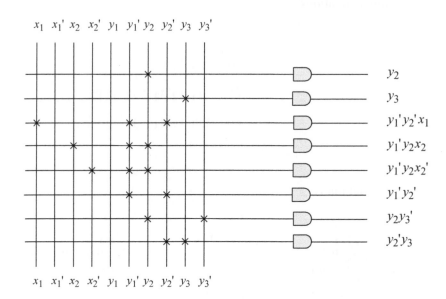

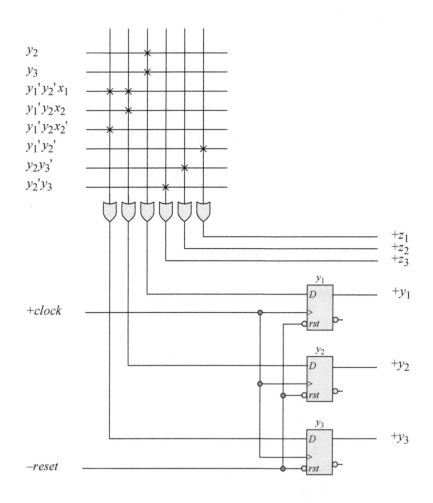

Chapter 10   Digital and Analog Conversion

10.4   Using the operational amplifier shown below,

(a) Determine V_{out} for the following conditions:
$A_v = 0.533$
$V_{in} = 3.75$ volts
$A_v = V_{out} / V_{in}$
$V_{out} = A_v \times V_{in} = 0.533 \times 3.75 = -2$ volts

(b) Determine V_{out} for the following conditions:
$V_{in} = 1.5$ volts
$R_{in} = 5$ KΩ
$R_F = 10$ KΩ
$A_v = R_F / R_{in} = 10$ K / 5 K = 2
$V_{out} = A_v \times V_{in} = 2 \times 1.5 = -3$ volts

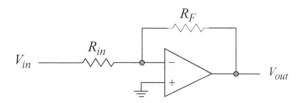

10.8   A D/A converter used a binary-weighted resistor network of 150 KΩ, 75 KΩ, 37.5 KΩ, and 18.75 KΩ (high order). The feedback resistor $R_F = 20$ KΩ. The input voltage $V_{in} = +3$ volts. Determine the voltage increment in V_{out} that is caused by a change in the low-order bit of the binary input.

When all switches are open, $V_{in} = 0$ and $V_{out} = 0$. The voltage gain when the low-order switch is closed is $A = R_F / R_{in} = (20 \times 10^3) / (150 \times 10^3) = 0.133$

Also, $A_v = V_{out} / V_{in}$
$V_{out} = A_v \times V_{in} = 0.133 \times 3$ volts $= -0.4$ volts

Alternative method:
$V_{out} = -(V R_F / 8R) (8b_3 + 4b_2 + 2b_1 + b_0)$
If the switches are (0000), then $V_{out} = 0$
If the switches are (0001), then
$V_{out} = -(V R_F / 8R)(1) = -3(20 \times 10^3) / 8(18.75 \times 10^3) = -60 / 150$
$V_{out} = -0.4$ volts

10.16   A binary-coded decimal (BCD) $R - 2R$ D/A converter has eight inputs that represent a voltage whose maximum value is 99 volts. The inputs are labeled as follows:

$$b_{30}\ b_{20}\ b_{10}\ b_{00} \qquad b_3\ b_2\ b_1\ b_0$$

where b_{30} and b_0 are the low-order bit positions of the tens and units digits, respectively. Write the equation for V_{out}.

$V_{out} = -(V R_F / 16R) [10(8b_{30} + 4b_{20} + 2b_{10} + b_{00}) + (8b_3 + 4b_2 + 2b_1 + b_0)]$, where $(V R_F / 16R)$ is the scaling factor.

There are two D/A converters: one for the tens decade and one for the units decade, as shown below.

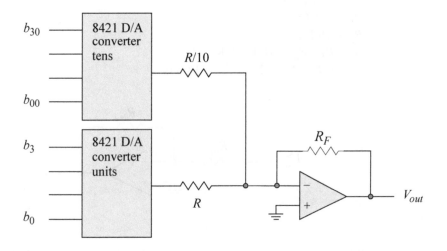

10.21   Show all possible output binary words of a simultaneous/flash comparator analog-to-digital converter that uses eight comparators *comp1* through *comp8*, where *comp8* is the low-order bit.

V_{out1}	V_{out2}	V_{out3}	V_{out4}	V_{out5}	V_{out6}	V_{out7}	V_{out8}
0	0	0	0	0	0	0	0
0	0	0	0	0	0	0	1
0	0	0	0	0	0	1	1
0	0	0	0	0	1	1	1
0	0	0	0	1	1	1	1
0	0	0	1	1	1	1	1
0	0	1	1	1	1	1	1
0	1	1	1	1	1	1	1
1	1	1	1	1	1	1	1

Chapter 11 Magnetic Recording Fundamentals

11.1 Indicate the data bit sequence for the read voltage waveforms shown below.

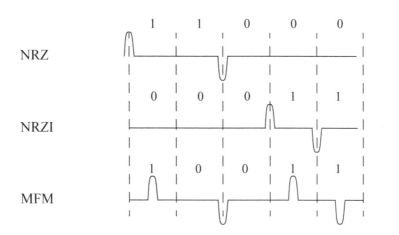

11.7 Given the bit pattern 11100101, convert the byte of data to the GCR format, then draw the write current waveform for the GCR code using the NRZI encoding technique. Begin the waveform with a +I level.

GCR = 01110 10101

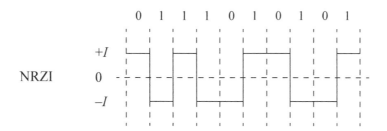

11.9 Each of the data shown below is obtained from two 4-bit segments using the GCR translation table. Indicate which data is valid or invalid for GCR encoding.

10011	00110	Valid ___	Invalid _x_
00111	11010	Valid ___	Invalid _x_
10100	11001	Valid ___	Invalid _x_
01010	10001	Valid ___	Invalid _x_
10110	00110	Valid ___	Invalid _x_

11.15   Each of the data shown below represents 1 byte, which was obtained by translating two 4-bit segments into two 5-bit segments. Indicate which data is valid or invalid for GCR encoding.

10011	00110	Valid ____	Invalid _x_
00111	11010	Valid ____	Invalid _x_
10100	11001	Valid ____	Invalid _x_
01010	10001	Valid ____	Invalid _x_
10110	00110	Valid ____	Invalid _x_
01100	11001	Valid ____	Invalid _x_
01001	01101	Valid _x_	Invalid ____
10010	01001	Valid _x_	Invalid ____
01101	00110	Valid ____	Invalid _x_
11100	11111	Valid ____	Invalid _x_

Chapter 12   Additional Topics in Digital Design

12.4   Given the following function, use functional decomposition to implement the function using a 4:1 multiplexer:

$$z_1(x_1, x_2, x_3, x_4) = x_1'x_4' + x_1x_3$$

x_1x_2 \ x_3x_4	0 0	0 1	1 1	1 0
0 0	1 (0)	0 (1)	0 (3)	1 (2)
0 1	1 (4)	0 (5)	0 (7)	1 (6)
1 1	0 (12)	0 (13)	1 (15)	1 (14)
1 0	0 (8)	0 (9)	1 (11)	1 (10)

z_1

$$z_1(x_1, x_2, x_3, x_4) = [(x_3, x_4), f(x_1, x_2)]$$
The map can be used as is or redrawn as shown below.

x_3x_4 \ x_1x_2	0 0	0 1	1 1	1 0	
0 0	1 (0)	1 (4)	0 (12)	0 (8)	$f(x_1, x_2) = x_1'$
0 1	0 (1)	0 (5)	0 (13)	0 (9)	0
1 1	0 (3)	0 (7)	1 (15)	1 (11)	$f'(x_1, x_2) = x_1$
1 0	1 (2)	1 (6)	1 (14)	1 (10)	1

z_1

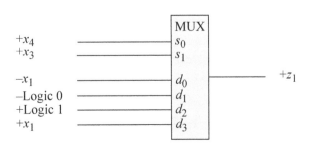

12.7   Design a parity checker using an iterative network to determine the parity of a 4-bit word x_1, x_2, x_3, x_4. If the parity of the word is odd, then the output z_1 will be asserted high; otherwise, $z_1 = 0$. Design the network using dataflow modeling. Obtain the design module, the test bench module, the outputs, and the waveforms.

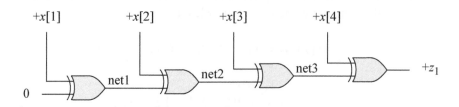

```
//dataflow odd parity checker
module iterative_parity (x1, x2, x3, x4, z1);

//define inputs and output
input x1, x2, x3, x4;
output z1;

//define inputs and output as wire
wire x1, x2, x3, x4;
wire z1;

//define internal nets
wire net1, net2, net3;

//design logic using continuous assignment
assign    net1 = x1 ^ 1'b0,
          net2 = x2 ^ net1,
          net3 = x3 ^ net2,
          z1   = x4 ^ net3;

endmodule
```

```
//dataflow odd parity checker
module iterative_parity (x1, x2, x3, x4, z1);

//define inputs and output
input x1, x2, x3, x4;
output z1;

//define inputs and output as wire
wire x1, x2, x3, x4;
wire z1;

//continued on next page
```

```
//define internal nets
wire net1, net2, net3;

//design logic using continuous assignment
assign   net1 = x1 ^ 1'b0,
         net2 = x2 ^ net1,
         net3 = x3 ^ net2,
         z1   = x4 ^ net3;

endmodule
```

```
x1 x2 x3 x4 = 0000, z1 = 0
x1 x2 x3 x4 = 0001, z1 = 1
x1 x2 x3 x4 = 0010, z1 = 1
x1 x2 x3 x4 = 0011, z1 = 0
x1 x2 x3 x4 = 0100, z1 = 1
x1 x2 x3 x4 = 0101, z1 = 0
x1 x2 x3 x4 = 0110, z1 = 0
x1 x2 x3 x4 = 0111, z1 = 1
x1 x2 x3 x4 = 1000, z1 = 1
x1 x2 x3 x4 = 1001, z1 = 0
x1 x2 x3 x4 = 1010, z1 = 0
x1 x2 x3 x4 = 1011, z1 = 1
x1 x2 x3 x4 = 1100, z1 = 0
x1 x2 x3 x4 = 1101, z1 = 1
x1 x2 x3 x4 = 1110, z1 = 1
x1 x2 x3 x4 = 1111, z1 = 0
```

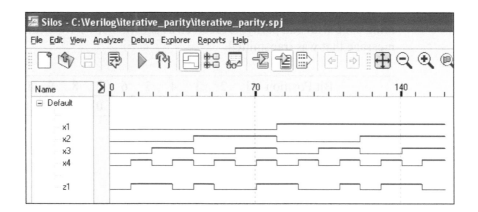

12.11   Three code words, each containing a message of 4 bits which are encoded using the Hamming code are received as shown below. Determine the correct 4-bit messages that were transmitted using odd parity.

	Received Code Words							Correct 4-bit
Bit Position	1	2	3	4	5	6	7	Message
(a)	0	1	0	1	0	1	0	0 0 1 1
(b)	1	1	0	0	1	1	0	0 1 1 1
(c)	0	0	1	0	1	1	1	1 1 1 1

(a) E_1 = 1, 3, 5, 7 = 0 0 0 0 = error = 1
    E_2 = 2, 3, 6, 7 = 1 0 1 0 = error = 1
    E_4 = 4, 5, 6, 7 = 1 0 1 0 = error = 1   Bit 7 is in error
    Correct message = 0011

(b) E_1 = 1, 3, 5, 7 = 1 0 1 0 = error = 1
    E_2 = 2, 3, 6, 7 = 1 0 1 0 = error = 1
    E_4 = 4, 5, 6, 7 = 0 1 1 0 = error = 1   Bit 7 is in error
    Correct message = 0111

(c) E_1 = 1, 3, 5, 7 = 0 1 1 1 = no error = 0
    E_2 = 2, 3, 6, 7 = 0 1 1 1 = no error = 0
    E_4 = 4, 5, 6, 7 = 0 1 1 1 = no error = 0   No error
    Correct message = 1111

12.15   A code word containing one 8-bit message, which is encoded using the Hamming code with odd parity, is received as shown below. Determine the 8-bit message that was transmitted.

Bit Position	1	2	3	4	5	6	7	8	9	10	11	12
Received Code Word	0	1	0	1	1	0	0	1	0	0	1	1

E_1 = 1, 3, 5, 7 ,9, 11   = 0 0 1 0 0 1   = error   = 1
E_2 = 2, 3, 6, 7, 10, 11   = 1 0 0 0 0 1   = error   = 1
E_4 = 4, 5, 6, 7, 12   = 1 1 0 0 1   = no error = 0
E_8 = 8, 9, 10, 11, 12   = 1 0 0 1 1   = no error = 0

Bit 3 is in error.   Correct 8-bit message = 1 1 0 0 0 0 1 1

12.21   Perform residue checking on the following BCD numbers:

    1000 0111 0101
    0110 0100 0010

                                                    Residue of
                                                    segments

         1000   0111   0101   2   1   2   → 2
         0110   0100   0010   0   1   2   → 0   → mod-3 = 2
                        0   ←   0111
                 1   ←   1011
    1 ←   1111   0110
         0110   0001
         0101   No error

      ↓   ↓   ↓   ↓
    0001 0101   0001   0111   1   2   1   1   →   → mod-3 = 2

12.24   Perform parity prediction on the following operands with a carry-in = 0:

        1010
        1010

             a =   1   0   1   0 P_a =   1
    +) b =   1   0   1   0 P_b =   1   ⊕ = 0
         Carries =   0   1   0   0   ← (c_{in}) P_{cy} =   0
            Sum =   0   1   0   0 P_{sum} = 0

12.28   Using behavioral modeling, design a memory unit that consists of sixteen 8-bit words. Store the bytes in memory by assigning a value to each address location individually. Beginning with address 0, store the following hexadecimal bytes: 00H, 01H, 02H, 03H, 14H, 15H, 16H, 17H, 28H, 29H, 2AH, 2BH, 3CH, 3DH, 3EH, 3FH. Obtain the design module, the test bench module, the outputs, and the waveforms.

```verilog
//behavioral memory load
module mem_load3 (addr, data_out);

//list input and output
input [3:0] addr;
output [7:0] data_out;

wire [3:0] addr;
reg [7:0] data_out;

//define memory size of data_cache
//an array of sixteen 8-bit registers
reg [7:0] data_cache [0:15];

//define data cache contents
initial
begin
   data_cache [0]  = 8'b0000_0000;
   data_cache [1]  = 8'b0000_0001;
   data_cache [2]  = 8'b0000_0010;
   data_cache [3]  = 8'b0000_0011;
   data_cache [4]  = 8'b0001_0100;
   data_cache [5]  = 8'b0001_0101;
   data_cache [6]  = 8'b0001_0110;
   data_cache [7]  = 8'b0001_0111;
   data_cache [8]  = 8'b0010_1000;
   data_cache [9]  = 8'b0010_1001;
   data_cache [10] = 8'b0010_1010;
   data_cache [11] = 8'b0010_1011;
   data_cache [12] = 8'b0011_1100;
   data_cache [13] = 8'b0011_1101;
   data_cache [14] = 8'b0011_1110;
   data_cache [15] = 8'b0011_1111;

end

//address the data cache
always @ (addr)
begin
   data_out = data_cache [addr];
end

endmodule
```

```verilog
//mem_load3 test bench
module mem_load3_tb;

reg [3:0] addr;
wire [7:0] data_out;

//assign values to the address
initial
begin
    #0     addr = 4'b0000;
    #10    addr = 4'b0001;
    #10    addr = 4'b0010;
    #10    addr = 4'b0011;
    #10    addr = 4'b0100;
    #10    addr = 4'b0101;
    #10    addr = 4'b0110;
    #10    addr = 4'b0111;
    #10    addr = 4'b1000;
    #10    addr = 4'b1001;
    #10    addr = 4'b1010;
    #10    addr = 4'b1011;
    #10    addr = 4'b1100;
    #10    addr = 4'b1101;
    #10    addr = 4'b1110;
    #10    addr = 4'b1111;
    #15    $stop;
end

//apply addresses and display data
initial
begin: apply_stimulus
    reg [3:0] invect;
    for (invect = 0; invect < 16; invect = invect + 1)
        begin
            addr = invect [3:0];
            #10 $display ("address = %b, data = %b",
                            addr, data_out);
        end
end

//instantiate the module into the test bench
mem_load3 inst1 (
    .addr(addr),
    .data_out(data_out)
    );

endmodule
```

```
address = 0000, data = 00000000
address = 0010, data = 00000001
address = 0010, data = 00000010
address = 0100, data = 00000011
address = 0100, data = 00010100
address = 0110, data = 00010101
address = 0110, data = 00010110
address = 1000, data = 00010111
address = 1000, data = 00101000
address = 1010, data = 00101001
address = 1010, data = 00101010
address = 1100, data = 00101011
address = 1100, data = 00111100
address = 1110, data = 00111101
address = 1110, data = 00111110
address = 1111, data = 00111111
```

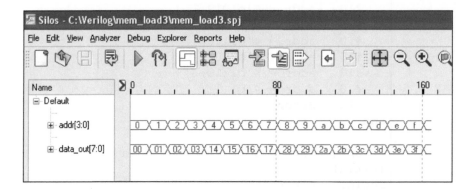

INDEX

T

T - #0333 - 101024 - C0 - 254/178/62 [64] - CB - 9781420074154 - Gloss Lamination